CW00621107

# MARE Publication Series

Volume 14

The MARE Publication Series is an initiative of the Centre for Maritime Research (MARE). MARE is an interdisciplinary social-science network devoted to studying the use and management of marine resources. It is based jointly at the University of Amsterdam and Wageningen University (www.marecentre.nl).

The MARE Publication Series addresses topics of contemporary relevance in the wide field of 'people and the sea'. It has a global scope and includes contributions from a wide range of social science disciplines as well as from applied sciences. Topics range from fisheries, to integrated management, coastal tourism, and environmental conservation. The series was previously hosted by Amsterdam University Press and joined Springer in 2011.

The MARE Publication Series is complemented by the Journal of Maritime Studies (MAST) and the biennial People and the Sea Conferences in Amsterdam.

More information about this series at http://www.springer.com/series/10413

Svein Jentoft • Ratana Chuenpagdee
María José Barragán-Paladines • Nicole Franz
Editors

# The Small-Scale Fisheries Guidelines

## Global Implementation

*Editors*
Svein Jentoft
Norwegian College of Fishery Science
UiT – The Arctic University of Norway
Tromsø, Norway

María José Barragán-Paladines
Development and Knowledge Sociology Working Group
Leibniz Centre for Tropical Marine Research – ZMT
Bremen, Germany

Ratana Chuenpagdee
Department of Geography
Memorial University of Newfoundland
St. John's, Newfoundland and Labrador
Canada

Nicole Franz
Fisheries and Aquaculture Department
Food and Agriculture Organization of the United Nations (FAO)
Rome, Italy

Too Big To Ignore (TBTI; toobigtoignore.net) is a global research network and knowledge mobilization partnership, funded by the Social Sciences and Humanities Research Council of Canada, and supported by 15 partner organizations and over 300 members from around the world. The network aims at elevating the profile of small-scale fisheries, arguing against their marginalization in national and international policies, and developing research and governance capacity to address global fisheries challenges.

ISSN 2212-6260 ISSN 2212-6279 (electronic)
MARE Publication Series
ISBN 978-3-319-55073-2 ISBN 978-3-319-55074-9 (eBook)
DOI 10.1007/978-3-319-55074-9

Library of Congress Control Number: 2017942813



*Cover Illustration*: A typical morning in Hann Bay, a landing site near Dakar, Senegal, with purse seiners returning from an overnight fishing trip, while hook and line fishers in smaller crafts are getting ready to go to sea. Photo credit: Aliou Sall, July 2014.

Printed on acid-free paper

This Springer imprint is published by Springer Nature
The registered company is Springer International Publishing AG
The registered company address is: Gewerbestrasse 11, 6330 Cham, Switzerland

# Foreword

At 9:30 am, on May 20, 2013, in the Green Room at FAO Headquarters, started the Technical Consultation to draft the Voluntary Guidelines for Securing Sustainable Small-Scale Fisheries in the Context of Food Security and Poverty Eradication (SSF Guidelines). On that spring morning, in Rome, the tide of the small-scale fisheries in the world, hopefully, started to change for the better. A little more than a year later, on June 9, 2014, the FAO Committee on Fisheries (COFI) endorsed the first internationally negotiated document explicitly devoted to small-scale fisheries. Despite its comprehensiveness and density, with more than a hundred paragraphs, and a very broad and bold scope, a consensus text was reached after only two 1-week meetings, in addition to side negotiations during COFI. The building process, however, had started much earlier, with several meetings and conferences, including the 2008 Global Conference on Securing Sustainable Small-Scale Fisheries, co-organized by the FAO and the Royal Government of Thailand. Therefore, the so-called zero draft that served as the basis for the negotiations of the Technical Consultation was already the result of a very broad, open, transparent and participatory consultation with thousands of stakeholders. The same spirit of broad participation and openness also guided the entire negotiation of the text during the Technical Consultation, with an unprecedented level of engagement and participation from civil society. It could not have, of course, been different, regarding a fisheries sector that accounts for the vast majority of fishworkers worldwide.

From the very early stages of the negotiation, which I had the honour and the privilege to chair, it became very clear for all delegations that we were not discussing the fate of an economic activity, but of livelihoods and communities – that small-scale fisheries are not about an economic sector but they are about families, culture and tradition. Considering the importance of small-scale fisheries for food security, nutrition, livelihoods, rural development and poverty and hunger eradication, it becomes clear that the adoption of a human rights-based approach in the SSF Guidelines, therefore, was much more a consequence than a choice. Or, as put by Chandrika Sharma, the executive secretary of the ICSF – to whom the Guidelines were dedicated – to adopt a human rights approach for improving the life and livelihoods of fishing communities was not really a matter of choice, but an obligation.

Since its official adoption by the FAO Conference, the SSF Guidelines have become a beacon to guide national and international policy, aimed at the sustainable development of small-scale fisheries and fishing communities, to establish political hierarchies and to elevate small-scale fisheries in the agendas of governments and international organizations. The document itself, however, is worthless unless the words and provisions it contains are able to find their way into the real world. The place of the SSF Guidelines is not in the shelves of public offices or ministerial departments, but at the beaches, aboard the canoes, in the hands of the fishers, by the sea. This is the challenge now lying ahead of us: to make it actually happen and to ensure its implementation at the local, national, regional and international levels.

In this context, the role and responsibility of the academic community cannot be overestimated. The present book, produced by the global research network 'Too Big To Ignore' (TBTI), is a very important step into the right direction. Its 37 chapters, from authors of so many different countries and regions of the globe, clearly show that implementation is already happening. The SSF Guidelines have been born and, beyond the talking, they are now already starting to walk by themselves. Let's hope this book will help guide the steps of this young but brave toddler, so that it may grow into a strong and energetic adult. To walk through the path ahead, nevertheless, it will be very important to understand, and never forget, that 'science' cannot prosper without 'experience' and that we will get nowhere unless scientific knowledge walks hand in hand with traditional knowledge. To empower fishing communities is not the best way to ensure the sustainability of small-scale fisheries, but the only one. And small-scale fisheries are not a problem to be solved, but a solution to be unfolded.

Recife, Brazil Fábio Hissa Vieira Hazin

# Preface

Since its inception, the peer-reviewed MARE Publication Series, hosted by Springer, has devoted much attention to fisheries. Of the 13 volumes realized at the time of writing, nine have actually dealt with fisheries-related issues. Within the fisheries realm, we have been particularly interested in the fate of small-scale fishers and their communities. This is also a consequence of the cooperation between the Centre for Maritime Research (MARE), of which this Series is part, and the Too Big To Ignore (TBTI) project, which strives to elevate the profile of small-scale fisheries around the world. The first volume dedicated specifically to the condition of this subsector was monumental – *Interactive Governance for Small-Scale Fisheries: Global Reflections* (2015). This book provides an authoritative overview of the trends prevailing small-scale fisheries governance around the globe.

The present volume continues where the previous book left off, now focusing on the most ambitious international policy instrument ever to have been developed for the benefit of this subsector: the Voluntary Guidelines for Securing Sustainable Small-Scale Fisheries (SSF Guidelines). These Guidelines, passed by FAO's membership in 2014, offer a holistic perspective on the needs of small-scale fishers and their ways forward to address these needs.

While the SSF Guidelines are slowly percolating to national and subnational levels, assisted by government agencies and a large number of civil society organizations, this book provides an extra impetus and a contribution from the academic and research community. Detailing the experiences and challenges faced during the application of sections of the SSF Guidelines in both Southern and Northern Hemispheres, the book supplies a relevant baseline for reflection as well as action. As one of the Series editors, I am more than proud to be hosting the volume and wish it a wide readership. Importantly, it is our hope that the book will help promote and support the implementation of the SSF Guidelines, leading therefore to sustainable small-scale fisheries around the world.

Amsterdam, The Netherlands Maarten Bavinck

# Acknowledgements

This book would not have been possible without the effort of the 95 authors from around the world. Thank you so much! Several other people have helped us with the production of the book. Ajit Menon and Brennan Lowery provided excellent language editing. Vesna Kereži and Mirella Leis, with support from David Bishop, were instrumental in communicating with the authors and in the formatting of chapters. Thanks also to Delphine Rocklin for producing the map for Chap. 1. Joseph Daniel and Fritz Schmuhl, as well as the Springer Books team, took care of the publishing process. We are enormously grateful for the effort of all the 65 reviewers, whose names are listed in Appendix 1. We are indebted to Maarten Bavinck and Fabio Hazin for writing the preface and the foreword, respectively. Professor Hazin played a key role as a chairperson of the FAO Technical Consultation on the SSF Guidelines, leading the 2 weeks of deliberations towards consensus. He was also the chair of the 32nd Session of the FAO Committee on Fisheries, when the strategies for the implementation of the SSF Guidelines were discussed.

Finally, we extend our sincere thanks to the Food and Agriculture Organization of the United Nations (FAO) for the opportunities to observe the development of the SSF Guidelines and contribute to their implementation. FAO colleagues, especially members of the Small-Scale Fisheries Task Force, have also helped with the production of this book, acting as reviewers of several chapters and connecting us to key organizations and information sources.

The book is a product of the Too Big To Ignore: Global Partnership for Small-Scale Fisheries Research (TBTI), funded by the Social Sciences and Humanities Research Council of Canada (grant number 895-2011-1011), headquartered at the Memorial University of Newfoundland, St. John's, Canada, and directed by Ratana Chuenpagdee. Most of the contributors to the book are members of TBTI. Svein Jentoft wishes to acknowledge the Centre for Sami Studies, UiT – The Arctic University of Norway, for the logistical support while working on this book.

Tromsø, St. John's, Bremen and Rome December 2016 — The Editors

# Contents

# Contributors

**Shehu Latunji Akintola** Fisheries Department, Lagos State University, Ojo, Lagos State, Nigeria

**Oscar Amarasinghe** University of Ruhuna, Matara, Sri Lanka

**Milena Arias-Schreiber** Department of Philosophy, Linguistics and Theory of Science, Gothenburg University, Gothenburg, Sweden

**María José Barragán-Paladines** Development and Knowledge Sociology Working Group, Leibniz Centre for Tropical Marine Research – ZMT, Bremen, Germany

**Maarten Bavinck** University of Amsterdam, Amsterdam, The Netherlands

University of Tromsø, Tromsø, Norway

**Ramachandra Bhatta** Indian Council of Agricultural Research Emeritus Scientist, College of Fisheries, Mangalore, India

**David Bishop** Too Big To Ignore Global Partnership for Small-Scale Fisheries Research, Memorial University of Newfoundland, St. John's, NL, Canada

**Meike Brauer** Nordic Master Aquatic Food Production – Quality and Safety, Norwegian University of Life Science, Ås, Akershus, Norway

**Joachim Carolsfeld** World Fisheries Trust – WFT, Victoria, BC, Canada

**Afrina Choudhury** WorldFish, Dhaka, Bangladesh

**Ratana Chuenpagdee** Department of Geography, Memorial University of Newfoundland, St. John's, Newfoundland and Labrador, Canada

**Philippa J. Cohen** WorldFish, Honiara, Solomon Islands

Australian Research Council Centre of Excellence for Coral Reef Studies, James Cook University, Townsville, Australia

**Steven M. Cole** WorldFish, Lusaka, Zambia

**Gustavo Henrique G. da Silva** Universidade Federal do Semiárido – UFERSA, Mossoró, Brazil

**Moslem Daliri** Fisheries Department, Faculty of Marine and Atmospheric Sciences, Hormozgan University, Bandar Abbas, Iran

**Maricela de la Torre-Castro** Department of Physical Geography, Stockholm University, Stockholm, Sweden

**Sérgio Macedo G. de Mattos** Ministry of Planning, Brasilia, Brazil

**Alyne Delaney** Innovative Fisheries Management, Aalborg University, Aalborg, Denmark

**María José Espinosa-Romero** Comunidad y Biodiversidad, A.C, Guaymas, SO, Mexico

**Kafayat Adetoun Fakoya** Fisheries Department, Lagos State University, Ojo, Lagos State, Nigeria

**Katia Frangoudes** Université de Bretagne Occidentale, Brest, France

**Nicole Franz** Fisheries and Aquaculture Department, Food and Agriculture Organization of the United Nations (FAO), Rome, Italy

**Carlos Fuentevilla** Food and Agriculture Organization of the United Nations (FAO), Sub-regional Office for the Caribbean, Bridgetown, Barbados

**Stuart Fulton** Comunidad y Biodiversidad, A.C, Guaymas, SO, Mexico

**Charlie J. Gardner** Blue Ventures Conservation, Omnibus Business Centre, London, UK

Durrell Institute of Conservation and Ecology (DICE), University of Kent, Canterbury, UK

**Sílvia Gómez Mestres** Faculty of Humanities, Department of Social and Cultural Anthropology, Autonomous University of Barcelona, Catalonia, Spain

**Miguel González** Department of Social Science, York University, Toronto, ON, Canada

**Charlotte Gough** Blue Ventures Conservation, Omnibus Business Centre, London, UK

**Surathkal Gunakar** Pompei College, Mangalore, India

**Alasdair Harris** Blue Ventures Conservation, Omnibus Business Centre, London, UK

**Johan Hultman** Department of Service Management and Service Studies, Lund University, Lund, Sweden

**Mohammad Mahmudul Islam** Department of Coastal and Marine Fisheries, Sylhet Agricultural University, Sylhet, Bangladesh

**Rikke Becker Jacobsen** Innovative Fisheries Management, Department of Planning, Aalborg University, Aalborg, Denmark

**Adam Jadhav** University of California at Berkeley, Berkeley, CA, USA

Panchabhuta Conservation Foundation, Kagal, India

Dakshin Foundation, Bangalore, India

**Svein Jentoft** Norwegian College of Fishery Science, UiT – The Arctic University of Norway, Tromsø, Norway

**Olufemi Olabode Joseph** Agriculture Department, Adeniran Ogunsanya College of Education, Ijanikin, Lagos State, Nigeria

**Kungwan Juntarashote** Department of Fisheries Management, Kasetsart University, Bangkok, Thailand

**Ehsan Kamrani** Fisheries Department, Faculty of Marine and Atmospheric Sciences, Hormozgan University, Bandar Abbas, Iran

**Vesna Kereži** Too Big To Ignore Global Partnership for Small-Scale Fisheries Research, Memorial University of Newfoundland, St. John's, NL, Canada

**Kate Kincaid** Department of Biology, Memorial University of Newfoundland, St. John's, NL, Canada

Cape Eleuthera Institute, Eleuthera, Bahamas

**Danika Kleiber** Pacific Island Fisheries Science Centre, Joint Institute for Marine and Atmospheric Research, Honolulu, HI, USA

Sociology Department, Memorial University Newfoundland, St. John's, Canada

**Mitchell Lay** Caribbean Network of Fisherfolk Organisations, St. John's, Antigua and Barbuda

**Adrian Levrel** Blue Ventures Conservation, Omnibus Business Centre, London, UK

**Lars Lindström** Department of Political Science, Stockholm University, Stockholm, Sweden

**Sebastian Linke** School of Global Studies, Gothenburg University, Gothenburg, Sweden

**Josep Lloret** Institute of Aquatic Ecology, University of Girona, Catalonia, Spain

**Joseph Luomba** Tanzania Fisheries Research Institute (TAFIRI), Mwanza, Tanzania

**Alison Elisabeth Macnaughton** World Fisheries Trust – WFT, Victoria, BC, Canada

**Allyssandra Maria Lima R. Maia** Universidade Federal do Semiárido – UFERSA, Mossoró, Brazil

**Patrick McConney** Centre for Resource Management and Environmental Studies, The University of the West Indies Cave Hill Campus, St. Michael, Barbados

**Cynthia McDougall** WorldFish, Batu Maung, Malaysia

**Thomas F. McInerney** FAO Consultant, Rome, Italy

**Tiffany H. Morrison** Australian Research Council Centre of Excellence for Coral Reef Studies, James Cook University, Townsville, Australia

**Cornelia E. Nauen** Mundus maris – Sciences and Arts for Sustainability, Brussels, Belgium

**Prateep Kumar Nayak** Faculty of Environment, School of Environment, Enterprise and Development, University of Waterloo, Waterloo, ON, Canada

**Nadine Nembhard** Caribbean Network of Fisherfolk Organisations, Belize City, Belize

**Zahidah Afrin Nisa** Coastal Resources Center, University of Rhode Island, Kingston, RI, USA

**Kim Olson** Government of Newfoundland and Labrador, St. John's, NL, Canada

**Paul Onyango** Department of Aquatic Sciences and Fisheries Technology, University of Dar-es Salaam, Dar-es Salaam, Tanzania

**Hajnalka Petrics** Dartmouth School of Graduate & Advanced Studies, Hanover, New Hampshire, USA

**Terrence Phillips** Coastal and Marine Livelihoods and Governance Programme, Caribbean Natural Resources Institute, Laventille, Trinidad and Tobago

**Cristina Pita** University of Aveiro, Aveiro, Portugal

**Joonas Plaan** Too Big To Ignore Global Partnership for Small-Scale Fisheries Research, Memorial University of Newfoundland, St. John's, NL, Canada

**Marilyn Porter** Sociology Department, Memorial University Newfoundland, St. John's, Canada

**Sarah Pötter** West Nordic Studies, University of Akureyri, Akureyri, Iceland

University of the Faroe Islands, Thorshavn, Faroe Islands

**James Prescott** Australian Fisheries Management Authority, Darwin, NT, Australia

**Steve Rocliffe** Blue Ventures Conservation, Omnibus Business Centre, London, UK

**Victoria Rogers** Too Big To Ignore Global Partnership for Small-Scale Fisheries Research, Victoria, BC, Canada

**Lina María Saavedra-Díaz** Programas de Biología e Ingeniería Pesquera, Universidad del Magdalena, Santa Marta, Magdalena, Colombia

**Gabriela Sabau** School of Science and the Environment, Memorial University of Newfoundland, Grenfell Campus, Corner Brook, NL, Canada

**Alicia Said** Durrell Institute for Conservation and Ecology, School of Anthropology and Conservation, University of Kent, Canterbury, UK

**Aliou Sall** Mundus maris – Sciences and Arts for Sustainability, Dakar, Senegal

**Anna Santos** National Marine Fisheries Service, Alaska Fisheries Science Center, Seattle, Washington, USA

**Suvaluck Satumanatpan** Faculty of Environment and Resources Studies, Mahidol University, Bangkok, Thailand

**Filippa Säwe** Department of Service Management and Service Studies, Lund University, Lund, Sweden

**Rebecca L. Singleton** Blue Ventures Conservation, Omnibus Business Centre, London, UK

**Hunter T. Snyder** Food and Agriculture Organization of the United Nations, Rome, Italy

**Lisa K. Soares** University of Warwick, Warwick, UK

**Kumi Soejima** National Fisheries University, Shimonoseki, Japan

**Francisco Javier Vergara Solana** Centro Interdisciplinario de Ciencias Marinas, Instituto Politécnico Nacional, La Paz, B.C.S, Mexico

**Andrew M. Song** WorldFish, Honiara, Solomon Islands

Australian Research Council Centre of Excellence for Coral Reef Studies, James Cook University, Townsville, Australia

**Siri Ulfsdatter Søreng** Agriculture Agency, Alta, Norway

**Dirk J. Steenbergen** Research Institute for Environment and Livelihoods, Charles Darwin University, Darwin, NT, Australia

**Wichin Suebpala** Faculty of Sciences, Ramkhamhaeng University, Bangkok, Thailand

**Jackie Sunde** Department of Environmental and Geographical Science, University of Cape Town, Cape Town, South Africa

International Collective in Support of Fishworkers (ICSF), Chennai, India

**Makamas Sutthacheep** Faculty of Sciences, Ramkhamhaeng University, Bangkok, Thailand

**Jorge Torre** Comunidad y Biodiversidad, A.C, Guaymas, SO, Mexico

**Xavier Vincke** Blue Ventures Conservation, Omnibus Business Centre, London, UK

**Lena Westlund** FAO Consultant, Rome, Italy

FAO Consultant, Stockholm, Sweden

**Rolf Willmann** Independent Expert, Kressbronn, Germany

**Matias John Wojciechowski** World Fisheries Trust – WFT, Victoria, BC, Canada

**Nobuyuki Yagi** Global Agricultural Sciences, The University of Tokyo, Tokyo, Japan

**Thamasak Yeemin** Faculty of Sciences, Ramkhamhaeng University, Bangkok, Thailand

**José Alberto Zepeda** Centro Interdisciplinario de Ciencias Marinas, Instituto Politécnico Nacional, La Paz, B.C.S, Mexico

Independent Consultant, Prolongación Baja California s/n, Ampliación Centenario, La Paz, B.C.S, Mexico

# Part I
# Vision and Ambition

In June 2014, FAO member states endorsed the Voluntary Guidelines for Securing Sustainable Small-Scale Fisheries in the Context of Food Security and Poverty Eradication (SSF Guidelines). For millions of small-scale fisheries people around, this was a historic event. With their broad agenda founded within a human rights-based approach, the SSF Guidelines are breaking new ground. However, given that the SSF Guidelines address a range of issues that are complex and politically contentious, there are reasons to expect that their implementation will be challenging. The first chapter of this book (Chap. 1) by the editors (Svein Jentoft, Ratana Chuenpagdee, María José Barragán-Paladines, and Nicole Franz) introduces the topic and the contexts for this major endeavor, while also presenting its content in broad terms. Chapter 2, by Rolf Willmann, Nicole Franz, Carlos Fuentevilla, Thomas McInerney, and Lena Westlund, discusses the human rights-based approach and what it implies in the context of the SSF Guidelines. They also examine critical views from social scientists on this approach while drawing on concrete examples on how human rights advocacy and human rights law have actually supported fishing communities in defending their rights to subsistence, livelihood, and culture. Chapter 3, by Nicole Franz and María José Barragán-Paladines, provides an account of the developments that have taken place since the SSF Guidelines were endorsed, occurring at various levels in countries and regions around the world. The implementation of the SSF Guidelines is now underway, and the chapter provides examples of concrete actions being taken to facilitate their uptake.

# Chapter 1
# Implementing the Voluntary Guidelines for Securing Small-Scale Fisheries

**Svein Jentoft, Ratana Chuenpagdee, Nicole Franz, and María José Barragán-Paladines**

**Abstract** On June 9, 2014 the Committee of Fisheries (COFI) of FAO endorsed the Voluntary Guidelines for Securing Sustainable Small-Scale Fisheries in the Context of Food Security and Poverty Eradication (SSF Guidelines). For millions of small-scale fisheries people around the world who are poor and marginalized, this was a historic moment and a potential turning point. The SSF Guidelines are the first instrument of its kind particularly aimed at promoting the sustainability of this sector. As the SSF Guidelines address a range of issues that are complex and politically contentious, there are reasons to expect that their implementation will be challenging and far from straight forward. In fact, one may assume that the SSF Guidelines will meet resistance as they are brought from the international level to local communities where fishing people live and work. This book examines the extent to which the SSF Guidelines' implementation is being initiated around the world and the limitations and opportunities involved in their contextualization and operationalization. It draws on case studies from more than 30 countries in which small-scale fisheries play an important role for food security and community well-being. What

S. Jentoft (✉)
Norwegian College of Fishery Science, UiT – The Arctic University of Norway, Tromsø, Norway
e-mail: svein.jentoft@uit.no

R. Chuenpagdee
Department of Geography, Memorial University of Newfoundland, St. John's, Newfoundland and Labrador, Canada
e-mail: ratanac@mun.ca

N. Franz
Fisheries and Aquaculture Department, Food and Agriculture Organization of the United Nations (FAO), Rome, Italy
e-mail: Nicole.Franz@fao.org

M.J. Barragán-Paladines
Development and Knowledge Sociology Working Group, Leibniz Centre for Tropical Marine Research – ZMT, Bremen, Germany
e-mail: mariaj.barraganp@leibniz-ztm.de



S. Jentoft et al. (eds.), *The Small-Scale Fisheries Guidelines*, MARE Publication Series 14, DOI 10.1007/978-3-319-55074-9_1

can the SSF Guidelines do to promote food security, alleviate poverty, and secure human rights, while at the same time empower fishing communities to take control of their future?

**Keywords** FAO • Small-scale fisheries guidelines • Human rights approach • Implementation • Empowerment • Governance

## Introduction

No one really knows exactly how many small-scale fishers there are in the world. By certain estimates, for instance by the United Nation's Food and Agriculture Organization (FAO), about half of the world's 51 million fishers are small-scale and that most of them live in developing countries.[1] In addition, hundreds of millions of people depend on fisheries for their livelihood throughout the value chain. Likewise, it is difficult to know how much small-scale fisheries produce in terms of catches. According to the Sea Around Us project, about one quarter of the world's catches originate from small-scale fisheries.[2] In all likelihood, the majority of small-scale fisheries catches is consumed in the fishing household or distributed to local markets, thus supporting local food security. Despite these estimates, most small-scale fisheries and communities are often not recognized or are overlooked in national, regional, and global decision- and policy-making processes. Despite the uncertainty about the actual figures due to inadequate statistical information, these approximations certainly imply that the small-scale fisheries sector is 'too big to ignore'.

Small-scale fisheries are now high on the research agenda, as championed by the global research network 'Too Big To Ignore' (TBTI),[3] which produced this book. The TBTI initiative and the work conducted by its members coincides with the development and implementation of the Voluntary Guidelines for Securing Sustainable Small-Scale Fisheries in the Context of Food Security and Poverty Eradication (SSF Guidelines; http://www.fao.org/fishery/ssf/guidelines/en), facilitated by FAO. The SSF Guidelines, which is a consensus document, resulted from extensive consultation with governmental bodies, small-scale fisheries through their organizations, civil society organizations (CSOs), practitioners, non-governmental organizations (NGOs), and other stakeholders, including the research community, culminating in intense negotiations by FAO member states. On June 9, 2014, the Committee of Fisheries (COFI) of FAO endorsed the SSF Guidelines, marking a historical moment for millions of small-scale fishing people around the world. Never before has this sector received such global recognition. Indeed, what the member states supported was remarkable.

[1] For more information refer to http://www.fao.org/fishery/ssf/people/en

[2] For more information refer to www.seaaroundus.org

[3] For more information refer to www.toobigtoignore.net

Expectations are now that the SSF Guidelines will make a big difference for small-scale fishing people around the world. Positive developments have already been observed in some places, but the full impact of the SSF Guidelines will undoubtedly take years, if not decades, to unfold. The voluntary nature of the SSF Guidelines implies that even if FAO member states have endorsed them, their implementation is not guaranteed. The SSF Guidelines call for major policy initiatives and governance reforms, which may involve legal and social innovation. The SSF Guidelines will not always meet fertile ground, as they befall in existing governing systems and their human rights and equity-based principles challenge and interfere with power relations. The more they challenge the *status quo*, the greater the likelihood that they will meet resistance, both at the governmental and large-scale industry level, especially when they call for reforms that involve the redistribution of resources and preferential treatment of small-scale fisheries. In some instances, small-scale fisheries may already be prominent on the political agenda, and the SSF Guidelines therefore will reinforce their status. In other instances, small-scale fisheries may have been forgotten, and great effort would therefore be needed in order to implement the SSF Guidelines.

Neither the worldwide stakeholder consultations nor the negotiations among state delegates about the SSF Guidelines were straightforward. Consequently, it would be naïve to assume that there will only be tailwind from now on. It would not be the first time that international agreements are shelved. The anticipated opposition on the home front may well have motivated the tough stance of some country delegates during the negotiations. Nevertheless, the SSF Guidelines are here to stay and it is to everyone's benefit that the conditions, factors, limitations, and opportunities for their implementation are closely examined, even on a case-by-case basis.

This book is about what lies ahead as far as the implementation of the SSF Guidelines is concerned. It aims to highlight challenges and opportunities as the SSF Guidelines land on the ground. How receptive are stakeholders to the SSF Guidelines? Will they agree with the many principles and propositions within the hundred paragraphs of the document? Will diverse stakeholders see their concerns addressed, their interests protected and their rights secured, or will some of them feel threatened? The SSF Guidelines stress the need to create a more level playing field, where small-scale fishing people have an active role in governance. However, as the SSF Guidelines emphasize the need for small-scale fisheries empowerment, it is easy to imagine that this will imply a zero-sum game, where empowerment will happen at the expense of the powerful stakeholders, who are not likely to remain passive.

Since it is too early to expect major changes to have happened on the ground with the recent adoption of the SSF Guidelines, it is nevertheless possible to juxtapose the SSF Guidelines with policies and governance systems that exist at various scales. How do, for instance, the basic principles of the SSF Guidelines match with those that already inform fisheries policy and governance in a particular country? In other words, is there a gap between the political reality and the social situation in small-scale fisheries in a given country in relation to the SSF Guidelines? If so, what explains it, and what policy and institutional reform will be needed to fill the gap?

How ready are governments, CSOs and other stakeholders to recognize and take ownership of the principles and perspectives that the SSF Guidelines advocate, such as the human rights based approach? These are examples of research questions which have inspired authors of this book.

It is premature to evaluate the success or failure of the SSF Guidelines. Such an evaluation must come at a later stage. The kind of transformation that the SSF Guidelines aspire to does not happen overnight, not only because of the institutional changes that must follow, but also because it requires a change of mindset among policy-makers and stakeholders. But it is already possible to address what challenges lie ahead and why they are there. By knowing the small-scale fisheries as they exist in a particular community, region, or country, and by knowing the institutional set up and governance mechanisms under which they operate, it is possible to perform an *ex ante* analysis of what the SSF Guidelines would imply in those contexts. This is also what the authors of the more than 30 case studies from around the world do in this book. In comparing what the SSF Guidelines set out to achieve and what it would require in terms of intervention and change in the system, the authors contribute their research experience to the implementation process.

The book is a product of the TBTI research cluster devoted to the study of the implementation of the SSF Guidelines. Small-scale fisheries have long been a focus of several academic disciplines which cover a broad range of issues. The stock of research-based knowledge is considerable, with a rich body of literature that has also informed the development of the SSF Guidelines. Thus, the SSF Guidelines invite researchers with an interest in, and a heart for, small-scale fisheries also to become engaged in the very process of realizing the governance principles of the SSF Guidelines, by exploring what difference they will make for the people they mean to serve, and how. It is hoped that this book will inspire the academic community to take initiative in studying how the SSF Guidelines will be received, and what impacts they will have at all levels of governance.

## The Relevance of the SSF Guidelines

Quoting from the document preface, the SSF Guidelines intend "to support the visibility, recognition and enhancement of the already important role of small-scale fisheries and to contribute to global and national efforts towards the eradication of hunger and poverty."[4] The expectation is that they will lead to policy change in the interest of current and future generations of small-scale fishers and fish workers and related activities. The SSF Guidelines will also:

> …be in support of national, regional and international initiatives for poverty alleviation and equitable social and economic development, for improving governance of fisheries and promoting sustainable resource utilization. Their objective is to provide advice and recommendations on implementation, establish principles and criteria, and information to assist

[4] For more information refer to http://www.fao.org/fishery/ssf/guidelines/en

States and stakeholders to achieve secure and sustainable small-scale fisheries and related livelihoods.[5]

As the full title indicates, the SSF Guidelines are voluntary. States may therefore choose whether to support or ignore them, either partially or entirely. However, the SSF Guidelines, as a consensus document resulted from an extensive, participatory, and transparent process, where small-scale fishers, their organizations, state governments and other stakeholders were involved. Even if states and CSOs cannot be held legally accountable, they can at least be held morally responsible for their operationalization. States shall also report to FAO what they have done to implement them. Thus, if not legally, they are formally accountable.

A challenge for the SSF Guidelines' implementation is the enormous diversity and complexity that characterize small-scale fisheries globally. They differ ecologically, organizationally, economically, culturally, and technologically, not just from one region to the next but often also from one type of fishery to another. They exhibit attributes that are often unique to a particular fishery or locality, and which must be taken into account when implementing the SSF Guidelines. Another important factor, which explains the broad focus of the SSF Guidelines, is that small-scale fisheries are rarely a distinct sector. They do not operate in isolation from the rest of the fishing industry, from other sectors or from society as a whole. Rather, small-scale fisheries are part of a larger social and ecological system – embedded as a 'system within systems', interwoven with economic, social, and cultural life in local communities.

The SSF Guidelines recognize that the well-being of people involved in small-scale fisheries relates more broadly to how they live and thrive in communities and how they are involved in decision-making on issues that affect them. Securing a healthy ecosystem is an important condition but only a step towards sustainable livelihoods and the general well-being of communities. This broad perspective is essential for the operationalization of the human rights-based approach that the SSF Guidelines promote, which includes, but is not limited to, the concept of fishing rights or tenure rights. Consequently, the SSF Guidelines also speak to other government departments that do not specialize in fisheries, as well as regional organizations, the private sector and CSOs, whose work impacts small-scale fisheries.

Small-scale fisheries must be understood in relation to their large-scale counterpart, as the two often interact. When there is conflict, keeping the two fisheries spatially apart is a solution but also a major challenge. The clash is not just between different ways of fishing and different economic rationalities; it is also a power relationship, where small-scale operators are generally the weaker party.

Small-scale fisheries form a complex system whose boundaries are permeable. What is happening inside small-scale fisheries is often due to what happens outside of them, which means that the problems that small-scale fisheries are confronted with are not necessarily of their own making. Instead, small-scale fisheries actors often find themselves at the receiving end of a string of causal factors, as when

[5] Ibid.

poverty in other sectors make people take up fishing and thereby contributes to more pressure on the resource base.

A prevalent image of small-scale fisheries is that they are traditional and thus lag behind in the modernization process. Therefore, in the long run, small-scale fisheries are bound to lose out, and become supplanted by more efficient, capital-intensive harvesting technologies of a larger scale. According to this view, the community-based owner-operator is a thing of the past; the take-over by big corporations is the future. This is regarded as a natural process, which should be left to run its own course. When looking at current trends in global fisheries, and in other sectors of society, it is easy to deem this as unavoidable. However, it would be erroneous not to consider the economic, social, and political drivers behind it. Governments are among these drivers, often providing economic incentives, such as subsidies that favor large-scale fisheries. Small-scale fisheries never received the same attention or support.

Small-scale fisheries technology and practice are often well adapted to the particular ecological and social circumstances within which they must operate, often developed through a long-term learning process of trial and error. One cannot therefore deem small-scale fisheries technology as outmoded, just by the look of it. Small-scale fisheries can also be sophisticated in the way they communicate, organize, and serve markets. They often have a complex technological dimension that is linked to the development of potential strategies that may help them run a more economically efficient fishery without increasing the scale of their operation and overexploiting resources. Thus, small-scale fisheries are not necessarily stuck in the past, but are part of dynamic value chains, undergoing change that the SSF Guidelines may help to spur. This corroborates why the SSF Guidelines emphasize the relevance of the contribution of small-scale fisheries to food security and poverty alleviation, and why they deserve attention not just for their problems, but also for their capabilities and potentials.

Small-scale fisheries do more than just provide society at large with a 'service'. They are important in and of themselves. Not only do millions of people depend on them for food, livelihoods, and well-being, small-scale fisheries also represent cultural heritage, a way of life, social cohesion, identity, and a lifestyle. They are not always 'an occupation of last resort', as they are frequently perceived, but provide an attractive livelihood and a meaningful and preferred lyfestyle.

The lack of a precise definition of what small-scale fisheries are in the SSF Guidelines is justified by the extreme diversity of small-scale fisheries globally. Their multi-faceted nature makes definitions complex and rich. The only way to define them is to employ 'thick description', as the case studies reported in this book provide. Since definitions of small-scale fisheries refer to particular situations for which they apply, they often differ from region to region. For this reason, small-scale fisheries go by different names from country to country, with terms such as inshore, coastal, artisanal, subsistence, small-boat, municipal, and community-based fisheries, to name some. Rather than providing a standard definition, the SSF Guidelines leave it to the countries to determine what small-scale fisheries are, in accordance with their own context. Thus, the implementation of the SSF Guidelines cannot follow a standard approach. Instead, the approach must be customized to the specific traits and circumstances of small-scale fisheries as they exist around the world.

Because of the emphasis on human rights and dignity, respect of cultures, non-discrimination, social justice, gender equality, and equity, the SSF Guidelines are universally applicable. They lead to the principles of 'good governance,' such as transparency, participation, and rule of law, which are all part of the SSF Guidelines' guiding principles, and therefore can reinforce other instruments and policies that governments have already embraced. Although still voluntary, by endorsing the SSF Guidelines, FAO member states have confirmed and at least morally committed themselves to implement them. How they exactly do that in their particular context of small-scale fisheries is a matter of empirical investigation for years to come.

## About This Book

It is always important to understand the context within which social events take place, as human life, and the communities and institutions they build, are always contextual. The SSF Guidelines are to be implemented in concrete situations, and must be sensitive to what limitations and opportunities exist. If not, they are likely to be ignored or resisted, as local stakeholders, be they government or fishers and fish workers, will not be able to see their relevance. Case studies are a well-suited method to illustrate this challenge. It is also important to stress that case studies are not simply in-depth, empirical descriptions of small-scale fisheries as they appear in concrete locations. They also serve to demonstrate the complexity of the challenges of implementing policy reforms, like those promoted by the SSF Guidelines. What is essential with case studies is what they are *a case of* because it allows us to draw general lessons from them that are beyond the particular case. Thus, as we learn about the case, we also learn about the issue, in this instance, about the implementation of the SSF Guidelines.

The book has 37 chapters, of which 32 are case studies (see Fig. 1.1). In addition to this first chapter, there is a chapter that lays out the human rights-based approach, which is foundational to the SSF Guidelines. This perspective is beyond the mere 'rights-based approach', which is basically about fishing and property rights, often associated with market-based approaches and privatization, something the SSF Guidelines do not promote as they are generally perceived to be detrimental to small-scale fisheries' communities and culture. The SSF Guidelines have a strong focus on customary rights and rights of tenure, without which small-scale fishing people would be insecure, but the human rights-based approach also includes social development and decent work, gender equality, and basic civil and political rights, which are as essential to small-scale fishing people as to anyone else. Chapter 3 summarizes what has been happening around the world since their endorsement in 2014. Many state governments and civil society organizations are currently in the process of actually implementing the SSF Guidelines, whereas others have yet to

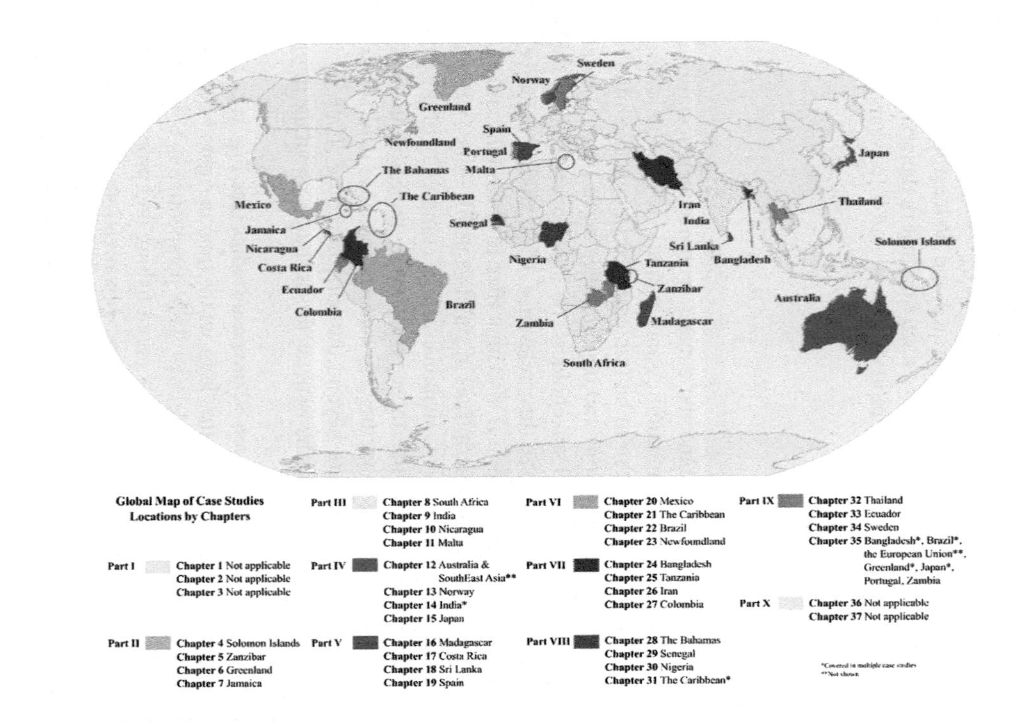


**Fig. 1.1** Global map of case study locations by chapter

act on them. The latter group may find inspiration in exploring what other countries and organizations are doing, as far as the SSF Guidelines are concerned. Chapters that follow will be introduced individually for the ten parts of the book. They all report on specific countries or fisheries and explore what challenges and responses the SSF Guidelines meet on the ground, and why the implementation process is still going slow in some instances. This is particularly the case in the chapters that form Part II of the book: *Politics of transformation.* In this part, the reader is introduced to developments in the Solomon Islands, Zanzibar, Greenland and Jamaica as the SSF Guidelines are about to be introduced there. The chapters are all about the extent to which there is coherence between current fisheries policies and the kind of developments that the SSF Guidelines are advocating. In some situations, there is considerable overlap, whereas in others the two are in conflict. The greater the distance between what is and what should be, according to the SSF Guidelines, the more fundamental reform will be required. The chapters demonstrate how the SSF Guidelines must relate to an existing social and ecological system, already governed within an institutional and political framework, which may or may not be conducive to their implementation. Integral to the existing institutional framework are rules about who has secure access to fisheries resources and how such entitlements are governed, and by whom. Therefore, the SSF Guidelines devote a whole section to this issue under the heading of 'Responsible Governance of Tenure'. States are encouraged to make sure that tenure rights and customary governance arrangements are recognized and respected. In Part III, *Securing tenure rights*, the four chapters all focus on this issue, starting with South Africa, and followed by India and Bay of Bengal lagoon, Nicaragua, and finally Malta. In all situations, customary tenure rights are under threat, and governments have a way to go in order to make them secure, even if there is a statutory or customary legal framework that is supposed to back them up. If governments are serious about supporting poor and marginalized small-scale fishing communities, securing their tenure rights would be an obvious entry point.

Appropriate management systems and practices would be required to make small-scale fisheries sustainable, as the resource base must remain healthy. In some instances, that effort would imply the restoration of damaged marine and inland ecosystems and fisheries, while in other instances mechanisms must be installed to safeguard their productivity. This is a challenge discussed by authors in Part IV: *Strengthening the resource base,* which takes the reader to Australia and Southeast Asia, Japan, India, and Norway. What kind of management institutions and approaches would serve small-scale fisheries? All authors have something to say about what would be needed to make small-scale fisheries sustainable while making sure that marine and inland ecosystems remain healthy and productive.

There are obvious reasons why small-scale fisheries are marginalized and vulnerable: small-scale fishing people often lack the capacities and capabilities for bringing them out of the trap they are in. For tenure rights to be secure, and management systems to work for small-scale fisheries, collective action and the empowerment of people are essential. Those two priorities are interlinked, but they do not always happen spontaneously. Sometimes they need initiatives led by actors who are centrally or externally situated, like CSOs. These issues are explored and illustrated in Part V of the book: *Empowerment and collective action*. Again, there are great geographical leaps being made, from Madagascar to Costa Rica to Sri Lanka to Spain. The chapters are all good illustrations of the importance of focusing on existing power-relations and how they can be rearranged through institutional value chain reforms, to benefit small-scale fisheries more than they currently do and as the SSF Guidelines suggest should be done.

Part VI, *Broadening participation*, is largely a follow-up on the previous theme. All chapters are situated in the Western hemisphere - Mexico, the Caribbean region, Brazil, and Canada - but could have included case studies from other parts of the world, as the issue is the same: how to involve people in decision-making that affects their lives and livelihoods. This is simply a good governance principle. As the SSF Guidelines emphasize, small-scale fisheries actors have a right to be heard, and their voices are just as important as other voices that are often louder and can be heard more easily by government. Yet participation also has functional merits. Organizations of fishers and other stakeholders within or outside the fishing industry may bring concerns, knowledge and interests into the decision-making process which could help to produce better and more legitimate outcomes. In this part of the book, the reader can obtain insights about what this can mean in practice.

Participation is also about building capacity for self-governance that makes small-scale fisheries communities more robust, resilient, and capable of pursuing both proactive and reactive strategies. This is the theme of Part VII, *Managing threats,* which speaks to the mitigation of risks to small-scale fisheries, which may be both internal and external to the community. Case studies from Bangladesh, Iran, Tanzania, and Colombia show that these risks can be different, ranging from natural hazards and climate change, the dissolution of the moral fabric of the community which erodes the ability to enforce locally observed rules, incompatible values leading to non-compliance of rules and regulations, to armed conflict and violence, which are all topics covered by the SSF Guidelines.

The SSF Guidelines stress the need to mobilize the capacity of small-scale fisheries by drawing on the local social and ecological knowledge that people possess, but which scientists and managers often regard as unreliable and anecdotal. However, it is obvious that treating people with respect, which is one of the guiding principles of the SSF Guidelines, is about acknowledging the relevance of people's experiences and knowledge. Research may help to support the capacities for self-governance, as data are usually in demand, but this research should be developed along with the integration of the knowledge that local people already have. Notably, it must not be introduced in a top-down fashion, as argued in chapters that are included in Part VIII of the book: *Building capacity*. This part makes similar visits around the world, from the Bahamas, to Senegal, Nigeria, and back to the Caribbean and their many island states, where CSOs and the academic community played an import role in the development of the SSF Guidelines.

As mentioned, one of the remarkable features of the SSF Guidelines is their solid foundation on human rights and other governance principles. They illustrate that poverty reduction and the end of marginalization require more than technical remedies. They also need sound ethical and normative underpinnings, which must be thought through and deliberated in public discourse, also by involving small-scale fishers and fishworkers. When this does not happen, small-scale fisheries actors feel alienated and excluded. People have their own values, norms, and principles that should inform the way fisheries governance is carried out and how the SSF Guidelines are implemented. This is the topic of Part IX, *Governing from principles,* which deals with the visions behind legislative reforms and policy paradigms in Thailand and Ecuador and Sweden, and the issue of gender relations and women rights in small-scale fisheries and the SSF Guidelines, as exemplified from multiple locations around the world.

The final section, *Moving forward*, again picks up the discussion about the human rights-based approach. Given its broad application and multiple dimensions, human rights thinking runs through the entire SSF Guidelines document. This approach requires implementing the SSF Guidelines in the full, which should also be the ambition and the test against which their effect should eventually be evaluated. It is also in this perspective that this book should be read. The final chapter summarizes how collectively we, the editors and authors of this book, see the challenges of implementing the SSF Guidelines and what futures lies ahead for people and communities who not only base their existence on these resources, but also form identity and relationships around them.

# Chapter 2
# A Human Rights-Based Approach to Securing Small-Scale Fisheries: A Quest for Development as Freedom

**Rolf Willmann, Nicole Franz, Carlos Fuentevilla, Thomas F. McInerney, and Lena Westlund**

**Abstract** Fishers, fishworker organizations, and supporting civil society organizations have played a critical role in the recent agreement by the international community of an international soft-law instrument that explicitly calls for the adoption of a human rights-based approach (HRBA) in small-scale fisheries development: the Voluntary Guidelines on Securing Sustainable Small-Scale Fisheries in the Context of Food Security and Poverty Eradication (SSF Guidelines). This chapter reviews some of the controversial views that have been expressed by social scientists about pursuing a HRBA in general and in small-scale fisheries specifically. While the experience of applying a HRBA in small-scale fisheries is still very limited, some concrete cases are presented where human rights advocacy and human rights law have helped fishing communities in defending their rights to livelihood, food, and culture.

---

The text of this chapter reflects the views of the authors and should not be attributed in any form to their current or former employer.

R. Willmann (✉)
Independent Expert, Kressbronn, Germany
e-mail: rolf.willmann@gmail.com

N. Franz
Fisheries and Aquaculture Department, Food and Agriculture Organization of the United Nations (FAO), Rome, Italy
e-mail: Nicole.Franz@fao.org

C. Fuentevilla
Food and Agriculture Organization of the United Nations (FAO) Sub-regional Office for the Caribbean, Bridgetown, Barbados
e-mail: carlos.fuentevilla@fao.org

T.F. McInerney
FAO Consultant, Rome, Italy
e-mail: tfmcinerney@me.com

L. Westlund
FAO Consultant, Rome, Italy

FAO Consultant, Stockholm, Sweden
e-mail: lena.m.westlund@telia.com



S. Jentoft et al. (eds.), *The Small-Scale Fisheries Guidelines*, MARE Publication Series 14, DOI 10.1007/978-3-319-55074-9_2

**Keywords** Small-Scale fisheries • Human rights-based approach • Neoliberal policy • Tenure rights

## Introduction

Since the mid-2000s, fishers, fishworker organizations, and supporting civil society organizations have demanded the strengthening of human rights principles and the adoption of a human rights based framework in small-scale fisheries development (Jaffner and Sunde 2006; ICSF 2007; FAO 2009; Sharma 2008, 2011; Allison 2011; Allison et al. 2011, 2012). They have played a leading role in shaping an agreement by the international community of an international soft-law instrument that explicitly calls for the adoption of a human rights-based approach (HRBA) in small-scale fisheries development: the Voluntary Guidelines on Securing Sustainable Small-Scale Fisheries in the Context of Food Security and Poverty Eradication (SSF Guidelines) (FAO 2015). Jentoft (2014), Franz et al. (2015), and Willmann et al. (2017, this volume Chap. 36), have highlighted the substantial new opportunities as well as challenges of implementing the SSF Guidelines. In this chapter, we review some of the reservations that have been expressed by social scientists about pursuing a HRBA. Critics make a number of points. First, they argue that human rights reflect Western ethical values that could undermine valuable traditional norms, cultural practices, and collective customary arrangements for resource management and benefit sharing. Second, they suggest that the emphasis on rights in a HRBA can be misused to promote the privatization of the commons as part of a neoliberal fisheries policy. Third, critics have pointed to the lack of empirical evidence supporting the efficacy of a HRBA in fisheries. We argue in this chapter that the SSF Guidelines are grounded in human rights principles that are universally shared and in a human development approach that seeks to expand fundamental freedoms.

Since the adoption of the Universal Declaration on Human Rights (UDHR) in 1948, a large body of international human rights law has been developed. The inclusion of human rights into international law and many national constitutions has been heralded as one of the greatest moral achievements of humanity. Since the turn of the millennium, international policy commitments and Plans of Action expressly include commitments to upholding international human rights (e.g. 2000 Millennium Declaration and Development Goals (UN General Assembly 2000), 2005 World Summit Outcome (UN General Assembly 2005), Rio+20 The Future We Want (UN General Assembly 2012), 2015 Sustainable Development Goals (UN General Assembly 2015)).

The widespread support for these policy statements at the highest level of global governance stands in contrast to the continued failure to implement them at the national level and the continuation of widespread abuses of human rights in many of the states that have formally ratified all or most international human rights conventions (Posner 2014). This discrepancy between words and deeds is one of the reasons why a number of legal, political and social scientists have expressed doubts about the reach and effectiveness of human rights. Another reason for a reserved, if not opposing position, is that some scholars argue that – intentionally or inadvertently – human rights rooted in Western moral philosophy constitute a form of cultural imperialism over nations and communities whose ethical values, culture, and political systems and institutions are not of Western origin (e.g. Brown 1997; Kennedy 2004; Uvin 2007; Mutua 2008; Langlois 2012; Golder 2014).

Whether there can be universality of human rights given the great diversity of human existences, traditions, cultures, and political settings continues to be the subject of academic and political debate (e.g. Donnelly 2007, 2013; Sen 2004; Ignatieff 2001). Controversies include the compatibility of universal human rights with cultural and religious traditions and values and the relative importance of individual versus societal well-being. Leaders of authoritarian regimes often argue that individual sacrifices for the common good and restrictions on political freedoms are needed in order to maintain societal peace and accelerate economic progress. This view has been challenged by economist, philosopher, and Nobel Laureate Amartya Sen in his widely cited article on 'Human Rights and Asian Values' (Sen 1997). He argues strongly that values underpinning fundamental freedoms have universal roots which can be traced through any culture. He also points out the great variations in culture within nations, across nations, and overtime. During long historical periods Western nations acted quite differently from their current ethical values (Sen 1997, 2004).

According to Donnelly "...human rights ideas and practices arose not from any deep Western cultural roots but from the social, economic, and political transformations of modernity" such as the French and American revolutions (Donnelly 2007, 287). Universal human rights, when properly understood, "...leave considerable space for national, regional, cultural particularity and other forms of diversity and relativity" (Donnelly 2007, 281).

The critique of the body of international human rights having a Western bias has been expressed from the time of the drafting of the Universal Declaration of Human Rights (UDHR). According to Glendon this is due to one of the most common and unfortunate misunderstandings of the UDHR. She emphasizes that the UDHR did not seek to impose a single model of right conduct but rather "...provide a common standard that can be brought to life in different cultures in a legitimate variety of ways (Glendon 2002, xviii)." The members of the drafting group had Chinese, Middle Eastern Christian, Marxist, Hindu, South and North American, and Islamic backgrounds and "... saw their task not as a simple ratification of Western convictions but as an attempt to delimit a range of moral universals from within their very different religious, political, ethnic, and philosophical backgrounds (Ignatieff 2001, 106)."

The 1993 Vienna Declaration on Human Rights states that the universal nature of human rights is 'beyond question' (UN General Assembly 1993). Notable are the large, and still increasing, number of ratifications of international human rights conventions. The ICESCR has been ratified by 164 states and the ICCPR by 168 states while the Convention on the Rights of the Child (CRC), has been ratified by 196 states (April 2016). Therefore, a large body of human rights has become international law and been incorporated into national law by many countries.

## Human Rights and Development

Human rights are not only consistent with economic development objectives but serve to facilitate their achievement. Human rights seek to guarantee freedoms that are fundamental for human development in procedural and outcome terms. Political freedoms in the form of free speech and democratic processes and institutions help to promote economic security. The rights to food, health, and education enhance human capabilities and economic participation. Human development in a human rights-based framework is a process of expanding the real freedoms that people enjoy, with economic growth constituting a means and not an end in itself (Sen 1999, 2009). The concern with the well-being of all people emphasizes equity as a major policy objective, requiring monitoring not only through national averages, but also via measures of deprivation and distribution (UNDP 2000; Fukuda-Parr 2003). The SSF Guidelines themselves recognize this HRBA emphasis on removing inequity and focusing development efforts on vulnerable and marginalized people. In contrast, neoliberal policy focuses on free market-led economic growth and efficiency.[1]

A HRBA is sensitive in several ways to the process of *how* certain human development outcomes can be achieved. Human rights establish limits to the losses that individuals can be made to bear, even where noble social goals are at stake. In fisheries, for example, fisheries management and conservation measures may be needed to restore depleted fish stocks and conserve ecosystems for the benefit of current and future generations. Such measures would often entail reduced fish harvesting opportunities in the short and medium term and could lead, if taken in isolation and without adequate consultation and safeguards, to a further marginalization and deprivation of small-scale fishing communities. A HRBA would help ensure that the

[1] The critique of neoliberal policy is not confined to its tendency to heighten maldistribution of incomes and wealth. Even in terms of strengthening economic growth, the view of Harvard economist Dani Rodrik is increasingly recognized among policy-makers. Based on thorough research he concluded that economic growth is not so much triggered by a long list of reforms as expressed in the so-called 'Washington Consensus' but often happens as a result of 'eclectic solutions' that combine the roles of the market and government and is often triggered by one or a few changes. Thus, there are many pathways to growth, not a single set of institutions and reforms that are necessary for growth (Loungani 2016).

burden of effort reduction does not fall disproportionately on poor and marginalized people. Fisheries management and conservation measures that have an impact on the livelihoods of fishery-dependent communities thus would need to be accompanied by adequate social safety measures and ensure that those communities have a say in developing diversified livelihoods.

A HRBA does not mean that the evaluative instruments of alternative development options are being ignored. On the contrary solid analyses remain of utmost importance to assess possible trade-offs such as, for example, between economic efficiency and employment or between resource conservation and the protection of the livelihoods of vulnerable sections of the population. A HRBA is not about an ideological battle but about development that prioritizes the betterment of marginalized and vulnerable people, those whose human rights remain denied or unfulfilled. It should be informed, inter alia, by practical approaches such as rigorous experimental methods that have recently been advanced in the struggle against poverty (e.g. Banerjee and Duflo 2011).

Assessments of human rights fulfillment should not only capture progress made but also assess the vulnerability and security of rights against potential threats (UNDP 2000). In the context of small-scale fisheries, social protection, occupational safety, and disaster risk management measures, for example, are critical because of the high vulnerability of the fishing occupation to accidents, disasters, and climate change impacts.

Human development in a human rights framework seeks to enhance certain functionings[2] and capabilities – i.e. the range of things a person can do and be in leading a life. Capabilities encompass basic freedoms such as the ability to avoid starvation and undernourishment, or to escape preventable morbidity or premature mortality. They also include the enabling opportunities provided by schooling or the liberty and the economic means to move freely and choose one's abode. Included are also important 'social' freedoms, such as the capability to participate in the life of the community, join in public discussion, participate in political decision making, and even the elementary ability 'to appear in public without shame,' a freedom whose importance was discussed by Adam Smith in The Wealth of Nations (quoted in Sen 1999; UNDP 2000). All these freedoms rarely apply as a matter of course in small-scale fisheries. Fishing communities are often 'outliers' of development with lower than average indicators of well-being, even in situations where pro-poor policies have had notable success, such as in the Indian state of Kerala (Kurien 1995).

When effectively applied to marginalized small-scale fishing communities, a HRBA strengthens a community's awareness of their rights and allows members to demand and claim these rights. However, awareness is not enough to ensure success. The ability to assert their rights and participate in decision-making is often compromised by weak organizations and representation of small-scale fishers and

[2] In Sen's capability approach, functionings are the states and activities constitutive of a person's being, i.e. being healthy, working in a good job, having self-respect, and being happy. Capability entails the freedom to achieve valuable functionings (Sen 1999).

fishworkers, frequently reinforced by low access to education in fishing communities.

A HRBA also enhances the ability and accountability of human rights duty-bearers. The main duty-bearer is the state and its institutions and agents, including all parts of government (e.g. ministries and departments, local authorities, courts, police, public health staff, and all those delivering a service on behalf of the state). In the case of small-scale fishing communities, duty-bearers are in the main fisheries agencies and entities such as ministries of education, health, public works, and the environment. While fisheries agencies might argue that human rights are beyond their scope of work, they maintain a critical responsibility because they are most knowledgeable about the sector and have, or should have, close day to day interaction with fishing communities (Sharma 2011). However, fisheries agencies themselves are not always able to influence policies in areas such as education, health, and public security given that they often have limited access and/or capacity to engage in intersectoral coordination and are generally weak in cabinet-level decision-making. As a consequence, fishery sector interests rarely receive the attention they deserve in national development policies.

This attention vacuum requires special efforts to enhance participation of fishing communities in the decision-making processes that impact their lives. These efforts might include targeted advocacy to increase fishing community political assertiveness and continued strengthening of sectoral organizations. On the other side, those with responsibility for fulfilling rights -including fisheries agencies – must recognize and learn how to respect those rights, and make sure they can satisfactorily exercise their duties as rights duty-bearers.

The duty to respect human rights is universal and not confined to the state and its agents: "All non-state actors including business enterprises related to or affecting small-scale fisheries have a responsibility to respect human rights. States should, however, regulate the scope of activities in relation to small-scale fisheries of non-state actors to ensure their compliance with international human rights standards" (FAO 2015, paragraph 3.1.1).[3]

Invoking of duties raises a bundle of related concerns, such as accountability, culpability, and responsibility. Accountability can be a powerful tool in seeking remedies. Accountability leads to an analysis of responsibilities of different actors when rights are violated or go unfulfilled (UNDP 2000). It will seek to establish, for example, why a certain government service such as basic health care or primary education is inadequate or not available at all in a fishing community, and who carries responsibility for such neglect.

[3]The UN Human Rights Council has endorsed in June 2011 the Guiding Principles on Business and Human Rights which should be implemented in a non-discriminatory manner, with particular attention to the rights and needs of … individuals from groups or populations that may be at heightened risk of becoming vulnerable or marginalized, and with due regard to the different risks that may be faced by women and men (UN 2011).

## Neoliberalism and Small-Scale Fisheries

Davis and Ruddle (2012) and Ruddle and Davis (2013) argue that co-management discourses and governance and human rights literature in respect to small-scale fisheries (e.g. Jentoft 1989; Pomeroy and Berkes 1997; Charles 2011; Chuenpagdee 2011; Allison 2011; Allison et al. 2011, 2012) ignore class structures, presuppose a benevolent state, and project a neoliberal market-dominated agenda onto small-scale fisheries thereby "betray[ing] resource harvesters by undermining family life and cultural systems and destroying the local social organization of production (Davis and Ruddle 2012, 244)." They hold that "... most new governance proposals would deepen the penetration of neoliberal values and, by so doing, further define and advance social class formation and differentiation in families and local societies (Davis and Ruddle 2012, 251)." They also argue that conventional co-management and human rights options rarely examine the concrete conditions on the ground, and therefore apply a generic or ideological approach that ignores "...the need for, the requirements of, and the methods by which to empower "voice" so as to achieve real and substantial powers enabling key aspects of the "fit" between local priorities and the attributes of resource governance (Ibid)."

In an attempt to accommodate the critique by Ruddle and Davis and shield small-scale fishing communities from a potentially damaging neoliberal approach and agenda, Song (2015) proposes human dignity as a nuanced alternative to a HRBA. Human dignity would be a less Europe-centric concept than a HRBA and could accommodate a wider spectrum of existing cultural norms and traditions with less emphasis on individual liberties and self-realization.

However, replacing the HRBA with the broader conception of human dignity would come with a significant drawback. The term human dignity does not provide a universalistic, principled basis for judicial decision-making because there is little common understanding of what human dignity requires substantively within or across jurisdictions (McCrudden 2008). Human rights, on the other hand, are conceptualized and defined in a number of international and national instruments. McCrudden acknowledges, however, that the context-specific meaning of dignity "...plays an important role in the development of human rights adjudication, *not in providing an agreed content to human rights* but in contributing to particular methods of human rights interpretation and adjudication (McCrudden 2008, 655 [emphasis added])."

Rather than promoting neoliberalism, a HRBA can be seen as a tool to stem the potentially negative influence of neoliberal policies.[4] Because of undesirable social

[4] Neoliberalism commonly refers to market-oriented reform policies such as removing price controls, deregulating capital markets, lowering trade barriers, and curtailing state influence on the economy, especially through privatization and austerity policies. In fisheries, neoliberal policies are often associated with the introduction of individual transferable quota management systems that establish quasi private property rights which can have a number of adverse impacts such as concentrating ownership, transfer of ownership to outsiders and investors or processing companies, blocking the entry of young fishers because of high quota prices, and others. A detailed

impacts that neoliberal policies can have (for example, growing income disparity or reduced social services), some scholars have argued that the parallel spread of neoliberalism and the discourse of human rights are antagonistic processes. This is due to a response to increasing needs as the welfare state is left behind (Speed 2007; Donnelly 2013) and to growing resistance by civil society organizations to the effects of economic globalization grounded in human rights discourse and doctrine. "Human rights has gone global not because it serves the interests of the powerful but primarily because it has advanced the interests of the powerless (Ignatieff 2001, 290)."

The use of advocacy and legal recourse to struggle for human rights realization has also been observed in small-scale fisheries. Individual fishers, fishworkers' organizations, and supporting NGOs have invoked human rights for food and livelihood claims and the human right of non-discrimination to defend their traditional access and use of fishery resources. In South Africa, for example, with the help of NGOs such as Masifundise Development Trust, a group of about 5000 small-scale fishers launched a class action in the High Court and in the Equality Court in Cape Town claiming that the Fisheries Minister had failed to provide them with just access to fishing rights, and sought an order giving them equitable access to marine resources. Fishers argued that the implementation of the Marine Living Resources Act (MLRA) of 1998 violated their right to food, a right that is protected in the South African Constitution of 1996. In 2007, the Minister issued a decision, granting traditional fishers the right to catch and sell West Coast Rock Lobster for commercial purposes. This decision followed an order by the Equality Court in May 2007-issued after an agreement among the parties-that small-scale fishers were entitled to some form of interim relief through fishing until the government had finalized its new subsistence fishing policy. Although the decision was not compatible with the long-term sustainable use of the resource, the Minister still authorized the interim relief, based on the argument that the fishers depended on the resource for their survival (Jaffer and Sunde 2006; Skonhoft and Gobena 2009).

Livelihood issues also figured in the case of Erlingur Sveinn Haraldsson and Irn Snaevar Sveinsson v. Iceland. Submitted to the UN Human Rights Committee under the Optional Protocol to the International Covenant on Civil and Political Rights (ICCPR), the case involved a challenge to the Icelandic fisheries management system which was based on individual transferable quotas.[5] Under this system, quotas were issued to regulate the catch of a variety of fish and shellfish. A portion of the permits were issued gratis to persons who had previously fished specified types of fish during a specified period. However, not all fishers received quotas and the petitioners did not receive that original quota. Those who received quotas were entitled

review of the influence of neoliberal policies in North American small-scale fisheries is provided by Pinkerton and Davis (2015) and other authors in the same issue of Marine Policy.

[5]Einarsson (2011) discusses in detail how the Icelandic ITQ policy allocated collective wealth unfairly to individuals and violated basic principles of human rights. According to him the de facto privatization of fishery resources in Iceland through the ITQ management regime was an important factor of exposing Iceland to the vagaries of international financial markets and the 2008 financial crisis that took a heavy toll on its economy.

to trade or sell them, whereby they effectively became private property. The basis of the petitioners' challenge was thus that the permit and quota system essentially gave the quota recipients public property for free and therefore discriminated against non-recipients, who would need to pay their fellow citizens to obtain quotas. In ruling in favor of the petitioners, the Committee found that the quota system constituted discrimination in violation of Article 26 of the ICCPR. Although the case turned particularly on the peculiarities of the quota system in question, the Committee implicitly recognized that practices which discriminate against persons' abilities to earn livelihoods are violations of human rights (U.N. Doc. CCPR/C/91/D/1306/2004 (2004)).

These two examples show the interplay and mutually supporting relationship between human rights protection and livelihood concerns, as contemplated in Sen's perspective of human development as freedom. In the South African case, the constitutional protection of the human right to food was instrumental in securing fishers' livelihoods on an interim basis and subsequently to the development of a dedicated policy for small-scale fisheries. In the Icelandic case, the design of the management system discriminated against people's abilities to earn livelihoods.

We argue that both the process by which the SSF Guidelines were developed and their contents are antithetical to the promotion of neoliberal policies in small-scale fisheries. The SSF Guidelines were developed through a wide-spread bottom-up process of consultation encompassing many different cultural, regional, and social settings. The HRBA was by no means imposed by governments or corporate entities onto fishing communities and their organizations. On the contrary, the small-scale fishers and fishworkers who participated in this process demanded it, recognizing its importance for safeguarding their livelihoods, securing their access to resources and services, promoting their values, and protecting their identity and culture.

Free or weakly regulated markets, expanding trade and the rapid penetration of IT and the internet into all spheres of life mark today's globalized world. Fisheries, both large-scale and small-scale, are caught in the midst of this increasingly interconnected and interdependent world and influenced by its ramifications in positive and negative ways. Among the positive sides of globalization are the ease of access to various kinds of valuable information on market and weather conditions, improved market access, the ability to network and organize for advocacy and collective action purposes, and to lobby for and benefit from greater transparency and accountability in policy-making at local and national levels.

Being a highly perishable food item that by its own is insufficient for a balanced diet (although highly nutritious and providing essential protein and micro-nutrients), fish has been a highly traded commodity for centuries if not millennia. With improved preservation and transportation methods, the variety of products and range of trade have greatly expanded during the last half century. In value terms, fish[6] has become the most internationally traded food item and is the leading foreign exchange earner among agricultural products. Expansion of international fish trade has given rise to concerns about whether fish exports might undermine local and

[6] For this chapter 'fish' includes all non-mammalian aquatic living organisms.

national food security and violate the human right to food. While available evidence does not allow for a definite statement on the impact of fish trade on food security and the right to food (FAO 2003; Kurien 2005; HLPE 2014; Béné et al. 2015), this concern is reflected in the SSF Guidelines: "States should ensure that promotion of international fish trade and export production do not adversely affect the nutritional needs of people for whom fish is critical to a nutritious diet, their health and well-being and for whom other comparable sources of food are not readily available or affordable (FAO 2015, paragraph 7.7)." Another concern is that fish trade may encourage overfishing which is also addressed in the SSF Guidelines: "States should ensure that effective fisheries management systems are in place to prevent overexploitation driven by market demand that can threaten the sustainability of fisheries resources, food security and nutrition. Such fisheries management systems should include responsible post-harvest practices, policies and actions to enable export income to benefit small-scale fishers and others in an equitable manner throughout the value chain (FAO 2015, paragraph 7.8)."

The SSF Guidelines thus explicitly support market-oriented – but not unfettered market-led – small-scale fisheries development. The state, private sector, and civil society organizations (CSOs) are called upon to avoid negative consequences for food security, nutrition, and resource sustainability so that small-scale fishers benefit from fish trade in an equitable and fair manner. This requires an active state to put in place safeguards against potentially damaging trade impacts and establish effective fisheries management and conservation systems. In case adverse impacts of international fish trade arise on the environment, small-scale fisheries culture, livelihoods and special needs related to food security, the SSF Guidelines call on states to adopt policies and procedures in consultation with concerned stakeholders to equitably address them (FAO 2015, paragraph 7.9) This provision is arguably fairly vague and could open the possibility of the state responding too late and in a manner that does not promptly curtail adverse trade impacts on small-scale fisheries. An active and vocal civil society is important in this regard.

The SSF Guidelines recognize that fishing and its associated activities in pre- and post-harvest are more than merely economic production activities. For many fishers and their communities they are a way of life characterized by a unique cultural richness comprising of traditions and norms anchored in local communities and supporting social cohesion (FAO 2015, 4). Where fishing becomes a dominant livelihood it usually extends to market-oriented production. As the fishery becomes more commercialized, market power may set in that make fishers dependent on merchants and their ability to advance credits on the requirement of the fisher to sell her/his fish at lower than the prevailing market price. The credit – indebtedness – price-taker mechanism so often observed in small-scale fisheries may then make the fisher perpetually dependent on the merchant and lead to a corresponding loss of autonomy and freedom to act.

Regarding markets and trade, Section 7 of the SSF Guidelines recognizes that market interactions are not necessarily equitable. In this regard, they require support for vulnerable and marginalized groups whose development is stunted due to unequal power relationships between value chain actors (FAO 2015, paragraph 7.1).

However, The SSF Guidelines stay silent on how such unequal power can be addressed. There are several successful examples of fishers getting collectively organized in cooperatives to ensure, for example, that auctions are run transparently and fairly. Examples include the South Indian Federation of Fisheries Societies (SIFFS) and the Japanese Fisheries Cooperatives. The experiences of fishworkers' organizations including fishery cooperatives from other parts of the world have been reviewed recently in a FAO workshop (FAO 2016).

## Sustainable Fisheries and Tenure Arrangements

Nowhere is the controversy among fisheries scientists as divided as on the question of the right course of action to achieve sustainable, efficient, and equitable fisheries. According to the Sustainable Development Goals (SDGs), a fishery would qualify as sustainable only if it is concurrently considered sustainable in environmental, economic, and social terms. To be economically sustainable in a market economy requires a minimum level of economic efficiency but not necessarily the maximization of economic rent which a fishery economist might postulate as the superior objective of a well-managed fishery. In environmental terms, if fisheries are considered purely in a fish stock specific manner, maintaining the size of the fish stock at a level that produces the maximum sustainable yield could be considered the most appropriate objective. Properly done, fishing at up to maximum sustainable yield allows nature to adjust to a new steady state, without compromising future harvests.[7] Not so, says Hilborn (2005), who considers this view as naive, because constancy is not an attribute of marine ecosystems. Instead, he argues for a sustainability concept that seeks to preserve intergenerational equity. This would acknowledge natural fluctuations and regard as unsustainable only practices which damage the genetic structure, destroy habitats, or deplete stock levels to an extent where rebuilding requires more than a single generation.

While there is no universally accepted definition of social sustainability, it is commonly associated with themes such as social justice and equity, community well-being and diversity, social security and support, human and labor rights, and social responsibility. The principles of the SSF Guidelines explicitly refer to economic, social, and environmental sustainability and to social responsibility. The Guidelines promote the application of the precautionary approach and risk management, encourage community solidarity and collective and corporate responsibility, and the fostering of an environment that promotes collaboration among stakeholders.

An alternative sustainability approach is to consider a biological, social, and economic system as a whole and assess concurrently the health of the human ecosystem

[7] To marry economic and environmental sustainability fishing at maximum economic yield is where fishery resource rent is maximized and where, theoretically in most cases, stock size is kept at a level above that which produces maximum sustainable yield.

as well as the marine ecosystem. The SSF Guidelines promote holistic and integrated approaches and recognize the ecosystem approach to fisheries (EAF) as a guiding principle (FAO 2015, paragraph 3.1). EAF is "to plan, develop and manage fisheries in a manner that addresses the multiple needs and desires of societies, without jeopardizing the options for future generations to benefit from the full range of goods and services provided by marine ecosystems (FAO 2003, 14)." It "strives to balance diverse societal objectives, by taking account of the knowledge and uncertainties of biotic, abiotic and human components of ecosystems and their interactions and applying an integrated approach to fisheries within ecologically meaningful boundaries" (Ibid).

Views differ about the primacy of environmental, economic, and social objectives in fisheries development and management. While ideally these should be attained concurrently and in a balanced fashion, environmental activists emphasize conservation of natural resources and biodiversity, economists economic efficiency and growth, and social scientists employment, equitable income distribution, and maintenance of cultural identity and diversity.

Well-developed and functioning democratic institutions and structures and market-based economies and social support systems that give effect to an extensive system of economic and social rights that remedy and compensate for inequitable outcomes of market transactions are the characteristics of societal arrangements that Donnelly calls liberalism, which is compatible with the model of the UDHR and whose leading practical example is the European welfare state (Donnelly 2013).

The tenor of the SSF Guidelines comprising the guiding principles and the thematic and procedural sections resonate well with this kind of UDHR compatible liberalism. They do not promote the privatization of fishing rights, reduction of state interference, or abolishment of customary rules. Instead, emphasis is given to the need to combine fisheries management with social and economic development, and promote equity and gender equality. The SSF Guidelines explicitly call on states to adopt measures to facilitate equitable access to fishery resources by small-scale fisheries operators, including, as appropriate, redistributive reform (FAO 2015, paragraph 5.8). Where gender equality would come in conflict with custom, the SSF Guidelines draw on the provision of the Voluntary Guidelines on the Responsible Governance of Tenure of Land, Fisheries and Forests in the Context of National Food Security (VG Tenure) (FAO 2012, paragraph 9.6): "Where constitutional or legal reforms strengthen the rights of women and place them in conflict with custom, all parties should cooperate to accommodate such changes in the customary tenure systems (FAO 2015, paragraph 5.4)." Thus the objective of equity and equitable resource access is central to the SSF Guidelines. Equity is also part of the SSF Guidelines' principles and thus in line with HRBA.

The SSF Guidelines envision the state as playing the main role in ensuring equitable and socially and culturally appropriate tenure rights to fishery resources (marine and inland) and small-scale fishing areas and adjacent land (5.1), granting of preferential access to fishery resources by small-scale fishers (paragraph 5.7), recognizing the role of small-scale fishing communities and indigenous peoples to restore, conserve, protect, and co-manage local aquatic and coastal ecosystems (5.5), recognizing and safeguarding publically owned resources that are collectively used and managed, in particular by small-scale fishing communities (5.6), facilitating, training, and supporting small-scale fishing communities to participate in and take responsibility for, …, the management of resources on which they depend for their well-being and that are traditionally used for their livelihoods (5.15), and many other things.

## *Collective and Communal Approaches to Tenure Rights*

The SSF Guidelines promote co-management as one of several participatory management systems (para 5.15) but not as a neoliberal means of rationalizing fisheries. The roles and responsibilities within co-management arrangements should be clarified and agreed through participatory and legally supported processes. The institutional arrangements should be such that "small-scale fisheries are represented in relevant local and national professional associations and fisheries bodies and actively take part in relevant decision-making and fisheries policy-making processes (FAO 2015, paragraph 5.17)." These provisions of the SSF Guidelines do not appear to leave much space for neoliberal policies.[8]

Widely dispersed and numerous small-scale fisheries are generally unsuited to management by centralized governance structures. This rules out a command and control based governance framework for the management of small-scale fisheries. More sensible are the various forms of co-management encompassing all spaces between a purely state-based management approach and a purely community-based self-governance approach. The conditions under which self-governance by resource users themselves can successfully and sustainably manage common pool resources such as fisheries have been well-researched and identified by various scholars (National Research Council 1986; Berkes et al. 1989; Ostrom 1990; Baland and Platteau 1996; Christy 2000; Kurien 2007).

Since the 1980s, there has been an upsurge of demands from small-scale fishing communities and their organizations and supporters worldwide for recognition of their customary and traditional fishing rights (Kurien and Willmann 2009; ICSF 2007). Community and group rights to inshore fisheries, often based on customary tenure rights, have led to successful systems of group and community-based management regimes (e.g. Berkes 1986; FAO 1993; Ruddle 1994; Christy 2000; Kurien 2000; Willmann 2000; Platteau and Seki 2001; Kuemlangan 2004; Platteau and Gaspart 2007; Townsend et al. 2008). A still significant impediment for such self-governance regimes is that current laws give little scope for conferring exclusive rights to communities. But with the increasing trend in decentralization of governance to local entities in many developing countries the scope for resource management by village groups is increasing.

Co-management systems rather than pure forms of self-governance are more typically found in small-scale fisheries. In these arrangements, the state (as well as

[8] Parallels can be drawn between the SSF Guidelines and the Rochdale Principles, a set of ideals for the operation of cooperatives first laid down in 1844 and on which, with two revisions in 1966 and 1995, co-operatives around the world continue to operate.

support organizations from civil society) and communities or groups of fishers come together in 'partnering' for fisheries management – taking a share of both the rights and responsibilities in these efforts (Jentoft 1989; Pinkerton 1989; Pomeroy and Williams 1994). These arrangements can often be more readily accommodated within existing legal frameworks and take into account customary norms and rules which may exist, and which have considerable social legitimacy (Wilson et al. 2003; Kurien and Willmann 2009). Co-management in combination with a HRBA that allows for meaningful and active participation by resource users and which is based on secure, equitable, and culturally appropriate tenure arrangements seems to be the best bet to avoid sacrificing the interests of small-scale fishers in favor of neoliberal agendas purely focused on economic efficiencies.

The notion of collective tenure rights is perfectly coherent and a valid legal proposition in a human rights-based conception. This is amply reflected in the frequent references to traditional and customary tenure systems – many of which are collective arrangements – in the VG Tenure and the SSF Guidelines. In the section on nature and scope, the VG Tenure explicitly include "… the governance of all forms of tenure, including public, private, communal, collective, indigenous and customary (VG Tenure para. 2.4)."

As is shown in cases involving natural resources and human rights, courts and international tribunals have shown consistent willingness to accept collective notions of tenure rights. A case decided by the Constitutional Court of Indonesia involved a challenge to the 2007 Management of Coastal Areas and Small Islands Act by a group of CSOs. Among other things, the law had authorized the granting of concessions for aquaculture, tourism, and mining in coastal waters and small islands. Among the concerns these concessions raised was the threat of privatizing existing customary rights of fishers, indigenous, and coastal communities. The coalition that brought the action included CSOs, leaders of fishworkers' organizations, academic experts, and representatives of artisanal fishers and indigenous communities. The Constitutional Court found the law to violate the rights of traditional communities. Specifically, it found that the state was required to consider pre-existing rights of traditional groups, whether collective or individual in nature. Such rights could not be revoked as long as the community existed. The Court also found the permitting process to impinge on the government's duty to promote the general welfare of all citizens because the procedures established were burdensome to traditional communities, which lacked working capital, technology, and knowledge of private sector actors. Coastal development strategies had only included the regional governments and the private sector, thus treating traditional communities unequally (Damanik 2013).

A number of indigenous peoples have used international human rights treaties to secure access to aquatic resources. Of these, perhaps the most successful and one that established international human rights law interpretative precedent was the decision of the Human Rights Committee on *Apirana Mahuika v. New Zealand*. The suitors, a group of Maori communities, claimed that the 1992 Treaty of Waitangi Settlement Act denied them their right to freely pursue their economic, social, and cultural development. In its decision, the Human Rights Committee established that the Treaty violated Article 27 of the International Covenant on Civil and Political Rights given that fisheries were an essential element of Maori culture. The Committee did not find the authorities in breach of the Convention given the large participatory process prior to the signing of the Treaty and the substantial support from the majority of Maori representatives. Nevertheless, the decision firmly established that fishing, when considered an essential element of the culture of the community, must be a guaranteed economic activity under the ICCPR (Smith and Dodson 2010).[9]

In the case of *Pueblo Indigena Kichwa de Sarayaku vs. Ecuador*,[10] a case that concerned land issues not fisheries, the Inter-American Court of Human Rights found that the Government of Ecuador had violated the rights of the Sarayaku indigenous community when it granted an oil concession without having consulted with or obtained their consent.[11] The significance of this case in the context of the SSF Guidelines is because in explaining the grounds of its decision the Court elaborated standards for proper consultations. These included the requirement that they be: (1) undertaken in good faith, (2) through culturally adequate procedures, (3) with the aim of reaching an agreement, and (4) prior, informed, and culturally appropriate. Further it held that such processes must be carried out by the state and not be delegated to third parties. The Court ruled that Ecuador had violated the right to prior informed consultation, community property rights, and the right to cultural identity.

Other indigenous peoples have also benefited from the application of human rights to achieve better security over fishing based livelihoods. The Canadian Mi'kmaq and Norwegian Saami (Davis and Jentoft 2001) as well as the Tagbanua peoples of the Philippines (Capistrano 2010) have made significant progress in terms of rights due to the national fisheries policy frameworks in their countries. While the cases and national policies cited vary in their specifics, they are all based on the recognition of fisheries as both an economic and social activity that should be protected on the basis of community and cultural identity. This is reflected in the UN Declaration on the Rights of Indigenous Peoples (UNDRIP) which provides that indigenous peoples are entitled to own, use, develop, and control the lands, territories, and resources that they possess by reason of traditional ownership, occupation, or use (UN General Assembly 2007). The UNDRIP principle of free, prior, and informed consent requires that any changes in use and access rights need the consent of the concerned indigenous peoples.

What these cases illustrate in particular is that judicial institutions in developed and developing countries and international and regional tribunals have come to see the rights of traditional and indigenous communities as deserving of particular attention. This is also recognized in the SSF Guidelines which specifically refer to the UNDRIP. On this emerging understanding, the relationship between basic legality, equality, and due process rights, considerations of livelihoods and traditional

[9] Text taken from Franz et al. 2015.

[10] Inter-American Court of Human Rights, 2012-6-27.

[11] Summary available at: http://www.escr-net.org/node/364959

cultural practices and development must be approached in a holistic fashion (McInerney 2013).

## Conclusions

The SSF Guidelines are based on internationally accepted human rights standards and are to be implemented in accordance with those standards. Their objectives are to be met through the promotion of a HRBA which offers an integrated, participatory, and holistic framework to securing sustainable small-scale fisheries. It does not replace other approaches to fisheries management and development but complements approaches such as EAF by adding a human rights perspective to the many challenges that small-scale fisheries face around the world including gradual loss of access to fisheries and land resources, environmental degradation, inadequate access to markets, social services and infrastructure, and persistent poverty of vulnerable and marginalized sections of fishing communities. A HRBA is about empowerment, recognition of rights and duties, accountability, transparency, and human dignity. It is not about privatization and fisheries reform efforts on the back of vulnerable fishing communities.

While there have already been successful cases of the use of human rights law and advocacy by fishers and their communities, especially in relation to resource access and the protection of their right to food and livelihoods, there is still very little documented experience in applying a HRBA to the sustainable development of small-scale fisheries. This innovative feature of the SSF Guidelines is an opportunity and a challenge. The opportunity is not just given by the now much greater global recognition of the economic, social, and cultural importance of small-scale fisheries and their critical roles in nutrition, food security, and poverty eradication, especially in developing countries. It is also given by the highly participatory way in which the SSF Guidelines have been developed. The participatory approach taken to the SSF Guidelines has given them legitimacy among fishers and their organizations and supporters and established them as an instrument of mobilization, advocacy, and policy guidance both globally and at national and local levels.

The challenge is to implement the SSF Guidelines and give practical meaning to their grounding in a human rights-based development perspective. This places demands on the way small-scale fisheries management and development is being planned, decisions are taken and implemented, and outcomes will be evaluated. These demands are generically spelled out in the SSF Guidelines' guiding principles and thematic areas. While their concrete requirements may vary with the specific context and situation of small-scale fisheries in a country or locality, they would improve not reduce access to social services such as education and health, respect cultures, traditional knowledge and practices, and existing forms of organization, and not undermine them, give special attention to marginalized and vulnerable sections of small-scale fishing communities, and not worsen inequity and inequality, and provide more secure rights to access and use of natural resources that fishing communities depend on, and not dispossess them of their means of livelihoods and cultural identity. Staff in governments, regional and international organizations, CSOs, fishworkers' organizations and others concerned with small-scale fisheries are called upon to prioritize the implementation of the SSF Guidelines in their work plans and activities and actively pursue a HRBA as this is the best way to expand fundamental freedoms of fishing communities, secure sustainable small-scale fisheries, and enhance their contribution to food security and poverty eradication.

## References


Allison, E. H. (2011). Should states and international organizations adopt a human rights approach to fisheries policy? *Maritime Studies, 10*(2), 95–116.

Allison, E. H., Åsgård, B., & Willmann, R. (2011). Human rights approaches to governing fisheries ed. *Maritime Studies, 10*(2), 5–13.

Allison, E. H., Ratner, B. D., Åsgård, B., Willmann, R., Pomeroy, R., & Kurien, J. (2012). Rights-based fisheries governance: From fishing rights to human rights. *Fish and Fisheries, 13*, 14–29.

Baland, J-M., & Platteau, J-P. (1996). Halting degradation of natural resources. Is there a role for local communities? Published for FAO by Clarendon Press, Oxford University.

Banerjee, V. A., & Duflo, E. (2011). Poor economics: A radical rethinking of the way to fight global poverty. Public Affairs.

Béné, C., Barange, M., Subasinghe, R., Pinstrup-Andersen, P., Merino, G., Hemre, G. I., & Williams, M. (2015). Feeding 9 billion by 2050 – Putting fish back on the menu. *Food Security, 7*, 261. doi:10.1007/s12571-015-0427-z.

Berkes, F. (1986). Local level management and the commons problem: A comparative study of Turkish coastal fisheries. *Marine Policy, 10*(3), 21–229.

Berkes, F., Feeny, D., McCay, B. J., & Acheson, J. M. (1989). The benefits of the commons. *Nature, 340*, 91–93.

Brown, C. (1997). Universal human rights: A critique. *The International Journal of Human Rights, 1*(2), 41–65.

Capistrano, R. C. G. (2010). Reclaiming the ancestral waters of indigenous peoples in the Philippines: The Tagbanua experience with fishing rights and indigenous rights. *Marine Policy, 34*, 453–460.

Charles, A. (2011). Small-scale fisheries: On rights, trade and subsidies. *Maritime Studies, 10*(2), 85–94.

Christy, F. C. (2000). Common property rights: An alternative to ITQs. Paper prepared for the FAO/Western Australia fish rights 99 conference on the use of property rights in fisheries management, Fremantle, Australia, 11–19 November 1999. In: Shotton, R. (Ed.), *Use of property rights in fisheries management*. Rome: Food and Agriculture Organization.

Chuenpagdee, R. (Ed.). (2011). *World small-scale fisheries: Contemporary visions*. Delft: Eburon Academic Publishers.

Damanik, R. M. (2013, May–August). An analysis of the constitutional court ruling on the annulment of the provisions on coastal water concessions (HP-3). *Indonesia Law Review, 2*, 163–172.

Davis, A., & Jentoft, S. (2001). The challenge and the promise of indigenous peoples' fishing rights – From dependency to agency. *Marine Policy, 25*, 223–237.

Davis, A., & Ruddle, K. (2012). Massaging the misery: Recent approaches to fisheries governance and the betrayal of small-scale fisheries. *Human Organization, 71*(3), 244–254.

Donnelly, J. (2007). The relative universality of human rights. *Human Rights Quarterly, 29*(2), 281–306.
Donnelly, J. (2013). *Universal human rights in theory and practice* (3d ed.). Ithaca: Cornell University Press.
Einarsson, N. (2011). Fisheries governance and social discourse in post-crisis Iceland: Responses to the UN human rights committee's views in case 1306/2004. *TheYearbook of Polar Law Online, 3*(1), 479–515.
FAO. (1993). *FAO/Japan expert consultation on the development of community-based coastal fishery management systems for Asia and the Pacific.* Papers presented at the FAO/Japan expert consultation on the development of community-based coastal fishery management systems for Asia and the Pacific. Kobe, Japan, 8–12 June 1992. FAO Fisheries Report. No. 474, Suppl., Vol. 1 and 2, Rome, FAO.
FAO. (2003). *The ecosystem approach to fisheries*. FAO technical guidelines for responsible fisheries No. 4, Suppl. 2. Rome, FAO.
FAO. (2009). *Report of the global conference on small-scale fisheries – Securing sustainable small-scale fisheries: Bringing together responsible fisheries and social development.* Bangkok, Thailand, 13–17 October 2008, No. 911. FAO. 2009.
FAO. (2012). *Voluntary guidelines on the responsible governance of tenure of land, fisheries and forests in the context of national food security*. Rome: FAO.
FAO. (2015). *Voluntary guidelines for securing sustainable small-scale fisheries in the context of food security and poverty eradication*. Rome: FAO.
FAO. (2016). Strengthening organizations and collective action in fisheries: Towards the formulation of a capacity development programme. In S. V. Siar, & D. C. Kalikoski (Eds.), *Workshop report and case studies*, 4–6 November 2014, Barbados. FAO Fisheries and Aquaculture Proceedings No. 41, Rome, Italy.
Franz, N., Fuentevilla, C., Westlund, L., & Willmann, R. (2015). A human rights-based approach to securing livelihoods depending on inland fisheries. In J. F. Craig (Ed.), *Freshwater fisheries ecology*. Chichester/Hoboken: Wiley-Blackwell.
Fukuda-Parr, S. (2003). The human development paradigm: Operationalizing Sen's ideas on capabilities. *Feminist Economics, 9*(2–3), 301–317.
Glendon, M. A. (2002). *A world made new. Eleanor Roosevelt and the universal declaration of human rights*. New York: Random House.
Golder, B. (2014). Beyond redemption? Problematizing the critique of human rights in contemporary international legal thought. *London Review of International Law, 2*(1), 77–114.
Hilborn, R. (2005). Are sustainable fisheries achievable? In E. A. Norse & L. B. Crowder (Eds.), *Marine conservation biology. The science of maintaining the sea's biodiversity*. Washington: Island Press.
HLPE. (2014). *Sustainable fisheries and aquaculture for food security and nutrition*. Committee on world food security (CFS), Rome 2014.
ICSF. (2007). Asserting rights, defining responsibilities: Perspectives from small-scale fishing communities on coastal and fisheries management in Asia. *Reports the proceedings of the Asian workshop and symposium held in Siem Reap, Cambodia*. Chennai.
Ignatieff, M. (2001). Human rights as politics and as idolatry. *The Tanner lectures on human values*. Delivered at Princeton University, April 4–7, 2000. http://tannerlectures.utah.edu/_documents/a-to-z/i/Ignatieff_01.pdf. Accessed 5 Apr 2016.
Jaffer, N., & Sunde, J. (2006). South Africa: Fisheries management – Fishing rights v. human rights? *SAMUDRA Report 44*, 83–86. ICSF, Chennai.
Jentoft, S. (1989). Fisheries co-management – Delegating government responsibility to fishermen's organisations. *Marine Policy, 13*(2), 137–154.
Jentoft, S. (2014). Walking the talk: Implementing the international voluntary guidelines for securing sustainable small-scale fisheries. *Maritime Studies, 13*, 16.
Kennedy, D. (2004). *The dark sides of virtue: Reassessing international humanitarianism*. Princeton: Princeton University Press.

Kuemlangan, B. (2004). *Creating legal space for community-based fisheries and customary marine tenure in the Pacific: Issues and opportunities* (Fish Code Review. No. 7 (En)). Rome: FAO.
Kurien, J. (1995). The Kerala model: Its central tendency and the outlier. *Social Scientist, 23*(1/3), 70–90.
Kurien, J. (2000). Community property rights: Re-establishing them for a secure future for small-scale fisheries. In Shotton, R. (Ed.), *Use of property rights in fisheries management. Proceedings of the FishRights99 conference*. Fremantle, Western Australia, 11–19 November 1999, pp. 288–294. Mini-course lectures and core conference presentations. FAO Fisheries Technical Paper. No. 404/1. Rome: FAO. 342p.
Kurien, J. (2005). *Responsible fish trade and food security* (FAO Fisheries Technical Paper No. 456). Rome: FAO.
Kurien, J. (2007). The blessing of the commons: Small-scale fisheries, community property rights, and coastal natural assets. In J. Boyce et al. (Eds.), *Reclaiming nature: Environmental justice and ecological restoration*. London/New York: Anthem Press.
Kurien, J., & Willmann, R. (2009). Special considerations for small-scale fisheries management in developing countries. In K. Cochrane & S. Garcia (Eds.), *A fishery manager's guidebook* (2nd ed.). Chichester/Ames: FAO and Wiley-Blackwell.
Langlois, A. J. (2012). Human rights in crisis? A critical polemic against polemical critics. *Journal of Human Rights, 11*(4), 558–570.
Loungani, P. (2016). Profile of Dani Rodrik: Rebel with a cause. *Finance & development*, June 2016, 2–5. Washington: International Monetary Fund.
McCrudden, C. (2008). Human dignity and judicial interpretation of human rights. *The European Journal of International Law, 19*(4), 655–724.
McInerney, T. F. (2013). *Report on addressing human rights and legal empowerment in small-scale fisheries*. Prepared for the UN Food and Agriculture Organization. Rome.
Mutua, M. (2008). *Human rights: A political and cultural critique*. Philadelphia: University of Pennsylvania Press.
National Research Council. (1986). *Proceedings of the conference on common property resource management*. Washington, DC: National Academy Press.
Ostrom, E. (1990). *Governing the commons: The evolution of institutions for collective action*. New York: Cambridge University Press.
Pinkerton, E. (Ed.). (1989). *Cooperative management of local fisheries: New directions for improved management and community development*. Vancouver: University of British Columbia Press.
Pinkerton, E., & Davis, R. (2015). Neoliberalism and the politics of enclosure in North American small-scale fisheries. *Marine Policy, 61*, 303–312.
Platteau, J.-P., & Gaspart, F. (2007). Heterogeneity and collective action for effort regulation: Lessons from the Senegalese small-scale fisheries. In J. M. Baland, P. Bardhan, & S. Bowles (Eds.), *Inequality, cooperation and environmental sustainability* (pp. 159–204). New York: Princeton University Press.
Platteau, J.-P., & Seki, E. (2001). Community arrangements to overcome market failures: Pooling groups in Japanese fisheries. In M. Aoki & Y. Hayami (Eds.), *Communities and markets in economic development* (pp. 344–402). New York: Oxford University Press.
Pomeroy, R. S., & Berkes, F. (1997). Two to tango: The role of government in fisheries co-management. *Marine Policy, 21*, 465–480.
Pomeroy, R., & Williams, M. (1994). Fisheries co-management and small-scale fisheries: A policy brief. International Centre for Living Aquatic Resource Management.
Posner, E. (2014, December 4). The case against human rights. *The Guardian*. https://www.theguardian.com/news/2014/dec/04/-sp-case-against-human-rights. Accessed 8 Sept 2016.
Ruddle, K. (1994). *A guide to the literature on traditional community-based fishery management in the Asia-Pacific tropics* (FAO Fisheries Circular. No. 869). Rome: FAO.
Ruddle, K., & Davis, A. (2013). Human rights and neo-liberalism in small-scale fisheries: Conjoined priorities and processes. *Marine Policy, 39*, 87–93.

Sen, A. K. (1997). Human rights and Asian values. *The New Republic*, July 14 & 21, 33–40.
Sen, A. K. (1999). *Development as freedom*. Oxford: Oxford University Press.
Sen, A. K. (2004). Elements of a theory of human rights. *Philosophy and Public Affairs, 32*(4), 315–356.
Sen, A. (2009). *The idea of justice*. The Belknap Press of Harvard University Press.
Sharma, S. (2008). *Securing economic, social and cultural rights of fishworkers and fishing communities*. Paper presented at the global conference on small-scale fisheries – Securing sustainable small-scale fisheries: Bringing together responsible fisheries and social development. Bangkok, Thailand, 13–17 October 2008.
Sharma, C. (2011). Securing economic, social and cultural rights of small-scale and artisanal fisherworkers and fishing communities. *Maritime Studies, 10*(2), 41–62.
Skonhoft, A., & Gobena, A. (2009). *Fisheries and the right to food. Implementing the right to food in national fisheries legislation*. Rome: FAO.
Smith, C., & Dodson, M. (2010). *Report on indigenous fishing rights in the seas with case studies from Australia and Norway*. Report prepared for the ninth session of the permanent forum on indigenous issues, New York, 19–30 April 2010. United Nations Economic and Social Council, document symbol E/C.19/2010/2.
Song, A. (2015). Human dignity: A fundamental guiding value for a human rights approach to fisheries? *Marine Policy, 61*, 164–170.
Speed, S. (2007). Exercising rights and reconfiguring resistance in Zapatista Juntas de Buen Gobierno. In M. Goodale & S. E. Merry (Eds.), *The practice of human rights: Tracking law between the global and the local*. Cambridge/New York: Cambridge University Press.
Townsend, R., Shotton, R., & Uchida, H. (Eds.). (2008). *Case studies in fisheries self-governance*. FAO Fisheries Technical Paper. No. 504. Rome: FAO.
UN. (2011). Office of the UN High Commissioner of human rights. *Guiding principles on business and human rights*.
UN General Assembly. (1993). *Vienna declaration and programme of action*, 12 July 1993, A/CONF.157/23.
UN General Assembly. (2000). *United Nations millennium declaration*. UNGA A/RES/55/.
UN General Assembly. (2005). *2005 world summit outcome*. UNGA A/RES/60/1
UN General Assembly. (2007). *United Nations declaration on the rights of indigenous peoples*. UNGA A/61/L.67 & Add.1
UN General Assembly. (2012). *The future we want*. UNGA A/RES/66/288.
UN General Assembly. (2015). *Transforming our world: The 2030 agenda for sustainable development*. A/RES/70/1.
UNDP. (2000). *Human development report. Human rights and human development*. Published for UNDP by Oxford University Press.
Uvin, P. (2007). From the right to development to the rights-based approach: How 'human rights' entered development. *Development in Practice, 17*(4-5), 597–606.
Willmann, R. (2000). Group and community-based fishing rights. In R. Shotton (Ed.), *Use of property rights in fisheries management. Proceedings of the FishRights99 conference* (pp. 51–57). Fremantle, Western Australia, 11–19 November 1999. FAO Fisheries Technical Paper. No. 404/1. Rome:FAO.
Willmann, R., Franz, N., Fuentevilla, C., McInerney, T., & Westlund, L. (2017). A human rights-based approach in small-scale fisheries – Evolution and challenges in implementation. In S. Jentoft, R. Chuenpagdee, M. J. Barragán-Paladines, & N. Franz (Eds.), *The small-scale fisheries guidelines – Global implementation*. Dordrecht: Springer.
Wilson, D. C., Nielsen, J. R., & Degnbol, P. (2003). *The fisheries co-management experience: Accomplishments, challenges and prospects* (Fish and Fisheries Series 26). Dordrecht/Boston: Kluwer Academic Publishers.

# Chapter 3
# The Small-Scale Fisheries Guidelines: A First Account of Developments Since Their Endorsement in 2014

**Nicole Franz and María José Barragán-Paladines**

**Abstract** The process for the implementation of the Voluntary Guidelines for Securing Sustainable Small-Scale Fisheries in the Context of Food Security and Poverty Eradication (SSF Guidelines) has unfolded since the very early stages if their development, even before their endorsement by the FAO Committee on Fisheries in June 2014. Their implementation, however, cannot be described as a linear-unidirectional process. On the contrary, there is abundant evidence for multi-directional, multi-scalar, and multi-temporal processes where individual and collective actions have been taken in order to achieve aims that are aligned with the SSF Guidelines. Globally, the process of implementing the SSF Guidelines has been discussed at various events which have detailed actions that should be taken towards their execution. Regionally and nationally, there have been various concrete illustrations about where and how the implementation process has started. The chapter is informed by an overview of key events and developments in relation to the implementation of the SSF Guidelines worldwide since 2014. It also draws on published and gray literature that has been made available, and that document events taking place globally, regionally or nationally, which are directly linked to the implementation of the SSF Guidelines. Examples are provided to illustrate concrete actions being taken to raise awareness about the existence and opportunities that the SSF Guidelines have to offer. Examples are also provided to demonstrate the importance of the institutionalization of their implementation as the active pull-and-push agent that can support this process in the long-term. Some key observations that emerged at the national, regional, and global scale are summarized as critical aspects to be taken into account, and to be further explored when addressing implementation in the future.

**Keywords** SSF Guidelines • Small-scale fisheries • FAO • COFI

---

N. Franz (✉)
Fisheries and Aquaculture Department, Food and Agriculture Organization of the United Nations (FAO), Rome, Italy
e-mail: Nicole.Franz@fao.org

M.J. Barragán-Paladines
Development and Knowledge Sociology Working Group, Leibniz Centre for Tropical Marine Research – ZMT, Bremen, Germany
e-mail: mariaj.barraganp@leibniz-ztm.de



S. Jentoft et al. (eds.), *The Small-Scale Fisheries Guidelines*, MARE Publication Series 14, DOI 10.1007/978-3-319-55074-9_3

## The Committee on Fisheries (COFI) and the SSF Guidelines

The year of 2014 marked a milestone event for small-scale fisheries worldwide. In June of that year, the 31st Session of the Committee on Fisheries (COFI) (FAO 2015a) of the Food and Agriculture Organization of the United Nations (FAO) endorsed the Voluntary Guidelines for Securing Sustainable Small-Scale Fisheries in the Context of Food Security and Poverty Eradication (SSF Guidelines).

COFI is the only global inter-governmental forum dealing specifically with fisheries and aquaculture. In addition to reviewing FAO's work on fisheries and aquaculture, COFI has the capacity to negotiate globally binding and voluntary instruments addressing governments, regional fisheries bodies, producer organizations, civil society organizations (CSOs), non-governmental organizations (NGOs), and the international community, as such. COFI membership is open to any FAO Member or non-Member State eligible to be an observer of the Organization. In fact, COFI observers include other representatives of the United Nations (UN) at large, including UN bodies and UN specialized agencies, regional fisheries bodies, CSOs, national and international NGOs who take part in the debate, but do not have the right to vote (FAO 2016c).

Despite the voluntary nature of the SSF Guidelines, the global scale of action at which COFI operates grants a special weight to the endorsement of this instrument. It confers a certain level of legitimacy and a moral obligation for compliance. The SSF Guidelines are the result of a long and participatory development process that took place between 2011 and 2014, which directly involved over 4000 stakeholders from 120 countries. The entire process generated a high degree of stakeholder ownership, in particular within the CSO community, as well as related expectations for action.

## 2014: Not an End, But a New Beginning for the SSF Guidelines

There is increasing recognition that a purely sectoral approach to small-scale fisheries has proven to be insufficient in addressing challenges which often lie outside the sector itself. The SSF Guidelines are the expression of this paradigm shift in how small-scale fisheries governance and development are addressed. The SSF Guidelines go beyond fisheries-specific issues, recognizing that small-scale fisheries do not exist in isolation. Indeed, they belong to highly complex systems embedded in specific historic, socio-economic, and institutional contexts that influence the behavior and capabilities of small-scale fisheries actors. In 2014, COFI 'noted the critical role of the adopted SSF Guidelines in improving the social, economic and cultural status of small-scale fisheries. It also reiterated the importance of the guiding principles of the SSF Guidelines, in particular the human-rights based approach' (FAO 2015j, 2).

The endorsement of the SSF Guidelines in 2014 was the end of the SSF Guidelines development phase, and represented at the same time the formal starting point of the even more important phase of implementation (Metzner and Franz 2015). This phase must be seen as a continuous process which began *de facto* even before the official endorsement of the SSF Guidelines. We claim that only when applied on the ground will the SSF Guidelines become an effective tool for change towards securing small-scale fisheries sustainability and viability.

As stressed during an FAO expert workshop in 2014, the SSF Guidelines are characterized by their holistic scope. This wide framework requires cross-sectoral collaboration and a balanced and equitable partnership approach to support their implementation. Throughout the SSF Guidelines development process, and in the 2014 FAO expert workshop, the following key actors were identified as critical for the implementation process (FAO 2015b):

*Governments,* who have a key responsibility for implementation at the national, local, and even regional level, including the incorporation of the SSF Guidelines principles into relevant policies, strategies, and actions – not only for fisheries but for overall socio-economic development. Political engagement at the highest level, together with investments in capacity and participatory decision-making processes, will be required in order to realize the new vision for small-scale fisheries.

*Fishers and fish workers*, through their organizations, are the main drivers of change and play a major role in 'bottom-up' processes. Collective action from the grass roots level is needed to ensure that small-scale fisheries are mainstreamed into relevant policies, strategies, and actions at the local, national, and regional level, and to ensure the implementation of these policies, strategies, and actions.

*Academia and research bodies, regional organizations, NGOs and other CSOs* should be at the interface of this dual 'top-down- bottom-up' strategy, with a function of connecting, supplementing, documenting, and strengthening the above-mentioned efforts.

During the SSF Guidelines development process itself, a number of key actions for implementation emerged as particularly relevant (FAO 2015c):

*Raising awareness* Efforts must be taken to raise awareness about the relevance of the small-scale fisheries sector: The SSF Guidelines can only be implemented if parties with the possibility to make a difference are aware of their existence and perceive that they relate to their area of interest and responsibility. Thus, there is a need for targeted communication materials, including implementation guides, translations into local languages, and promotion and awareness-raising activities.

*Strengthening the science-policy interface* There is need for a strengthened knowledge base that informs policy reforms and leads to increased integration of sustainable resource management with social and economic development within a human rights context. Good practices need to be identified and shared. Improved collaboration and exchange of experiences between relevant research initiatives is needed. Case studies, technical support, and assistance for reviews and revisions of policy and legal frameworks creating an enabling environment for SSF Guidelines implementation are also required.

*Empowering stakeholders* Organizational structures which ensure effective participation and fair representation in local, national, and regional processes, as well as effective public institutions, represent the key building blocks for continuous improvement towards secure and sustainable small-scale fisheries governance and development. Accordingly, capacity development should be the backbone of the implementation of the SSF Guidelines and is required at different levels, for different stakeholders, and with respect to different abilities (e.g. technical skills, organizations skills, and business development).

*Collaboration and monitoring* New and strengthened partnerships and experience sharing, as well as a monitoring system to track progress of the SSF implementation at the national, regional, and international levels, contribute to a more effective and coherent implementation of the SSF Guidelines.

FAO has been confirmed as a natural custodian of the SSF Guidelines implementation process. In that light, in 2015 FAO established an Umbrella Programme for the Promotion and Application of the SSF Guidelines based on the before-mentioned four key areas through which it collaborates with resource partners. In addition, the need for broader participation within facilitating mechanisms to guide the SSF Guidelines implementation has been confirmed by further consultations. In 2016, COFI recognized this need and welcomed the development by FAO of a global strategic framework on small-scale fisheries. This framework will serve as a mechanism that will further strengthen collaboration among the various stakeholders, facilitate sharing of experiences, and support monitoring, to the benefit of small-scale fishing communities all over the world, including in the Global North.

In fact, the strong sense of ownership developed by stakeholders during the SSF Guidelines development process has already borne fruit. During the 32nd session of the FAO COFI held in June 2016, the first signs of progress were reported of initial implementation of the SSF Guidelines at international, regional, and national scales. The main milestones of implementation are summarized in the following sections and a summary overview of the main recommendations from various events discussing the implementation of the SSF Guidelines is provided in Table 3.1.

## Mainstreaming the SSF Guidelines at the International Level

While small-scale fisheries are a sector that is mainly rooted in the local and national scales, there is a need to ensure an enabling policy environment and political will for small-scale fisheries at the highest international level as well. This requires that various relevant policy debates and stakeholders organize around broader thematic issues, from fisheries-specific issues to social development, post-harvest issues, food security and nutrition, poverty, and climate change.

**Table 3.1** Main recommendations from workshops and meetings discussing the implementation of the SSF Guidelines (varied sources)

| Issue addressed | Challenges/Strategies | Source |
|---|---|---|
| Governance of tenure in SSF and resources management (chapter 5 of SSF Guidelines) | Improve current arrangements for access to fishery resources for SSF | FAO (2015f) Summary conclusions and recommendations of the South East Asia Regional Consultation Workshop on the Implementation of the Voluntary Guidelines for Securing Sustainable Small-Scale Fisheries in the Context of Food Security and Poverty Eradication. Bali, Indonesia, 24–27 August 2015 |
| | Review existing tenure rights systems (for fisheries and land) to protect SSF | |
| | Follow an ecosystem approach to fisheries (EAF) and apply a human rights based approach (HRBA) | |
| | Ensure equitable participation of SSF in co-management and other initiatives and frameworks | |
| | Ensure that appropriate fora, including regional human rights and legal mechanisms, exist to address transboundary issues | |
| | Include SSF – and not only fisheries in general – in national and regional climate change adaptation and disaster risk management legislation and plans | |
| Social development, employment and decent work and gender equality (chapter 6 + 8 of SSF Guidelines) | Empower small-scale fishing communities through an integrated ecosystem/holistic approach for small-scale fisheries development | |
| | Address tensions generated by transboundary and trans-border issues to support an environment for small-scale fisheries communities that have decent work and living conditions | |
| | Enable access to education for all to achieve informed and educated coastal communities | |
| | Improve living and working conditions and social protection in small-scale fisheries to contribute to ensuring decent work in the region | |
| | Actively promote and realize gender equality and equity in small-scale fisheries through the development and implementation of gender-sensitive legal, regulatory and policy frameworks. | |
| | Ensure effective climate change adaptation, emergency response and disaster risk management in small-scale fisheries by including fisheries and fishing communities, including indigenous people, in related national policies and plans at all levels | |

(continued)

**Table 3.1** (continued)

| Issue addressed | Challenges/Strategies | Source |
|---|---|---|
| Value chains, post-harvest and trade (chapter 7 of the SSF Guidelines) | SSF meeting local food security and human development needs, participate as partners in domestic, regional and global value chains and get a fair share of the benefits | |
| | Reduction of fish losses and ensure quality of the product to increase fishers' income and support sustainable fisheries management | |
| | Develop a conducive policy and business environment to encourage investment in infrastructure appropriate to SSF: | |
| | Establish transparent market information systems for local and international market and trade, facilitate networking between SSF and end users, and promote better access to information | |
| | Organize SSF associations, facilitate their evolution and strengthening to encourage fair and inclusive environment, improve their bargaining positions through an inclusive legal framework, and promote community-based resource management combining local wisdom and scientific knowledge | |
| Governance of tenure in SSF and resources management (Chapter 5 of the SSF Guidelines) | Strength representation of small-scale fisheries actors, including women and marginalized groups, in decision making needs | FAO (2015h) Summary conclusions and recommendations of the South Asia FAO-BOBLME Regional Consultation on the Implementation of the Voluntary Guidelines for Securing Sustainable Small-Scale Fisheries in the Context of Food Security Colombo, Sri Lanka, November 2015 |
| | Initiate multi-tier platforms for joint management at the regional level for management of shared resources | |
| | Implement mechanisms for effective and meaningful consultations with communities, including consultative Committees | |
| | Promote the legalization of legitimate customary tenure rights, both to fishery resources and land | |
| | Identify, document and address Human Rights violations through collaboration including fisheries sector stakeholders and national human rights commissions | |
| Social development, employment and decent work and gender equality (Chapters 6 + 8 of the SSF Guidelines) | Develop capacities of small-scale fisheries stakeholders at all levels | |
| | Improve welfare schemes to address the high level of vulnerability of fishing communities often aggravated by the lack of (sector-specific) structures | |
| | Strength of effective community and/or fisherfolk organizations empowered for responsible fisheries Management | |
| | Promote equal access to opportunities and a safe and fair source of income, in particular for women and in inland fisheries | |

(continued)

**Table 3.1** (continued)

| Issue addressed | Challenges/Strategies | Source |
|---|---|---|
| Value chains, post-harvest and trade (Chapter 7 of the SSF Guidelines) | Raise awareness and enhance technical support programmes to assist women in setting up women's cooperatives/societies | |
| | Ensure fair distribution of benefits from fish trade | |
| | Searching for better return from fish and fishery products through support to post harvest infrastructures, processing technology and capacity development | |
| | Promote safety at sea to ensure the safety of small-scale fishers, the efficiency of their operations and the likeliness of fish to reach the market | |
| Governance of tenure in small-scale fisheries and resources management (Chapter 5 of the SSF Guidelines) | Provide regulatory framework to small-scale fisheries who lacks of specific small-scale fisheries areas, about preferential access rights | FAO (2015g) Summary conclusions and recommendations of the Near East and North Africa Regional Consultation Workshop: Towards the implementation of the Voluntary Guidelines for Securing Sustainable Small-Scale Fisheries in the context of Food Security and Poverty Eradication. Muscat, Oman, December 2015 |
| | Enhance participation of all relevant small-scale fisheries actors through the existing institutional frameworks which do not always enable their enhancement towards the achieve sustainable management | |
| | Provide/generate data and information necessary to support sustainable management of small-scale fisheries | |
| Social development, employment and decent work and gender equality (Chapters 6 + 8 of the SSF Guidelines) | Strength the organization of small-scale fisheries actors in the region, particularly of deprived categories, women and migrants, who lack of structures to actively participate in fisheries management and policies. | |
| | Provide access to social security protection to small-scale fishers and fish workers, in particular women and deprived groups | |
| | Promote safety at sea and other decent working conditions, including for women, which are currently insufficient in small-scale fisheries | |
| | Contribute with integrated approaches that reconcile environmental, social and economic development in order to enable small-scale fisheries to be a driver for development | |
| | Provide/improve access to education and professional development opportunities, in particular for children/women within the small-scale fisheries sector | |

(continued)

**Table 3.1** (continued)

| Issue addressed | Challenges/Strategies | Source |
|---|---|---|
| Value chains, post-harvest and trade (Chapter 7 of the SSF Guidelines) Climate change and disaster risks | Empower small-scale fishers actors to ripe more of the benefits of and income from the sales of their produce | |
| | Better understand and consideration of the links between trade (demand) and production | |
| | Strength the organizational structures of small-scale fisheries actors along the value chain to enhance their negotiating power | |
| | Improve the availability of trade related information facilitating the access to domestic, regional and international markets | |
| | Provide adequate conditions and controls to ensure the quality and prices of fishery products | |
| | Enhance investments for small-scale fisheries in appropriate infrastructures and equipment, marketing facilities, financial support | |
| | Include the context of food security and poverty eradication as essential | |
| | Consider the negative effects of climate change and disasters over the small-scale fisheries in the region | |
| Governance of Tenure in SSF and resources management (chapter 5 of the SSF Guidelines) | Secure tenure for small-scale fishing communities with regard to fishery resources and land needs to be ensured | (2015i) Conclusions of the East Africa Consultation Workshop on improving small-scale fisheries in the context of food security and poverty eradication FAO Sub-Regional Office for Eastern Africa, Addis Ababa, Ethiopia 15–18 September 2015 |
| | Protect existing zones and preferential access arrangements for small-scale fisheries need to be protected | |
| | Strength the capacity and organizations of small-scale fisheries actors | |
| | Harmonize policy frameworks and fishery regulations on shared water bodies and for shared fishery resources | |
| | Promote inter-ministerial collaboration – as well as coordination with other actors –on small-scale fisheries governance and development | |
| | Promote the usage of the ecosystem approach to fisheries (EAF) as a model for developing small-scale fisheries management | |

(continued)

**Table 3.1** (continued)

| Issue addressed | Challenges/Strategies | Source |
|---|---|---|
| Social development, employment and decent work and gender equality (chapter 6 + 8 of the SSF Guidelines) | Increase the access to amenities, facilities and services for small-scale fishing communities | |
| | Shift the current focus of fisheries management on the resource to a more people-focused approach to small-scale fisheries governance | |
| | Enhance the availability of financial services and insurance schemes for small-scale fisheries actors | |
| | Address the poor standard of living, lack of decent working conditions and discriminatory policies in small-scale fisheries | |
| | Efforts should be made to build entrepreneurial capacity for alternative and complementary livelihood opportunities to help reduce the vulnerability of small-scale fisheries actors | |
| Value chains, post-harvest and trade (chapter 7 of the SSF Guidelines) | Improve landing, processing and marketing infrastructures (including access roads) and enhanced data collection and information systems | |
| | Active involvement of fish value chain actors at decision making processes and representative fora | |
| | Fully engagement of women, vulnerable and marginalized groups in a dignified and respected manner | |
| | Improve infrastructures for small-scale fisheries, reduce post-harvest losses to a minimum and enhance added value. | |
| | Enabling regulations, guidelines and harmonized fish product quality standards | |
| Disaster risks and climate change (chapter 9 of the SSF Guidelines) | Leverage existing strategies to address climate change by small-scale fisheries actors within their countries, improve ability to access to funds and to insurance coverage for climate change adaptation | |
| 2nd World Congress for Small-Scale Fisheries (TBTI) | Consider SSF Guidelines as voluntary and necessary self-regulatory instrument which must be built on values and principles. Ensure their accessibility at national scale and within different cultural and indigenous settings | TBTI (2015). Too Big To Ignore Congress Report Number 08.1/2015 http://toobigtoignore.net/wp-content/uploads/2015/01/GAP-recommendations-on-SSF-Guidelines_Report-from-the-2WSFC-session.pdf |

(continued)

**Table 3.1** (continued)

| Issue addressed | Challenges/Strategies | Source |
|---|---|---|
| Policy communication | Improve information channels through National Workshops, legislation, policies and notifications about national and subnational fisheries, biodiversity, labour, coastal, marine and inland resource use as well as human development and rights in benefit of fishers, fishworkers and fishing communities. | ICSF (2016) National Workshop on Capacity-building for the Implementation of the Voluntary Guidelines for Securing Sustainable Small-scale Fisheries, March 2016, New Delhi |
| Implementing the FAO Voluntary Guidelines for Securing Sustainable Small-scale Fisheries in the Context of Food Security and Poverty Eradication (SSF Guidelines) | Improve resources governance by effectively implementing / providing appropriate laws. Recognize practical strategies as means to improve the fisheries performance (e.g., freezer on fishing vessels, monitoring of motors and fuel usage, etc.). Identify existing gaps in policy and action and bring them to the notice of his Ministry. Governance is a continuous process and the responsibility of drawing out the road map must also lay with the communities, and CSOs. Ensure the active participation of the government in the strategic paths | Report of the International Collective in Support of Fishworkers (ICSF)-Bay of Bengal Large Marine Ecosystem (BOBLME) India (East Coast) Workshop. Report prepared by Seema Shenoy International Collective in Support of Fishworkers www.icsf.net |

The SSF Guidelines have already been successfully mainstreamed into a number of instruments covering a wide range of topics developed through the Committee on World Food Security (CFS). These include the Security Framework for Action for Food Security and Nutrition in Protracted Crises (CFS-FFA) adopted in 2016, the Principles for Responsible Investment in Agriculture and Food Systems adopted in 2014, and policy recommendation on Water for Food Security and Nutrition. In addition, the debate spurred by the development of the SSF Guidelines also influenced the formulation of Target 14b of the United Nations Sustainable Development Goals (SDGs): "provide access for small-scale artisanal fishers to marine resources and markets" (SDG 2015).

The uptake of the SSF Guidelines is equally important within the fisheries sector itself. One encouraging development has been the production of *The Rome Declaration: Ten Steps to Responsible Inland Fisheries* document that was derived from contributions and interventions at the Global Conference on Inland Fisheries: Freshwater, Fish, and the Future, convened at FAO Headquarters in Rome, Italy, on 26–28 January 2015. This declaration specifically builds on the principles of the SSF Guidelines. Given that inland fisheries are almost entirely small-scale, this is an important signal to support the mainstreaming the SSF Guidelines in inland-fisheries related developments.

In addition to awareness-raising about and mainstreaming of the SSF Guidelines in global policy process, the empowerment of small-scale fisheries actors is another key aspect of implementation. In this context, the International Fund for Agricultural Development (IFAD) is an important player. The Fifth and Sixth Global Farmers' Forum, convened by IFAD in 2014 and 2016, respectively, included sessions dedicated to small-scale fisheries issues. IFAD has also initiated mainstreaming the SSF Guidelines in its fisheries projects and is providing funding support to the International Planning Committee (IPC) Fisheries Working Group[1] with a focus on facilitating the engagement and strengthening of small-scale fisheries actors and their organizations (NFF and ICSF 2011). This is an important contribution to enable CSOs to continue playing a major role also in the implementation of the SSF Guidelines. In fact, the commitment of CSOs, including fish worker organizations in particular, remains critical in order to ensure the uptake of the SSF Guidelines at all levels. Only if small-scale fishing communities themselves recognize the value of the SSF Guidelines and use them as a tool to call for and initiate positive change their goals can be achieved. The empowerment of small-scale fishing communities to know and realize their rights is at the heart of the SSF Guidelines themselves to enable them to contribute to food security and nutrition, poverty reduction, and sustainable resource use.

Shortly after the endorsement of the SSF Guidelines in June 2014, the International Collective in Support of Fish Workers (ICSF) organized a workshop for CSOs and fish worker organizations entitled 'Towards Socially Just and Sustainable Fisheries: Workshop on implementing the FAO Voluntary Guidelines for Securing Sustainable Small-Scale Fisheries in the Context of Food Security and Poverty Eradication (SSF Guidelines)' in July 2014 in Puducherry, India (ICSF 2014). This CSO event was followed by discussions on the SSF Guidelines implementation during the 6th General Assembly of World Forum of Fisher Peoples (WFFP) in South Africa in September 2014 (FFFP 2014). On that occasion, the General Assembly decided to further strengthen the cooperation between the WFFP and FAO and to strengthen the work of WFFP and WFFP members in regards to the SSF Guidelines.

Other stakeholder groups have also taken action to promote the application of the principles of the SSF Guidelines. Focusing on the research community, but with the declared intention to bridge the research community, fishing communities and policy makers, the 2nd World Small-Scale Fisheries Congress organized in Mérida, Mexico in October 2014, organized by the Too Big To Ignore (TBTI) global partnership network for small-scale fisheries research, discussed the implementation of the SSF Guidelines (TBTI 2015). The Congress confirmed that, even if the SSF Guidelines are voluntary, small-scale fishing communities consider them necessary regulations. The discussions confirmed that research can play a role in enabling

[1]The International Planning Committee for Food Sovereignty (IPC) Fisheries Working Group is composed of the World Forum of Fishers People (WFFP), the World Forum of Fish Harvesters and Fish Workers (WFF) and the International Collective in Support of Fishworkers (ICSF). Crocevia operates the Rome based IPC secretariat.

access to the SSF Guidelines in various cultural and national contexts to support implementation. Additional initiatives have been taking place under the TBTI umbrella. One of them is the TBTI-monthly webinar series, which focuses on specific topics about small-scale fisheries worldwide, addressed from different perspectives. The webinar conducted on March 1st, 2016 was specifically dedicated to the SSF Guidelines implementation. Another interesting initiative is the Learning Circles for small-scale fisheries, a virtual venue where specialists in small-scale fisheries-related issues have participated in a discussion panel, covering a diverse scope of thematic disciplines.[2]

In addition to the above mentioned events aiming at raising awareness about the SSF Guidelines and initiating discussions about implementation, a number of events with a more thematic focus have taken place. For example, as reported to COFI in June 2016 (FAO 2016a), the meeting of FAO and indigenous peoples on Indigenous Food Systems, Agroecology and the Voluntary Guidelines on Tenure held in Rome, Italy, in February 2015, developed recommendations in relation to the implementation of the SSF Guidelines, including on the need for capacity development for indigenous peoples and dissemination of materials at the local level (FAO 2015e). As a follow-up, FAO is currently collaborating with the *Fondo para el Desarrollo de los Pueblos Indígenas de América Latina y El Caribe* to deliver a capacity development training on the use of the SSF Guidelines to fisheries-dependent Indigenous Peoples representatives in Central America in April 2017.

At the IUCN World Parks Congress in Sydney, Australia, held in November 2014, a side event was held entitled 'Connecting the dots: Marine Protected Areas (MPAs) and sustainable small-scale fisheries', which explored the importance of the SSF Guidelines in the context of MPAs. This discussion allowed to highlight the importance of involving small-scale fishing communities in the design and management of such areas to ensure their sustainability without compromising the livelihoods of coastal communities which depend on aquatic resources.

In addition, the SSF Guidelines are complementary to, and supportive of, national, regional, and international initiatives that address human rights, responsible fisheries, and sustainable development. These initiatives include the Code of Conduct for Responsible Fisheries, the Voluntary Guidelines to support the Progressive Realization of the Right to Adequate Food in the Context of National Food Security, and the Voluntary Guidelines on the Responsible Governance of Tenure of Land, Fisheries, and Forests in the Context of National Food Security (Tenure Guidelines). While these linkages are usually pointed out at awareness-raising events, the complementary nature of these instruments and their concurrent implementation, where and whenever possible and appropriate, remains an important operational dimension to develop to enhance their implementation. One example for awareness raising efforts in this direction are given by a side event on 'Human rights, food security and nutrition and small-scale fisheries' organized during the 43rd Session of the Committee on Global Food Security in October 2016[3] in which

---

[2] http://www.marineresourcecentre.ca/small-scale-fisheries-learning-circles-project/

[3] For more information: www.fao.org/cfs/cfs-home/plenary/cfs43/side-events/15/en/

the UN Office of the High Commissioner for Human Rights joined forces with FAO, the International Fund for Agricultural Development (IFAD), and the International Planning Committee (IPC)- Fisheries Working Group to explain the fundamental links between these issues.

Within the context of the before-mentioned Umbrella Programme, FAO organized a multi-stakeholder workshop in October 2016 on exploring the human rights-based approach in the context of implementation and monitoring of the SSF Guidelines. The purpose of the event was to explore what the human rights-based approach (HRBA) means within the context of small-scale fisheries in general and the thematic areas covered by the SSF Guidelines in particular; to discuss what HRBA entails in terms of the conduct of the various state and non-state actors to whom the SSF Guidelines are addressed; and to better understand the needs of different stakeholders in various policy areas with a view to develop guidance materials for the application of HRBA in implementing and monitoring the SSF Guidelines. The workshop confirmed that this approach still has to be further understood in the context of small-scale fisheries, including through the collection and documentation of more evidence and good practices. Related to this, in November 2016, FAO will hold an expert workshop on gender-equitable small-scale fisheries which will inform the preparation of a practical guide on the subject in support of the SSF Guidelines implementation.

## Taking the SSF Guidelines to the Regional Level

Many fisheries, both in marine and inland waters, cope with boundaries, borders, and frontiers which conflate different scales of complexity. At a regional scale, the management organizations and institutions operating at this level therefore play a crucial role in fisheries management and development. Many regional organizations that have actively participated in the SSF Guidelines development continue to engage in the implementation process. In fact, their early involvement in the process greatly facilitated the integration of this instrument into regional policies, strategies, and initiatives, as illustrated by the following examples.

In November 2013, the General Fisheries Commission for the Mediterranean (GFCM) together with partners organized the First Regional Symposium on Sustainable Small-Scale Fisheries in the Mediterranean and Black Sea in Malta. This symposium included a session on ‘Setting up a regional platform to promote the implementation of SSF Guidelines’. During this session, members of the Maghreb Platform of Artisanal Fishing – established with FAO support – joined forces with MedArtNet–a small-scale fisheries platform established with support from the World Wildlife Fund in order to enhance their collaboration regarding small-scale fisheries development in the Mediterranean region. The symposium also informed the development of the first regional program on small-scale fisheries for the GFCM region (FAO 2015d).

As a follow-up to this symposium, the GFCM co-organized a Regional Conference on 'Building a future for sustainable small-scale fisheries in the Mediterranean and the Black Sea' in Algeria in March 2016 in collaboration with partners. One of the outputs of this conference was the recommendation to the GFCM to establish a permanent GFCM working group on small-scale fisheries in order to facilitate the implementation of the SSF Guidelines in this region. This recommendation was taken up during the 40th Session of the GFCM (FAO 2016f) held in Malta in May/June 2016. In the same session, a resolution in support of small-scale fisheries and a proposal for General Fisheries Commission for the Mediterranean (GFCM 2016) mid-term strategy towards the fisheries' sustainability of the Mediterranean and the Black Sea were also adopted. Both refer specifically to the support of the implementation of the SSF Guidelines in the GFCM region.

A second example of the early integration of the SSF Guidelines into regional strategies is the African Union Commission and NEPAD Planning and Coordinating Agency Policy Framework and Reform Strategy for Fisheries and Aquaculture in Africa (AUC-NEPAD 2014) which was adopted at ministerial level in May 2014. It could be argued that this effort at integration even preceded the final endorsement of the SSF Guidelines. The expected outcomes of this regional strategy include that "Provisions of the FAO led International Guidelines for Securing Sustainable Small-scale Fisheries be widely applied across Member States".

In this context, a consultative Think Tank meeting on enhancing the governance of small-scale fisheries in Africa was jointly organized by the African Union-Inter-African Bureau for Animal Resources (AU-IBAR), the New Partnership for Africa's Development (NEPAD)-Planning and Coordinating Agency (NPCA), and the Government of Senegal in Senegal in January 2016. This consultative meeting produced important results to support the implementation of this pan-African Policy Framework and Reform Strategy, including the collection and promotion of lessons and best practices on governance of small-scale fisheries in Africa, the identification of practical constraints (institutional and technical) and possible solutions for ecosystem-based approaches and co-management in small-scale fisheries in Africa, as well as related priority actions to improve/enhance the contribution of small-scale fisheries to the agricultural transformation agenda of the African Union. Another significant African event was the meeting of the Fishery Committee for the Eastern Central Atlantic (CECAF) held in Tenerife in October 2015, which revived its artisanal fisheries working group through the adoption of revised terms of reference where the SSF Guidelines are explicitly mentioned (FAO 2016b). Another example for 'earlier adopters' include the Western Central Atlantic Fishery Commission (WECAFC) resolution WECAFC/15/2014/8 on Promoting the implementation of the SSF Guidelines and the Tenure Guidelines which was adopted in 2014.

In Central America, the Central America Fisheries and Aquaculture Organization (OSPESCA) adopted a new strategy for 2015–25 which explicitly refers to the SSF Guidelines (OSPESCA 2015). In June 2016, OSPESCA, along with the regional fish workers' organization CONFEPESCA, organized a regional workshop on the implementation of the SSF Guidelines in Nicaragua. During that event, OSPESCA agreed to establish a working group to support the implementation of the SSF

Guidelines in the Central American context and to establish a protocol of intent with FAO to collaborate on the SSF Guidelines' implementation in this region.

In the Pacific, the 9th Heads of Fisheries Meeting of the Secretariat of the Pacific Community (SPC), which took place in New Caledonia in March 2015, welcomed the endorsement by COFI of the SSF Guidelines and recognized the high degree of concordance with 'A New Song for coastal fisheries – pathways to change', also known as the Noumea Strategy (SPC 2015).

In Asia, the Southeast Asian Fisheries Development Center (SEAFDEC) supported the organization of the Southeast Asia Regional Consultation Workshop on the Implementation of the SSF Guidelines in Bali, Indonesia, in August 2015 (FAO 2015f). Following this workshop, SEAFDEC organized a Regional Technical Consultation on a Regional Approach to the Implementation of the SSF Guidelines in June 2016 in Bangkok. Ultimately, this process will be integrated into the related initiatives of the Association of South-East Asian Nations (ASEAN). The Consultation developed specific recommendations for both marine and inland fisheries, including provisions in relation to the thematic issues included in Part 2 of the SSF Guidelines. The recommendations included, among others, the establishment of regional CSO networks and inventories of fishers and fish worker organizations, gender-sensitive policy planning, and capacity-building for the adaptation to climate change and disaster preparedness.

In 2015 and 2016, together with partners, FAO organized four other regional workshops to discuss the implementation of the SSF Guidelines, namely for Eastern Africa (in Ethiopia in September 2015), Latin America (in Peru in September 2015), South Asia (in Sri Lanka in November 2015), and the Near East and North Africa region (in Oman in December 2015). Key summary recommendations of these workshops are provided in Table 3.1. Another regional workshop for the Southern Africa Development Community (SADC) countries is planned for December 2016 in Mauritius.

In relation to CSO-led events it is worth mentioning World Fisheries Day on 21 November. In Africa it was celebrated in 2015 in Morocco (ICSF 2016), and it will take place in Togo in 2016 under the lead of the African Confederation of Artisanal Fisheries Professional Organizations (CAOPA). In both cases, the SSF Guidelines implementation is part of the celebrations agenda. In relation to regional research community initiatives, TBTI organized the First Symposium for Small-Scale Fisheries and Global Linkages at the European scale through its Europe-based Working Group in Tenerife, Canary Islands, in June–July 2016. One of the main aspects addressed by the different panels and round tables was the SSF Guidelines' implementation at the national and regional scale.

Furthermore, the TBTI symposium on small-scale fisheries in the Asia-Pacific region, held on 7–9 August 2016 in Thailand, focused on inland fisheries and on fish as food, with the stated intention to provide an opportunity to develop a research agenda and capacity development program for promoting sustainable small-scale fisheries and for the implementation of the SFF Guidelines in the region and elsewhere. The symposium identified a few issues to follow up on in this regard: (1) hidden or under-valued fisheries; (2) privatization of the commons in inland fisheries; (3) ecosystem indicators; (4) comparison of marginalization in inland and marine fisheries; (5) migration; and (6) how to include inland fisheries into water management discussions.

## Implementing the SSF Guidelines Nationally: Back to the Roots to Achieve Sustainable Small-Scale Fisheries

The national level is the most crucial level for the implementation of the SSF Guidelines. This scale is where change will ultimately have to take place. A number of countries have taken steps to implement the SSF Guidelines. The following sections provide some examples of these initiatives.

- Since 2012, FAO has been assisting the fisheries administration in Cambodia in relation to small-scale fisheries development. This support includes the assessment of the legal realm, the policy instruments, and the institutional framework as well as the strengthening of the community fisheries in the context of the implementation of the SSF Guidelines.
- In Algeria, the national fisheries sector strategy launched in 2014 includes a 'Chartre' for sustainable fisheries and aquaculture which specifically refers to the SSF Guidelines. Steps have been taken since to advance the implementation of this instrument through the implementation of the national strategy.
- In Mauritania, the NGO Mauritania 2000 organized a national workshop to discuss the implementation of the SSF Guidelines in March 2015.
- In August 2015, Costa Rica enacted an executive decree about the official application of the SSF Guidelines, which is being operationalized in the country with FAO support.
- Since 2015, Indonesia has been developing a national plan of action for small-scale fisheries.
- In Sierra Leone in 2015, FAO provided support regarding governance of tenure within the context of the Voluntary Guidelines for the Responsible Governance of Tenure of Land, Fisheries and Forests in the Context of National Food Security (VGGT), including for small-scale fisheries, as called for under Chapter 5 of the SSF Guidelines (FAO 2016d, 2016e; ICSF et al. 2016).
- South Africa has requested FAO collaboration on, among other things, the implementation of a national small-scale fisheries policy with a view to ensure its consistency with the SSF Guidelines principles and provisions. This initiative will be supported in 2016.
- Members of the IPC Fisheries Working Group have organized workshops on the implementation of the SSF Guidelines in India, Thailand, and Myanmar in the context of the Bay of Bengal Large Marine Ecosystem (BOBLME) project in 2015. This working group has continued this process of taking back the SSF Guidelines to the national level by developing national workshops supported through an IFAD grant in: Brazil (June 2016), India (March 2015 and 2016), Pakistan (August 2016), Tanzania (August/September 2016), and Ecuador (September 2016) (MAGAP 2016). A workshop in Myanmar has also been planned (ICSF 2014, 2015; ICSF et al. 2016).

## Conclusion

Translating the principles of the SSF Guidelines into concrete action is critical in securing sustainable small-scale fisheries. Progress made in this regard since 2014 is encouraging, but much more remains to be done. As illustrated in this chapter, efforts thus far have concentrated primarily on raising awareness and bringing different players together to initiate discussions on how to apply the SSF Guidelines in various contexts. In some cases, these efforts also included the identification of priorities, potential actions, and related roles and responsibilities.

There are two types of key actors in this context: governments and small-scale fishing communities and their representations at various levels. The latter have demonstrated their role as a driving force in the promotion of the SSF Guidelines. Without the commitment, active engagement, and collaboration of these two groups, the implementation of the SSF Guidelines in their true spirit is impossible. On the other hand, other actors, including FAO and other UN organizations, regional organizations, and academia/research, continue to play a key role in catalyzing, supporting, and connecting the efforts of the two main actors. The further strengthening of the collaboration among the different actors in full respect of the guiding principles of the SSF Guidelines is therefore crucial to ensure that the awareness and implementation planning processes will bear fruits in the form of concrete change at national level in the future. This will require further opportunities for dialogue and exchange to follow up on the implementation processes which have started, while also initiating others.

The importance of institutions in implementation is also a major need emphasized in this chapter, illustrated through the examples of initiatives at global, national, and regional levels. Thus, the institutionalization of the SSF Guidelines' implementation in national, regional, and international policies, initiatives, and, ideally, legal frameworks is thus critical as a pull-and-push agent to support this process in the long-term. Institutions can also serve as the engine that mobilizes the actors towards the successful achievement of the common aim of small-scale fisheries sustainability.

The uptake of the SSF Guidelines in a number of global and regional policy processes, initiatives, and strategies is a key achievement which contributes towards an enabling environment for their implementation. Regional organizations have a catalytic role to play in this regard, in particular in the challenging task of applying those policies, initiatives, and strategies in their respective areas of competency. Small-scale fisheries organizations, through their regional and international networks, also need to remain key actors in these developments, and have already demonstrated their ability to convene and constructively discuss the SSF Guidelines at both regional and national levels.

Interesting questions that arise from the experiences thus far, which should be explored in the future, include exploring what has enabled or hindered implementation from an institutional perspective at national and regional levels. It appears that the self-organization, self-awareness, empowerment, and capacity development of small-scale fishing communities have played in favor of initiating implementation efforts. Another interesting aspect to explore further is the potential of connecting small-scale fisheries to other policy issues to foster the institutionalization of the SSF Guidelines. Food security, for example, is receiving increased attention, in large part due to the newly agreed-upon United Nations Sustainable Development Goals (SDGs). The uptake of the SSF Guidelines in a number of products of the CFS seems to illustrate that there is an openness to recognize the contribution of small-scale fisheries to food security and nutrition, a key SDG. Other entry points could be for example processes in relation to indigenous peoples or gender.

Finally, the human rights-based approach upon which the SSF Guidelines are founded is often still controversial, in particular regarding the rights-based approach to fisheries in a narrower sense, which focus on access and user rights. Often related to this controversy is the question of the role of the private sector, in the sense of large-scale enterprises within and beyond the fisheries sector, in relation to small-scale fisheries. More often, scepticism to the human rights-based approach stems from the lack of a common understanding and practical evidence of application of this approach in small-scale fisheries. An important task for both academia as well as international organizations like FAO and the UNOHCHR is to fill this gap, for example through the collection, documentation, and analysis of cases where the human rights-based approach has resulted in improved small-scale fisheries sustainability.

In conclusion, it is essential to build on the initial awareness-raising and implementation planning efforts which have taken place so far surrounding the SSF Guidelines. This effort requires serious political will, the full participation of small-scale fisheries representatives, and support from all other relevant actors in order to generate tangible change at the local and national level. This process also requires tools to facilitate planning and action, as well as a monitoring framework to assess progress and capture lessons learned from these early adapter experiences, which in turn could contribute to an inspiring global learning process. This will ultimately enable the small-scale fisheries sector to contribute fully to food security and poverty eradication.

## References

AUC-NEPAD. (2014). *The policy framework and reform strategy for fisheries and aquaculture in Africa*. African Union Commission and NEPAD Planning and Coordinating Agency (NEPAD Agency) 2014.

FAO. (2015a). *Towards the implementation of the SSF Guidelines: Proceedings of the workshop on the development of a global assistance programme in support of the implementation of the voluntary guidelines for securing sustainable small-scale fisheries in the context of food security and poverty eradication*. Workshop proceedings, 8 December 2014, Rome, Italy.

FAO. (2015b). *Report of the thirty-first session of the Committee on Fisheries. Rome, 9–13 June 2014*. FAO Fisheries and Aquaculture Report No. 1101 Rome: FAO.

FAO. (2015c). *Towards the implementation of the SSF Guidelines. Proceedings of the workshop on the development of a global assistance programme in support of the implementation of the voluntary guidelines for securing sustainable small-scale fisheries in the context of food security and poverty eradication*. Workshop proceedings, 8–11 December 2014, Rome, Italy. FAO Fisheries and Aquaculture Proceedings No. 40.

FAO. (2015d). *Voluntary guidelines for securing sustainable small-scale fisheries in the context of food security and poverty eradication: At a glance*. Rome: FAO.

FAO. (2015e). *First regional symposium on sustainable small-scale fisheries in the Mediterranean and Black Sea*, A. Srour, N. Ferri, D. Bourdenet, D. Fezzardi, A. Nastasi (Eds.), Workshop proceedings, 27–30 November 2013, Saint Julian's, Malta. FAO Fisheries and Aquaculture Proceedings No. 39.

FAO. (2015f). *Indigenous food systems, agroecology and the voluntary guidelines on tenure: A meeting between indigenous peoples and FAO*. Rome: FAO.

FAO. (2015g). *Towards the implementation of the SSF guidelines in the Southeast Asia region. Summary conclusions and recommendations of the South East Asia Regional Consultation*. Proceedings of the Southeast Asia Regional Consultation Workshop on the implementation of the voluntary guidelines for securing sustainable small-scale fisheries in the context of food security and poverty eradication, 24–27 August 2015, Bali, Indonesia.

FAO. (2015h). *Near summary conclusions and recommendations of the Near East and North Africa Regional Consultation Workshop: Towards the implementation of the voluntary guidelines for securing sustainable small-scale fisheries in the context of food security and poverty eradication*. Workshop proceedings, 7–10 December 2015, Muscat, Oman.

FAO. (2015i). *Summary conclusions and recommendations of the South Asia FAO-BOBLME Regional Consultation on the implementation of the voluntary guidelines for securing sustainable small-scale fisheries in the context of food security and poverty eradication*. Workshop proceedings, 23–26 November 2015, Colombo, Sri Lanka.

FAO. (2015j). *Conclusions of the East Africa consultation workshop on improving small-scale fisheries in the context of food security and poverty eradication*. Workshop proceedings, 15–18 September 2015, Addis Ababa, Ethiopia.

FAO. (2016a). *Securing sustainable small-scale fisheries: towards implementation of the voluntary guidelines for securing sustainable small-scale fisheries in the context of food security and poverty eradication (SSF Guidelines)*. Report of the thirty-second session of the Committee on Fisheries (COFI/2016/7), 11–15 July 2016, Rome, Italy.

FAO. (2016b). *Fishery committee for the eastern central atlantic*. Report of the seventh session of the Scientific Sub-Committee, 14–16 October 2015, Tenerife, Spain. FAO Fisheries and Aquaculture Report No. 1128.

FAO. (2016c). Committee on Fisheries (COFI) – Fisheries and Aquaculture Department. http://www.fao.org/fishery/about/cofi/en.

FAO. (2016d). Committee on world food security. Making a difference in food security and nutrition. Major products. http://www.fao.org/cfs/cfs-home/products/it/. Accessed 18 Nov 2016.

FAO. (2016e). *Freshwater, fish and the future: Cross-sectoral approaches to sustain livelihoods, food security, and aquatic ecosystems*. The Rome declaration: Ten steps to responsible inland fisheries. FAO, Michigan State University. http://www.fao.org/3/ai5735e.pdf. Accessed 18 Nov 2016.

FAO. (2016f). *Report of the fortieth session of the General Fisheries Commission for the Mediterranean (GFCM)*, St. Julian's, Malta, 30 May – 3 June 2016. GFCM Report No. 40. Rome, Italy. http://www.fao.org/gfcm/reports/statutory-meetings/detail/en/c/423828/

FFFP. (2014). *Assert our rights, restore our dignity*. Report of the 6th General assembly of the World Forum of Fisher Peoples. 6th General Assembly. 1–5 September 2014, Cape Town, South Africa. http://worldfishers.org/wpcontent/uploads/2015/04/WFFP_GA6_REPORT.pdf. Accessed 18 Nov 2016.

GFCM. (2016). *Resolution GFCM/40/2016/3 on sustainable small-scale fisheries in the GFCM area of application*. Report on the 40th session. http://www.fao.org/gfcm/reports/statutory-meetings/detail/en/c/423828/. Accessed 18 Nov 2016.

ICSF. (2014). *International workshop on: Towards socially just and sustainable fisheries. ICSF workshop on implementing the FAO voluntary guidelines for securing sustainable small-scale fisheries in the context of food security and poverty eradication (SSF Guidelines)*. Workshop proceedings, Chennai, India.

ICSF. (2015). *Workshop to introduce the FAO voluntary guidelines for securing sustainable small-scale fisheries in the context of food security and poverty eradication (VG-SSF) in Tanzania*. Workshop proceedings, 17–18 August 2015, Bagamoyo, Tanzania.

ICSF. (2016). *National workshop on capacity-building for the implementation of the voluntary guidelines for securing sustainable small-scale fisheries (SSF Guidelines)*. Workshop proceedings, 21–22 March 2016, New Delhi, India.

ICSF, CROCEVIA, & IFAD. (2016). *Implementation of the voluntary guidelines for securing sustainable small-scale fisheries in the context of food security and poverty eradication (SSF Guidelines)*. https://sites.google.com/site/ssfguidelines/. Accessed 22 Nov 2016.

MAGAP. (2016). FAO, Federación Nacional de Cooperativas Pesqueras de Ecuador. (2016). *Informe taller nacional sobre la implementación de las Directrices Voluntarias para lograr la sostenibilidad de la pesca en pequeña escala, en el contexto de la seguridad alimentaria y la erradicación de la pobreza*. Workshop proceedings, 21–22 September 2016, Guayaquil, Ecuador.

Metzner, R., & Franz, N. (2015). SSF guidelines: Vital momentum for small-scale fishers. *Rural 21: The international journal for rural development, 49*(3), 20–21.

NFF & ICSF. (2011). *Sustainable small-scale fisheries: Towards FAO Guidelines on marine and inland small-scale fisheries*. Workshop proceedings, 19–21 September, 2011, Kolkata, West Bengal, India.

OSPESCA. (2015). *Política de integración de pesca y acuicultura 2015–2025*. El Salvador: OSPESCA.

SDG. (2015). United Nations: Sustainable development goals. http://www.un.org/sustainabledevelopment/. Accessed 18 Nov 2016.

Shenoy, S. (2015). *Report of the International Collective in Support of Fishworkers (ICSF)-Bay of Bengal Large Marine Ecosystem (BOBLME) India (East Coast) workshop: implementing the FAO Voluntary guidelines for securing sustainable small-scale fisheries in the context of food security and poverty eradication (SSF Guidelines)*. Workshop proceedings, 6–7 March, 2015, Chennai, India.

SPC. (2015). *A new song for coastal fisheries pathways to change: The Noumea strategy*. New Caledonia: Secretariat of the Pacific Community (SPC).

TBTI. (2015). *Compilation of ideas from the TBTI-FAO session on the voluntary guidelines for securing sustainable small-scale fisheries (SSF Guidelines)*. Too Big To Ignore Research Report Number 12.1/2014. http://toobigtoignore.net/wp-content/uploads/2015/01/GAP-recommendations-on-SSF-Guidelines_Report-from-the-2WSFC-session.pdf. Accessed 28 Nov 2016.

# Part II
# Politics of Transformation

Even if it is premature to measure the success or failure of the implementation of the SSF Guidelines, it is possible to say something about the distance that the implementation would need to go. This is what the chapters in this section do, given that they compare what the SSF Guidelines advocate with the current fisheries policies and practices in different countries and regions. It is fair to assume that the greater the distance is between what is and what should be from the perspective of the SSF Guidelines, the more cumbersome their implementation will be. In Chap. 4, Philippa Cohen, Andrew Song, and Tiffany Morrison take us to the Pacific Solomon Islands. They argue that here the SSF Guidelines are entering a very complex situation, where small-scale fisheries policy has already been established. Although they identify large overlaps between existing policies and the SSF Guidelines, they also find gaps. The implementation of the SSF Guidelines would therefore largely be about filling these gaps. In Chap. 5, Lars Lindström and Maricela de la Torre-Castro observe a similar situation in Zanzibar, an archipelago off the coast of Tanzania. Here, the authorities were not aware of the new Guidelines until the authors brought them to their attention. When comparing the SSF Guidelines with the existing policy framework, they not only find gaps, but also contradictions between a new fisheries policy which is underpinned by a market-liberal paradigm and that of the Guidelines, which they think will complicate the implementation. They argue that conflicts between the two may possibly hinder the SSF Guidelines from being implemented. Whereas the SSF Guidelines do not come with donor funding, the new fisheries policy does. In Chap. 6, Hunter T. Snyder, Rikke Jacobsen, and Alyne Delaney observe that Greenland is well positioned to implement the SSF Guidelines, given existing policies on human rights, food security, gender equity, and a political history of supporting small-scale fisheries. They argue, however, that implementing the SSF Guidelines requires re-harmonizing Greenland's small-scale fisheries policy design together with local, national, and international objectives. Chapter 7 takes us to the Caribbean for a case study in Jamaica. Here, Lisa Soares discusses to what extent the SSF Guidelines can take root, and whether or not there is an enabling environment for them to do so. She argues that the implementation of the SSF Guidelines would need to be coherent with national and local specificities, with an eye for nuances that exist at various levels.

# Chapter 4
# Policy Coherence with the Small-Scale Fisheries Guidelines: Analysing Across Scales of Governance in Pacific Small-Scale Fisheries

**Philippa J. Cohen, Andrew M. Song, and Tiffany H. Morrison**

**Abstract** Concerns about the sustainability of small-scale fisheries, and the equitable distribution of fisheries benefits, are wide-spread within government agencies, non-government organizations, and rural fishing communities throughout Pacific Island Countries and Territories. Addressing these concerns was given renewed impetus in recent years with the completion and adoption of the Voluntary Guidelines for Securing Sustainable Small-Scale Fisheries (SSF Guidelines). This global document enters a complex policy landscape within the Pacific region. In anticipation of its region-wide implementation, this chapter focuses on policy coherence; using Solomon Islands as a case we investigate the potential interplay of the SSF Guidelines with priority policies at the regional, national, and sub-national levels. We first examine the SSF Guidelines to identify 22 dominant themes, including human rights, adaptive capacity, and tenure rights. We then focus in on 11 on policy instruments known to directly influence small-scale fisheries governance; we examine to what extent and in which direction the small-scale fisheries themes are represented in these 11 regional, national, and sub-national policies. We find areas of incoherence in addition to nine themes that are relatively poorly represented ('gaps') in the current policy landscape. More positively, however, we also observe a large-scale overlap on many of the key themes. While our analysis is specific in its application to Solomon Islands, our approach to diagnose areas of incoherence and gaps is easily applicable to other countries. This type of policy-based analysis is a useful first step to understanding priorities and strategies for implementation, and in particular opportunities for the SSF Guidelines to prompt adjustment and transformation of existing policies.

P.J. Cohen (✉) • A.M. Song
WorldFish, Honiara, Solomon Islands

Australian Research Council Centre of Excellence for Coral Reef Studies, James Cook University, Townsville, Australia
e-mail: p.cohen@cgiar.org; andrew.song@jcu.edu.au

T.H. Morrison
Australian Research Council Centre of Excellence for Coral Reef Studies, James Cook University, Townsville, Australia
e-mail: Tiffany.Morrison@jcu.edu.au



S. Jentoft et al. (eds.), *The Small-Scale Fisheries Guidelines*, MARE Publication Series 14, DOI 10.1007/978-3-319-55074-9_4

**Keywords** Policy coherence • Multi-scale governance • Solomon Islands • Oceania • Co-management • Implementation

## Introduction

Small-scale fisheries provide food, income, and a way of life for a high proportion of the largely coastal dwelling populations of the 22 Pacific Island Countries and Territories.[1] Although cash-based economies are expanding, in many Pacific Island countries where human development is low, the subsistence economy plays an important role in maintaining self-sufficiency and human well-being (Bell et al. 2009; Gillett 2009; Adams 2012). Coastal marine resource use, including small-scale fisheries are a major contributor to many national economies. Small-scale fisheries are typically diverse, but particularly so in the Pacific due to the region's exceptionally high political and cultural diversity and marine biodiversity (Veron et al. 2009).

Concerns about the sustainability of the Pacific's marine resource use arise in light of rapid population growth, increased connectedness to global markets, intensifying interactions with commercial enterprise, and projected effects of climate change (Gillett and Cartwright 2010). These concerns have generated policy responses at higher scales, such as the emergence of the Coral Triangle Initiative on Coral Reefs, Fisheries, and Food Security (CTI) which defines a region that includes Pacific countries. The CTI subsequently pushed for the development of national plans of action within each of the six member states (Malaysia, Philippines, Indonesia, Timor-Leste, Papua New Guinea, and Solomon Islands) that intended to consolidate and increase efforts of fisheries and environmentally focussed government agencies and NGOs (CTI Secretariat 2009). More recently the governments of Pacific Island countries convened to discuss the future of inshore and coastal fisheries. The deliberations led to the formation and commitment to a series of recommendations related to coastal fisheries captured in a policy document referred to as "A new song for coastal fisheries – pathways to change" (Secretariat of the Pacific Community 2015). In sum, the commitment made in this document was to increase regional and national level focus on improving coastal fisheries governance.

Pacific nation states are responsible for the management of small-scale fisheries, but most fishing activities occur hundreds to thousands of kilometers away from urban centers, beyond the reach of management and enforcement capacity of centralized governments. In many countries, there have been calls for devolution and decentralization of governance authority, for example to the provincial governments

[1] 22 Pacific Island Countries and Territories commonly considered to comprise the Pacific region and are served by the Secretariat of the Pacific Community, http://www.spc.int/en/about-spc/members.html; American Samoa, Cook Islands, Federated States of Micronesia, Fiji, French Polynesia, Guam, Kiribati, Marshall Islands, Nauru, New Caledonia, Niue, Northern Mariana Islands, Palau, Papua New Guinea, Pitcairn Islands, Samoa, Solomon Islands, Tokelau, Tonga, Tuvalu, Vanuatu, Wallis and Futuna.

in Solomon Islands (Lane 2006). In many cases, provincial governments have responded with the formation of their own development strategies and fisheries legislation. In parallel, in many Pacific Island countries (including Solomon Islands) there are legally recognized systems of customary tenure and traditional management that are commonly promoted in policy, and invoked in practice, for the governance of small-scale fisheries (Ruddle 1998; Govan et al. 2009). The small-scale fisheries attributes of diversity, complexity, dynamism, interactions across scales and sub-systems of governance, all offer substantial challenges and opportunities for securing the benefits of small-scale fisheries.

Adding to the varied policy backdrop of the region are the recently published Voluntary Guidelines for Securing Sustainable Small-Scale Fisheries (SSF Guidelines). The SSF Guidelines are a global document whose formulation was led by the Food and Agriculture Organization and adopted by its 143 member states. In support of national implementation of the SSF Guidelines, *policy coherence* has been identified as one of the main themes to be pursued by all relevant governance actors. The SSF Guidelines document itself lays the framework for this by highlighting the need for policy coherence in the multi-level structure of policy instruments, *inter alia*: national legislation and international documents, including those relating to the fields of human rights, economic development, environmental protection, and other fisheries sectors (Part 3, Chapter 10.1). In addition, the SSF Guidelines highlight the importance of integrating and harmonizing policies (10.2 and 10.3), basing the coherence on a long-term vision set out for sustainable small-scale fisheries and poverty eradication (10.4) as well as promoting institutional linkages (local-national-regional-global) necessary for achieving policy coherence and cross-sectoral collaboration (10.5).

Policy coherence is, in fact, a long enduring policy challenge. Across all policy fields, the aspiration to coherence has been widespread (Jordan and Harpin 2006). It is generally believed that increased coherence is associated with greater policy stability and more substantive policy delivery. Likewise, incoherence can engender fragmentation, coordination problems, and implementation gaps (Jordan and Harpin 2006; May et al. 2006). Policy coherence has thus been referred to as "the synergic and systematic support towards the achievement of common objectives within and across individual policies" (den Hertog and Stroß 2013, 377) or, in a more straightforward sense, an "overall state of mutual consistency among different policies" (di Francesco 2001, 8; OECD 1996). In essence, coherence is about eliminating conflict and promoting synergy (Nilsson et al. 2012).

Initially a concern at the nation-state level, in which two or more domestic policies may push in different directions, Carbone (2008, 325) reports that "with globalization, not only has the distinction between internal and external policies become blurred, but the interplay between different policies also involves the regional and global level". The recognition of an increasingly dense policy arena and the need to understand the degree of coordination between instruments has seen international organizations in the environmental and natural resource field, such as the International Institute for Sustainable Development (IISD) and the Food and Agriculture Organization of the United Nations (FAO), taking on the goals of "development and environment policy coherence at the international level" and "coherence between agriculture and trade policies" (Nilsson et al. 2012, 396).

Likewise, the newly promoted SSF Guidelines do not enter a policy vacuum. In the multiple policy framings that pervade the Pacific Islands small-scale fisheries, the SSF Guidelines may be competing for political attention vis-à-vis existing policies at the national, regional, and global levels. The policy goals might also be at odds with the dominant governing visions, thus stalling the acceptance of its key messages and ultimately hindering implementation. On the other hand, the implementation of the SSF Guidelines may prove more immediately governable than expected if high consistency with existing policy objectives is observed. It follows that understanding and exploring policy coherence as we anticipate the implementation of the SSF Guidelines could be a useful undertaking that provides an early outlook into implementation hurdles and synergies, and a chance to strategize practical actions. In this regard, this chapter aims to understand coherence of the SSF Guidelines with the current policy landscape into which the SSF Guidelines will enter. We focus on Solomon Islands as a case study within the Pacific region. Specifically, three research questions are pursued.

1. What is the policy state of play for implementing the SSF Guidelines in Solomon Islands?
2. Where are the areas of coherence, incoherence, and gaps between the SSF Guidelines and policies in Solomon Islands?
3. What mechanisms and arenas are appropriate for resolving incoherence? Do these currently function in Solomon Islands?

## Methods

Our assessment of policy coherence was informed by a three-step framework used in Nilsson et al. (2012). We first created an inventory of key policies relevant to small-scale fisheries in Solomon Islands; second proceeded with a 'screening' comparison matrix, and, third, conducted an in-depth analysis of key interactions.

We constructed an inventory of legal and policy instruments that relate directly to coastal resource management and conservation in Solomon Islands from three sources: (1) an analysis of regional and national instruments that influence community-based management (Govan et al. 2009); (2) data from interviews where respondents were asked to free-list policies and legislation that influence their activities on coastal resource related food security, biological conservation, and climate change adaptation within Solomon Islands (unpublished data); and (3) a Solomon Islands fisheries, marine, and coastal policy gap analysis (Healy and Hauirae 2006). Once this list was established, we categorized each instrument as global, regional (Pacific-wide), sub-regional (Coral Triangle Region, or Melanesia), national, or sub-national (provincial). We selected a subset of 11 documents which, for the purpose of this analysis, included policies whose scope or responsible agency was most directly related to the governance and management of small-scale fisheries (see Table 4.1). We consider our analysis to be illustrative, but preliminary, and note that there would be also value in an analysis of a larger sample of policy instruments.

The second step involved the thematic comparison of these policy documents with the SSF Guidelines by way of coding. We used this method of comparing the frequency and extent to which different themes were detailed within policy documents as an *indicator* of policy coherence. We commenced by using a general inductive approach to coding Part 2 of the SSF Guidelines given its focus on w*hat* was to be implemented. NVivo 11 was used to facilitate coding and comparison of the codes. Coding was independently carried out by two of the authors. Codes were then compared using the NVivo coding comparison function. Amendments to codes or their definition and discrepancies were discussed and resolved between the two authors; from this, 22 dominant themes and their definitions were agreed upon (see Table 4.2). We then used these themes for coding the content of other policy documents and Part 3 of the Guidelines. We coded according to interpretation of the text against the definitions the 22 themes (Table 4.2).

Finally, an in-depth reading of key comparisons was sought to understand the degree and details of coherence or incoherence between the SSF Guidelines and the reviewed policy documents. This step focused on a subset of codes that our analysis had indicated to be a high priority within the SSF Guidelines, and simultaneously demonstrated high coherence or major gaps with other policies, thus warranting a more detailed examination.

## Results and Discussion

### *What Is the Policy State of Play for Implementing the SSF Guidelines?*

We identified a total of 45 policies, conventions, and strategies from the (a) analysis of national and regional instruments that influence community-based management, (b) fisheries, marine and coastal gap analysis review, and (c) interview data on influential policies on coastal resource related food security, climate change, and conservation. These policies spanned themes of biodiversity conservation, climate change development, environmental (e.g., marine pollution), national and rural community development, and those directly related to the management and development of fisheries. Of those related to fisheries, 13 were primarily concerned with larger scale, commercial pelagic fisheries (e.g., including international access agreements for migratory stocks); we did not examine these given that, in Solomon Islands, these were only indirectly related to the majority of small-scale fisheries activities. Table 4.1 contains a brief profile of each of the 11 policy instruments that we considered, based on their descriptions from the three sources from which they were

**Table 4.1** The SSF Guidelines (the First Item) and policy instruments directly influencing small-scale fisheries governance in Solomon Islands

| Policy Instrument (abbreviations in bold) | Year | Administrative body | Scale | Summary of vision or objectives |
|---|---|---|---|---|
| **SSF Guidelines**<br>Voluntary guidelines for securing sustainable small-scale fisheries in the context of food security and poverty alleviation | 2014 | Food and Agriculture Organisation | Global | "to enhance the contribution of small-scale fisheries to global food security and nutrition and to support the progressive realization of the right to adequate food; contribute to the equitable development of small-scale fishing communities and poverty eradication; to achieve the sustainable utilization; to promote the contribution of small-scale fisheries to a sustainable future; to provide guidance for the development and implementation of ecosystem friendly and participatory policies; to enhance public awareness and promote the advancement of knowledge". To promote a "human rights-based approach, by empowering small-scale fishing communities" |
| **New Song**<br>A new song for coastal fisheries – pathways to change. The Noumea Strategy | 2015 | Secretariat of the Pacific Community | Regional | Vision "Sustainable well-managed inshore fisheries, underpinned by community-based approaches that provides food security, and long-term economic, social and ecological benefits to our communities". Eight overarching outcomes; empowered communities with clear rights, information to support management, political commitment to coastal fisheries, good governance in agencies, legal and policy framework strengthened, collaboration and coordination, equitable benefits within communities, diverse livelihoods |
| **MSG Roadmap**<br>Roadmap for inshore fisheries management and sustainable development | 2014 | Melanesian Spearhead Group | Regional | Vision "sustainable inshore fisheries, well managed using community-based approached that provide long-term, social, ecological and food security benefits to our communities". To build capacity to develop and manage coastal resources, raise profile of important of management of inshore fisheries, manage and restore fish stocks for coastal community benefit |
| **SI Constitution**<br>The Constitution of Solomon Islands | 1978 | Government of Solomon Islands | National | To provide the supreme law for Solomon Islands; related to fisheries prescribes adherence to the 'protection of fundamental rights and freedoms of the individual', and Constitution, 1978 – The Constitution recognizes the right of landowners to exercise control over their lands and resources. Also, the Solomon Islands shall "cherish and promote the different cultural traditions", make provision to apply customary laws (e.g., related to customary tenure), consider customary law as part of the national law |
| **SI NPOA**<br>Solomon Islands National Plan of Action; Coral Triangle Initiative on coral reefs, fisheries & food security | 2010 | Ministry of Environment, Climate, & Disaster Management | National | Vision: "Marine and coastal resources are sustainably managed and utilized to secure the long term improvement of the livelihoods of Solomon Islanders", Goal: "Solomon Islands sustainably manages marine and coastal resources to ensure food security, sustainable economic development, biodiversity conservation and adaptation to emerging threats through community based resource management approaches supported by government agencies and other partners" |

| | | | | |
|---|---|---|---|---|
| **Fisheries Management Act** | 2015 | Ministry of Fisheries & Marine Resources | National | "to ensure the long-term management, conservation, development and sustainable use of Solomon Islands fisheries and marine ecosystems for the benefit of the people of Solomon Islands" |
| **MFMR Corp Plan**<br>Ministry of Fisheries and Marine Resources Corporate Plan | 2014 | Ministry of Fisheries and Marine Resources | National | Vision a "fisheries sector that generates an economically viable and equitable distribution of benefits for all Solomon Islanders from a biologically and economically sustainably managed marine ecosystem", and lead management and development of offshore and coastal fisheries, actively promote CBRM. To "promote government inter-agency cooperation and act as the focal point for national capacity building, research and development within the sector". Six points of strategic focus; improve market access for rural fishers, grow livelihoods through sustainable aquaculture development; improve health of fisheries and marine resources; grow economy through sustainable fisheries investments; effective enforcement; increase skills and knowledge of partners in fisheries development |
| **SI Dev Strategy**<br>National Development Strategy | 2011 | Ministry of Development Planning and Aid Coordination | National | "create a modern, united and vibrant Solomon Islands founded on mutual respect, trust and peaceful co-existence in a diverse yet secure and prosperous community where tolerance and gender equality are encouraged and natural resources are sustainably managed; and enable all Solomon Islanders to achieve better quality of life and standard of living" |
| **CBRM Principles**<br>Principles for Best Practice of CBRM in Solomon Islands | 2011 | NGO & Government management advisors | National | To encourage some consistency and response to lessons on the way partner organisations interact with communities to facilitate community-based resource management, further clarify what communities can expect from partners and donors |
| **SI NAPA**<br>National Adaptation Programmes of Action | 2008 | Ministry of Environment, Climate, & Disaster Management | National | To "communicate priority activities addressing the urgent needs and concerns of Solomon Islands, relating to adaptation to the adverse effects of climate change", and ultimately [directly related to fisheries] to "ensure that increased agriculture, livestock and fisheries productivity contributes to sustainably enhanced food security and improved livelihoods of the people of the Solomon Islands" |
| **Western Ordinance**<br>Western Provincial Fisheries Ordinance | 2015 | Western Province Government | Provincial | To support management of fishery and marine resources in a manner which; is consistent with national policy; acknowledges the economic significance of fishing; promotes environmental sustainability; upholds the place of customary fisheries rights and practices |
| **Malaita Ordinance**<br>Malaita Provincial Fisheries Ordinance | 2015 | Malaita Provincial Government | Provincial | "to regulate fisheries in Malaita Province" |

**Table 4.2** Dominant themes in Part 2 of the SSF Guidelines

| Thematic code | Description |
|---|---|
| Adaptive capacity | Refers to adaptive capacity of small-scale fisheries communities (to unspecified or specified drivers of change or shocks), but also adaptive management |
| Capacity building | Refers to specific training activities or capacities to general calls for investments in building capacity of communities, NGO or government managers |
| Climate change | Any specific mention of climate change |
| Co-management | Includes specific mention of community-based approaches to management, and co-management and also general references to State, fishers and communities working collaboratively to address management. Includes some mentions of specific management measures implemented under these approaches |
| Compliance | Refers to enforcement and compliance strategies, and also sanctions |
| Equitable access and distribution | Refers to the distribution of benefits within fishing communities, but also include distribution of "project" geographically |
| Fisher participation | Include participation and representation of fishers in management efforts to policy forums |
| Gender | Calls for special attention and differentiated strategies for women and men |
| Human rights | Direct references to human rights, or references to respecting freedom, social justice etc |
| Human social development | Calls for local to higher level social development efforts i.e., simultaneous to management efforts, or as a specific objective of fisheries reform/management etc |
| Institutional coordination and strengthening | Includes general calls for institutional coordination and also details specific mechanisms to achieve coordination or coherence – includes building legal institutions to back co-management and community-based management for example. Also includes cross-sectoral and cross-scale interactions – see also integrated approaches |
| Integrated approaches | Includes inter-sectoral community development efforts (i.e., addressing health, education alongside fisheries), but also ecosystem approach to fisheries management |
| International fish trade | Specific reference to trade across national borders |
| Management for sustainability | Refers to the objectives of ecological sustainability or sustainability in broader sense. May include term conservation. May refer to specific measures (reduction of efforts, catch limits) where they are applied to promote ecological sustainability |
| Migration | Reference to migrants (rights etc.) or migration |
| Monitoring, research, information | Includes calls for improved data management, data collection and research. Also includes calls for integration of multiple knowledge sources (e.g., contemporary science and local knowledge). Also includes calls for "awareness raising" |
| Political recognition and will | Calls to increase the profile and recognition of small-scale fisheries and fishers and associated concerns |
| Post-harvest economic development | Specific calls for investment in post-harvest developments, including food safety |
| Resource competition | Mentions of competition for resources between small-scale fishers, with commercial fisheries and with other sectors |
| Safety at sea | Investments or concerns about safety at sea |
| Tenure rights | Specific mention of tenure as an instrument, tenure rights, and the interpretation of tenure rights |
| Transboundary fishing | Includes reference to fishing or fisheries that span international borders |

identified, to most directly relate to the governance of small-scale fisheries in Solomon Islands.

An insight into the policy state of play was initiated through a comparison of the SSF Guidelines with the other 11 documents. A comparison matrix (Table 4.3) displays the presence or absence of each key theme from the SSF Guidelines in the existing policy documents. First, we can observe nine themes with reasonably good representation over the 11 policy documents. They are reiterated here in decreasing order of frequency of mention in the Guidelines:

- Monitoring research information / awareness raising
- Fisher participation
- Institutional cooperation and strengthening
- Tenure rights
- Human and social development
- Management for sustainability
- Co-management
- Integrated approaches
- Compliance

The presence of these themes in most of the policies we examined should not come as a surprise since they are, arguably, among some of the most salient and widely-discussed topics in the contemporary fisheries governance literature. A description of the themes in the context of small-scale fisheries is provided in Table 4.2. What begs greater attention is perhaps the themes that are poorly represented, relatively speaking, in the existing regional, national, or sub-national policy documents, nine of which are listed below:

- Gender
- Human rights
- Post-harvest economic development
- International fish trade
- Equitable access and distribution
- Resource competition
- Safety at sea
- Transboundary fishing
- Migration

Some topics (e.g., post-harvest economic development, safety at sea, transboundary fishing, and migration) are arguably more specific in scope to commercial, offshore fisheries. Therefore, the fact that they received little attention is more a reflection of the sub-sample of policies we analyzed. However, the relatively light treatment given to gender, human rights, and equitable access and distribution, for example, suggests that these issues have, to date, been viewed as of relatively low pertinence or urgency. Despite commitments to these themes appearing in more global policy instruments, e.g., the *Convention for Human Rights*, the global fisheries community has perceived that their importance warrants reiteration in fisheries

**Table 4.3** The 22 most cited SSF Guidelines themes (arranged from the most frequently cited on the left, and in descending order towards the right), and their presence (indicated by shaded areas) in the 11 policy instruments (see abbreviated names in Table 4.2 for full description) impacting upon small-scale fisheries governance in Solomon Islands

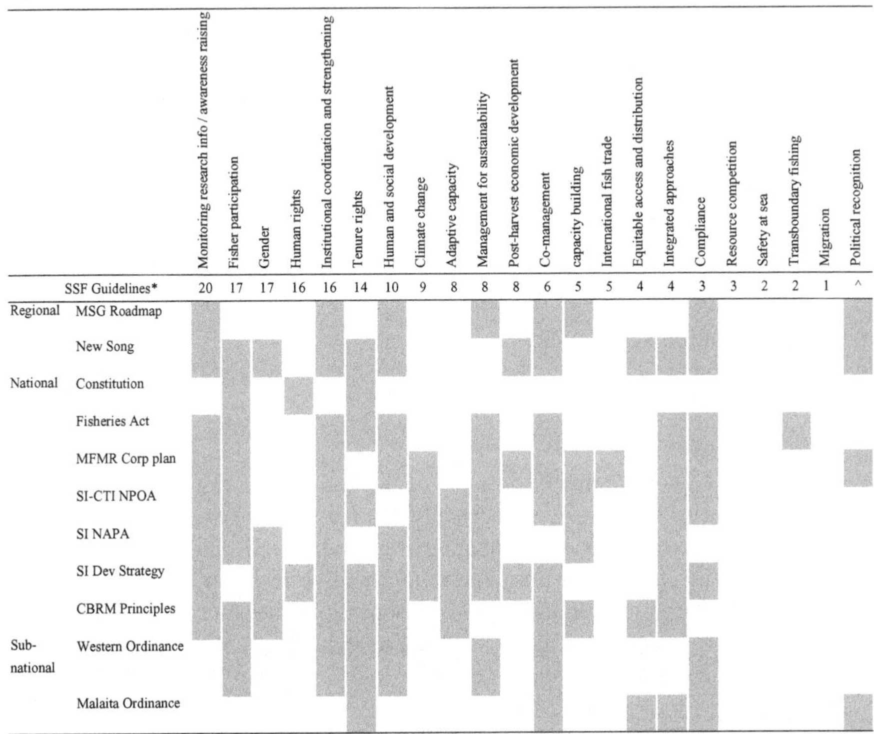

| | | Monitoring research info / awareness raising | Fisher participation | Gender | Human rights | Institutional coordination and strengthening | Tenure rights | Human and social development | Climate change | Adaptive capacity | Management for sustainability | Post-harvest economic development | Co-management | capacity building | International fish trade | Equitable access and distribution | Integrated approaches | Compliance | Resource competition | Safety at sea | Transboundary fishing | Migration | Political recognition |
|---|---|---|---|---|---|---|---|---|---|---|---|---|---|---|---|---|---|---|---|---|---|---|---|
| | SSF Guidelines* | 20 | 17 | 17 | 16 | 16 | 14 | 10 | 9 | 8 | 8 | 8 | 6 | 5 | 5 | 4 | 4 | 3 | 3 | 2 | 2 | 1 | ^ |
| Regional | MSG Roadmap | ■ | | | | ■ | | ■ | | | ■ | | ■ | ■ | | | | ■ | | | | | ■ |
| | New Song | ■ | ■ | ■ | | ■ | ■ | ■ | | | | ■ | ■ | | | ■ | ■ | ■ | | | | | ■ |
| National | Constitution | ■ | ■ | | ■ | | ■ | | | | | | | | | | | | | | | | |
| | Fisheries Act | ■ | ■ | | | ■ | ■ | ■ | | | ■ | | ■ | | | | ■ | ■ | | | ■ | | |
| | MFMR Corp plan | ■ | ■ | | | ■ | | ■ | ■ | | ■ | ■ | ■ | ■ | ■ | | ■ | ■ | | | | | ■ |
| | SI-CTI NPOA | ■ | ■ | | | ■ | ■ | | ■ | ■ | ■ | | ■ | ■ | | | ■ | ■ | | | | | |
| | SI NAPA | ■ | ■ | ■ | | ■ | | ■ | ■ | ■ | ■ | | | ■ | | | ■ | | | | | | |
| | SI Dev Strategy | ■ | | ■ | ■ | ■ | ■ | ■ | ■ | ■ | ■ | ■ | ■ | | | | ■ | ■ | | | | | |
| | CBRM Principles | ■ | ■ | ■ | | ■ | ■ | ■ | | ■ | | | ■ | ■ | | ■ | ■ | | | | | | |
| Sub-national | Western Ordinance | | ■ | | | ■ | ■ | ■ | | | ■ | | ■ | | | | | ■ | | | | | |
| | Malaita Ordinance | | | | | | ■ | | | | | | ■ | | | ■ | ■ | ■ | | | | | ■ |

*coding frequency; ^*political recognition* theme is consistently expressed throughout the SSF Guidelines making frequency redundant (in urging States to recognize all these themes and take action in favour of small-scale fisheries).

specific commitments (i.e., the SSF Guidelines). As part of the in-depth analysis of key comparisons, we discuss them in greater detail in the following section.

Based on the observed patterns of thematic coverage (Table 4.3), we anticipate that the degree of coherence between the SSF Guidelines and the 11 policy documents that we analyzed is at least moderate. If we were to expand our analysis to policies that primarily dealt with economic growth, economic development, and/or trade, we may find higher degrees or more fundamental inconsistencies. We limit the scope of our analysis to understand the points of incoherence that are faced most directly by small-scale fisheries managers as they seek to navigate their commitments within the existing policy landscape, and in terms of implementing the SSF Guidelines. Only two policy instruments – the *New Song*, and the *Fisheries Management Act* – were endorsed subsequent to the endorsement of the SSF Guidelines and, unsurprisingly, we find only one direct reference to the SSF Guidelines themselves within the policies we analyze. Therefore, we would suggest that the 'implementation of the SSF Guidelines' involves, for the most part, continuing and strengthening the implementation of existing policies, conventions, and strategies. The gaps and points of incoherence we identify highlight where policy reform may be necessary or desirable to aid implementation of the commitments as they are laid out in the SSF Guidelines.

## *Where Are the Areas of Coherence, Incoherence, and Gaps Between the SSF Guidelines and Policies in Solomon Islands?*

### Areas Indicative of Coherence

The SSF Guidelines place particular emphasis on promoting participatory management approaches such as *co-management*. Fisheries co-management is defined by relationships between a resource-user group (e.g., local fishers) and another entity (e.g., a government agency or non-government organization) in which management responsibilities and authority are shared (Pomeroy and Berkes 1997; Evans et al. 2011). In practice, co-management arrangements vary according to the degree of authority and influence the resource users have over management, relative to partners, with some arrangements relatively tokenistic and others truly empowering (Sen and Nielsen 1996). The SSF Guidelines are not specific about what point on this spectrum small-scale fisheries co-management should take. In contrast, regional (the New Song), sub-regional (the MSG Roadmap, CTI NPOA), national (the Corporate Plan and Fisheries Management Act) and sub-national (Provincial ordinances) settings are very specific and almost exclusively promote collaborative, community-based forms of co-management (CBRM). A common view expressed within these policies is that communities hold the rights and responsibilities for management largely through customary tenure and governance systems. For example, the CTI NPOA (p. 18) states: "Solomon Islands focuses on people-centred approaches where communities will be the primary drivers, as well as beneficiaries, of sustainable resource management".

Policies assert that communities will be responsible for ensuring their management plans are coherent with provincial and national law. In a similar vein, non-government organizations working with communities to implement CBRM are instructed to consult with the government "to ensure their activities align with current national strategies" (CBRM principles, 8) and a 'minimum standard' management model (SI NPOA, 17). Regional policies indicate that the government supports community-based efforts and that fisheries agencies will be "transparent, accountable, and adequately resourced, supporting coastal fisheries management and sustainable development, underpinned by CBR" (Secretariat of the Pacific Community 2015, 10) and "conducting effective CBRM activities" (p. 13). These prescriptions for the role of government remain quite broad. The commonly articulated specific role was that the national and provincial government responsibilities were to 'raise awareness' and to provide information to community managers. Further specifics include "establishment of, and strengthening of the CBRM unit within the Ministry of Fisheries and Marine Resources" (MFMR Corp Plan, 12) and that the National government must prepare and enforce laws and regulations. In sum, the policies we examine tend to place substantial management responsibility, for dealing with a range of issues (i.e., food security, adaptive capacity to climate change and other pressures, conservation of target or threatened species and habitats), onto communities. Whilst the values and 'empowerment' potential of community-based approaches are promoted in the region, there are also clear critiques of devolving the burden of management responsibility and risk to communities (e.g., Davis and Ruddle 2012). In contrast, the SSF Guidelines place substantial responsibilities onto the state, with 64 of the 89 commitments laid out in Part 2 and 3 directed firstly to states. The balance of roles, responsibility, and authority is one that requires careful negotiation and open discussion, as it would have a major bearing on how the CBRM-based activities play out in implementation.

The SSF Guidelines provide nuanced and multiple recommendations for *fisher representation and participation* along multiple points of the value chain. For example, calling for small-scale fisheries to be "represented in relevant local and national professional associations and fisheries bodies and actively take part in relevant decision-making and fisheries policymaking processes" and "[states] and small-scale fisheries actors should encourage and support the role and involvement of both men and women, whether engaged in pre-harvest, harvest or post-harvest operations" (p. 7). Solomon Islands' higher level policies (the Constitution and the Fisheries Management Act, respectively) call broadly for fisher participation and representation, too, with language to "ensure that participation of our people in the governance of their affairs and provide within the framework of our national unity for the decentralisation of power" (p. 12), and "the interests of artisanal and subsistence fishers shall be taken into account, including their participation in management of their respective fisheries" (p. 31). Whereas the more operational national fisheries and environment policies have an explicit and relatively limited emphasis on 'participation' at the level of 'community', within Solomon Islands policies, there are relatively few calls to address the participation of women or vulnerable and marginalized groups explicitly. Further, we found no mention of professional associations, fisheries bodies, or equivalents; although the intent to establish a 'Fisheries Advisory Council' may be such a forum where this is potentially addressed (although details are not provided). In sum, the policies we examined had a much narrower view of participation and representation than that articulated in the SSF Guidelines and related consultations (e.g., Food and Agrictulture Organisation 2013).

The SSF Guidelines demonstrate a strong call for *institutional coordination* where states specifically "should establish and promote the institutional structures and linkages – including local–national–regional–global linkages and networks – necessary for achieving policy coherence, cross-sectoral collaboration and the implementation of holistic and inclusive ecosystem approach" (p. 15). The requirement for between-organization, cross-sector, and cross-level institutional coordination is echoed in Pacific and Solomon Islands policies, and is perhaps one of the most dominant themes. Within these policies, the objectives of institutional coordination were stated as to align levels of fisheries governance, to implement integrated management and development approaches, and to foster capacity building and learning between organizations. The policies we examined do not yet reflect details about the national-level platforms that states will utilise and facilitate "with cross-sectoral representation and with strong representation of civil society organizations, to oversee implementation of the Guidelines" (p. 18). Three multi-stakeholder forums are briefly mentioned in Solomon Islands policy: the National Coordinating Committee for the Coral Triangle Initiative (described as the mechanism designated to coordinate and promote implementation of the CTI Plan of Action, specifically), the Solomon Islands Locally Managed Marine Area network (policies call for support of this network, and for the network to provide reports to the Ministry of Fisheries and Marine Resources), and 'a functioning Fisheries Advisory Council' (the intent to form this council is expressed). In contrast to the degree of value and importance given to institutional coordination, there is in fact very little treatment of how this will be achieved in practice and through which forums. We discuss this further later on in this chapter.

## Areas Indicative of Incoherence

Prospects for policy incoherence are observed in two themes: *equitable access and distribution* and *tenure rights*. First, the SSF Guidelines places a considerable emphasis on equitable access to resources and just distribution of fishery benefits. It states: "States should where appropriate grant preferential access of small-scale fisheries to fish in waters under national jurisdiction, with a view to achieving equitable outcomes for different groups of people, in particular vulnerable groups… States should adopt measures to facilitate equitable access to fishery resources for small-scale fishing communities" (p. 6). These recommendations converge on ensuring an equitable outcome for small-scale fisheries and their vulnerable subgroups as a whole in recognition of the lop-sided political economy that has historically disadvantaged the small-scale sector (Bailey 1988; Pinkerton and Davis 2015). Ensuing marginalization of small-scale fisheries as implicated in the 'sectoral politics' has been identified a key problem around the world (Jacquet and Pauly 2008; Chuenpagdee 2011; Song and Chuenpagdee 2015). This, unfortunately, is reflected in Solomon Islands, and in much of the Pacific as well, in that coastal and small-scale fishers receive little policy attention and resources for management, development, and research relative to commercial tuna fisheries, in particular.

While *equitable access and distribution* has received relatively scarce attention in the existing policy field, appearing in only three out of 11 documents (see Table 4.3), when it is mentioned, it is shown to highlight another important aspect about equity. For example, the New Song includes phrases such as: "More equitable access to benefits and decision making within communities, including women, youth and marginalised groups" (p. 10), "equitable access to the resource and benefits from coastal fisheries within communities" (p. 14), and "[management] plans take account of equity issues, especially those involving gender and youth" (p. 14).

In these instances, emphasis is placed in the equity issues that arise within communities among different individuals or sub-groups (i.e., local-level interactions). This is indeed a highly pertinent issue in everyday fisheries management affairs upon which community functioning and individual well-being hinge, for instance, through elite capture (Béné et al. 2009; Cohen and Steenbergen 2015). However, what is missing here is the equity considerations of coastal small-scale fisheries in relation to larger-scale industrial fleets. Hence, we see that the SSF Guidelines and the New Song are, in fact, discussing two different dimensions of small-scale fisheries equity issues. Without fully recognizing and integrating these two perspectives, what may transpire is the prioritization of small-scale fisheries over large-scale fisheries, leading to the persistence of inequitable relations in the community (a bigger pie but unfairly split pieces), or achieving just mechanisms to share resource wealth but with few resources to distribute in the first place (fair splitting of a pie that remains small). As competition for coastal resources intensify and as global markets reach further into Solomon Islands marine resources, the potential for both forms of inequity may well increase.

Secondly, policy incoherence may also surface from the *tenure rights* theme. Tenure rights are a popular and well-circulated notion recently elevated through the Voluntary Guidelines on the Responsible Governance of Tenure of Land, Fisheries and Forests in the Context of National Food Security (Food and Agrictulture Organisation 2012). In it, tenure systems are defined as formal law or informal arrangements that regulate who gains access to natural resources, for how long and under what conditions. Its significance in relation to vulnerability, food security, and poverty is deservedly reflected in the SSF Guidelines. Customary tenure is also referred to in most policy documents we reviewed (see Table 4.3). While this consistency should be regarded as a positive step towards securing tenure rights for those working in the small-scale fishery, upon an in-depth look, there may be important discrepancies that could erode any perceived policy coherence.

Customary marine tenure in Solomon Islands (and many other Pacific Island Countries) has a long history ingrained in the cultural and socio-economic makeup of inshore fishing communities (Ruddle et al. 1992). It is effectively inseparable from community functioning and coastal villagers' life-world (Siriwardane-de Zoysa and Hornidge 2016). Still forming the backbone of rural governance structures, the emphasis on customary tenure is evident in policy documents. For instance, the Fisheries Management Act of Solomon Islands defines customary fishing as "fishing by indigenous Solomon Islanders, in waters where they are entitled by custom to fish, where – the fish are taken in a manner that, having regard to the boat, the equipment and the method used, is substantially in accordance with the indigenous Solomon Islanders' customary traditions…" (p. 16), and refers to the 'customary rights area' as "the areas within Solomon Islands waters that communities of indigenous Solomon Islanders own, use or occupy according to current customary usage" (p. 42). Similarly, the Western Province Fisheries Ordinance specifies that "Customary Fishing Rights refers to the rights that certain indigenous groups of people from the Western Province are able to establish over certain fishing areas by virtue of historical use and association with such areas of water and through acknowledgement of such rights by traditional leaders" (p. 7).

The difference is that the SSF Guidelines offer a view of tenure rights that is broader, that is to include also the rights of non-indigenous and non-traditional people insofar as they form a part of a small-scale fishing community, as it states "Local norms and practices, as well as customary or otherwise preferential access to fishery resources and land by small-scale fishing communities including indigenous peoples and ethnic minorities, should be recognized, respected and protected in ways that are consistent with international human rights law" (FAO 105, 5). While this subtle variation in the scope of how tenure holders are perceived may not appear particularly contradictory, incoherence becomes apparent if one traces its economic ramifications. The SSF Guidelines are unequivocal that "Small-scale fishing communities need to have secure tenure rights to the resources that form the basis for their social and cultural well-being, their livelihoods and their sustainable development" (FAO 2015, 5). In other words, tenure rights are central to the realization of food security, sustainable livelihoods, and economic growth (also see 5.2 of the SSF Guidelines). On the contrary, customary tenure as stipulated by the Fisheries Management Act encompasses only those activities regarded to be consistent with the emblematic image of customary fishing in Solomon Islands. For example, it declares that customary fishing is only recognizable where "any boat used is small scale, individually operated and if motorised does not have more than one motor; the fish are taken primarily for household consumption, barter or customary social or ceremonial purposes; and the fish are not taken or used for commercial purposes" (p. 16). What this means, however, is that such a view (if applied in practice) would restrict the full range of potential economic benefits derived from a tenure-protected fishery including fishing for commercial exchanges, thereby placing the idea and practice of tenure rights in contradiction with what is envisioned in the SSF Guidelines. Broadly, it seems that tenure rights are, in Solomon Islands, seen as instrumental for protecting the cultural and subsistence functions of small-scale fisheries, but (at least in these policy documents) not as an instrument to enable profit and economic growth.

## Gaps

*Gender* and *Human rights* themes are conspicuously featured in the SSF Guidelines. They appear to hold an important position as a normative and deep-seated foundation that can enable an expression of other concerns (such as those expressed by the themes of *human and social development*, *fisher participation*, *tenure rights*, *capacity building*, and *post-harvest economic development*). We deal firstly with gender. The low frequency with which gender (as shown in Table 4.3) is mentioned in the policy documents we analyzed is therefore a noteworthy outcome. Does the current absence mean that gender is to be overlooked and policy incoherence is unavoidable? We cannot be certain. Despite the relative lack of attention to the gender aspect, when gender sensitivities are expressed, they appear quite consistent with the provisions in the SSF Guidelines. For example, the 2015 completed New Song is unequivocal in promoting gender balance: 'Gender relations have a significant effect on the course of development and so the voice of women and youth must be heard and acted upon effectively in all future CBRM strategies. In addition to playing a greater role in decision-making, women and youth must have more equitable access to the benefits flowing from coastal fisheries' (p. 6). Nonetheless, to be in line with the high emphasis placed on gender equity in the SSF Guidelines, there is a need for wider inclusion of gender priorities in the existing and any newly-forming policies.

Secondly, the SSF Guidelines have clearly stipulated that small-scale fisheries are to be treated with international human rights standards such as universality and inalienability, non-discrimination and equality, participation and inclusion, accountability, and the rule of law. Despite being inscribed in the Solomon Islands' national constitution, we find that the high-level ideals of human rights have yet to percolate down to domestic and regional policy documents. This deficit may be because it has so far lacked a superior policy impetus for this translation to occur, which the SSF Guidelines are poised to provide, or because their reiteration is seen as unnecessary. This likely represents the biggest and most significant gap between the Guidelines and the existing instruments (together with the gender aspect). There is ample room that formulation of implementation strategies for themes such as human social development, tenure rights and capacity building explicitly incorporates the idea of human rights. At minimum, there would need to be a re-examination of the existing plans to ensure their compliance with such principles.

## *What Mechanisms and Arenas Are Appropriate to Promote Policy Coherence? Do These Currently Function in Solomon Islands?*

Based on this initial look at the potential interplay of policies that pertain to coastal, small-scale fisheries in Solomon Islands, we can be reasonably confident that policy coherence would generally prevail across scales when the SSF Guidelines enters the implementation phase. Many key policy documents already contain provisions that delve into the themes championed in the SSF Guidelines, as shown in Table 4.3. There were themes that were yet to be adequately elaborated, such as gender and human rights. In addition, important themes such as equitable access and tenure rights, upon closer reading, may incite potential inconsistencies as we observed some contradictory elements. This is no reason to be discouraged, however, as a certain degree of incoherence is deemed inevitable in all pluralist societies (Carbone 2008). But to increase the chances of successful implementation we can, and should, strive to achieve greater coherence. The findings of this study highlight some differences and discrepancies between policies that could be initial points of focus to move the SSF Guidelines from intentions to implementation.

One way that even a very crowded policy field can still cohere is through 'integrative properties' or 'policy glue' that connect issues and interests, such as a clear set of goals, a compelling policy image, and the strong involvement of the institutions that perform as the executive agency (May et al. 2006; Jordan and Harpin 2006, 23; Carbone 2008, 327). Who controls the governing vision or images that form the basis of policy formulation and how to administer them to influence other actors and sustain their interests are key meta-concerns. For instance, Forster and Stokke (1999) submits that coherence depends on the capacity and commitment of the political and administrative leadership at the centre, whether the national government at the domestic level or a supra-national organization at the regional or global level. In this sense, the SSF Guidelines (along with their explicitly articulated guiding principles) can be understood as serving a meta-governance function, providing the high-order rationale to the lower-scale policies that can help align various policy objectives (Song et al. 2013; Morrison 2014). The meta-governance approach (Kooiman 2003) of the SSF Guidelines would be an indication of both the political importance and cross-cutting nature of small-scale fisheries, and an acknowledgment of the national sovereignty of the FAO member states as well as the semi-autonomous interests of regional institutions. Meta-governance thus represents this tacit and often behind-the-scenes steering of policies. Similarly, within these nation-states, adoption of the SSF Guidelines provides a strong political signal that small-scale fisheries functions and governance in the context of food security and poverty eradication needs to be integrated and taken seriously across all sectors (Morrison and Lane 2005). At the same time, local champions are needed to mainstream the SSF Guidelines at the on-the-ground level (Cuevas et al. 2015).

Practically, multiple cross-agency, cross-scale, and cross-sector learning and governance networks could act as mechanisms or arenas to negotiate and promote policy coherence. In Solomon Islands governance networks with concerns over coastal (including fisheries) resources include a national branch of the Asia-Pacific 'LMMA' network (Cohen et al. 2012) and the National Coordinating Committee for the CTI (unpublished data). These networks bring together government and non-government organizations, with mandates spanning social and economic development, environmental management, and human health and well-being. They provide a forum for deliberation between multiple objectives and actors, as well as a point at which the design of actions (to address regional to national policies) can be tailored to fit the Solomon Islands context. In some cases, they also promote representation of small-scale fishers and managers in higher level policy forums. Most directly, they offer an avenue through which to pursue the 'institutional coordination and strengthening' objectives, and it has also been suggested that these networks may be logical candidates to mainstream and implement the SSF Guidelines (Nisa 2014). Whilst these networks have had successes in improving local level representation and increasing coordination between agencies, they also face substantial logistical, financial, and political challenges in effectively acting as arenas and mechanisms for multi-stakeholder dialogue and coordination (e.g., Ratner et al. 2013). Improving policy coherence, addressing policy gaps, and continually strengthening institutional coordination will require time and resources, the distribution of which may affect more grounded efforts of implementation such as managing resources or building the capacity of managers.

## Conclusion

The SSF Guidelines explicitly call for policy coherence to promote substantive policy delivery; conversely, incoherence will likely engender coordination problems, implementation gaps, and conflicts between regimes (May et al. 2006; Jordan and Harpin 2006). The policies we examined are guiding the way in which marine resource conservation, coastal and rural development, and small-scale fisheries governance are tackled within Solomon Islands. Amid the dense small-scale fisheries policy field observed in Solomon Islands, we found numerous areas of coherence, which suggest that the SSF Guidelines enter a policy and implementation landscape that is relatively consistent with the intentions laid out in the Guidelines. As a result, we submit that the implementation of the SSF Guidelines will largely be a case of continuing to strengthen the implementation of these policies. Yet, the policy gaps and more nuanced areas of incoherence we identify suggest early hurdles for implementation, and these gaps in fact highlight areas in which the SSF Guidelines can potentially have the greatest impact on the current policy landscape within Solomon Islands.

It is important to recognize that the policies we examined sit within a political landscape that has the potential to be more influential on small-scale fisheries than the intensions formally laid out in policy. Within the sector priorities may shift between supporting the welfare function of small-scale fisheries (Béné et al. 2010), to promoting economic goals and the wealth generating functions of small-scale fisheries (sensu Cunningham et al. 2009). The relative weight given to these priorities, or indeed any of the themes that are analyzed, will be heavily influenced by the dominant development vision of the nation which can shift, for instance, with changing political cycles or international donor policies. For example, emphasis on commercial industry and economic growth would impact upon small-scale fisheries resources and fishers differently to emphasis on improving access to services in rural areas. Understanding the influence of this broader political landscape on the implementation of the SSF Guidelines is beyond the scope of this analysis, but is an important area for future research.

While our analysis has been specific in its application to a Solomon Islands case situated in the Pacific regional setting, our approach is easily applicable to other countries and provides a way to diagnose areas of incoherence and gaps, and also opportunities for the SSF Guidelines to prompt adjustment and transformation of existing policies. We suggest that this type of policy-based analysis is a useful first step to understanding priorities and strategies for implementation. Further, we hope that our findings are illustrative of how and why policy coherence can be an influential concept for informing multi-scalar implementation of the SSF Guidelines. Nonetheless, our analysis and findings remain restricted to the policy level. A suite of further challenges exist in the space between policy intent and actual implementation; while this is beyond the scope of this study, the challenges and opportunities that rest in this space are expected to be substantial. As the national governments begin to take on this massive implementation task, *coherent policy aims across scales* should, however, give us renewed optimism that the small-scale fisheries policy interplay can be a governable process, providing a sensible platform for organizing implementation efforts and ultimately boosting our chance of realizing the noble visions laid out in the SSF Guidelines.

## References


Adams, T. (2012). The characteristics of Pacific Island small-scale fisheries. *SPC Fisheries Newsletter, 138*(May/August), 37–43.

Bailey, C. (1988). The political economy of marine fisheries development in Indonesia. *Indonesia, 46*, 25–38.

Bell, J., Kronen, M., Vunisea, A., Nash, W. J., Keeble, G., Demmke, A., Pontifex, S., & Andrefouet, S. (2009). Planning the use of fish for food security in the Pacific. *Marine Policy, 33*, 64–76.

Béné, C., Belal, E., Baba, M. O., Ovie, S., Raji, A., Malasha, I., & Njaya, F. (2009). Power struggle, dispute and alliance over local resources: Analyzing 'democratic' decentralization of natural resources through the lenses of Africa inland fisheries. *World Development, 37*(12), 1935–1950. doi:10.1016/j.worlddev.2009.05.003.

Béné, C., Hersoug, B., & Allison, E. H. (2010). Not by rent alone: Analyzing the pro-poor functions of small-scale fisheries in developing countries. *Development Policy Review, 28*(3), 325–358.

Carbone, M. (2008). Mission impossible: The European Union and policy coherence for development. *Journal of European Integration, 30*(3), 323–342.

Chuenpagdee, R. (2011). *World small-scale fisheries: Contemporary visions*. Delft: Eburon Academic Publishers.

Cohen, P., & Steenbergen, D. (2015). Social dimensions of local fisheries co-management in the coral triangle. *Environmental Conservation, 42*(3), 278–288. doi:10.1017/S0376892914000423.

Cohen, P., Evans, L., & Mills, M. (2012). Social networks supporting governance of coastal ecosystems in Solomon Islands. *Conservation Letters, 5*, 376–386. doi:10.1111/j.1755-263X.2012.00255.x.

CTI Secretariat. (2009). *Regional plan of action; coral triangle initiative on coral reefs, fisheries and food security (CTI-CFF)* (p. 42). Manado: Interim Regional CTI Secretariat.

Cuevas, S. C., Peterson, A., Robinson, C., & Morrison, T. H. (2016). Institutional capacity for long-term climate change adaptation: Evidence from land use planning in Albay, Philippines. *Regional Environmental Change, 16*(7), 2045–2058.

Cunningham, S., Neiland, A. E., Arbuckle, M., & Bostock, T. (2009). Wealth-based fisheries management: Using fisheries health to orchestrate sound fisheries policy in practice. *Marine Resource Economics, 24*(3), 271–287.

Davis, A., & Ruddle, K. (2012). Massaging the misery: Recent approaches to fisheries governance and the betrayal of small-scale fisheries. *Human Organization, 71*(3), 244–254.

den Hertog, L., & Stroß, S. (2013). Coherence in EU external relations: Concepts and legal rooting of an ambiguous term. *European Foreign Affairs Review, 18*(3), 373–388.

di Francesco, M. (2001). Process not outcomes in new public management? 'Policy coherence' in Australian government. *The Drawing Board: An Australian Review of Public Affairs, 1*(3), 103–116.

Evans, L., Cherrett, N., & Pemsl, D. (2011). Assessing the impact of fisheries co-management interventions in developing countries: A meta-analysis. *Journal of Environmental Management, 92*(8), 1938–1949. doi:10.1016/j.jenvman.2011.03.010.

FAO. (2015). *Voluntary guidelines for securing sustainable small-scale fisheries in the context of food security and poverty eradication*. Rome: Food and Agriculture Organization of the United Nations.

Forster, J., & Stokke, O. (1999). Coherence of policies towards developing countries: Approaching the problematique. In J. Forster & O. Stokke (Eds.), *Policy coherence in development co-operation* (pp. 16–57). London: Frank Cass.
Gillett, R. (2009). *Fisheries in the economies of Pacific Island countries and territories*. Mandaluyong City: Pacific Studies Series.
Gillett, R., & Cartwright, I. (2010). *The future of Pacific Island fisheries*. Noumea: Secretariat of the pacific community.
Govan, H., Tawake, A., Tabunakawai, K., Jenkins, A., Lasgorceix, A., Schwarz, A.-M., Aalbersberg, B., Manele, B., Vieux, C., Notere, D., Afzal, D., Techera, E., Rasalato, E. T., Sykes, H., Walton, H., Tafea, H., Korovulavula, I., Comley, J., Kinch, J., Feehely, J., Petit, J., Heaps, L., Anderson, P., Cohen, P., Ifopo, P., Vave, R., Hills, R., Tawakelevu, S., Alefaio, S., Meo, S., Troniak, S., Malimali, S., Kukuian, S., George, S., Tauaefa, T., Obed, T. (2009). *Status and potential of locally-managed marine areas in the South Pacific: meeting nature conservation and sustainable livelihood targets through wide-spread implementation of LMMAs* (p. 95). Suva.
Healy, J., & Hauirae, J. (2006). *Bismarck Solomon seas ecoregion: Solomon islands' fisheries, marine and coastal legislation and policy gap analysis* (p. 67). Honiara.
Jacquet, J., & Pauly, D. (2008). Funding priorities: Big barriers to small-scale fisheries. *Conservation Biology, 22*(4), 832–835.
Jordan, G., & Harpin, D. (2006). The political costs of policy coherence: Constructing a rural policy for Scotland. *Journal of Public Policy, 26*(1), 21–41.
Kooiman, J. (2003). *Governing as governance*. London: Sage.
Lane, M. B. (2006). Towards integrated coastal management in Solomon Islands: Identifying strategic issues for governance reform. *Ocean & Coastal Management, 49*(7–8), 421–441.
May, P. J., Sapotichne, J., & Workman, S. (2006). Policy coherence and policy domains. *The Policy Studies Journal, 34*(3), 381–403.
Morrison, T. H. (2014). Developing a regional governance index: The institutional potential of rural regions. *Journal of Rural Studies, 35*, 101–111.
Morrison, T., & Lane, M. (2005). What 'whole-of-government' means for environmental policy and management: An analysis of the connecting government initiative. *Australasian Journal of Environmental Management, 12*(1), 47–54.
Nilsson, M., Zamparutti, T., Petersen, J. E., Nykvist, B., Rudberg, P., & McGuinn, J. (2012). Understanding policy coherence: Analytical framework and examples of sector–environment policy interactions in the EU. *Environmental Policy and Governance, 22*, 395–423.
Nisa, Z. (2014). Sharing the responsbility of the implementation of the new International voluntary guidelines for small-scale fisheries via regional and national networks. University of Rhode Island.
OECD. (1996). *Building policy coherence: Tools and tensions*. Paris: OECD.
Pinkerton, E., & Davis, R. (2015). Neoliberalism and the politics of enclosure in north American small-scale fisheries. *Marine Policy, 61*, 303–312.
Pomeroy, R. S., & Berkes, F. (1997). Two to tango: The role of government in fisheries co-management. *Marine Policy, 21*(5), 465–480. doi:10.1016/s0308-597x(97)00017-1.
Ratner, B. D., Cohen, P., Barman, B., Mam, K., Nagoli, J., & Allison, E. H. (2013). Governance of aquatic agricultural systems; analyzing representation, power and accountability. *Ecology and Society, 18*(4), 59.
Ruddle, K. (1998). The context of policy design for existing community-based fisheries management systems in the Pacific Islands. *Ocean & Coastal Management, 40*(2–3), 105–126.
Ruddle, K., Hviding, E., & Johannes, R. (1992). Marine resources management in the context of customary tenure. *Marine Resource Economics, 7*, 249–273.
Secretariat of the Pacific Community. (2015). *A new song for coastal fisheries – pathways to change: The Noumea strategy* (p. 14). Noumea.
Sen, S., & Nielsen, J. R. (1996). Fisheries co-management: A comparative analysis. *Marine Policy, 20*(5), 405–418.

Siriwardane-de Zoysa, R., & Hornidge, A.-K. (2016). Putting lifeworlds at sea: Studying meaning-making in marine research. *Frontiers in Marine Science, 3*(197). doi:10.3389/fmars.2016.00197.

Song, A. M., & Chuenpagdee, R. (2015). A principle-based analysis of multilevel policy areas on inshore fisheries in Newfoundland and Labrador, Canada. In S. Jentoft & R. Chuenpagdee (Eds.), *Interactive governance for small-scale fisheries: Global reflections* (pp. 435–456). Dordrecht: Springer.

Song, A. M., Chuenpagdee, R., & Jentoft, S. (2013). Values, images and principles: What they represent and how they may improve fisheries governance. *Marine Policy, 40*, 167–175.

Veron, J., Devantier, L. M., Turak, E., Green, A. L., Kininmonth, S., Stafford-Smith, M., & Perterson, N. (2009). Delineating the coral triangle. Galaxea. *Journal of Coral Reef Studies, 11*(2), 91–100.

# Chapter 5
# Tuna or Tasi? Fishing for Policy Coherence in Zanzibar's Small-Scale Fisheries Sector

**Lars Lindström and Maricela de la Torre-Castro**

**Abstract** Zanzibar in 1964 merged with Tanganyika to become the United Republic of Tanzania (URT). Zanzibar enjoys autonomy in the governance of marine resources having adverse effects on the implementation of the *Voluntary Guidelines for Securing Sustainable Small-Scale Fisheries in the Context of Food Security and Poverty Eradication* (SSF Guidelines) as Zanzibar is not a member of the FAO as a unit on its own, but only as a part of the URT. While the Guidelines were still unknown to Zanzibar a new fisheries policy was formulated complicating their implementation, as the Guidelines clashes with the new fisheries policy. We examine this clash using the concept policy coherence defined as the coherence between (a) development and other policies, and (b) development policies of different donors. We downscale it to apply to policies within *one* sector, small-scale fisheries, by comparing the fisheries policy which is grounded in liberal ideas like commercialization and capitalization, with the SSF Guidelines which ideationally are based in human rights and a view of fishing as also culture and not just any economic activity subject to economic laws. We argue that conflicts between the two may result in failure to implement the SSF Guidelines as they do not come with World Bank and other external funding as the new fishery policy does. Choosing between conflicting policy elements the choice will likely be the fishery policy if the implementation of SSF Guidelines comes with a cost.

**Keywords** Policy coherence • Institutional fit • Small-scale fisheries governance • Zanzibar

L. Lindström (✉)
Department of Political Science, Stockholm University, SE-106 91, Stockholm, Sweden
e-mail: lasse.lindstrom@statsvet.su.se

M. de la Torre-Castro
Department of Physical Geography, Stockholm University, Stockholm, Sweden
e-mail: maricela@natgeo.su.se



S. Jentoft et al. (eds.), *The Small-Scale Fisheries Guidelines*, MARE Publication Series 14, DOI 10.1007/978-3-319-55074-9_5

## Introduction

In June 2014, a historical event in the advancement of the valuation of global fisheries took place: the endorsement of the *Voluntary Guidelines for Securing Sustainable Small-Scale Fisheries in the Context of Food Security and Poverty Eradication* (hereafter SSF Guidelines) by 147 FAO member countries. This document represents the first real acknowledgement of the important role of small-scale fisheries for not only food but also income security for the world population, contributing as it does to about half of global fish catches (FAO 2015, ix).

More importantly, though, they are targeted at the equitable development of small-scale fishing communities and consider its contribution to poverty eradication. The SSF Guidelines do so by emphasizing participatory policies and the rule of law, as well as by raising public awareness of the broader cultural and social role played by small-scale fisheries.

What remains now for those countries that have adopted the SSF Guidelines is to implement them, or to 'walk the talk' (Jentoft 2014). The implementation gap - the distance between stated goals and objectives and the realization of them - is often wide, particularly in developing countries. Corruption, lack of human and material resources, and unstable political environments are only a few of the factors that impede implementation in these contexts.

However, we will analyze a different implementation problem: the one of 'double talk', or when policies are not coherent, which can create synergies, but rather are in conflict with one another, which turns the walk into limping. The concept of 'policy coherence' has gained increasing weight ever since the Paris Declaration on Aid Effectiveness in 2005, and is now part of Organization for Economic Cooperation and Development (OECD) and European Union aid frameworks. It is also part of the SSF Guidelines' instruments.

In this chapter, we turn our attention to the case study of Zanzibar. The former sultanate of Zanzibar is an archipelago off Tanzania's coast that merged with former Tanganyika in 1964 to become the United Republic of Tanzania (URT). Zanzibar enjoys autonomy in non-union matters, one of which is the governance of marine resources. Thus, the URT and Zanzibar have their own institutions, both political and administrative, for the governance of small-scale fisheries. As a consequence of the peculiar institutional design of the URT, which resembles a federation without a federal level, Zanzibar is not a member of the FAO as a unit of its own, but through its inclusion within the URT state. Zanzibar was, therefore, not invited to the FAO negotiation process that began in 2010 and ended in the adoption of the SSF Guidelines in 2014. Zanzibar neither participated on its own terms nor was invited to the Tanzanian representation. Thus, the talk was unheard by Zanzibar's government! Zanzibar authorities had no previous knowledge, or opportunities to be part, of the negotiations, consultation process, or the drafting of the final document.

It seems unlikely, though, that the architects behind the new fisheries policy (FP), SWIOfish and SMARTfish (see below), were unaware of this process, which started in 2010 and involved more than 4000 representatives of governments, small-scale fishers, fish workers and their organizations, researchers, development partners, and other relevant stakeholders from more than 120 countries in six regions and more than 20 civil society organization-led national consultative meetings, leading up to the endorsement of the SSF Guidelines in June 2014 (FAO 2015, v).

The Zanzibar elections in October 2015 were annulled because of alleged violations of the electoral laws and the political situation remained in limbo until new elections were held in March 2016. It was not until early summer 2016 that the new administration had taken over.

In addition to the lack of knowledge about the FAO negotiations by key authorities involved in marine resources, the SSF Guidelines were also unknown to other members of Zanzibar society. Interviews with representatives of civil society organizations and individual fishers in July–August 2016 showed a complete unawareness of even the existence of the SSF Guidelines.

As we will demonstrate below, these factors will have adverse effects on the prospects for the effective implementation of the SSF Guidelines in Zanzibar. Furthermore, since a new FP was being developed while negotiations in the context of FAO were taking place, policy coherence with the new Guidelines under development was not on the agenda in creation of the new FP.

In what follows, we will present Zanzibar's small-scale fisheries context, then present the policy framework and, finally, examine the inherent conflicts between the new FP and the SSF Guidelines. At the end of the chapter, we interpret our findings against the broad key objectives in the SSF Guidelines.

## Small-Scale Fisheries in Zanzibar

Small-scale fisheries in Zanzibar are *de facto* operated as an open access system, mainly taking place inshore in intertidal areas with less than 20–30 m depth around both major islands, Pemba and Unguja. The fishers use traditional vessels like small wooden boats, canoes, and outrigger canoes chiseled out from huge tree trunks, and use a variety of fishing gears including different types of nets (drag-nets, gill nets, ring nets, beach-seines, etc.), basket traps (*demas*), longlines, as well as hook and line. Fishing takes place along the whole seascape comprising different ecosystems - coral reefs, mangroves, sandbanks, and seagrass beds - while a few fishers reach deeper waters (de la Torre-Castro et al. 2014). A large variety of fin-fish and invertebrate species is targeted, and it has been considered that the fishery in Tanzania (including Zanzibar) is in decline and in urgent need of management reform (Jiddawi and Öhman 2002; Eriksson et al. 2010). There are about 35,000 fishers using approximately 8500 vessels, of which 10–15% are equipped with outboard engines (ZFC 2007, 2010; RGZ 2016).

It is difficult to estimate the importance of fish production and related activities to the national economy, since large parts of the sector are predominantly unregulated. Although monitoring agents (*Bwana dikos*) take catch data, the statistics are error prone (de la Torre-Castro 2006). In addition, the coastal population depends almost entirely on small-scale fisheries for income and food security. With a population of about 1.3 million and an average family size of seven persons – a staggering 250,000 persons, or about 20% of Zanzibar's total population – are directly dependent on small-scale fisheries.

Moreover, small-scale fisheries are by and large the most important livelihood sector in the coastal villages in Unguja. Based on information in diaries, the small-scale fisheries sector (including not only harvest, but 21 different livelihood activities) is the most important for the coastal people. It provides both the highest income and the highest frequency of performed activities. Although male-dominated, small-scale fisheries in Zanzibar are performed by, and are important for, both men and women. There is a gendered differentiation of activities, ecosystems used, and species targeted, but the importance of small-scale fisheries are high for all coastal inhabitants. In this context, even small children at pre-school age fish in their villages and start interacting with markets, active fishers, and traders as early as before five years of age.

Markets are mainly local, but some valuable products, such as sea cucumbers, have IUU (Illegal, Unregulated, and Unreported) characteristics and reach international markets through clandestine commercialization that takes place parallel to the established legal market (Eriksson et al. 2012). The local markets present a high degree of complexity, with a myriad of actors varying from auctioneers, middlemen, credit providers, sidewalk fishers, women fish traders, fish cleaning children, and many other actors. The market chains are also complex, with distribution of fish products to local people, restaurants, hotels, and larger markets in Zanzibar town (e.g. Darajani and Mwanakwerekwe). Market dynamics are not thoroughly studied, but it is clear that the inclusion of female fish traders in management would be a positive management action, as well as a deeper understanding of the ecological effects of market exploitation which targets different functional groups and all maturity stages of fish (Fröcklin et al. 2013; Thyresson et al. 2013). The high fishing pressure, as well as the complex market in which the diverse array of market agents currently buys up all that is offered, seem to create a situation in which there is no refuge for fish (and invertebrates) to complete their life-history undisturbed.

Small-scale fisheries in Zanzibar have a management history similar to many other sites in developing countries. They have passed through periods of local traditional management marked by sustainable practices and robust local institutions. Like in many other countries in post-war and/or post-colonial periods, fisheries in Zanzibar experienced radical systemic changes characterized by external management interventions and the disruption of historical well-functioning management practices by donor-funded projects (de la Torre-Castro and Lindström 2010; de la Torre-Castro 2012). The introduction of extensive Marine Protected Areas (MPAs) with the support of the World Bank between 2005 and 2011 has radically changed the national management path, moving Zanzibar's fisheries into traditional Western management. Small-scale fisheries in Zanzibar are clearly in need of better governance. Catch decline, destruction of ecosystems, inadequate fishing gears, gender inequality, and lack of proper policy and management are major problems threatening the sustainability of the overall fisheries governance system. The problem of destructive gear use and the unsuccessful gear exchange programs are particularly important negative factors (Wallner-Hahn et al. 2016). Also, the selective fishing of key species, such as sea-urchin predators, contributes to cascade effects which are detrimental to both coral and seagrass ecosystems (Wallner-Hahn et al. 2015).

## *The Institutional Setup*

Zanzibar is a semi-autonomous part of the URT, and RGZ and its Ministry of Agriculture, Natural Resources, Livestock, and Fisheries (MANLF) enjoy full rights to regulate marine resource use. Thus, RGZ exercises sovereignty over fisheries governance. Fisheries are formally regulated by mainly the *Fisheries Law* of 1993 and the *Fisheries Act* of 2010, but also informally by the daily praxis of fishers and other stakeholders (RGZ 1993, 2010a).

In addition to the legislative assembly, the government agencies engaged in the governance of fisheries are MANLF at the central level, a number of District Fisheries Offices at the regional level, and *Bwana Diko* (fishing monitoring agents), who operate at the majority of landing sites. At the local level, there is a variety of informal institutions established by fishers themselves, as well as the fishers' committees that were formalized by the previously called Department of Fisheries during the last decade (Lindström and de la Torre-Castro 2015).

The *Bwana Dikos* have a key role at the local level as they record catches on a regular basis and report these to higher levels of government, as well as issue fishing licenses and are supposed to report law breakers. They are, however, caught in a number of dilemmas. Their remuneration is low and issues related to kinship, multitasking, poverty, and control complicate their monitoring and sanctioning roles (de la Torre-Castro 2006).

## *Policy Coherence*

In the following analysis, we will make use of the concept of *policy coherence* that has become one of the pillars in OECD and EU development policy, as well as in FAO documents like the one on trade and agricultural policy and as a policy objective in the SSF Guidelines (FAO 2006, 2015, 10.1; EC 2013; OECD 2016). The concept refers to the reduction of conflicts and the promotion of synergies between, both vertically and horizontally, and within different policy areas to achieve the desired outcomes (Nilsson et al. 2012). We apply the concept to Zanzibar's fisheries sector by analyzing horizontal conflicts within the fisheries sector, showcased in the comparison of the new FP announced in 2014 and the SSF Guidelines (RGZ 2014, 2016; FAO 2015). More precisely, we will conduct a policy analysis by comparing the policy objectives and instruments in the two policies in terms of conflicting and incompatible objectives and instruments.

We argue that the less coherence between the two policies, the more difficult it will be for government agencies to implement the SSF Guidelines. This challenge is heightened by the fact that the new FP comes with international donor funding. Struggling with meagre finances makes it tempting to choose that the policy option which comes with funding.

## *Policy Framework*

Tanzania has been a major recipient of foreign aid for decades and the coastal management sector, including fisheries, are no exemption. The World Bank has been a major actor in shaping the governance and management trajectories of small-scale fisheries. Since the early 1980s, there have been efforts and donations directed at commercialization and the increase of fish catches (World Bank 1984). In addition, from the early 2000s, the Bank's presence has been critical in changing the management system through the Marine and Coastal Environment Management Project (MACEMP) (World Bank 2012). The project was made possible with an approximate $63 million USD World Bank loan and a grant from the Global Environment Facility (World Bank 2013). It was designed and implemented in cooperation with the Union Government and the RGZ. The purpose of MACEMP was to 'improve management of coastal and marine resources, to enhance the contribution of these resources to economic growth, to reduce poverty, and to develop the scientific understanding of the marine and coastal resources and major threats to them' (World Bank 2012). However, during the period of intervention (2005–2011), the project mainly focused on creating a series of MPAs and, although the objective was to achieve conservation and development at the same time, the positive results of the project are limited and difficult to see by the local people, especially since many issues of procedural and distributive injustices have been reported (Gustavsson et al. 2014). A new project supported by the World Bank after MACEMP is now in effect, the 'First South West Indian Ocean Fisheries Governance and Shared Project Growth' (SWIOFish). This project comprises a larger geographical area, targeting South West Indian Ocean countries and addressing fisheries as a key sector in which regional collaboration, improved management, and governance, as well as economic benefits, are needed. The investment for this project consists $75.5 million USD from the International Development Association (IDA) and $15.5 USD from the Global Environmental Facility. Each country included in the project elaborates its own fisheries policy. Tanzania will receive a sum of $36 million USD to continue the work for better governance and management, improve livelihoods, and increase private sector participation in fishing industry activities. This new World Bank project is projected to operate until 2021 (World Bank 2016).

Other guiding policy documents are *The Zanzibar Vision 2020* (RGZ 2002) and *The Zanzibar Strategy for Growth and Reduction of Poverty,* MKUZAII (RGZ 2010b). The former articulates the overall development goals for Zanzibar as the eradication of absolute poverty and the attainment of sustainable human development. The Vision's policy on fisheries is to strengthen the management of marine and coastal resources to support sustainable tourism development while conserving the richness of the environment. The Vision also recognizes the key role played by the fisheries sector in the social and economic development of the country.

*The Zanzibar Strategy for Growth and Reduction of Poverty*, MKUZAII, recognizes that fisheries are of great importance to the economy of Zanzibar. It also stresses that recent government efforts have been directed towards the conservation of marine and coastal environments and that this has largely contributed to significant increases in fish catch (however, there are no scientific catch data to substantiate this assertion). Despite this positive performance, MKUZAII highlights that marine resources further offshore in deeper waters are still underutilized, as most fisheries activities are done in inshore waters which are unsustainably overexploited. It also emphasizes that there is a great potential on the part of domestic fishers for offshore fishery expansion in Zanzibar.

## *Double Talk*

The new FP was announced on December 1, 2014, with a first draft published in June 2014 (RGZ 2014) and a second draft in July 2016 (RGZ 2016). The first draft had as its main objective the protection of coral reef ecosystems in order to promote tourism, a clearly non-fishing and non-ecological objective. This was removed from the second draft. Both drafts have been developed within the context of SWIOfish, financed by the World Bank, and SmartFish, funded by the EU. The second draft has three main objectives, namely to:

(i) Enhance sustainability of fishery resources through preserving biodiversity of marine ecosystems
(ii) Enhance the social and economic performances of the fishery sector through improvement fisheries management
(iii) Increase fish production and improve quality of fish and fish products in line with food security and safety requirements in compliance to international standards (RGZ 2016, 14).

These objectives are indicated to be achieved through a set of guiding principles: sustainability, conservation, research, equity, poverty reduction, gender equity, and good governance principles like participation, transparency, and accountability, as well as a number of policy instruments (RGZ 2016, 14–15). In the following section, we will analyze the policy objectives and instruments with bearing on small-scale fisheries, or artisanal which is the term used in the policy documents, in a comparison with the SSF Guidelines.

## From Tasi[1] to Tuna

A main objective in the second draft of the new FP is to 'promote responsible and sustainable development of artisanal fisheries further offshore in deeper waters with an aim to increase the contribution of the fishery sector to economic growth and food security. Meanwhile *to facilitate the removal of fishing capacity in shallow waters* for improved fisheries management in coastal zones' (RGZ 2016, 21).[2]

This goal, however, depends upon the replacement of simple wooden inshore vessels with large fiberglass ones equipped with engines which would enable fishers to go offshore. Through the MACEMP program, a few vessels like these, along with sizeable gill nets, were distributed freely among the villages to be controlled by the Village Fisheries Committees. However, our study shows that this introduction of technology contributed to distributional injustices, as only a few fishers were granted access to these vessels, predominantly the upper ranks of the Committees

**Fig. 5.1** *Tasi*, rabbit fish (*Siganus sutor*), at the local fish market. *Tasi* is key for small-scale fisheries in Zanzibar providing animal protein and monetary income (Photo credit: M. de la Torre-Castro)

[1] Tasi is the Kiswahili word for the predominantly seagrass-associated Rabbit fish *Siganus spp.* (see Fig. 5.1). Tasi is mainly fished inshore and is a valuable species particularly for food security and often targeted using basket traps *(demas)*.

[2] Emphasis added by authors.

(Gustavsson et al. 2014). In the new FP, fishers are expected to buy fiberglass boats, engines, and fishing nets by taking up loans on the commercial market (RGZ 2016, 22). Given that most fishers have an income slightly above the poverty line the policy will likely contribute to a segregation of the fishery into a few 'haves' and the many 'have nots'. Those that cannot afford the technical upgrading will be forced to remain inshore and intertidal areas, where they most probably will be squeezed by the extension and establishment of MPAs and other Marine Conservation Areas (MCAs). There are also risks that the poorest fishers may enter disadvantageous relationships with market traders, seeking loans and pressed by their wish to retain their livelihoods and fishing as a cultural life style.

This kind of stratification runs contrary to what the SSF Guidelines state: "States should also adopt measures to facilitate *equitable access* to fishery resources for small-scale fishing communities" (FAO 2015, 6).[3] Those remaining inshore will also struggle to compete on the market with their less valuable species (there are signs that intertidal invertebrates are decreasing and fish juveniles fetch lower value in the market). Thus, the policy may contribute to Zanzibar's export earnings but it is difficult to perceive how it will secure or improve food and income security of small-scale fishers or communities outside of a small elite group.

In addition, offshore fisheries will benefit from "technological (fishing and navigation techniques, type and localization of Fish Aggregating Devices, etc." (RGZ 2016, 22), as well as value addition like fish preservation on-board, fish storage, processing and marketing, further aggravating the situation for those left to remain inshore. It also implies that the vessels required will be far larger than, for example, simple 25 ft. fibreglass boats with outboard engines, thus requiring an even larger investment that few small-scale fishers can afford. Wallner-Hanh et al. (2016) demonstrated that the economic constraints to change gears among Zanzibarian fishers are huge and difficult to overcome. It is expected that, when it comes to investments in boats and all extra inputs needed for such a radical change in the way of fishing, the economic constraints will increase disproportionally for lower-income harvesters. The strategy of fishing offshore, rather, privileges large commercial vessels owned by pavement fishers employing a captain and a small crew manning 25 ft. wooden boats mainly engaged in inshore fishing, an already fairly common arrangement.

However, these changes are all in line with the objective of removing fishing capacity in shallow waters, which stands in conflict with the SSF Guidelines' principles of poverty reduction and equity (FAO 2015; RGZ 2016). The equity principle, as formulated in the new FP, is restricted to encouraging "local and international researchers into the equitable use, distribution, and conservation of marine resources in sustainable ways that limits in transboundary oceanic fisheries. Meanwhile, the Government will establish transparent and equitable rules and frameworks for assessing and distributing conservation burdens for all fisheries stakeholders" (RGZ 2016, 15). Indeed, this is a strange conceptualization of equity. That rules and

[3] Emphasis added by authors.

**Fig. 5.2** Back from work after one tidal cycle in the ocean, small-scale fishers leave their catch to particular 'assistants' to sell through auction (Photo credit: M. de la Torre-Castro)

frameworks are transparent and equitable is far from the same as the burdens of conservation being equitable distributed.

The SSF Guidelines speak differently. While the draft of the FP refers to *regulation* being equitable, the SSF Guidelines discusses equitable *distribution* of "the benefits yielded from responsible management of fisheries and ecosystems" (FAO 2015, 5).

The new FP puts much emphasis on leaving inshore fishing and instead develop an industrial tuna fleet together with "Zanzibar potential investors" (RGZ 2016: xx). It is interesting to note the emphasis given to include private actors in the new FP. In addition, the development of aquaculture activities in the form of seaweed, pearls, and fish is promoted (RGZ 2016, 19–25). Some of these priorities are, at least is on the surface, in line with the SSF Guidelines, which also looks to the creation of complementary and alternative income generating activities (FAO 2015, 9). What is problematic, however, is that fishers cannot realistically turn to seaweed aquaculture as this is predominantly, at least in Zanzibar, is a business of women and is one of the already existing complementary livelihoods for coastal families (Fröcklin et al. 2012). Men in the household are engaged in fin fishing, and women in seaweed farming (Fig. 5.2).

It is also questionable if seaweed farming is a sustainable venture, with common occurrences of declining crop yields and falling prices. The price of *Euchema den-*

*ticulatum,* locally known as *spinosum* seaweed, was previously around 700 Tanzanian shillings per kg, but is now selling at 300 shillings per kg, while the price for *Kappaphycus alvarezii,* locally known as *cottonii*, the high-quality variety of seaweed, has plummeted from around 1100 shillings per kg to 700 (*Daily Mail*, 1 May 2016). Seaweed farming is a major livelihood only for a few, while it is complementary activity for the overwhelming majority of the female farmers. Most of the seaweed farmers report that working with seaweed alone is not sufficient for a decent income and, thus, women seaweed farmers are forced to perform other activities. In fact, during the last 10–15 years an abandonment of the activity has been reported and observed. The new FP's push to cement a poor income generating activity such as seaweed farming will result in the reduction of income security, running contrary to one of the main objectives of the SSF Guidelines.

Together with the economic problems faced by the activity, climate change poses a larger threat to seaweed farming. Increasing sea surface temperature in the intertidal zone, where seaweed is grown, has negative effects and diseases have been reported, causing a substantial decrease in production (Lindström and de la Torre Castro 2016). The production of seaweed also shows several negative health effects for women (Fröcklin et al. 2012). Another concern related to climate change is shifting distribution patterns of target species in the overall fishery. Increasing sea temperatures in the wake of climate change result in the migration towards the poles of many fish species, as is also stated in the new FP (RGZ 2016, 26). This change will, however, also affect fish cultivations whether in ponds or in the sea itself and how this in the long run affects income and food security is not properly addressed.

When introduced, seaweed farming was targeted as an alternative livelihood for men to decrease fishing pressure. However, as soon as men realized the lack of profits, they abandoned the activity and women took over. Why do women perform such an activity which negatively affects their health and provides so little income? It has been argued that the advantage of seaweed farming is that it allows women to continue with activities of production and reproduction since they are able to take care of the household duties while earning some cash (de la Torre-Castro et al. in press). Given the historical context, the new proposition of male fishers turning to seaweed farming may completely fail and, if it succeeds, the shift will force women out from the business, effectively violating gender equity principles in an already patriarchal division of labour.

## Which Talk to Walk: Coherent Policies?

The ideas expressed in the different fisheries policy documents in Zanzibar clearly expose the wish to solve a perceived problem of an overcrowded and overexploited fishery in near coastal waters. The ideas are explicit and there is no doubt about the wish to promote radical changes. Two key policy objectives from this policy are troubling in particular:

1. "The Government will play a catalytic role in the gradual development of artisanal fisheries further offshore in deeper waters through promoting the formulation and the implementation of an Integrated FAD Fisheries Development and Management Programme and through improving efficiency and safety at sea" (RGZ 2016, 22).
2. "To promote the development of a domestic offshore industrial fishing fleet targeting tuna and tuna-like species in Tanzania EEZ thorough examining all possibilities that could provide incentives for potential local investors" (RGZ 2016, 22).

The offshore solution is portrayed as a viable alternative since this part of the Exclusive Economic Zone (EEZ) is perceived as low or non-exploited by national Tanzanians, while foreign vessels are harvesting the economic benefits by landing in neighboring countries (RGZ 2016). Our analysis. as well as our previous research, show that the idea of making the proposed radical changes to the way of fishing in the area, and in the coastal activities to be supported, are both in direct contradiction to the SSF Guidelines and are highly prone to failure.

We have identified incoherencies between small-scale fishers' activities and strong conservation approaches; between small-scale fishers and the economic capacity needed to change their fishing techniques; between small-scale fishers and spatial issues (by removing them from inshore areas and promoting fishing offshore); between small-scale fishers' production and new activities, mainly aquaculture, that have their own social and economic repercussions.

The proposed new FP does not convincingly provide for a secure situation for small-scale fishers in Zanzibar. To the contrary, their livelihoods, lifestyle, and way of fishing will be jeopardized and face the threat of either disappearing or experiencing radically change. Due to these implications, the proposed new FP is in incoherent with the SSF Guidelines. Livelihood security including economic security, gender equity, minimum rights to design one's own management regimes and long term sustainability, and tenure rights over the fishing resources and ecosystems are all weak. A tailored and already decided solution is presented at hand. The following table (Table 5.1) illustrates our findings regarding the governance implications of the proposed new FP in reference to the general objectives stated in the SSF Guidelines.

If these incoherencies remain, it is not difficult to foresee that the new FP has the winning hand, given its attachment to extensive donor funding. The SSF Guidelines are not confined to responsible fisheries governance, but also to furthering social development and decent work, gender equality, and basic civil and political rights, which implies costs without funding. In the context of poor budgetary resources, Zanzibar, like most other developing nations, will find it difficult to implement the bulk of the principles in the SSF Guidelines. While the new FP seems founded in common ideas of expansion, capitalization, and commercialization as problem solvers, the SSF Guidelines' ideational foundation is in human rights. This forms the most fundamental conflict between the two.

**Table 5.1** Policy Incoherence between the SSF Guidelines (FAO 2015) and the new Zanzibar Fisheries Policy (RGZ 2016)

| SSF guidelines' broad objectives | Incoherencies with the draft new fisheries policy |
|---|---|
| To enhance the contribution of small-scale fisheries to global food security and nutrition and to support the progressive realization of the right to adequate food | Radical changes in fishing ways and techniques that promote export and commercialization rather than enhancement of production for local consumption. Relocation of SSF and promotion of industrial fleet which jeopardizes lifestyle and might reduce access to fish protein for locals |
| To contribute to the equitable development of small-scale fishing communities and poverty eradication and to improve the socio-economic situation of fishers and fish workers within the context of sustainable fisheries management | Low gender equity as men and women activities are put in competition |
| | Key actors in the fishery system such as women fish traders are not addressed in the FP |
| | Low overall economic equity as actors which are economically strong will be able to fish offshore while poor ones are not |
| | Poor conditions on coastal people health as promotion of aquaculture activities with detrimental health effects are encouraged and proposed |
| | No convincing efforts to manage near coastal fisheries in a sustainable way, since much emphasis is put on 'Western solutions' such as strengthening MPAs, MCAs, and a regulations and control approach. Emphasis to 'move' the problem to offshore waters |
| To achieve the sustainable utilization, prudent and responsible management and conservation of fisheries resources consistent with the Code of Conduct for responsible Fisheries (the Code) and related instruments | No explicit synergies between the new proposed FP and this FAO code |
| To promote the contribution of small-scale fisheries to an economically, socially and environmentally sustainable future for the planet and its people | In the new FP, there is an aspiration towards this broad objective, however, the proposed policy oversees the complexity of the SSF situation, the local ecological knowledge of the fishers and the scientific research produced in the area |
| To provide guidance that could be considered by States and stakeholders for the development and implementation of ecosystem friendly and participatory policies, strategies and legal frameworks for the enhancement of responsible and sustainable small-scale fisheries | Although well-functioning and healthy ecosystems are named a number of times in the policy document, it is unclear how this will be achieved in practical terms and how fishers will contribute |
| To enhance public awareness and promote the advancement of knowledge on the culture, role, contribution and potential of small-scale fisheries, considering ancestral and traditional knowledge, and their related constraints and opportunities | Weak, vague and with a discourse towards "technological advancements and improvements" rather than supporting ancestral and traditional aspects |
| | Traditional knowledge and traditional fisheries management practices are used sometimes in the policy document but mainly linked to complement conservation approaches, e.g. traditional knowledge is useful when it informs for example closed fishing seasons or can be part of the Marine Conservation initiatives |

It is interesting to note how the Zanzibar management path differs from the global patterns of fishery management history. While an emphasis on industrial fisheries and offshore exploitation was increased globally during the 1950–1970s, small-scale fisheries were seen as backwards and underdeveloped. However, overcapitalization and extensive fishery collapse has turned attention worldwide to potential solutions, including traditional management and small-scale fisheries (Carvalho et al. 2011). In contrast, Zanzibar goes the other way, probably since they have not reached the degree of industrialization of other fishing nations. The question here is if they necessarily have to embrace and experience that phase, and the harmful mistakes that come with it, or learn from world experiences and jump directly into a fisheries governance regime that emphasizes and protects the sustainability of small-scale fisheries.

Another aspect worth mentioning is that other cases which address tuna fisheries dynamics have shown that competition between industrial fleets and small-scale fishers is present and can exacerbate conflicts and resource depletion (SPC 2013; Leroy et al. 2016); in contrast, this dynamic has so far been unforeseen by Zanzibarian policy makers. Industrial fishing is not necessarily deemed to constitute unsustainable practices. When the industry is responsible, supportive of ecosystem-based management, have working systems for resource allocation and conflict resolution as well as real commitment to long-term work, they can be a positive and viable economic alternative (Bodal 2013). However, it is difficult to see that the proposed offshore tuna fleet adheres to those conditions. The potential for conflict is especially high considering the sensitive issue of potential conflicts between foreign vessels already fishing, small-scale fishers, and new actors entering the system financed by private capital.

It is apparent that authorities in Zanzibar are struggling to write the Fisheries Policy drafts from a position in which they are squeezed between differing interests, economic constraints, and strong global actors. The SSF Guidelines were adopted in 2014, implying that a very short period of time for reflection on the implementation and the previous lack of knowledge about this new policy instrument have without a doubt been important factors contributing to their lack of consideration in the proposed new FP. What we may hope for is that an eventual third draft of the new FP will be released which pays close attention to this and address coherence issues. Hopefully, in doing so, most of the principles and objectives in the SSF Guidelines will be included and reinforced in the ultimate final fisheries policy, and hopefully also implemented in practice.

**Acknowledgements** We are grateful to the Department of Fisheries Development in Zanzibar for always being friendly with us, and for sharing information and documentation. Thanks to two anonymous reviewers for appreciated reviews of previous versions of the manuscript. To TBTI for continued energy injections to work with small-scale fisheries. A very special thanks to two individuals for their invaluable assistance in the field: Narriman S. Jiddawi and Tesfaye Ayele Woubshet. This work was partly financed by the Swedish Research Council VR project diarienr. 344-2011-5448.

## References

Bodal, B. O. (2013). Incorporating ecosystem considerations into fisheries management: Large-scale industry perspectives. In M. Sinclairand & G. Valdimarsson (Eds.), *Conference on responsible fisheries in the marine ecosystem* (pp. 41–46). Rome: FAO.

Carvalho, N., Edwards-Jones, G., & Isidro, E. (2011). Defining scale in fisheries: Small versus large-scale fishing operations in the Azores. *Fisheries Research, 109*, 360–369.

Daily Mail. (2016, May 1). *Seaweed farmers in hot water as Zanzibar struggles*. http://www.dailymail.co.uk/wires/afp/article-3567780/Seaweed-farmers-hot-water-Zanzibar-struggles.html. Accessed 12 June 2016.

de la Torre-Castro, M. (2006). Beyond regulations in fisheries management: The dilemmas of the 'beach recorders' Bwana dikos in Zanzibar, Tanzania. *Ecology and Society, 11*, 4.

de la Torre-Castro, M. (2012). Governance for sustainability: Insights from marine resource use in a tropical setting in the Western Indian Ocean. *Coastal Management, 40*, 612–633.

de la Torre-Castro, M., & Lindström, L. (2010). Fishing institutions: Addressing regulative, normative and cultural-cognitive elements to enhance fisheries management. *Marine Policy, 34*, 77–84.

de la Torre-Castro, M., Di Carlo, G., & Jiddawi, N. S. (2014). Seagrass importance for a small-scale fishery in the tropics: The need for seascape management. *Marine Pollution Bulletin, 83*(2), 398–407.

EC. (2013). *A decent life for all: Ending poverty and giving the world a sustainable future*. Brussels: European Commission.

Eriksson, B. H., de la Torre-Castro, M., Eklöf, J. S., & Jiddawi, N. S. (2010). Resource degradation of the sea cucumber fishery in Zanzibar, Tanzania: A need for management reform. *Aquatic Living Resources, 23*(4), 387–398.

Eriksson, H., de la Torre-Castro, M., & Olsson, P. (2012). Mobility, expansion, and management of a multi-species scuba diving fishery in East Africa. *PloS One, 7*(4), e35504.

FAO. (2006). *The state of food and agriculture: Food aid for food security?* Rome: FAO.

FAO. (2015). *Voluntary guidelines for securing sustainable small-scale fisheries in the context of food security and poverty eradication*. Rome: FAO.

Fröcklin, S., de la Torre-Castro, M., Lindström, L., & de la Torre-Castro, M. (2012). Seaweed mariculture as a development project in Zanzibar, East Africa: A price too high to pay? *Aquaculture, 356*, 30–39.

Fröcklin, S., de la Torre-Castro, M., Lindström, L., & Jiddawi, N. S. (2013). Fish traders as key actors in fisheries: Gender and adaptive management. *Ambio, 42*(8), 951–962.

Gustavsson, M., Lindström, L., Jiddawi, N. S., & de la Torre-Castro, M. (2014). Procedural and distributive justice in a community-based managed Marine Protected Area in Zanzibar, Tanzania. *Marine Policy, 46*, 91–100.

Jentoft, S. (2014). Walking the talk: Implementing the international voluntary guidelines for securing sustainable small-scale fisheries. *Maritime studies, 13*, 16.

Jiddawi, N. S., & Öhman, M. C. (2002). Marine fisheries in Tanzania. *Ambio, 31*(7–8), 518–527.

Leroy, B., Peatman, T., Usu, T., Caillot, S., Moore, B., Williams, A., & Nicol, S. (2016). Interactions between artisanal and industrial tuna fisheries: Insights from a decade of tagging experiments. *Marine Policy, 65*, 11–19.

Lindström, L., & de la Torre-Castro, M. (2015). Promoting governability in small-scale fisheries in Zanzibar, Tanzania: From self-governance to co-governance. In S. Jerntoft & R. Chuenpagdee (Eds.), *Interactive governance for small-scale fisheries: Global reflections* (pp. 671–686). London: Springer MARE Publication Series.

Lindström, L., & de la Torre-Castro, M. (2016). *Does the heat hit the same. Gender, agency and climate change in coastal communities*. Manuscript. Stockholm: Department of Political Science, Stockholm University.

Nilsson, M., Zamparutti, T., Petersen, J. E., Nykvist, B., Rudberg, P., & McGuinn, J. (2012). Understanding policy coherence: Analytical framework and examples of sector–environment policy interactions in the EU. *Environmental policy and governance, 22*, 395–423.

OECD. (2016). *Policy coherence for development.* www.oecd.org/pcd/. Accessed 24 July 2016.

RGZ. (1993). *Fisheries law*. Zanzibar Town: Revolutionary Government of Zanzibar.

RGZ. (2002). *The Zanzibar vision 2020*. Zanzibar Town: Zanzibar Revolutionary. Government of Zanzibar.

RGZ. (2010a). *Fisheries act*. Zanzibar Town: Zanzibar Revolutionary Government.

RGZ. (2010b). *MKUZAII. Zanzibar strategy for growth and reduction of poverty 2010–2015.* Zanzibar Town: Revolutionary Government of Zanzibar.

RGZ. (2014). *Zanzibar fisheries policy*. First draft prepared with the support of Smartfish. Zanzibar Town: Revolutionary Government of Zanzibar.

RGZ. (2016). *Zanzibar fisheries policy*. Second Draft prepared with the support of SmartFish. Zanzibar Town: Revolutionary Government of Zanzibar.

SPC. (2013). *Balancing the needs: Industrial versus artisanal tuna fisheries. SPC policy brief 22/2013*. Noumea/New Caledonia: Secretariat of the Pacific community.

Thyresson, M., Crona, B., Nystrom, M., de la Torre-Castro, M., & Jiddawi, N. S. (2013). Tracing value chains to understand effects of trade on coral reef fish in Zanzibar, Tanzania. *Marine Policy, 38*, 246–256.

Wallner-Hahn, S., de la Torre-Castro, M., Eklöf, J. S., Gullström, M., Muthiga, N. A., & Uku, J. (2015). Cascade effects and sea-urchin overgrazing: An analysis of drivers behind the exploitation of sea urchin predators for management improvement. *Ocean & Coastal Management, 107*, 16–27.

Wallner-Hahn, S., Molander, F., Gallardo, G., Villasante, S., Eklöf, J. S., Jiddawi, N. S., & de la Torre-Castro, M. (2016). Destructive gear use in a tropical fishery: Institutional factors influencing the willingness-and capacity to change. *Marine Policy, 72*, 199–210.

World Bank (1984). *Project completion report: Tanzania fisheries development project (Credit 652-TA)* (Report No. 5136). Washington, DC: The World Bank.

World Bank. (2012). *Tanzania - marine and coastal environment management project: Restructuring* (Vol. 1 of 2). Main report (English). Washington, DC: The World Bank.

World Bank. (2013). *Implementation, completion and results report (IDA-41060, TF-52440, TF-55580)* (Report No. ICR 2754). Washington, DC: The World Bank.

World Bank. (2016). *Implementation, status and results report AFCC 2/R1. First South West Indian Ocean governance and shared growth project (P132123)*. Washington, DC: The World Bank.

ZFC. (2007). *Zanzibar fishery census*. Zanzibar Town: Department of Fisheries and Marine Resources.

ZFC. (2010). *Zanzibar fishery census*. Zanzibar Town: Department of Fisheries and Marine Resources.

# Chapter 6
# Pernicious Harmony: Greenland and the Small-Scale Fisheries Guidelines

**Hunter T. Snyder, Rikke Becker Jacobsen, and Alyne Delaney**

**Abstract** Greenland appears well-positioned to implement the Voluntary Guidelines for Securing Sustainable Small-Scale Fisheries (SSF Guidelines). Its existing national policies focused on human rights, food security, gender equity, and fisheries align with the objectives of the SSF Guidelines. Further, Greenland's history of political support for small-scale fisheries gives reason that implementation is feasible. However, Greenland's economic growth objectives via natural resource exploitation are in opposition to the SSF Guidelines and Greenland's history of political support of small-scale fishers and fish workers. We show that small-scale fishers and fish workers in Qeqertarsuatsiaat, Greenland are in pernicious harmony with large-scale buyers. Through a representative survey among small-scale fishers and fish workers (N = 21), we find that small-scale fishers do not recognize large-scale buyers as resource competitors. Beyond recognition, such a configuration between government, small-scale fishers and the large–scale sector underestimates the small-scale sector's capacity for innovation, hinders profit margin growth and market diversification, and perpetuates inequality between the large- and small-scale stakeholders. Our results demonstrate that, while Greenland may have a track record of implementing progressive human rights policy, implementing the SSF Guidelines requires reconciling competitiveness and understandings thereof, in Greenland's fisheries. By highlighting blind spots in Greenland's small-scale fisheries governance and management, we anticipate our study will help serve as a starting point for re-harmonizing Greenland's small-scale fisheries policy design together with local, national and international objectives.

**Keywords** Small-scale fisheries • Greenland • Markets • Competition • Governance

H.T. Snyder (✉)
School of Environmental Studies, Dartmouth College, Hanover, NH, USA
e-mail: h.snyder@oxon.org

R.B. Jacobsen • A. Delaney
Innovative Fisheries Management, Department of Planning, Aalborg University, Aalborg, Denmark
e-mail: rbj@ifm.aau.dk; ad@ifm.aau.dk



S. Jentoft et al. (eds.), *The Small-Scale Fisheries Guidelines*, MARE Publication Series 14, DOI 10.1007/978-3-319-55074-9_6

## Introduction

Greenland, a former colony of Denmark, appears well-positioned to implement the Voluntary Guidelines for Securing Sustainable Small-Scale Fisheries (SSF Guidelines) put forward by the Food and Agriculture Organization of the United Nations, or FAO (FAO 2015). A country with progressive human rights policy and an historic national emphasis on small-scale fishers and fish workers' livelihoods, there is reason to expect it to be feasible if not straightforward to implement the SSF Guidelines. However, Greenland's growth objectives through natural resource exploitation are in partial opposition to the historical, harmonious, and interdependent relationship that small-scale fishers have maintained with a critical economic engine: Greenland's large-scale fisheries sector. Growing opportunities on the world market call the viability of this relationship into question and test the resilience of the country's small-scale fishing livelihoods (Stockholm Resilience Center 2014).[1]

The introduction of select neoliberal fisheries management throughout the Arctic is not new and Greenland's growth ambitions are not singular; it is a Northern country whose movements stage an array of regionally particular but globally parallel concerns for the recognition and development of small-scale fisheries (Thornton and Hebert 2015). Drastic, global shifts in small-scale fisheries are precisely what FAO seeks to address by Member States implementing the SSF Guidelines. Our study shows that implementation, even in politically progressive scenarios, is not without its challenges. Importantly, we also show that considering the readiness to implement highlights tenuous relationships between fisheries stakeholders. While such relationships facilitate harmony in a host of contexts, they can limit the overarching objectives of fisheries policy and national economic development.

As Greenland has not formally endorsed the SSF Guidelines, we demonstrate how Greenland should be well-poised to implement them. With empirical evidence, we illustrate a common relationship between small-scale fishers and large-scale fishing and industrial operations. Our findings show how the SSF Guidelines may secure not just political recognition for small-scale fisheries, but may also call into question what we describe as a state of pernicious harmony between the three actors: government, small-scale fishers and fish workers, and the large-scale fishing sector.

Our study investigates the political economy of Greenland's small-scale fisheries in advance of implementing the SSF Guidelines (Forsyth 2003). Focusing on one settlement with economic, cultural, and infrastructural characteristics akin to others in Greenland in isolated regions and throughout the Arctic, we marshal empirical evidence of how Greenland's small-scale fisheries function and how the SSF Guidelines may highlight the relational complexities between government,

[1] Our study draws upon the Stockholm Resilience Centre's definition of resilience, which is 'the capacity of a system, be it an individual, a forest, a city or an economy, to deal with change and continue to develop' (Stockholm Resilience Center 2014, 3).

small-scale fishers and fish workers, and the large-scale sector. As such, we establish two basic hypotheses:

> **$H_1$**: If Greenland's economy has historically integrated small-scale fishing traditions into the national economy and designs policy that upholds human rights, gender equality and small-scale fisheries, then implementing the SSF Guidelines in Greenland will be programmatic and feasible.
>
> **$H_2$**: If the SSF Guidelines implore member states to recognize and improve the competitive capacity of small-scale fishers and Greenland's small-scale fishers have historically worked harmoniously with the large-scale sector and national government, then Greenland will not struggle with the power inequalities that are common among small-scale fishers and large-scale actors.

We test our hypotheses successively, initially focusing on how small-scale fisheries governance in Greenland is legislated nationally and situated globally. Our empirical study of small-scale fisheries in Qeqertarsuatsiaat, Greenland grounds national and international fisheries governance and questions the extent and complexity of harmony between government, small-scale fishers and fish workers, and the large-scale fishing sector. We intend to stir discussion on how Member States could implement the SSF Guidelines as they reconcile competing political, cultural and economic interests. If the Guidelines are designed to recognize and support small-scale fisheries globally, then we find it necessary to understand how they are currently situated, how they function in relation to other actors and, ultimately, of what utility the SSF Guidelines may be not just for small-scale fishers, but for government and the large-scale sector within and outside of Greenland.

## Greenland's Small-Scale Fisheries

From an economic and cultural standpoint, fishing has become significantly important in the colonial and post-colonial era. In addition to the $570 million block grant from Denmark, fishing helps to underpin Greenland's economy, accounting for more than half of the total value of the country's traded commodities (FAO 2014). In addition to the employment base of the public sector, small-scale fishing is also *the* economic mainstay of several settlements throughout the country, with few exceptions, such as limited agriculture in South Greenland as well as hunting in North and East Greenland (Nielsen 2010; Dzik 2014). As the SSF Guidelines reaffirm, fishing is of social and cultural significance for many worldwide, and the same holds true in Greenland and throughout the Arctic (Condon et al. 1995; Nuttall 2009; Poppel and Kruse 2009). A fundamental basis for fishing and other livelihoods in Greenland is the health and wellbeing of the country's aquatic ecosystem biodiversity, which is heavily emphasized in both the SSF Guidelines and in Greenland's natural resources policy.

In addition to a weak labor market, Greenland's limited but expanding coastal infrastructure sets the backdrop for Greenland's economic growth challenges. Greenland's shipping monopoly, Royal Arctic Line (RAL), ensures that settlements have access to shipped goods. However, the responsibility to serve all settlements places extraordinary pressure on the government-owned company's bottom line and, in turn, bars external shipping companies from transporting goods between Greenland and nearby ports in Iceland, Maine, Newfoundland, and elsewhere in Western Europe.

Another government-owned entity, Greenland's largest seafood supplier, Royal Greenland, also has lifeline significance as a direct and indirect employer in several settlements. As we test in the case study of Qeqertarsuatsisaat, their longstanding interdependence blurs the relational complexities between small and large-scale fishers and buyers.

## Small-Scale Fisheries Governance in Greenland

The SSF Guidelines are comprehensive and deeply ambitious. Set in the context of human rights, poverty alleviation, and hunger eradication, and emphasizing the importance and recognition of women, the SSF Guidelines are attracting interest from several UN Member States, research consortia, and the public. Greenland is a country that is technically listed as 'developed', due to its position under the realm of the Kingdom of Denmark. However, Greenland has conditions often seen in countries that are typically defined as 'developing countries'. It receives an annual block grant from Denmark worth USD $570 million and has a nascent self-rule government, limited economic growth, and poor but improving infrastructure. Despite its controversial status as a developed country, the SSF Guidelines appear to be a possible mechanism for achieving the country's goals of economic growth that would aid the current ruling political party's hopes of eventual independence from Denmark.

Greenland is included in the United Nations as a member of the Kingdom of Denmark. In 2009, Greenlanders voted in a national referendum in favor of further independence in the form of Greenlandic Self-Government. This new arrangement transferred to Greenland the full rights to its underground resources as well as the possibility to 'take home', whenever the Greenland Self-Government decide, some of the remaining fields of responsibility that are currently administered by the Danish state. Other fields of responsibility can only be transferred when Greenland decides to be an autonomous state (Civilstyrelsen 2009).

The relationship between Denmark and Greenland concerning Greenland's ratification and implementation of international agreements is defined by the Self-Government Act of 2009. When Denmark ratifies a treaty or convention, as a general rule it includes all of the territories within the Kingdom, which includes Greenland and the Faroe Islands. However, Denmark can decide to include a territorial reservation whereby the treaty will not include Greenland. The Self-Rule Act stipulates that the Danish Government shall inform the Government of Greenland (Nalakkersuisut) before initiating negotiations about international agreements relevant to Greenland. Deference to Greenland is given so that Nalakkersuisut can express its views. Before any agreement is ratified or cancelled, it is submitted to Nalakkersuisut for comments (H. R. C. O. Greenland 2014).

Concerning the most recent FAO agreements, in 2012, the Greenland Parliament (Inatsisartut) discussed and ratified the FAO Port State Measures Agreement of 2009 and subsequently sent their comments to Denmark to ratify the agreement on behalf of Greenland (G. O. Greenland 2012b).[2] Before the establishment of the Self-Rule Government of Greenland in 2009, the practice of ratifying FAO agreements was similar, inasmuch as the Home-Rule Government of Greenland would debate and formulate its own comments. For example, the Home-Rule Government of Greenland debated its comments to the FAO Compliance Agreement (1993) and the FAO Code of Conduct (1995) in 1999 (G. O. Greenland 1999). As of 2016, the Self-Rule Government of Greenland has not endorsed the SSF Guidelines. Furthermore, there is no evidence at all that the SSF Guidelines have entered the ministerial agenda in Greenland. A first and critical step for implementation may so far be missing: that they first be identified and considered relevant by at least some of the actors involved in Greenland's fisheries policy design. The researchers note that a range of pertinent fishery issues (e.g. general quota and license management, climate change, and the development of new pelagic fisheries) compete for the limited resources of the Ministry of Fishing, Hunting, and Agriculture. Furthermore, we note that there are already a range of other political mechanisms in place to address the concerns of small-scale fisheries and coastal communities within Greenland's Self-Government decision-making (Jacobsen and Raakjær 2012; Delaney 2016). For the SSF Guidelines to be implemented in Greenland, they first need to be brought to the decision-making table by those who find them to be a relevant means to achieve their political goals.

The design of the SSF Guidelines are to 'spur new legislation' among Member States (Jentoft 2014). But who is the relevant Member State to spur or approve legislation, considering Greenland's position as part of the Kingdom of Denmark? The SSF Guidelines cover an array of policy areas, particularly fisheries policy, but also issues such as gender equality and equity. However some concerns, most notably those over human rights, are the shared responsibility of Denmark and Greenland (H. R. C. O. Greenland 2014). Regarding fisheries policy and, in particular, the coastal fishery within Greenland's exclusive economic zone (EEZ), Greenland is best described as a sovereign state and its history of protecting its EEZ is especially telling of the national importance of fishing.

The entrance of Greenland into the European Union (EU) in 1973 strengthened the emerging push for Greenlandic independence from Denmark. Greenland achieved Home-Rule in 1979 and decided to withdraw from the EU after a 1985 national referendum. Since then, Greenland has managed its fisheries independent of both Denmark and the EU. That Greenland explicitly withdrew from the EU to gain total control of fishing activities within its EEZ demonstrates the national, historical, and economic importance of fishing. The country's departure from the EU

[2] Greenland has had a home-rule government since 1979, and a self-rule government (Nalakkersuisut) and Parliament (Inatsisartut) since 2009.

occurred when the Greenlandic fishery was small to medium in scale and in the wake of the intense international cod fishery battle from 1950 to 1975, which was followed by the stock's collapse in West Greenland (Hovgård and Wieland 2008; Lilly et al. 2008).

Although not part of the EU, Greenland can in practice be represented by Denmark from within EU fisheries policy should Denmark decide to act on behalf of the entire Kingdom of Denmark.[3] Continuing its long-term interest in Greenland, the EU made a financing agreement that focuses on improving education in country (including vocational training) for 2014–2020 (Commission 2014). The program highlights the need to improve educational quality and opportunities, including language skills in particular, which are a critical part of a developing country's integration with global markets.

Greenland has since developed a state-of-the-art, extremely efficient offshore fishery and large-scale seafood companies that operate internationally with supply chains in both developed and developing countries. Today, Greenland has its own national 'Institute of Natural Resources', Pinngortitaleriffik, and represents itself independently in all relevant international and bilateral fishery organizations, including ICES, NAFO, NEAFC, and NASCO.[4]

The SSF Guidelines align well with Greenland's fisheries governance framework, though some of the principle themes, such as gender equity and labor conditions, are more appropriately addressed outside of the Ministry of Fishing, Hunting, and Agriculture on account of those policies being under the remit of other ministries. Because the SSF Guidelines' themes may be best distributed throughout the nine ministries, implementing the Guidelines would need parliamentary support and would require coordination across departments. Other aspects of the SSF Guidelines would require resources from Denmark, in particular the themes that concern human rights.[5]The most pernicious policy-induced challenges for small-scale fisheries stem from domestic fisheries management and domestic development policies, as opposed to bilateral policy (Delaney et al. 2012; Jacobsen n.d.; Jacobsen and Raakjær 2014; Delaney 2016). The assessment of the SSF Guidelines and their suitability for implementation would take place through policymaking at the Government of Greenland, preferably with 'an eye to the interactive governance processes' as suggested by Jentoft including, in particular, the diverse interests and influences of different Greenlandic stakeholders (Jentoft 2014, 8). Against this governmental backdrop, we stage our first hypothesis:

---

[3] Recently, in the context of mackerel disputes, Denmark found itself in a dilemma of this sort in relation to EU sanctions against another territory with a similar status, namely the Faroese Islands. http://www.dr.dk/nyheder/penge/faeroesk-makrel-krig-saetter-danmark-i-penibel-situation

[4] International Council for the Exploration of the Sea (ICES); Northwest Atlantic Fisheries Organization (NAFO); North East Atlantic Fisheries Commission (NEAFC); The North Atlantic Salmon Conservation Organization (NASCO).

[5] Justice and law enforcement are two of the most prominent and expensive fields of responsibility that can be transferred whenever Greenland self-government decides. Military and defense and a couple of other key fields of responsibility can only be transferred if Greenland decides to become an autonomous state.

> **H$_1$**: If Greenland's economy has historically integrated small-scale fishing traditions into the national economy and designs policy that upholds human rights, gender equality and small-scale fisheries, then implementing the SSF Guidelines in Greenland will be programmatic and feasible.

The hypothesis is supported by the fact that small-scale fishing traditions have remained an integrated part the national economy so far. Whether this has occurred due to or in spite of the chosen fisheiesy policies is a relevant question. Greenlands's fisheries policy is balancing a range of competing concerns for regional and national development and there have been moments where the chosen policy has, intentionally as well as unintentionally, impeded the development of small-scale fisheries as a viable economic sector (Rasmussen 1998; Jacobsen 2013). Considering the changing political agendas for the fishery, one cannot expect that small-scale fisheries will also be immediately prioritized. On the other hand, human well-being in the Arctic region is greatly tied to land-based activities like fishing, hunting, and gathering. However, such perspectives are not explicitly included in the management of the commercial fisheries. Due to the broad scope of the SSF Guidelines, they could serve to mainstream issues of human well-being into Greenland's fisheries governance.

Stemming from an understanding of the political institutions that shape fisheries policies on a local, national and international level, we turn our attention to how small-scale fisheries stakeholders organize themselves in practice through a case study of the small-scale Atlantic cod (*Gadus morhua)* fishery of Qeqertarsuatsiaat, Greenland.

Since commercial fishing began in Greenland in the late nineteenth and early twentieth centuries, customary fishing practices in small dinghies have persisted. In fact, there are estimated to be five times more small-scale fishing vessels than there are registered, large-scale vessels (FAO 2004). Small-scale fishers are not only too big to ignore in Greenland (Chuenpagdee 2011); they also grapple with several principle concerns set forth in the SSF Guidelines, including: the transfer of traditional ecological knowledge and local management (Point 2, 5.18 and 11.4); remoteness and disadvantaged access to markets (Point 7.3); limited alternative livelihoods (Point 6.7); and competition from large-scale fishing and industrialization (Points 5.9, 5.10) (FAO 2015).

Beyond a focus of several of the key areas of SSF Guidelines, our study builds upon studies that detail the conditions and challenges in which small-scale fishers exist, several of which we find relevant to Greenland. We espouse Jentoft (2014)'s definition of small-scale fishing, especially his conjecture that they exhibit attributes that are 'unique to localities, that they are often family enterprises that include women and children, and that they are not stuck in the past but instead undergo constant change' (Jentoft 2014, 3). We audit these and other conjectures, including the argument that fishing is not always an occupation of last resort (Onyango and Jentoft 2011) and that privatization *merely* tends to lead to the exclusion of poor and vulnerable small-scale fisher groups (Jentoft 2014).

## Qeqertarsuatsiaat

Qeqertarsuatsiaat is a settlement that was initially founded in the mid-eighteenth century as a trading post during the Danish colonial era. It has historically been a place of special significance for the cod fishing industry, owing in part to its relatively close proximity to Nuuk, the capital of Greenland, and cod fishing grounds that are once again thriving in the wake of the stock collapse in the middle of the twentieth century. Qeqertarsuatsiaat's population has remained more or less static over the last 5 years (2010–2015), with a 2015 population of 213 residents. The overwhelming majority of residents are Kalaallit Inuit who primarily speak Kalallisut and, to a lesser extent, Danish. The settlement has limited infrastructure, including a school, municipal office, general store, kiosk, a fish factory that provides additional employment opportunities on an as-needed basis, an elder's home, recreation center, church, medical outpost, diesel power plant, incinerator, a municipal workshop, and a small port.

Regarding life in the settlement, findings from the 2015 Survey of Living Conditions in the Arctic (SLiCA) indicate that when comparing the general satisfaction of living conditions between 2005 and 2015, residents are as happy as they were then and furthermore that they engage in more fishing and hunting activities than they did a decade ago (Snyder and Poppel, in press). Among all residents, there has also been a steady increase in the percentage of residents who own fishing gear, including fishing nets (which rose from 45% in 2005 to 71% in 2015) and outboard engines (from 49% to 65%), which is more than those who have personal computers (41.7%).

While Qeqertarsuatsiaat has received attention for the study of Arctic living conditions (and we will discuss this in more detail below), the settlement has also received unprecedented interest in both national and international geopolitical stages because of its sub-soil resource potential. Qeqertarsuatsiaat and nearby areas are geologically diverse and have for decades been regularly combed for gem-quality precious stones. Rubies are the most commonly sought after stone and can be found with relative ease in brittle ores common to the area's geology. Just under an hour sail from the settlement transports one to a high-concentration deposit of rubies. Although the country has had a tumultuous history of mineral exploration and exploitation, Qeqertarsuatsiaat is currently the only settlement in Greenland that hosts a nearby ruby mine. The commercial production of rubies began only 4 months after the completion of this study and 2 weeks after the SLiCA reassessment. Policymakers and the public alike are watching the development of the ruby mine in relation to the settlement with great attention to detail, particularly as the track record of their relationship could serve as a harbinger for hopes of forthcoming expansion of Greenland's industrial sector in relative close proximity to other settlements. Industrialization on this scale also bears relevance for small-scale fisheries. The impact benefit agreement that sets out the goals of this ruby mine mentions not only an additional buyer of fish, but also the potential for alternative livelihoods. There is also the potential to increase shipping, access new markets and in turn augment the flow of goods and services to the otherwise remote region. Initial findings indicate that nearly three quarters (74%) of local residents were not employed or involved in the construction activities of the mine and that only one fisher supplied fish to the mine (Snyder and Poppel, in press).

Greenlandic fisheries are managed by the Greenland Self-Government in consultation with the formal and well-established associations of Greenlandic Fishers and Hunters (KNAPK) as well as the larger-scale fishing industry, represented by Greenlandic Employers (GE). The commercial coastal fisheries of Greenland, in order of greatest marine capture production, include Greenland halibut (*Reinhardtius hippoglossoides*), Atlantic cod (*Gadus morhua*), capelin (*Mallotus villosus*), Atlantic herring (*Clupea harengus*), lumpfish (*Cyclopterus lumpus*), Redfish (*Sebastes marinus*) Queen crab (*Chionoecetes opilio*) and Iceland scallops (*Chlamys islandica)* (FAO 2017). Qeqertasuatsiaat's fisheries, which consist primarily of cod, are managed by a total allowable catch (TAC) quota system. Recreational or subsistence fishing is generally not subjected to formal management by the Greenlandic Self-Government. Regulation for Atlantic salmon (*Salmo salar*) is, however, robust, stringent and enforced. Further, Arctic char (*Salvelinus alpinus*) and Atlantic salmon fishing areas are often retained over generations, negotiated informally and respected by others who fish for these and other finfishes. Both commercial and subsistence hunting for terrestrial and marine mammals is governed by Greenland Self-government, though decisions on quota distribution are also transferred to the municipal level and municipal hunting committees (Jacobsen and Raakjær 2012).

Beyond Qeqertarsuatsiaat's geopolitical relevance for Greenland's mineral resource future, the settlement bears similar demographic, infrastructural, labor market conditions and social-ecological characteristics to not just other settlements in Greenland, but to other Arctic communities. By extension and for the purposes of discussing the SSF Guidelines, Qeqertarsuatsiaat may also bring to bear observations relevant to small coastal communities below the Arctic circle. We find Qeqertarsuatsiaat an ideal site, not just because it circumscribes the challenges of implementing the SSF Guidelines, but also due to its larger political economy of marine and sub-soil resources, both of which play an important part in how we understand Greenland's social-ecological systems.

## Methods

In August 2015, the lead author and a small research support team carried out fieldwork that focused on several key themes of the SSF Guidelines. In addition to participant observation and unstructured interviews, a brief questionnaire was issued among stakeholders in the small-scale fishing sector. Stakeholders are defined as fishers, fish workers, and fish buyers who reside in Qeqertarsuatsiaat. Several themes from the SSF Guidelines guided the design and organization of the questionnaire. Questions concerned:

1. the transfer of traditional ecological knowledge and local management (Point 2, 5.18 and 11.4)
2. remoteness and disadvantaged access to markets (Point 7.3)
3. limited alternative livelihoods (Point 6.7)
4. competition from large-scale fishing and industrialization (Points 5.9, 5.10)

Although the questionnaire was drafted independently, it is informed by conversations with section heads of the Ministry of Fishing, Hunting, and Agriculture in the Government of Greenland, the Hunters and Fishers Association of Greenland (KNAPK), and other small-scale fisheries stakeholders. In addition to the 12 months of ethnographic fieldwork on Greenlandic small-scale fisheries that preceded this study, these conversations helped to isolate key concerns, gauge interest and to ensure transparency before carrying out fieldwork in Qeqertarsuatsiaat. The questionnaire, which consists of possible categorical responses, was drafted in English and translated to Kalaallisut, after which the translation was verified to be clear and accurate through two audits undertaken by multilingual (Kalaallisut and English) project research assistants. The questionnaire was issued in Kalaallisut to respondents in confidence and the results were anonymized to protect identities.

Of the 213 residents, 29 work as either fishers, fish workers or fish buyers, and the team sought to elicit responses from all (defined hereon as SSF stakeholders). Caribou hunting and summer travel are common when temperatures and ocean conditions permit safer travel, which left 21 SSF stakeholders available for participation. The research team attained a 72% participation rate among all SSF stakeholders and 100% among those who were not hunting or travelling. The remaining eight stakeholders did not participate for a variety of reasons, including but not limited to illness, travel or conflicting participation in the caribou hunting season. Because we met with all available SSF stakeholders who comprise a wide variety of subcategories (i.e. fish workers, buyers, including junior, senior and retired fishers) and none refused to participate, we assert that our sample is representative of the stakeholders of the small-scale fishing sector in Qeqertarsuatsiaat.

Of those who participated, 24% of the sample were fish workers or buyers and 76% were small-scale fishers. All SSF stakeholders who participated were male.[6] The research team interviewed respondents at home, over lunch breaks in the fish factory with management approval but without their oversight, in the municipal workshop or assembly hall, and when necessary, aboard fishing boats. The results that follow are described in frequencies. Because the responses are categorical in nature and the sample size small, regressions and/or more detailed descriptive statistics are not available.

[6] We recognize that women play a significant role in the household economies of small-scale fisheries. Please see **Chap.35** for a case study of the significance of women within the small scale fishing sector.

## Results

To establish a baseline understanding of respondents' commitment to the sector, we asked if stakeholders were interested in small-scale fishing, which yielded a 90% positive response. To establish whether the SSF Guidelines had been introduced formally or informally to Qeqertarsuatsiaat's small-scale fishers, we asked if they were familiar with them, to which we received a two-thirds response of no. Those who said otherwise admitted confusion with other national or international fisheries management initiatives or chose not to the answer the question. This result emphasizes the need for disseminating and explaining the SSF Guidelines to the stakeholders for whom they are intended to benefit.

### *Local Management (Points 2, 5.18 and 11.4)*

Despite confusion over ongoing fisheries management plans in effect, more than 70% indicated that there was no resistance to the management of fishing activities within and around the community. Our informal conversations with fishers and community members corroborate approval but also the practical limitations of enforcing management measures. Although small-scale fishers are part of a commercial fishery, fishers and community members regularly engaged in hunting, fishing, and gathering activities for subsistence purposes. Through an informal subsistence economy, they enact locally-understood and respected norms of environmental stewardship, the results of which are akin to the kind of marine cultivation strategies seen elsewhere (Thornton et al. 2015). In one example, a community member cited community-wide disagreement with the ban on the hunting of eider ducks, but remarked that the community had refrained from hunting them, despite the community referring to them as both delicious and observing their abundance. The same can also be said in the case of the Atlantic salmon season, to which residents abide by the regulated season and, among other conservation measures, limit catches to projected need for the following year.

According to residents, waterfowl, terrestrial species, some plants and especially fish are a cornerstone of year-round food security. While most fish that are caught in Qeqertarsuatsiaat are sold, much remains shared. Food sharing is a staple activity in both town and particularly settlement life in Greenland and Qeqertarsuatsiaat is no exception. More than half of small-scale fishers (57%) share fish that they catch, primarily among friends and family and through informal social networks (43% and 43% respectively). Regularly engagement by the majority of residents in hunting, fishing and gathering activities is related to not just the utilization of natural resources, but also an interest in their management. More than 80% of respondents indicated that they wanted to be more involved in small-scale fisheries management, and more than 70% wanted to learn more about the SSF Guidelines. Access to several common-pool resources (CPRs) not only affirms the existence of environmental stewardship in Qeqertarsuatsiaat. It also goes hand-in-hand with a concerted interest and a track record of stakeholder involvement in the local, but not national, management of living marine natural resources. In accordance with H1, we recognize that even though national policies recognize small-scale fisheries, Qeqertarsuatsiaat illustrates that small-scale fishers do not directly engage in the formal governance of the resources upon which they depend. Our findings begin to suggest there to be reason to implement the SSF Guidelines on these grounds.

While access to CPRs, particularly fish, are important for all SSF stakeholders, so too is their access to markets. Because the Arctic cod stock upon which the settlement's economy is based is plentiful, fishers sail only several hundred meters from the harbor to fish, and do so in very close proximity to each other. Longline and jig fishers alike in the Disko Bay and the Nuuk Fjord of Greenland also fish in close proximity with little conflict over access to fishing grounds. Fishers with small dinghy boats less than 9 m length overall (or LOA) comprise the largest fleet category in Qeqertarsuatsiaat. Local fishers with larger boats (10–24 m LOA) that are capable of landing larger catches and hiring a small crew of one-to-three sail further from the settlement and typically return within 24–36 h. Because Nuuk is an area where buyers compete with each other to offer a premium to fishers, fishers receive a higher price per kilo of landed Atlantic cod, which compels larger vessels to land most fish in Nuuk instead of in Qeqertarsuatsiaat.[7]

## *Remoteness and Disadvantaged Access to Markets (Point 7.3)*

Remote locations are directly linked to small-scale fishers' disadvantaged access to markets. While the SSF Guidelines set out to highlight and call for lessening disadvantaged access to markets, it is important to highlight how small-scale fishers are mutually dependent on the large-scale seafood buyers who are offer low prices for fish. In a form remarkably similar to colonial systems of trade and market access that were and still are common in Greenland, fishers in settlements have limited options as to whom and for how much they can sell fish. As a result of the regular presence and reliability of a buyer located in the settlement, small-scale fishers in Qeqertarsuatsiaat indicate that they can always bring their product to market. More than two-thirds of respondents indicated that they can always sell fish. Furthermore, fishers clarified that they sell fish every day, especially in season, and if not every day then every other day. Such market conditions create harmony between sellers and buyers.

The current buyer located in Qeqertarsuatsiaat is responsible for the trading of the majority of Greenlandic seafood within national and international export markets. The buyer, Royal Greenland, is a descendant of the Danish colonial trade (Den

[7] A landing obligation is in effect in Greenland to promote employment in settlements. Large-scale vessels must land at least 25% of their catch in Greenland, and coastal vessels are obliged to land 100% of their catches on land.

Kongelige Grønlandske Handel) and is now the government-owned seafood company with headquarters in Nuuk and international offices worldwide.

Despite remoteness, Qeqertarsuatsiaat small-scale fishers are part of a global supply chain. Cod caught in the waters at Qeqertarsuatsiaat are cleaned, landed with the heads off and block frozen, after which they travel by ship to Denmark and ultimately to China where they are then thawed and filleted. In fact, filleted Atlantic cod that travel to China can be purchased in the settlement grocery store. Though they have access to a global market, they do not wield any control over the price of their products nor possess the capacity or timing to singularly satisfy supply. Market access is global, but narrow.

The introduction of competition is obstructed by Greenland's limited infrastructure. Also a colonial byproduct like Royal Greenland, Royal Arctic Line is a government-owned shipping monopoly built upon a history of serving the country's most remote locations, however much it may cost. While subsidized shipping rates allow goods to enter the market where and when they would not otherwise, restrictions on competitors limits the ability to assess how small-scale fishers could respond to alternative supply chains. In an attempt to maintain harmony in settlements by stabilizing fish prices and employment, Greenland has hitherto staged a fight-or-flight opportunity for increasing access to the markets and competition in the fishing sector.

That small-scale fishers' access to the market is disadvantaged is *not*, however, by sole virtue of the sole presence and leverage of a single-operating fish buyer in Qeqertarsuatsiaat. The settlement's small-scale fishing activities take place within a complex system of regulatory, infrastructural, and post-colonial institutions that shape market access, competition and, in this first instance, the readiness to implement the SSF Guidelines.

## *Limited Alternative Livelihoods (Point 6.8)*

Although Greenland's economy is bolstered by the block grant (Auchet 2011; Commission 2014), a vision put forth by the Government of Greenland focuses on expanding three main areas of the economy and the labor market: fishing, tourism, and resource exploitation (G. O. Greenland 2011). While all three pillars of economic activity take place in and around Qeqertarsuatsiaat, employment within those specific sectors remains limited. Jentoft's argument that small-scale fishing is a family enterprise holds true in Qeqertarsuatsiaat (Jentoft 2014). Over 81% of respondents described their fishing activities as built around the family. Small-scale fishers indicated that they fished with their sons, daughters, extended family, wives and siblings. When asked whether they felt as if fishing was an occupation of last resort, only 14% agreed. Instead, the majority felt it was better to describe fishing as a way of life, a heritage activity and/or in support of cultural cohesion (43%, 26%, and 17%, respectively). Concerning employment opportunities elsewhere, the majority felt that the nearby mine would offer more jobs and income opportunities (54%).

Reforms in the coastal halibut fishery have shifted the expectations of new entrants, especially among young men (Delaney et al. 2012). Historically, young men who leave school at the age of 16 plan to enter the coastal fishery. However, when the new regulation entered into force in 2012, fewer were able to join the commercial fishery, which led to a paradigm shift in what a proportion of young men expect to pursue for employment (G. O. Greenland 2012a). However, the reform took place amid optimism that Greenlandic economy would soon prosper from mineral exploitation and mega projects that would in turn reduce deficits in national employment. Thus, the fishery would be free to function on free market conditions without subsidies and employment priorities (G. O. Greenland 2009). However, extractive industry employment has been non-existent, due in part to the 2014 commodities crash, which wiped several prospective mining projects off of Greenland's drawing board. In July of 2014, the limitation on new entrants was rescinded, enabling anyone, including young men, to join the fishery.

Some in Greenland derive hope for a subsoil resource future based upon its historical significance. Mines have opened and closed over the past 120 years and several new exploration projects are ongoing (Auchet 2011). However, while Greenland has the geological capacity for numerous mining projects, it lacks the robust management and regulatory know-how, as well as an historical track record of successful projects necessary for investor support and experience sharing of the benefits of such projects (Hansen 2014). Of importance for Qeqertarsuatisiaat and its labor market is how the Government of Greenland could build off the progress that the nearby ruby mine project has had in satisfying the employment goals of its IBA by ensuring that nearby residents have equitable, if not preferential, access to job opportunities in the mine.

## *Competition from Large-Scale Fishing and Industrialization (Paragraphs 5.9, 5.10)*

Competition is prominently featured in the SSF Guidelines. We observed competitive behavior is conceptualized with a degree of peculiarity in Qeqertarsuatsiaat. Indeed, it forms the basis of what we refer to as pernicious harmony. As the SSF Guidelines outline, one of the challenges that face small-scale fishers is the presence of competition from large-scale fishing operations and other large-scale industrial projects. The SSF Guidelines recognize that unequal power relations are found in small-scale fisheries and that communities often suffer as a result. Qeqertarsuatsiaat's nearby waters support large-scale fishing as well as the nearby industrial project. However, small-scale fishers do not directly compete with large-scale fishers on spatial grounds (i.e. there is no overlap of fishing grounds). For some species, however, such as West Greenland halibut (*Reinhardtius hippoglossoides*) and Atlantic cod, both small and large-scale fishers supply the market, even if they do not fish in the same areas. One common scenario for indirect competition is when large-scale fishers land large quantities and augment supply, which drives down the prices for all sellers, including for small-scale fishers. High volumes of landed fish push out small-scale fishers first, for they do not benefit from the efficiencies that large-scale fishing boats possess.

Although competition between sectors appears a non-issue, one area of concern that also sits at the core of the readiness to implement the SSF Guidelines is how competition has hitherto been conceptualized and enacted among both small and large-scale fishers. In this section, we seek to test our second hypothesis:

> **$H_2$**: If the SSF Guidelines implore member states to recognize and improve the competitive capacity of small-scale fishers and Greenland's small-scale fishers have historically worked harmoniously with the large-scale sector and national government, then Greenland will not struggle with the power inequalities that are common among small-scale fishers and large-scale actors.

With the insignificant exception of the nearby ruby mine, Royal Greenland is the sole commercial buyer and offers fishers of all vessel sizes the minimum price per kilo for cod when landing cod at the fish factory in Qeqertarsuatsiaat. Small- and medium-scale fishers' landings keep the fish factory operating and offers limited seasonal employment for fish workers. Small-scale fishers earn their income from their sales to Royal Greenland. Because Royal Greenland's large-scale fishing vessels both buy and fish for cod, they are both a buyer and a competitor.

However, our findings show that a majority of small-scale fishers in Qeqertarsuatsiaat consider competitors differently from how they are. Although all are competitors, the category indicated with the greatest frequency as a competitor was large-scale fishers (38%), followed by vessels from outside of Qeqertarsuatsiaat (32%), Royal Greenland (14%) and fellow fishers (10%). The fact that small-scale fishers do not think of the sole seafood buyer as a competitor is a paradox, insofar as they not only compete for the same resource, but that the economic fate of small-scale fishers is directly dependent upon the purchasing capacity of Royal Greenland. Concerns over competition are therefore shrouded within the relative stability and dependability of the sole seafood buyer. How small-scale fishers interact with small-scale providers not only highlights the utility of Point 5.9 of the SSF Guidelines, but it also shows the danger of development that is akin to the status quo. For these reasons, the status quo furthers a pernicious harmony between buyers and sellers that would have to be reformed if Greenland were to implement the SSF Guidelines.

## Analysis

Our findings raise uncertainties about the fate of small-scale fishing activities not just in Qeqertarsuatsiaat, but within localities above and below the Arctic circle with similar characteristics. In the first instance, fishing is not always an occupation of last resort (Onyango and Jentoft 2011). Instead, it is part of a diverse livelihood portfolio that people in Qeqertarsuatsiaat maintain. The SFF Guidelines would fortify the viability of small-scale fishing as a livelihood option. Alternative livelihoods can be expected if market access improves, particularly vis-à-vis shipping. Such an opening up could lead to the diversification of fishing activities and other ancillary industries. Fishing is more than just a means to an end; it is a way of life – a heritage activity – and remains important for cultural cohesion.

In our investigation of the relationships between small-scale fishers and Royal Greenland, the government-owned seafood company, we find that small-scale fishers work interdependently and harmoniously with Royal Greenland and are not excluded from income-generating activities. Small-scale fishers buy and sell product to Royal Greenland in a way that is not manipulative, though not as mutually beneficial as it could be. Although a paucity of competition benefits some, it comes at the cost of lower-tier stakeholders, who, in this case, are the small-scale fishers of Qeqertarsuatsiaat. The harmony produced through this relationship is pernicious because: firstly, it underestimates the small-scale sector's capacity for innovation; secondly, it hinders profit margin growth and market diversification; and, thirdly, because it perpetuates inequality between large- and small-scale stakeholders. It is also a relationship that stands in opposition to the principles set forth in the SSF Guidelines. Above all, it is a relationship that could be reconfigured to achieve Greenland's national fisheries objectives and to uphold a notable history of support for small-scale fishers and fish workers.

Even though Greenland's national policies emphasize indigenous peoples' rights and human rights as well as gender equity and equality principles, they are not spelled out in relation to management of the fishery sector. Nor do Greenland's fisheries policies explicitly include UNDRIP or other international human rights declarations. The SSF Guidelines could provide Greenland's fisheries stakeholders with an instrument to govern, respect, and fortify small-scale fishing activities amid trends toward ultra-efficient, large-scale fishing activity. While our findings indicate that implementing the SSF Guidelines in Greenland may be feasible, it will require policy makers to reconcile the power and position of government-owned and supported companies in relation to available living marine resources, tradition and national economic ambitions. If supporting small-scale fisheries comes at the cost of the current development of the large-scale sector, then we may see the strength of the small-scale sector falter.

Competition is an indicator of any healthy economy and the absence of it can have deleterious effects on the productivity and resilience of social-ecological systems. Because large-scale companies such as Royal Greenland are moving away from small-scale fisheries, future work should scrutinize interdependences and assess the risk that would come to small-scale fishers, fish workers, and communities in the event of the dissolution of this single market. There is, however, promise to build upon the pre-existing strengths between small- and large-scale fishers and buyers. Doing so could improve the economic productivity of the two sectors and achieve national growth agendas. We also expect that implementing the SSF Guidelines would recognize the net contributions that small-scale fishing activities make to society, including but not limited to local food security, stable, independent, and flexible employment, and the mixed cash and subsistence economies that have hitherto thrived in Greenland.

## Closing Remarks

Despite adversity, Greenland's small-scale fisheries are well-positioned to implement the SSF Guidelines. It stands ahead of other countries on several key issues facing small-scale fisheries globally. For example, Greenland does not have chronic issues with unhealthy or unsafe working conditions, nor with forced labor. As a signatory to the Convention on the Elimination of all Forms of Discrimination Against Women (CEDAW) in 1983 and in ratifying its own Gender Equality Act in 2013, Greenland's gender policies are advanced and aligned with the SSF Guidelines' emphasis on gender equality.[8] Despite Greenland's social welfare state supporting tuition-free education at all levels and other mechanisms that would otherwise support upward labor mobility, Greenlanders, especially among those who fish and/or who are in settlements, have limited but improving livelihood alternatives. These and other factors serve as a legitimate basis for not only considering implementation of the SSF Guidelines, but for also having the political velocity to adopt them.

Although our findings are particular to Qeqertarsuatsiaat, they are not total. What we outline here does, however, highlight stark disparities between the aims of the SSF Guidelines and the realities in which they would be implemented not just in Qeqertarsuatsiaat or Greenland, but elsewhere as well. These data also suggest that how small-scale fishers think about and act on several key concerns circumscribes the organization and behavior of the social-ecological system.[9] They also point to how the SSF Guidelines would demand reconfiguring the relationships between government, small-scale fishers and fish workers, and large-scale fishers. Current relations between small-scale fisheries actors in Qeqertarsuatsiaat and in other settlements are dubious, particularly when the option exists to implement a novel management instrument like the SSF Guidelines. When small-scale fishers and fish workers do not recognize their lack of control, all small-scale stakeholders and first-order beneficiaries lose out. Though it is in large part up to the people who call Qeqertarsuatsiaat home to decide, it is necessary to highlight where blind spots could become political sticking points as well as the areas of promise if or when Greenland implements the SSF Guidelines.

---

[8] Please refer to **Chap. 35**, which includes a case study by Snyder on Greenland's Gender and Fisheries Policy.

[9] Key concerns include: (1) the transfer of traditional ecological knowledge and local management (Point 2, 5.18 and 11.4); (2) remoteness and disadvantaged access to markets (Point 7.3); (3) limited alternative livelihoods (Point 6.7); (4) competition from large-scale fishing and industrialization (Points 5.9, 5.10).

**Acknowledgements** We thank Qeqertarsuatsiaat's SSF stakeholders for the openness and willingness with which they spoke with us. Structural inspiration for the questionnaire is credited to the Survey of Living Conditions in the Arctic (Poppel 2015) and the Pacific Coast Groundfish Fishery Social Survey produced by Suzanne Russell in the Human Dimensions Program of the Northwest Fisheries Science Center at NOAA (Russell 2012). Institutional support was provided by the Ilisimatusarfik, University of Greenland during the fieldwork stages of this project. Special thanks is also given to colleagues in the Environmental Studies Department, Dartmouth College, for their support during the writing stages of this project. We also thank the Greenland Climate Research Centre for providing institutional support and the Ministry of Fishing, Hunting, and Agriculture for helpful dialogue.

**Funding** The National Geographic Society provided Hunter Snyder a Young Explorer Grant (GEFNEY74–14) to carry out the case study upon which this chapter is based. Additional fieldwork in Qeqertarsuatsiaat, Greenland was supported by Statistics Norway under the ECONOR III, *Economies of the North* project. The writing of the article has been further supported by research funds of the Greenland Self-Government (Grønlands Selvstyres midler til forskningsfremme).

## References

Auchet, M. (2011). Greenland at the crossroads: What strategy for the Arctic? *International Journal, 66*(4), 957–970. doi:10.2307/23104404.

Chuenpagdee, R. (Ed.). (2011). *World small-scale fisheries: Contemporary visions*. Delft: Eburon.

Civilstyrelsen. (2009). *Lov om Grønlands Selvstyre* (pp. 1–5).

Condon, R. G., Collings, P., & Wenzel, G. (1995). The best part of life: Subsistence hunting, ethnicity, and economic adaptation among young adult Inuit males. *Arctic, 48*(1), 31–46.

Delaney, A. E. (2016). The neoliberal reorganization of the Greenlandic Coastal Greenland halibut fishery in an era of climate and governance change. *Society for Applied Anthropology, 75*, 193–203. doi:10.17730/1938–3525-75.3.193.

Delaney, A., Kaare, H., & Jacobsen, R. B. (2012). *Greenland halibut in Upernavik: A preliminary study of the importance of the stock for the fishing populace*. Deliverable to the Ministry of Fishing, Hunting, and Agriculture.

Dzik, A. J. (2014). Kangerlussuaq: Evolution and maturation of a cultural landscape in *Greenland. Bulletin of geography. Socio-Economic Series*. doi:10.2478/bog-2014-0014

European Commission. (2014). *Programming document for the sustainable development of Greenland 2014–2020*. Brussels: European Commission.

FAO. (2004). *Fishery and aquaculture profile: Greenland*. Rome: FAO.

FAO. (2014). *The state of world fisheries and aquaculture 2014*. Rome: FAO.

FAO. (2015). *Voluntary guidelines for securing sustainable small-scale fisheries in the context of food security and poverty eradication*. Rome: FAO.

FAO. (2017). *Fishery and aquaculture country profiles* (G. H. Snyder, Ed.). Rome. Available at: http://www.fao.org/fishery/facp/GRL/en

Forsyth, T. (2003). *Critical political ecology*. New York: Routledge. doi:10.4324/9780203017562.

Greenland, G. O. (1999, October 26). *Meeting minutes Tuesday October 26 1999*. Nuuk: Government of Greenland.

Greenland, G. O. (2009). *Fiskerikommissionens betænkning*. Nuuk: Government of Greenland.

Greenland, G. O. (2011). *Vores velstand og velfærd – kræver handling nu*. Nuuk: Government of Greenland.

Greenland, G. O. (2012a). *Selvstyrets bekendtgørelse nr. 2, af 2. februar 2012 om kystnært fiskeri efter hellefisk.*http://lovgivning.gl/lov?rid={A898F2BA-CABB-45E8-AA87-392B29B26975}. Accessed 9 Dec 2016.

Greenland, G. O. (2012b, August 16). *Meeting minutes August 16, 2012.* http://naalakkersuisut.gl/da/Naalakkersuisut/Moeder-for-Naalakkersuisut/M%C3%B8dereferater/2012/08/16_08_2012. Accessed 9 Dec 2016.

Greenland, H. R. C. O. (2014). *Menneskerettigheder i Grønland – Status 2014* (pp. 1–103). Nuuk: Government of Greenland.

Hansen, A. M. (2014). *Community impacts: Public participation, culture and democracy.* Background paper for the Committee for Greenlandic Mineral Resources to the Benefit of Society.

Hovgård, H., & Wieland, K. (2008). Fishery and environmental aspects relevant for the emergence and decline of Atlantic Cod (*Gadus morhua*) in West Greenland waters. In G. H. Kruse, K. Drinkwater, J. N. Ianelli, J. S. Link, D. L. Stram, V. Wespestad, & D. Woodby (Eds.), *Resiliency of Gadid stocks to fishing and climate change* (pp. 1–22). Proceedings of the symposium Resiliency of Gadid stocks to fishing and climate change, 31 October–3 November 2006, Anchorage, Alaska, USA.

Jacobsen, R. B. (2013). Small-scale fisheries in Greenlandic planning – The becoming of a governance problem. *Maritime studies, 12*(1), 2. doi:10.1186/2212-9790-12-2.

Jacobsen, R. B. (n.d.). *Power and participation in Greenlandic fisheries governance: The becoming of problems, selves and others in the everyday politics of meaning.* vbn.aau.dk. Accessed 9 Dec 2016.

Jacobsen, R. B., & Raakjær, J. (2012). A case of Greenlandic fisheries co-politics: Power and participation in total allowable catch policy-making. *Human Ecology, 40*(2), 175–184. doi:10.1007/s10745-012-9458-7.

Jacobsen, R. B., & Raakjær, J. (2014). *Who defines the need for fishery reform? Actors, discourses and alliances in the reform of the Greenlandic fishery.* Cambridge: Cambridge University Press.

Jentoft, S. (2014). Walking the talk: Implementing the international voluntary guidelines for securing sustainable small-scale fisheries. *Maritime Studies, 13*(1), 1–15. doi:10.1186/s40152--014-0016-3.

Lilly, G. R., Wieland, K., Rothschild, B. J., Sundby, S., Drinkwater, K. F., Brander, K., Ottersen, G., Carscadden, J. E., Stenson, G. B., Chouindard, G. A., Swain, D. P., Daan, N., Enberg, K., Hammill, M. O., Rosing-Asvid, A., Svedäng, H., & Vázquez, A. (2008). Decline and recovery of Atlantic Cod (Gadus morhua) stocks throughout the North Atlantic. In G. H. Kruse, K. Drinkwater, J. N. Ianelli, J. S. Link, D. L. Stram, V. Wespestad, & D. Woodby (Eds.), *Resiliency of Gadid stocks to fishing and climate change* (pp. 1–28). Proceedings of the symposium Resiliency of Gadid stocks to fishing and climate change, 31 October–3 November 2006, Anchorage, Alaska, USA.

Nielsen, A. B. (2010). *Present conditions in Greenland and the Kangerlussuaq area* (Posive Working Paper 2010–07). Olkiluoto: Posiva Oy.

Nuttall, M. (2009). Living in a world of movement: Human resilience to environmental instability in Greenland. In S. A. Crate & M. Nuttall (Eds.), *Anthropology and climate change: From encounters to actions* (pp. 292–310). Walnut Creek: Left Coast Press.

Onyango, P. O., & Jentoft, S. (2011). Climbing the hill: Poverty alleviation, gender relationships, and women's social entrepreneurship in Lake Victoria, Tanzania. *Maritime Studies, 10*(2), 117–140.

Poppel, B. (2015). SLiCA: *Arctic living conditions.* Nordic Council of Ministers. doi:10.6027/TN2015-501.

Poppel, B., & Kruse, J. (2009). The importance of a mixed cash- and harvest herding based economy to living in the Arctic: An analysis on the survey of living conditions in the Arctic (SLiCA). In V. Møller & D. Huschka (Eds.), *Quality of life and the millennium challenge: Advances in quality-of-life studies, theory, and research* (pp. 27–42). New York: Springer.

Rasmussen, R. O. (1998). *Managing resources in the Arctic: Problems in the development of fisheries* (Working Paper 135). Roskilde: Roskilde University, Department of Geography and Development Studies.

Russell, S. (2012). *Pacific coast groundfish fishery social survey post catch shares 2012* (Version 206).

Stockholm Resilience Center. (2014). *What is resilience?*http://www.stockholmresilience.org/. Accessed 9 Dec 2016.

Thornton, T. F., & Hebert, J. (2015). Neoliberal and neo-communal herring fisheries in Southeast Alaska: Reframing sustainability in marine ecosystems. *Marine Policy, 61*, 366–375. doi:10.1016/j.marpol.2014.11.015.

Thornton, T., Deur, D., & Kitka, H. (2015). Cultivation of salmon and other marine resources on the Northwest coast of North America. *Human Ecology, 43*(2), 189–199. doi:10.1007/s10745-015-9747-z.

# Chapter 7
# Walking the Talk of the Small-Scale Fisheries Guidelines in Jamaica

**Lisa K. Soares**

**Abstract** Conventional rights based regimes designed to streamline small-scale fisheries governance have slowly moved towards embedding policy considerations that aim to bridge fishing rights and human rights (including social and economic rights). This chapter discusses the extent to which the Voluntary Guidelines for Securing Sustainable Small-Scale Fisheries in the context of Food Security and Poverty Eradication (SSF Guidelines) can take root within the Commonwealth Caribbean state of Jamaica. Particularly, this chapter explores to what extent Jamaica's enabling environment facilitates the implementation of Section 10(7) of the SSF Guidelines: Policy Coherence, Institutional Coordination and Collaboration. In order to implement the current SSF Guidelines effectively, they should be in tune with national and local specificities and nuances.

**Keywords** Jamaica • Small-scale fisheries • SSF Guidelines • Fisheries politics • Fisheries policy • Governance

## Introduction

Jamaica's small-scale fisheries historically represent one of if not the most socio-economically and politically undervalued of the island's natural resources. Recently this oversight has been addressed with the endorsement of the Voluntary Guidelines for Securing Sustainable Small-Scale Fisheries in the context of Food Security and Poverty Eradication (hereafter the 'SSF Guidelines') in 2014. The SSF Guidelines are considered to be the most recent normative packaging designed to influence global fisheries governance dynamics, particularly the rights of small-scale fisheries and fisherfolk in developing countries (see also Jentoft 2014).

Jamaica was as host in 2012 at the center of the the Caribbean consultations of the SSF Guidelines (FAO 2013). Given the island's centrality in the region (i.e. it is

L.K. Soares (✉)
University of Warwick, Warwick, UK
e-mail: L.K.Soares@warwick.ac.uk



S. Jentoft et al. (eds.), *The Small-Scale Fisheries Guidelines*, MARE Publication Series 14, DOI 10.1007/978-3-319-55074-9_7

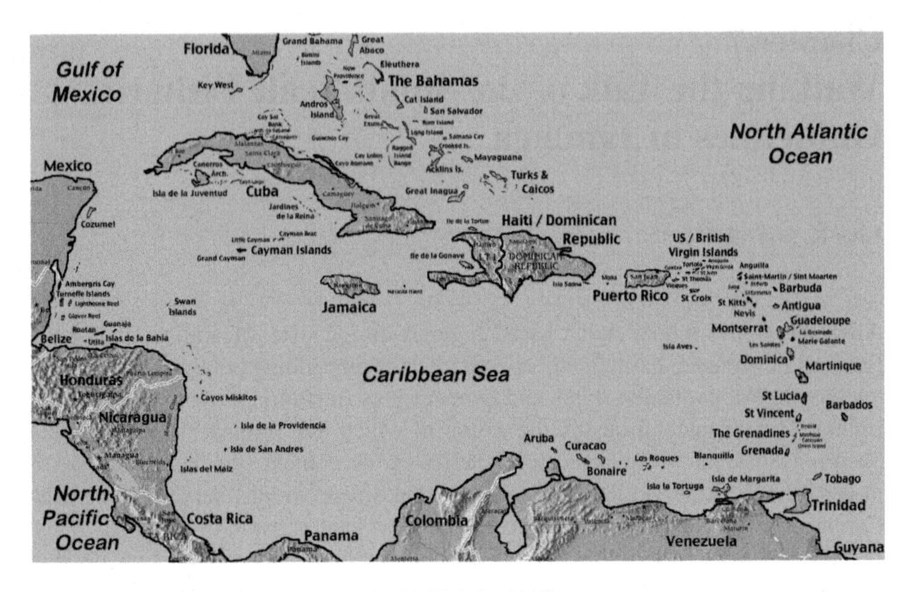


**Fig. 7.1** Map of Caribbean (Source: The World Atlas 2016)

the third largest of the Greater Antillean islands, and the largest English-speaking island – see Fig. 7.1), it is important to contextualize whether or not the country has a sound enabling environment to implement Section 10(7) of the SSF Guidelines. Section 10(7) emphasizes that

> States should recognize, and promote as appropriate, that local governance structures may contribute to an effective management of small-scale fisheries, taking into account the ecosystem approach and in accordance with national law (FAO 2015).

Section 10 of the Guidelines highlights three important concepts that should be defined, as they are integral to successfully implementing the Guidelines: Policy Coherence, Institutional Coordination, and Collaboration. The OECD (2003, 2) defines policy coherence as the "systematic promotion of mutually reinforcing policy actions across government departments and agencies creating synergies towards achieving the agreed objectives." Mattessich et al. (2001, 39) explain that institutional coordination 'involves a low level of joint planning, sharing of resources, defining of compatible roles, and interdependent communication channels,' and that collaboration involves "a mutually beneficial and well-defined relationship entered into by two or more organizations to achieve common goals."

The SSF Guidelines clearly call for a stable enabling environment embedded in local governance structures that ought to be constructed in accordance with national law. Therefore, the role of the state is emphasized in promoting policy coherence, institutional coordination, and collaboration. Equally important are the roles and capacities of non-state actors, who alongside local government, are part of the local governance structures needed to implement the Guidelines. The aim of this chapter is to identify and evaluate how state and non-state actors interact with the socio-economic, political, legal and regulatory factors, which might make possible or prohibit the production of a stable enabling environment for implementation of the Guidelines. By stable enabling environment, I mean an environment that achieves synergy between: (a) the state and status of Jamaica's small-scale fisheries and (b) the political, legal, and regulatory frameworks, i.e. fisheries laws, policies, and management actions which define and govern Jamaica's fisheries as well as the environment in which Jamaica's small-scale fisheries ought to function. The issues under consideration therefore, aim to explore the viability of synergies existing between local governance and national law, structures ultimately needed to shore up implementation of the SSF Guidelines in Jamaica.

The conclusions I offer are based upon a review of secondary data and reports, as well as insights gained from key informants (KIs) in the course of fieldwork between December 2015 and January 2016. The chapter is divided into five sections. Section "Conceptual and analytical groundings" describes the conceptual and analytical groundings that underpin the analysis. Section "The state and status of Jamaica's small-scale fisheries and the SSF guidelines" provides an overview of the state and status of Jamaica's small-scale fisheries and fishers. In Section "The enabling environment for the SSF guidelines in Jamaica: political, legal, and regulatory frameworks and management actions", I chart out the political, legal, and regulatory frameworks that govern Jamaica's small-scale fisheries, and consequently influence the enabling environment for the SSF Guidelines in Jamaica. Section "An enabling environment for the SSF guidelines in Jamaica: the existence of synergies?" evaluates the extent to which synergies exist between Jamaica's small-scale fisheries and the political, legal, and regulatory frameworks needed to produce a stable enabling environment. Finally, Section "Policy recommendations: how can the SSF guidelines be implemented in Jamaica?" offers policy recommendations that might influence or constructively shape the enabling environment needed to harmonize efforts to implement the SSF Guidelines in Jamaica with the needs of actors who have an interest in Jamaica's fisheries.

## Conceptual and Analytical Groundings

### *Conceptual Groundings*

Similar to a majority of Commonwealth Caribbean nations, Jamaica perpetually struggles with finding adequate capacity, resources, and political will at both national and local scales to implement evolving global and regional normative directives (Bravo 2006; Commonwealth Fisheries 2009). Indeed, modelling frameworks for fisheries governance also suffer from a lack of contextual and material translatability (Espeut 1993). The SSF Guidelines represent a call to engage in such policy transfer from global to local scales. Dolowtiz (2003, 103) suggests that when examining what motivates an individual or the state to engage in the policy transfer process, it is important to acknowledge that the power to authorize and legitimize a policy is not necessarily located within the domestic political system. Often the pressure to adopt a policy position results from decisions made at the international level or emerges out of treaty or multi-lateral membership obligations. Of course, this raises questions with regard to the feasibility of the trickle down/bottom up framework successfully structuring the SSF Guidelines policy transfer process from the global to local and vice-versa, as was agreed to by state representatives at the 31st Session of the FAO Committee on Fisheries (COFI 2014).

### *Analytical Groundings*

Some practitioners contend that the SSF Guidelines ought to be implemented in accordance with country-specific legal systems and their institutions, i.e. small-scale fisheries should have access to justice and effective remedies guaranteed by both national and international law. This new 'norm' has been met with both support and criticism because of the perception that the SSF Guidelines are linked to hard international law such as the United Nations Declaration on Human Rights, and therefore are not as voluntary as they seem (Jentoft 2014). While the question of voluntariness remains debatable, the more relevant challenge revealed by the research conducted is adherence. Even if one concedes that the new global norms posed by the SSF Guidelines are inevitably going to be influenced by more powerful governing bodies, compliance and efficacy become extremely tricky when the viability of the enabling environment for local governance structures are not taken into account.

So, can Jamaica walk the talk of the SSF Guidelines (Jentoft 2014)?

## The State and Status of Jamaica's Small-Scale Fisheries and the SSF Guidelines

In order to explore the viability of the enabling environment for the SSF Guidelines in Jamaica, it is important to first understand the Jamaican fishing industry and as such, the benefactors of the Guidelines.

### *The Jamaican Small-Scale Fishing Industry*

The Government of Jamaica (GOJ) classifies the Jamaican fishing industry into five operational categories: industrial fisheries, offshore artisanal, mainland artisanal, inland aquaculture and sports fisheries (Government of Jamaica 2008). Here the subject of focus is the island's marine capture small-scale fisheries. The island's main marine capture fisheries comprise: coral reef fishes, spiny lobsters, queen conch, small coastal pelagic finfish, and large offshore pelagic finfish (Aiken and Kong 2000, 31).

Fish is a staple in the Jamaican diet (Espeut 1993). Fish consumption in the island is estimated at 14.74 kg per capita, close to the global average of over 20 kg per capita (Fisheries Division and STATIN 2011a; FAO 2016). Two major fleet components exist: (1) the small-scale fisheries that harvest mostly coral reef fishes for domestic consumption and (2) the queen conch and lobster industrial fisheries (ECOST 2007). The latter are dominated by a handful of private entities with significant capital investments. Here, labor inputs provided by Jamaica's small-scale fisheries normally feed harvesting efforts centralized around larger 'mother-ship' vessels (Grant and Lewis, personal communication, 17, December, 2015). With the exception of the industrial conch and lobster fisheries, all other fisheries operate on an open-access basis (Espeut 1992; Kong 2003; Fisheries Division 2016).

Sustaining Jamaica's valuable fisheries has not been without its pressures. Jamaica's coastal waters are classed among the most overfished in the Caribbean and the world as a whole (Aiken and Kong 2000; Figueroa 2009a; Waite et al. 2011; P. Espeut, personal communication, 15, March, 2016). Dating as far back as 1945, Jamaica's fisheries were deemed to be a case of "…too many men trying to catch too few fish" (Thompson 1945). In reality, many of Jamaica's small-scale fishers have few alternative sources of income, creating a high level of dependence on Jamaica's near-shore reef fisheries (Fisheries Division 1997; Aiken and Kong 2000). With Jamaica's near-shore fishing grounds depleted, fishing pressure has increased in Jamaica's offshore fisheries, mainly in the Pedro Bank located 80 km off the mainland (Haughton and Kong 2006).

## *Economic and Livelihoods Contributions*

Considering the economic importance of fisheries in Jamaica is especially paramount, given that between 1998 and 2007, the industry contributed an average of 0.6% to the island's Gross Domestic Product (GDP) (Fisheries Division and STATIN 2011a). The livelihoods contributions of Jamaica's small-scale fisheries are equally important to consider as they provide Jamaican coastal communities with an important 'safety net' for food and employment (ECOST 2007). Although there are conflicting data on the ultimate contribution of fisheries to the island, fisheries activities are estimated to support the livelihoods of more than 200,000 Jamaicans (Van Riel and Wijkstrom 2005; ECOST 2007; Waite et al. 2011). Fish is a vital protein source that is available all year round and which occurs in sufficiently diverse forms to be available both to the rich and poor. There is the view (see for example, Kong 2003; Waite et al. 2011) that many who participate in the sector are the poorest, most vulnerable, and marginalized in Jamaican society. Indeed, the perception exists that small-scale fisheries throughout the world are frequently characterized as 'the occupation of last resort' and fishers as the 'poorest of the poor.' This characterization, however, often leads to inaccurate development approaches for small-scale fisheries (see Allison and Ellis 2001). International agencies and development directives throughout the world tend to overemphasize the concepts and extent of poverty, marginalization, and vulnerability in the fishing industry. In the context of the SSF Guidelines securing the rights of small-scale fishers, it is important to highlight that many small-scale fishers in Jamaica depend on the fishing industry to establish a viable means of income to ensure the future for their second and third generations. For most coastal dwellers who reside on small-island developing nations like Jamaica, fishing is the occupation of 'first resort'—as a cultural and social standing—on which generations are able to support self and family, thereby casting a net to secure well-being with otherwise limited means.

It is important to mention also that small-scale fishers who participate in the fishing industry are a large part of the informal economy. Secondary data indicates a significant number of unlicensed fishers, up to 10,000 unlicensed fishers at one time (Kong 1990; Espeut 1992; Halcrow 1998). This undoubtedly affects the accuracy of the fishing industry's contributions to the island's overall GDP. Despite these pressures, Jamaica's fisheries continue to provide valuable jobs and revenue for the country (Waite et al. 2011). There is also significant foreign exchange revenue generated through the export of mostly conch and lobster (Kong 2003). It is true that due to inadequate formal education, Jamaican small-scale fishers often have limited alternate opportunities for income. Moreover, given Jamaica's harsh and unpredictable economic environment, many people from other sectors often turn to fisheries seasonally, temporarily, or permanently when faced with unemployment and poverty. The most recent public data available indicates that, in 2008, there were approximately 18,000 registered fishers (Fisheries Division 2011b). It is important to clarify here who can register as a fisher. According to the Fisheries Division, only boat owners/operators and harvest-fishers—the latter otherwise referenced as 'fishermen'—are required to register (Fisheries Division 2011f). Although the data indicates that a mere 6% of those actively involved in the fishing industry are women (Fisheries Division 2011b), it is a fact that the inputs of women in Caribbean fisheries have largely been ignored in both the policy and academic literature (Grant 2004; Centre for Resource Management and Environmental Studies 2016). Fishmongers, vendors, and '*higglers' or informal traders* are not required to register as active participants in Jamaica's fishing industry (Fisheries Division 2011c)— post-harvest marketing and distribution roles dominated by women (Grant 2004). In other words, women occupy critical positions in Jamaica's small-scale fisheries (see Kong 2003)—potentially giving them considerable yet undocumented power in the industry.

## *Current Realities*

Jamaica's small-scale fisheries and its fishers are in trouble. The island's waters are severely over-fished. In fact, due in large part to decreasing Jamaican fish stocks, *the country now has to import most of the fish eaten on the island* (Waite et al. 2011, 7, emphasis mine). This is evidenced from the most recent public data available that points out that total fishery imports in 2007 were of the value of $USD 63.1 million as compared to $USD 667,000 worth of exports (Fisheries Division 2011d). Equally staggering for the state and status of Jamaica's small-scale fisheries are threats borne by the gross underestimation of their economic and livelihood contributions to the island. As mentioned, these underestimations include small-scale fisheries contributions to GDP, social welfare, value-add components through the informal economy, as well as women's employment. If the state and status of Jamaica's small-scale fisheries continue to be largely un-acknowledged, it may be difficult to establish synergy with the political, legal, and regulatory frameworks needed to create a stable enabling environment for the Guidelines. There are reports, however, of positive steps taken by the Government of Jamaica to improve the enabling environment for local governance structures in accordance with national law (see for example, Waite et al. 2011). These provisions include:

- Jamaica's Charter of Rights;
- the establishment of the 1975 Fisheries Act;
- the 2008 Draft Fisheries Policy;
- the application of the ecosystem approach to fisheries (EAF); and
- EAF management through the introduction of special fishing conservation areas in 2009.

Equally important to consider are the contributions of fisher's organizations and non-governmental organizations in influencing, implementing, and facilitating efforts to sustain Jamaica's small-scale fisheries. The next section explores these developments and assesses their alignment to the SSF Guidelines to help determine efficacy.

## The Enabling Environment for the SSF Guidelines in Jamaica: Political, Legal, and Regulatory Frameworks and Management Actions

Jamaica has authorized more than 50 pieces of legislation that deal with the management of the environment including fisheries. The key pieces of national legislation are: The Morant and Pedro Cays Act (1945), aspects of the Wildlife Protection Act (1945), The Fishing Industry Act (1975), The Natural Resources Conservation Agency Act (1991), the Beach Control Act (1956), the Exclusive Economic Zone Act (1996), and the Maritime Areas Act (1997) (NEPA 2003). Jamaica has also endorsed and is signatory to several regional and international agreements as well as mechanisms that deal specifically with fisheries ecosystems and resources. Some of the most significant agreements Jamaica is party to include: the Caribbean Regional Fisheries Mechanism (2003), the United Nations Convention on the Law of the Sea (UNCLOS III) (1982), the Convention on Biological Diversity (1992), and the FAO Code of Conduct for Responsible Fisheries (1995) (Fisheries Division

2011g). Jamaica also served as host for the Caribbean regional consultation on the SSF Guidelines in 2012 (FAO 2013). Jamaica's apparent commitment to seeing the implementation of the Guidelines go through was supported by the Caribbean Regional Fisheries Mechanism's endorsement of the Guidelines by COFI Member States at its 31st session in June 2014. So, what is Jamaica's capacity to facilitate a stable enabling environment for local governance structures in accordance with national law and in support of the implementation of the SSF Guidelines?

## *Jamaica's Charter of Rights and the SSF Guidelines*

There are some practitioners (see for example, Espeut 1992; D. McCaulay, personal communication, 15, December, 2015) who suggest that the role of the state in fisheries governance is establishing policy coherence, institutional coordination, and collaboration. The state is the trustee of the natural environment and has the duty to ensure that resources are not exploited beyond the limits of sustainability. The state, however, should respect the rights of resource users to have a say in how resources are exploited and utilized. It is noteworthy that Article (13,1) of Jamaica's 'Charter of Fundamental Rights and Freedoms' aligns to the SSF Guidelines' focus on people-centered perspectives. This is through specific provisions for patrons of the state regarding "the right to enjoy a healthy and productive environment free from the threat of injury or damage from environmental abuse and degradation of ecological heritage." Moreover, the Charter stipulates that the "state has an obligation to promote universal respect for, and observance of, human rights and freedoms" (Government of Jamaica 2011; Jamaica Environment Trust 2013). Given that the underlying aims of the SSF Guidelines are explicitly upheld in Jamaica's Charter of Rights, it is even more critical to contextualise their efficacy alongside national and local political, legal, and regulatory engagement with the fishing industry.

## *An Institutional Framework for the SSF Guidelines: The Fisheries Division, Fisheries Act, and Policy*

The primary fisheries legislation in Jamaica is the Fishing Industry Act (1975). The 'Act' came into effect in 1976, and made the Fisheries Division of the Ministry of Agriculture the primary agency responsible for fishing. The Division is responsible for enforcing the 1975 Fisheries Act, which includes regulating closed areas and closed seasons, and the banning of destructive gears, licensing, etc. (CARICOM Fisheries Unit 2000).

Like in many other fisheries throughout the world, Jamaica's approach to fisheries can be characterized as utilizing conventional top-down fisheries governance and management policies (see Allison and Ellis 2001). However, since 2003, the

GOJ has been developing a new draft National Fisheries and Aquaculture Policy (Waite et al. 2011) based on a co-management approach. The development of the Policy, first conceived in 2008, was followed by the development of a 'new' Fisheries Act in 2009 (Jamaica Observer 2012). In 2008, the GOJ also founded the National Fisheries Advisory Board (FAB), comprised of stakeholder representatives from the commercial fishing industry, sport fishers, small-scale fishers, and marine ecologists (Jamaica Gleaner 2008). According to the Fisheries Division, extensive public consultations on the new Fisheries Policy and Act were held within the sector as the result of these developments—signalling a change in the Division's ethos to one reflecting a more consultative and participatory approach to fisheries governance (Jamaica Information Service 2008).

These developments have indeed mirrored the call for people-centered approaches to global fisheries governance like those called for in the SSF Guidelines. For the GOJ, this represented a distinct move away from the largely failed top-down governance and managerial approaches towards approaches that emphasized local governance guided by co-management and community-based fisheries resource management (Waite et al. 2011). Despite a direct reference to the SSF Guidelines, the drafts of Jamaica's new Fisheries Act and Fisheries Policy also replicate concepts found in the SSF Guidelines—namely co-management approaches to fisheries governance. Moreover, key informants (KIs) reported that since their first drafts both the Fisheries Act and Policy have undergone several revisions to enhance the state's consultative and participatory apparatuses in theory, the most recent revision being in 2015 (Government of Jamaica 2015a, b).

## *The SSF Guidelines and an Eco-System Approach to Fisheries Management via Special Fishing Conservation Areas*

Section 10(7) of the SSF Guidelines also emphasizes the need to adopt an ecosystem approach to fisheries (EAF) (FAO 2003). It is important to highlight, therefore, that the GOJ has attempted to incorporate the EAF into its fisheries governance agenda. This was accomplished with the development and expansion of the country's fish sanctuaries and marine protected areas (MPAs), now classified as special fishing conservation areas (SFCAs), from two to eleven between 2009 and 2010 (Waite et al. 2011; D. Newell, personal communication, 10, December, 2015). A few of these areas are the Oracabessa (St. Mary), Bluefields (Westmoreland), and Galleon (St. Elizabeth) fishing sanctuaries (CaribSave 2015).

As a matter of national and local policy transfer, the emphasis on EAF through the provisioning of SFCAs also suggests the GOJ's intention to align policy with the SSF Guidelines (Aiken and Kong 2000; Waite et al. 2011). According to the Fisheries Division (2011a), SFCAs are managed through a cooperative arrangement between the government and community organizations. A memorandum of agreement (MOA) formalizes each partnership between the ministry and the collaborating organization. Under the MOA, the GOJ provides the resources for partner NGOs to undertake day-to-day operations. Earnings from taxes on the exports of Jamaica's queen conch, also known as the 'conch-cess' largely funds these efforts.

## *Fishers' Organizations, NGOs and the SSF Guidelines*

Boding well for the SSF Guidelines in Jamaica is the fact that there is a great deal of community cohesion among fishers, especially in rural areas. 'Jamaican fishers do belong to organizations and those organizations are part of a country wide network' (E. Figueroa, personal communication, 9 January, 2016). Despite their existence, however, Jamaican fishers' organizations often struggle to unite as a strong political unit that pursues and establishes consensus positions—bringing into focus the importance of fishers' organizational capacity and attitudes towards working together. As Figueroa (personal communication, 9 January, 2016) explains, ultimately Jamaican fishers, though willing to come together for a particular reason or cause, prefer to operate within their own respective agencies and not be subsumed under someone else's control, or a larger umbrella structure. They prefer freedom over functionality.

It is true that globalization has weakened wholly state-reliant traditional governance structures. NGOS and other civil society groups have become not only critical stakeholders in governance, but also the driving force behind greater cooperation through their active mobilization of public support for evolving normative global directives (Gemmill and Bamidele-Izu 2002). With regard to mobilizing public support, there are several conservation NGOs, local and international, that have also attempted to play a role in small-scale fisheries governance in Jamaica. Some of the more active include: CaribSave (now defunct), The Nature Conservancy, Jamaica Environment Trust, Yardie Conserve, and the Caribbean Coastal Area Management Foundation (C-CAM). Although their efforts have led to varied levels of success (Waite et al. 2011; E. Figueroa, personal communication, 7, December 2015; L. Meggs, personal communication, 15, December, 2015; McCaulay, personal communication, 15, December, 2015; P. Espeut, personal communication, 15, March, 2016), these NGOs too could be strategic partners in working towards the streamlining of the SSF Guidelines across national and local spheres. They could be partners through the promotion of collaboration and dialogue between different actors who have interest in Jamaica's small-scale fisheries. Consequently, NGOs and other civil society actors should continue to seek out ways to engage all those with a stake in Jamaica's fisheries in matters pertaining to:

- information dissemination;
- policy development consultations;
- policy implementation and training;
- assessment and monitoring; and
- advocacy for environmental justice (see Gemmill and Bamidele-Izu 2002).

### *Current Realities*

There have been concerns (see for example, Espeut 1992; P. Espeut, personal communication, 15, March, 2016), that existing governance structures view fishers as 'part of the problem' rather than integral to the solution. This perception often leads to conflict between fishers, lawmakers, and environmentalists. "The latter tend to enforce the laws they have engendered, while the former fight for survival against those they perceive trying to erode their traditional rights. What these approaches lack is an appreciation of the political and cultural dynamics of fishing and the socio-economic parameters within which fisherfolk operate" (Espeut 1992, 3). Treating fisheries solely in biological/positivist terms means that governance approaches focus on the idea of managing fish rather than fishers. Espeut (1992, 2) points out that such a dehumanized approach fails to adequately acknowledge power relations and structures of dominance that influence notions of empowerment and community democracy. The embedded lack of emphasis on power relations and structures of dominance are long-standing concerns with fisheries governance that the SSF Guidelines attempt to address. Empowerment and community democracy are certainly integral concepts of governance and development, and arguably, the underlying tenets needed to enable the synergistic conditions between local governance structures and national law to implement the SSF Guidelines in Jamaica.

## An Enabling Environment for the SSF Guidelines in Jamaica: The Existence of Synergies?

It may be worth exploring at this point, if the statement proffered by Victor and Raustiala (1989, 660) that "often, a country adopts an international accord without a clear plan for putting commitments into practice' has merit." Given Jamaica's high dependence on fish and the predominance of small-scale fisheries within the industrial landscape (Halcrow 1998; Aiken and Kong 2000), it is plausible to conclude that a loose and protracted enabling environment will affect the island's small-scale fishers and thus the implementation of the SSF Guidelines. Documented in detail below are a number of critical observations.

### *Institutional Capacity for Policy Coherence: The Fisheries Division*

It is a fact that the need for improved fisheries governance and management is critical to Jamaica. Jamaica's small-scale fishers have to contend with competition with both small-scale and industrial commercial counterparts, as well as conflicting development interests and perennial maritime boundary disputes.

However, research indicates that the GOJ's Fisheries Division, the agency in charge of managing such mandates, is severely handicapped by limited capacity and funding (see for example, Waite et al. 2011). Recent interviews with KIs, as well as the review of the literature undertaken, suggest that this lack of institutional capacity is on-going and long standing. Consequently, the Division suffers from a limited presence and effectiveness in the field (Espeut 1993; Halcrow 1998).

To mitigate these pressures, in 2008, under the leadership of the then Minister of Agriculture and Fisheries, the GOJ underwent a restructuring from a regulatory division to an executive agency to be headed by a Chief Executive Officer. This was done to ensure increased effectiveness and "promote transparency, accountability, and efficiency and involvement of all stakeholders in the management of the sector" (Jamaica Information Service 2008). As mentioned earlier, in 2008, the GOJ also founded the National Fisheries Advisory Board (Jamaica Gleaner 2008), signaling its commitment to move towards more participatory and people-centered governance approaches like the SSF Guidelines. However, KIs indicate and secondary data suggests that these developments have stalled, creating further frustration within the Fisheries Division, and among its partners (see for example, Waite et al. 2011, 12–13). KIs also question whether the restructuring of the Fisheries Division into an executive agency with much power vested in the Minister and ministerial appointees, is a real departure from top-down management and governance—an approach that the SSF Guidelines strives to correct.

Furthermore, if one conceives of the importance of situating a viable enabling environment for local governance structures to implement the SSF Guidelines, it is certainly concerning studies indicate that the GOJ's Fisheries Division lacks the capacity to implement initiatives that would influence or affect the island's widely dispersed, small-scale fisheries (Halcrow 1998). Efforts regarding Jamaica's small-scale fisheries have been mainly in the area of collecting data on landings (Grant et al. 1996). This situation has not altered in the recent past, and despite several efforts at restructuring, the Fisheries Division still lacks the capacity to interact viably with Jamaica's small-scale fisheries (Auditor General's Department 2009). However, what of the legislative and regulatory framework needed to implement the SSF Guidelines?

## *Institutional Collaboration and Coordination: The Fisheries Act and Policy*

Historically, the development and enforcement of Jamaica's fisheries laws and regulations have been long, tedious, and weak (Waite et al. 2011). Arguably, this has also resulted in a poor enabling environment for local governance structures to implement the SSF Guidelines.

Several KIs reported being frustrated that both the new Fisheries Policy is yet to be adopted by the Government of Jamaica. Moreover, despite changing times, current and up to date legislation, i.e. the new Fisheries Act, has still not been passed or implemented (twenty-five plus years—1975 and on-going) (see also, Jamaica Observer 2012). KIs also raised concerns that the development of the Policy is leading the development of the law, i.e. the Act, as opposed to vice versa. They see this inversion as a problem because history illustrates the necessity of having a cogent and established legislative environment from which you can then develop and apply policy (Dolowitz 2003). The concerns put forward by the KIs with regard to the development of the new Fisheries Act on the back of the new Fisheries Policy are certainly critical; especially when one considers that, the development of a cohesive enabling policy environment which emphasizes 'Policy Coherence, Institutional Collaboration and Coordination' is what is needed to implement the SSF Guidelines.

One of the primary problems with the Fishing Industry Act is that the fines for breaches to provisions in the Act are too low to constitute a deterrent (see Jamaica Gleaner 2015). KIs report that the sanction for an infringement, no matter how large or damaging (for e.g. a restaurateur harvesting over 200 lbs. of lobster during closed season), is $1200 JMD (less than $10 USD). For a large restaurateur, such a fine surely holds little to no disincentive to break the law, and thus fails miserably to address overfishing—one of many issues that the Act seeks to rectify. KIs also lamented that the process to develop the 'Act' and 'Policy' was not as consultative and inclusive as the GOJ had claimed. Environmental perspectives tended not to be included, or were buried in the GOJ's development initiatives. Consequently, there is a perception of "environment at the mercy of development." Indeed, there are studies (for example, Waite et al. 2011, 7) which indicate that there is currently little political will to tackle environmental issues in Jamaica, as the environment is perceived to be of lower importance than economic priorities such as job creation, GDP growth, and public debt. Levels of funding for coastal and fisheries management are likewise low. A clear signal was sent when the newly elected (January 2016) Government of Jamaica did not include a dedicated Ministry of the Environment among its ministry portfolios, and also put the once dedicated Ministry of Agriculture and Fisheries under the Ministry of Industry and Commerce (Mundle 2016; RJR News On-line 2016).

The concerns expressed by several KIs that after twenty-five plus years the Fisheries Policy had yet to be reconstituted and the new fisheries legislation (i.e. the Act) was still just on the books seem justifiable and warranted. To present the legislation for official authorization and not pass it was taken as an indication of failure—that the GOJ was either not serious about the environment, its fisheries, or simply did not see the importance of the resource for livelihoods and food security in the country. The lack of genuine interest to create a sound legislative and regulatory environment for the industry certainly has dire implications for the enabling environment needed to shore up the implementation of the SSF Guidelines in Jamaica.

## *Local Capacity: Fishers' Organizations and NGOs*

As NGOs and fishers' organizations are required to implement the SSF Guidelines, it is necessary to take stock of them. According to the Jamaican Fishermen's Cooperative Union (JFCU's) website, JFCU consists of 17 registered member organizations (JFCU 2009). It is not clear how many of JFCU's member organizations are active. KIs reported that some fishing cooperatives appear to be struggling and do not see themselves as being empowered to play a significant role in small-scale fisheries management and governance. This may be because conventional management and governance models used for fisheries position fishers as socially atomistic actors who often are in conflict with each other. The result has been a continued dependence on the state to oversee fishery management and decision making (see Jentoft 1989; McConney 1995).

Notwithstanding the state's dominant role, the fact that there is union representation indicates that there is a national platform of local governance structures for Jamaica's small-scale fisheries (Gardner 2016). However, KIs reported that the capacity of fishers' organizations and NGOs to assist in crafting an enabling policy environment for the SSF Guidelines is severely hampered by an 'us' vs. 'them' mentality; "the big man will always get more than the little man," and, "money talks." The SSF Guidelines addresses this problem of how to synergize the competing interests of the state and fishers—politics will be key, particularly on national and local scales.

## *The Loss of Traditional Fishing and Foreign and Domestic Poaching*

Due to the GOJ pursuing large development plans, a stable enabling environment for the SSF Guidelines in Jamaica has not been there. In recent years, it has been common for Jamaican coastal dwellers to perceive themselves to be at a disadvantage concerning their access to beaches and thus fish (Figueroa 2009b; McCaulay, personal communication, 15, December 2015). Even though the island's small-scale fisheries are mostly classified as open access and hence, in theory, open to small-scale fishers as well, in practice there is no such space for them (CVM TV 2014).

Another problem that has prevented achieving synergy between Jamaica's small-scale fisheries and the existing political, legal, and regulatory frameworks is foreign and domestic poaching, also known as illegal, unreported and unregulated (IUU) fishing. Negotiations surrounding the delimitation of Jamaica's maritime boundaries are ongoing. The island currently has only two delimited maritime boundaries with Columbia and Cuba (US Bureau of Oceans and International Environmental and Scientific Affairs 2004). Given Jamaica's large exclusive economic zone (EEZ) of 263,283 $km^2$, it is easy to see why poaching is so prevalent. Aiken et al. (2006) document how the threat of foreign poaching is highest during the closed seasons for lobster and conch, Jamaica's only regulated yet most lucrative fisheries. Further, they suggest that this is the norm not the exception.

The former Minister of Agriculture and Fisheries' proclamations regarding Jamaica's poaching problem are noteworthy. Referencing an event where two registered Jamaican vessels had been caught fishing illegally in Nicaraguan waters, the Minister noted that 'the deviants were instructed to pay US$35,000 each in fines to the Nicaraguan authorities to retrieve their vessels. Contrast this with a maximum fine of US $2.30 if Nicaraguan vessels were caught in Jamaican waters.' The Minister thus acknowledged that the existing fines were not a deterrent. Despite the Minister's proclamation, he somewhat contradicted his stand about the incident when speaking to fishers. He said that 'as government, we can create all the laws necessary, but it will never be enough if you do not abide by these rules and help to enforce them as necessary' (Ministry of Agriculture and Fisheries 2011). Perhaps the question is how can the SSF Guidelines be implemented within structures of local governance where fishers are tapped to be self-appointed guardians of Jamaica's marine space, considering that, the Minister himself, has acknowledged the weak state of the enabling environment?

KIs also reported incidents of domestic poaching of reef and other finfish in the special fishery conservation areas (SCFAs). Moreover, there have been recent legislative reforms to reduce the tax on exports of conch, i.e. the conch-cess used to fund sanctuaries has been drastically lowered (Bryan 2016). These events are *lessening the legitimacy of existing regulations* and, in fact weakening local governance structures needed to shore up the SSF Guidelines.

## *Current Realities*

The above observations bear serious considerations, particularly while assessing the viability of achieving synergy between the state and status of Jamaica's small-scale fisheries and the political, legal, and regulatory frameworks that ought to shore them up. The binding elements of Section 10(7) of the SSF Guidelines are grounded on people-centered perspectives that depend on a sound enabling environment for local governance structures facilitated by national law. However, if the perceptions of several of the KIs are true—namely that Jamaican small-scale fishers often prove to be individualistic, apathetic, and not worth the capital outlay they receive from the government due to not achieving optimal outcomes; and, if there is indeed a weak enabling environment in place for local governance structures to be shored up by national law, Jamaica's small-scale fisheries will continue to be marginalized. The question then remains will Jamaica walk the talk of the Guidelines? Based on the findings put forward, concrete policy recommendations are offered in the concluding section to answer this question.

## Policy Recommendations: How Can the SSF Guidelines Be Implemented in Jamaica?

Demerrit (1996) contends that norms are crafted as scientific knowledge based on local construction(s) that are dependent upon local practices. Seen like this, norms and values cannot be generalized into laws and theories as the deductive method would suggest. Perhaps this is the problem with the applicability of the SSF Guidelines to Jamaica. The grounded realities that could facilitate a stable enabling environment for Jamaica's small-scale fisheries are nuanced and highly complex and seem to be disconnected from the Guidelines. If the SSF Guidelines are to be viably implemented, there remains a considerable amount of work to be done to overcome the challenges related to creating and sustaining a stable enabling environment for local governance structures that is supported by national law. Having said that perhaps the time is right to synergize efforts between state and non-state actors to achieve the policy coherence and institutional coordination and collaboration needed to create an enabling environment for Jamaica's small-scale fisheries and implement the Guidelines.

It is with this hope that I offer the following recommendations to improve: (1) the state and status of Jamaica's small-scale fisheries and (2) the political, legal, and regulatory frameworks that shore up Jamaica's small-scale fisheries (Table 7.1). These recommendations are meant as policy advice to decision makers, Jamaica's fishers, and other actors with an interest in Jamaica's fisheries, including NGOs and civil society. They are in tune with the SSF Guidelines.

It is true that the state and its agents both have significant roles to play in securing the future of Jamaica's small-scale fisheries and the implementation of the Guidelines. However, on both sea and land, industrial interests and community commerce can clash with small-scale fisheries. This means that leadership within the highest levels of the Jamaican government, as well as the environmental and business communities—in consultation with fishers, need to push for a viable enabling environment through sound legislation, regulations, and capacity building in the private and public sectors as well as at the local level. This will be critical to the success of implementing the Guidelines. Furthermore, political will is also necessary if these institutions and regulations are to serve their function of improving the condition of Jamaica's small-scale fisheries, and therefore creating the stable enabling environment needed to secure the implementation of the SSF Guidelines. If there is indeed a proper balancing of the socio-political, legal, regulatory and cultural dimensions that shape fisheries governance in Jamaica, then the SSF Guidelines can have a real impact on fisheries in the future. It will also ensure that Jamaica's small-scale fisheries make an even greater contribution toward the country's economy and to the well-being of its people than it does at present.

**Table 7.1** Policy advice to improve the political, legal and regulatory frameworks for Jamaica's small-scale fisheries for implementation of the SSF guidelines

| **Policy advice to improve the state and status of Jamaica's small-scale fisheries for implementation of the SSF guidelines:** | |
|---|---|
| **Issue** | **How achieved?** |
| (a) Promote greater awareness of the contributions and roles of small-scale fisheries and fisherfolk to the fabric of Jamaican society | The state should work in conjunction with the private sector, and established fisher organizations, NGOs and civil society to promote the contributions and efforts of Jamaica's SSF through joint public awareness campaigns. These campaigns could focus on small-scale fisheries contributions to economic stability, livelihoods, social welfare, food security, and, the role of women. Funding for this initiative should be provided by the private sector as well as the state. |
| (b) Establish greater capacity for fisherfolk organizations | Fishers should seek to work together around common political issues to ensure the credibility and legitimacy of their collective voice. This will mean that fisherfolk have to discard the cloak of social atomism. Efforts may include: building greater institutional capacity within their national and local organizations; gaining better recognition with regional and international partners, as well as improving external relationships with key policy partners. Regional networks such as the Caribbean Network of Fisherfolk Organizations and the Caribbean Regional Fisheries Mechanism of which Jamaica is a part may prove useful partners in these efforts. |
| (c) Improve the institutional capacity of the Fisheries Division | The powers that be in the state ought to acknowledge the importance of a having a sound institutional unit for the island's fisheries and small-scale fishers. |
| (d) Promote the eco-system approach to fisheries management | The state should continue its efforts to implement the eco-system approach to fisheries management as evidenced with the establishment of several special fisheries conservation areas (SCFA's). The state should also make a concerted effort to ensure that the mediums for participatory and consultative co-management exist between them, fisher organizations, NGO's and civil society. This includes ensuring sustainable funding mechanisms for these efforts, for e.g. the conch-cess (i.e. the tax on exports of conch used to fund SCFAs). |
| **Policy Advice to Improve the Political, Legal and Regulatory Frameworks for Jamaica's Small-Scale Fisheries for Implementation of the SSF Guidelines:** | |
| **Issue** | **How achieved?** |
| (a) Promote greater awareness of the SSF Guidelines | The majority of the KIs interviewed had either not heard of the Guidelines or if they had heard of the Guidelines they had not read them. It is perhaps incumbent upon surrogates of the Guidelines to find ways to disseminate the value of the Guidelines from a grassroots perspective. This might include a targeted strategic partnership with private sector actors, as well as NGOs and civil society actors that promote the welfare of Jamaica's fisheries. The state should also play a role in this campaign, by facilitating participatory consultations with key stakeholders around the implementation of the Guidelines. |

(continued)

**Table 7.1** (continued)

| | |
|---|---|
| (b) Update and implement the dated 1975 Fisheries Act and the 2008 Draft Fisheries Policy | Government and private sector stakeholders need to get serious about their political will to implement the legal and regulatory mechanisms for Jamaica's fisheries. NGO and civil society actors need to continue to hold the government accountable to these matters. The Fisheries Advisory Board should also reinstate itself as a political conduit representing, as best possible, all stakeholder interests, but especially small-scale fishers. The Act and Policy cannot be seen as laying the groundwork for the SSF Guidelines or seen as a example of the SSF Guidelines at work, if these implementation issues are not addressed. There should therefore be a focus on shared governance. |
| (c) Improve the political, legal, and regulatory capacity of the Fisheries Division | The powers that be in the State ought to acknowledge the importance of a having a sound institutional and regulatory government unit for the island's fisheries. This includes positioning the Division with: (1) proper management capacity – staff; (2) adequate funding for monitoring and oversight to address issues related to – domestic poaching, illegal unregulated fishing; and (3) adequate enforcement mechanisms for: special fisheries conservation areas and breaches to fisheries regulations. The Fisheries Division should also improve mechanisms of engaging with Jamaica's small-scale fishers to promote shared governance. This may include the facilitation of training workshops with Jamaica's small-scale fishers and private sector fisheries actors. |
| (d) Promote, in general, the proper management and governance of Jamaica's fisheries, as well as the improved welfare of vulnerable and marginalised coastal dwellers and others whose livelihoods depend on Jamaica's small-scale fisheries | The state in partnership with the private sector, fisherfolk organizations and fisherfolk, NGOS, civil society as well as the community at large should work together to promote the general management of Jamaica's fisheries. There is evidence of such efforts existing, e.g. the parrot fish ambassador program – which strives to discourage the harvesting of over-exploited Jamaican parrot fish. Moreover, all stakeholders in Jamaica's fisheries need to hold each other accountable on efforts to institute a stable enabling environment for Jamaica's small-scale fisheries. This could be accomplished with participatory and consultative stakeholder and community meetings to address issues that are affecting Jamaica's small-scale fisheries and fisheries in general. |

## References

Aiken, K., & Kong, A. (2000). The marine fisheries of Jamaica. *The ICLARM Quarterly, 23*(1), 29–35.

Aiken, K., Kong, A., Smikle, S., Appledorn, R., & Warner, G. (2006). *Managing Jamaica's queen conch resources*. Mona/Kingston/Mayaguez: Ocean and Coastal Management.

Allison, E. H., & Ellis, F. (2001). The livelihoods approach and management of small-scale fisheries. *Marine Policy, 25*(July), 377–388.

Auditor General's Department. (2009). *Auditor general's performance audit report on the fisheries division of the ministry of agriculture*. Kingston: Auditor General's Department.

Bravo, K. E. (2006). CARICOM, the myth of sovereignty, and aspirational economic integration. *North Carolina Journal of International Law and Commercial Regulation, 31*, 146–206.
Bryan, C. (2016, May 7). ConchCess reduced by $0.25 US Cents. *Jamaica Information Service*. http://jis.gov.jm/conch-cess-reduced-us0-25-cents/. Accessed 7 May 2016.
CaribSave. (2015). *IntaSave Caribbean – Our projects*.http://intasave.org.www37.cpt4.host-h.net/our_projects/. Accessed 5 May 2016.
CARICOM Fisheries Unit. (2000). *Jamaica national marine fisheries atlas*. Belize City: CARICOM Fisheries Unit.
Centre for Resource Management and Environmental Studies. (2016). *Caribbean gender in fisheries team*.http://www.cavehill.uwi.edu/cermes/projects/gift/overview.aspx. Accessed 9 July 2016.
COFI. (2014). *Securing sustainable small-scale fisheries: Update on the development of the voluntary guidelines for securing sustainable small-scale fisheries in the context of food security and poverty eradication (SSF guidelines)*. Rome: COFI.
Commonwealth Fisheries. (2009). *Commonwealth fisheries programme: Report of Caribbean study tour to St. Lucia, Trinidad and Tobago and Belize, 24 January – 4 February 2009*. London. http://www.commonwealthfisheries.org/admin/downloads/docs/Caribbean%20Study%20Tour%20Report.pdf. Accessed 6 Jan 2016.
CVM TV. (2014). *Beach access in Jamaica part 1 of 3*. Kingston: CVM TV.
Demerrit, D. (1996). Social theory and the reconstruction of science and geography. *Transaction of the Institute of British Geographers, 21*, 484–503.
Dolowitz, D. P. (2003). A policy-maker's guide to policy transfer. *The Political Quarterly, 74*(1), 101–108. doi:10.1111/1467-923X.t01-1-00517.
ECOST. (2007). *Jamaica case study – Institut de Recherche pour le Developpement France*.http://www.ird.fr/ecostproject/doku.php?id=case_study_2_jamaica_c_m_s_centre_for_marine_sciences. Accessed 8 Feb 2016.
Espeut, P. (1992). Managing the fisheries of Jamaica and Belize: The argument or a co-operative approach. *The third annual conference of the International Association for the Study of Common Property (IASCP)*, 1–32. Washington, DC.
Espeut, P. (1993). Managing the fisheries of Jamaica: Is co-management a viable option? *XXVIII annual conference of the Caribbean studies association*, 1–15. Ocho Rios.
FAO. (2003). The ecosystem approach to fisheries. *FAO technical guidelines for responsible fisheries, 4*(2). Rome. http://www.fao.org/docrep/005/y4470e/y4470e02.htm#TopOfPage. Accessed 4 Mar 2016.
FAO. (2013). *FAO fisheries and aquaculture report No. 1033 report of the FAO/CRFM/WECAFC Caribbean regional consultation on the development of international guidelines for securing sustainable small-scale fisheries*. Kingston, Jamaica for Rome, Italy. http://www.fao.org/docrep/017/i3207e/i3207e.pdf. Accessed 5 Mar 2016.
FAO. (2015). *Voluntary guidelines for securing sustainable small-scale fisheries*. Rome: FAO.
FAO. (2016). Global per capita fish consumption rises above 20 kilograms a year, *FAO*. http://www.fao.org/news/story/en/item/421871/icode/. Accessed 10 Mar 2016.
Figueroa, E. (2009a). *Massa god fish can done*. The Nature Conservancy.
Figueroa, E. (2009b). *Jamaica for sale*. Vagabond Media, Jamaica Environment Trust.
Fisheries Division. (1997). *Fish production survey for 1996*. Kingston: Ministry of Agriculture and Fisheries. http://moa.gov.jm/Fisheries/. Accessed 14 Mar 2016.
Fisheries Division. (2011a). Special fishery conservation areas. Kingston: Ministry of Agriculture and Fisheries. http://moa.gov.jm/Fisheries/. Accessed 14 Mar 2016.
Fisheries Division. (2011b). *Number and percentage of registered fishers by parish*. Kingston: Ministry of Agriculture and Fisheries. http://moa.gov.jm/Fisheries/. Accessed 14 Mar 2016.
Fisheries Division. (2011c). *Number and percentage of registered vehicles by vessel type*. Kingston: Ministry of Agriculture and Fisheries. http://moa.gov.jm/Fisheries/. Accessed 14 Mar 2016.
Fisheries Division. (2011d). *Imports, exports and re-exports of fish and fishery products along with the total value 2004–2007*. Kingston: Ministry of Agriculture and Fisheries. http://moa.gov.jm/Fisheries/. Accessed 14 Mar 2016.

Fisheries Division. (2011e). *Source and consumption of fish in Jamaica, 2001–2007*. Kingston: Ministry of Agriculture and Fisheries. http://moa.gov.jm/Fisheries/. Accessed 14 Mar 2016.
Fisheries Division. (2011f). *Licensing and registration of fishers and vessels*. Kingston: Ministry of Agriculture and Fisheries. http://moa.gov.jm/Fisheries/boat_inspection.php. Accessed 29 Oct 2016.
Fisheries Division. (2011g). *Legislation*. Kingston: Ministry of Agriculture and Fisheries. http://moa.gov.jm/Fisheries/. Accessed 29 Oct 2016.
Fisheries Division. (2016). *Fisheries and aquaculture*. Kingston: Ministry of Agriculture and Fisheries. http://www.moa.gov.jm/Fisheries/index.php. Accessed 14 Mar 2016.
Fisheries Division and STATIN. (2011a). *GDP of fishing industry 1998–2007 (at constant prices) $millions*. Kingston: Ministry of Agriculture. http://moa.gov.jm/Fisheries/. Accessed 14 Mar 2016.
Gardner, C. (2016, March 29). Negril stakeholders blast NEPA. *Jamaica gleaner*.http://jamaica-gleaner.com/article/western-focus/20160329/negril-stakeholders-blast-nepa. Accessed 2 Apr 2016.
Gemmill, B., & Bamidele-Izu, A. (2002). The role of NGOs and civil society in global environmental governance. In D. Esty & I. Maria (Eds.), *Global environmental governance: Options and opportunities* (pp. 77–101). New Haven: Yale School of Forestry and Environmental Studies.
Government of Jamaica. (2008). *DRAFT fisheries policy*. Kingston: Ministry of Agriculture and Fisheries Division.
Government of Jamaica. (2011). *An act to amend the constitution of Jamaica's charter of fundamental rights and freedoms*. Kingston: Joint Select Committee of Parliament.
Government of Jamaica. (2015a). *DRAFT fisheries act*. Kingston: Jamaican Parliament.
Government of Jamaica. (2015b). *Final draft national fisheries and aquaculture policy*. Kingston: Ministry of Agriculture and Fisheries.
Grant, S. (2004). Caribbean women in fishing. In *Proceeedings of the fifty fifth annual Gulf and Caribbean fisheries institute* (pp. 69–76). Fort Pierce: Gulf and Caribbean Fisheries Institute.
Grant, S., Smikle, S., & Galbraith, A. (1996). *Jamaica fisheries sampling plan*. Kingston: Fisheries Division, Ministry of Agriculture.
Halcrow, W. (1998). *South Coast sustainable development study*. Kingston: Marine Resources.
Haughton, M., & Kong, A. (2006). Socio-economic indicators in integrated coastal zone and community-based fisheries management case studies from the Caribbean. *FAO Fisheries technical paper 491*. Rome.
Jamaica Environment Trust. (2013). *Save goat islands: Take a stand for our island*.http://savegoat-islands.org/photos-videos/videos/. Accessed 5 Apr 2016.
Jamaica Gleaner. (2008, January 24). Tufton appoints new Fisheries Advisory Board. *Jamaica Gleaner*. http://new.jamaica-gleaner.com/power/717. Accessed 5 Mar 2016.
Jamaica Gleaner. (2015, June 6). *Massive increases fines and penalties proposed for fisheries breaches*. *Jamaica Gleaner Online*. http://jamaica-gleaner.com/article/lead-stories/20150615/. Accessed 5 Apr 2016.
Jamaica Information Service. (2008, November 26). Fisheries division being restructured to ensure greater efficiency. http://jis.gov.jm/fisheries-division-being-restructured-to-ensure-greater-efficiency/. Accessed 5 Apr 2016.
Jamaica Observer. (2012, August 29). *New fisheries law limps along*. Jamaica Observer. http://www.jamaicaobserver.com/environment/New-fisheries-law-limps-along_12349301. Accessed 1 Mar 2016.
Jentoft, S. (1989). Fisheries co-management: Delegating government responsibility to fishermen's organizations. *Marine Policy, 13*, 137–154.
Jentoft, S. (2014). Walking the talk: Implementing the international voluntary guidelines for securing sustainable small-scale fisheries. *Maritime Studies, 13*(16), 1–15.
JFCU. (2009). Jamaica fishermen's cooperative union. *Members*. http://www.ja-fishermen.com/home/about-us/members. Accessed 6 Apr 2016.

Kong, A. (1990). *The Jamaican fishing industry: Its structure and major problems*. Kingston: Fisheries Division, Ministry of Agriculture.

Kong, A. (2003). *The Jamaica fishing industry: Brief notes on its structure, socio-economic importance and some critical management Issues*. Kingston.

Mattessich, P., Murray-Close, M., & Monsey, B. R. (2001). *Collaboration: What makes it work* (2nd ed.). St. Paul: Amherst H. Wilder Foundation.

McConney, P. (1995). *Fishery planning in Barbados: The implications of social strategies for coping with uncertainty* (Doctoral dissertation). The University of British Colombia.

Ministry of Agriculture and Fisheries. (2011, June 9). Jamaica's Fishing Laws do not deter illegal activities on the high seas- they're the least severe in the region. *MOAF – News*. http://www.moa.gov.jm/News/2011-06-09_Jamaica_fishing_laws.php. Accessed 10 Apr 2016.

Mundle, T. (2016, March 10). *Environment portfolio should not be hidden. Jamaica Observer Online*. http://www.jamaicaobserver.com/latestnews/Environment-portfolio-should-t-be--hidden---Pickersgill-argues. Accessed 10 Apr 2016.

NEPA. (2003). *National strategy and action plan on biological diversity in Jamaica*. Kingston: NEPA.

OECD Observer. (2003). *Policy coherence: Vital for global development, Policy brief*.www.oecd.org/pcd/20202515.pdf. Accessed 21 Oct 2016.

RJR News On-line. (2016, March 7). Prime Minister announces new cabinet. *Local News*. http://www.jamaicaobserver.com/latestnews/Environment-portfolio-should-t-be--hidden---Pickersgill-argues. Accessed 7 Mar 2016.

The World Atlas. (2016). Map of Jamaica in Caribbean. *Jamaica*. http://www.worldatlas.com/img/areamap/fe82a41adeba42e2e891633dbb223b7d.gif. Accessed 1 Mar 2016.

Thompson, E. (1945). *The fisheries of Jamaica. Development and welfare in the West Indies Bulletin*, No. *18*. Barbados.

US Bureau of Oceans and International Environmental and Scientific Affairs. (2004). *Jamaica's maritime claims and boundaries, No. 125*. Washington, DC.

Van Riel, W., & Wijkstrom, U. (2005). *FAO/Ministry of agriculture fisheries division economic study of Jamaican fishing industry governance and macro-policy*. Rome.

Victor, D. G., & Raustiala, K. (1989). Conclusions. In D. G. Victor, K. Raustiala, & B. Skolikoff (Eds.), *The implementation and effectiveness of international environment commitments: Theory and practice*. Cambridge, MA: MIT Press.

Waite, R., Cooper, E., Zenny, N., & Burke, L. (2011). *Coastal capital: Jamaica. The economic value of Jamaica's coral reef-related fisheries*. Washington, DC.

# Part III
# Securing Tenure Rights

Respect for human rights and dignity constitutes a foundational principle, with broad application throughout the SSF Guidelines. Consistent with this principle, the Guidelines devote a whole section on the rights to tenure as a way to secure sustainable livelihoods. Without their secure access to, and control over, the resources on which small-scale fishing people rely, these resources and the people that use them are both vulnerable. In Chap. 8, Jackie Sunde discusses the experiences of customary communities and expressions of the tenure of indigenous peoples in South Africa. She argues that it is essential to understand the plurality of tenure systems based on different legal concepts and norms that exist, because they create a challenging and potentially conflictual context for the implementation of the SSF Guidelines. Sunde issues a warning that a particular interpretation of tenure may contravene the implementation of the SSF Guidelines. In Chap. 9, Prateep Kumar Nayak takes us to the Bay of Bengal, India, to look at tenure issues related to small-scale lagoon fisheries. Like Sunde, he emphasizes the need for small-scale fishing communities to have secure tenure rights to resources that form the basis for their social, economic, and cultural well-being, something the state should recognize, as instructed by the SSF Guidelines. His chapter offer lessons for responsible governance of lagoon tenure, including insights on possible institutional arrangements. In Chap. 10, Miguel González describes policy actions taken by the government of Nicaragua that interfere with indigenous peoples' titled customary lands and aquatic tenure systems in relation to the planned Interoceanic Canal. He identifies gaps in the process of implementation and explores the implications for the human rights of indigenous and Afro-descendant peoples. Chapter 11 moves the focus to the Mediterranean, with a case study in Malta by Alicia Said. Due to a number or reasons, including the industrialization of the Bluefin tuna fisheries, small-scale fisheries have become marginalized to a point where leaving the fishery has become a necessary livelihood strategy. As explained in the chapter, the provision of tenure rights would be needed as an empowerment strategy in the neoliberal era.

# Chapter 8
# Expressions of Tenure in South Africa in the Context of the Small-Scale Fisheries Guidelines

**Jackie Sunde**

**Abstract** Tenure relations lie at the heart of the livelihoods of small-scale fishing communities who depend on their access to and control of fisheries and other natural resources in order to realise their right to food as well as a range of other human rights. In recognition of this, the *Voluntary Guidelines for Securing Sustainable Small-Scale Fisheries in the Context of Food Security and Poverty Eradication* (SSF Guidelines) identify the responsible governance of tenure as central to the realization of the human rights of small-scale fishers. Drawing on the experience of customary communities in South Africa, this chapter explores expressions of tenure and their implications for the implementation of the SSF Guidelines. An exploration of tenure relations within the customary systems of indigenous peoples and local communities suggests that these forms of tenure are embedded in epistemologies and ontologies that are foundationally different to most statutory tenure systems. Contrary to the individual, market-orientated conception of rights within the neoliberal property rights paradigm dominating state fisheries management, these tenure systems reference an alternative conception of rights and tenure governance. This plurality of tenure systems, embodied in different systems of law, creates a challenging, potentially conflictual context in which the objective of the SSF Guidelines to develop responsible governance of tenure will be achieved. Recognition and accommodation of the plurality of tenure systems is imperative if the transformative potential of the SSF Guidelines to achieve equitable and sustainable fisheries is to be realized.

**Keywords** Small-scale fisheries guidelines • Tenure • Responsible governance • Human rights • Customary law • Legal pluralism

---

J. Sunde (✉)
Department of Environmental and Geographical Science, University of Cape Town, Cape Town, South Africa

International Collective in Support of Fishworkers (ICSF), Chennai, India
e-mail: jsunde@telkomsa.net



S. Jentoft et al. (eds.), *The Small-Scale Fisheries Guidelines*, MARE Publication Series 14, DOI 10.1007/978-3-319-55074-9_8

## Introduction

Tenure in fisheries refers to the relationships between people, either individuals or groups of people, that determine who has the power to decide who may access what fisheries resources and under what conditions (FAO 2016). Tenure thus plays a critical role in the lives of small-scale fishing communities who depend on their access to and control of fisheries and other natural resources for their livelihoods.

In recognition of this link between tenure and issues such as food security and poverty, the *Voluntary Guidelines for Securing Sustainable Small-Scale Fisheries in the context of Food Security and Poverty Eradication* (SSF Guidelines) identify the responsible governance of tenure as central to the realization of the human rights of small-scale fishers (FAO 2015, 5). The Guidelines urge all parties to ensure that small-scale fishers, fish workers and their communities have "secure, equitable, and socially and culturally appropriate tenure rights to fishery resources (marine and inland) and small-scale fishing areas and adjacent land, with a special attention paid to women with respect to tenure rights" (FAO 2015, 55).[1] In addition, states and all other parties, in line with their national legislation, should "recognize, respect and protect all forms of legitimate tenure rights, taking into account, where appropriate, customary rights" (FAO 2015, 5).

This focus on tenure in fisheries reflects a global strengthening of interest in tenure across natural resource governance over the past decade as states and other actors seek to develop mechanisms to secure the sustainable use of the earth's natural resources (FAO 2016). This has received impetus through the adoption of the *Voluntary Guidelines on the Responsible Governance of Tenure of Land, Fisheries and Forests in the Context of National Food Security (Tenure Guidelines)* (FAO 2012). These Guidelines have been supported by the G20, Rio+20 and the United Nations General Assembly (FAO 2016). The Tenure Guidelines establish the principle that tenure systems include statutory systems which may be written down, as well as those that have their source in local law but might not be recorded but are considered legitimate by the customary communities who observe them (FAO 2016). The SSF Guidelines draw extensively on the Tenure Guidelines on issues related to tenure governance.

Whilst the need for 'secure tenure' is being propounded across diverse stakeholders, from small-scale fishing communities to large multi-national institutions involved in small-scale fisheries governance, the interpretation of 'secure tenure' and the arguments motivating the need to promote secure tenure appear to differ. A narrow, neoliberal conceptualization of tenure predominates amongst certain fisheries actors, proposing the introduction of various forms of property 'rights-based' tenure mechanisms, largely divorced from the human rights-based approach

[1] The SSF Guidelines use the terms "all parties" and 'small-scale fisheries actors" to refer to the broad range of government and non-government stakeholders and rights holders working in small-scale fisheries to whom the Guidelines apply. This includes the business sector.

underpinning the SSF Guidelines.[2] The concept of 'secure tenure' envisaged in these instances focuses predominantly on user rights and assumes a market-orientated rationality, underpinned by a distinctive neoliberal conception of individual rights, a significant role for the private sector and its handmaiden, the enabling state. It neglects the social relations and the multiplicity of meaning and values that are woven into tenure relations in different legal systems and across diverse cultural and socio-ecological systems.

This approach towards tenure in policy on natural resource governance is contested however and international social movements have raised the alarm with regard to how tenure is framed in neoliberal, multinational governance policies and programmes (IDI 2015; Baarbesgaard 2016; FAO 2016). In contrast to this neoliberal perspective, indigenous peoples and local communities have advocated for recognition of their collective rights to natural resources, including fisheries, inextricably linked to their cultures, their knowledge systems and their governance systems, in the context of their right to self-development.[3] Scholarship on tenure relations within the customary systems of indigenous peoples and local communities supports the claim that their tenure relations are, in many instances, embedded in epistemologies and ontologies that are foundationally different to statutory tenure systems (Davis and Ruddle 2012; Sunde 2014; Almeida et al. 2015). Contrary to interpretations of ownership and property within the property rights-orientated paradigm informing state policy and management in many Western industrialized nations and post-colonial contexts, these customary tenure systems reference an alternative conception of rights and responsibilities in the governance of tenure. This plurality of tenure systems, embodied in different systems of law, creates a challenging, potentially conflictual context in which the objective of developing responsible governance of tenure in the SSF Guidelines must be implemented.

Based on evidence from the expressions of tenure in two customary communities in South Africa, coupled with a review of the submissions to FAO made by indigenous and local small-scale fishing communities during the development of the SSF Guidelines, this chapter explores implications of interpretations of tenure for responsible governance of tenure.[4] The chapter concludes that recognition and accommodation of the plurality of tenure systems is imperative if the transformative potential of the SSF Guidelines to achieve equitable and sustainable fisheries is to be realized.

---

[2] See the Global Partnership for Oceans (GPO), (2014) and Environmental Defence Fund (EDF) et al. (2012) as key examples of this approach.
This submission was developed in the form of a synthesis document for the CSO (Sowman et al. 2012).

[3] The right to self-development is recognised explicitly in the UN Declaration on the Rights of Indigenous Peoples but draws on existing human rights instruments that recognise this right.

[4] This submission was developed in the form of a synthesis document for the CSO Co-ordinating Committee by Sowman et al. (2012) but is referred to as 'CSO Submission on SSF Guidelines' for the purposes of this document.

The following section introduces and explores the concept of tenure in fisheries and how this concept has evolved in fisheries governance. It highlights the distinctive interpretation of tenure rights that indigenous peoples and small-scale fishing communities expressed through the Civil Society Organisation (CSO) submissions on the development of the SSF Guidelines (Sowman et al. 2012; CSO 2013).

Section "International human rights standards guiding the interpretation of tenure and 'responsible governance' in the SSF Guidelines" identifies the approach and principles shaping the SSF Guidelines. It explores the international human rights instruments underpinning the provisions on collective tenure and customary governance of tenure.

Section "Customary expressions of tenure along the coastline of South Africa" identifies the distinctive characteristics of customary tenure systems in South Africa. These systems challenge dominant interpretations of the nature and scope of tenure relations and tenure rights. In line with the concept of tenure articulated by the international fisher social movements, they offer an alternative interpretation of responsible governance of tenure.

Section "Interpreting and implementing the SSF Guidelines" analyzes and discusses the implications of an interpretation of tenure through the small-scale fisheries policy in South Africa. It links this to a neo-liberal interpretation of the SSF Guidelines, made possible through the current political economy dominating fisheries not only in South Africa, but worldwide. It concludes with a discussion on the need for all actors to be vigilant in the process of implementing the Guidelines to ensure that implementation reflects the original vision of responsible governance of tenure that inspired the Guidelines.

## Changing Tenure Narratives Over Time

The term 'tenure' is derived from the Classical Latin verb *tĕnĕo* and refers to the 'holding' of land and other natural resources. During the Classical period, this verb expressed not only the action of 'holding', but also that of 'directing', 'attaining' 'understanding, knowing', 'occupying, dwelling' and 'conserving'.[5]

It is recognized that, in many parts of the world, communities have devised local practices and rules to manage their interactions with nature and with each other (Almeida et al. 2015). This includes how they hold, use, and manage terrestrial and marine resources and who has the right to do so. The term 'customary marine tenure' is now used regularly in fisheries governance and management to refer to a broad range of customary practices and systems of access, use, and management that have persisted whereby communities "perceive, define, delimit, own and defend their rights" (Ruddle and Akimichi 1984, cited in Aswani 2005, 289).

These customary marine resource use and governance practices and the spiritual, cultural and livelihood values embedded in them are evidenced in a range of spatial,

[5] These terms are translated from the Latin (Gaffiot and Flobert 2000, 1579–1582).

temporal, technical, social, and legal norms and rules. In most instances, these practices are embedded in local systems of customary law (Bavinck and Gupta 2014; Jentoft and Bavinck 2014). These systems of customary law reflect particular communities' cosmologies and cultural identities, expressing the unique interplay of the ecological, social, and cultural aspects of their lives and livelihoods. They are informed by finely tuned systems of indigenous and local ecological knowledge. These customary systems thus play an important role in the maintenance of biocultural diversity, itself inseparable from the protection of biodiversity (UNEP 2009; Almeida et al. 2015). It is estimated that communities own, control, or otherwise claim under customary ownership up to 6.8 billion hectares or about 52% of the global land area and collective customary tenure systems covering land and marine areas regulate the lives of at least 1.5 billion people around the world (Almeida et al. 2015, 3).

Customary marine tenure has been impacted in many parts of the world when statutory laws and related fisheries management regimes have been introduced and enforced with varying strength in different localities, alongside existing customary law (Bavinck and Gupta 2014). As a result, in these contexts a situation of legal pluralism exists in fisheries governance to varying degrees, "leaving layers of different legal systems intact" (Bavinck and Gupta 2014, 80). Legal pluralism exists where "different legal systems apply to identical situations" (Bavinck and Gupta 2014, 80).

The imposition of statutory systems of law and management has been most keenly felt in colonial contexts where the most common approach of colonizing regimes was to subject indigenous communities to the law of the conquering power (Bennett 2006, cited in Sunde 2014, 26). In many of these contexts the dominance of the colonial state-centric approach to law held sway over the local law of indigenous communities. One consequence of this is that the interpretation of ownership of marine resources has been influenced by colonial interpretations of property and ownership, particularly those embedded in Western law, in which marine resources are considered *res nullius*, or property that is not owned by any person. In many such jurisdictions, state law has therefore affirmed the authority of the state over these resources, ignoring communities who have lived according to customary law and who regard marine and coastal resources as being collectively owned by the community (FAO 2013). This situation has been exacerbated by the dominance of neo-classical economic theories influenced by Scott Gordon's theory of common property resources (Gordon 1954), which led to calls for the introduction of property rights as a means addressing fisheries over-exploitation. It is now recognized that the theoretical basis of neo-classical economics emerging in the late 1950s, with its emphasis on economic growth, contributed towards the subsequent rise of neoliberalism in the 1980s and, more specifically with regard to fisheries, to the influence of neoliberalism in fisheries discourse (Mansfield 2007, cited in Pinkerton and Davis 2015).

Drawing on these earlier influences but developing in response to post-war shifts, neoliberalism has emerged as a dominant global economic ideology since the 1980s (Harvey 2005). It refers to a range of political and economic policies that centre on the core belief that development, freedom, and prosperity are best fostered through "the optimizing efforts of self-interested entrepreneurs efficiently coordinated by self-regulating markets" (Hartwick and Peet 2003, cited in Pinkerton and Davis 2015, 304). It is characterized by 'principles privileging narrow conceptions of individual self-interest and economic efficiency' (Foley et al. 2015, 391). Most importantly, Pinkerton and Davis (2015) note that it is important to recognize that 'neoliberalism is not a singular, monolithic entity. The form that policies inspired by neoliberal ideals take in practice can vary widely from place to place in response to any number of local factors' (Pinkerton and Davis 2015, 308).[6] Mansfield (2004, 315) argues that neo-liberalism has taken a particular historical turn in ocean governance due to the centrality accorded notions of property rights in debates about common property resources, environmental degradation and economic efficiency. Notwithstanding localized expressions of neoliberalism, trends have been documented across a range of socio-ecological and geographical contexts that highlight the distinctive tendencies of neoliberal fisheries management policies to emphasise privatised conceptions of property and ownership, strengthen the power of the market and facilitate an expansion of individual rights, responsibilities and risks to the private sector (Ruddle and Davis 2013; Pinkerton and Davis 2015). The tendency of neoliberalism to support class formation (Harvey 2005) is reflected in the way in which power operates in neoliberal fisheries contexts to reinforce class priorities (Ruddle and Davis 2013, 89).

Despite evidence of the existence of elaborate customary tenure systems based on collective governance of common-property regimes in many jurisdictions, fisheries management approaches involving the statutory introduction of neo-liberal regimes in these contexts has continued (Ruddle and Davis 2013). Policy approaches have been orientated towards supporting commercial fisheries at the expense of near-shore artisanal and small-scale fisheries. The introduction of Individual Transferable Quotas (ITQs) as a key private property management tool, together with other individual rights-orientated management measures reflecting various degrees of privatization and commodification of fisheries rights became popular in the late 1970s and garnered the term 'rights-based' measures due to their emphasis on property rights. A wide range of formal, or statutory, rights-based measures now exist ranging from quotas which allocate harvesting rights, effort-based rights which regulate the effort that might be used, and a range of additional specific licensing measures that include various effort or catch regulations. These can all include individual rights as well as community-based rights such as spatial or area-based systems based on a local communities' territory. One such example is the territorial user rights in fisheries system (TURFS) (FAO 2013, 13). A distinction is usually made in statutory tenure management measures between 'use rights', also referred to as 'access rights', and 'management rights' (Charles 2011; FAO 2013). In this interpretation of rights, 'use rights' are conceptualized as separate from but parallel to 'management rights' (Charles 2011). Use rights do not automatically confer

[6] See the Special Edition of Marine Policy (2015) in which the impact of neoliberalism on a range of fisheries governance and management issues in North America are explored.

management rights. As will be discussed below, in contrast, in many customary systems of law, use or access rights are indivisible from the right and responsibility to collectively manage resources.

## Claiming Tenure Rights

In response to the emphasis on individual, market-orientated property rights conceptualized within a neoliberal paradigm of ownership and control, small-scale fishing communities and their supporters have advocated internationally for recognition of their collective tenure systems and the interdependence of their tenure rights and other rights such as the right to their own culture and, most importantly, their right to determine their own systems of management (WFFP et al. 2008). At the first FAO Small-Scale Fisheries Conference held in Bangkok in 2008, the international small-scale fishers' organizations released the Bangkok Statement, in which they articulated the indivisibility of their tenure rights and their broader human rights. The fishers' movement demanded that States should, amongst several other actions:

- Guarantee access rights of small-scale and indigenous fishing communities to territories, lands, and waters on which they have traditionally depended for their lives and livelihoods
- Protect the cultural identities, dignity, and traditional rights of fishing communities and indigenous peoples
- Ensure the integration of traditional and indigenous knowledge and customary law in fisheries management decision making (WFFP et al. 2008)

In 2011, these fishers' organizations established a CSO platform for the purpose of advocating for the development of the SSF Guidelines.[7] They organized and coordinated 22 consultative national level and two regional level workshops, collectively enabling over 2300 representatives of fisher communities to participate in the development of the SSF Guidelines. In these workshops, small-scale fisher communities from around the world articulated the distinctive nature of their tenure systems, the fact that these systems formed the material basis of their culture and their systems of traditional and local knowledge. Their claims to their right to manage their own fisheries as part of their right to self-determination were clear. These demands were synthesized into a common document which was submitted to the FAO and which was used as the basis for the CSO submission on the Zero Draft of the SSF Guidelines (Sowman et al. 2012).[8] An extensive range of very strong

---

[7] The CSO coordination group included the World Forum of Fishworkers and Fish Harvesters (WFF), the World Forum of Fisher Peoples (WFFP), the International Collective in Support of Fishworkers (ICSF), and the International Planning Committee on Food Sovereignty (IPC).

[8] See CSO Submission Document (Sowman et al. 2012) and CSO submission on the Zero Draft (CSO 2013).

demands in reference to tenure rights was included. These included, among others, the following:

*2.1.1.* All parties must recognize, respect, and promote the exercise and enjoyment by small-scale fishers, fishworkers, and fishing communities of their human rights: most notably their right to dignity, to freedom, to civil and political rights.... with respect for the cultural and territorial rights of such fishing communities and their right to choose their own path of development. Their right to manage fisheries resources is inherent in these related rights.

*2.1.2.* States and other institutions and organizations should take into account the heterogeneity, diversity, and complexity of artisanal and small-scale fisheries from a human rights perspective. The distinctive cultural, customary, and ancestral rights and legal systems of native, indigenous, and customary communities should be respected insofar as they are compliant with international human rights principles.

*2.1.3.* States should consider the rights of small-scale fishing communities to territory, on land and water, as a collective human right (Sowman et al. 2012, 9–20).

A considerable portion of this text proposed by the CSO platform (CSO 2013) was incorporated into the final text of the Guidelines, albeit with slightly different language. Most significantly from the perspective of the CSO platform, the Guidelines affirmed a human rights-based approach and are based on international human rights standards (FAO 2015, 2). However, some subtle references in the proposed CSO text were not included. The above demands echo the right to self-development and self-governance inherent in the international recognition of customary law, but these were not supported by States in the negotiations and were not included in the final text. Indeed the term ‘territories’ was removed entirely from the final text. Reference to ‘customary law’ in the section on responsible governance of tenure proposed by the CSO negotiating team and initially incorporated into the first draft was subsequently also removed. The implicit references to customary law in the SSF Guidelines are now reliant on detailed knowledge and interpretation of ‘international human rights standards’ in order to be visible.[9]

## International Human Rights Standards Guiding the Interpretation of Tenure and ‘Responsible Governance’ in the SSF Guidelines

The SSF Guidelines include the principles upon which ‘responsible governance of tenure’ is based and can be assessed. These principles resonate closely with the principles of the Code of Conduct for Responsible Fisheries, but elaborate on

[9] In particular, Canada and the United States lobbied successfully for the weakening of language on Free, Prior Informed Consent which replaced consent with consultation (C. Sharma, February 7, 2014, personal communication).

**Table 8.1** The Principles of responsible governance in the SSF Guidelines

| |
|---|
| 1. Respect for human rights and dignity |
| 2. Respect of cultures |
| 3. Non-discrimination |
| 4. Gender equality and equity |
| 5. Equity and equality |
| 6. Consultation and participation which should be based on human rights standards including the UN Declaration of the Rights of Indigenous Peoples |
| 7. Rule of law |
| 8. Transparency |
| 9. Accountability |
| 10. Economic, social, and environmental sustainability |
| 11. Holistic and integrated practices |
| 12. Social responsibility |
| 13. Feasibility and social and economic viability (FAO 2015, 2–3). |

principles that have become internationally accepted as principles of 'responsible governance' since the development of the Code (FAO 1995). In the FAO context, governance includes 'the formal and informal rules, organisations, and processes through which public and private actors articulate their interests, frame and prioritise issues; make, implement, monitor and enforce decisions' (FAO 2016, 6). These principles apply to both state and non-state actors, as shown in Table 8.1 above.

As noted in section "Changing tenure narratives over time", the SSF Guidelines urge all parties to recognize all legitimate forms of tenure, taking into consideration, "where appropriate, customary rights" (FAO 2015, 5). Significantly, the Tenure Guidelines have stronger language than the SSF Guidelines, recognizing customary tenure systems. Articles 4, 8, and 9 specifically advise States to recognize to the "legitimate tenure rights of indigenous peoples and other communities with customary tenure systems" (FAO 2012, 6–14).

Section 17 (1) of the UN Declaration of Human Rights recognizes that "[everyone] has the right to own property alone as well as in association with others" (UN General Assembly 1948). In addition to this general protection of the right to collective ownership, the legal protection in international law addressing customary tenure specifically is derived from human rights instruments such as the International Covenant on Civil and Political Rights (ICCPR), the Convention on Biological Diversity (CBD), the International Labour Organization (ILO) Convention concerning Indigenous and Tribal Peoples in Independent Countries (No. 169), and the Declaration on the Rights of Indigenous Peoples (UNDRIP) (UN General Assembly 2007). Most importantly, UNDRIP recognizes and affirms that "indigenous individuals are entitled without discrimination to all human rights recognized in international law, and that indigenous peoples possess collective rights which are indispensable for their existence, well-being and integral development as people" (Preambular paragraph 21, UN General Assembly 2007). UNDRIP Article 26 recognizes indigenous peoples' right 'to own, use, develop and control the lands, territories and resources that they possess' and this approach recognising the centrality of their access to and control over their natural resources permeates the declaration (UN General Assembly 2007, 10). The importance of customary systems of natural resource use and traditional knowledge in securing sustainable use and protecting biodiversity is recognized in CBD Article 8 (j) and 10 (c) (UNEP CBD 2009).

Although some states may not have ratified the above mentioned instruments, the collective body of international human rights law and regional laws 'create an extensive body of state practice recognising Indigenous peoples' rights to apply their own customary laws to protect their resource and knowledge rights' and "[support] claims that these rights of indigenous peoples and state obligations have become norms of customary international law, which are applicable to all states whether or not they are parties to relevant international human rights instruments" (Tobin 2013, 155–156). Emerging jurisprudence on international law and general tenets of African customary law point to the inclusion of other principles and procedural rights such as recognition and integration of indigenous and local knowledge, recognition of customary institutions and practices, and free and prior informed consent (FIPC) when changes to tenure rights are being proposed (Sunde et al. 2013; Tobin 2013; Almeida et al. 2015). Wicomb and Smith (2011), drawing on the 2010 Endorois decision of the African Human Rights Court, argue that this judgement provides a basis which affirms custom as culture when considered in the context of Constitutional jurisprudence in South Africa which recognizes that customary tenure is often the central expression of a community's culture. Significantly, this judgement paved the way for local communities that have customary tenure systems and depend on access to natural resources to then assert their right to culture and protect their customary tenure rights without having to do so in terms of their aboriginality or indigeneity (Wicomb and Smith 2011, 446).

## Customary Expressions of Tenure Along the Coastline of South Africa

The very clear requirements in the SSF Guidelines for "secure, equitable, and socially and culturally appropriate tenure rights to fishery resources" (FAO 2015, 5), coupled with the provisions in international human rights law, provide the backdrop to the recognition of the SSF Guidelines in South Africa and the standards upon which the reform of the small-scale fisheries policy in South Africa is accordingly considered.

As a result of its history of colonialism and apartheid, South Africa has inherited an extremely complex legacy of interacting systems of law and governance. From the late 1890s onwards the capital intensive, commercial fisheries on the Western seaboard of the country have influenced the governance of fisheries resources.

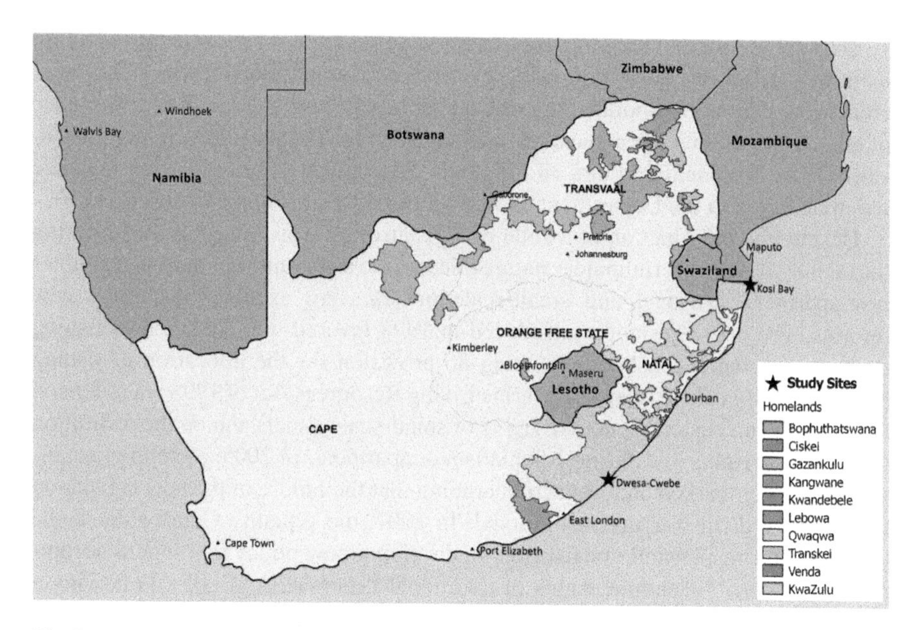


**Fig. 8.1** Case study sites located within the former apartheid homelands

Initially managed at the provincial level, the authority to manage fisheries shifted from the provinces to the federal government in 1930 in the state's efforts to gain a measure of control over the lucrative and rapidly expanding commercial fishing sector. The state embarked on the progressive introduction of legislative and policy mechanisms that favored white industrial fishing interests (van Sittert 1992). A series of regulations placed increasing restrictions on African and colored subsistence and artisanal fishers and brought them under the control of the industrial sector, steadily eroding the customary access and use rights of local fishers (van Sittert 2002).

Over the course of the past century and a half, the imposition of a statutory system of fisheries governance by the colonial authorities and then the apartheid state has over-shadowed customary systems of governance along the coast in South Africa (Sunde et al. 2013; Bavinck et al. 2014; Sunde 2014). These customary systems of tenure have continued, despite the state implementing various restrictions on small-scale fishing. This is most evident in the former apartheid homelands, known as the Bantustan areas of the country on the eastern seaboard, where many African communities live on communally owned land, according to customary law and varying levels of traditional authority (see Fig. 8.1).[10]

[10] Bantustan is the term used to refer to the homeland areas established by the Apartheid regime for the settlement of African peoples.

During the Apartheid regime, these areas were set aside for the separate development of African residents. Following the first democratic elections in 1994, these areas were then reincorporated into the Republic of South Africa. The governance of all marine resources became a national mandate for the state department responsible for environmental affairs and tourism. Subsequently, in 2009, this mandate was transferred to the Department of Agriculture, Fisheries and Forestry (DAFF).

Despite the introduction of a suite of legislative reforms aimed at transforming the racially based, discriminatory nature of access to and control of marine resources post-apartheid, artisanal and small-scale fishers were excluded from the new reforms. The new legislation introduced in 1998 favored the allocation of fishing rights to commercial fisheries, making no provision for the allocation of fishing rights to small-scale fishers. The Marine Living Resources Act of 1998 was silent on the pre-existing customary tenure rights of small-scale fishers and on the restitution of any tenure rights lost during colonialism or apartheid. In 2005, a group of small-scale fishers embarked on legal action arguing that the failure of the new legislation to recognize them was unconstitutional. In 2007, the Equality Court ordered the then Minister responsible for fisheries to develop a new policy that would accommodate the socio-economic rights of traditional fishers (EC 1/2005). Following a process of consultation that included the provinces of Eastern Cape and KwaZulu Natal, where customary systems of tenure predominated, a new Policy for Small-scale Fisheries was gazetted in 2012 (DAFF 2012). This policy recognized that many indigenous and local communities had been dispossessed of their customary tenure systems, however in some instances these customary systems have continued (DAFF 2012).

The governance arrangements in these areas where customary systems of tenure continue to exist are generally extremely complex, with multiple layers of law and authority. This is particularly so in areas where marine protected areas (MPAs) and adjacent terrestrial reserves were declared by the Apartheid state on what was communally owned land. In many of these areas, African residents were dispossessed of their land and access to natural resources, forced to relocate beyond the borders of the reserve or MPA. Post-apartheid, these communities have launched restitution claims to these areas. In most instances the communities' land claims have been recognized but the conservation status of the area has been retained and the community has not been permitted to resettle the land. Instead, settlement agreements include the right to co-manage the area, to benefit from the protected area through eco-tourism and regulated, sustainable use of resources. A very complex, plural system of tenure governance thus exists in reality, shaping the context in which the SSF Guidelines and the national Policy on SFF will be implemented. The tenure system of the Tembe-Thonga peoples of Kosi Bay Lakes and that of the Dwesa-Cwebe communities are illustrative of this tenure complexity.

**Fig. 8.2** A Tembe-Thonga fisherman in his fish trap in Kosi Bay

## *The Tembe-Thonga People of the Kosi Bay Lakes Region*[11]

In the far Northeast of the country (Fig. 8.1), the Tembe-Thonga people of the Kosi Bay Lakes region have practiced a customary system of fisheries tenure for over 600 years. They depend on fish and the harvesting of inter-tidal resources as a key component of their source of food.

The Tembe-Thonga recall that their ancestors who settled along the coastal peninsula between the lake and the sea harvested both marine and fresh water species in the lakes. Kosi fishers report that their ancestors initially used grass woven fish funnels in the estuary and lakes. Subsequently fish traps, known locally as *utshwayelo,* were developed (see Fig. 8.2).

[11] The information on the Tembe-Thonga community is drawn from research conducted by the author (Sunde 2013).

**Table 8.2** Customary norms and rules of the trap fishing communities of Kosi Bay

| |
|---|
| The larger community, comprising all of the Thembe-Thonga clans, holds communal ownership of the total area of the lakes |
| Families that are members of these clans hold family rights to the lake within this communal system |
| These rights are transferred within families from one generation to another through the male line. The community is aware of this gender discrimination and this practice is changing slowly with a few women being allocated traps in recent years |
| Individuals (usually male) within these families have individual rights, nested within a family right that in turn is part of the communal right |
| Only members of the families that comprise the community are permitted to own a trap |
| No outsiders may own or use a trap however, under certain circumstances, where someone moves into the community, permission may be granted from the community committee and the local headman is then informed |
| The construction of a new *utshwayelo* must be done after discussion with the owner of the neighbouring *utshwayelo* |
| No rights to any area or *utshwayelo* may be sold |
| Rights may be leased but the original individual and his or her family retain ownership of the right |
| Where there is no male within a family to inherit the ownership of the *utshwayelo,* use of the *utshwayelo* may be given to another male member of the clan or extended family but the original family retains ownership |
| In the above instance, the female partner of the original owner may assume the role of 'supervisor' of the trap and the new user may be required to provide fish to the family owning the trap |
| Where the male owner of a trap is deceased his wife may assume control over the trap in certain circumstances and this trap may be inherited by her sons rather than the passing back to the brothers of her deceased husband |
| Decisions-making is managed by a local customary committee. All households are members of this committee. This committee is male dominated but this is changing as men increasingly recognise the Constitutional requirement that women must be allowed to participate |
| A dispute is first managed at the level of the individuals impacted by the dispute. If this is not resolved it is referred to the committee and then, in turn, to the local headman |
| There is a social obligation to share one's catch with neighbours or others in the community if they are in need. This is a custom that continues to the present time (Sunde 2013) |

The communities using the lake describe a number of shared norms and rules relating to access, ownership, use, and management of *utshwayelo.* They have a customary system of decision-making and dispute resolution as well as shared cultural rituals that reinforce their distinctive culture and customary system. Rights to access, own, and/or use a portion of the lake for *utshwayelo* are derived from membership of a system of customary law that is common to the families descended from Tembe-Thonga clans. The following customary norms and rules are evident (Table 8.2).

As can be seen from the above norms and rules, the tenure system reflects a complex set of social relations and governance arrangements. The Tembe-Thonga believe that the source of their tenure is located in their relationship with their ancestors and is transferable to the next generation.

> Both the land and the sea, we own. We grew up here – it is a God given asset. You know where your ancestors lived and it's your heritage...this is our *isiko (custom).* The right is conferred from generation to generation and it is a thing that is known in the village...it is *injalo* (on-going) (KB 1 in Sunde 2013).

In the words of another informant "it is part of our estate...it is our heritage, our '*lifa*'.... It is our property... so its automatically transferred from one to another – we regard it as property. It is like a cow is part of a kraal. If someone comes in and confiscates your kraal with your cattle in it ...Your kraal is your right that you have enjoyed for ages and comes from your grandparents – it is *lefa lethu* – (our heritage) – and that's why we say "*ungateki lifa dranga'* (do not take my heritage)" (KB 2 in Sunde 2013).

## *The Dwesa-Cwebe Communities of the Eastern Cape*[12]

The Dwesa-Cwebe communities live adjacent to the Dwesa-Cwebe coastal forest reserve and marine protected area (MPA), situated on the Eastern Cape coast of South Africa (Fig. 8.1). Like the Thembe-Thonga peoples, the residents of Dwesa-Cwebe trace the source of their land and marine tenure rights to their ancestors. The land comprising the reserve was settled by the ancestors of the current residents several centuries ago. The seven communities who lived on this land regard the land, the forest and the coastline and associated natural resources as their common property, upon which they depend for their livelihoods. Their relationship with this land and the sea, derived through their relationship with their ancestors, is reflected in their culture and their customary tenure system. Over the course of the past century these communities were dispossessed of their land, their forest and marine resources in the name of nature conservation (Fay et al. 2002; Sunde 2014).

As with the residents of Kosi Bay, the entitlement to use resources derives from one's ancestors and from having been born in this space and hence belonging there.

Participant: "It was a norm that the land does not belong to the chief, it belongs to the people who are the first inhabitants. We grew up with this nature, we love this nature. The nature was created by God for us, and you cannot deny us what is rightfully ours."

Researcher: "Is there a word for this entitlement, this sense of what is 'rightfully' yours...?"

Participant: "I don't really remember the isiXhosa word but what I know, people say this is our land....*umhlaba wethu* (our land), *imithi*, (our forest), *nezihlanzi* (our fish), *nezimbaza* (our mussels)'...When a person grows up we grow together with these things... How can you divorce us now? We were born together with

[12] This case study is based on the author's PhD research (Sunde 2014).

> these things… we grow up, we are born as if you are one with the same mother. The mussel or fish is a brother to a human being that was born and grew up next to the sea. (When I was born) the first thing to see was my mother, the second thing to see was my father, then the sea with the mussels to eat as well as whatever was our tradition … *imveli yethu* (our tradition), we were born with these things." (H 41 in Sunde 2014).

Common to both the Tembe-Thonga of Kosi Bay and the Dwesa-Cwebe communities is the expression of a relational ontology that cites a distinctive relationship with nature that has shaped the development of their customary law and traditional knowledge system and, within this, the source of their rights to use natural resources.

There are several local idioms to describe this social-ecological connectivity. However, when translated into English these terms lose their meaning. The English word 'belonging' is used by several Dwesa-Cwebe respondents when describing their relationship to the land and resources. In addition to *indalo yethu* (our nature), residents frequently use the phrase '*umhlaba wethu*'. While '*umhlaba wethu*' literally means 'our land', according to local isiXhosa speakers, it refers to their collective relationship of belonging to their land, not just their land belonging to them. *Umhlaba wethu* signifies the on-going relationship of the ancestors to that land and to the community. It gives expression to the belief in the continuing presence of the ancestors and their on-going interactions with members of the community (Sunde 2014). It has a metaphysical, temporal and spatial component that transcends the meaning usually ascribed to it in English. This embedded notion of belonging to the land is very different to the relationship of ownership inherent in dominant interpretations of property ownership in statutory rights-based tenure systems (Fig. 8.3).

Viewed through the lens of customary tenure, tenure rights might be held individually or by families but these rights are nested within communal rights. In both of these systems, the layered nature of rights gives rise to a similarly layered system of institutions for decision-making and dispute resolution. Dispute resolution processes are vested at the local village level. They are deliberative and create an opportunity for the socio-cultural logic in the context of the conflict to be understood in order to make decisions or facilitate restoration of relationship between parties in dispute (Sunde 2014). Outcomes are guided by an ethic that prioritizes the restoration of social relationships.

In both the Kosi and Dwesa-Cwebe customary tenure systems, rights to resources are linked to responsibilities to feed one's family and those who are in need in one's community. Bennett has observed that in African customary law, in general, 'rights' are largely expressed as duties or obligations rather than as rights. This sense of duty to one's family is so strong in African customary law that it has been incorporated into Article 27 of the African Charter on Human Rights (Bennett 2012, 32).

This sense of duty in the customary systems of law of both these cases can be traced to the foundational philosophical principle of interconnectedness between humans, and between humans and nature underpinning these two systems (Sunde 2013, 2014). The ontological mode of being interconnected to nature and each other

**Fig. 8.3** Dwesa-Cwebe customary fishing community shares local knowledge with conservation officials

shapes the subsequent symbolic ordering of structures and rules in the customary system of law. This foundational principle of interconnectedness, as expressed through the ethic of *uBuntu*, shape how access to and use of marine resources is ordered and shapes interaction between members of the group. Translated literally, *uBuntu* means "a human being is a human being because of other human beings" (Mokgoro 2012, 317). *UBuntu* is simultaneously referred to as a foundational African value and legal principle (Mokgoro 2011, 1) and "an ancient principle of traditional African methods of government" (Froneman 2010, cited in Bennett 2011, 6). It is interpreted as "a web of values that informs conduct, and fosters group solidarity – the knit between an individual and his or her community; and the interconnectedness of individuals within their communities" (Mokgoro 2011, 1).

These distinctive ethical expressions inherent in these examples of customary law call for an approach to the responsible governance of tenure that is able to accommodate this alternative, more collectively-orientated, relational logic within which tenure is embedded.

In South Africa detractors of customary systems are quick to point out the examples where power relations within customary systems have skewed resources in favor of elites, or have not achieved environmental sustainability. In some contexts, the very patriarchal nature of customary systems is in direct violation of human rights standards. Evidence from elsewhere suggests that in many contexts customary marine tenure systems have struggled to adapt to demographic changes, technological developments, and market penetration, amongst other changes (Cinner and Aswani 2007). Customary systems may lack the capacity to respond to increasingly complex governance challenges that are present on a broader scale (Cinner et al. 2012). Whilst these governance challenges are real, they are not peculiar to customary tenure systems but have to be managed by all community-based tenure systems. As in the case of South Africa, these governance challenges are often exacerbated by the manner in which customary systems are undermined by statutory systems. In Dwesa-Cwebe and Kosi Bay, the colonial and then apartheid regimes imposed an interpretation of chiefly power onto the customary system, distorting the authority of the chiefs. Subsequently the dispossession of these two communities from their land and the imposition of strict restrictions imposed on their access to resources have meant that these customary governance systems have not been able to flourish, adapt to external changes and achieve ideals like *Ubuntu* 'in an ideal way'. Nonetheless, these principles have been persistent and remain integral to the customary systems of governance in these communities and hence retain the potential to be reinvigorated if their rights to practice their customary systems of tenure governance are recognized.

## *Legislative and Policy Frameworks for the Recognition of Customary Marine Tenure in South Africa*

The South African Constitution recognizes and protects customary law as an independent source of law that must be recognized and respected alongside common law and statutory law (Constitution of South Africa 1996). This means that where a community has had occupation, use, and/or access to coastal lands and marine resources according to the customary laws of their community, this customary system must be recognized as valid as long as its practice conforms to the Bill of Rights. Customary systems can be regulated however, and their regulation is subject to the balancing of these rights with other human rights according to the provisions of the Constitution (Sunde 2014). The Constitution recognizes the right to culture and the right of communities dispossessed of their property rights due to racial discrimination to restitution or equitable redress (Constitution of South Africa).[13]

Despite a strong human rights-based Constitutional framework, post-apartheid policy reforms adopted in 1998 failed to recognize small-scale fishing communities in South Africa (Isaacs 2011; Sowman et al. 2014). As noted above, in 2007 the Equality Court ordered the Minister responsible for fisheries to develop a new policy that would recognize the socio-economic rights of traditional, small-scale fishing communities (Kenneth George EC/105). Through a participatory policy development process, a new Policy for Small-scale Fisheries was finally gazetted in

[13] Section 26 of the Constitution states that property is not limited to land.

2012 (DAFF 2012). This policy, developed alongside the international negotiations on the Tenure Guidelines and the SSF Guidelines, incorporates the principles and human rights based standards of the SSF Guidelines.

Yet, notwithstanding this human rights-based framework and the South African government's commitment to the SSF Guidelines, DAFF continues to ignore customary tenure by imposing a statutory regulatory system onto these communities (Sunde 2014, 2016). The communities are required to apply for individual fishing exemptions, their customary institutions of governance are not recognized, and there is no co-management system. Instead, they are regulated top-down through a state-centric system that devises permit conditions based on an individual permit. The DAFF continues to do this despite a court ruling in which the Magistrate confirmed that the Dwesa-Cwebe community had a system of customary fishing rights (State vs. Gongqose and Others E382/10).[14]

In early 2016, DAFF released a set of regulations to give effect to the new Policy on Small-Scale Fisheries (DAFF 2016). In line with the policy, these regulations make provision for a community-based system of tenure. However, contrary to the provision in the Guidelines that states should 'identify, record and respect legitimate tenure right holders and their rights" including customary rights and ensure "secure, equitable and socially and culturally appropriate tenure rights to fisheries' (FAO 2015, 5), the DAFF regulations propose a neoliberal 'one size fits all' approach to tenure that is firmly orientated towards promoting market-based relations. All small-scale fishing communities will be required to establish and register a cooperative in order to qualify for fishing rights for which they must apply to the Minister (DAFF 2016). The market orientation of the regulations is evident in the way in which the regulations give the authority to decide which species will be harvested for commercial purposes and which for own consumption to the Minister. Although he or she is obliged to consult communities in making such a determination, the regulations prohibit the use of species identified for commercial purposes being used for personal consumption (DAFF 2016). In this way, the department is able to ensure that high value species are available for commercial purposes. This is significant in the South African political economy of fisheries where the existing large industrial companies control the segment of the value chain that processes and markets resources. As the bulk of the value adding benefits come from the export of most high value species, control over the sale of commercial species benefits these companies.

The state has refused to give specific recognition to customary systems of tenure (C. Smith, June 10 2015, personal communication). No steps have been taken to identify communities living according to customary law. Contrary to the Constitution and the SSF Guidelines, the regulations make no provision for redress and restitution of the territories and tenure rights dispossessed under colonialism or Apartheid (FAO 2015, 6). There is no means of accommodating the bundle of different rights

[14]Subsequently this was also confirmed by the High Court in 2016 (*State v Gongqose and others*).

within systems such as Kosi Bay and Dwesa-Cwebe communities. The regulations do not make provision for the customary management rights of indigenous and local communities or for their right to self-governance. Despite the SSF Guidelines urging states to take "existing power imbalances between different parties into consideration" (FAO 2015, 3) and to "recognise and safeguard publicly owned resources that are collectively used and managed by small-scale fishing communities" (FAO 2015, 5–6), small-scale fishers in South Africa report increasing inequities in the sector and accuse the state of continuing to give preferential treatment to the commercial sector (Sunde 2016). Long-term rights allocations to the commercial sector continue with no reference to the needs and rights of the small-scale fishers. Whilst the proposed reforms and new regulations aim to introduce a community-based rights system, this new model of tenure is also conceptualized within the same neoliberal property rights-based paradigm as earlier reforms and fails to address the underlying political economy of the fisheries governance model which favors the industrial fisheries sector's access to and control over the means of production.

## Interpreting and Implementing the SSF Guidelines

An analysis of the interpretation of 'tenure' in the context of customary communities in South Africa highlights the disjuncture between this meaning and the understanding of tenure being used to inform state policy. It is clear that tenure in the context of state policy fails to accommodate the values, ethics, and norms and social relations embedded in the customary systems of the Kosi Bay and Dwesa-Cwebe communities. It fails to respond to their right to self-determination of their norms and ethics of social responsibility, and to their right to control the resources they own collectively. In this regard, it discriminates against these communities and fails the standards of 'responsible governance' in the SSF Guidelines. Awareness of this mismatch between the approach to tenure reform in South Africa and the SSF Guidelines prompts a re-examination of the CSO submission on tenure rights in the development of the SSF Guidelines. As in the case studies cited above, the CSO submission included the articulation of a very particular concept of tenure. It linked tenure to the demand for a perspective that recognized the distinctive customary laws, cultures, and territories of indigenous peoples and local communities, in the context of UNDRIP. It was based on their right to control the natural resources which are the material basis of their culture and of their production systems. This has been recognized in the UNDRIP and in the subsequent development of human rights jurisprudence (Perry 2011; Tobin 2013).

At the time of developing these demands, activists within the movement were aware of the dangers of supporting a normative human rights framework which, in the context of the dominance of neoliberalism, might interpret human rights as individual rights. This did not resonate closely with communities' expression of their collective rights and the interpretation of their rights in the UNDRIP. The reliance on the human rights-based approach in fisheries advocacy was critiqued for enabling neoliberal tendencies that promote a market-based, individualized notion of property rights and neglect socially embedded collective rights (Ruddle and Davis 2013). However, this approach was defended at the time of the development of the SSF Guidelines text by key actors involved in the SSF Guidelines process (Jentoft 2014). Whilst considerable scholarship is now emerging on this issue of both the possible contradictions and synergies between human rights standards and ethics and principles underlying indigenous peoples and local communities' systems of law, this issue has yet to receive more detailed discussion amongst the international fishworkers movement itself (Engle 2011; Ruddle and Davis 2013; Song 2015). At the time of the negotiations, the attraction of a common set of international governance norms that reached beyond the national experience of many fisher communities was compelling. The shared demand for the human rights of fishers' to be recognized provided a common platform upon which the alliance across international fisher movements was developed and could be articulated in international fora with donors and the UN agencies amongst whom the human rights-based approach was recognized. It is significant to note that the synthesis document observed that CSO national level workshop reports "emphasise the need for all parties to recognise, respect and promote a human-rights based approach whilst simultaneously having respect for indigenous peoples and customary fishing communities" own systems of law and governance' (Sowman et al. 2012, 8). This indicates that CSOs were not unaware of the dangers that a universalizing, individualized, potentially neoliberal interpretation of the human rights framework would subsume the interpretation of collective rights arising from their customary systems of law and governance.

Perhaps a naïve, untested faith in the ability of the human rights standards emerging internationally and in several jurisdictions that recognize collective rights resulted in fisher leaders hoping their customary law would be protected by the fact that the SSF Guidelines were underpinned by these evolving and developing international human rights standards and instruments? It is now apparent that the standpoint and world view of indigenous peoples and local communities on their territories was rendered largely invisible in the text, despite the reference to the UNDRIP. As a consequence, and contrary to international human rights standards, there is a risk that whilst legal pluralism and the presence of customary practices may be acknowledged in a particular national jurisdiction, such as is the case in South Africa, customary law will not recognized as a legitimate system of law in the *de facto* tenure governance reforms that flow from the SSF Guidelines.

Fears concerning the appropriation and misuse of human rights language in the text of the Tenure Guidelines have now been voiced at the FAO User Rights 15 Conference in Cambodia and by an international coalition of CSOs who participated in the development of these Guidelines (International Alliance Statement on Tenure Guidelines 2015; FAO 2016). It would appear that as implementation processes gather momentum around the world, states and other governance actors will be forced to confront the contradiction inherent in their professed support for the SSF Guidelines and yet their concomitant discrimination against indigenous fisher peoples and local communities with customary systems of tenure.

## Conclusion

It is increasingly apparent that a gap exists between the concept of tenure as expressed by the two customary communities in South Africa presented above, the Constitution of South Africa and the SSF Guidelines. This gap has become visible through the way in which the SSF Guidelines are being interpreted and implemented through regulations and national policy in South Africa. As is evident in this context, the exclusion of subtle but important terms that framed the interpretation of tenure by the social movements in their submission on the development of the SSF Guidelines may render collective tenure and customary governance in general vulnerable to neoliberal tendencies, and powerless in many contexts. In commentating on conflicting positions during the negotiations of the text, Jentoft observed that 'the most controversial issues tend to be phrased in ways that allow interpretation flexibility' (Jentoft 2014, 7). This flexibility may prove to be a useful weapon of power for all actors across the spectrum of interests represented in fisheries. However, for small-scale fishing communities the litmus test will surely be the extent to which professed supporters work with them to secure their tenure governance rights in the context of their expression of their rights and not just 'secure tenure'. This will require vigilance to ensure that implementation programs recognize indigenous people and local customary communities' customary law with all that this implies in terms of respect for their distinctive ontologies and world views. Most importantly, implicit in this interpretation of 'responsible' tenure is recognition of their rights to their systems of customary governance and collective tenure. It does not exempt them from the responsibility of ensuring sustainability and equity in the governance of their resources, but it enables them to determine their own pathway to achieving the objectives of the Guidelines. This recognition and accommodation of the customary law and associated customary tenure systems of indigenous peoples and local communities is imperative in order that implementation remains true to the letter and spirit of the full spectrum of rights underpinning the SSF Guidelines.

## References

Almeida, F., Borrini-Feyerabend, G., Garnett, S., Jonas, H. C., Kothari, A., Lee, E., Lockwood, M., Nelson, F., & Stevens, S. (2015). *Collective land tenure and community conservation: Exploring the linkages between collective tenure rights and the existence and effectiveness of territories and areas conserved by indigenous peoples and local communities (ICCAs)*. Tehran: ICCA Consortium in collaboration with Maliasili Initiatives and Cenesta.

Aswani, A. (2005). Customary sea tenure in Oceania as a case of rights-based fishery management: Does it work? *Reviews in Fish Biology and Fisheries, 15*, 285–307.

Baarbesgaard, M. (2016, February 4–5). *Blue growth: Saviour or ocean grabber?* Paper presented at International Colloquium: Global governance/politics, climate justice & agrarian/social justice: linkages and challenges.

Bavinck, M., & Gupta, J. (2014). Legal pluralism in aquatic regimes: A challenge for governance. *Current Opinion in Environmental Sustainability, 11*, 78–85.

Bavinck, M., Sowman, M., & Menon, A. (2014). Theorizing participatory governance in contexts of legal pluralism – a conceptual reconnaissance of fishing conflicts and their resolution. In M. Bavinck, L. Pellegrini, & E. Mostert (Eds.), *Conflict over natural resources in the global south – conceptual approaches* (pp. 147–171). Leiden: CRC Press/Taylor and Francis.

Bennett, T. W. (2006). Comparative law and African customary law. In M. Reimann & R. Zimmermann (Eds.), *The Oxford handbook of comparative law* (pp. 641–674). Oxford: Oxford University Press.

Bennett, T. W. (2011). Ubuntu: An African equity. In F. Diedrich (Ed.), *Ubuntu, good faith and equity: Flexible legal principles in developing a contemporary jurisprudence* (pp. 3–23). Claremont: Juta.

Bennett, T. W. (2012). Access to justice and human rights in the traditional courts of sub-Saharan Africa. In T. Bennett, E. Brems, G. Corradi, L. Nijzank, & M. Schotsmans (Eds.), *African perspectives on tradition and justice* (pp. 19–40). Cambridge: Intersentia Publishing Ltd.

Charles, A. (2011). Human rights and fishery rights in small-scale fisheries management. In R. S. Pomeroy & N. L. Andrew (Eds.), *Small-scale fisheries management: Frameworks and approaches in the developing world* (pp. 59–74). Oxfordshire: CABI.

Cinner, J. E., & Aswani, S. (2007). Integrating customary management into marine conservation. *Biological Conservation, 140*, 201–216.

Cinner, J. E., Basurto, X., Fidelman, P., Kuange, J., Lahari, R., & Mukminin, A. (2012). Institutional designs of customary fisheries management arrangements in Indonesia, Papua New Guinea and Mexico. *Marine Policy, 36*, 278–285.

CSO. (2013). *Submission to FAO on the zero draft of the small-scale fisheries guidelines*. Unpublished internal document. International collective in support of fishworkers (ICSF). Accessed 14 Feb 2013.

DAFF. (2012). *Policy for the small-scale fisheries sector in South Africa*. Pretoria: Government Gazette.

DAFF. (2016). *Regulations relating to small-scale fishing*. Pretoria: Government Gazette.

Davis, A., & Ruddle, K. (2012). Massaging the misery: Recent approaches to fisheries governance and the betrayal of small-scale fisheries. *Human Organization, 71*(3), 244–254.

EDF, The Prince of Wales International Sustainability Unit. (2012). *Towards investment in sustainable fisheries: A framework for financing transition* (Discussion document). https://www.edf.org/sites/default/files/content/towards-investment-in-sustainable-fisheries.pdf. Accessed 10 May 2016.

Engel, K. (2011). On Fragile architecture: The UN declaration on the rights of indigenous peoples in the context of human rights. *The European Journal of International Law, 22*, 1 (141–163).

FAO. (1995). *Code of conduct for responsible fisheries*. Rome: FAO.

FAO. (2012). *Voluntary guidelines on the responsible governance of tenure of land, fisheries and forests in the context of national food security*. Rome: FAO Committee on World Food Security.

FAO. (2013). *Implementing improved tenure governance in fisheries – A technical guide to support the implementation of the voluntary guidelines on the responsible governance of tenure of land, fisheries, and forests in the context of national food security* (Preliminary version). Rome: FAO.

FAO. (2015). *Voluntary guidelines for securing sustainable small- scale fisheries in the context of food security and poverty eradication*. Rome: FAO.

FAO. (2016). *Applying the voluntary guidelines on the responsible governance of tenure of land, fisheries and forests: A technical guide for fisheries* (Draft version). Rome: FAO.

Fay, D., Timmermans, H., & Palmer, R. (2002). Closing the forests: Segregation, exclusion, and their consequences from 1936 to 1994. In R. Palmer, H. Timmermans, & D. Fay (Eds.), *From conflict to negotiation: Nature-based development on South Africa's wild coast* (pp. 78–110). Pretoria: Human Sciences Research Council.

Foley, P., Mather, C., & Neis, B. (2015). Governing enclosure for coastal communities: Social embeddedness in a Canadian shrimp fishery. *Marine Policy, 61*, 390–400.

Froneman, J. (2010). Albutt v Centre for the Study of Violence and Reconciliation, and Others. 2010. (3). SA 293 (CC).

Gaffiot, F., & Flobbert, P. (2000). Le Grand Gaffiot: Dictionnaire latin-français. Paris: Hachette-Livre. http://www.dualjuridik.org/uk/Etymology/tenure.htm#_ftn5. Accessed 25 Feb 2016.

Global Partnership for Oceans (2014). *Global document for a global partnership*.http://www.worldbank.org/en/topic/environment/brief/global-partnership-for-oceans-gpo. Accessed 10 Oct 2016.

Gordon, H. S. (1954). The Economic theory of common property resources: The fishery. *Journal of Political Economy, 62*, 124–142.

Government of South Africa. *Constitution of South Africa* Act. (108 of 1996). Pretoria: Government Publications.

Hartwick, E., & Peet, R. (2003). Neoliberalism and nature: The case of the WTO. *The Annals of the American Academy of Political Social Science, 590*, 188–211

Harvey, D. (2005). *A brief history of Neoliberalism*. Oxford: Oxford University Press.

IDI (Inclusive Development International). (2015). Campaign to reform the World Bank's policies and practices on land and human rights. http://www.inclusivedevelopment.net/world-bank-safeguards-campaign/. Accessed 25 Feb 2016.

International Alliance Statement on Tenure Guidelines. (2015). *The guidelines on the responsible governance of tenure at a crossroads*. International Statement. http://www.fian.org/fileadmin/media/publications/2015_TG_Statement. Accessed 17 Dec 2015.

Isaacs, M. (2011). Governance reforms to develop a small-scale fisheries policy for South Africa. In R. Chuenpagdee (Ed.), *Contemporary visions for world small-scale fisheries* (pp. 221–233). Amsterdam: Eburon.

Jentoft, S. (2014). Walking the talk: Implementing the international voluntary guidelines for securing sustainable small-scale fisheries. *Maritime Studies, 13*, 1–16.

Jentoft, S., & Bavinck, M. J. (2014). Interactive governance for sustainable fisheries: Dealing with legal pluralism. *Current Opinion in Environmental Sustainability, 11*, 71–77.

Mansfield, B. (2004). Neoliberalism in the oceans: 'Rationalization', property rights, and the commons question. *Geoforum, 35*, 313–326.

Mansfield, B. (2007). Privatization: Property and the remaking of nature–society relations. *Antipode, 39*(3), 393–405.

Mokgoro, Y. (2011). *Ubuntu* as a legal principle in an ever-changing world. In F. Diedrich (Ed.), *Ubuntu, good faith and equity: Flexible legal principles in developing a contemporary jurisprudence* (pp. 1–2). Claremont: Juta.

Mokgoro, Y. (2012). Ubuntu and the law in South Africa. In D. Cornell & N. Muvangua (Eds.), *uBuntu and the law: African ideals and post-apartheid jurisprudence* (pp. 317–323). New York: Fordham University Press.

Perry, R. (2011). Balancing rights or building rights? reconciling the right to use customary systems of law with competing human rights in pursuit of Indigenous sovereignty. *Harvard Human Rights Journal, 24*, 71–114.

Pinkerton, E., & Davis, R. (2015). Neo-liberalism and the politics of enclosure in North American small-scale fisheries. *Marine Policy, 61*, 303–312.

Ruddle, K., & Akimichi, T. (1984). Sea tenure in Japan and the South Western Ryukyus. In J. C. Cordell (Ed.), *A sea of small boats: Customary law and territoriality in the world of inshore fishing* (pp. 1–9). Stanford: Stanford University Press.

Ruddle, K., & Davis, A. (2013). Human rights and neo-liberalism in small-scale fisheries: Conjoined priorities and processes. *Marine Policy, 39*, 87–93.

Song, A. M. (2015). Human dignity: A fundamental guiding value for a human rights approach to fisheries? *Marine Policy, 61*, 164–170.

Sowman, M., Sunde, J., Raemaekers, S., & Mbatha, P. (2012). *Synthesis document developed for the civil society coordinating committee*. South Africa: University of Cape Town.

Sowman, M., Raemaerkers, S., & Sunde, J. (2014). Shifting gear: A new governance framework for small-scale fisheries in South Africa. In M. Sowman & R. Wynberg (Eds.), *Governance for justice and environmental sustainability. Lessons across natural resources sectors in Sub-Saharan Africa*. London: Earthscan/Taylor and Francis.

Sunde, J. (2013). *Living customary law along the South African coastline: Securing the rights of small-scale fishing communities. (Unpublished research report).* Cape Town: Legal Resources Centre.

Sunde, J. (2014). *Customary governance and expressions of living customary law at Dwesa-Cwebe: Contributions towards small-scale fisheries governance in South Africa.* Doctoral dissertation, University of Cape Town, Cape Town, South Africa.

Sunde, J. (2016). *'Caught in a net': Social relations and dynamics shaping the implementation of the voluntary guidelines on small-scale fisheries (VG SSF) in South Africa.* Monograph, International collective in support of fishworkers (ICSF), Chennai, India.

Sunde, J., Sowman, M., Smith, H., & Wicomb, W. (2013). Emerging proposals for tenure governance in small-scale fisheries in South Africa. *Land Tenure Journal,* (1–13), 117–146. Rome: FAO.

Tobin, B. (2013). Bridging the Nagoya compliance gap: The fundamental role of customary law in protection of indigenous peoples' resource and knowledge rights. *Law, Environment and Development Journal, 9*(2), 142–161.

UNEP (United Nations Environment Programme). (2009). *Ad hoc open-ended inter-sessional working group on Article 8 (j) and related provisions of the convention on biological diversity.* Sixth meeting. Montreal, 2009. Item 7. UNEP/CBD/WG8J/6/2/Add.1. https://www.cbd.int/doc/meetings/tk/wg8j-06/official/wg8j-06-01-add1-rev1-en.pdf. Accessed 10 Oct 2016.

United Nations General Assembly. (1948). *Universal declaration of human rights.* 10 December 1948. http://www.un.org/en/universal-declaration-human-rights/. Accessed 10 Oct 2016.

United Nations General Assembly. (2007). *United Nations Declaration on the Rights of Indigenous Peoples: Resolution/adopted by the General Assembly.* (A/RES/61/295). http://www.un.org/esa/socdev/unpfii/documents/DRIPS_en.pdf. Accessed 10 Oct 2016.

van Sittert, L. (1992). *Labour, capital and the state in the St. Helena Bay fisheries c. 1856- c. 1956.* Doctoral thesis, University of Cape Town, Cape Town, South Africa.

van Sittert, L. (2002). Those who cannot remember the past are condemned to repeat it: Comparing fisheries reforms in South Africa. *Marine Policy, 26*, 295–305.

Wicomb, W., & Smith, H. (2011). Customary communities as 'peoples' and their customary tenure as 'culture': What we can do with the *Endorois* decision. *African Journal of Human Rights Law, 11*, 422–446.

WFFP (World Forum of Fisher Peoples), WFF (World Forum of Fish Harvesters), & ICSF (International Collective in Support of Fishworkers). (2008). *Bangkok Statement.* Statement made at the FAO Small-scale Fisheries Conference, Bangkok, October 2008.

# Chapter 9
# Conditions for Governance of Tenure in Lagoon-Based Small-Scale Fisheries, India

**Prateep Kumar Nayak**

**Abstract** This chapter begins by confirming that issues around tenure within lagoon-based small-scale fisheries context have largely been neglected. Despite a growing body of literature on lagoon commons and property rights systems, existing literature on marine and terrestrial tenure tend to subsume tenure issues of coastal lagoons. Lack of specific attention to lagoon tenure can potentially affect their long-term sustainability and further marginalize small-scale fishers that have depended on them for generations. This chapter identifies important challenges associated with lagoon tenure in relation to the implementation of the Voluntary Guidelines for Securing Sustainable Small-Scale Fisheries in the Context of Food Security and Poverty Eradication (SSF Guidelines), particularly focusing on its provisions for responsible governance of tenure. The tenure provisions in the SSF Guidelines highlight that small-scale fishing communities need to have secure tenure rights to resources that form the basis for their social, economic, and cultural wellbeing, and that the state should recognize and ensure such rights. To this effect, the chapter sets forth some of the key conditions for governance of tenure in the context of lagoon small-scale fisheries social-ecological systems through an extensive treatment of a broad range of fishers' rights and multi actor responsibilities. Fisher experiences with the impacts of ongoing rapid social-ecological changes on lagoon tenure and community responses in Chilika Lagoon, Bay of Bengal, India region is used as a case. Data analyzed in this chapter comes from a series of workshops, meetings, and consultations with small-scale fishers and other stakeholders in Chilika. The chapter offers important lessons for governance of lagoon tenure by highlighting its key connections with resource systems, resource sectors, and user-level dynamics, to offer insights on possible institutional and governance arrangements around secure lagoon tenure. Further, it provides suggestions and reflections on the specific characteristics of lagoon small-scale fisheries tenure and possible future directions for governance. Despite its specific focus on lagoon systems, the main learnings about the key conditions, characteristics, and governance directions of small-scale fisheries tenure provides crucial insights on successful implementation of the SSF Guidelines, especially its tenure provisions.

P.K. Nayak (✉)
Faculty of Environment, School of Environment, Enterprise and Development, University of Waterloo, Waterloo, ON, Canada
e-mail: pnayak@uwaterloo.ca



S. Jentoft et al. (eds.), *The Small-Scale Fisheries Guidelines*, MARE Publication Series 14, DOI 10.1007/978-3-319-55074-9_9

**Keywords** Tenure • Social-ecological system • Lagoon • Governance • Rights • Fisher perception • Resource system • Resource units • Bay of Bengal

## Introduction

Small-scale fisheries are a large rural sector in many parts of the world, especially Asia and India. It provides jobs, livelihoods, food security, and cultural identity to a significant number of people (Béné et al. 2007; Kurien and Willmann 2009). However, the small-scale fisheries sector throughout the world is experiencing rapid changes because of impacts from both environmental and human drivers (e.g. climate change and globalization along with local, national, regional development, and policy processes), resulting in large-scale dispossession of fishers from their livelihoods, property rights, and cultural identity (MEA 2005; Allison et al. 2009; Nayak and Berkes 2010). This is particularly true in the case of coastal lagoons[1] which constitute about 13 % of the world's shoreline (Barnes 1980) and host millions of small-scale fishers by supporting their livelihoods, and offering essential habitat for a range of coastal, marine, and lagoon species. However, many lagoon systems of the world are undergoing a peculiar 'identity' crisis, partly because of the ongoing processes of change in their social-ecological attributes, but particularly because of a lack of recognition of their unique position (Coulthard 2008; Nayak 2011). In many parts of the world lagoons are considered neither 'marine' nor 'inland,' but a mixture of both and, as such, risk being overlooked by fisheries, and other, policies. Lagoons have been referred to as 'grey zones' with regard to policy making and implementation (Rana et al. 1998). In this context, this chapter looks at conditions for lagoon tenure with a view that tenure within a lagoon small-scale fishery social-ecological context has largely remained a neglected area (See introduction chapter of the book for tenure related definitions). Existing work on tenure almost entirely directs its attention to the discussion of sea/marine and land/forest tenure. Even though there is a growing body of literature concerning lagoon property rights systems with attention to commons' arrangements, existing literature on marine and terrestrial tenure tend to subsume tenure issues of coastal lagoons (Seixas 2002; Almudi and Kalikoski 2010; Coulthard 2011; Huong and Berkes 2011; Nayak and Berkes 2011; Benessaiah and Sengupta 2014). Lack of specific attention to lagoon tenure can potentially affect their long-term sustainability and further marginalize fishers that have depended on them for generations.

[1] Lagoons are shallow coastal bodies of water separated from the ocean by a series of barrier islands which lie parallel to the shoreline. Inlets, either natural or man-made, cut through barrier islands and permit tidal currents to transport water into and out of the lagoons. Because lagoons are characteristically shallow, they are strongly influenced by precipitation and evaporation, which results in fluctuating water temperature and salinity. Lagoons can also be fragile ecosystems susceptible to pollution effects from municipal, industrial, and agricultural runoff (Hill 2001).

This chapter identifies important challenges associated with lagoon tenure in relation to the implementation of the Voluntary Guidelines for Securing Sustainable Small-Scale Fisheries in the Context of Food Security and Poverty Eradication (FAO 2015) (SSF Guidelines, particularly focusing on its provisions for responsible governance of tenure (See the introduction chapter in this volume). The tenure provisions in the SSF Guidelines highlight that small-scale fishing communities need to have secure tenure rights to resources that form the basis for their social, economic, and cultural wellbeing, and the state should recognize and ensure such rights. To this effect, the chapter sets forth some key conditions for governance of tenure in the context of lagoon small-scale fisheries social-ecological systems through an extensive treatment of a broad range of fishers' rights and multi actor responsibilities. Despite the chapter's specific focus on lagoon tenure systems, the main learnings (e.g. key conditions of tenure, characteristics of small-scale fisheries tenure, important challenges and issues shaping tenure, and possible future directions for governance) will provide crucial insights into successful implementation of the SSF Guidelines, especially its tenure provisions.

Fisher experiences with the impacts of ongoing rapid social-ecological changes in lagoon tenure and community responses in the Chilika Lagoon, Bay of Bengal, India region is used as a case. Data analyzed in this chapter comes from a series of workshops, meetings, and consolations with fishers including: (1) a 2009 community workshop on fishers' rights to Chilika lagoon (17 participants from fisher villages and their regional network); (2) a 2013 policy workshop on the topic of 'human-environment relationship in Chilika Lagoon, India' (35 participants from fishing communities of Chilika, state, and national level NGOs); (3) a 2015 regional consultation on the SSF Guidelines with small-scale fishers of the Bay of Bengal, Odisha coast, India (45 participants from lagoon and marine fishing communities, NGOs, regional fisheries research institute, and government representatives); (4) a 2016 state level fisher workshop on the 'future of Chilika Lagoon' (37 fisher community and Regional Fisher Federation representatives).

The section following the introduction describes the context and overarching conditions for framing and implementing lagoon tenure, and offers an understanding of the key conceptual underpinnings around governance of lagoon-based small-scale fisheries tenure. I then turn to local perceptions of tenure in lagoon-based small-scale fisheries systems, analysing various rights, responsibilities, and enabling conditions as articulated by fishers themselves. Based on the findings reported in this section, important lessons for governance of lagoon tenure are distilled in the discussion section to highlight key connections of tenure with resource systems, resource sectors, and user-level dynamics, to offer insights on possible institutional and governance arrangements around secure lagoon tenure. The chapter concludes with some suggestions and reflections on the specific characteristics of lagoon small-scale fisheries tenure and possible future directions for governance.

## Conceptual Orientation: Defining Context and Overarching Conditions for Tenure

Conceptually, success in implementing governance of tenure in lagoon small-scale fisheries systems largely depends on three key conditions. First, lagoon-based fisheries systems are unique and different than other fisheries despite many similarities. Coastal lagoons are distinctively located at the interface of the sea and the land (includes freshwater systems). They not only epitomize both marine and terrestrial systems but also maintain their unique disposition by acting as a link between the two. Here, both tenure and its governance become somewhat tricky. While laws and practices of the sea apply on one side cultures, norms, and regulations associated with the terrestrial resource systems offer influence from the other (Nayak 2011). Thus, the 'in-betweenness' character of the lagoons becomes a determinant factor for understanding governance of tenure. It suggests, as a starting point, that lagoon tenure is not the same as land or forest tenure, nor the same as marine or sea tenure. Thus, the unique character of coastal lagoons, in relation to marine-terrestrial systems, becomes a fundamental basis for a discussion on lagoon tenure and its governance.

Second, there is a need to understand coastal lagoons as complex social-ecological systems. Social-ecological systems are systems that recognize the integration of humans in nature and stress that the delineation between social and ecological is artificial and arbitrary (Berkes and Folke 1998). Addressing only the social dimension of resource management without an understanding of resource and ecosystem dynamics will not be sufficient to guide society towards sustainable outcomes (Folke et al. 2005). This understanding implies that both social and ecological processes define and shape the nature of coastal lagoons where social outcomes remain contingent upon ecological dynamics and vice-versa. Similar to other coastal systems, coastal lagoons have many drivers, an array of impacts, unpredictable ways in which drivers act, uncertain system dynamics, and two-way feedback interaction between human and biophysical systems, all of which make them complex (Nayak 2014). Characterizing costal lagoons as complex social-ecological systems has important implications for how tenure can be defined and governed.

Third, using the social-ecological system context, it is possible to extend our understanding of coastal lagoons as highly interconnected systems of humans and environment and see them also as coupled human-environment systems (Turner et al. 2003). The term human-environment system emphasizes that the two parts (human system and environmental/biophysical system) are equally important, and that they function as a coupled, interdependent, and co-evolutionary system. Human actions affect biophysical systems, biophysical factors affect human well-being, and humans in turn respond to these factors (Berkes 2011). This understanding helps to explore how governance of tenure in coastal lagoons can be studied within an understanding of these areas being complex human-environment systems, with particular attention to interactions, relationships, and connections between the two (Kates et al. 2001; MEA 2005; Nayak and Berkes 2014). Consequently, any disconnect in the interaction and relationship between people and their lagoon environments may adversely impact tenurial arrangements, and this could be seen as a two-way process.

## Governance of Tenure in Lagoon-Based Small-Scale Fisheries

### *The Chilika Case*

Connected to the Bay of Bengal on the south, with the Eastern Ghats Mountain ranges forming most of its catchment on the north and the west, Chilika is a Ramsar site of international conservation importance and a biodiversity hotspot in India (Fig. 9.1). Rare, vulnerable, and endangered species inhabit the lagoon. It is the largest wintering ground for migratory waterfowl found anywhere on the Indian subcontinent and home to Irrawaddy dolphins and the Barkudia limbless skink. The total number of fish species is reported to be more than 225. According to a survey by the Zoological Survey of India in 1985–87, in addition to a variety of phytoplankton, algae, and aquatic plants, the lagoon region also supports over 350 species of

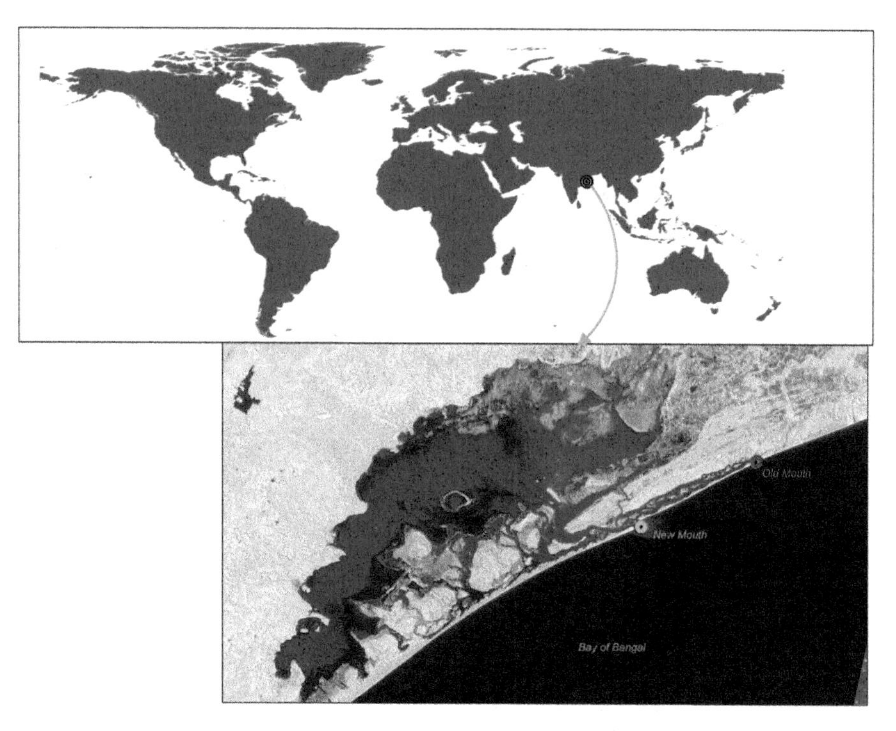


**Fig. 9.1** Map of Chilika Lagoon, India – The site of the case study

nonaquatic plants and over 800 fauna species. This represents the ecological subsystem of the lagoon and offers a solid ecological foundation to the lagoon small-scale fisheries system. Chilika's biodiversity is an integral part of sustaining the culture and livelihoods of the roughly 400,000 fishers and their families, who live in more than 150 villages. People in these villages have been engaging in customary fishing occupations for generations. The fisheries consist of traditional fisher groups whose vocation is identified by their membership in certain Hindu castes: there are seven different types of fisher castes and five sub-castes in Chilika (Nayak and Berkes 2011). The lagoon ecosystem also indirectly supports 0.8 million non-fisher higher caste villagers (e.g. Brahmins, Karans, Khandayat, and Khetriyas) in the watershed areas, who traditionally engaged in farming, forestry, and other livelihood occupations.

The lagoon historically provided multi-species and small-scale capture fisheries. However, in the 1980s, the sudden boost in the international shrimp markets and increase in export prices made shrimp aquaculture a major driver of change in the lagoon. Powerful local elites encroached customary capture fishery sources to use as aquaculture farms which led to resource conflicts. In 2001 a second important driver emerged. The state government created an artificial sea mouth with the Bay of Bengal through a hydrological intervention to deal with persisting siltation problems in the lagoon. The creation of an artificial sea mouth backfired as it furthered the ecological crisis by increasing the intensity of daily water inflow and outflow, and altering the saltwater-freshwater balance. The social-ecological system of the lagoon came under stress from the adverse impacts of the two drivers acting synergistically. Ecologically, habitats of most key species of fish, crab, and shrimps, along with associated species such as Irrawaddy dolphins (*Orcaella brevirostris*) and migratory birds, were reportedly damaged. Fluctuations in the main biophysical processes led to a change in species composition and altered food webs in Chilika Lagoon, indicating an ecological crisis gradually pushing the system towards a major crisis. There were corresponding impacts on the social subsystem as well. Fish production plunged dramatically and the small-scale fish economy including its management and institutional structures began to collapse. There have been corresponding changes in the rights of fishers to the fisheries resources of Chilika and the customary tenure arrangements that were recognized through annual leases by government to village fisheries cooperatives. Policy and civil society responses to the ongoing crisis have not yielded desired results and there are unresolved issues and complex uncertainties looming large.

## Fisher Perceptions About Tenure: Rights, Responsibilities, and Enabling Conditions

This section presents findings obtained from a series of community workshops, conducted over a number of years, involving Chilika lagoon fishers and representatives from marine fisheries, NGOs, government, and fisheries research institutions.

The emphasis is on three related aspects of tenure as it applies to small-scale fisheries in the context of coastal lagoons, but with ample relevance to other types of fisheries systems. The three aspects of tenure include: tenure as a set of rights, key responsibilities in relation to tenure, and enabling conditions that facilitate governance of tenure.

## *Rights Define the Nature of Tenure*

Small-scale fishers of Chilika Lagoon speak about issues around tenure using the lens of rights. In other words, tenure is understood as the combination of a set of specific rights that connect the fishers with various aspects of the fisheries and puts the control of their own tenure in their hands (through fisher institutions). These rights, as expressed by the fishers, span social, ecological, economic, and political aspects of tenure, and help provide directions to moving toward effective governance. Several rights were listed in three community/state level workshops (2009, 2013, and 2016) and one regional consultation meeting on the SSF Guidelines (2015) that helped gain a fuller understanding of the small-scale fisheries tenure system in Chilika and its governance challenges (Table 9.1).

The list of rights outlined by fishers clarifies two important aspects of tenure in Chilika Lagoon's small-scale fisheries system. First, tenure can be characterized by numerous rights, the status of which indicate the robustness of the tenure system. In ascribing meanings to each of the rights (Table 9.1, column 2) fishers chose to elaborate upon some of the major deficiencies by which these rights have been affected with ultimate impact on the status of tenure security (Table 9.1, column 3). Fishers argued that a simple approach to strengthen tenure in the small-scale fisheries system of Chilika is to repair the various rights that govern their relationship with the Lagoon. A subgroup at the regional consultation on SSF Guidelines stated, "there is no effective tenure without a basket of rights that are active and, in return, a strong tenure system will provide basic protection to the same rights." Second, the diversity of rights signify key aspects of small-scale fisheries related tenure that span economic/livelihood, political/legal, environmental, and social/cultural boundaries (Table 9.1, column 1) Explaining tenure using the lens of rights helps to understand the multifaceted nature of tenure. It further suggests that no one aspect will be enough to achieve tenure security without simultaneous emphasis on all other aspects as identified by the fishers through the enlisting of various rights. It is pertinent to highlight that an understanding of tenure through the lens of rights positively informs the implementation of tenure provisions of the SSF Guidelines.

Based on the explanations of various rights provided by fishers (Table 9.1), rights can be grouped into five categories (as below). It is important to note that these categories are not exclusive in nature but open to inclusion of other rights depending on the small-scale fisheries context under consideration. Moreover, these categories do not attempt to list completely new rights; instead they combine various rights from Table 9.1 to create the five categories.

**Table 9.1** Fisher perceptions of rights and their contribution to tenure security

| Nature of rights | Meaning | Link to tenure |
|---|---|---|
| Right to proper demarcation of fishing areas | No survey to demarcate and recognize fishing area boundaries after the 1950s (early-independence period), resulting in conflicting claims, lack of clarity on fishing rights, and increased number of legal disputes in courts. | Proper demarcation of fishing area boundary necessary to settle issues of fishing rights and allocate tenure to fishing villages. |
| Right to livelihoods | Fishers strongly believe that alternate livelihood is a misnomer: 'Once you have displaced a group of people from their original source of living there is nothing that could replace it as an alternative. Rather, it leads to a situation of fishers' livelihood being compromised.' | Since fishing alone is not sufficient for many fisher households there should be 'additional' livelihood support provided to supplement fishing-based incomes, which form the basis of the rights to livelihoods. |
| Right to fishing related loans | In the absence of fishing related loan provisions from the government or the banks, fishers become vulnerable to being exploited by fish traders or middlemen often paying interest rates of about 200% per annum. | Right to fishing related loans will make fishers financially self-sufficient, provide protection from exploitation, and make them better able to exercise their tenure rights. |
| Right to proper price | In the absence of any arrangement for government regulation of price and provision of minimum support price, fishers are left to the mercy of traders who control the price of fish in Chilika. It is important to mention that there are district level price fixation committees and minimum support prices for agricultural and forest products already in place. | Right to proper price will ensure the economic value of fish, protect fishers from artificial market price fluctuations, and ensure steady income. |
| Right to institution | Dissolution of the Central Fishermen Cooperative Marketing Society (CFCMS) and the imposition of a state level institution like the FISHFED in 1991 was a gross violation of fishers' rights to have their own institutions. In addition, other state institutions (e.g. Chilika Development Authority) have worked to suppress the voices of community fishery institutions. | Institutions provide continuity to tenure rights, which are primarily vested in the village level fisheries cooperatives. Rights to institutions are seen as a basic protection to the collective rights of fishers and their political voice. |
| Right to manage Chilika | Management of the lagoon through institutional and biophysical means must rest primarily with fishers. Additional management provisions to work collaboratively and within a respectful partnership arrangement for lagoon management need to be created. | Management rights put fishers in the forefront of deciding what is good for the lagoon. It provides an opportunity to make management interventions appropriate to ensure ecosystem services and fisher wellbeing. |

(continued)

**Table 9.1** (continued)

| Nature of rights | Meaning | Link to tenure |
|---|---|---|
| Right to speech | Implementing tenure requires deliberative, collaborative, and cooperative processes with ample freedom for fishers to engage in the governance system. Without the right to speak, fishers will be voiceless and less able to influence decision-making around lagoon management. | Right to speak provides the opportunity for fishers to add meaning/expressions to tenure. It makes them capable of advocating for their tenure rights. |
| Right to decide ones' own occupation ('Bruti') | Various government interventions in recent years have forced many fishers out of their customary occupation of fishing. One key example is the artificial sea mouth which has resulted in low fish production, leading to loss of customary occupation by fishers. Ecotourism is another example which has the potential of displacing traditional fishers from fishing occupations and converting them into petty service providers in tourism industry or pushing them to out-migrate. | Right to decide one's own occupation will ensure that the decision to be or not to be a fisher is left to the fishers alone without any interference. It will also include the state ensuring the right to one's occupation. |
| Right to make decisions | 'What kind of boats will run, where should the sea mouth have been created, when, how much and what type of fish to catch, how much should the lease fee be, and who should be the leaders are important decisions pertaining to fisheries and they should be left mainly to the fishers,' fisher leaders said. | Right to make decisions is crucial in putting the fishers at the center of fishery decision-making and can be a crucial step in successfully operationalizing tenure arrangements. |
| Right to information | Currently there are no formal arrangements for providing information on policies and markets to the fishers of Chilika. The only network through which information, mostly on markets, reaches the fishers is that of the fish traders and their agents. The absence of formal channels of information makes fishers exceptionally prone to exploitation. | Right to information will ensure government provides information on not only policies, regulations, domestic fish market, fish prices, etc. but also on regional and international market trends, including shrimp price fluctuations, and timely analyzed data on the merits and demerits of various fishing related practices. |
| Right to the rule of law (right to have a fishery – Chilika – law) | Absence of a law to govern Chilika exposes the lagoon and its fishers to multiple threats and the often contradictory policies of numerous government departments. Overall, it creates barriers to the full realization of fishing tenure. | Right to the rule of law will provide legal protection to tenure arrangements and other rights of fishers, and strengthen the legal and political environment influencing fisheries. |

1. **Rights that ensure the physical integrity of the resource to which tenure is allocated.** The foremost right in this category is the right to proper demarcation of the fisheries area boundary in order to avoid growing instances of resource conflict and many legal disputes over tenure in courts. "A better clarity on what area belongs to whom will provide long-term security to hold and implement tenure rights," explained Krushna Jena, fisher leader from Berhampur village.
2. **Rights that safeguard the economic (e.g. livelihood and subsistence) aspects of fishers' engagement in small-scale fishing and ensure that fishing occupations provide adequate economic incentives and becomes financially rewarding.** In this category, fishers talk about their right to have a fishing-based livelihood and putting in place response mechanisms to avoid situations in which their livelihoods may be compromised. A related right is to ensure that each fisher and fisher cooperatives receive appropriate remunerations for their fish and fish products in a fair manner. One of the ways proposed by fishers to ensure this is for the government to set minimum support prices for fish and fish products. This could be done through the formation of a district/state level price fixation committee comprising of private, public, and community representatives (a model that already exists in the case of pricing forest and agricultural products). Making available institutional loans to fishers will enhance their entrepreneurial capacity and financial stability to engage in a seasonally fluctuating and highly variable occupation like fishing.
3. **Rights that build and strengthen the institutional foundation of secure tenure and help create opportunities for better governance of tenurial arrangements.** Fishers in Chilika have experienced the loss of many community fisheries related institutions in the recent past. Prominent among the lost institutions are the Central Fishermen Cooperative Marketing Society (CFCMS), many Primary Fishermen Cooperative Societies (PFCS) at village level, and the weakening of the caste assembly (*Jati panchayats*) that have contributed to serious weakening of fishers control over fish resources. The right to have their own institution is a demand that results from this particular experience. Moreover, fisher institutions often create necessary rules and mechanisms for fishers to engage in fisheries management.
4. **Rights that strengthen the voices of the fishers, both individual fishers as well as their communities, and strengthen their political standing and decision-making capacity.** Fishers of Chilika, like elsewhere, are politically powerless and voiceless. Their subordinate position in the hierarchical Indian caste system adds to this disadvantage. "Right to speak will allow us to voice our concerns against the atrocities we face from the government as well as the non-fisher caste elites, and seek appropriate protection when necessary, in order to freely express our freedom and rights," some fishers said in a fisher sub-group at the 2015 regional consultation. Such a right will lead to fisheries related decision-making powers resting with community institutions. Since, information is power, fishers recommended that regular flow of information on market prices and fluctuations, policies and laws, rules and procedures, and knowledge and skills must be made available as a basic right. The most important right in this category is the right to decide one's own occupation which the fishers think has integral links with them being fishers by caste. A number of fishers think that "being a

fisher by occupation is our caste identity, and if we are not fishers (and/or not fishing for a living) then we are not complete humans." In recent years, significant occupational displacement has taken place in Chilika, resulting in caste-based fishers shifting towards non-fishing jobs or even outmigrating (Nayak and Berkes 2010). Fishers demanded that their right to have a fishing occupation needs to be fully protected by the state.

5. **Rights that provide a stronger legal basis and policy recognition to current and future tenure arrangements.** This is suggested by the fishers as an overarching right that protects all other rights listed in the above categories. A fishers' sub-group at the regional consultation (2015) emphasized that "there is chaos and foul play in the absence of the rule of law, and Chilika continues to suffer from the lack of a Lagoon law made to protect it and the fishers". Other fishers complained that "Asia's largest lagoon does not have a law to govern it and that violates our right to have the rule (protection) of law." Fishers further assert that without the rule of law tenure arrangements are bound to fail with serious adverse consequences for the small-scale fisheries and fishing people.

## Rights Come with Responsibilities

In the views of fishers, rights are always connected with responsibilities (Table 9.2). Fisher expressions about responsibilities ranged from their own duties to the lagoon to duties to be performed by the government/state. There were also areas identified

**Table 9.2** Various responsibilities articulated by fishers

| Responsibilities | Details |
|---|---|
| Fishers | Maintain a clear channel for navigation by not blocking it |
| | Do not catch juvenile fish |
| | Maintain seasonality of fishing |
| | Do not fish in breeding areas |
| | Follow catch and release principle |
| Government | Create awareness about all aspects of the lagoon |
| | Provide quality and timely information |
| | Catch the spun mafia instead of engaging in corruption (e.g. bribes) |
| | Monitor ecological conditions of the lagoon |
| | Take steps for increasing fish productivity |
| | Keep Chilika free from shrimp farming/aquaculture |
| Joint (fisher and government) | Find out why fish production has reduced and take immediate steps to address the problem |
| | Policy making with regard to the lagoon |
| | Management responsibilities for the bird sanctuary, dolphin census, and ecotourism |
| | Share enforcement responsibilities and power to deliver it. |
| | Enforcement should be carried out in consultation between fishers' institutions and government. |

which require joint responsibilities to be taken by fishers and the government. Main responsibilities of the fishers include their role in maintaining the ecological integrity of the lagoon by facilitating and not disrupting some of the natural processes associated with lagoon fisheries (e.g. keeping the numerous lagoon channels clear to allow regular flow of water and movement of fish, tracking fishing seasonality to aid breeding during the season, not interfering with fish breeding areas or fish habitats, and following village institution rules such as the policy of catch and release to ensure long-term sustainability of the lagoon).

If all these responsibilities are taken seriously, it would positively to governance of tenure. Further, fisher level responsibilities are not enough if the government does not adhere to its own responsibilities with regard to the lagoon. Table 9.2 lists a number of the government's responsibilities including its role in providing timely information to raise awareness involving both fishers and visitors (such as tourists), providing scientific support through knowledge and management interventions, regulating illegal interventions in the lagoon to minimize threats such as shrimp aquaculture and collection of larval fish/juveniles, and implementing programs for increasing productivity in the lagoon.

Fishers identified a number of areas where interventions are required as a collaborative venture between themselves and the government (Table 9.2). Such collaborative ventures are particularly important because it might be too challenging for fishers alone to handle some of the issues given the political and power dynamics prevalent in the lagoon (e.g. providing local level protection from threats of encroachment by higher caste non-fishers). Similarly, the government also requires support from the community in dealing with certain priorities (e.g. enforcement and management). Fishers asked, "why should the government only have responsibility of enforcement? Why not local enforcement by fishers themselves in which the government joins in as a partner?"

## Enabling Conditions for Tenure Rights

The exercise of tenure rights can only be possible if certain conditions are meaningfully met. These conditions offer the much required social, ecological, and political environment for the operationalization of tenure rights, and necessary security and protection against tenure rights violations. According to fishers (Table 9.1), some rights are better positioned to create enabling conditions for small-scale fisheries tenure. For example, fishers' right to have their own institutions creates a collective platform for everyone to work together to protect tenure and engage in advocacy efforts and negotiations with the government if and when tenure is under threat. Similarly, right to rule of law will ensure basic legal conditions for unhindered implementation of tenure. It will also recognize fishers' tenure of fisheries as a legal right and help see them as integral to the rule of law. Therefore, tenure rights have the ability to function both as specific rights pertaining to the use, access, and management of small-scale fisheries systems as well as provide enabling conditions under which these rights can thrive.

**Table 9.3** Key enabling conditions for secure tenure rights

| Enabling conditions |
|---|
| Maintain ecological condition |
| Increase fish production |
| Continuous protection of the lagoon |
| Rights without unhealthy restrictions |
| Timely survey and demarcation |
| Active peoples' movement |
| Priority to village institutions |

Based on views expressed by fishers, Table 9.3 outlines a number of enabling conditions and how they might contribute to the governance of tenure in small-scale fisheries. Key enabling conditions can be categorized as environmental, economic/rate of production, legal and policy instruments and physical and social dimensions. Maintaining the ecological health of the lagoon is a top priority for tenure. Fishers said that "Our rights to the lagoon and its fish will make sense only if the natural conditions of Chilika are good. It is hard to imagine what we will do with the secure tenure being granted to us by the government if there is no fish left." This concern clarifies that tenure is contingent upon healthy biophysical and resource conditions in a small-scale fishery system, and, therefore, acts as a key determinant of secure tenure.

There is a general agreement amongst fishers that there has been significant loss of fish stock in the lagoon primarily due to the impacts of the artificial sea mouth and intensive aquaculture. While this view has been substantiated by recent studies (Pattanaik 2007; Nayak 2014), the state government continues to deny it. Urgent strategies are required to increase fish production that in turn will strengthen fishers' tenure rights. As Sadashiv Jena, President of Eastern Chilika Fishers' Federation, explained: "An increase in fish production will automatically bring solutions to most of the problems we face in the lagoon today." Other fishers added: "We will benefit from having enough fish in Chilika only if they are protected from illegal capture along with the Lagoon areas that act as home to the fish." Obviously, the reference here is to the large customary fishing areas of the Lagoon that are already under elite capture for tiger shrimp aquaculture. Unofficial estimates highlight that over 60% of customary fishing areas are encroached in Chilika (Nayak and Berkes 2010 quoting Sea Food News 2005). Fishers strongly believe that without continuous protection of the Lagoon and its resources (including fish) it will not be possible to create a conducive environment for successful implementation of tenure rights.

Fishers of Chilika have witnessed many undue restrictions on their fishing rights and activities in the last three decades and more. These restrictions have curtailed their rights and put existing customary tenure arrangements into disarray. Some of the main restrictions reported by fishers include: (1) denial of access to customary fishing grounds as more than 60% of the Lagoon is currently under direct and indirect control of aquaculture owners (e.g. social caste and economic elites), (2) adverse impacts of new sea mouth (e.g. barnacle and sand infestation, excessive shallowing or deepening of fishing areas, speed of water inflow and outflow with high and low tides, and excessive salinity levels) that make fishing in the Lagoon waters risky, (3) changes in fishing area lease policy (e.g. imposition of a 27%

annual increase in lease fees, hassle of applying for annual leases instead of three-year leases as was earlier the case, (4) shifting the locus of institutional control from the village fisher cooperatives and regional fishers' organizations to district and state level statutory institutions (e.g. gradual weakening of the Primary Fishermen Cooperative Society (PFCS), dissolution of the Central Fishermen Cooperative Marketing Society (CFCMS), creation of FISHFED - a state level apex institution - which now controls fishing area leases in Chilika, and setting up of Chilika Development Authority (CDA) as a centralized state body to regulate lagoon management).

Our Chilika-wide village-level survey (2008–2009) showed that the number of past and ongoing resource conflicts in Chilika exceeded 100, more than 80% of which were fishing area boundary disputes. At a state level consultation (2015), fishers called for conducting up-to-date surveys and demarcating of fishing areas in Chilika as important steps to ensure tenure security. The last survey of fishing area boundaries and rights was done in the late 1800s, followed by another survey in the early 1950s. However, these surveys have not been fully used by the state government to legally settle fisher rights and boundaries of customary fishing areas despite prolonged demand by the fishers. There has also been significant blurring of the customary boundaries over the past decades. The old surveys have become somewhat dated leading to several conflicting boundary claims, resulting in court cases and fierce fights in Chilika. Boundary conflicts, along with the absence of proper boundary survey and settlement of fishery rights, have resulted in adverse impacts on tenure governance.

Chilika has been known in the past for its strong social and environmental movements. These movements have been organised by various fisher institutions (Pattanaik 2003a). The *Chilika Bachao Andolan* (Save the Chilika) movement was effective in uprooting the giant Tata company which initiated industrial shrimp aquaculture in Chilika after the state government signed a MOU and transferred 6000 hectares of lagoon area to the company in the 1990s (Mishra 1996; Pattanaik 2003b). Another movement has been ongoing since 2001 against the Chilika Bill, which, if passed into law, promises to reserve 30% of the lagoon area for non-fishers (which the fishers fear might also include industries and corporate houses engaging in large-scale aquaculture). This area might include significant customary fishing areas currently with caste-based fishers and involve them being transferred to non-fishers. "If that happens then it will be a final blow to us in terms of our ability to fish in the lagoon, and violate our identity as caste-based fishers," Biranchi Behera, a fisher participant in the regional consultation (2015), stated. Another fisher participant, Tapana Behera added, 'it is because of these protests and movements we have been able to hang on to our rights in Chilika. We will all die if the movement dies.' Nayak (2011) has used fishers' metaphors, such as "for the poor, when hunger becomes unbearable, movement becomes our last resort" to discuss the importance of fisher social movements in protecting their rights. In both the community workshop on rights to Chilika (2009) and the regional consultation (2015), fisher representatives emphasized the role of social movements as key to ensuring governance of tenure rights.

## Lessons for Governance of Tenure in Lagoon Social-Ecological Systems

In the above section I discussed the views of fishers regarding tenure in Chilika. I address them further in this section in relation to lessons that could be applied at the broader level, especially in the context of implementing the SSF Guidelines. Discussion on fishers' rights (Table 9.1), responsibilities (Table 9.2), and enabling conditions (Table 9.3) for successful small-scale fisheries tenure, and analysis of relevant published work offer a number of key leanings for governance of lagoon small-scale fisheries tenure. I discuss the main learnings in four categories that provide key insights for effective operationalization of the tenure provisions in the SSF Guidelines.

### *Attention to the Resource System*

Given the complexity of lagoon social-ecological systems, the starting point for establishing and recognising tenure should be the resource system (i.e. the lagoon itself). This is particularly important because resource system traits vary given their ecological construct, functions, and services, and their implications for related social and economic aspects. Even though there are system interconnectivities that exist, attention to a particular resource system within the web of interconnected resource systems (e.g. lagoon, marine, terrestrial) is the foremost basis for determining tenure arrangements. Keeping this in mind, I have already highlighted that lagoons are different from land (including forests) and sea, and, therefore, need to be considered as a distinct resource system for the purpose of administering tenure. This view was further confirmed through the articulation of tenure rights, responsibilities, and enabling conditions by fishers who chose to use their specific experiences with the lagoon as a basis of their explanations.

Within a lagoon, different resource sectors (e.g. capture fisheries, aquaculture, ecotourism, sand mining, and salt pans) need different tenure arrangements and specific attention to their governance. The context defines which resource sector gets priority (and which ones should not be allowed at all) and how to maintain links between sectors from a tenure point of view. In other words, stronger tenure in one particular resource sector may adversely affect tenure security in another sector within the same resource system. For example, the government effort in 1991 to legalise shrimp aquaculture at the expense of existing traditions of customary capture fisheries in Chilika Lagoon caused serious conflicts between fishers and non-fishers, and led to a gross violation of fishers' access and use rights. Here, the history of property rights, nature of user dependence, and ecological threshold levels of the lagoon are key considerations in the determination of tenure. An important function of tenure in the context of multi-resource sector social-ecological systems, such as coastal lagoons, is to create spaces and conditions for positive interactions between resource sectors, between people (users) who depend on those sectors, and between resource sectors and its users. As pointed out earlier, tenure is not only about relationships between people who are users (and non-users) but also about interactions between those users and their respective resources.

The overall size of the resource system matters as it clarifies the extent to which allocation of tenure rights can be accommodated given the number of resource sectors and resource users, and the general scope of governance requirements. Boundaries of both resource systems and resource sectors are useful. Resource system boundaries are important for clarity of use and management and, as in the case of Chilika, interdepartmental coordination. Specific attention to demarcating customary fishing boundaries is a top priority and a foremost activity in tenure arrangements. Experiences in Chilika suggest that demarcation does not necessarily entail a physical boundary as long as it is properly defined and contains some elements of tangibility. Absence of clarity of resource boundaries leads to tenurial confusion and may result in prolonged conflicts and legal disputes. While boundaries of the lagoon resource system are often defined naturally, boundaries of resource sectors are based on the history of lagoon use, customary and emerging resources practices and rights, and ecological and biophysical composition of the resource. For tenure, clear and recognized boundaries are preferred to fluctuating and contested boundaries. Detailed surveys and settlement of customary rights to fisheries have been suggested as mechanisms to address issues around boundaries in Chilika.

Resource system productivity helps maintain human-environment connections because it takes care of livelihood and food security concerns (Nayak et al. 2014). When productivity goes down, tenure including institutional arrangements tends to break. This happens partly because of economic and livelihood concerns but also because it leads to contestations and conflicts over resource extraction that may involve violation of agreed upon tenure rules. Moreover, when one type of productivity is lost other stakeholders, including the government, are often forthcoming to convert the resource system or parts of it into distinctly different resource sectors using loss of productivity as the reason for making such shifts. Similar situations were recorded in Chilika where a decrease in fish production was used as a plea to justify the opening of an artificial sea mouth that caused ecological and economic disaster (Pattanaik 2007; Dujovny 2009; Nayak et al. 2016). Tam Giang lagoon in Vietnam had a comparable experience where official promotion of aquaculture replaced several resource sectors (e.g. capture fisheries and mangroves) on similar grounds. Relevant to this analysis is the consequence of those government actions which worked against existing tenurial arrangements, mostly related to customary tenure of capture fishery sources. The productivity logic is also used by many economists and fisheries scientists who see subsistence and local fish production systems as not contributing to increasing global food demands. They, therefore, argue in favour of converting these systems into intensive production units. Such views, and any actions resulting from them, bear significant adverse consequences for tenure rights of small-scale fishers.

## *Understanding Resource Unit Dynamics*

A resource unit (e.g. fish) can be seen as one of the central bases for allocation of tenure as it helps to answer 'tenure of what?' It links a unit of specific resource area to a unit of defined users, both of which are essential components of any tenurial arrangement. It signifies a unit of physical resource area (i.e. village fishing ground) that is stationary and a unit of resource (i.e. fish) contained within it which may be highly mobile. In a lagoon situation this combination of 'mobile fish on stationary fishing grounds' creates an extremely dynamic and complex situation with significant implications for governance of tenure. Fishing behaviour of users at one location can potentially influence the availability of the mobile resource units at another location. Questions regarding tenure then would include: who (users) catches what (type of mobile resource unit), where (location) and how much (volume), when (seasonality), and how (fishing methods and practice). Governance of tenure will face the challenge of 'how to manage mobile resources as mobile and stationary resource units as stationary' requiring complex negotiations and conflict resolution on a constant basis. Similar questions on tenure and its governance may arise at places where open water fishing systems are in practice as 'mobile fishers (users) chasing mobile resource units (fish)' can become a source of resource conflict and threat to existing lagoon tenure.

The dynamics related to mobile and stationary resource units lead to unsustainable levels of resource extraction, a factor that underscores the importance of resource unit health from the perspective of tenure, among other things. Related aspects include interaction among resource units which is critical for sustenance of the lagoon social-ecological system. Healthy interaction among lagoon resource units has a foundational role in the sustenance of many biophysical processes within a lagoon. For example, larval shrimp hibernates in one resource unit (habitat), moves out to other resource units soon after reaching post-larval stage, and often gets caught in yet another resource unit when mature. Resource unit interaction related problems are prominent in the case of aquaculture and capture fishery linkages, as in the case of Chilika. Even in marine ecosystems of Atlantic Canada, there is growing tension between the lobster industry and salmon aquaculture as two leading resource sectors (Wiber et al. 2012). Can the governance of lagoon tenure address these critical interactions among resource units? Improper tenure can create gaps between resource units and block channels of interaction among them.

## *Users: Who Gets Tenure and Why*

A well demarcated lagoon resource unit in isolation is not of much use without an equally defined and recognized user unit in whose favor tenure will be allocated. Tenure defines and distinguishes between "who is a user and, therefore, has rights in the resource, and who does not." A tourist or an urban consumer of lagoon fish may be a 'user' but may not qualify for tenure. In spite of best efforts there would always be a few 'latecomers' (people who are late to join fishing), 'incomers' (people who come into fishing from outside) and 'out goers' (people who move out of fishing to other occupations or even migrate elsewhere) so far as allocation of tenure rights is concerned. Thus, governance of tenure needs to be flexible to allow the user units to define and redefine themselves as they develop and mature over time (Nayak 2003).

The number, location, and socio-economic attributes of users are important requisites. For tenure, both the total number of users in the whole resource system and specific number of users in the local resource unit matters (Meinzen-Dick 2007). While a comparatively small, defined, and regulated size of the user unit (homogenous) is preferred over a large, undefined and unregulated (heterogeneous) number of users, what matters most is whether the user unit is manageable or unmanageable with regard to tenure. Additionally, location in terms of proximity of the users to the resource unit or their residence relative to the lagoon can have significant bearing on long-term security of tenure. A large number of fisher villages on the north-west end of Chilika Lagoon have lost their customary fishing grounds through encroachment mainly because of their distant location on the south-east end of the lagoon. Socio-economic-cultural attributes, such as wealth, heterogeneity, land tenure, stability (Meinzen-Dick 2007), class, and caste (Nayak 2011), of the group impacts resource use and practices. For example, in Chilika, groups of refugee fishers have a record of engaging in invasive fishing practices; *Nolias* (sea going fishers who also fish in the lagoon) are known for using fishing gears that are not lagoon friendly; seasonal out-migration is a growing phenomenon, all of which are affecting the fishing behaviour in local communities. A related factor is the history of resource use by the group which allows distinguishing between customary users vs. new users in deciding tenure. This is particularly important in lagoons where new resource sectors are emerging that introduce groups of users without a history of involvement in the lagoon.

## Governance System: An Institutional Basis for Tenure

Institutions are vehicles of tenure and it is not a surprise that rights to institutions emerged as one of the important tenure rights in fisher workshops. Building stronger institutions can be seen as a precursor (Berkes 2004; Ostrom 2005) to successful governance of tenure. Institutions connect the resource to the people (users) and formulate norms and rules to regulate their behaviour vis-a-vis the resource. They provide means to the users to exercise their tenure rights and responsibilities as a collective rather than as individuals. Institutions that can create, hold, and govern tenure arrangements are necessary. The strength of institutions lies in their ability for renewal and reorganization, learning and adaptation, and in dealing with change (Holling 2001; Berkes et al. 2003). The 2009 community workshop on tenure rights and 2016 regional consultation on SSF Guidelines brought up several considerations with regard to an institutional basis for governance of lagoon tenure (Table 9.1).

Existing literature on the commons, institutions, and community-based conservation/resource management support the insights provided by fishers.

First, user institutions (e.g. village fish cooperatives in Chilika), not users (e.g. fishers) themselves, create conditions for successful governance of tenure (in the sense that Berkes et al. 2003 define it). Thus, institutions, not users, should hold tenure. Of course, there may be exceptions to this such as individual transferable quota systems and other forms of customary individual or family-based tenure arrangements as prevalent in many places (Johannes 1978; Coulthard 2011; Nayak and Berkes 2011).

Second, institutions at all levels are required but the local (user) level is important as a starting point (Berkes 2007) for better tenure governance. Higher level institutions are necessary but not at the cost of the local level. In Chilika, fishers' experiences with the removal of community institutions through creation of state level government institutions (i.e. replacing CFCMS with FISHFED) is a good example in this regard. Failure to recognize local level institutions along with bottom-up arrangements, and giving them the political space to continue may lead to collapse of tenure governance (Agrawal 2002; Nayak and Berkes 2008).

Third, the extent to which institutions are able to connect to each other transcending multiple geographical, administrative, political, and social-ethnic boundaries determines tenure success. For example, fishers in Chilika think that tenure rights are reciprocal in nature, whereby success of tenure in one particular village fishery system depends on the tenure arrangement in neighbouring villages, and such linkages are crucial for stronger tenure. Tenure governance should recognize the need to consider multiple levels of management (Ostrom et al. 1999; Young 2002; Adger 2003; MEA 2003) with appropriate cross-scale linkages among institutions (Adger 2001; Cash et al. 2006). Recent work in this area has identified 'missing institutions' and 'missing linkages' as factors for looming multi-scale governance failures (Walker et al. 2009; Almudi and Berkes 2010) which applies to tenure and related institutions (Nayak 2015).

Fourth, from a tenure perspective, multi-level institutional linkages are important in a globalized world (Lebel et al. 2005; Adger et al. 2006; Berkes 2007) but it requires significant attention to cooptation (Lele 2000; Gelcich et al. 2006; Nayak and Berkes 2008). In practice, better linkages to global shrimp markets may bring short-term economic gains to Chilika fishers, but may lead to large-scale changes in the lagoon ecosystem and jeopardize existing livelihood systems. Therefore, there is a need to have institutional linkages that help avoid cooptation.

Fifth, problems of cross-scale governance and cooptation can be addressed through institutional arrangements that provide an arena for trust building, sense making, learning, knowledge co-production, vertical and horizontal collaboration, and conflict resolution (Hahn et al. 2006; Berkes 2009), roles that can be taken up by 'bridging organizations' (Brown 1991; Cash 2001; Folke et al. 2005) or 'boundary organizations' (Guston 1999; Cash and Moser 2000; Berkes 2009). The role of bridging organizations in effective governance of tenure includes creating effective local institutions, horizontal linkages across sectors, and vertical linkages that enable grassroots influence on national policy making (Brown 1991).

Sixth, there is a need to recognize the challenges to governance of tenure from the perspective of institutional plurality, i.e. influence from different institutions with overlapping jurisdictions and capabilities (Ostrom 2005; Andersson and Ostrom 2008). The cross-influence of multiple hierarchical institutions can be effectively managed through polycentric governance arrangements that helps build relationships among these multiple authorities with overlapping jurisdictions (Oakerson 1999; McGinnis 2000; Ostrom 2005; Andersson and Ostrom 2008; Brewer 2010). This has the ability to create inclusive foundations for effective implementation and governance of tenure in small-scale fishery systems.

Seventh, the nature of property rights determines the extent of tenure security. Tenure may be ambiguous if property rights are not appropriately considered. In recent decades, Chilika Lagoon has witnessed diversification of property types (e.g. privatization for aquaculture, declaration of wildlife sanctuaries as state property, customary commons and some open access resources) leading to multiple or mixed property rights regimes which mutually contradict each other (Huong and Berkes 2011). Existing commons are being decommonized (Nayak and Berkes 2011) through state interventions with protected areas and even by private interests (such as aquaculture owners in Chilika) encroaching upon customary fishing grounds. Governance of tenure in a mix or multiple property rights scenario becomes complicated. Moreover, policies for actual implementation of property rights to stationary and mobile resource units require progressive negotiations on a constant basis.

## Conclusions

The chapter considered fisher perspectives to shed light on issues of rights, responsibilities, and enabling conditions for governance of tenure with specific reference to lagoon small-scale fishery systems. A key focus was on linking the information and analysis in the chapter with questions around successful implementation of the 2015 SSF Guidelines, especially the section on successful governance of tenure. Three broad conditions are emphasized as a precursor to lagoon tenure and its governance. First, the distinctive location of coastal lagoons at the interface of marine-terrestrial systems (in-between sea and land) puts them in a unique position (Nayak 2011, 2014). This in itself constitutes a fundamental basis for lagoon tenure and a reminder that lagoon tenure is not the same as land or sea tenure. Second, lagoons are complex social-ecological systems which imply that social (human) and ecological (biophysical) processes, interconnections and cross-influences among social-ecological system attributes, and the extent of system complexity influence how tenure is defined and governed. Third, lagoons can be seen as coupled, interdependent, and co-evolutionary human-environment systems (Turner et al. 2003; Berkes 2011) that stress on relationships (MEA 2005), interactions (Kates et al. 2001), and connections (Nayak 2014) between people and their environment, and each of these has implications for tenure. Thus, disconnect between fishers and the lagoon environment is detrimental to tenurial arrangements. This, moreover, is a two-way process.

The use of these three conditions extends our understanding of tenure as a 'relationship between people' (see FAO work on governance of tenure) and leads to a more inclusive definition that values 'relationships (also interactions and connections) between people and the environment (includes resource)' to which tenure is being sought. Governance of tenure is then about the manner in which the host of relationships, interactions, and connections are addressed and promoted. Additionally, dealing with complexity is a key task of tenure and its related governance system.

Given this understanding, analysis of various perspectives provided by fishers through a series of workshops and interactions highlighted specific rights, responsibilities, and enabling conditions that either promote or hinder tenurial security and related governance arrangements. Analysis of fisher perspectives of lagoon small-scale fishery tenure rights indicates that change is a common phenomenon, which not only offers an overall context for understanding and implementing lagoon tenure but also signifies specific conditions and challenges for governance success. However, they do not constitute a 'blue-print' for success and, therefore, should not be considered as panaceas (in the way that Ostrom et al. 2007 sees them) to deal with tenure related problems. Success and failure are inherent to lagoon social-ecological dynamics.

This analysis calls for a relook at the existing definitions of lagoon tenure, both legal and otherwise. The possibility of going beyond a legal definition of lagoon tenure is yet to be completely explored. There are several examples of customary lagoon tenure, property rights arrangements, and traditional knowledge in different parts of the world that precede legal tenurial arrangements and could potentially provide some directions (Johannes 1978, 1981). "Who should define tenure" and "whose realities count" for designing appropriate institutional and governance arrangements could be an important consideration in this regard? Exploring "how people define or what do people understand by tenure" (in the sense of Narayan et al. 2000; Nayak and Berkes 2010; Andrachuk and Armitage 2015) in different lagoon contexts can be a starting point. In Chilika, fishers verbalized their relationships with the lagoon using four metaphors (Nayak and Berkes 2010), each of which is significant for defining tenure in that specific context. Exploring similar illustrations elsewhere can only further this understanding.

Tenure in the context of complex lagoon small-scale fisheries social-ecological systems is not a static concept and, therefore, can be best understood as a process (see Nayak and Berkes 2011, in the context of lagoon commons) and its governance as continuous. Such a perspective, along with the three broad conceptual considerations and a host of specific rights, responsibilities, and enabling conditions (based on fisher perspectives), has the potential to further our understanding of good practices in governance of lagoon tenure and help sustain appropriate tenurial arrangements in the face of multilevel challenges.

# References


Adger, N. (2001). Scales of governance and environmental justice for adaptation and mitigation of climate change. *Journal of International Development, 13*, 921–931.

Adger, W. N. (2003). Building resilience to promote sustainability. *International Human Dimensions Programme IHDP Update, 2*, 1–3. Update 02/2003:1–3, Bonn, Germany.

Adger, W. N., Brown, K., & Tompkins, E. L. (2006). The political economy of cross-scale networks in resource co-management. *Ecology and Society, 10*(2), 9. http://www.ecologyandsociety.org/vol10/iss2/*art9/*. Accessed 9 Dec 2016.

Agrawal, A. (2002). Common resources and institutional sustainability. In O. E. T. Dietz, N. Dolsak, P. C. Stern, S. Stonich, & E. U. Weber (Eds.), *The drama of the commons* (pp. 41–86). Washington, DC: National Academy Press.

Allison, E., Perry, A. L., Badjeck, M. C., Adger, W. N., Brown, K., Conway, D., Halls, A. S., Pilling, G. M., Reynolds, J. D., Andrew, N. L., & Dulvy, N. K. (2009). Vulnerability of national economies to the impacts of climate change on fisheries. *Fish and Fisheries, 10*(2), 173–196.

Almudi, T., & Berkes, F. (2010). Barriers to empowerment: Fighting eviction for conservation in a southern Brazilian protected area. *Local Environment, 15*(3), 217–232.

Almudi, T., & Kalikoski, D. (2010). Traditional fisherfolk and no-take protected areas: The Peixe Lagoon National Park dilemma. *Ocean and Coastal Management, 53*, 225–233.

Andersson, K. P., & Ostrom, E. (2008). Analyzing decentralized resource regimes from a polycentric perspective. *Political Science, 41*, 71–93.

Andrachuk, M., & Armitage, D. (2015). Understanding social-ecological change and transformation through community perceptions of system identity. *Ecology and Society, 20*(4), 26. doi:10.5751/ES-07759-200426.

Barnes, R. S. K. (1980). *Coastal lagoons* (p. 106). Cambridge: Cambridge University Press.

Béné, C., Macfadyen, G., & Allison, E.H. (2007). *Increasing the contribution of small-scale fisheries to poverty alleviation and food security* (Fisheries Technical Paper No. 481, 124 pp). Rome: FAO.

Benessaiah, K., & Sengupta, R. (2014). How is shrimp aquaculture transforming coastal livelihoods and lagoons in Estero Real, Nicaragua? The need to integrate social–ecological research and ecosystem-based approaches. *Environmental Management, 54*(2), 162–179.

Berkes, F. (2004). Rethinking community-based conservation. *Conservation Biology, 18*(3), 621–630.

Berkes, F. (2007). Community-based conservation in a globalized world. *Proceedings of the National Academy of Sciences, 104*, 15188–15193.

Berkes, F. (2009). Evolution of co-management: Role of knowledge generation, bridging organizations and social learning. *Journal of Environmental Management, 90*, 1692–1702.

Berkes, F. (2011). Restoring unity: The concept of marine social-ecological systems. In R. Ommer, I. Perry, P. Cury, & K. Cochrane (Eds.), *World fisheries: A social-ecological analysis* (pp. 9–28). Oxford: Blackwell Publishing.

Berkes, F., & Folke, C. (Eds.). (1998). *Linking social and ecological systems: Management practices and social mechanisms for building resilience*. Cambridge: Cambridge University Press.

Berkes, F., Colding, J., & Folke, C. (Eds.). (2003). *Navigating social-ecological systems: Building resilience for complexity and change*. Cambridge: Cambridge University Press.

Brewer, J. E. (2010). Polycentrism and flux in spatialised management: Evidence from Maine's lobster (*Homarus americanus*) fishery. *Bulletin of Marine Science, 86*(2), 287–302.

Brown, D. L. (1991). Bridging organisations and sustainable development. *Human Relations; Studies Towards the Integration of the Social Sciences, 44*(8), 807–831.

Cash, D. W. (2001). In order to aid in diffusing useful and practical information: Agricultural extension and boundary organisations. *Science, Technology & Human Values, 26*, 431–453.

Cash, D. W., & Moser, S. C. (2000). Linking global and local scales: Designing dynamic assessment and management processes. *Global Environmental Change, 10*, 109–120.

Cash, D. W., Adger, W., Berkes, F., Garden, P., Lebel, L., Olsson, P., Pritchard, L., & Young, O. (2006). Scale and cross-scale dynamics: Governance and information in a multilevel world. *Ecology and Society, 11*(2), 8. http://www.ecologyandsociety.org/vol11/iss2/art8/. Accessed 9 Dec 2016.

Coulthard, S. (2008). Adapting to environmental change in artisanal fisheries – Insights from a South Indian lagoon. *Global Environmental Change, 18*(3), 479–489.

Coulthard, S. (2011). More than just access to fish: The pros and cons of fisher participation in a customary marine tenure (Padu) system under pressure. *Marine Policy, 35*(3), 405–412. doi:10.1016/j.marpol.2010.11.006.

Dujovny, E. (2009). The deepest cut: Political ecology in the dredging of a new sea mouth in Chilika Lake, Orissa, India. *Conservation and Society, 7*, 192–204.

FAO. (2015). *Voluntary guidelines for securing sustainable small-scale fisheries in the context of food security and poverty eradication*. Rome: FAO.

Folke, C., Hahn, T., Olsson, P., & Norberg, J. (2005). Adaptive governance of social-ecological systems. *Annual Review of Environment and Resources, 30*, 441–473.

Gelcich, S., Edward-Jones, G., Kaiser, M. J., & Castilla, J. C. (2006). Co-management policy can reduce resilience in traditionally managed marine ecosystems. *Ecosystems, 9*, 951–966.

Guston, D. H. (1999). Stabilizing the boundary between politics and science: The role of the office of technology transfer as a boundary organization. *Social Studies of Science, 29*, 87–112.

Hahn, T., Olsson, P., Folke, C., & Johansson, K. (2006). Trust-building, knowledge generation and organizational innovations: The role of a bridging organization for adaptive co-management of a wetland landscape around Kristianstad, Sweden. *Human Ecology, 34*, 573–592.

Hill, K. (2001). *The Indian River Lagoon species inventory: What is a lagoon?* Smithsonian Marine Station, USA. http://www.sms.si.edu/irlspec/Whatsa_lagoon.htm. Accessed 9 Dec 2016.

Holling, C. S. (2001). Understanding the complexity of economic, ecological, and social systems. *Ecosystems, 4*, 390–405.

Huong, T. T. T., & Berkes, F. (2011). Diversity of resource use and property rights in Tam Giang Lagoon, Vietnam. *International Journal of the Commons, 5*(1), 130–149.

Johannes, R. E. (1978). Reef and lagoon tenure systems in the Pacific islands. *South Pacific Bulletin, 4*, 31–33.

Johannes, R. E. (1981). *Words of the lagoon: Fishing and marine lore in the Palau District of Micronesia*. Berkeley: University of California Press.

Kates, R., Clark, W. C., Corell, R., Hall, J. M., Jaeger, C. C., Lowe, I., McCarthy, J. J., Schellnhuber, H. J., Bolin, B., Dickson, N. M., Faucheux, S., Gallopin, G. C., Grübler, A., Huntley, B., Jäger, J., Jodah, N. S., Kasperson, R. E., Mabogunje, A., Matson, P., Mooney, H., Moore, B., O'Riordan, T., & Svedin, U. (2001). Sustainability science. *Science, 292*, 641–642.

Kurien, J., & Willmann, R. (2009). Special considerations for small-scale fisheries management in developing countries. In K. L. Cochrane & S. M. Garcia (Eds.), *A fishery manager's handbook* (2nd ed., pp. 404–424). Chichester: FAO/Wiley-Blackwell.

Lebel, L., Garden, P., & Imamura, M. (2005). The politics of scale, position, and place in the governance of water resources in the Mekong region. *Ecology and Society, 10*(2), 18. http://www.ecologyandsociety.org/vol10/iss2/art18/. Accessed 9 Dec 2016.

Lele, S. (2000). Godsend, sleight of hand, or just muddling through: Joint water and forest management in India. In *ODI natural resource perspectives* (Vol. 53, pp. 1–6). London: Overseas Development Institute.

McGinnis, M. D. (Ed.). (2000). *Polycentric games and institutions: Readings from the workshop in political theory and policy analysis*. Ann Arbor: University of Michigan Press.

MEA (Millennium Ecosystem Assessment). (2003). Ecosystems and human wellbeing: A framework for assessment. Chapter 5: Dealing with Scale. In *Millennium ecosystem assessment*. Washington, DC: Island Press.

MEA (Millennium Ecosystem Assessment). (2005). Ecosystems and human well-being: General synthesis. In *Millennium ecosystem assessment*. Chicago: Island Press. http://www.Millenniumassessment.org/en/Synthesis.aspx. Accessed 9 Dec 2016.

Meinzen-Dick, R. (2007). Beyond panaceas in water institutions. *Proceedings of the National Academy of Science, 104*(39), 15200–15205.
Mishra, B. K. (1996). *Reframing protest: The politics of livelihood and ecology in two environmental movements in India*. Doctoral thesis, Cornell University. 270 pp.
Narayan, D., Chambers, R., Shah, M. K., & Petesch, P. (2000). *Voices of the poor: Crying out for change*. New York: Published for the World Bank, Oxford University Press.
Nayak, P. K. (2003). Community-based forest management in India: The significance of tenure. *Forests, Trees and Livelihoods, 13*, 135–160.
Nayak, P. K. (2011). *Change and marginalisation: Livelihoods, commons institutions and environmental justice in Chilika Lagoon, India*. Doctoral thesis, University of Manitoba, Canada.
Nayak, P. K. (2014). The Chilika Lagoon social-ecological system: an historical analysis. *Ecology and Society,* 19(1), 1. doi:http://dx.doi.org/10.5751/ES-05978-190101.
Nayak, P. K. (2015). Institutional Pluralism, Multilevel Arrangements and Polycentricism: The Case of Chilika Lagoon, India. In M. Bavinck & A. Jyotishi (Eds.), *Conflict, negotiations and natural resource management: A legal pluralism perspective from India* (pp. 148–177). London: Routledge.
Nayak, P. K., & Berkes, F. (2008). Politics of cooptation: Self-organized community forest management and joint forest management in Orissa, India. *Environmental Management, 41*, 707–718.
Nayak, P. K., & Berkes, F. (2010). Whose marginalisation? Politics around environmental injustices in India's Chilika Lagoon. *Local Environment, 15*(6), 553–567.
Nayak, P. K., & Berkes, F. (2011). Commonisation and decommonisation: Understanding the processes of change in Chilika Lagoon, India. *Conservation and Society, 9*(2), 132–145. doi:10.4103/0972-4923.83723.
Nayak, P. K., & Berkes, F. (2014). Linking global drivers with local and regional change: A social-ecological system approach in Chilika Lagoon, Bay of Bengal. *Regional Environmental Change, 14*(6), 2067–2078. doi:10.1007/s10113-012-0369-3.
Nayak, P. K., Oliveira, L. E., & Berkes, F. (2014). Resource degradation, marginalization, and poverty in small-scale fisheries: Threats to social-ecological resilience in India and Brazil. *Ecology and Society, 19*(2), 73. doi:http://dx.doi.org/10.5751/ES-06656-190273.
Nayak, P. K., Armitage, D., & Andrachuk, M. (2016). Power and politics of social–ecological regime shifts in the Chilika lagoon, India and Tam Giang lagoon, Vietnam. *Regional Environmental Change, 16*(2), 325–339. doi:10.1007/s10113-015-0775-4.
Oakerson, R. (1999). *Governing local public economies: Creating the civic metropolis*. Oakland: ICS Press.
Ostrom, E. (2005). *Understanding institutional diversities*. Princeton: Princeton University Press.
Ostrom, E., Burger, J., Field, C. B., Norgaard, R. B., & Policansky, D. (1999). Revisiting the commons: Local lessons, global challenges. *Science, 284*, 278–282.
Ostrom, E., Janssen, M. A., & Anderies, J. M. (2007). Going beyond panaceas. *Proceedings of the National Academy of Science, 104*(39), 15176–15178.
Pattanaik, S. (2003a). Globalization and a grassroots environmental movement: The case of Chilika Bachao Andolan (CBA). *Journal of the Indian Anthropological Society, 38*(1).
Pattanaik, S. (2003b). Development, globalization and the rise of a grassroots environmental movement: The case of Chilika Bachao Andolan (CBA) in eastern India. *The Indian Journal of Public Administration, XLIX*, 55–65.
Pattanaik, S. (2007). Conservation of environment and protection of marginalized fishing communities of Lake Chilika in Orissa, India. *Journal of Human Ecology, 22*(4), 291–302.
Rana, K. J., Grainger, R., & Crispoldi-Hotta, A. (1998). Current methods and constraints for monitoring production from inland capture fisheries and aquaculture. In T. Peter (Ed.), *Inland fishery enhancements* (FAO Fisheries Technical Paper T374). Papers presented at the FAO/DFID Expert Consultation on Inland Fishery Enhancements, Dhaka, Bangladesh, 7–11 Apr 1997.

Seixas, C. S. (2002). *Social ecological dynamics in management systems: Investigating a coastal lagoon fishery in Southern Brazil* (265 pp). Doctoral thesis, University of Manitoba, Winnipeg, Canada.

Turner, B. L., Matson, P. A., McCarthy, J. J., Corell, R. W., Christensen, L., Eckley, N., Hovelsrud-Broda, G. K., Kasperson, J. X., Kasperson, R. E., Luers, A., Martello, M. L., Mathiesen, S., Naylor, R., Polsky, C., Pulsipher, A., Schiller, A., Selin, H., & Tyler, N. (2003). Science and technology for sustainable development special feature: Illustrating the coupled human-environment systems for vulnerability analysis. Three case studies. *Proceedings of Nation Academy of Science U S A, 100*, 808085.

Walker, B., Barrett, S., Polasky, S., Galaz, V., Folke, C., Engström, G., Ackerman, F., Arrow, K., Carpenter, S., Chopra, K., Daily, G., Ehrlich, P., Hughes, T., Kautsky, N., Levin, S., Mäler, K., Shogren, J., Vincent, J., Xepapadeas, T., & Zeeuw, A. (2009). Looming global-scale failures and missing institutions. *Science, 325*(5946), 1345–1346.

Wiber, M. G., Young, S., & Wilson, L. (2012). Impact of aquaculture on commercial fisheries: Fishermen's local ecological knowledge. *Human Ecology, 40*(1), 29–40. doi:10.1007/s10745-011-9450-7.

Young, O. (2002). *The institutional dimensions of environmental change: Fit, interplay and scale*. Cambridge: MIT Press.

# Chapter 10
# Beyond the Small-Scale Fisheries Guidelines: Tenure Rights and Informed Consent in Indigenous Fisheries of Nicaragua

**Miguel González**

**Abstract** This contribution seeks to provide an overview of policy actions taken by the government of Nicaragua in relation to critical aspects of the *Voluntary Guidelines on the Responsible Governance of Tenure* (FAO 2012) as well as the *Voluntary Guidelines for Securing Sustainable Small-Scale Fisheries* (SSF Guidelines) (FAO 2015), which are concerned with the governance of Indigenous peoples' customary tenure systems, including the rights to aquatic resources. The chapter devotes attention to the relationship between provisions to land and aquatic rights, assesses the impact of recent programs of land titling on Indigenous collective property rights and access to fisheries, and identifies gaps in the process of implementation in the Rama-Kriol territory. It also explores the implications for the human rights of Indigenous and Afro-descendant peoples emerging from the prospective construction of an Interoceanic Canal, a large-scale infrastructure project vigorously endorsed by the Nicaraguan government. This chapter suggests that, in this context, the SSF Guidelines alone hold little practical value in shifting the balance towards protecting the rights of Indigenous fisheries. Therefore, the implementation of the Guidelines should consider their potential synergies in tandem with other instruments in national and international law.

**Keywords** Nicaragua • Indigenous fisheries • Interoceanic canal • Small-scale fisheries • Rama-Kriol territory

M. González (✉)
Department of Social Science, York University, Toronto, ON, Canada
e-mail: migon@yorku.ca



S. Jentoft et al. (eds.), *The Small-Scale Fisheries Guidelines*, MARE Publication Series 14, DOI 10.1007/978-3-319-55074-9_10

## Introduction

> We are not against development or the Grand Canal. However, we would like to see that things are done in good faith. We would like to see the government putting in our hands every single information related to the Canal, including the possible impact on fisheries, and under this basis our communities shall freely decide without pressure. (Alison May, Kriol leader of Monkey Point, Bluefields, July 7 2016)

> Local norms and practices, as well as customary or otherwise preferential access to fishery resources and land by small-scale fishing communities including Indigenous peoples and ethnic minorities, should be recognized, respected and protected in ways that are consistent with international human rights law. (SSF Guidelines, Section 5.4)

In June 2013, after with minimal consultation or debate, the Nicaraguan National Assembly passed a law which granted a multibillion, 50-years concession to a Chinese consortium (the Hong Kong Nicaraguan Development Corporation, HKND Group) to build an Interoceanic Canal linking both the Pacific and the Caribbean coasts of the country (Huete-Pérez et al. 2016). Its proponents have presented the megaproject as a unique opportunity for Nicaragua – the second poorest country in the Western hemisphere – to be lifted up from endemic poverty and underdevelopment. The Canal project was received with optimism by the country's private sector, while peasants and Indigenous Afro-descendant peoples whose lands might be affected by the proposed route have expressed their opposition by actively mobilizing against the Canal (Wade 2016).

The Canal project is particularly troubling for the sustainability of small-scale fisheries in the country, particularly through its potential negative impact on Lake Nicaragua, which is the largest freshwater tropical lake in Americas (Campos 2013; Huete-Pérez et al. 2015). Researchers have commented on the possible impact of the megaproject on unique species of freshwater fish that inhabit Lake Nicaragua and its tributaries (Meyer and Huete-Pérez 2014). In addition, Indigenous and Afro-descendant peoples on the Eastern Coast – also known as the Caribbean Coast – whose collectively-held lands are located in the proposed Canal route, have been very active in expressing their concerns about the potential impact of the project on their tenure rights and access to aquatic resources (Goett 2016).

This chapter devotes attention to the contested dynamic that ensued once the Canal project was first launched by the Nicaraguan government in 2013, the local responses the initiative elicited from the Rama-Kriol Territorial Government (GTRK), the legal governing body of the Afro-Indigenous territory on the Caribbean Coast, and the mounting tensions that exist on the Eastern Coast of Nicaragua. I explore the above case study through the normative lenses of the *Voluntary Guidelines for Securing Sustainable Small-Scale Fisheries* (SSF Guidelines), which were endorsed by FAO member states in 2014 and provide substantial directives for states to 'contribute to equitable development of small-scale fishing communities and poverty eradication and to improve the socio-economic situation of fishers and fish workers within the context of sustainable fisheries management' (FAO 2015, 1). The Guidelines also state that "Small-scale fishing communities need to have secure tenure rights to the resources that form the basis for their social and cultural well-being, their livelihoods and their sustainable development." In addition, "states, in accordance with their national legislation should ensure that small-scale fishers, fish workers and their communities have secure, equitable, and socially and culturally appropriate tenure rights to fishery resources (marine and inland) and small-fishing areas and adjacent land" (FAO 2015, 5, sections 5.1, 5.3).

I contend that the Canal project in Nicaragua illustrates some of the most pressing challenges regarding the country's efforts in securing customary tenure rights for Indigenous peoples. It appears that Nicaragua is now pursuing a large-scale infrastructure project that threatens to dismantle the accomplishments of progressive legislation passed to protect Indigenous and Afro-descendant collective rights to land and aquatic resources that were realized in the last two decades. Therefore, this chapter examines the complex relationships in which the SSF Guidelines are embedded in the case of Nicaragua and, more particularly, in a context where the collective rights of Indigenous peoples constitute a fundamental avenue for the protection of small-scale fisheries and the promotion of tenure rights. From a more general analytical point of view, this case illustrates the complicated relationships and dilemmas of small-scale fisheries governance from where lessons on the normative utility of the SSF Guidelines can be drawn. Due to their voluntary nature, the SSF Guidelines have had a very limited practical value in halting the progress of the Canal project (which is still in its preparatory phase) in a context in which both the government and global investors have decided – almost at any cost – to carry out their vision of development. In light of these circumstances, I contend that the SSF Guidelines can play a role in challenging the asymmetries of power involved in large-scale infrastructure project as long as they become embodied along with other legally binding instruments which, as a whole, would help protect the fundamental human rights of Indigenous peoples.

The chapter also identifies important challenges to policy coherence with regard to the implementation of the SSF Guidelines. In the case of Indigenous tenure rights, these challenges constitute a critical barrier to better governance. Policy coherence, in this case, involves the interaction of multiple jurisdictions (local, regional, and national) which are responsible for providing a supportive and enabling environment for the development of small-scale fisheries.

The research that informs this chapter was conducted over a period of 3 months in the summer of 2016. It involved multiple ethnographic techniques such as a field visit to the Rama – Kriol territory, participant observation, in-depth and group interviews with 25 Indigenous and Afro-descendant community members of the territory, including authorities, leaders, fishers, and their legal and political representatives in regional governing bodies. Research participants were selected through purposive sampling on the basis of their roles within these communities, leadership and authority positions, and overall knowledge on small-scale fisheries, tenure rights, and the negotiation process with government authorities on the Canal project. Interviews were conducted both in English and Spanish. The author also participated in one assembly in the community of Suumu Kaat, a Rama village that belongs to the wider Rama-Kriol territory. Access to the communities and research participants was granted by members of the Rama-Kriol Territorial Government, who invited me to participate in a field trip to Rama Cay and other communities in the territory. They also allowed me to participate in communal assemblies and introduced me to individuals I was interested in interviewing in Bluefields. The methodology also included literature review and face-to-face interviews with regional and national authorities, including senior officials in the Ministry of Fisheries, the Regional Council, and its executive branch. All interviewees have been given pseudonyms to protect their anonymity.

## The Context of the Eastern Coast

The Eastern (Caribbean) Coast of Nicaragua is populated by Indigenous peoples (Miskitu, Ramas, and Sumu-Mayangnas) as well as non-Indigenous peoples (Afro-descendants and mestizos), who inhabit diverse ecosystems and reside in approximately 250 scattered and impoverished communities along the Coast and inland (PNUD 2005).[1] The fishers amongst them target several species such as shrimp, lobster, and various finfish, but they also capture sea turtles, mainly for subsistence (Christie 2000; Hostetler 2005).

Information about small-scale fisheries in Nicaragua is dated and few systematic efforts have been made to fully assess the contribution of SSF to economic growth and food security. Current Nicaraguan fisheries policies are aimed largely at the industrial fishing sector, virtually ignoring small-scale fisheries, while international development initiatives targeted toward small-scale fisheries have often lacked continuity (Salas et al. 2011; Hostetler 2005). This situation contrasts with the substantial contribution of small-scale fisheries – particularly Indigenous-fisheries – to the conservation of natural resources, food security, and global seafood exports (Béné 2003; Jentoft and Chuenpagdee 2009; Jentoft and Eide 2011). This is also true for Nicaragua, where Indigenous artisanal lobster fisheries, for example, which is conducted mostly through diving, contributed 37% of the 5.5-million-pound total catch registered on the Caribbean Coast in 2015 (INPESCA 2015). The contribution of artisanal Indigenous fisheries is also substantial with regard to other species, which have an important commercial value, such as the sea cucumber (INPESCA 2015).

Indigenous coastal and inland fisheries have also been recognized as one of the key factors which positively impacts the region's food security and overall human development (PNUD 2005; United Nations Development Program 2011; Williamson 2016). However, researchers have also pointed out the marginalizing conditions under which Indigenous fishers operate due to precarious labour conditions, lack of social security, and inadequate health services (Acosta et al. 2002; Daw 2008).

[1]According to the most recent census the population of the Coast is comprised by 79% mestizo, 17% Miskitu, 1.06% Mayangna, 0.02% Ulwa, 0.22% Rama; 2.8% Kriol and 0.2% Garifuna. See Williamson and Fonseca (2007).

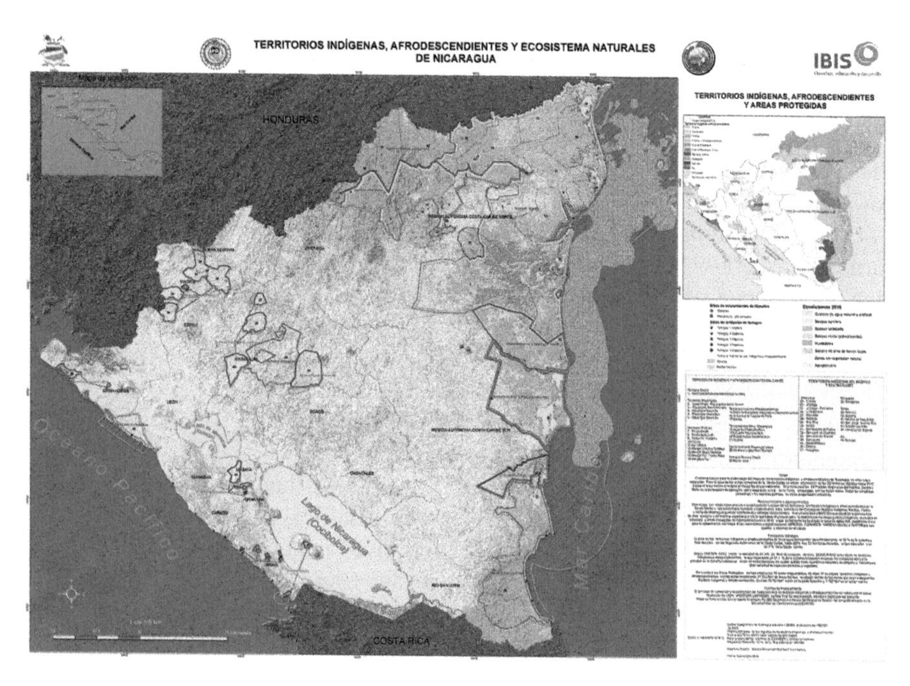


Fig. 10.1 Indigenous Territories of the Nicaraguan Caribbean Coast (Source: Mendoza 2016)

Globally, small-scale fisheries are for the most part marginalized from policy making, which tends to favour large scale, industrial fisheries (Chuenpagdee 2011a, b).

This state of affairs is of particular relevance in the Nicaraguan case, given that 94% of historically claimed Indigenous territorial areas in Nicaragua have been titled to coastal and inland communities (CONADETI 2013). This tenure system leaves Indigenous rights to aquatic resources for the most part unaddressed. Consequently, outstanding claims over who controls ownership over and access to fishing areas in coastal and inland communities have not received proper attention in policy-making or in national development plans (Fig. 10.1).

Furthermore, national authorities have issued programs to protect endangered species, establish coastal and marine protected areas, and expand seasonal closures for overexploited species. However, these initiatives have met with scepticism, while enforcement has been inconsistent (Lagueux and Campbell 2005). The disconnect in public policy between collective rights granted over communal lands and aquatic rights to secure a sustainable resource base for Indigenous communities, the limited functionality of established marine protected areas, and an overall disregard or inadequacy of policies aimed at poverty reduction, are widespread along the Eastern Coast (González and Jentoft 2010; González 2011). This has resulted in weak governance and a lax regulatory environment for Indigenous small-scale fisheries.

### *The Rama-Kriol Territory: Emerging Governance of Tenure Rights to Land*

The Rama-Kriol – an Indigenous and Afro-descendant territory – was granted a full ownership title in 2009. The territory comprises 406,849 ha of terrestrial lands and 44,308 ha of marine areas, including 22 cays, for the purpose of "development of artisanal fisheries" (Asamblea Nacional de Nicaragua 2016). Comparatively speaking, the Rama-Kriol Territory is the second largest territory that has been granted under collective ownership to the Indigenous and Afro-descendant peoples of the Eastern Coast of Nicaragua, after *Prinzu Awala*, with 4149.5 km$^2$. The territory consists of six Rama Indigenous communities: Rama Cay, Wiring Cay, Bang Kukuk Tai, Tiktik Kaanu, Suumu Kaat, and Indian River; and three Kriol (mostly coastal) communities – Monkey Point, Corn River, and Greytown. The Rama is the second smallest Indigenous people on the coast, with three thousand inhabitants. The Kriols, an Afro-descendant people that has deep historical and cultural relationship with the Rama, including inter-marriage, have lived in coastal communities along the Rama territory, particularly Monkey Point and Greytown (see Fig. 10.2 below).

The multi-communal territorial government (GTRK) comprises six representatives of each community and is a relatively new governing body, inaugurated in 2004 in the context of Law 445 for the Demarcation of Indigenous Territories (Acosta 2007).

Indigenous fisheries for the Rama and Kriols are diverse, targeting multiple species, as well as seasonal and highly adaptive to coastal, marine, and inland ecosystems (Roe 2006). For all nine communities in the Rama-Kriol territory, fishing and subsistence agriculture equally play a fundamental role in their social, cultural, and economic well-being. Fishing activities also rely on a deep-rooted local traditional knowledge of coastal and inland ecosystems, a dimension that has also been observed for Indigenous fisheries more generally (Neiland et al. 2005; Tress et al. 2005; Deepananda et al. 2015). The Rama and Kriol fish both at sea and in inland waters, as well as in brackish lagoons in the surrounding coastal and marine ecosystems. They target multiple species, both for subsistence and for commercial purposes. Subsistence fisheries in particular are highly adaptive, as they are mostly dependent on the availability of target species in lagoons and inland rivers (Nietschmann 1973). A few species are targeted for commercial purposes, particularly oysters and fish that are traded in the regional capital city of Bluefields.

The Kriols living in the coastal communities of Greytown and Monkey Point have been actively involved in subsistence, commercial, and recreational fishing (for tourism), which are conducted in relatively small boats at sea and inland. Over the last 5 years, government incentives in the form of equipment and stock centers have been provided to small-scale fishers organized in cooperatives in Monkey Point and Greytown. However, access to regional markets has been a bottleneck issue for the capitalization of cooperatives, while fluctuation in prices for fish has discouraged fishers to make a living off the sea (G. Enriquez, personal communication, 4 July 2016).

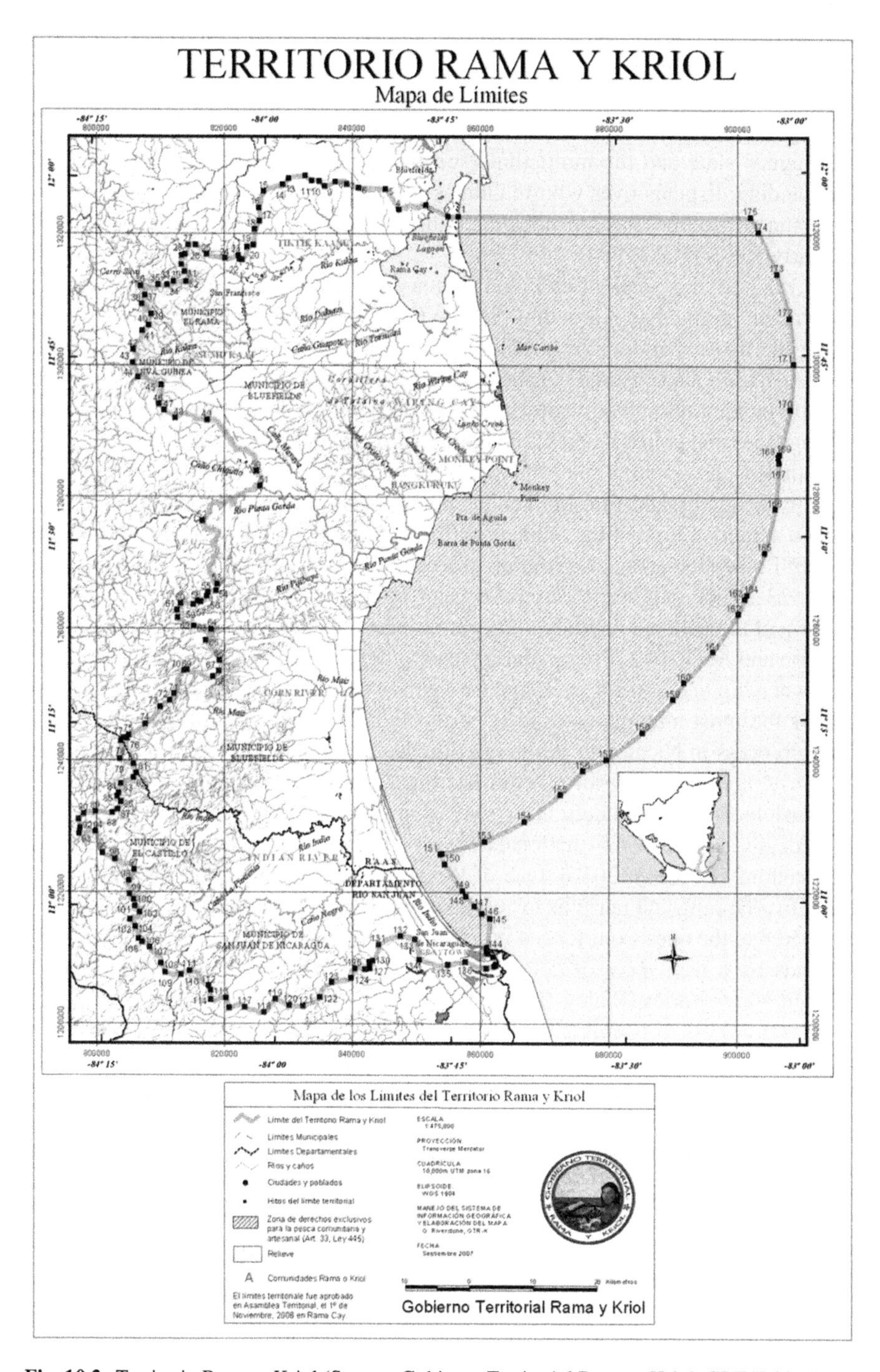


Fig. 10.2 Territorio Rama y Kriol (Source: Gobierno Territorial Rama y Kriol, GRT-K 2017)

The current state of Indigenous fisheries in the Rama-Kriol territory cannot be fully appreciated without considering the significance of the struggle for land and access to aquatic rights in which they have engaged in recent history. Struggles for collective land have been a defining element in the relationship between the Nicaraguan state and the multiethnic society of the Coast (Frühling et al. 2007). Outstanding disputes over who maintains ultimate control and ownership of land and natural resources have resulted in conflict, including a military confrontation in the early 1980s (Hale 1994).

In 1987, after 5 years of civil and military unrest on the Coast, the Nicaraguan government granted an Autonomy Statute (Law 28) to the Coastal region, inaugurating the first experience of a multiethnic territorial autonomy in Latin America. The legislation also ensured constitutional protection of ancestral lands historically held by Indigenous and Afro-descendant peoples, as well as the cultural protection of languages and political and social participation in regional governing institutions. Nonetheless, it was not until 2002 that the Nicaraguan state passed legislation specifically to survey and title Indigenous lands. This process began in 2006 when a second administration of the Sandinista Front of National Liberation (Spanish acronym FSLN) endorsed a new program to advance land demarcation. Until November 2016, 23 Indigenous territories have been titled on the Coast, which represents 31.16% of the national territory and approximately 54% of the autonomous regions (Larson and Soto 2012; Procuraduría General de la República 2016).

Ownership and control over aquatic rights – sea and coastal tenure rights – are one of the most ambiguous aspects of the Indigenous and Afro-descendant land titling process in Nicaragua. Although title deeds include both maritime as well as terrestrial areas, the overall process has been biased toward surveying terrestrial areas while defining and delimiting spaces on maritime areas have not received the same degree of attention by national or regional authorities (J. Lewis, personal communication, 18 August 2016). One of the implicit reasons for this is that disputes about overlapping tenure rights in coastal or marine areas are relatively less pressing compared to the ones experienced in most inland areas where the illegal occupation of lands by a new wave of colonist peasants have produced violent conflicts in Indigenous territories (Finley-Brooks 2011).

In fact, access to fishing grounds in recently established maritime, coastal, and inland areas within Indigenous territories has remained for the most part under an open access system for non-holder of tenure rights (both Indigenous and non-Indigenous peoples who live within or in neighboring areas). Thus, *de facto* open access regimes in a context of emerging and overlapping claims to land and aquatic resources by multiple users have, in turn, resulted in frequent conflicts, particularly during the last decade. This situation is deeply felt in the Rama-Kriol territory and has been manifested through new colonist occupations (including into a protected area, the Indio-Maiz Reserve), encroachment by illegal fishing, and state-sponsored agribusiness and infrastructure initiatives in various communal areas that are legally owned by the GTRK.

## *The Canal Project*

The Interoceanic Canal project – although still in the planning stage – has a long history in the nationalist narratives of the Nicaraguan governing elites (López 2014). In the early twentieth century, the Canal project – which was considered in Nicaragua but later built in Panama – was a critical part of US-Nicaragua relations, involving tense negotiations and the meddling of the US in Nicaraguan politics. Similarly, multiple plans for building transoceanic infrastructure have been part of the history of the Rama and Kriol communities, given that initiatives to build infrastructure were often authorized by the national governments without consultation of the inhabitants of Coastal territories (González et al. 2006). Some of these projects were started but never completed, including a railway, a pipeline, as well as others.

Once Nicaragua approved progressive legislation protecting Indigenous rights to land and aquatic resources in mid 1980s, communities on the Coast could invoke constitutional rights to preserve the integrity of the ancestral territories and tenure rights to land and marine areas. However, the Nicaraguan courts, for the most part, have not been supportive of upholding Indigenous rights, resulting in two groundbreaking cases brought by Indigenous peoples of Nicaragua to the Inter-American human rights system over the last two decades. In these two cases – *Awastingni Vs. Nicaragua* (2001) which related to Indigenous collective rights to land, and *YATAMA Vs. Nicaragua* (2005) that centered around political participation of Indigenous peoples – the rulings favoured Indigenous claimants and demanded that the state of Nicaragua institute protections and reparations of violated Indigenous rights to land tenure systems and political participation (Anaya and Grossman 2002).

Interestingly, the Interoceanic Canal project was proposed by a national administration that – until 2013 – had been politically committed to supporting Indigenous land titling and demarcation on the Coast (González 2016). From the perspective of dissenting members of the Rama-Kriol Territorial Government (GTRK) the official endorsement of the Grand Canal Project means "granting the land to the weakest of the Indigenous peoples of the Coast, just to disguise the official dispossession of the same recognized land" (T. Bill, personal communication, 5 July 2016).

There are legitimate reasons why the Rama and Kriol are hesitant about the ultimate goal the FSLN administration is pursuing with its endorsement of the Canal project. Article 12 of the Canal legislation (Law 840), states that it "is in the public interest of the Republic of Nicaragua to expropriate any property or right that is reasonably necessary to complete all or part of the project, therein the 'required property' whether private property, communal lands of the autonomous regions, or those belonging to the Indigenous communities, or property in control of government entities" (Asamblea Nacional de Nicaragua 2013).[2] This section of Law 840 was perceived as especially troubling for the integrity of Indigenous territories, as existing norms – in particular Laws 28 on Autonomy, and Law 445 on Communal Land Rights – state that communal lands cannot be sold, they are imprescriptible

[2] Translation by the author.

and cannot be used as collateral. Moreover, article 89 of the Nicaraguan Political Constitution mandates that "the state recognizes the communal forms of property of the land that belong to the communities of the Caribbean Coast" (Corte Suprema de Justicia 2016).

Intuitively, members of the GTRK have interpreted article 12 of Law 840 as a threat to their tenure rights to land and inland fisheries. They responded to Law 840 by demanding that the Nicaraguan Supreme Court repeal the legislation, a petition that was denied on the basis of the government not knowing whether the final route of the Canal would affect Indigenous lands. The Court ruling required the government to "initiate the consultations with the aboriginal peoples of the Caribbean Coast until the final route has been defined; this is, until is known which authorities must be consulted" (Corte Suprema de Justicia 2013).

## *Stakeholder Perspectives on Tenure Rights*

The conditions above – uncertainties about the capacity of legal mechanisms to protect Indigenous tenure rights to fisheries, and safeguards against state-sponsored development initiatives that threaten Indigenous peoples' access to fisheries livelihoods – speak directly to the kind of critical issues the SSF Guidelines seek to address. In fact, the Guidelines provides a policy-relevant framework with regard to the governance of tenure rights of small-scale fisheries as it is linked to states' responsibilities in providing the enabling environment under which Indigenous rights can be protected. Of particular importance is Article 5.9, which states: "States should ensure that small-scale fishing communities are not arbitrarily evicted and that their legitimate tenure rights are not otherwise extinguished or infringed." Moreover, Article 5.4 emphasizes that:

> [states], in accordance with their legislation, and all other parties should recognize, respect and protect all forms of legitimate tenure rights, taking into account, where appropriate, customary rights to aquatic resources and land and small-scale fishing areas enjoyed by small-scale fishing communities. When necessary, in order to protect various forms of legitimate tenure rights, legislation to this effect should be provided. States should take appropriate measures to identify, record and respect legitimate tenure right holders and their rights. (FAO 2015, 5)

In an interview about the seeming contradiction between SSF Guidelines and domestic and international legally binding norms, the executive director of the South Caribbean Autonomous Government, which is the main governing authority in the region, claimed that:

> The Canal project should be seen as the only possible protection that might be available to the Ramas and Kriols in order to protect their lands and fisheries. Colonist invasion to communal lands is so prominent that only an infrastructure of this scale can deter more colonists from entering into Indigenous lands. A beneficial effect of the Canal is to ensure the survival of the Rama people. (Carlos Moni, personal communication, Bluefields, July 10)

This same view resonated at the national ministerial level on issues related to small-scale fisheries. Currently, the government is gathering data to launch a national strategy to promote the small-scale fisheries sector. Some of the priorities include: diversifying fishery activities through the promotion of a value change approach; strengthening fishers' associations and collective agency; contributing to innovation in fishing techniques; processing of marine products; and encouraging domestic consumption (ProNicaragua 2016, 7). In an interview about this plan and its relationship to the SSF Guidelines, a senior officer within Ministry of Fisheries (INPESCA) pondered that:

> ...the FAO SSF Guidelines should not be regarded as being more important than national development plans with regard to promotion of small-scale fisheries. This point of view, which is sometimes promoted within international organizations including FAO, does not allow for a better appropriation of the FAO's principles into national priorities – as defined by the current FSLN administration. The FAO Guidelines are not an overarching supranational scheme to do things in relation to fisheries. If anything, it should complement national efforts. (C. Trico, personal communication, 15 July 2016)

This same government official suggested that the governance of tenure rights – as relevant to the concerned Indigenous communities of the Rama-Kriol territory with regard to the Canal project – fell outside of the jurisdiction of the Ministry of Fisheries. Contradictorily, the Ministry of Fisheries is very much involved in granting fishing permissions to the industrial fleet that fish on the marine areas that rightfully belong to the Rama-Kriol Territory (M. Queens, personal communication, 16 July 2016). The GTRK is not usually informed about the amount of licences being granted every year, but it does receive monthly cash transfers from the Ministry of Finance originating from annual licences as per provisions included in Law 445. The fact that this practice exists does not mean that the GTRK is content with its continuation. A GTRK member – representing one the Coastal communities comprising the territory–explained that:

> We would like to see more transparency and inclusion in the way decisions are made at the level of the Ministry of Fisheries. We don't know how many ships are authorized to fish in our waters. We get to know because we see them [fishing boats] coming close to the shoreline, sometimes fishing all-night long. These are our territorial waters, the government should respect that. (G. Enriquez, personal communication, 4 July 2016)

Thus, both overlapping spatial and environmental conflicts exist within the tenure rights being granted to the Rama and Kriol Territory. These conflicts highlight critical gaps in regulatory mechanisms involving competing governing jurisdictions that claim various degrees of authority with regard to the management of natural resources of the territory, particularly over Indigenous fisheries. These competing actors include the Indigenous territorial government, the regional autonomous council, and the national government, through is various ministerial agencies. The interests of the investors have now being added to these complex relationships of contending forces and interests.

## *Summary of Governance Issues*

In synthesis, the Canal Project has been envisioned by official authorities as a matter of public national interest and, therefore, its realization seems to stand over and above any precedent constitutional rights recognized to Indigenous and Afro-descendant communities, particularly with regard to tenure rights to lands and marine areas. Therefore, legally binding domestic and international norms – although formally invoked – as a matter of fact are considered as secondary principles to be observed in light of a project of 'national interests.' With regard to the SSF Guidelines, the perception amongst some government officials is that they constitute a referential normative framework that could support national priorities, but they are not to determine the country's primary goals with respect to promoting small-scale fisheries. These views should not be taken as representing the official position of the Nicaraguan government as a whole with regard to the SSF Guidelines. On the contrary, it may well be the case – as some of my interviewees cautiously articulated – that difference in opinion exists within the various branches of the Nicaraguan government (regional and national) with regard to the prospects of the Canal project and its overall impact on Indigenous collective rights. For instance, when questioned about the possibility for dispossessing Indigenous peoples from their lands contained in chapter 12 of Law 840, Carlos Moni, the Executive Director of the Regional Government, said:

> I do not know how that chapter was passed in the National Assembly. It might have been an oversight. But what I can tell you is that confiscation of Indigenous property wouldn't work. We went through confiscation in the past [during the first FSLN administration in the 1980s]. It will not happen again. Our regional government will be the first one to step in and oppose the possibility that Indigenous people are confiscated. (C. Moni, personal communication, 10 July 2016).

Another prominent view is that the SSF Guidelines do not recognize the institutional diversity under which overlapping jurisdictions interact with regard to the governance of tenure rights in terrestrial and marine areas. The consequence of this is that governance of tenure rights – which is a fundamental piece for the governance of the entire system due to its stabilizing effects – are dealt with at different governing scales (local, regional, and national), competing jurisdictions (the justice system, the national government), and the influence of international processes and dynamics (such as the HKND group).

A much richer understandings of the complexities entailing the interaction of multiple and often conflicting governing logics (and decision-making processes) under which small-scale fisheries operate require a scalar, interactive governance approach. Interactive governance theory has suggested the need for a comprehensive understanding of small-scale fisheries to better design policies which able to respond to the interaction of complex social systems (Kooiman et al. 2005). The Nicaraguan Canal project and its potential to undermine Indigenous tenure rights to fisheries makes this question of complexity and the interactions of multi-scalar networks of governance particularly evident, given that decision-making processes related to the project are not confined within domestic actors, norms, or national institutions. Key global actors – such as international investors – are also playing a role, which may override legal and policy instruments that protect the rights of Indigenous peoples. The same might be said about the Inter-American Human Rights System, which has been instrumental in upholding Indigenous rights in light of reluctant governments that have sought to undermine these rights in the name of national development. Therefore, this chapter makes the case that in Nicaragua securing tenure rights has not been and end in itself but the initiation of a more complicated process in which the applicability of domestic and international norms – both binding and non-biding – aimed at protecting the rights of Indigenous peoples are being tested.

## Malicious Consent

Early in May 2016, the government of Nicaragua announced it had secured 'prior, free and informed consent' from the GTRK for the "due implementation of the [Interoceanic Canal] Project in approximately 263 square kilometers of its territory, which includes both terrestrial and marine areas" (Asamblea Nacional de Nicaragua 2016). Acknowledging the inalienable status of communal lands, the "Covenant of prior, free and informed consent for the Implementation of the Development Project of the Grand Interoceanic Canal' subscribed by both parties state that they will sign an '*indefinite contract lease*, which will determine the annual royalties, payable in American dollars due at the time when the lease is signed by the parties" (Asamblea Nacional de Nicaragua 2016).[3] The 'Covenant' was signed into law and published by the official newsletter of the National Assembly (Asamblea Nacional de Nicaragua 2016).

Hector Thomas, the president of the GTRK, was quoted by the official media as saying:

> On behalf of the nine communities that make up the Rama and Kriol territory, we give our vote of confidence to the National Commission of the Grand Interoceanic Canal of Nicaragua because we are sure that the construction of this project will contribute to the human development of all nations. For this reason, we will continue to move forward with faith in God and the hope that our peoples will eventually be able to achieve their longed desire for development and out of poverty. (El 19 Digital, May 3 2016)

Almost immediately, dissenting members of the GTRK who were also elected authorities to the territorial government, demanded in national courts the Covenant should be declared illegal, claiming that the government had crafted it out of pressure of participating community representatives to the GTRK. In fact, the claimants argued, Communal Assemblies – the ultimate decision-making bodies within each of the nine communities that comprise the Rama-Kriol Territory – were not adequately and timely informed about the consent the government claimed it had

[3] Emphasis added by author.

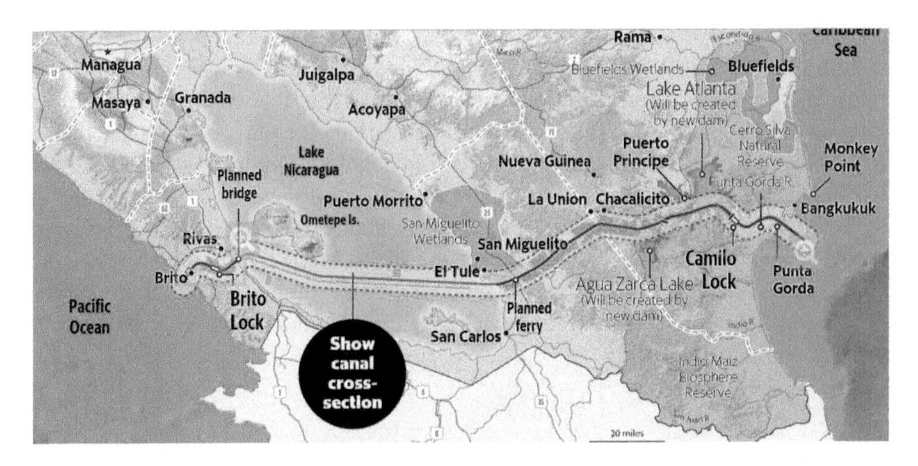


**Fig. 10.3** The Proposed Grand Canal Route (Source: Johnson 2015)

secured. The dissenting members of the GTRK pointed out the Government did not follow the *Lineamientos* (Guidelines) the GTRK had especially designed in December 2014 for the consultation of the Rama and Kriol communities in relation to the Interoceanic Canal (Gobierno Territorial Rama y Kriol, GRT-K 2014) (Fig. 10.3).

The purpose of the interviews conducted in this study was to find out whether community members were informed about the possible consequences and impact of the Canal Project over their tenure rights, particularly rights to inland and marine fisheries. Almost unanimously important gaps of information were noticeable in two main areas: the possible impact of the project over communal lands in general, and on the prospective effects over marine and inland fisheries. Particularly telling was the intervention of an elderly woman from the community of Suumu Kaat who posed the following question to the community assembly:

> I don't understand how our land will be leased for eternity to the Canal. I have asked the government if our land can be returned at some point, later in time, for instance if the project is never completed. But they did not respond. (M. Clair, personal communication, 7 July 2016)[4]

Similarly, Carlos Brigg, a community leader to the GTRK representing Greytown who decided not to sign the government-sponsored Covenant, posed the same concern:

> We've been pressured to sign the consent for the whole project, not only for the Canal. This means, I believe, if the Canal is never built, we will still be dispossessed from our land. This is a point of no return. (C. Brigg, personal communication, 18 July 2016)

The same concern was expressed by another representative from Rama Cay, a Rama community, with regard to marine and inland fisheries:

[4] Marina Clair is a member of the Sumu Kaat community assembly.

> We do seasonal fisheries all the way down from Rama Cay to Punta de Aguila [also known as Bang Kukuk). With the Canal crossing over our land and going out to the sea, we don't know if we will be able to continue this traditional practice which we've been doing for years and years. (L. Carlos, personal communication, 4 July 2016)

For the vast majority of people interviewed and those who participated in communal assemblies the notion of 'indefinite' lease was completely unknown as there is no precedent for the GTRK to lease communal lands beyond a 10-year term, as specified in the national Civil Code (Asamblea Nacional de Nicaragua 1904, *Código Civil*, article 2820). It should also be noticed that the initial draft of the Covenant between the Government and the GTRK proposed the notion of 'perpetual lease', but the Rama-Kriol found that to lease something in 'perpetuity' was a subject matter which "only God could decide upon", not a decision that could be taken by human beings. The final document introduced the term 'indefinite', which for disagreeing members of the Rama and Kriol communities was synonymous with legalizing the dispossession of their lands.

More importantly, I discovered, was the ambiguous and contradictory information available to community members of the Rama-Kriol Territory with regard to the proposed route of the Canal in the marine area included in the 263 square kilometers granted under the indefinite lease. The proponents of the Canal have not clearly identified the area, while community members remained uninformed about potential changes in fishing practices and access to their traditional fishing grounds. It is unlikely that the proposed Canal route will not conflict with customary fishing grounds of the Rama and Kriol coastal communities. The Environmental and Social Impact Assessment (ESIA) identifies substantial consequences resulting from the construction of the access channel on both coasts (the Pacific and the Caribbean). More concretely, the ESIA foresees that a 14.1 km channel in length on the Caribbean, as well as the construction of port infrastructure, would "remove or disturb nearshore marine habitats and beaches that are used by five species of globally vulnerable, endangered, or critically endangered marine turtles" (HKND and ERM 2015, 47). With regard to inland (riparian) ecosystems, the ESIA indicates that "roughly, 1650 km of riverine habitat would be lost by Project construction and flows in many portions of the remaining channels would be severely reduced or eliminated entirely especially in the remnant portions of the main river channels" (HKND and ERM 2015, 48). The highly technical information contained the ESIA is not fully known or accessible to the Rama and Kriol communities. Accessibility to these documents has been an issue, considering the fact that the ESIA executive summary alone is 123 pages, while the full report including appendices totals 7000 pages. When the government held community meetings in order to seek consent, these documents were not studied in full detail (M. Queens, personal communication, 16 July 2016).

Therefore, there are legitimate concerns arising from various communities comprising the Rama-Kriol Territory: timely information was not delivered by the proponents of the Canal; specific data about the prospective route of the Canal over marine areas was not openly shared; and pressure to sign off the Covenant was exerted by government representatives under the promise that signing the 'Consent' would mean a rainfall of American dollars for the Rama people (Centro de Asistencia Legal a los Pueblos Indígena de la Costa Atlántica 2016). Surprisingly for many community members was the fact that according to Provision 9.2 of the agreement, the full presentation of the Socio-Economic Impact Assessment of the Canal project (completed in 2015) would only be delivered to communities once the Covenant had been signed. For one community member, this 'after-the-fact manoeuvre' was manufactured by the government for seeking consent, and threw the legitimacy of the consent itself into question:

> If we have not been given the chance to get to know in detail how this project will affect our ways of life and culture, our economic activities including fisheries, what have been given consent to? We can only give true consent until we consent to the project, and after all the information has been brought and discuss with us. (M. Lister, personal communication, 5 July 2016)

## Conflicting Demands: Land and Aquatic Tenure System

The implementation of the SSF Guidelines in the context of Indigenous fisheries on the Nicaraguan Eastern Coast is complicated by the conflicting demands that exist between users and legitimated holders of land and aquatic rights, and competing actors – both individuals and large infrastructure corporations that seek to ascertain control over Indigenous lands. On the other hand, present perceptions of Nicaraguan officials, both at the local and at national levels, are often informed by the mainstream development narrative that superimposes the 'public interest' over legitimate demands appealing for free, prior, and informed consent. This was evident as information gathered among research participants about the potential impact of the Canal project on marine and inland fisheries was scattered, incomplete, and often contradictory.

On the other hand, the Covenant signed by the Nicaraguan government and members of the GTRK for the 'implementation' of the Interoceanic Canal is especially problematic due to its lack of critical provisions to guarantee protections for Indigenous livelihoods, particularly with regard to securing tenure rights and on the possibility of reverting possible adverse effects over the sustainability of resources. This failing is highlighted by Article 5.11 of the SSF Guidelines, which states that:

> States should provide small-scale fishing communities and individuals, including vulnerable and marginalized people, access through impartial and competent judicial and administrative bodies to timely, affordable and effective means of resolving disputes over tenure rights in accordance with national legislation, including alternative means of resolving such disputes, and should provide effective remedies, which may include an entitlement to appeal, as appropriate. (FAO 2015, 6)

In fact, the Covenant has prevented the possibility of modifying the terms of the agreement for 20 years, thus leaving the GTRK with limited options to appeal a massive infrastructure project. Such a project, for example the Interoceanic Canal project, if built, would likely have a detrimental impact on the survival of Indigenous peoples in the region.

In June 2016, lacking trust in the Nicaraguan justice system, legal representatives of the GTRK brought their case to the Inter-American Human Rights Commission, which is now following proceedings with the Nicaraguan government. It is worth mentioning that the dissenting members of the GTRK have not publicly opposed the Interoceanic Canal project, but instead have demanded that the consultation be completely legitimately under internationally recognized standards (including the 169 ILO Convention) and according to their especially-designed Guidelines for informed consent (Gobierno Territorial Rama y Kriol, GRT-K 2014). This document, as stated earlier, was overlooked during the state-sanctioned consultation process for securing consent, which instead opted to gain approval from selective and supportive representatives of the Rama and Kriol communities.

## Conclusions

The SSF Guidelines represent an important milestone in the global effort towards recognizing the contribution of small-scale fisheries and its long-term sustainability, particularly in developing countries. Of critical importance is the recognition that the Guidelines make with regard to the tenure rights of Indigenous peoples to terrestrial and marine areas, especially in the context of the implementation of large-scale development projects. Article 5.10 of the Guidelines stipulate that "States and other parties should, prior to the implementation of large-scale development projects that might impact small-scale fishing communities, consider the social, economic and environmental impacts through impact studies, and hold effective and meaningful consultations with these communities, in accordance with national legislation" (p. 6).

The discussion I have presented in this chapter illustrate some of the challenges for the implementation of the SSF Guidelines with respect to Indigenous and Afro-descendants' tenure rights to land and marine areas in the context in which a large-scale infrastructure project has been endorsed by Nicaraguan national authorities. From their initial enthusiastic support of advancing land titling and Indigenous rights, the Nicaraguan government has transitioned to actively crafting a questionable consent from the GTRK for securing support for the Interoceanic Canal project. This consent involves the 'indefinite lease' of a significant territorial and marine swath that would eventually be used for the Canal and its associated projects. In manufacturing this dubious consultative process, the Nicaraguan state has shown little consideration to observing domestic and international standards pertaining to the rights of Indigenous peoples with regard to the free, prior, and informed consent.

In fact, Indigenous tenure rights over marine and inland fishery resources, which have long been ancestral practices of the Rama and Kriol communities, are now being threatened by the proposed Canal route through a disguised form of a state-sponsored, land privatization scheme (Mansfield 2004). Particularly troubling is the fact that limited information exists among community members on the possible changes to fishing practices and access to traditional fishing grounds as a consequence of building the Interoceanic Canal over Indigenous lands. On the other hand, regional and national authorities see 'public interest' development priorities as preeminent to minority rights, thus this narrative has overshadowed the legitimate concerns expressed by the Rama and Kriol communities. I suggest that these views are not unanimously held across the all branches of government, which in fact reflect different perspectives and interests of the various actors at play relative to the Canal project, its perceived potential benefits compared to its negative consequences.

Moreover, the scenario is very much complex due to the dilemmas and contradictory relations involved in the plans to build the Canal vis-à-vis the protection of Indigenous rights and the overall governance of small-scale fisheries. These dilemmas emerged out of the conflicting relations and interplay that exist between the following actors and processes: firstly, the interests of the global investors, the HKND Group; secondly, the constitutional obligation of the Nicaraguan state to safeguard the rights of Indigenous peoples, particularly the ownership to collective property to land and maritime areas; thirdly, the objectives of the current FSLN administration that has championed the Canal as a matter of strategic 'public' interest; fourthly, the views and interest of regional authorities, as elected, representative bodies of the autonomous regions; and fifthly, the concerns and interests of the GTRK at the legitimate body representing the Rama and Kriols communities, whose members are also deeply divided regarding the consent given to the Canal project.

The SSF Guidelines should be evaluated against the above-mentioned dynamics to determine what their practical effect could be in implementation. If the SSF Guidelines are to have a substantive impact – and any other progressive legislation on Indigenous rights at the domestic or international level, for that matter – then a clearer and more explicit political commitment to implementation must be demanded from member states. One way of moving forward is for policy makers to better integrate the principles of the Guidelines as guiding mechanisms for the promotion of small-scale fisheries, particularly with regard to decisions that may have an irreversible impact on Indigenous human rights. From the perspective of Indigenous peoples, the GTRK – which has been torn apart due to the divisive political dynamics involved in the government's efforts to seek consent – the SSF Guidelines as a stand-alone instrument seem to hold little practical value in reversing the imbalance of power that threaten the livelihoods of these groups. The relative importance of the SSF Guidelines arise from their normative embeddeness in other norms – norms which are supranational and legally binding – related to the rights of Indigenous peoples. These are also norms against which states and global investors can be held accountable.

With respect to tenure rights to terrestrial and marine areas, both the Tenure Guidelines and the SSF Guidelines provide substantial policy-pertinent guidance to states in their effort to produce better outcomes with regard to sustainable Indigenous fisheries. The Rama-Kriol experience in securing the integrity of their ancestral

**Fig. 10.4** Rama Fishers in the Bluefields Lagoon (Source: Miguel Gonzalez)

lands, even after being granted full ownership, has been further complicated by the state's vested interest in dispossessing collective ownership for building the Interoceanic Canal through dubious legal means, therefore potentially privatizing a significant portion of Indigenous lands and aquatic resources. It is then evident that the only possible deterrent for states and other external and powerful actors from encroaching into Indigenous tenure rights relies upon a combination of legal actions, political mobilization, and social support at multiple scales.

In light of this reality, the SSF Guidelines still constitute an important moral support for the causes of Indigenous peoples in securing sustainable fisheries. Most importantly, the Nicaraguan case shows that these efforts must be done through the strategic articulation with domestic and international legally binding norms and standards for the protection of the rights of Indigenous peoples (Fig. 10.4).

## References

Acosta, M. L. (2007). La política del estado de Nicaragua sobre las tierras indígenas de las regiones autónomas de la costa atlántica. *Wani, 33*, 35–48.

Acosta, M. L., Moreno, E., & Weil, D. (2002). *Condiciones laborales de los buzos miskitos en la costa atlántica de Nicaragua*. San José: Oficina Internacional del Trabajo.

Anaya, S. J., & Grossman, C. (2002). The case of Awas Tingni v Nicaragua: A new step in the international law of Indigenous peoples. *Journal of International and Comparative Law, 19*(1), 1–15.

Asamblea Nacional de Nicaragua. (1904). *Código civil de la república de Nicaragua, Managua, Nicaragua*. https://www.oas.org/dil/esp/Codigo_Civil_Nicaragua.pdf. Accessed 1 Dec 2016.
Asamblea Nacional de Nicaragua. (1987). *Estatuto de autonomía de las regiones de la costa atlántica de Nicaragua* (La Gaceta No. 238). Managua: Asamblea Nacional de Nicaragua.
Asamblea Nacional de Nicaragua. (2013). *Ley no. 840. Ley especial para el desarrollo de infraestructura y transporte Nicaraguense atingente al canal, zonas de libre comercio e infraestructuras asociadas* (La Gaceta, CXVII, June 14 2013). Managua: Asamblea Nacional de Nicaragua.
Asamblea Nacional de Nicaragua. (2016). *Convenio de Consentimiento Previo, Libre e Informado para la Implementacion del Proyecto de Desarrollo del Gran Canal Interoceanico de Nicaragua* (La Gaceta No. 85, September 5 2016). Managua: Asamblea Nacional de Nicaragua.
Béné, C. (2003). When fishery rhymes with poverty: A first step beyond the old paradigm on poverty in small-scale fisheries. *World Development, 31*(6), 949–975.
Campos, V. (2013). The canal will irreversibly damage Lake Cocibolca. Envío, 384. http://www.envio.org.ni/articulo/4726. Accessed 1 Nov 2016.
Centro de Asistencia Legal a los Pueblos Indígena de la Costa Atlántica. (2016). We do not consent [Documentary Film]. https://drive.google.com/file/d/0B1FGs5LsBqBHbi0tcFUyZUN5SXc/view. Accessed 25 Nov 2016.
Christie, P. (2000). The people and natural resources of Pearl Lagoon. In P. Christie et al. (Eds.), *Taking care of what we have. Participatory natural resource management on the Caribbean Coast of Nicaragua* (pp. 17–46). Ottawa: CIDCA-IDRC.
Chuenpagdee, R. (Ed.). (2011a). *World small-scale fisheries. Contemporary visions*. Delft: Eburon.
Chuenpagdee, R. (2011b). Thinking big on small-scale fisheries. In V. Christensen & J. Maclean (Eds.), *Ecosystem approaches to fisheries: A global perspective* (pp. 226–240). Cambridge: Cambridge University Press.
CONADETI. (2013). Informe ejecutivo Conadeti y Cid. Bilwi, Nicaragua CONADETI. http://www.poderjudicial.gob.ni/pjupload/costacaribe/pdf/informe_costacaribe3006.pdf. Accessed 1 Dec 2016.
Corte Suprema de Justicia. (2013). *Sentencia número 30-2013*. Managua: Corte Suprema de Justicia.
Corte Suprema de Justicia. (2016). *La constitución política y sus reformas*. Managua: Biblioteca Jurídica.
Daw, T. M. (2008). Spatial distribution of effort by artisanal fishers: Exploring economic factors affecting the lobster fisheries of the Corn Islands, Nicaragua. *Fisheries Research, 90*, 17–25.
Deepananda, K., Ashoka, H. M., Amarasinghe, U. S., & Jayasinghe-Mudalige, U. K. (2015). Indigenous knowledge in the beach seine fisheries in Sri Lanka: An indispensable factor in community-based fisheries management. *Marine Policy, 57*(Complete), 69–77. doi:10.1016/j.marpol.2015.03.028.
El 19 Digital. (2016, May 3). Firman acuerdo de consentimiento para uso de tierras del Gobierno Territorial Rama y Kriol.' https://www.el19digital.com/articulos/ver/titulo:41492-firman-acuerdo-de-consentimiento-para-uso-de-tierras-del-gobierno-territorial-rama-y-kriol. Accessed 25 Nov 2016.
FAO. (2012). *Voluntary guidelines on the responsible governance of tenure of land, fisheries and forest in the context of national food security*. Rome: Food and Agriculture Organization of the United Nations.
FAO. (2015). *Voluntary guidelines for securing sustainable small-scale fisheries in the context of food security and poverty eradication*. Rome: Food and Agriculture Organization of the United Nations.
Finley-Brook, M. (2011). 'We are the owners': Autonomy and natural resources in northeastern Nicaragua. In L. Baracco (Ed.), *National integration and contested autonomy. The Caribbean coast of Nicaragua* (pp. 309–335). New York: Algora.

Frühling, P., González, M., & Buvollen, H. P. (2007). *Etnicidad y nación. El desarrollo de la autonomía de la Costa Atlántica de Nicaragua: 1987–2007*. Guatemala City: Guatemala: F and G Editores.

Gobierno Territorial Rama y Kriol, GRT-K. (2014). *Lineamientos para realizar un proceso de consulta en el territorio Rama y Kriol en relación al Proyecto Gran Canal Interoceanico y sub proyectos asociados*. Bluefields: GRT-K.

Gobierno Territorial Rama y Kriol, GRT-K. (2017). *Diagnostico del Territorio Rama-Kriol*. Bluefields: GRT-K.

Goett, J. (2016, May 20). In Nicaragua, the latest zombie megaproject. *Nacla*. http://nacla.org/news/2016/05/20/nicaragua-latest-zombie-megaproject. Accessed 10 Nov 2016.

González, M. (2011). To make a life in Marshall Point: Community empowerment and small-scale fishery. In S. Jentoft & A. Eide (Eds.), *Poverty mosaics. A better future in small-scale fisheries* (pp. 275–308). New York: Springer.

González, M. (2016). The unmaking of self-determination: Twenty-five years of regional autonomy in Nicaragua. *Bulletin of Latin America Research (BLAR), 33*(3), 306–321.

González, C., & Jentoft, S. (2010). MPA in labor: Securing the Pearl Cays of Nicaragua. *Environmental Management, 47*(4), 617–629.

González, M., Koskinen, A., Jentoft, S., & Lopez, D. (2006). *The rama people: Struggling for land and culture*. Managua: URACCAN and University of Tromso-Norway.

Hale, C. R. (1994). *Resistance and contradiction. Mískitu Indians and the Nicaraguan state, 1894-1987*. Stanford: Stanford University Press.

HKND Group & ERM. (2015). *Canal de Nicaragua: Environmental and social impact assessment*. Managua: ESIA, HKND Group. http://hknd-group.com/portal.php?mod=viewandaid=243. Accessed 27 Nov 2016.

Hostetler, M. (2005). *Enhancing local livelihood options: Capacity development and participatory project monitoring in Caribbean Nicaragua*. Doctoral dissertation, York University, Toronto.

Huete-Pérez, J., Alvarez, P. J., Schnoor, J. L., Rittmann, B. E., Clayton, A., Acosta, M. L., Bicudo, C. E., Arroyo, M. T., Brett, M. T., Campos, V. M., Chaimovich, H., Jimenez-Cisneros, B., Covich, A., Lacerda, L. D., Maes, J. M., Miranda, J. C., Montenegro-Guillén, S., Ortega-Hegg, M., Urquhart, G. R., Vammen, K., & Zambrano, L. (2015). Scientists raise alarms about fast tracking of transoceanic canal through Nicaragua. *Environmental Science & Technology, 49*, 3989–3996.

Huete-Pérez, J., Meyer, A., & Alvarez, P. J. (2016). Rethink the Nicaraguan Canal. *Science, 347*(6220), 354–355.

INPESCA. (2015). *Anuario Pesquero y Acuicola 2015*. Managua: INPESCA.

Jentoft, S., & Chuenpagdee, R. (2009). Fisheries and coastal governance as a wicked problem. *Marine Policy, 33*, 553–560.

Jentoft, S., & Eide, A. (Eds.). (2011). *Poverty mosaics: Realities and prospects in small-scale fisheries*. New York: Springer.

Johnson, T. (2015). Remaking world trade. McClatchy DC. http://media.mcclatchydc.com/static/features/NicaCanal/. Accessed 6 Dec 2016.

Kooiman, J., Bavinck, M., Jentoft, S., & Pullin, R. S. V. (Eds.). (2005). *Fish for life: Interactive governance for fisheries*. Amsterdam: Amsterdam University Press.

Lagueux, C. J., & Campbell, C. L. (2005). Marine turtle nesting and conservation needs on the south-east coast of Nicaragua. *Oryx, 39*(4), 398–405.

Larson, A. M., & Soto, F. (2012). *Territorialidad y gobernanza: Tejiendo retos en los territorios Indígenas de la RAAN*. Managua: Nitlapan-UCA.

López, M. (2014). Truths about the canal concession all Nicaraguans should know. *Envío, 390*. http://www.envio.org.ni/articulo/4805. Accessed 1 Nov 2016.

Mansfield, B. (2004). Neoliberalism in the oceans: 'Rationalization', property rights, and the commons question. *Geoforum, 35*(3), 313–326.

Mendoza, L. J. (2016). *Gobernanza de la propiedad comunal: Institucionalidad para la administración de la propiedad de la tierra y los recursos naturales*. Paper presented at the Seminar Encuentro de Saberes, Palma Real, Peru, August 14 2016.

Meyer, A., & Huete-Pérez, J. (2014). Nicaragua could wreak environmental ruin. *Nature, 505*, 288–289.

Neiland, A., Madakan, S., & Béné, C. (2005). Traditional management systems, poverty and change in the arid zone fisheries of northern Nigeria. *Journal of Agrarian Change, 5*(1), 117–148.

Nietschmann, B. (1973). *Between land and water: The subsistence ecology of the Miskito Indians, eastern Nicaragua*. New York: Seminar Press.

PNUD. (2005). *Informe de Desarrollo Humano 2005: Las regiones autónomas de la costa Caribe*. Nicaragua asume su diversidad. Managua: PNUD.

Procuraduría General de la República. (2016). *Títulos entregados por el GRUN*. http://www.pgr.gob.ni/index.php/entregados-por-el-grun. Accessed 1 Nov 2016.

Pro-Nicaragua. (2016). *Políticas y proyectos de desarrollo para potenciar la inversión 2017–2021*. www.PRONicaragua.gob.ni. Accessed 1 Nov 2016.

Roe, K. H. (2006). Indigenous ecological knowledge and marine resource management: Perceptions within a Rama community. In M. González, A. Koskinen, S. Jentoft, & D. Lopez (Eds.), *The Rama: Struggling for land and culture* (pp. 107–123). Managua: URACCAN and University of Tromso-Norway.

Salas, S., Chuenpagdee, R., Charles, A., & Seijo, J. C., (Eds.) (2011). *Coastal fisheries of Latin America and the Caribbean.* (FAO Fisheries and Aquaculture Technical Paper 544). Rome: Food and Agriculture Organization of the United Nations.

Tress, B. Tress, G., & Fry, G. (2005). *Defining concepts and the process of knowledge production in integrative research.* Published online at Learning for Sustainability (LfS), http://learningforsustainability.net/research/interdisciplinary.php#integration. Accessed 1 Aug 2016.

United Nations Development Program. (2011). *Experiences of titling of Indigenous territories. Main breakthroughs*. UNDP-Caribbean Coast Program Preparatory Meeting of the Conclave of Latin America and the Caribbean, Managua, April 4–6 2011.

Wade, L. (2016). A nation divided. *Science, 351*(6270), 220–223.

Williamson, D. (2016). Contribución de las regiones autónomas al Producto Interno Bruto de Nicaragua. *Revista Electrónica de Investigación en Ciencias Económicas, 4*(7), 1–11.

Williamson, D., & Fonseca, G. (2007). *Compendio estadístico de las regions autónomas de la costa Caribe de Nicaragua*. Managua: CIDCA-UCA.

# Chapter 11
# Are the Small-Scale Fisheries Guidelines Sufficient to Halt the Fisheries Decline in Malta?

**Alicia Said**

**Abstract** The fishing sector in Malta has always been one of a small-scale nature with a long history of fishers engaging in traditional small-scale fishing practices. However, this image has undergone a radical shift in the past decade since Malta's accession to the EU in 2004. With the industrialization of the Bluefin tuna fishery and the increase in the number of industrial trawlers, small-scale fishers are facing multi-faceted deprivation to a point where exiting is the only option, a reality evident by the declining number of small-scale fishers engaged in the sector. This case study demonstrates that the problems small-scale fishers are facing are the result of ineffective governance systems which do not cater to the needs of the small-scale fisheries sector and thus the establishment of imminent protective strategies for small-scale fishers are needed. I argue that the Voluntary Guidelines for Securing Sustainable Small-Scale Fisheries (SSF Guidelines) should be the starting point for the regeneration of the small-scale fisheries sector in Malta. In line with the scope of the SSF Guidelines, fishers can benefit from enriched stability through the provision of tenure rights and the formation of fisheries local action groups (FLAGs). This way, small-scale fishers, who represent the relics of sustainable fishing in Malta, can become empowered and proactively get involved in designing a long-term vision that restores the image of the small-scale fisheries sector in the neoliberal era.

**Keywords** Fisheries governance • SSF Guidelines • Mediterranean • FLAGs • Rationalization • Livelihood struggles

---

A. Said (✉)
Durrell Institute for Conservation and Ecology, School of Anthropology and Conservation, University of Kent, Canterbury, UK
e-mail: as946@kent.ac.uk; alicia.said87@gmail.com



S. Jentoft et al. (eds.), *The Small-Scale Fisheries Guidelines*, MARE Publication Series 14, DOI 10.1007/978-3-319-55074-9_11

## Introduction

The geographical location of Malta – an island with high accessibility to the sea – has always been a crucial element to human beings' link to fish resources. The small-scale fisheries sector, more than any other, represents this link, one that dates back hundreds of years. The legacy of fishing has been sustained for centuries with traditional know-how and fishing vessels being passed from one generation to the next. These fisheries have always been characterized by small family enterprises engaging in traditional low-impact small-scale fishing methods to produce small volumes of high-value products (Dimech et al. 2009).

Altogether small-scale fishers represent around 1% of the Maltese working population, and their catches comprise just around 0.2% of the national GDP (Dimech et al. 2009). Due to this insignificant direct contribution to the national economy, the small-scale fisheries sector is less important economically than it is socially and culturally (Dimech et al. 2009). It is important socially and culturally for a number of reasons: these include the supply of local and fresh fish to authentic fish markets and traditional restaurants; the reproduction of the social fabric that solidifies community networks and stability; and the creation of an emblematic cultural representation of fishing sought by tourists, especially in the main fishing villages of Marsaxlokk (Fig. 11.1) and Mġarr (Gozo) (Fig. 11.2) where local craftsmanship of vessel-and-gear making is still vibrant.

The sustainability of small-scale fisheries is questionable, however, since, as this chapter will show, the small-scale fisheries sector has been in constant decline for the past 10 years. The disintegration of the small-scale fisheries sector is mostly due to major changes in traditional fishing norms and values which were heavily changed-with the advent of new supranational regimes of policy and trade, and the concomitant modus operandi of the governing system at the national level. This case study borrows important concepts from the governance theory to illustrate how Maltese small-scale fisheries have become increasingly vulnerable as a result of socio-political governance processes that favoured the endurance of a small but powerful group of large-scale, predominantly industrial, fishing companies, which managed to strive and prosper at the expense of the small-scale fisheries sector (Said et al. 2016).

**Fig. 11.1** Known as Malta's largest fishing village, Marsaxlokk hosts the largest number of artisanal fishermen. The multi-coloured vessels in the picture are the Maltese traditional *luzzu* (Source: Author, 15th May, 2015)

**Fig. 11.2** An artisanal fisherman from the island of Gozo making a traditional bogue trap. This authentic local craftsmanship is slowly dying out (Source: Author, 25th April, 2015)

We show how these developments are diluting the importance of small-scale fisheries to the extent that they have become marginal.

Following an assessment of the main changes and challenges faced by small-scale fisheries, in Section 'The End of the Small-Scale Fishing Sector: A Governance Problem', I argue that the small-scale fisheries sector would benefit from a governance structure that truly represents the realities and needs of the sector through policy frameworks that promote the sustainability of small-scale fishing communities.

By drawing on the framework of the SSF Guidelines, I provide recommendations that aim at highlighting trajectories for the renaissance of the Maltese small-scale fisheries sector. The SSF Guidelines, which are the first international instrument dedicated entirely to the "often-neglected" small-scale fisheries sector (FAO 2015, v), were adopted by the Food and Agriculture Organization of the United Nations (FAO) in 2014 with the aim of providing guidance to promote secure and sustainable small-scale fisheries across the world. The guidelines provide a resourceful roadmap in situations, such as that in Malta, where the small-scale fisheries sector is sinking under the pressure of multi-faceted challenges emanating from policy shocks, processes of elite capture as well as ineffective governance structures.

Being an international tool that applies to various contexts, the SSF Guidelines provide pragmatic routes that can be enacted to support the sustainability of small-scale fisheries through good governance principles based on "equality and non-discrimination, participation and inclusion, accountability and the rule of law" (FAO 2015, xi). In this chapter, I suggest that the SSF Guidelines may in fact "spur new legislation" (Jentoft 2014, 7) that could improve the fisheries' social, economic and ecological pillars, for the benefit of current and future generations. As an example, I discuss the potential advancements that small-scale fisheries can obtain through the establishment of fisheries local action groups (FLAGs), supported by an EU-funded scheme aimed at promoting endogenous community-based management. Ultimately, I argue, however, that although the SSF Guidelines provide promising guidance on the future of small-scale fisheries, their voluntary nature might be the main challenge especially in cases such as Malta, where herculean efforts are needed to overhaul the status quo that favours the large-scale sector.

As a case study, this chapter seeks to explain the main drivers that have entrapped small-scale fisheries in a cycle of vulnerability by providing a detailed description of the state of affairs at the local level as gauged through 15-months of fieldwork between May 2014 and August 2015 within the main fishing villages. The narrative, which details the pressures faced by fishers in the context of dynamic changes that are constantly occurring in the fishing landscape, seeks to provide recommendations based on the SSF Guidelines. Recommendations are given around two main research questions, namely:

1. What are the main challenges faced by the small-scale fisheries sector?
2. How can the implementation of the SSF Guidelines halt the decline of the sector?

## Methodology

A mixed-method approach was implemented and encompassed various qualitative data collection techniques including in-depth interviews ($n = 50$), participant observation including fishing trips ($n \approx 100$) and opportunistic conversations ($n \approx 150$) with various fishers, their families, and other social actors from the community.

Ethical considerations were taken into account throughout the data collection process, especially during happenstance encounters that elicited rather tacit and sensitive information.[1] The use of gatekeepers was essential for accessing different networks of fishers through purposeful snowballing. Rapport with the fishers was well-established to an extent that they themselves made direct requests to be interviewed towards the end of the first quarter of fieldwork. The primary data was triangulated with other sources including online forums, media articles, local and regional legislations, and formal national statements, including ministerial speeches. Detailed information about the fleet vessel registry and the métier-based (fishing gear) data were kindly provided by the Department of Fisheries and Aquaculture (DFA), within the Ministry for Sustainable Development, the Environment and Climate Change (MSDEC). Following the compilation of the various datasets, an open-ended comparative technique was implemented to triangulate the data so as to ensure that the findings were consistent, valid, and reliable.

## The Fishing Sector in Malta

Fishing in the Maltese islands has always formed an important socio-economic component for a number of coastal communities (Camilleri 2002), and along with agriculture, in the past it played a central role in rural household economies. With the increase of job opportunities in manufacturing and tertiary industries, dependency on fishing started decreasing. However, up until the beginning of the 2000s, Malta still had a full-fledged fleet consisting of 1850 commercial fishing vessels. These included 314 full-time fisher vessels which represented family-based entities whose livelihood was predominantly dependent on fishing; 266 market-fisher vessels, which represented part-time fishers who were obligated to register their fish catches at the main fish market; and 1270 part-time fisher vessels which represented fishers who fish on part-time basis but were not required to declare their landings at the fish market (Grupetta 2002).

This fleet classification system was phased out in 2004 and commercial fishing vessels are now divided into two main segments, namely: full-time and part-time. It is generally considered that full-time vessels encompass fishers whose main income is earned from fishing whilst part-time vessels include fishers who engage in commercial fishing for a supplementary income, however in reality, the fishing fleet system is much more complex. In terms of catches, part-time vessels produce landings that amount to approximately 10% of what full-time vessels produce by weight and value, and in terms of fishing activity, part-timers conduct fewer and shorter trips than the commercial counterpart (Dimech et al. 2009). At the time of writing, the total number of vessels in the part-time cohort was 601, whilst the full-time counterpart hosts 401 vessels. The vessels differ by type, and some of the most

[1] The data collection process was aligned to the code of ethics of the American Anthropological Association (AAA).

**Table 11.1** This data, which was supplied by the DFA in 2015, shows that the commercial fleet is majorly composed of vessels smaller than 12 m, and thus the Maltese fleet is predominantly of a small-scale nature

| The Maltese commercial fleet | | | | | | |
|---|---|---|---|---|---|---|
| Size of vessel (meters) | <6 | 6–<12 | 12–<18 | 18–<24 | 24–<40 | Total |
| Total Number of Vessels | 504 | 431 | 24 | 33 | 10 | 1002 |
| Percentage of fleet | 50.3% | 43% | 2.4% | 3.3% | 1% | 100% |

Source: Data supplied by DFA 2014

traditional ones have their origins in the early nineteenth century when the most traditional boats were manually operated by wooden oars. Some traditional names of these vessels are *'kajjiek'*, *'luzzu'*, *'bimbu'*, *'firilla'*, and *'frejgatina'* (NSO 2011).[2] All the vessels today are operated by motorized propellers, and the traditional wooden vessels are gradually giving way to fiber-based vessels.

Over time and through technological advancement, the small-scale fleet has become better equipped with navigation systems and fish finding equipment that facilitate the search for good fishing grounds, as well as with motorized winches that make fishing relatively less strenuous. Despite becoming more efficient in terms of fishing activity, small-scale fisheries have retained their artisanal and low-impact fishing practices. The small-scale fisheries sector has, however, as a segment started co-existing and competing for resources with a number of small trawlers since the 1980s when the government issued a number of trawling permits as a response to greater local demand for fish to feed the local population (Camilleri 2002). After 2004, increased access to larger trawlers as well as the introduction of purse seine licenses has seen the rise of industrial operations within the Maltese fleet. Nonetheless, in terms of number of vessels the small-scale fleet has always outnumbered its industrial counterpart. In 2015, small-scale vessels accounted for 98% of the commercial sector. Multi-purpose vessels comprise the largest number within the small-scale cohort, followed by the '*kajjiek*' and the '*luzzu*'. In terms of vessel length (LOA), 50% of the vessels are smaller than 6 m, and overall, 93% of the fleet is composed of vessels which are smaller than 12 m (Table 11.1). Hence in line with the European Union (EU) definition of small-scale fisheries, 93% of Maltese commercial vessels are small-scale in nature.

The Maltese small-scale fishing fleet is engaged in different types of fishing activities throughout the year which range from offshore to coastal fisheries. Offshore fisheries, which mostly take place within the high seas (outside the 12-mile-territorial-waters), and at times extend as far out as 120 miles from the coast, include fishing for offshore pelagic species such as Bluefin tuna (*Thunnus thynnus*), swordfish (*Xiphias gladius*), and dolphinfish/mahi-mahi (*Coryphaena hippurus*) (Gatt et al. 2015). On average, offshore trips last between two to ten days depending on the fishery, the weather, and the catches. Fisheries are seasonal: Bluefin tuna long-line fishing takes place between April and June until the allowable

[2] The difference in the vessels is mainly a structural one as their operational purpose in terms of fishing is relatively similar.

quotas are caught as stipulated in the EU Bluefin tuna recovery plan (EC 302/2009); swordfish long-lining is open all year except for the months of March, October and November in line with EU law; and dolphinfish fishing, which is practised through the use of floating aggregate devices and surrounding nets, is permitted from 15th August up until the end of December (EC 1967/2006 Art. 27).

Coastal small-scale demersal fishing is a widely-practised activity that can take place throughout the year since there are no specific seasonal regulations other than for the use of the hand-held seine targeting white bait, which according to national law is permitted only between 25th June and 15th August (GOM 2013a). Fishing with small-scale fishing gear is very versatile and dynamic since fishers might fish with a trammel net one day and deploy a bottom long-line the next day. Basically, this multi-gear system which includes the use of trammel nets, gillnets, pots, traps, hand-held seines, bottom and surface long-lines, trolling, and pole lines, is very typical both in Malta and in other Mediterranean countries (Leiva et al. 1998; Tzanatos et al. 2006; Battaglia et al. 2010). Locally, these fishing gears are majorly deployed within the 12 nautical mile zone, although fishing within the 25 nautical mile zone, especially with set bottom long-lines is very common also. Fishing grounds closer to the mainland within the 3 nautical mile circumference are also popular amongst small-scale fishers (Stelzenmüller et al. 2008).

On average, the coastal demersal trips with small-scale fishing gear within the 3, 12 or 25 nautical mile zones take between half-a-day to two days maximum, and the actual geographical location of the fishing is mostly determined by the richness of the fishing grounds. Fishers explain that demersal fisheries depend on the depth of the water and the sea bathymetry, and that there are different fishing grounds which are good for specific fish during different seasons. This type of traditional ecological knowledge about Maltese fisheries exists across small-scale fishers as they have gained it through years of experience; however, no records of such knowledge is available, mainly because no research has actually been carried out on the ethnoichthyology of Maltese small-scale fishers to date.[3] The only data that is collected about traditional fishing activities is in line with the EU Data Collection Framework. As an annual dataset it provides information on the small-scale fisher fleet in terms of usage of gear vis-a-vis the number of days at sea, the total catches made, and the value of the catches. Interestingly, this data shows that small-scale fisheries account for the dominant fishing activity in terms of days at sea (MSDEC 2013c), indicating that they encompass the largest chunk of operations of the Maltese fishing sector, mostly because there are no restrictions on them.

The use of trammel nets, gillnets, pots and traps, bottom long-lines and trolling lines is open to all commercial fishers with no restrictions on fishing effort within these fisheries. In other words, those owning a commercial fishing license (full-time and part-time) can utilize these fishing gears without specific permits. On the other hand, offshore fisheries do not host a large number of fishers since the Bluefin tuna fishery is restricted by a fishing quota and a permit system that recognizes only 63

[3] This is a field within anthropology that examines the knowledge that humans have about fish, their use and their importance in different societies.

vessels (Said et al. 2016) whilst the dolphinfish fishery is restricted to 130 vessels (EC 1967/2006 Art. 27). In fact it is not uncommon to find small-scale fishers who only work with coastal fishing gear and do not target offshore fisheries for they do not hold the necessary permits to do so. In contrast, most small-scale fishers who target offshore fisheries with vessels larger than 12 m, also own a small-scale vessel so as to be able to use coastal small-scale fisheries gear especially in the winter season (Grupetta 2002; Dimech et al. 2009).

In terms of fish marketing, the majority of the catches of the small-scale fisheries sector was, and still is, sold mainly through the central fish marketplace (*pixkerija*) and are destined to the local market through middlemen, although some fishers sell their catches to the consumer themselves. Consumers consist of hotels, restaurants and individuals who buy fish at the famous fish market in Marsaxlokk on Sundays. Some of the catches, especially of large pelagic fish, are also exported. The export of Bluefin tuna and swordfish to foreign lucrative markets has been taking place for around 30 years (Farrugia et al. 2004) and has always been considered as necessary because the small-and-fragile local market gets saturated.[4] The fluctuating fish prices in the local and foreign markets, along with other challenges that fishermen are confronted with on an everyday basis during their fishing activity, shape the nature of fishers' livelihoods.

Small-scale fishers pointed out that it is not unusual to not catch a single fish in a day, or to lose gear at sea and incur losses as opposed to make profits. When fishing in the high seas, clashes with foreign vessels have become the norm, and at times these fights have escalated to 'wars', especially during the tuna (Vella 2002) and dolphinfish season (Bilocca 2006). Fishing comes with a lot of challenges but fishers narrated that despite these challenges, they still continued to fish as they consider fishing a way of life. It seems then that fishers are resilient and persist through various cycles of changes. What is less understood, however, is how fishers have experienced changes as a result of Malta becoming part of the EU. Prior to accession to the EU Malta, as an island state, had sole jurisdiction over fisheries.

## The Major Drivers of Change

As already mentioned, Maltese fishers have endured various patterns of change throughout their fishing lives, which have strengthened their endogenous ability to respond to ecological variability. In his work on small-scale fisheries in Europe, Symes (2014) explains that fishers are more predisposed to adjust to changes which they understand and can make sense of. On the other hand, when changes occur that test their local systems of knowledge and are detached from their realities, these can potentially erode the fishers' resilience (Coulthard 2012; Hadjimichael et al. 2013).

[4] In fact, before Malta joined the EU a 'tuna fish importation restriction order' to control the amount of tuna entering the Maltese market was also in place to curb economic impacts on local fishers (Cap 425.03).

Exogenous processes that have been noted to threaten the functional endogenous systems of small-scale fishers include globalization in the form of trade networks (Frank et al. 2007), as well as supranational policy formulations which deregulate patterns of local norms and traditions (Symes et al. 2015).

In the Maltese case, various changes have happened within the fishing sector, the most notable of them since 2004. The period 2004–2014 was the first 10 years after EU accession and the period in which the fishing sector was exposed to supranational processes that now shape most fisheries management and market trajectories. These include the ratification of the Common Fisheries Policy (CFP) with regard to internal policy, fleet restructuring, fisheries management, monitoring and controls, fleet regulations, fishing opportunities; and the integration into the EU Common Market Organization which brought major changes in fish marketing standards, and the import and export policy (European Commission 2003).

I argue that these changes, some of which are listed in Table 11.2, have had different levels of negative impact on small-scale fisheries due to a transformation of traditional patterns of fishing; and in the market case, due to the overhauling of the national legislation that had previously shielded small-scale fishers from adverse economic impacts of fish imports (S.L.425.03 1965). The main impacts dealt with in detail in this chapter include EU fisheries management policies that have been framed around the concept of sustainable exploitation of fisheries through reduced fishing effort and capacity at the Member State level. As highlighted in Table 11.2, these include the restriction on national fishing capacity, the legislations pertaining to trawling that deal with circumnavigating the 25 nautical mile zone open-access demersal fisheries, and the Bluefin tuna fishery policy that has developed in line with the EU Bluefin tuna recovery plan (EC 302/2009).

These policies were aimed at reducing the overall fishing capacity and effort at the Member State level. The EU, by introducing specific capping in the Maltese fishing industry, ensured that the fisheries did not remain open-access in nature. By placing a cap on total fishing capacity and by reducing the total allowable catch of Bluefin tuna at the national level, the EU has created a situation of regulated scarcity at the Member State level. Regulated scarcity refers to a situation where a resource is scarce due to a regulation, and one in which its distribution is controlled by those in power. In Gezelius' (2002, 64) words, regulated scarcity is "a political decision which limits the citizens' access to a good". Thus, in this case, the government, following EU restrictions, became the entity, at the national-level, responsible for determining who gets what in terms of access to fishing quotas and licenses. So whereas the open-access nature of the fleet prior to EU accession enabled fishers to partake freely within the fleet, top-down restrictions imposed by the EU have overhauled this 'flexibility' and empowered the national government to distribute 'scarce' resources as deemed 'proper'. I show that this 'proper' way of distribution was mostly aimed at rationalizing the fleet to make it more competitive in the global market; however, this has triggered a series of derailing consequences in terms of the stability of the small-scale fisheries sector.

One example of this is the Bluefin tuna fishery, which was a profitable niche for small-scale fishers, but which is increasingly becoming the property of industrial

**Table 11.2** This table represents the various legislative changes in the Maltese fisheries policy since EU accession in 2004

| Fisheries legislative changes following EU accession | | | | |
|---|---|---|---|---|
| Fishing and Marketing | PRE-EU | POST-EU | Post-EU National Implementation | Implication on Small-Scale Fisheries sector |
| Vessel Register | Open-access | Closed | Fleet restructuration through capping system | Concentration of capacity in large-scale vessels |
| 25 NM Fisheries Management Zone (FMZ) | Maltese Territorial Waters since 1971 | Territoriality reduced to 12 NM and retained as FMZ | Implementation of FMZ measures on flag vessels | Increased competition by foreign vessels outside 12NMZ. |
| Demersal Fishery | Open-access for artisanal vessels and restriction on trawlers | Restriction on 12m boats within FMZ & Derogation for Trawlers | Artisanal Vessels larger than 12m relocated to outsize FMZ and trawlers operating within | Discrimination of artisanal vessels vs trawlers operating in 25NMZ. |
| Bluefin Tuna Fishery | Open-access in an ICCAT national quota | Drastic reduction of national TAC | Implementation of Individual Transferable Quotas | Concentration of quotas amongst large-scale operations mainly fish farms |
| Swordfish Fishery | No regulation | Swordfish Conservation Policy | Implementation of technical measures and minimum size | Reduced ability to catch swordfish |
| Trawling Zones | Prohibited fishing within the 3NMZ (since 1990s) | Re-opening of coastal zones for trawling | Increase in the number of trawlers and relocation of the activity | Increased fishing competition and reduced fishing grounds for artisanal segment |
| Fish Imports Regulation | Restriction on imports of fish | Single market | Ratified local law on restrictions | Increased competition from imported products |

fleets as a result of government providing the latter with enabling policies (Said et al. 2016). This has happened primarily through the introduction of a national purse seine tuna license system in 2005. This national policy overhauled a long-standing precautionary policy that had prohibited the licensing of purse seiners for catching large pelagic species and avoiding monopolization of catch at the expense of small-scale fishers fishing with Bluefin tuna long-lines (Vella 2002). Added to this, with the introduction of the EU Bluefin tuna recovery plan (EC 302/2009), which brought drastic reductions to national total allowable catch (TAC), specific clauses within the policy have enabled industrial purse seine fleets and concomitant tuna ranching industries to take over the fishery. Specifically, the introduction of individual transferable quotas (ITQs), which shifted fishing rights into a set of marketable property rights, enabled a gradual system of accumulation by dispossession which has undermined the continued sustainability of the small-scale fishing fleet engaged in the Bluefin tuna fishery (Said et al. 2016).

Another example that illustrates how supranational policies have been implemented to rationalize the fleet was the restructuration of the Maltese fleet in 2005. This restructuration, as illustrated in national statistics (NSO 2004, 2005) reduced the small-scale fleet by half and enabled the expansion in the number of larger-scale vessels. The number of trawling vessels between 2002 and 2010 increased by 43% (from 16 to 23 trawlers) (NSO 2010). This increase, in the name of rationalization, means that more trawlers are now able to operate within what has become, after EU accession, 'the 25 nautical mile fisheries management zone (FMZ)'. The main aim of the FMZ was to protect marine resources by limiting fishing effort to a minimum; however, this has worked against, rather than in favor of the artisanal fleet. Fishermen owning vessels larger-than-12-m engaging in artisanal long-lining and trammel netting were displaced from this zone (Dimech et al. 2009), whilst trawlers, since 2004, have been benefitting from a derogation that allows them to operate within the FMZ. Moreover, a number of trawlers are allowed to trawl within the inshore 3 nautical mile zone. These relaxation of rules for the trawling sector, which were primarily aimed at making the Maltese fleet (or powerful businesses) more competitive in the EU single market, overhauled pre-EU legislations that had restricted the number of trawlers that could operate in the 25 FMZ, and which had prohibited all forms of trawling within the 3 nautical mile zone due to the presence of spawning grounds and small-scale fishers in the coastal zone (Camilleri 2002). Hence, with these changes, small-scale fishers have become subjected to heightened competition with industrial fleets targeting demersal fisheries within the coastal waters (Dimech et al. 2009).

Simultaneous to this development in the demersal fisheries, there has been a parallel growth in trammel and gillnetting operations which too are now dominated by fishing companies who engage in relatively higher 'small-scale' fishing effort than the independent small-scale fishers. The latter are able to deploy 12–15 trammel nets per day due to human capital and time constraints (i.e. they need to ensure that they get their fish to the market in time) and as a result are in a constant competition with the companies that employ multiple foreign hands "to work around the clock" and deploy between 50 and 80 nets per day (Caruana 2015; Said et al. 2016). The engagement of companies in intensive small-scale fishing is also a sign of the growth of a capitalist fishing sector. Since there are no supranational policies that limit the use of trammel nets, the government has been enabling the expansion of such operations, even though these are being done at the detriment of the independently-owned small-scale fisher sector, and might be causing overexploitation of such fisheries – as argued by fishers (Said et al. 2016).

The aforementioned situation that characterizes the trammel net fishery, the trawling industry, and the Bluefin tuna fishery policy all demonstrate that the government has been pushing the fishing fleet towards increased rationalization so that decision-making has become narrowly founded on neoliberal modernization within the ambit of supranational policy compliance (Said et al. 2016). Neoliberalism, which has favored the expansion of capitalist operations through aggregation and

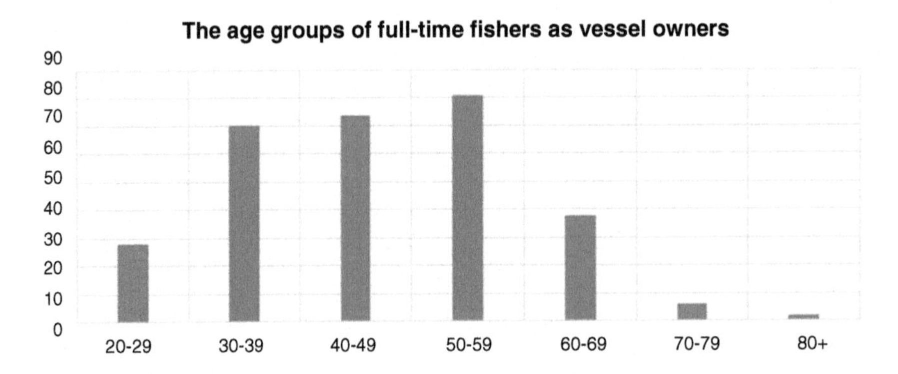


**Fig. 11.3** This data, provided by the DFA, shows the age distribution of full-time vessel owners, and illustrate that there is an inter-generational deficit which might be threatening the socio-economic renewal of the Maltese fishing sector (Source: Compiled by author)

ownership schemes, is creating a series of obstacles for the small-scale fishing community. For example, most small-scale fishers, who do not hold sufficient social, economic and political power, have been facing different forms of marginalization with regard to the allocation of fishing rights, access to fishing grounds, and other forms of restricted growth (e.g. vessel enlargement) (Said et al. 2016). In fact, only a few fishers have advanced within the sector. My fieldwork showed that these individuals were either more financially equipped or politically connected than most others in the fleet, and as a result have been able to benefit from different opportunities such as licenses, permits and quotas. With these distributive injustices, the majority of small-scale fishers have been unable to compete on a level playing field with the burgeoning capitalist class, which is now dominating the different fisheries that were once communally shared by the small-scale fisheries sector.

As a result, small-scale fisheries are threatened. One can observe that since EU accession there has been a gradual ongoing phasing out of small-scale fishers who have become less able to adapt to neoliberalism. Since 2004, the number of commercial vessels has decreased by 45%. There has also been a parallel decrease in the number of fishers. Moreover, the fishing community is not being rejuvenated as was the case in the past, since different exogenous forces have been pushing fishers out and simultaneously restricting the entrance of new fishers. This is leading to a situation where the fishing fleet will dissipate in numbers, especially if the present pattern of intra-generational deficit in the fleet persists, as indicated in Fig. 11.3. Overall, the age cohorts of the full-time Maltese fishers who own a vessel illustrate that there is not an equivalent inter-generational increase to cater for the socio-economic renewal of the fleet and this is leading to the greying of the sector.

The number of owners within the youngest age cohort (20–29), as shown in Fig. 11.3, is less than half of what it was in earlier generations. This is the outcome of two main drivers. First, in the past, it was the norm that sons inherited the boat from their fathers, and this legacy kept the sector somewhat sustainable. In contrast, nowadays families are discouraging their sons/daughters from taking over the business as they do not see fishing as a desirable livelihood any longer due to the uncertainties that it holds. Second, newcomers find it difficult to enter fishing as vessel permits have become scarce due to the closure of the fleet, and any available fishing licenses/rights are mostly bought by and affordable to corporate fishing companies which are expanding their operations through the ownership of multiple small-scale and large-scale vessels (Said et al. 2016).

With these changes, family-based enterprises have been on the decline, and while a few still remain they are mainly supported by foreign labor which has been on the rise over the past 8 years. Foreign employees have been gradually replacing family members and local deckhands who in the past were employed for most of the peak fishing seasons including that of the Bluefin tuna, swordfish and dolphinfish. The change of the vessel crew dimension has also changed the mode of production from a share-based system to one which is now mostly based on wage labor. The profits used to be shared amongst the crew according to a share-based ranking system, whereas now the foreign deckhands are on a payroll and paid wages. In the case of large companies who own multiple small-scale fishing vessels, the employment of three to four foreigners as deckhands has been part of their capitalist cyclical performance of expansion. On the other hand, independent small-scale fishers who employ one to two foreigners explained that foreign wage labor is their last resort to remain a fully-operational 'family-based' entity. Moreover, some independently-owned fishers who own a full-time fishing vessel but do not afford to pay foreigners, also have a part-time job on land since, as a fisherman explained, "fishing alone is not enough to feed the family these days", and they are aware that their situation cannot improve within the current policies.

These cases have shown how different policies have favored the most powerful interest groups, perhaps because they are likely to be more influential in steering national policy trajectories in their favour. However, we have noticed how these processes, especially those related to questions of distribution, have been justified on lines of rationalization, i.e. the allocation of resources and opportunities to the most efficient cohorts within a system (Davis 2015). In this regard, I argue that although most of these changes have their origin in EU supranational policies of sustainable resource use, the way they have been implemented at the national level within the ambit of conservation, efficiency and rationalization, has been underpinned by an implicit motive of accommodating specific interests of powerful/politically-connected groups that foster capitalist growth at the expense of small-scale entrepreneurship. The governance of the small-scale sector, therefore, has inevitably become enmeshed in a system of interplaying forces that favour the large-scale operators, and thus it became difficult for the small-scale sector to have their say.

## The End of the Small-Scale Fishing Sector: A Governance Problem

The realities of the Maltese small-scale fishing sector demonstrate that although fishers are subject to challenges emanating from policies that favored capitalist growth at the expense of their own existence, they are unable to challenge these policies since governance structures are controlled by capitalist enterprises. In the SSF Guidelines, this problem of unequal power relations has been highlighted as a shared phenomenon amongst many small-scale fishing communities who are unable to secure their rights mainly due to the lack of effective governance structures. Malta's case shows how fishers, who pre-2004 benefitted from relatively transparent governance systems that responded to the requirements of small-scale fisheries, are now victims of a governance structure that is not aligned to meeting the needs of the sector.

One of the major roots of the problem is the way fisheries co-operatives, which are the local governing institutions that ought to "work for the sustainable development of their communities through policies approved by their members" (CAP442 2001), have not been fulfilling their roles. Instead, co-operative representatives have used their legally-enshrined socio-political powers to make selfish decisions that have adversely affected the majority of small-scale fishers. As indicated in the study by Said et al. (2016), "politically-connected representatives within the Fisheries Co-operative allegedly used their legitimate power to benefit from the investment opportunities of tuna ranching" without the knowledge of small-scale fishers. The latter indicated that "they were unaware of the decisions that were being agreed on their behalf behind closed doors and perceive the institutional process to be high-handed and lacking transparency."

Having the 'ministers' ear', these representatives engage in various processes of elite capture such as opportunistic pursuits of licenses, permits, and financial aid. By investing in capitalist corporations, they have progressively shifted their operations from small-scale to capitalist frameworks, and thus have been unable to continue representing the real needs of small-scale fishers. Despite being rather insensitive to the realities of the small-scale fisheries sector, these individuals were still perceived by fishers as the 'bridge' to ministers. In other words, if fishers wished to speak to the Minister responsible for fisheries affairs, to obtain 'favours', their chance of doing so improved if cooperative leaders helped. As a result, the same leaders are continuously elected to power by small-scale fishers.

Another problem related to governance is the functioning of the national government which seems to be rather detached from fishers. Decisions are mainly taken in consultation with the co-operatives who are considered to be the spokespersons of fishers, and with large-scale operators (including foreign companies) who have invested in industrialized fisheries. In other words, the sector is now ruled by a system of plutocracy, wherein the power to take decisions and steer national trajectories lies in the hands of the wealthy. This could partly explain why the agenda of successive national governments in power since EU accession has not addressed the resilience of the sector. Catering for the survival of small-scale fisheries does not seem to have been a priority of governments. Rather, the neo-liberal urge to rationalize the sector so as to promote the interests of a few powerful agents in order to increase the foreign exchange earnings of Malta has been the main aim of successive governments. (Galea 2011; Said et al. 2016).

Indeed, it transpires how the government has hidden behind science and pilot scientific studies to, for example, legitimize the purse seiner segment within the fleet, in the process masking the fact that it is supporting the interests of powerful actors (Said et al. 2016). In doing so, it has simultaneously disempowered and disintegrated local communities from their livelihoods (Said et al. 2016). Rationalization of fisheries management is used to magnify the prospects of rapid-uptake entrepreneurs at the expense of 'less efficient' small-scale fishers (Gibbs 2009) who are unable to challenge the overarching structures that disempower them. Unlike the powerful few within the fishing sector, most small-scale fishers are unable to influence nationally-enacted policy trajectories, and hence have become subject to multiple stresses which have made them less resilient.

Against this backdrop, it is safe to assume that leading powers within the aforementioned national authorities are aggravating the powerlessness of small-scale fishers to maintain the hegemony of dominant groups within the sector. Through this process, small-scale fishers remain detached from the forces that are shaping their fishing landscape and livelihoods. Although participatory governance systems exist – such as political meetings at national clubs – they seem to be more tokenistic than inclusive. This is because most of these meetings, which are predominantly and purposely organized by the politicians close to the national elections, only serve as an avenue within which the politicians can fish for votes by promising fishermen a better future. History has shown how politicians also commit to do the impossible, and after the election, they state that such commitments cannot be fulfilled as they are, for example, not in line with the EU laws (Muscat 2015).

So, in other words, the government's pledge to protect and safeguard the sustainability of small-scale fisheries is largely rhetorical. Phrases such as 'safeguarding of small-scale fisheries' (Schembri 2010), and fisheries management plans for the 'sustainability of fishing fleets' (MSDEC 2013a, b) often do not translate into practice. Sustainability as a concept has become more of a dogma to justify the distribution of resources to the 'efficient' segments of the sector and legitimize the political and elite capture of opportunities, rather than a vision to truly implement sustainable pathways for small-scale fisheries. A ray of hope for small-scale fisheries surfaced in 2013 when the Labour party, through its electoral manifesto, promised a future of 'equality for all', and was consequently elected to power. Most fishers were optimistic for a better future. However, their hope faded away and in 2015 fishers publicly contended that the Labour government 'broke pre-electoral promises' with regard to policies aimed at the wider distribution of fishing opportunities (Muscat 2015). The future does not look bright given the fact that the government has also not fulfilled another promise regarding the establishment of better consultative platforms for fishers, a promise it said it would enact before 2018 in the form of a Fisheries Consultative Council (Barry 2013).

To date there has been no news about the formation of the Fisheries Consultative Council while an Agriculture Consultative Council has been formed and is operating in line with electoral pledges. By the look of things, it is likely the Fisheries Consultative Council will not be established and the pledge will not be fulfilled within the current political legislature which ends in 2018. Nonetheless, even if the Council is established and the fishers get to have their say within the new framework, the prospects of a redistribution of resources with a fair share for small-scale fishers seems near to impossible. This is because the neoliberal framework has established patterns of ownership, and it transpired that the government has a limited say over transactions that are now in the 'invisible' hands of the market. Thus, the establishment of the Consultative Council, and its embeddedness in the current governance framework, is likely to be an extension of the failing systems that are already in place.

Along with the disempowerment and oppression that small-scale fishers are facing at the local and national level, they are also detached from regional bodies that are meant to be the voice of fishers for the Mediterranean region as a whole. Most of the interviewed small-scale fishers are unaware of the existence of the Regional Advisory Council for the Mediterranean (MEDAC) and have never been able to partake in these forums. Rather, the Maltese representatives within MEDAC are co-operative representatives whose interests are, as argued above, more in line with large-scale fleets. Furthermore, unlike most small-scale fishers from other Mediterranean countries (within the EU), such as Spain, Greece, and France, Maltese small-scale fishers to date are not partners to the Low Impact Fishers of Europe (LIFE) organization.[5] They are thus not in a position to benefit from the recognition that this body gives in providing "a clear and coherent voice at EU level for the previously mainly silent majority of European fishermen, who are smaller scale and who use low impact fishing gears and methods, but have historically lacked dedicated and effective representation in Brussels and at Member State level."[6] This illustrates that Maltese small-scale fishers are being isolated from multi-scalar governance structures, making their prospects of changing their situation of powerlessness bleak.

This section has illustrated how small-scale fishers in Malta have remained underrepresented and their problems camouflaged by the actual policies of successive governments. It shows how policies have revolved around the neoliberal growth of the sector with the aim of making the fleet more efficient (and the wealthy wealthier). Rhetoric, on the other hand, has kept small-scale fishers hopeful of a secure and stable future through political promises (especially close to national elections). The actual and rhetorical appear to be at loggerheads with each other and reconciling them appears near to impossible, since it is the ideology of neoliberal growth that suffocates the adaptive capacity of the small-scale fisheries sector. The small-scale fisheries sector has become overly alienated and oppressed by the governing

[5]There are ongoing discussions on the potential membership of the Maltese small-scale fishing sector.

[6]http://lifeplatform.eu/

structures that restrict small-scale fishers' agency to the extent that they are unable to change their pathways. The recently-formulated SSF Guidelines seem to provide hope for a new dawn; however, their voluntary nature might make them rhetorical rather than real.

## SSF Guidelines as a New Dawn for the Small-Scale Fisheries Sector?

As illustrated in the preceding sections, it can be concluded that the fisheries governance framework in Malta has consistently disempowered small-scale fishers who in turn have been unable to challenge existing systems, which have reiteratively failed them. As a tool that serves for the protection and sustainability of fishing communities, the SSF Guidelines can potentially be the new dawn for the development of good governance based on principles of transparency, equity, and inclusivity. The SSF Guidelines provide straightforward recommendations for the implementation of 'governance of tenure in small-scale fisheries and resource management', such as legitimate tenure in the allocation of fishing rights (5.4) and in the recognition of territorial fishing grounds (5.7) that ought to enable small-scale fishers to practice fishing and maintain livelihoods, rather than be arbitrarily evicted from their fishing grounds to make space for new marine users/sectors (5.9). Furthermore, the SSF Guidelines give a direction to how the state may work towards empowering small-scale fishers to engage in the restoration, conservation, and protection of coastal ecosystems, which form the basis of fishers' livelihoods (5.5). Moreover, the SSF Guidelines provide guidance on how the sector could be given the necessary support so that small-scale fishers can participate in decision-making through participatory management systems (5.15) that enable them to be represented within various multi-scalar decision-making bodies (5.17).

To survive, small-scale fishers should be given the opportunity to benefit from transparent and effective governance which enables inclusivity that can improve the future of small-scale fisheries management and the resilience of small-scale fishers. Rather than remaining alienated from policy-frameworks which have jeopardized their livelihoods throughout the past decade, small-scale fishers should become empowered, as the SSF Guidelines attempt to do, so they can shape management frameworks that impact upon their traditional legacy of fishing. The possibility of reverting to the pre-2004 scenario is near to impossible, as globalizing forces have had significant influence in giving a new shape to the sector. Hence, the hope for the small-scale fisheries sector lies in the possibility of bridging the divide between small-scale fishing and the globalized future through feasible management strategies that enable the sector to flourish.

The establishment of a national Fisheries Local Action Group (FLAG), which as a social structure, is embedded within the principle of community-based governance could potentially be a step in this direction. The FLAG is a relatively recent development recognized through the EU CFP and funded through Axis 4 of EU funds (the European Fisheries Fund [EFF 2007–2013]) and the European Maritime and Fisheries Fund [EMFF 2014–2020]) to encourage "sustainable development and the improvement of the quality of life in areas with activities in the fisheries sector"[7] so as to promote "a balanced and inclusive territorial development of fisheries and aquaculture areas."[8] The establishment of FLAGs across Europe is the potential ground where seeds of resilience needed by small-scale fishing communities may take root (Symes 2014). Since 2007, in 80% of European Member States, a number of FLAGs have been established (Budzich-Tabor 2014). Malta remains an exception since EU funding for 2007–2013 has been more focused on other pillars of the sector such as reduction of fishing capacity through permanent demolition of fishing vessels and other modernization and infrastructural investments related to fishing ports, landing sites and the central fish market (GOM 2013b); and in the plans for EU funding for 2014–2020 FLAGs have not been recognized, even though, as I have explained, they are much needed for small-scale fishers.

As a concept, FLAGs merge well with the substance of the SSF Guidelines, and in the Maltese context its implementation ought to be considered as the "new solution to address local needs" (Budzich-Tabor 2014). Through it, fishers would be able to work collectively to seek opportunities and prospects that support small-scale fisheries and the local community. Fishers can, for example, pool in their indigenous, practical, and ecological knowledge to support grass-root development of robust and localised frameworks along with NGOs and other local entities. FLAGs can be used to provide the platform for partnership projects at both regional and national levels, and can empower fishers to participate within fisheries management, including in the implementation of marine protected areas and NATURA 2000 sites which are presently being implemented by the government (MEPA 2010). Fishers can then potentially explicate their needs through the development of marine and fisheries conservation strategies, and become stewards of the marine environment on which their livelihoods depend.

Furthermore, through FLAGs, small-scale fishers would be able to seek new ventures for diversification and alternative livelihoods that would boost the local economy and entice new entrants through job creation, such as for example, tourism-related activities. Fishing tourism has proved to be a very popular diversification strategy for coastal communities across the Mediterranean, including in Italy and Spain [2]. Malta, as a tourist hub with an annual influx of tourists that exceeds 1.5 million (MTA 2015), could exploit this lucrative economic niche. The eco-cultural product of colourful wooden boats and fish-gear mending is already a highly sought trademark that entices thousands of tourists annually to experience the craftsmanship and folklore that lies in fishing villages; however, at present, fishermen do not earn any additional income from the tourism industry. Thus, the establishment of a fishing-tourism market wherein fishers can take tourists onboard their vessels, show them the actual fishing activity and enable them to engage in the activity itself could

[7] Council Regulation 1198/2006, Article 4(f).

[8] Council Regulation 508/2014, Article 5(c).

be a valuable FLAG target since fishers can earn extra cash by tapping on the already-existing tourist visitation rates.

By using the FLAG framework, fishers can potentially invest in initial training to upgrade their capacity and become better equipped to undertake fishing-tourism ventures. Learning how to speak English and equipping their vessels with safety kits for tourist passengers would be very necessary, for example. In its full-fledged form, fishing tourism would reduce small-scale fishers' sole dependence on fish catches, maintain their link with the sea and fishing, and most importantly heighten their resilience within the fishing sector. Ultimately, through FLAGs, fishers can overturn the current image of the small-scale fisheries sector and convert a sector in terminal decline into a highly vibrant niche so as to regain the community's sense of social, economic and cultural identity. The question remains, however, are 'voluntary' guidelines sufficient to see these opportunities materialize?

## Voluntary Guidelines Might Not Work

The previous section demonstrated that there are possibilities and avenues for change and that the SSF Guidelines provide tangible targets that can be implemented to halt the decline of small-scale fisheries in Malta. However, the realities that currently shape the Maltese governance system trigger major questions of whether these opportunities can be actually mobilized and targets achieved. This is because the priorities of the governance system have been majorly embedded in the ideology of economic efficiency, and since the low-capital based nature of the small-scale fisheries sector is implicitly considered an obstacle to its rationalization, there have been no defined pathways aimed at making it resilient. In this regard, even though the SSF Guidelines provide the foundations for the development of new opportunities, the danger exists that securing the livelihoods of the sector remains at the level of rhetoric. To date, the Maltese authorities have not responded to these Guidelines and their implementation are not part of the Maltese Fisheries Strategy for upcoming years, probably because they are not of a binding nature, unlike the array of EU obligations that determine the political and administrative lines of action at the national level. Unless a governance overhaul takes place, therefore, it is unlikely that one will ever witness the country taking the path of 'walking the talk' of the Guidelines (Jentoft 2014).

Furthermore, the provisions for the empowerment of small-scale fisheries that lie within the Guidelines are not served well by the current governance system that is orchestrated by the powerful few who determine most paths of decision-making at the national level. Hence, the possibility of a top-down overhauling of the current governance system in a way that fits the SSF Guidelines is illusionary, to say the least. Rather, the ideal way forward is to dismantle current politically-biased governance structures that are suffocating the regeneration of the small-scale fleet, and make way for a framework that truly supports the small-scale fisheries sector. It seems that the only hope for marginalized fishers to benefit from the SSF Guidelines is through a bottom-up approach in which communities join forces and work to establish an association that represents the needs of the small-scale fisheries sector.

A representative body, that breaks away from current co-operative structures could result in more representativeness and counter the inequalities that have fragmented the sector over the past many years (Said et al. 2016). This neo-endogenous formation requires the mobilization of small-scale fisher agency. To get it started, fishers might benefit from the assistance of non-governmental organizations such as 'Friends of the Earth'[9] and 'Fish For Tomorrow'[10], which are two national bodies that both call for the protection of sustainable fishing livelihoods. At the regional level, fishers can benefit from the support of organizations such as the 'Low Impact Fishers of Europe' organization and the 'Too Big To Ignore'[11] since these bodies both have significant power vis-a-vis small-scale fisher concerns. With such support, fishers can become empowered, realize their potential, and gradually re-establish themselves within the fishing sector.

## Conclusion

This case study has highlighted how, in the past 10 years, the governing systems of Maltese fisheries have pushed the small-scale fisheries sector into a globalized scenario without providing sufficient ways for it to adapt to the new context. By promoting a neoliberal agenda of rationalization and focusing on efficiency-based distribution, the governing systems have implicitly renounced small-scale fishers' rights to important fisheries such as Bluefin tuna. Simultaneously, benign pathways have been created for those who had the ability to invest in industrial fishing. The industrialization of fisheries has adversely affected the resilience of small-scale fishers who are subject to constant struggles to retain their livelihoods as fishers. Many small-scale fishers have left the fleet altogether.

Evidently, the small-scale fisheries sector in Malta is in troubled waters. It is foreseen given the current circumstances that it will not be long before the small-scale fishing community will totally dissolve. A threat to the small-scale fisheries sector has both direct employment implications for fishers and their families and indirect repercussions for other socio-economic systems that are dependent on the small-scale fisheries sector, such as the local market economy and tourism. Adverse consequences are also there for the social fabric of the sector, namely a breakdown in social cohesion, community networks, and social stability. There are thus various reasons why the small-scale fisheries sector in Malta should be supported rather than neglected. It is time that a commitment is given and effort pooled to overturn the trajectories that are perpetuating the demise of the small-scale fisheries sector.

---

[9] http://www.foemalta.org/about.html

[10] http://fish4tomorrow.com/

[11] http://toobigtoignore.net/

The starting point for this could be the SSF Guidelines for these provide the right direction upon which the rebuilding of the small-scale fisheries can take shape.

In line with the SSF Guidelines, this case study highlights the need for incentives that raise human capital and empower small-scale fishers in decision-making processes in ways that promise a better and more resilient future for fishing communities. Recognition and investment in existing local communities through the concept of community-led local development should create the right platform for inclusive fisheries management, something that is not possible through existing co-operative structures. The challenge remains, however, to actually operationalize the SSF Guidelines as this is not yet foreseen in the Maltese context. Hence, real efforts need to be invested in doing so. This would be the way forward as it would acknowledge the significance, heterogeneity, and socio-cultural richness of the small-scale fisheries sector and allow it to rejuvenate.

## References


Barry, D. (2013, July 10). Fisheries consultative council to replace current board. *The Malta Independent.* http://www.independent.com.mt/articles/2013-07-10/news/fisheries-consultative-council-to-replace-current-board-2041708545/. Accessed 10 June 2016.

Battaglia, P., Romeo, T., Consoli, P., Scotti, G., & Andaloro, F. (2010). Characterization of the artisanal fishery and its socio-economic aspects in the central Mediterranean Sea (Aeolian Islands, Italy). *Fisheries Research, 102*(1–2), 87–97.

Bilocca, R. (2006). *Rationality, risk, and reckoning: A maritime anthropological study of a pressured way of life in Gozo* (Honours dissertation). University of Malta, Malta.

Budzich-Tabor, U. (2014). Area-based local development – A new opportunity for European fisheries areas. In J. Urquhart, T. G. A. Acott, D. Symes, & M. Zhao (Eds.), *Social issues in sustainable fisheries management* (pp. 183–197). Dordrecht: Springer.

Camilleri, M. (2002). An overview of the conservative management of Maltese fisheries. In *the code of conduct for responsible fisheries and its implementation in the Mediterranean*, (pp. 41–66). APS Seminar Proceedings.

Caruana, M. (2015, December 14). *Fascist rules in the fishing industry* (p. 3). *Malta Today.*

Coulthard, S. (2012). Can we be both resilient and well, and what choices do people have? Incorporating agency into the resilience debate from a fisheries perspective. *Ecology and Society, 17*(1), 4.

Davis, R. (2015). 'All in': Snow crab, capitalization, and the future of small-scale fisheries in Newfoundland. *Marine Policy, 61*, 323–330.

Dimech, M., Darmanin, M., Smith, I. P., Kaiser, M. J., & Schembri, P. J. (2009). Fishers' perception of a 35-year old exclusive fisheries management zone. *Biological Conservation, 142*(11), 2691–2702.

European Commission. (2003). *Comprehensive monitoring report on Malta's preparations for membership.* http://ec.europa.eu/enlargement/archives/pdf/key_documents/2003/cmr_pl_final_en.pdf. Accessed 10 May 2016.

FAO. (2015). *Voluntary guidelines for securing sustainable small-scale fisheries in the context of food security and poverty erradication.* Rome: FAO.

Farrugia, A. F., De Serna, J. M., & De Urbina, J. O. (2004). Description of swordfish by-catch made with bluefin tuna longliners near Malta during 2002. *Collective Volume of Scientific Papers ICCAT, 56*(3), 912–920.

Frank, K., Mueller, K., Krause, A., Taylor, W., & Leonard, N. (2007). The Intersection of global trade, social networks, and fisheries. In W. Taylor, M. Schecther, & L. Wolfson (Eds.), *Globalization: Effects on fisheries resources* (pp. 499–523). New York: Cambridge University Press.

Galea, M. (2011, March 12). Tuna wars in Maltese territory. *Times of Malta*, p. 1.

Gatt, M., Dimech, M., & Schembri, P. J. (2015). Age, growth and reproduction of Coryphaena Hippurus (Linnaeus, 1758) in Maltese waters, central Mediterranean. *Mediterranean Marine Science*, 334–345.

Gezelius, S. S. (2002). Environmental sustainability and political survival: A comparative analysis of the cod fisheries of Canada and Norway. *Environmental Politics, 11*(4), 63–82.

Gibbs, M. T. (2009). Resilience: What is it and what does it mean for marine policymakers? *Marine Policy, 33*(2), 322–331.

GOM (Government of Malta). (2013a). *Fisheries management plan, Tartarun.*

GOM (Government of Malta). (2013b). *Malta's national strategic plan for fisheries.*

Grupetta, A. (2002). An overview of the fishing industry in Malta. In *APS annual seminar on the developent of agriculture and fisheries in Malta*, (pp. 29–40).

Hadjimichael, M., Delaney, A., Kaiser, M. J., & Edwards-Jones, G. (2013). How resilient are Europe's inshore fishing communities to change? Differences between the north and the south. *Ambio, 42*(8), 1037–1046.

Jentoft, S. (2014). Walking the talk: Implementing the international voluntary guidelines for securing sustainable small-scale fisheries. *Maritime Studies, 13*(1), 16.

Leiva, I.D., Busuttil, C., Darmanin, M., & Camilleri, M. (1998). Artisanal fisheries in the western Mediterranean: Malta Fisheries. Copemed – FAO sub-regional project.

MEPA (Malta Environmental and Planning Authority). (2010, March 12). Four new marine protected areas. *Malta environmental and planning authority* 1. https://www.mepa.org.mt/outlook5-article2. Accessed 15 June 2016.

MSDEC (Ministry for Sustainable Development, the Environment and Climate Change). (2013a). *Fisheries management plan: Otter bottom trawl Lampara fishery*. https://stecf.jrc.ec.europa.eu/documents/43805/595618/Maltas+Fisheries+Management+Plan+-+Trawler+and+Lamapra.pdf. Accessed 23 June 2016.

MSDEC. (2013b). *Fishing effort adjustment plan for bluefin tuna*. https://agriculture.gov.mt/en/agric/Documents/EFF/documents/FEAP%20Tuna%20FINAL.pdf. Accessed 23 June 2016.

MSDEC. (2013c). *National data collection programme proposal*. https://datacollection.jrc.ec.europa.eu/np/2013. Accessed 10 May 2016.

MTA (Malta Tourism Authority). (2015). *Tourism in Malta*. http://www.micc.org.mt/malta-tourism-authority.html. Accessed 20 June 2016.

Muscat, C. (2015, January 26). At the end of the line we are no better off…Fishermen claim labour 'Broke pre-electoral promise.' *Times of Malta*. http://www.timesofmalta.com/articles/view/20150126/local/At-the-end-of-the-line-we-are-no-better-off-.553375. Accessed 19 July 2016.

NSO (National Statistics Office). (2004). *Agriculture and fisheries*. http://nso.gov.mt/en/publicatons/Publications_by_Unit/Pages/B3-Agricultural-and-Environment-Statistics.aspx. Accessed 20 May 2016.

NSO. (2005). *Agriculture and fisheries*. http://nso.gov.mt/en/publicatons/Publications_by_Unit/Pages/B3-Agricultural-and-Environment-Statistics.aspx. Accessed 20 May 2016.

NSO. (2010). *Agriculture and fisheries*. http://nso.gov.mt/en/publicatons/Publications_by_Unit/Pages/B3-Agricultural-and-Environment-Statistics.aspx. Accessed 20 May 2016.

NSO. (2011). *Agriculture and fisheries*. http://nso.gov.mt/en/publicatons/Publications_by_Unit/Pages/B3-Agricultural-and-Environment-Statistics.aspx. Accessed 20 May 2016.

S.L.425.03. (1965). *Tunny fish (Importation) Restriction order*.

Said, A., Tzanopoulos, J., & MacMillan, D. (2016). Bluefin tuna fishery policy in Malta: The plight of artisanal fishermen caught in the capitalist net. *Marine Policy, 73*, 27–34.

Schembri, D. (2010, October 15). *Minister calls for safeguarding of small-scale fisheries* (p. 5). *Times of Malta.*

Stelzenmüller, V., Maynou, F., Bernard, G., Cadiou, G., Camilleri, M., Crec'hriou, R., Criquet, G., Dimech, M., Esparza, O., Higgins, R., Lenfant, P., & Pérez-Ruzafa, A. (2008). Spatial assessment of fishing effort around European marine reserves: Implications for successful fisheries management. *Marine Pollution Bulletin, 56*(12), 2018–2026.

Symes, D. (2014).Finding solutions: Resilience theory and Europe's small-scale fisheries. In J. Urquhart, T. G. Acott, D. Symes, & M. Zhao (Eds.), *Social issues in sustainable fisheries management*. MARE Publication Series (Vol. 1, pp. 23–32). Dordrecht: Springer.

Symes, D., Phillipson, J., & Salmi, P. (2015). Europe's coastal fisheries: Instability and the impacts of fisheries policy. *Sociologia Ruralis, 55*(3), 245–247.

Tzanatos, E., Somarakis, S., Tserpes, G., & Koutsikopoulos, G. (2006). Identifying and classifying small-scale fisheries métiers in the Mediterranean: A case study in the Patraikos Gulf, Greece. *Fisheries Research, 81*(2), 158–168.

Vella, S. (2002). *Tuna wars in Malta.* B.A Anthropology, Thesis, University of Malta.

# Part IV
# Strengthening the Resource Base

The authors of the chapters included in this section address the SSF Guidelines from the viewpoint of viable management mechanisms for the small-scale fisheries sector. In Chap. 12, James Prescott and Dirk Steenbergen focus on the ecosystem-based approach to fisheries management, which are also proposed in SSF Guidelines. Writing from Australia and Southeast Asia, they argue that small-scale fisheries still lack the institutional and scientific foundation for implementing this approach. Instead they suffer the consequences of operating with little or no effective management. There seems to be too little capacity to manage the complexity associated with ecosystem-based approaches. They present six case studies where comparatively simple management models were applied, and suggest that management should primarily strive for better grounding and more realistic targets. Chapter 13 takes the reader far north, to the small-scale fisheries as practiced by the indigenous Sami inhabiting northern Norway. The authors, Svein Jentoft and Siri Ulfsdatter Søreng, discuss what inspiration the SSF Guidelines represent in establishing management institutions that operate in accordance with the rights that indigenous peoples have for sustainable livelihoods and self-governance. They demonstrate how the need for special treatment of small-scale fisheries may trigger controversy when implemented. They argue that the institutional management reforms that have been recently introduced go a long way to accommodate the livelihood and cultural rights that indigenous, small-scale fishing people have. Chapter 14, by Gunakar Surathkal, Adam Jadhav and Ramachandra Bhatta, is a case study of India's monsoon fishing ban. They posit that a critical perspective on Indian fisheries governance reveals a lack of compliance with the SSF Guidelines. Instead, the existing governance regime has generally benefitted large-scale, industrial, and semi-industrial fisheries. However, they see potential in the seasonal fishing ban in effect during the monsoon period, which could aid in implementation if framed as safeguard for small-scale fisheries as part of dynamic fisheries management to privilege and protect India's small-scale fisher communities. In Chap. 15, Alyne Delaney and Nobuyuki Yagi take the reader to Japan. Given the observation that Japan is an industrialized nation with a relatively low poverty rate and secure food supply, how can they benefit from the SSF Guidelines, and what lessons could Japan provide to other countries? Despite industrialization, Japanese fisheries are overwhelmingly small-scale and based in local communities with historic and strong cultural links to nearby coastal resources.

# Chapter 12
# Laying Foundations for Ecosystem-Based Fisheries Management with Small-Scale Fisheries Guidelines: Lessons from Australia and Southeast Asia

**James Prescott and Dirk J. Steenbergen**

**Abstract** Ecosystem approaches are increasingly mainstreamed in contemporary debate on small-scale fisheries management, however many small-scale fisheries lack solid institutional and scientific foundations on which to build such holistic and inherently more complex management systems. Most small-scale fisheries still operate with little or no effective management. Proponents of ecosystem approaches frequently malign single-species management models that placed less emphasis on wider ecosystem effects. However these 'simpler' approaches are responsible for significant management successes, even in contexts where fisheries were not strictly single species. We argue for incremental development of fisheries management more deeply rooted in successful past management systems. At this stage, there appears too little capacity to manage the complexity associated with a complete paradigm overhaul towards ecosystem-based approaches. The multi-dimensional importance of small-scale fisheries is highlighted in the Voluntary Guidelines for Securing Sustainable Small-Scale Fisheries, where ecosystem approaches are identified to guide holistic, integrated management, and facilitate cross institutional interactions. Its application is nuanced and connected with practical measures to ensure that principles of decency, equity, and responsibility, define management's fabric. We draw from this in problematizing the adoption of ecosystem approaches and examine the implications for small-scale fisheries management. We present six small-scale fisheries case studies; two in Australia where comparatively simple management models were applied, two operating in trans-boundary contexts with

J. Prescott (✉)
Australian Fisheries Management Authority, Darwin, NT, Australia
e-mail: jimdeb@bigpond.com

D.J. Steenbergen
Research Institute for Environment and Livelihoods, Charles Darwin University, Darwin, NT, Australia
e-mail: dirk.steenbergen@cdu.edu.au



S. Jentoft et al. (eds.), *The Small-Scale Fisheries Guidelines*, MARE Publication Series 14, DOI 10.1007/978-3-319-55074-9_12

Australia and two operating under very different social, political and economic conditions in the wider region of Indonesia. We suggest initial management approaches should primarily strive for better grounding and more realistic targets.

**Keywords** Ecosystem Approach to Fisheries (EAF) • Governance • Small-scale fisheries • Transboundary fisheries • Australia • Indonesia

## Introduction

Small-scale fisheries are recognized globally as important sources of food and employment, and frequently identified as viable channels for socio-economic development, or poverty alleviation (Barkin and DeSombre 2013; FAO 2015). With an estimated 90% of fishers and 50% of fisheries production globally accounted for through small-scale fisheries (FAO and WFC 2008; FAO 2015), it comes as no surprise that many small-scale fisheries attract substantial subsidies. These are often 'bad' or 'ugly' subsidies (see Khan et al. 2006), that bolster capacity through assisting access to fuel and fishing gear or, in some circumstances, providing complete fishing units (e.g. vessel and gear) (Prescott et al. 2015). In contrast, however, few small-scale fisheries attract equivalent government support to establish and maintain effective governance and management that industrial fisheries draw, for example. With such inconsistency, the Voluntary Guidelines for Securing Sustainable Small-Scale Fisheries in the Context of Food Security and Poverty Eradication (SSF Guidelines) form a timely response and potentially provides an international mechanism to address governance needs and human-ecology management of small-scale fisheries in the context poverty and food security (FAO 2015).

The ecosystem approach to fisheries (EAF) features strongly in these SSF Guidelines and has been widely embraced as the most appropriate vehicle to achieve goals of improving food security and alleviating poverty (FAO 2003; Pikitch et al. 2004; Murawski 2007). As in many parts of the world, in the Australasian region EAF has attracted strong support from international donors, particularly where small-scale fisheries approaches are embedded in broader marine conservation initiatives seeking also to address the human dimensions of fisheries management. In such contexts, financial and technical support tend to frame the urgency for small-scale fisheries management primarily to safeguard globally significant biological diversity, and secondarily because of small-scale fisheries' value in improving food security and alleviating poverty (Clifton 2009). Fisheries management experiences in the Australasian region, however, show diverse approaches to small-scale fisheries management being applied to comparable ecosystems, ranging from single-species management to holistic EAF approaches. This provides an important opportunity to critically examine experiences of contemporary fisheries under EAF approaches and other relevant fishery management approaches to assess whether investments to reform small-scale fisheries management towards EAF approaches is the most appropriate near-term strategy for all fisheries.

First, we discuss how and why the EAF is by default the standard approach for many influential actors, as evident from its prominent feature in the SSF Guidelines, but also from its role in guiding fisheries management under the auspices of regional marine conservation programs like the Coral Triangle Initiative (CTI) across insular Southeast Asia and the Western Pacific. We proceed to examine how this approach has thus far performed. Next, we set out six case studies of fisheries management involving small-scale fisheries in the region. Two Australian case studies are discussed first. These enterprises are typically family-owned and operated but have larger cash flows than would be expected in most small-scale fisheries in the nearby developing country contexts. We consider the important junctures in the management history of these fisheries and how these align or contrast with the EAF. We then describe two trans-boundary fisheries with Australia; one with Papua New Guinea and the other with Indonesia. Finally, we examine two fisheries from nearby eastern Indonesia. We make comparisons between the Australian management experience and regional fishery management to explore what can be learned and what lessons may be transferable across fisheries in different development and governance contexts. We draw on the SSF Guidelines in our analysis to maintain consistency in language and understanding to that developed in the SSF Guidelines.

## Ecosystem Approach to Fisheries (EAF)

EAF is defined by FAO as an approach to fisheries that:

> strives to balance diverse societal objectives, by taking into account the knowledge and uncertainties about biotic, abiotic and human components of ecosystems and their interactions and applying an integrated approach to fisheries within ecologically meaningful boundaries (FAO 2003, 6).

Related variants, like the ecosystem based fisheries management (EBFM), ecosystem based management (EBM) or the ecosystem approach to fisheries management (EAFM), follow the same principles but implement fisheries management with greater or less emphasis on addressing ecological scale and complexity (Patrick and Link 2015). EAF attempts to manage the interactions between the fishery and the ecosystem in which it operates while integrating current and future human interests.

Despite widespread recognition of the merits of EAF, there are few empirical examples that demonstrate its successful application as the complete holistic frame in which it is often theoretically conceptualized, particularly in less developed countries. In some jurisdictions, such as Australia, incrementally adaptive processes have moved numerous fisheries towards systems that might be recognizable as EAF, well before EAF was identified as an end goal. As elsewhere, for much of the twentieth century Australia sought to develop, rather than restrict, its fisheries. In the latter part of the century, however, Australia controlled its harvests by introducing limited entry licensing, gear limits, and seasonal access (Harris 1991). While these measures were partially successful, overcapacity necessitated further interventions in many fisheries. Fisheries data gradually accumulated and stock assessments progressively yielded better information to support science-based management, but generally at the species level, though there have been notable exceptions (see Smith et al. 2007). Reducing interactions with threatened, endangered and protected species and reducing by-catch became another management focus. Sensitive areas associated with the life history stages of target species and vulnerable marine ecosystems, including certain sea mounts, were protected. Large spatial scale marine biodiversity conservation zoning in national and state jurisdictions is the most recent overlay to which fisheries must adapt.

Given the importance of small-scale fisheries to millions of people while also understanding the excessive pressures that many small-scale fisheries put on resource bases, and thereby the sustainability of the associated human-ecological systems, implementing the SSF Guidelines successfully has tremendous significance. Inherently implied transitions to EAF management, however, require several considerations. For example, we find it unlikely that suggestions that EAF is no more complex than single-species management approaches (Patrick and Link 2015), given that one has to deal more fully with the human-ecological interactions next to the complexity of ecosystems' natural elements. More importantly, regardless of any differences in complexity, it is not necessarily the actual degree of complexity that is relevant in the context of 'doing management' but rather the complexity that is constructed by the broad range of actors implicit in EAF. Moreover, the human and financial resources required to create the operational platform for EAF to function are too often critically insufficient. This emphasizes the need for management efficiency, and urges the consideration of how limited resources are best utilized. This, we argue, may require interventions to apply scarce resources to address the most pressing and critical issues first in order to secure a system, from which point progressive processes of building adaptive and long-term management capacity towards effective EAF reform can be initiated.

## Results

The six fisheries case studies fall into three categories. The Queensland Reef-line Fishery (QRLF) and the Northern Demersal Scalefish Fishery (NDSF) are entirely Australian fisheries.[1] Of the trans-boundary fishery case studies, the Torres Strait Lobster Fishery is exploited by Papua New Guinea (PNG) and Australia, while the Scott Reef Sea Cucumber Fishery is exploited by Indonesian fishers in the Australian Fishing Zone. The final two fisheries case studies are Indonesian fisheries, representative of small-scale fisheries found in the broader Southeast Asian region: the Purse Seine Fishery around the island of Rote and the Flying Fish Roe Fishery around Tanimbar Kei Island (Table 12.1).

[1] The NDSF shares its northern boundary with Indonesian fisheries.

**Table 12.1** Annual production and fishery characteristics as assessed in 2015

| | Fishery | Annual catch | Vessel numbers | Management arrangement | Monitoring and research | MPAs – management relevance | EAF[a] practiced | Monitoring control and surveillance |
|---|---|---|---|---|---|---|---|---|
| *Australian* | Queensland Line Fish | 1458 t live wt. | ~225 but variable (not all licensed vessels participate) | Spawning closures | Logbooks | Very relevant | No | Quota monitoring |
| | *Multispecies* | | | TAC and ITQ | Catch sampling | | | Aerial surveillance |
| | | | | Gear limits | Long-term stock and fish assemblage monitoring | | | Limited at-sea inspections |
| | | | | Size limits | | | | |
| | | | | MAC | | | | |
| | Northern Demersal Scalefish | 1111 t live wt. | 6 | TAE | Logbooks | Weak relevance | No[b] | VMS effort-day monitoring |
| | *Multispecies* | | | Gear limits | Fishery dependent and independent catch sampling | | | At sea and in-port inspections |
| | | | | Size limits | | | | |
| | | | | Bycatch limits | | | | |
| | | | | MAC | | | | |

(continued)

**Table 12.1** (continued)

| | Fishery | Annual catch | Vessel numbers | Management arrangement | Monitoring and research | MPAs – management relevance | EAF[a] practiced | Monitoring control and surveillance |
|---|---|---|---|---|---|---|---|---|
| *Trans-boundary* | Torres Strait Tropical Rock Lobster (Australia and PNG) | 666 t live wt. | 12 MB/33 tenders | Seasonal closures | Logbooks on mother boats | Weak relevance | No[b] | At sea and in-port inspections |
| | | | 268 indigenous | Size limits Gear controlled | Catch landing monitoring | | | Controlled licensing |
| | | | 7 PNG MB/49 tenders | MAC | Fishery independent surveys | | | |
| | | | ~500 divers | | Routine stock assessments | | | |
| | Scott Rf Sea Cucumber | 50–250 t live wt. | Few to ~80 | Sail only vessels | Aerial vessel counts and ad hoc logbooks and catch sampling | Relevant at two reefs where fishing was excluded | No | At sea inspections |
| | Indonesia | | | Non-motorised gear | | | | |
| | *Multispecies* | | | | | | | |

| | | | | | | | | |
|---|---|---|---|---|---|---|---|---|
| *Indonesian* | Rotinese Purse Seine | Up to 3600 t live wt. | 43 | | Ad hoc catch sampling | Unestablished relevance | No | Controlled licensing |
| | *Multispecies* | | | | | | | |
| | Kei Flying Fish Roe | Est. 60-100 t dried wt. | Est 90–120 seasonally (up from ~9 during fishery's emergence in 2009) | Zone closures | Monitoring information on boat numbers and crew. | Very relevant | Fishery falls under a broader NRM program, but is managed autonomously | Regulations against destructive fishing |
| | *Single-species* | | | Controlled fisher access (vessel retribution system) | Sampled catch effort data | | | Monitoring registration |
| | | | | | Sampled harvest data | | | |

*MAC* a stakeholder committee established to provide management advice, *TAE* total allowable effort, *MB* mother boat, *MPAs* Marine Protected Areas
[a]EAF used synonymously with EBM, EBFM but not with ESD; [b]Not a full EAF approach but ecologically risk assessed

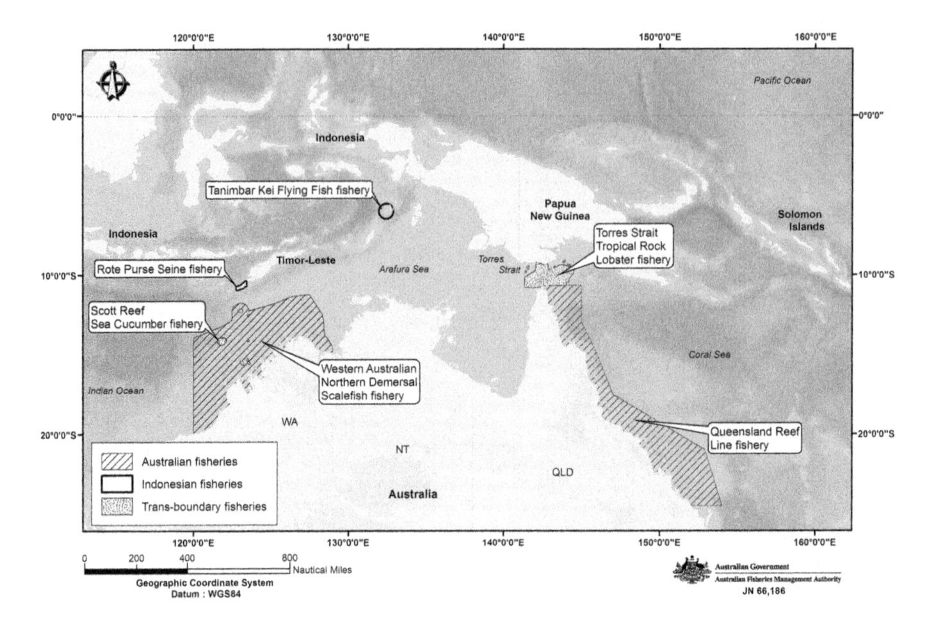


**Fig. 12.1** Locations of the six fisheries case studies (Note that the defined area of the Australia fisheries is much greater than the area actually exploited by them. Moreover, the PNG lobster fishery adjoins the Australian fishery in Torres Strait but is not shown)

## *Australian Case Studies*

The Australian case studies (Table 12.1 and Fig. 12.1) involve small-scale fisheries in the Australian context. They operate in tropical northern Australia where the environmental conditions are similar to other parts of the Indo-Pacific region and the same species are targeted as throughout the Asian-Pacific region.

### Queensland Reef-Line Fishery (Multispecies Tropical Fish)

The Queensland reef-line fishery (QRLF) targets coral trout (*Plectropomus spp.*) and redthroat emperor (*Lethrinus miniatus*) but also takes other 'mixed' tropical reef fish species. The QRLF handline fishery operates within the Great Barrier Reef Marine Park (GBRMP), which is managed by a commonwealth authority while the QRLF is managed by the state of Queensland under state legislation. A feature of this fishery is the rezoned GBRMP that expanded no-take zones to about 33% of the park. Minimum size limits were introduced for three species in 1976 and increased in 2003 when a maximum size for coral trout was also introduced. Quota management was introduced a year later with the total allowable catch (TAC) allocated as individual transferable quota (ITQ) for coral trout, red throat emperors, and a third ITQ for 'other species' (Andersen et al. 2005). Fishery closures to protect spawning

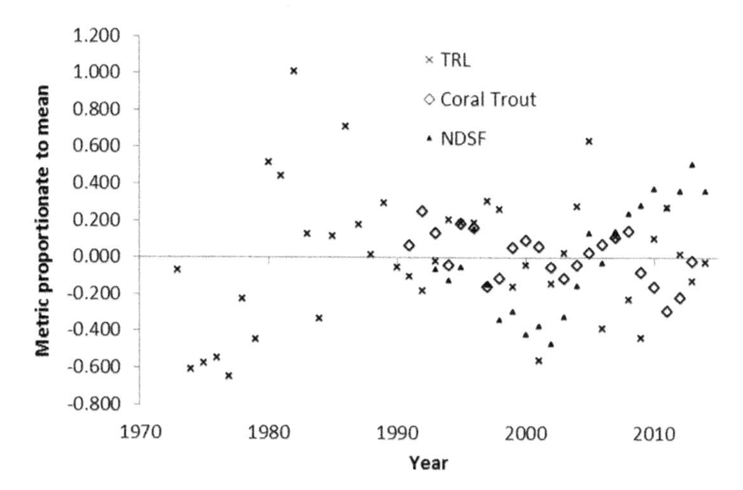


**Fig. 12.2** Catch rates (coral trout) and catches (NDSF and TSTRLF) are presented as deviations from the mean standardised catch rate or mean catches, respectively

aggregations were introduced at the same time as quotas. Since all fishing is done by handline, the fishery is not considered to cause serious 'collateral' environmental damage (Queensland Government 2012).

Like many Australian fisheries, multiple sectors harvest reef fish in the GBRMP, including commercial, recreational, tourists aboard charter boats, and indigenous fishers. Multiple sectors require a multi-part data collection system including commercial catch and effort logbooks, charter boat records, recreational surveys, and fishery independent transect surveys.

Coral trout catches are comprised of four species that are stock-assessed as a single group, while red-throat emperor undergo periodic, stand-alone assessments. Many other species are monitored but not formally stock-assessed. The QRLF's effects were assessed at specific experimental reefs by The Effects of Line Fishing project (Mapstone et al. 2004) and long-term monitoring programmes are carried out by Queensland and the Australian Institute of Marine Science. Standardized catch rates of the 20 years since 1993 are a crude, but ready indicator of the sustainability of the fishery and are presented in Fig. 12.2.

## Northern Demersal Scalefish Fishery (NDSF) (Multispecies Tropical Fish)

The Western Australian-managed NDSF off north-western Australia overlapped with but eventually replaced a large foreign trawl fishery. Australia declared its 200 mile exclusive economic zone in 1979, which eventually led to cessation of all foreign fishing in 1990 (Nowara and Newman 2001). The NDSF fishery traps goldband snapper (*Pristipomoides multidens*), red emperor (*Lutjanus sebae*), other snapper species (*Lutjanidae* spp.), emperors (*Lethrinus* spp.) and cod (*Epinephelus* spp.). There are 11 licensees and their fishing effort (in units of standard fishing days) is tightly regulated through individual allocations of Total Allowable Effort (TAE) via a vessel monitoring system (VMS). The TAE is adjusted as necessary to meet fishery reference points. Because this fishery operates well offshore from the very sparsely populated Kimberley coastline (Fig. 12.1) there is little or no recreational fishing and almost no reported illegal fishing. This indicates that there are no other significant anthropogenic sources of fishing mortality (Western Australian Department of Fisheries 2009). The passive trap fishing methods are relatively benign in terms of disturbance to the benthic habitat (ICES 2006, 2013) and fish can exit lost traps (Newman et al. 2012).

Two Commonwealth no-take marine protected areas exist within the NDSF boundaries and several recently proclaimed areas that have yet to be implemented. The current designated no-take zones have little bearing on the fishery because their habitats are little exploited. Fishery data are collected through logbooks, voluntary fishery-dependent sampling, and observer trips. The science program, delivered by the Western Australian Department of Fisheries, provides biological research (e.g. stock structure) and periodic stock assessments. This information supports a strong management framework that includes well developed and interactive governance through the NDSF Management Committee (Personal communication, S. Hinge, April 2016), although management decisions are ultimately made by the Minister of Fisheries.[2] Total annual catches from the NDSF are shown in Fig. 12.2 and demonstrate, if anything, increasing production from the fishery over the past two decades.

## Summary of Australian Case Studies

Both fisheries are assessed as ecologically sustainable through independent 'strategic assessment' against Guidelines for the Ecologically Sustainable Management of Fisheries by the Commonwealth Department of the Environment. Two critical elements are the capacity to control harvests using input and/or output controls and the means to enforce the management arrangements. Further, there must be a capability to assess, monitor, avoid, remedy, or mitigate adverse impacts on the wider marine ecosystem in which the fishery operates. These measures move these fisheries towards EAF in respect of the environmental attributes of the fishery.

Management success in these fisheries is arguably attributable to 'single-species' approaches, despite multiple species being harvested. Although each fishery interacts with the whole ecosystem, the management objectives are narrower than a fully-fledged EAF approach would be. While potentially less precautionary than the EAF prescription, provided precautionary decisions are made regarding the managed species and they are similarly vulnerable to the other unassessed species in the catch, it is reasonable to expect that the unassessed species will also remain sustainably exploited (see Hilborn et al. 2015). Trophic interactions are implicitly expected to be within acceptable bounds while stocks of the exploited species are maintained at 'sustainable levels'. Management is simplified because under this model fewer

[2] Steve Hinge, former licence holder in NDSF, discussion held January 2016.

stock assessments and management decisions are needed. Moreover, governance performance can be evaluated against fewer criteria than would be necessary under full-fledged EAF.

Both fisheries catch high-value fish and the fisheries are generally profitable (or very profitable) for the fishers engaged in them. This may reduce incentives for illegal fishing found in fisheries with marginal economics. Monitoring, Control, and Surveillance (MCS) is effective and non-compliance, when detected, is penalized appropriately. Access tenure, while not granted in perpetuity, is effectively ongoing, tradeable, and valuable.

Independent processes spanning several decades have gradually transitioned these fisheries from having little management control to operating effective quotas on effort or catch. Through this process, governance, management, and research have to varying degrees become more inclusive and interactive. When the GBRMP was rezoned for example, affected fishers (and other affected businesses) in the QRLF were eligible for compensation. While this process stopped short of officially recognizing historical tenure rights these were implicit in the process. Although often cited as a flawed process from the fishers' perspective (Lédée et al. 2012), it seems probable that it reinforced tenure in this fishery and more broadly in Australian fisheries.

## *Trans-Boundary Case Studies*

### Torres Strait Tropical Rock Lobster (Single-Species Tropical Lobster)

The Torres Strait Tropical Rock Lobster Fishery (TSTRLF) (Fig. 12.1) exclusively targets the ornate tropical rock lobster (*Panulirus ornatus*). The resource is shared by three jurisdictions: Australia's Torres Strait Protected Zone (TSPZ), PNG's area of the TSPZ and Gulf of Papua, and the State of Queensland (south of TSPZ). Within the Australian TSPZ, lobsters are fished by small-scale indigenous fishers operating from 5 m dinghies who generally return home after fishing each day, and non-indigenous fishers who also fish from dinghies but generally return to a mother vessel (Table 12.1). Operations are similar in PNG.

Management of the fishery began during the 1970s when, in PNG, it was demonstrated that lobsters migrated from Torres Strait across the Gulf of Papua to spawn and were exposed to intensive trawling during this important life-history phase (Moore and MacFarlane 1984). A trial quota constrained trawling in 1978 and trawling was temporarily banned in 1979. Australian prawn trawlers later targeted the migration before it entered PNG waters for several seasons. Trawling was banned permanently in both countries in 1984 which demonstrated early management cooperation in this trans-boundary fishery. The TSTRLF has remained a dive fishery since 1984 in which lobsters must be taken by hand or spear.

The Torres Strait Treaty, ratified in 1985, officially created the TSPZ, put in place catch sharing arrangements, and fostered small-scale fishing. Non-indigenous fishing licences were capped in Australia; however, licences for indigenous 'Islanders' remained uncapped. Minimum size limits, seasons, and periodic closures for diving with surface supplied air were introduced later. PNG's management plan constrains the number of mother boats and tenders, and creates complementary closed seasons, as is the case in Queensland. Latent (inactive) non-indigenous licences were removed by the Australian government to prevent expansion of fishing effort in 2003 and others were bought back to allow expansion of the indigenous sector in 2006. The fishery has effectively been managed by input controls without fishing effort being systematically controlled in either the indigenous Australian or PNG sectors. There are no closed areas in the Torres Strait fishery; however, waters beyond diving depth provide some refuge for breeding lobsters (Pitcher and Prescott 1991) in addition to the extensive closed areas in the adjacent Queensland fishery.

Annual fishery-independent visual transect surveys have provided estimates of pre-recruit and legal size lobster abundance since 1989. Logbooks are mandatory on all mother boats in both Australia and PNG and systems are in place to record landings in both countries. The history of catches in the fishery (Fig. 12.2) is displayed as deviations from the mean combined catch of PNG and Australia and suggests that the fishery has sustained production over several decades.

Fishery management and governance arrangements are well established in Australia where stakeholders engage through management and research committees and PNG representatives routinely participate in meetings by invitation. Decision making power is vested in the 'Protected Zone Joint Authority'.

While the stock has been sustained, meeting 'diverse societal objectives' has probably been less successful because of the disparity in indigenous and non-indigenous objectives, which has only recently been investigated by Plagányi et al. (2013). Aware of the differences, current management continues to work towards a solution.

## Traditional Fishery at Scott Reef (Multispecies Sea Cucumber and Reef Fish)

The trans-boundary Indonesian fishery at Scott Reef (Fig. 12.1), like the TSTRLF, is conducted under a formal bilateral arrangement (DFAT-Australia and Republic of Indonesia 1974), between Australia and Indonesia. It targets sea cucumbers and opportunistically exploits many other species of economic or nutritional value, including several vulnerable fish species like the humphead Maori wrasse (*Cheilinus undulatus*) and ecologically important bumphead parrotfish (*Bolbometapon muricatum*) (Bellwood et al. 2003). The 1974 Memorandum of Understanding (MOU) and 1989 'Practical Guidelines' require fishers to sail to and from Scott Reef and use only non-motorized equipment. The fishery is otherwise unregulated and, while only fishers with sailing vessels and skills access the fishery, with many such vessels and crews in Indonesia, the fishery is over-exploited.

Ashmore Reef, once the fishery's focal point, became a no-take marine reserve in 1988. Aside from reducing the fishing area and the exploitable biomass of sea cucumbers it is unknown what impact this reserve had on the fishery. Given the remoteness from the exploited reefs (> 100 nautical miles) it is unlikely that it contributes 'spill-over' recruits.

Prices paid to the fishers for their dried sea cucumbers have risen about 30% per year over a decade (Prescott et al. In Review) and the fishery remains profitable. Despite the fishery's profitability, it bears the hallmarks of unregulated fishing with severely depleted stocks of the valuable species and extraordinarily high exploitation rates in some years, and very low catch rates (Skewes et al. 1999; Prescott et al. 2013). The ecosystem effect of removing such a high proportion of the sea cucumber biomass from Scott Reef is unknown (Prescott et al. 2013). Governance arrangements for the fishery were undefined by the MOU. Moreover, the remoteness of the fishery from central authorities in Canberra and Jakarta means that governance interaction is practically difficult and costly. Fishers infrequently see fisheries staff and would scarcely be aware of their existence aside from the strong MCS presence in the fishery which exists as part of Australia's border protection program.

In the past, loans were made by the District of Rote Fisheries Department to some Rotenese crews to assist them meeting the financial burden of voyaging to Scott Reef. This was discontinued when loans were not repaid (Personal communication, J. Riwu, October 2015),[3] but in effect was a temporary government subsidy for the over-fished fishery. Subsidized safety and fishing equipment, provided respectively by Australia and Indonesia, and safety equipment from the FAO Regional Fisheries Livelihoods Program have promoted safety, which is positive in terms of safe work conditions but may have also encouraged greater participation. Research has been funded by the Australian government with contributions from Indonesia, while Australia entirely funds the compliance program, which costs many times the value of the fishery, but ensures compliance with requirements for non-mechanized propulsion and fishing equipment.

## Summary of Trans-Boundary Fisheries

Both of the trans-boundary case study fisheries operate under a bilateral treaty or MOU. Both fisheries target high-value species which are predominantly exported to China. Both fisheries have also remained small-scale amid measures that have strictly limited the vessels' size (Torres Strait) or propulsion (Scott Reef). There are also similarities between the fisheries in terms of many fishers working on boats owned by others who also often purchase and on-sell the products (patron-client). However, this is where the similarities end. Poverty is a feature of fishers' communities in Indonesia and PNG, but not in the indigenous Torres Strait communities of Australia. The Torres Strait Treaty explicitly has the objective of 'optimum sustainable yields' and formulas for sharing that yield that the MOU does not. By adopting an optimum sustainable yield management objective, the Treaty created a need for the science necessary to estimate it. This led to research programs and a time series

[3] James Riwu, District of Rote Department of Fisheries.

of fishery independent surveys and catch and effort data that never existed for the poorly documented Scott Reef fishery.

The TSTRLF compliance program, while less robust than some others in Australia because of the compromises made to protect traditional fishing rights, has integrity in terms of licensing, seasons, and size limits and has been adequate to prevent overfishing under the present fishing capacity constraints. Indigenous fishers in the Australian TSPZ have strong cultural and economic ties to the fishery, however, economic dependence is moderated through Australia's social welfare program(s). Indonesian fishers have no equivalent access to comparable programs and therefore may have a higher dependence on fishing which is likely to challenge management.

Governance in the TSTRLF is demonstrably interactive. Over the past three decades, stakeholders have become progressively more engaged in management through stakeholder advisory groups, and may have direct access to the decision making PZJA. Through this process, PNG, though not formally part of the Australian fisheries governance regime, has also been drawn into that system. In contrast the Scott Reef fishery governance is poorly defined and interaction is almost non-existent. Despite some similarities, the differences have left the sea cucumber fishery severely depleted while the TSTRLF continues to enjoy high catch rates, albeit fluctuating inter-annually.

## *Case Studies from Fisheries Adjacent to Australia*

### The Rotinese Purse Seine Fishery (Multispecies Fish and Squid)

The small-scale purse seine fishery around the island of Rote Ndao (Rote) in Nusa Tenggara Timur province is comparable to many small pelagic purse seine fisheries in eastern Indonesia. It is estimated that the catch of this fishery supplies as much as 90% of the fresh fish sold on the island (Personal communication, J. Riwu, April 2014), making it an important source of food and income. All purse seine vessels must be licensed, but licences are uncapped and fishing licence fees are a 'lucrative' source of revenue for the District as they are in other maritime districts, municipalities and provinces (Resosudarmo et al. 2009). However, subsidies, as fishing equipment including entire fishing units (e.g. hulls, engines, and nets) (Prescott et al. 2015), and national fuel subsidies which moderate fuel costs exceed the licensing revenue from the fishery.

Fishing effort is influenced by the seasonal winds. However, there are no formal closed seasons to limit effort or potentially protect fish stocks at spawning times. Fishers and fisheries officials anecdotally note that catch rates in the fishery have fallen precipitously despite the fishery using only manual labour.

Catch landing data are incomplete and there is no robust statistical procedure to estimate the uninspected catches. It is also likely that some catches leave the district waters without being first landed there. These conditions leave the fishery without a robust monitoring program.

Fisheries governance in Rote is strongly hierarchical and lacks mechanisms and intent to facilitate interaction between the decision makers and stakeholders, thus limiting opportunities for interactive governance. Importantly, there are no clear management objectives for the fishery which could assist governors to formulate policy and enact laws to sustain the stocks. MCS is weakened by the lack of precise laws, infrastructure with which to control fishing in the District, and trained fisheries compliance officers. The jurisdiction has porous borders thereby leaving small-pelagic stocks exposed to exploitation by external as well as internal actors. Complex jurisdictional arrangements further complicate the fishery with vessels notionally licensed by size to particular district, provincial, and national governments, often operating well beyond their jurisdiction for which they are licensed (Resosudarmo et al. 2009).

Waters around Rote have become part of the Savu Sea marine protected area (MPA) which is Indonesia's largest. However, the comparatively small no-take areas of the MPA appear unlikely to have any measurable impacts on the abundance and productivity of small-pelagic species harvested by the fishery since individual small protected areas are less effective than large areas because of boundary fishing (Walters 2000), and particularly so given the life history of small pelagic fishes (Fréon et al. 2005).

### The Kei Flying Fish-Roe Fishery (Multispecies Tropical Fish)

In eastern Indonesia, seafaring ethnic groups, particularly the Bugis and Butonese from south and southeast Sulawesi, have for generations collected flying-fish roe (Zerner 1987). In recent decades, the fishery has progressively moved to Indonesia's more eastern waters following the collapse of former fishing grounds under pressures from growing market demands and expanding per capita fishing capacity. Consequently, a dramatic increase in exploitation of what are thought to be relatively intact flying-fish stocks in the Kei Islands (Maluku province) has occurred over the last decade, primarily by external fishers. Localized response is evident in the establishment of a community-based roe fishery management regime around a small island community, Tanimbar Kei.

Since 2009, Tanimbar Kei has experienced intensive annual influxes of Bugis and Butonese fishers targeting spawning schools of flying fish between the months of April and October. The initial nine recorded vessels in 2009 expanded to the current 90–120 vessels that consistently return in season. Since 2012, local management and control interventions are being applied as part of a community fisheries management regime, which functions under a co-management arrangement with an NGO.

Fishing vessels, ranging between 10 and 15 m with crews of three to five men, operate around the island. On average, crews spend 6–14 h daily harvesting roe, involving deployment of locally made fishing aggregation devices (FADs) on which the flying fish deposit their roe. Crews operate from makeshift island bases where they prepare FADs for deployment and dry harvests of fish eggs. After drying, a majority of the roe is packed and sent to 'bosses', boat owners or middlemen in export collection centres (e.g. Galesong, South Sulawesi) where it is processed and shipped to overseas markets. Crews from time to time sell portions of harvest to local middlemen to cover immediate operating costs. Although roe of several species of flying fish (*Exocoetidae*) is collected, we treat this fishery as a single-species fishery since fishing effort is not geared towards one species over another. The collection of eggs is from a group of species that have similar ecology, behaviour and for which the market is relatively uniform.[4]

The Tanimbar Kei community has exclusive management and traditional ownership rights over their defined marine territory, which has allowed the establishment of the island's management regime. The national decentralization of government following Indonesia's political reform in 1999 devolved political and resource management authority to lower administrative levels. The passing of a national village autonomy act, first declared in 2004 and later refined and amended under Law No. 6/2014 concerning villages, allowed village level claims to territory to be passed based on proven traditional access. Such rights are furthermore echoed in context of fisheries under Law 31/2004 concerning fisheries, where Article 6 (2) states that "fisheries management for capture fisheries and fish culture must take into account any existing customary laws and local wisdom including the community participation". Law No. 27/2007 of Coastal and Small Islands Management, later revised by Law No. 1/2014, also acknowledges the traditional rights of communities to manage their own coastal waters. Building on these legislative opportunities, the island's management regime allows a defined number of boats to operate in the waters throughout the roe season, all of which must register and pay the seasonal retribution which gives fishers access to fresh water and storage space on the island, and provides them with mooring sites. Furthermore, fishing effort is monitored and basic catch data is collected throughout the season from a sample of fishing vessels.

Without comprehensive fisheries data available, management arrangements are designed to be broadly restrictive. Spatial management is achieved through an annual cycle whereby no take zones are allocated in Tanimbar Kei's territory. Fishing effort is controlled through setting seasonal quotas on how many vessels may operate. Although such restrictions are noted in the island fishery management plan, the capacity to enforce compliance with these measures is limited. External fishers' knowledge of these management measures appeared limited. Moreover, many crew claimed that they fished well beyond the defined island territory, thus arguing that their presence in Tanimbar Kei waters is only a mere fraction of their total coverage.

The management regime has arguably produced positive fisheries outcomes, including monitoring and control measures preventing complete open access to the

[4] Fishers identified different 'size classes' of eggs, which in turn have different value. Size variation is likely indication of different subspecies.

resource, societal benefits in the form of communal income, formulation of local enforceable fishery rules and regulations, and a sense of value and ownership over resources amongst recognized custodians. Benefits of the management thus feed largely back to the Tanimbar Kei community. However, critical assessments show lack of consistent control and inconsistent application of sanctioning and penalties due in part to socio-cultural relationships and accountabilities. Moreover, the lack of comprehensive serial data prevents robust fishery management measures from being developed and their performance from being evaluated. Although long term, annual catch monitoring and technically intensive stock assessments might help rectify this data gap, these are improbable in the context of this fishery. Therefore, responsive management is most applicable and manageable in this case.

Although the roe fishery management forms one component of a larger and more holistic resource management program (including various other aquaculture and fisheries components, like seaweed farming, coral reef fishing, etc.), the specific management interventions applied to the fishery are strongly focused on the species rather than the broader ecosystem. Since restrictive measures were generally perceived undesirable, seeing clear links between directed action and resulting outputs were highly important for 'new' locally unfamiliar management tools to gain support amongst local managers. As a result, consensus on which control and management measures were to be applied depended on how visible the direct connection to the fishery was and how intrusive the secondary effects would be on the broader community. Fisheries management therefore remains firmly focused on the particular fish stock in question rather than the broader ecosystem.

## Summary of Fisheries Adjacent to Australia

The adjacent fisheries case studies are characterized by weak engagement of state institutions in governance and weak control of trans-jurisdictional fishing capacity. The comparatively open access circumstances in these fisheries suggest that they are likely to be exploited to the point that they are no longer profitable, which almost always corresponds to stocks at lower than optimal ecological levels (Bjørndal and Conrad 1987). Fishers may respond by resorting to lower cost methods of fishing to maintain profitability, which often having higher destructive impacts, leading to what is sometimes referred to as Malthusian overfishing (McManus 1997). Various subsidies in these fisheries further reduce fishing costs and thereby encourage even more fishing. Although there are formal tenure arrangements in the Kei case, in neither case is the access tenure of participants effectively controlled. Informal tenure is moreover threatened by erosion of the resource base.

The fishery baselines are poorly defined in terms of catches, catch rates, or abundance of the stocks. Combined with the fact that the current data from the fisheries are not collected comprehensively and the future prospect for assessing the fisheries is remote, important management issues may go undetected and promoting better management is made more difficult. Under these circumstances, considering broader, ecosystem aspects of EAF appears overly ambitious.

## Discussion

In reviewing all six objectives formulated under the SSF Guidelines, it appears that giving effect to five of them arguably depends on the third objective being met first, namely "(c) to achieve sustainable utilisation, prudent and responsible management and conservation of fisheries resources consistent with the Code of Conduct for Responsible Fisheries" (FAO 2015, 1). Maintaining the productivity of the fishery resource base, or restoring it as necessary, is the foundation on which each of the other Guideline objectives rest. While multiple strategies may meet this critical objective, they are not all equal in terms of their efficiency. Technical attractiveness aside, each small-scale fishery requires practical and efficient strategies to reach sustainability – giving each the best chance to 'beat the odds', which empirical evidence suggests is stacked against them. This is effectively captured in the SSF Guidelines' 13th guiding principle, that notes how "policies, strategies, plans and actions for improving small-scale fisheries governance and development [need to be] socially and economically sound and rational" (Ibid, 3).

Although the two Australian fisheries case studies may not be viewed as equally ecologically sustainable by every stakeholder, both did meet the sustainability criteria specified in the Australian "guidelines for the ecologically sustainable management of fisheries". Yet it is hard to argue that these Australian fisheries have overtly taken an ecosystem approach to reach their respective sustainability goals. Robust data collection and research programs supported stock assessments carried out on single species or congeners. The stock assessments were then used to set TACs in the QRLF, or TAE in the NDSF. Fishing capacity was, by and large, matched with the stock to be caught. In the NDSF this was partially achieved by eliminating a large foreign industrial trawl sector.

Most harvested species are monitored but not formally stock assessed. The interactions of target and non-target species with the wider ecosystem (e.g. as predators, prey, or competitors) are not formally assessed or taken into account when setting TACs/TAE other than by ecological risk assessment. It is also hard to demonstrate that many Australian fisheries are managed by balancing 'diverse societal objectives'. In the TSTRLF the objectives of indigenous and non-indigenous sectors remain highly polarized. Plagányi et al. (2013), however, have more recently undertaken research to better understand social objectives, which is a step towards this goal.

MPAs are often a priority in EAF programs, particularly under conservation-oriented programs, however they have little importance in the management of two of the Australian fisheries while in the QRLF they have had importance in some heavily fished areas (Mapstone et al. 2004). Buxton et al. (2014) demonstrated, using a relatively simple model, that MPAs can have positive spill-over effects, particularly when fisheries are seriously overfished and when spill-over species have low rates of movement and high growth rates. The first of these conditions is probably met where most MPAs are proposed in conservation framed regions like the Coral Triangle. However, even if this and the other conditions are met, MPAs must still contend with ineffective governance, compliance, monitoring, and excess fishing capacity – the reasons why fisheries management failed and MPAs are needed.

We argue that the key factor in the sustainability of Australian fisheries were the effective systems of governance through which 'single-species' fisheries management was applied. Facilitating this was fisheries legislation with clear and coherent objectives written using precise language for administrative clarity which assists both the application of the laws as well as providing certainty for stakeholders (Winter 2009). However, we acknowledge that governing Australian fisheries has generally benefited from the favourable human conditions that exist in Australia. Against the World Bank governance indicators including 'control of corruption', 'rule of law', and 'regulatory quality' Australia ranks among the highest in the world and is clearly different from neighbouring states in the Asia-Pacific region (World Bank 2016). This contributes to Australian fishers' confidence that their tenure to the resources is secure and there are precedents for compensation for partial or complete loss of tenure where spatial management removes resources from their access (e.g. through the GBRMP rezoning exercise). Fishers' livelihoods are not threatened by serious, subtractable, unregulated, or non-compliant fishing by others because compliance programs, grounded on rule-of-law, are effective and compliance with the laws by fishers is generally high. Low population densities, low unemployment, and greater per capita wealth found in Australia also foster governability. Most, but not all, Australian fishers seem to perceive that they have benefited from the governance and management systems. In other fisheries where this is not the case attitudes towards exploiting the resource to get what one can before it is appropriated by someone else are a natural outcome.

In other contexts, such as those in our adjacent fisheries case studies, achieving sustainable fisheries may meet insurmountable obstacles and setting goals of more complex EAF will almost certainly compound the challenges. The case of the flying fish roe fishery indicates the importance of visibly consequential management, in order to gain legitimacy among local stakeholders for potentially restrictive interventions. Where a national system's governing infrastructure is weak and communicable law and legislation rarely functions well on the ground, immediate management interventions that have clear and favourable objectives have greater currency. In such contexts, single-species management may form a more logical departure point that local fishers and dependent households comprehend while other seemingly unrelated interventions that fit within broad complex EAF ambit but have imperceptible outcomes are unlikely to be similarly perceived. With the legislative arena in which the fishery operated (i.e. Tanimbar Kei territorial waters) being small, meant governability improved, at least locally. Although, the fishery's spatial domain is far larger than the 'local' area, suggesting that Tanimbar Kei's measures do not determine the state of health of the whole system. Nevertheless, if fisheries stakeholders are to comply and support management, outcomes such as at Kei emphasise the need for governance that is comprehensible and legitimate.

## *Recommendations to Promote Sustainability and Lay the Foundations for Progressive Improvement*

Successful adoption of the SSF Guidelines across a wide spectrum of small-scale fisheries will ultimately rest on their compatibility and proven utility under varying contexts. With this in mind we revisit our fisheries case studies in and near Australia, build on relevant lessons from the Australian fisheries and draw on the Guidelines to suggest practical ways to achieve more sustainable outcomes.

### Appropriate Modes of Governance

In many fisheries management discourses, addressing governance shortfalls appears secondary to achieving 'technical' management outcomes. For example, implementing a total allowable catch (TAC) may be seen as a high priority and an extremely important outcome. However, in many fisheries across the developing country context of the Asia-Pacific where this has been attempted, ineffective governance has hampered capacity to control the catch. Fisheries experts drawing from successful experiences in implementing a TAC often do so from cases where appropriate and effective governance systems are in place to support such fisheries management. In taking appropriate governance conditions for granted, small-scale fisheries management discourses are driven across broader contexts with insufficient acknowledgement of its importance. Alternatively, governance issues may simply be deemed too overwhelming. Either way, weak fisheries governance, which is "expressed in terms of biological overuse, loss of economic productivity, costly management, and inequitable processes and outcomes" (Hanna 1999, 284), seems consistent with, and a likely explanation for, the state of many fisheries in the Asia-Pacific. It thus remains the central problem for achieving sustainability.

Jentoft and Chuenpagdee (2009) argued that fisheries governance faces particularly 'wicked' problems, on the basis that they are complex and difficult merely to specify. Despite their wickedness, our Australian case studies indicate that opportunities to resolve wicked problems do exist, though solutions for wicked problems in other contexts are unlikely to be found as easily. Jentoft and Chuenpagdee (2015) suggested that moving from the typical hierarchal governance approach to co- or self-governance to enhance governability in small-scale fisheries may be necessary in certain contexts and this suggestion is consistent with the SSF Guidelines.

Among our case studies are examples which likely would do better under different modes of governance. The traditional fishery at Scott Reef may benefit most by drawing on Australia's existing system of governance even though the fishery is effectively Indonesian. Given Australia's experience in interactive governance and the fact that the fishery is trans-boundary in nature, takes place in Australian waters, and involves different ethnic groups competing for access, there are multiple advantages for this governance path. Such an approach would, however, require considerable commitment of resources to support as well as to learn from such an 'experiment'. Complementary governance arrangements in fishing communities, where decisions about who could participate and when, would be also needed since there is an imbalance between the number of fishers and the resource that requires less fishing to resolve.

The Rote purse seine fishery would seem to benefit from enhanced and more interactive governance at the level of the district which controls licensing and therefore fishing capacity, however it may benefit equally from more traditional complementary community governance. Given the nature of movements by small pelagic fish there is also a need for governance at a higher level with broader geographic reach.

The flying fish fishery relies on a mode of co-governance where local aspects of the fishery are governed by the community while the district and province must govern at broader geographic scales given the connectivity between the Kei fishery and those around it. Mediating stakeholders in the form of, for example co-management partners, becomes critical in facilitating a form of interactive governance between formal governing agencies and those on the ground (Cohen and Steenbergen 2015).

### Governance of Tenure

There is strong evidence from around the world that some form of tenure provided to fishery participants is essential (Allison et al. 2012), regardless of the prevailing mode of governance. This is verified in the flying fish roe fishery case in Kei. The SSF Guidelines similarly emphasize that secure tenure of access is "central for the realisation of human rights, food security, poverty eradication" (FAO 2015, 5). However, it is also clear that not all conventional wisdom on property rights resolving the problems of sustainability and, in particular, promoting economic efficiency in fisheries, is directly applicable to many small-scale fisheries contexts. As Allison et al. (2012) point out, possible casualties of ill-conceived property rights include people's fulfilment of the right to fish for food security and for distributive justice.

Creating secure tenure in fisheries suffering the effects of overfishing, where the levels of participation and fishing capacity exceed sustainable limits, is a particularly wicked problem, but nevertheless a critical one to overcome to achieve a satisfactory end. In such situations, catch and capacity must be reduced but, where there are high levels of dependency, support must be provided for any realistic prospect of meeting the governance, social, economic, and nutritional imperatives. Where fishing effort is too high but dependence is relatively lower, it may be possible to assign tenure through a rotational system which fairly re-distributes a reduced level of access. The beginnings of this are seen in the Kei flying fish fishery. It is likely that the traditional fishery at Scott Reef may also fit into this category.

Regardless of how tenure of access is created it is important that excess capacity is not displaced to other fisheries where access is more open. Promoting alternative livelihoods is a common approach to manage human capital to limit displacement by improving its (labour) mobility. However, much more effort needs to also be applied to managing capital tied up in vessels and fishing gear. After all, many small-scale fishers will be unable to rationalize limiting their fishing effort if it means their vessel sits idle, particularly given that is likely to be their single most valuable capital asset. However, even this entails risk as Clark et al. (2005) also demonstrated that buyback subsidies can have severe disadvantages in the longer term where they are anticipated.

## The Right to Share Knowledge

Fisheries governance and even single-species management are complex processes that require knowledge of them and of the social, economic, and other technical matters to which they apply. Traditionally, most 'accepted' knowledge resided with certain actors or groups of actors forming, to varying degrees, hierarchies of power. Governors and managers had knowledge of these systems while scientists had knowledge, usually far from complete, of the resources. Fishers have always had knowledge of the aspects of the human and natural systems that were relevant to them which today may or may not conflict with the dominant Western knowledge system that pervades EAF.

Open systems for knowledge exchange and information sharing are critical for interactive governance. In the Australian fisheries, we highlighted processes (granted imperfect ones) that promote information exchange and facilitate participation in the science and management of the fisheries, and effort to improve these continues including the broadening of stakeholders' inclusion. The SSF Guidelines state in 12.1 that "states and other parties should enhance the capacity of small-scale fishing communities in order to enable them to participate in decision making processes" (FAO 2015, 17). In addition, in 12.3 the SSF Guidelines go on to state that "all parties should recognize that capacity development should build on existing knowledge and skills and be a two way process of knowledge transfer", (Ibid, 17).

These are fundamental principles and the likelihood of achieving sustainable fisheries without giving effect to them is remote. We are confident that it is possible to build local capacity by employing a pedagogy relevant to small-scale fisheries and investing additional resources to build the capacity of institutional actors to be better 'teachers' and listeners. It is equally important and productive to create opportunities for fishers (and other stakeholders) to participate in the collection of the data as they have in the sea cucumber fishery at Scott Reef (Prescott et al. 2016), and which forms the basis for small-scale fisheries co-management in Tanimbar Kei (Steenbergen 2013, 2016). Fishers can inform assessments about observed patterns of abundance, stock movements, or other observable factors which, if not accounted for, usually bring the credibility of assessments into question. Among our case studies, fishers are contributing their knowledge in several assessments; however, this process is still poorly developed and poorly utilized, which goes back to capacity building of all actors as well as formally adopting institutional processes to support this.

### Creating Opportunities for Success

Jentoft and Buanes (2005) identified that avoiding 'negative' rather than meeting 'positive' goals that are unfamiliar to fishers may be a more productive discourse to develop. Although the cognitive baseline in fisheries is constantly shifting (Pauly 1995; Ainsworth et al. 2008) which makes it difficult for many fishers to fully accord with the resource losses, most small-scale fishers are likely to recall times when fisheries conditions were better and it is likely that they are concerned about further losses. There is anecdotal evidence for this in Rote, Scott Reef, and Kei. It is much less likely that fishers have experienced a stock recovering so a discourse based on improving sustainability may be completely unfamiliar and abstract. It is better to open discourses with familiar experiences and over time introduce concepts of improving sustainability and demonstrating this (as far as possible) through experiential capacity building.

With a few exceptions, fishing capacity is poorly controlled in Indonesian waters, which contributes to the serious levels of illegal fishing. However, the Indonesian Ministry of Marine Affairs and Fisheries, under the leadership of Minister Susi Pudjiastuti, has made significant progress in removing illegal foreign fishing vessels from its waters. This is tremendously important because unless illegal fishing is eliminated there is little prospect to control legal fishing capacity since it is unlikely legal fishers (at all scales) could reconcile reducing their capacity while witnessing illegal activities.

However, rather than economic incentives to control and reduce fishing capacity, many governments continue to subsidize fishing directly or indirectly. While the SSF Guidelines are silent on subsidies, as they are often applied, many subsidies remain a threat to long-term sustainability. Re-directing 'bad subsidies' (Khan et al. 2006) may be possible and could produce immediate impacts on both sustainability and providing new financial resources needed to implement many of the SSF Guidelines' recommendations. Unlike many aspects of fisheries, subsidies are one of the few institutionally controllable factors and, political objectives aside, could be readily adjusted.

We have sought to identify the most effective mode(s) of governance, understand and deal with the factors underlying governability, and then to apply simple management approaches to build a solid foundation which can be later improved rather than taking an immediate EAF path to sustainability. Others, like Patrick and Link (2015) have counter-argued that EAF is no more complex than the simple single-species management approaches we advocate. However, we find it unlikely that a system of management committed to dealing not only more fully with the complexity of ecosystems' natural elements but also with the human ecological interactions, is less or equally complex to one that focuses purely on the catch management of a single species. Regardless of any differences in complexity, it is not necessarily the actual degree of complexity that is relevant in the context of 'doing management' but rather the complexity that is constructed by the broad range of actors implicit in EAF. Applying fisheries management is usually painful for all actors and each additional issue becomes an opportunity to remain fixed in a relatively more comfortable position of stasis while the issues are debated and delay in making hard choices is legitimized. This trend is termed 'rational inaction' by Walters and Martell (2004). We therefore urge those engaged in implementing the SSF Guidelines to look beyond the technical differences between the various management approaches and consider what human constructions may be made of them.

The noble objective to 'balance diverse societal objectives' is one aspect of EAF that is likely to consume many resources and remain most challenging. Sumaila (2005) presents a view of societal objectives in the context of discount rates which, as he notes, reflect societies' relative weighting of benefits received versus at some time in the future, for example as a stream of sustainable catches. Poverty and indebtedness are conditions endemic in many parts of the world, including in the Asia-Pacific and are identified as two major factors shaping societies' views of the future. Furthermore, pursuit of EAF at wide geographic scales means that societies with very different wealth, health, and livelihood opportunity characteristics are likely to collide in programs attempting to operationalize EAF. For example, private and public donors, ENGOs, and aid agencies in wealthy countries fund EAF in less wealthy parts of the world. Aligning the objectives and implicit expectations of the donors with the objectives of the primary stakeholders under circumstances of extreme diversity effectively becomes a mathematical impossibility (Sumaila 2005). This situation emphasizes the need to create more fertile conditions for inter-societal communication and knowledge sharing which will assist implementation of the SSF Guidelines. However, this effort will always require walking a fine line between many competing and often conflicting objectives.

## Conclusions

Many 'sustainable' fisheries achieved this status through relatively simple single-species management approaches. However, reaching a sustainable level of harvest alone in a fishery falls well short of achieving all the objectives of the ecosystem approach to fisheries championed by the SSF Guidelines. In light of management experience and more information about fisheries' interactions with their ecosystems, there will be opportunities to adaptively and incrementally improve management and governance. Progress must be made soon, as it is likely that management of natural resources will become more difficult as human populations grow. For example, based on World Bank (2016) data, a growth rate 1.14% per year is implied, suggesting a corresponding 24 million additional people in coastal areas adding to the current population of 130 million people living in the Coral Triangle who derive their livelihoods directly from marine ecosystems (CTI-CFF 2013). Further, as fisheries resources diminish, they become less likely to recover (Neubauer et al. 2013). In addition, this recovery is even less likely to occur at time scales that matter to dependent and often marginalized fishing communities.

In the Australian context, and undoubtedly every other context where fisheries have been sustainably managed, the process has been enabled through effective governance and the promotion of local social, economic, and other factors that enhance governability. In Australia, governance has been interactive, with the result that governing systems have been better informed while also creating a sense of participation and ownership among stakeholders. These are both essential outcomes. Australia has also been assisted in managing its fisheries by precise management objectives and implementing laws. Nevertheless, agreeing on measures and applying them was not easy. Ray Hilborn[5] made the observation that in Australia's commonwealth fisheries, 'there was a lot of blood on the floor' when more restrictive measures were introduced, which was a reference to the many heated management meetings between governments and fishers. Escaping this confronting process is unlikely when 'doing fisheries management'.

The contexts in which fisheries are managed in the region adjacent to Australia are very different and we do not suggest Australia's approach can produce equivalent outcomes. However, some factors seem to be directly relevant and include: starting with the application of appropriate single-species approaches; developing an interactive governance system; and applying suitable management measures. The modes of governance through which these interventions are delivered may be very different than in Australia but must be inclusive, interactive, and adaptable as experience grows. Sustainability may be substantially and quickly improved by finding the most appropriate mode of governance whereby rule of law is strengthened because, for instance, communities still have strong affiliation to traditional law. Therefore, the means to implement the SSF Guidelines needs to be chosen carefully and the success of implementation judged within the full context in which they are implemented, which includes the multiple dimensions of space, time, culture, starting points, and capacity. The SSF Guidelines are there to help small-scale fisheries precisely because help is needed. We advocate that the help provided must be rooted in these contextual realities and not in theory.

## References

Ainsworth, C. H., Pitcher, T. J., & Rotinsulu, C. (2008). Evidence of fishery depletions and shifting cognitive baselines in Eastern Indonesia. *Biological Conservation, 141*(3), 848–859.

Allison, E. H., Ratner, B. D., Åsgård, B., Willmann, R., Pomeroy, R., & Kurien, J. (2012). Rights-based fisheries governance: From fishing rights to human rights. *Fish and Fisheries, 13*(1), 14–29. doi:10.1111/j.1467-2979.2011.00405.x.

Andersen, C., Clarke, K., Higgs, J., & Ryan, S. (2005). *Ecological assessment of the Queensland coral reef fin fish fishery*. A report to the Australian Government Department of Environment and Heritage on the ecologically sustainable management of a multi-species line fishery in a

[5] Ray Hilborn, Professor at the School of Aquatic and Fisheries Sciences, University of Washington; discussion held at Commonwealth Scientific and Industrial Research Organisation (CSIRO) in 2011.

coral reef environment. Brisbane: Department of Agriculture Forestry and Fisheries, Queensland Government.

Barkin, J. S., & DeSombre, E. R. (2013). *Saving global fisheries: Reducing fishing capacity to promote sustainability*. Cambridge, MA: MIT Press.

Bellwood, D. R., Hoey, A. S., & Choat, J. H. (2003). Limited functional redundancy in high diversity systems: Resilience and ecosystem function on coral reefs. *Ecology Letters, 6*(4), 281–285. doi:10.1046/j.1461-0248.2003.00432.x.

Bjørndal, T., & Conrad, J. M. (1987). The dynamics of an open access fishery. *The Canadian Journal of Economics/Revue canadienne d'Economique, 20*(1), 74–85. doi:10.2307/135232.

Buxton, C. D., Hartmann, K., Kearney, R., & Gardner, C. (2014). When is spillover from marine reserves likely to benefit fisheries? *PloS One, 9*(9), e107032. doi:10.1371/journal.pone.0107032.

Clark, C. W., Munro, G. R., & Sumaila, U. R. (2005). Subsidies, buybacks, and sustainable fisheries. *Journal of Environmental Economics and Management, 50*(1), 47–58. doi:10.1016/j.jeem.2004.11.002.

Clifton, J. (2009). Science, funding and participation: Key issues for marine protected area networks and the Coral Triangle Initiative. *Environmental Conservation, 36*(2), 91–96.

Cohen, P. J., & Steenbergen, D. J. (2015). Social dimensions of local fisheries co-management in the Coral Triangle. *Environmental Conservation, 42*(03), 278–288. doi:10.1017/S0376892914000423.

CTI-CFF. (2013). *Annex – 3 decision document on: EAFM working group and priority actions*. Paper presented at the The 9th CTI-CFF Senior Officials Meeting (SOM9), Manila, Philippines, 26–27 November 2013.

DFAT-Australia, & Republic of Indonesia. (1974). *Memorandum of understanding between the Government of Australia and the Government of the Republic of Indonesia regarding the operations of Indonesian traditional fishermen in areas of the Australian Exclusive Fishing Zone and Continental Shelf*. Jakarta: DFAT-Australia, & Republic of Indonesia.

FAO. (2003). *Fisheries management – 2. The ecosystem approach to fisheries* (Vol. 4, FAO Technical guidelines for responsible fisheries, Vol. 2). Rome: Food and Agriculture Organization of the United Nations.

FAO. (2015). *Voluntary guidelines for securing sustainable small-scale fisheries in the context of food security and poverty eradication*. Rome: Food and Agriculture Organization of the United Nations.

FAO & WFC. (2008). *Small-scale capture fisheries – A global overview with emphasis on developing countries: A preliminary report of the Big Numbers Project*. Rome: Food and Agriculture Organization of the United Nations (FAO) & World Fish Center (WFC).

Fréon, P., Cury, P., Shannon, L., & Roy, C. (2005). Sustainable exploitation of small pelagic fish stocks challenged by environmental and ecosystem changes: A review. *Bulletin of Marine Science, 76*(2), 385–462.

Hanna, S. S. (1999). Strengthening governance of ocean fishery resources. *Ecological Economics, 31*(2), 275–286. doi:10.1016/S0921-8009(99)00084-1.

Harris, A. J. (1991). The development of Australian fisheries management: A historical perspective 1800–1990. In *Ecologically sustainable development working group – Fisheries*. Canberra: Final Report AGPS.

Hilborn, R., Fulton, E. A., Green, B. S., Hartmann, K., Tracey, S. R., & Watson, R. A. (2015). When is a fishery sustainable? *Canadian Journal of Fisheries and Aquatic Sciences, 72*(9), 1433–1441. doi:10.1139/cjfas-2015-0062.

ICES. (2006). *Report of the working group on the ecosystem effects of fishing activities (WGECO)* (p. 117). Copenhagen: International Council for the Exploration of the Sea (ICES).

ICES. (2013). *Report of the working group on the ecosystem effects of fishing activities (WGECO)* (p. 117). Copenhagen: International Council for the Exploration of the Sea (ICES).

Jentoft, S., & Buanes, A. (2005). Challenges and myths in Norwegian coastal zone management. *Coastal Management, 33*(2), 151–165. doi:10.1080/08920750590.

Jentoft, S., & Chuenpagdee, R. (2009). Fisheries and coastal governance as a wicked problem. *Marine Policy, 33*(4), 553–560.

Jentoft, S., & Chuenpagdee, R. (Eds.). (2015). *Interactive governance for small-scale fisheries: Global reflections, Vol. 13, MARE Publication Series*. Cham/Heidelberg/New York/Dordrecht/London: Springer.

Khan, A. S., Sumaila, U. R., Watson, R., Munro, G., & Pauly, D. (2006). The nature and magnitude of global non-fuel fisheries subsidies. In *Fisheries Centre research reports* (Vol. Vol. 14). Vancouver: Fisheries Centre, University of British Columbia.

Lédée, E. J. I., Sutton, S. G., Tobin, R. C., & De Freitas, D. M. (2012). Responses and adaptation strategies of commercial and charter fishers to zoning changes in the Great Barrier Reef Marine Park. *Marine Policy, 36*(1), 226–234.

Mapstone, B. D., Davies, C. R., Little, L. R., Punt, A. E., Smith, A. D. M., Pantus, F., Lou, D. C., Williams, A. J., Jones, A., Ayling, A. M., Russ, G. R., & McDonald, A. D. (2004). *The effects of line fishing on the Great Barrier Reef and evaluations of alternative potential management strategies*. Townsville: CRC Reef Research Centre.

McManus, W. J. (1997). Tropical marine fisheries and the future of coral reefs: A brief review with emphasis on Southeast Asia. *Coral Reefs, 16*(1), S121–S127. doi:10.1007/s003380050248.

Moore, R., & MacFarlane, W. (1984). Migration of the ornate rock lobster, *Panulirus ornatus* (Fabricius), in Papua New Guinea. *Australian Journal of Marine and Freshwater Research, 35*, 197–212.

Murawski, S. A. (2007). Ten myths concerning ecosystem approaches to marine resource management. *Marine Policy, 31*(6), 681–690.

Neubauer, P., Jensen, O. P., Hutchings, J. A., & Baum, J. K. (2013). Resilience and recovery of overexploited marine populations. *Science, 340*(6130), 347–349. doi:10.1126/science.1230441.

Newman, S. J., Harvey, E. S., Rome, B. M., McLean, D. L., & Skepper, C. L. (2012). *Relative efficiency of fishing gears and investigation of resource availability in tropical demersal scalefish fisheries* (Vol. Final Report FRDC Project No. 2006/031, p. 72). Western Australia: Department of Fisheries.

Nowara, G. B., & Newman, S. J. (2001). *A history of foreign fishing activities and fishery- independent surveys of the demersal finfish resources in the Kimberley region of Western Australia*. Perth: Fisheries Western Australia.

Patrick, W. S., & Link, J. S. (2015). Myths that continue to impede progress in ecosystem-based fisheries management. *Fisheries, 40*(4), 155–160. doi:10.1080/03632415.2015.1024308.

Pauly, D. (1995). Anecdotes and the shifting baseline syndrome of fisheries. *Trends in Ecology & Evolution, 10*(10), 430.

Pikitch, E. K., Santora, C., Babcock, E. A., Bakun, A., Bonfil, R., Conover, D. O., Dayton, P., Doukakis, P., Fluharty, D., Heneman, B., Houde, E. D., Link, J., Livingston, P. A., Mangel, M., McAllister, M. K., Pope, J., & Sainsbury, K. J. (2004). Ecosystem-based fishery management. *Science, 305*(5682), 346–347. doi:10.1126/science.1098222.

Pitcher, C. R., & Prescott, J. (1991). Deep water survey for *Panulirus ornatus* in Papua New Guinea and Australia. *Lobster Newsletter, 4*(2), 8–9.

Plagányi, É. E., van Putten, I., Hutton, T., Deng, R. A., Dennis, D., Pascoe, S., Skewes, T., & Campbell, R. A. (2013). Integrating indigenous livelihood and lifestyle objectives in managing a natural resource. *Proceedings of the National Academy of Sciences, 110*(9), 3639–3644. doi:10.1073/pnas.1217822110.

Prescott, J., Vogel, C., Pollock, K., Hyson, S., Oktaviani, D., & Panggabean, A. S. (2013). Estimating sea cucumber abundance and exploitation rates using removal methods. *Marine and Freshwater Research, 64*(7), 599–608. doi:10.1071/MF12081.

Prescott, J., Riwu, J., Steenbergen, D. J., & Stacey, N. E. (2015). Governance and governability: The small-scale purse seine fishery in pulau rote, eastern Indonesia. In S. Jentoft & R. Chuenpagdee (Eds.), *Interactive governance for small-scale fisheries: Global reflections,*

*Vol. 13, MARE Publication series 13, pp. 61–84*. Cham/Heidelberg/New York/Dordrecht/ London: Springer.

Prescott, J., Riwu, J., Stacey, N. E., & Prasetyo, A. (2016). An unlikely partnership: fishers' participation in a small-scale fishery data collection program in the Timor Sea. *Reviews in Fish Biology and Fisheries, 25*(4), 679–692. doi:10.1007/s11160-015-9417-7.

Queensland Government. (2012). *Annual status report 2011: Coral reef fin fish fishery*. Brisbane: State of Queensland.

Resosudarmo, B. P., Napitupulu, L., & Campbell, D. (2009). Illegal fishing in the Arafura Sea. In F. Jotzo (Ed.), *Working with nature against poverty: Development, resources and the environment in Eastern Indonesia* (pp. 178–200). Singapore: Institute of Southeast Asian Studies (ISEAS).

Skewes, T. D., Dennis, D. M., Jacobs, D. R., Gordon, S. R., Taranto, T. J., Haywood, M., Pitcher, C. R., Smith, G. P., Milton, D., & Poiner, I. R. (1999). *Survey and stock size estimates of the shallow reef (0–15 m deep) and shoal area (15–50 m deep) marine resources and habitat mapping within the Timor Sea MOU74 box, Vol. 1, Stock estimates and stock status*. Brisbane: CSIRO Marine Research.

Smith, A. D. M., Fulton, E. J., Hobday, A. J., Smith, D. C., & Shoulder, P. (2007). Scientific tools to support the practical implementation of ecosystem-based fisheries management. *ICES Journal of Marine Science: Journal du Conseil, 64*(4), 633–639. doi:10.1093/icesjms/fsm041.

Steenbergen, D. J. (2013). *Negotiating the future of local 'backwaters': Participatory marine conservation on small islands in Eastern Indonesia*. Perth: Murdoch University.

Steenbergen, D. J. (2016). Strategic customary village leadership in the context of marine conservation and development in Southeast Maluku, Indonesia. *Human Ecology, 44*(3), 311–327. doi:10.1007/s10745-016-9829-6.

Sumaila, U. R. (2005). Differences in economic perspectives and implementation of ecosystem-based management of marine resources. *Marine Ecology Progress Series, 300*, 279–282.

Walters, C. (2000). Impacts of dispersal, ecological interactions, and fishing effort dynamics on efficacy of marine protected areas: How large should protected areas be? *Bulletin of Marine Science, 66*(3), 745–757.

Walters, C. J., & Martell, S. J. D. (2004). *Fisheries ecology and management*. Princeton: Princeton University Press.

Western Australian Department of Fisheries. (2009). *Application to the department of the environment, water heritage and the arts on the northern demersal scalefish managed fishery: Against the guidelines for the ecologically sustainable management of fisheries*. Perth: Western Australian Department of Fisheries.

Winter, G. (2009). *Towards sustainable fisheries law: A comparative analysis, IUCN environmental policy and law paper, Vol. 74*. Gland: IUCN, in Collaboration with IUCN Environmental Law Centre, Bonn, Germany.

World Bank. (2016). World Data Bank: Worldwide Governance Indicators. http://databank.worldbank.org/data/reports.aspx?source=worldwide-governance-indicators&preview=on. Accessed 18 Feb 2016.

Zerner, C. (1987). The flying fishermen of Mandar. *Cultural Survival, 11*(2), 18–22.

# Chapter 13
# Securing Sustainable Sami Small-Scale Fisheries in Norway: Implementing the Guidelines

**Svein Jentoft and Siri Ulfsdatter Søreng**

**Abstract** Emboldened by the UN's Declaration of the Rights of Indigenous Peoples and other international human rights instruments, the Sami of Norway have been criticizing fisheries authorities for being blind to their particular rights, interests, and concerns. With the Voluntary Guidelines for Securing Sustainable Small-Scale Fisheries (SSF Guidelines), they have yet another opportunity to raise their concerns, as the rights of indigenous peoples are addressed in many chapters and paragraphs throughout the SSF Guidelines. Sami claims to fishing rights have, however, been met with resistance from Norwegian fisheries authorities and non-Sami fisher organizations. The institutional reforms initiated in recent years to accommodate Sami demands have been limited and not eased tensions. If this experience of institutional reforms is anything to go by, there is reason to believe that the implementation of the SSF Guidelines will be far from smooth. This chapter addresses the obstacles that implementation is likely to meet and also what opportunities exist as far as Sami fisheries management and tenure rights are concerned. More particularly, the chapter describes the functioning of a new governing institution, the Fjord Fisheries Board (FFB), established in 2014 as a vehicle for securing the sustainability of Sami small-scale fisheries in the fjords of northern Norway. It is argued that any positive outcome as a result of the FFB's formation will depend on the its ability to integrate indigenous values, norms, and principles in its work and on the Norwegian government's willingness to back the FFB not just in principle but also in practice. For the Norwegian government, the FFB may become an important instrument to help Norway fulfill its commitment to the SSF Guidelines. The chapter also explores what lessons can be learned from the FFB experience as far as the implementation of the SSF Guidelines is concerned.

S. Jentoft (✉)
Norwegian College of Fishery Science, UiT – The Arctic University of Norway, Tromsø, Norway
e-mail: svein.jentoft@uit.no

S.U. Søreng
Agriculture Agency, Alta, Norway
e-mail: siriulfsdatter@hotmail.com



S. Jentoft et al. (eds.), *The Small-Scale Fisheries Guidelines*, MARE Publication Series 14, DOI 10.1007/978-3-319-55074-9_13

**Keywords** Small-scale fisheries • Sami • Indigenous rights • SSF Guidelines • Implementation • Legal pluralism

## Introduction[1]

In December 2013 a new institution, the Fjord Fishing Board (FFB), saw the light of day in Norway. The FFB was to play an advisory role in matters of particular interest to the indigenous Sami fishers who are mostly small-scale fishers and limited to the fjords of northern Norway. The FFB was the result of a long lasting struggle by the indigenous Sami to have their fishing rights acknowledged. This chapter explores the possibilities of the FFB succeeding in securing Sami small-scale fishing tenure in the fjord areas as a part of ensuring the future implementation of the Voluntary Guidelines for Securing Sustainable Small-Scale Fisheries (SSF Guidelines) in Norway. Whether the FFB represents the endpoint or is just a step on the way remains to be seen. What is clear is that Sami ambitions clearly go beyond this particular institution. The Norwegian government seems, however, unwilling to acknowledge fully Sami's rights claims. For instance, the government has rejected the idea that the Sami have historical rights to fish in the north.

In principle, Sami fishers have the same fishing rights as other small-scale fishers in Norway because the state has not traditionally differentiated between Sami and ethnic Norwegian fishers. In fact, the latter often lived side by side with the Sami the northern fjords. As long as Sami fishing practices and interests were not in direct conflict with the state's fisheries management system, there was no problem. However, a problem arose when a new quota system for cod was introduced in 1990.[2] It turned out that almost none of the Sami small-scale fishers qualified for a guaranteed vessel quota. The newly established Sami Parliament reacted to what they perceived was discrimination against Sami fishers (Davis and Jentoft 2001; Eythórsson 2008). Since that time, and emboldened by the UN Declaration for the Rights of Indigenous Peoples and other international human rights instruments, the Sami have put fishing rights issues on their political agenda. With the passing of the SSF Guidelines, which also talk about indigenous peoples, the Sami have yet another impetus and opportunity to advance their cause.

In response to legitimate Sami claims of common property rights and self-government, (and in conformity with the SSF Guidelines), Norway has modified its

[1] This chapter can be read as a follow-up to Søreng's (2013a) article 'Legal Pluralism in Norwegian Inshore Fisheries: Perceptions on Sami fishing Rights.' (*Maritime Studies* 12: 9 doi: 10.1186/2212-9790-12-9) and Jentoft's article: 'Governing Tenure in Norwegian/Sami Fisheries: From Common Pool to Common Property?" (*Land Tenure*, No.1, pp. 91–115 (2013).

[2] The cod quotas for individual fishing vessels, which replaced a total quota for all cod fishers north of the 62nd parallel, was introduced in response to a rapidly declining cod stock. Many small-scale fishers in northern Norway felt they were on the losing side of the policy, because the highest quotas seemed to be reserved for the biggest, most capital-intensive vessels (Eythórsson 2008).

fisheries governance system – with the FFB as the most conspicuous new institution. However, as we shall argue in this chapter, Norwegian fisheries authorities remain hesitant to recognize fully Sami historical tenure rights. Although established to support Sami small-scale fisheries, the FFB is an institutional compromise between Sami rights claims and Norwegian national interests, born in a context of ethno-political strife. The FFB may therefore be seen as a test case for what the implementation of the SSF Guidelines may involve, and what to expect when the SSF Guidelines meet practical and political realities on the ground. As would typically be the situation, new governing institutions for small-scale fisheries such as the FFB, are introduced into an already established governance structure, where power structures and paradigms are entrenched in a way that tend to work against small-scale fisheries and in favor of the status quo. Given the Norwegian fishing industry's outright opposition to make established rules more in tune with the needs of Sami fisheries, the FFB is likely to experience resistance.

The FFB could potentially be an important instrument for the Norwegian government to address the rights of indigenous Sami small-scale fisheries and communities in the context of the SSF Guidelines. It is likely that when the Norwegian government reports to the FAO (Food and Agriculture Organization of the United Nations) what it has done in terms of implementing the SSF Guidelines, the FFB will be mentioned. However, the FFB would not be alone in implementing the SSF Guidelines. The SSF Guidelines are not included in the FFB mandate although there is a clear overlap between the SSF Guidelines and the FFB mandate. The FFB could therefore also serve as a watchdog, monitoring what the Norwegian government is doing to fulfill its promises as far as the SSF Guidelines are concerned. The SSF Guidelines can also help ensure that the FFB gets the needed support and room of maneuver to fulfil its task.

The chapter draws on documents, public media reports, meeting attendances, and interviews with key actors associated with the FFB. It also uses information gathered by the authors over many years of following, and writing about, the governance of Sami small-scale fisheries and the political discourse surrounding it. The next section gives an account of the origin of the FFB, as well as why and how it was established. Thereafter, Section "Institutionalizing legal pluralism" outlines some analytical perspectives through which to interpret and reflect upon the SSF Guidelines. Section "The Fjord Fisheries Board (FFB)" summarizes the mandate, organizational design, and issues related to the working process during the first year of the FFB's existence. The discussion section focuses on the challenges and potential pitfalls that may determine the success of the FFB in securing Sami fisheries tenure and participatory decision making in Norway, while the conclusion summarizes some general lessons with regard to the implementation of the SSF Guidelines.

## Prelude: The Position of Sami Fisheries Rights

For the Sami settled along the coast and fjords of northern Norway, small-scale fishing has been a way of life and a source of subsistence and income for centuries (Paine 1965). Although the number of Sami fishers has declined in recent decades, as have the numbers of those employed in small-scale fisheries in Norway as a whole, small-scale fisheries are still important, but in need of institutional and economic support. Overall, the total population and number of Sami fishing communities have decreased significantly over the years (Broderstad et al. 2014). In 1990, in a report to the government, Smith characterized Sami fisheries as being in a 'five-to-twelve' situation, suggesting that time is running out for Sami small-scale fishing (Smith 1990). Between 2008 and 2013, the number of Sami employed in the fisheries sector dropped by 18% and today the future of the small-scale, indigenous Sami fjord fisheries looks uncertain (Brattland 2014). The FFB is considered an instrument to help secure what is left of Sami fisheries and their distinct indigenous culture. However, the task ahead is formidable.

The FFB resulted from a 2-year (2006–2008) investigation into Sami fisheries rights in Finnmark, the northernmost county of Norway. Finnmark is Norway's biggest county, with a land mass of the size of Denmark, and is home to the majority of the Sami population in the country. The Sami are better known for their reindeer pastoralism than for their fisheries. Traditionally, however, the Sami fjord economy comprised both small-scale fishing and small-scale agriculture (Paine 1965; Nilsen 2003; Eythórsson 2008).

By signing ILO (International Labor Organization) 169 Convention on Indigenous and Tribal Peoples, the UN Covenant on Civil and Political Rights and the UN's Declaration on the Rights of Indigenous Peoples, Norway committed herself to respecting the human rights of the Sami as an indigenous people, and securing the material basis for Sami culture, including fisheries. These commitments are also there in Norwegian law, most prominently in Paragraph 108 of Norway's Constitution, which says: "It is the responsibility of the authorities of the State to create conditions enabling the Sami people to preserve and develop their language, culture and way of life." This so-called 'Sami paragraph' was added in 1988, whereas the Constitution was enacted in 1814. An important incident that instigated the amendment of this paragraph was the damming of the Alta River in Finnmark, which triggered heated demonstrations and even hunger strikes in front of Norway's parliament in Oslo in 1979 and 1981. As a follow-up to the constitutional amendment, the Sami Parliament was established in 1989.[3] The government also initiated

[3]The Norwegian Sami Parliament (in Sami 'Samediggi') is the main political institution for strengthening the Sami's political, social, and cultural position. It is a democratically elected body comprised of 39 representatives elected from seven districts every 4 years. Only those listed in the Sami Electoral Register have the right to vote. The central government has transferred authority to the Sami Parliament in some areas, primarily those concerning preservation of Sami cultural heritage, education, language, and culture. The Sami Parliament is a mandatory body that has to be consulted on matters of special concern to the Sami population (www.samediggi.no).

an investigation into Sami indigenous land rights (Svensson 2002; Minde 2005) which eventually, in 2005, resulted in the so-called Finnmark Act that recognized Sami rights to ancestral land and terrestrial resources. It also transferred former state land in Finnmark to a new regional institution named the 'Finnmark Estate', for which the Sami Parliament and the County Assembly appoint the Board of Directors. However, what remained to be done after this new institution was established, was to investigate whether similar historical rights pertained to ocean space and marine resources. An inquiry into this started when the Coastal Fisheries Committee (CFC) was appointed in 2006, with a mandate to clarify "Sami and others [non-Sami residents'] rights to fish in the sea in Finnmark" (Brattland 2010; Jentoft 2013; Søreng 2013a).

In a series of local hearings (eighteen in total) in Finnmark that the second author monitored, the CFC made an effort to document historical usage and people's perceptions of traditional and inherited fishing rights (Søreng 2013a). Based on the documented information, and with reference to Norwegian and international law, the CFC concluded that fishers living in Finnmark indeed have a historical right to fish and that the Norwegian government should legally recognize and formally implement this right. The CFC's report included a complete proposal for a fisheries law for Finnmark (NOU 2008: 5) that also outlined how the regional management system should look.

For 3 years, the Ministry of Fisheries and Coastal Affairs[4] remained silent about the report while the media intensely debated the report's findings and recommendations. However, when the Ministry finally responded, it appeared to be largely negative towards the proposals. Indeed, the government argued that the existing Norwegian fisheries management system already sufficiently addressed indigenous rights. After consulting the Ministry of Law, the Ministry of Fisheries and Coastal Affairs decided not to support the idea of historical Sami fishing rights. Nor did it agree with the recommendation to create an autonomous co-management institution for Sami fisheries in Finnmark (Prop. 70 L (2011–2012).

As could be expected, the government's reaction met with criticism. The Sami Parliament and the chair of CFC (Prof. Carsten Smith) conveyed disappointment. The media was flooded with letters from angry small-scale fishers in Finnmark. However, the condemnation was not unanimous. The Norwegian Fishers' Association (NFA), which had been highly critical of the CFC report was, as could be expected, supportive of the government's conclusion. When the Sami Parliament, after passionate debate, finally voted on the Ministry's alternative proposal, which included an advisory function for the FFB plus a number of other measures pertaining to fishing and fish processing, a small majority of delegates of the Sami Parliament supported it. Those who voted for the Ministry's proposal argued that it was the best they could hope for now (Jentoft 2013).

[4] When the current government took office in 2012, it merged this ministry with another ministry. The new ministry is the Ministry of Trade, Industry, and Fisheries (https://www.regjeringen.no/en/dep/nfd/id709/). To simplify the term, ministry is used hereafter.

In 2013, the Norwegian Parliament passed a reformed version of the Marine Resource Act,[5] which confirmed the government's position on the Sami rights issue. The amendment included a new management institution, the FFB. As mentioned at the outset, this chapter aims to explore what this new creation may involve in practice, and the conditions under which it will advance. Will the FFB be able to live up to expectations, despite its limitations? In the context of the SSF Guidelines, how relevant is the FFB? Is it the answer to what the SSF Guidelines envisage and the way through which the Sami's aspiration for tenure rights can be met? After 2 years of operation, it is worth reflecting upon the experience of the FFB.

## Institutionalizing Legal Pluralism

In Norway, Sami use and property rights have traditionally been interpreted from a Norwegian rather than Sami perspective. This is partly due to 'translations problems' that have occurred between experts who investigated Sami rights and local Sami informants (Ween 2006). Legal experts, overwhelmingly ethnic Norwegians, lacked the cultural competence needed to identify and conceptualize the Sami's own justice principles. Sami is a language that few Norwegians speak. In fact, not even all Sami due to assimilation know the Sami language. This is particularly the case for Sami living in coastal areas where the assimilation process was harsher than in the inland (Minde 2005). Cultural barriers and geographical distance continue to hinder mutual communication and understanding between the Sami and the non-Sami. Consequently, Sami rights and the culture that underpin them have been underdeveloped in Norwegian courts and within the political establishment, including with regard to fisheries management.

In this respect, the situation for the Sami in Norway is not very different from that which exists in other post-colonial settings. The SSF Guidelines result from the recognition that small-scale fish workers are a marginalized group, politically, legally, and economically. This is even more the case for small-scale fishing people who are indigenous. Therefore, the SSF Guidelines document refers frequently to indigenous small-scale people and to the UN Declaration of the Rights of Indigenous Peoples. In Norway too, there is a need for legal empowerment of indigenous fishing people.

The legal pluralism literature argues for legal 'decolonization', meaning that established law should undergo reform by recognizing and integrating indigenous legal norms and principles or being replaced by them (Woodman 1995; Zips 2003). The Norwegian CFC investigations arrive at the same conclusion. Likewise, the new Sami paragraph in The Marine Resources Act of 2008 and the consequent establishment of the FFB are important steps in the same direction. In fact, one may see this as a *de facto* acknowledgement of the existence of 'legal pluralism' (Vanderlinden 1989) in small-scale fisheries in Norway, which involves the Sami.

[5] In Norwegian, it is 'Lov om forvaltning av viltlevande marine ressursar (havressurslova).'

Josefsen (2014) sees it as an opportunity for the indigenous Sami community more generally and the Sami Parliament particularly, not to 'break out' of the national governing system but to 'break in', i.e. gain formal access to fisheries governance decision-making in a way that was previously not possible. With a foot in the door, institutional innovation can happen, as was the case with the FFB. The following describes what this new institution is, what its mandate is, and how it is organized.

## The Fjord Fisheries Board (FFB)

*Organization* Although fisheries management institutions have existed at local and regional levels with considerable stakeholder involvement in Norway for a long time and have undergone reforms (Jentoft 1989; Jentoft and Mikalsen 1994; Søreng 2006, 2013b), never before has an institution like the FFB with a specific Sami mandate been established. The FFB covers the three northernmost counties in Norway: Finnmark, Troms, and Nordland, where most of the Sami fishers reside (Fig. 13.1). The FFB has a vast and heterogeneous area to look after, which includes numerous unique fjord ecosystems and fisheries.

**Fig. 13.1** Small-scale fishing vessels anchored up in Smørfjord (Smiervuotna), Porsanger, Finnmark (Photo credit: Svanhild Andersen)

The FFB board has six members appointed for 2 years. The Sami Parliament appoints three members and the three county administrations one each. The three county administrations and the Sami Parliament must consult each other when appointing new members. Members must have knowledge about coastal and fjord fisheries, and have experience of public affairs in general. Moreover, at least some of the members should be active coastal and fjord fishers. The FFB leader is chosen from the six members and serves for 2 years. If two members get an equal share of votes, they split the time between them. During the first year, the appointee of the Sami Parliament led the FFB. He regularly spoke to the press either to inform the public about the FFB, or to announce or defend decisions made. He also spoke out against the Ministry for not following the FFB's recommendations. After 1 year, due to the parity of votes, the member appointed by one of the county councils took over as leader. When taking the position, the new leader declared that the FFB represents a different fisheries management paradigm, one that privileges small-scale fishers.

*Mandate* According to the mandate (§ 1), formally given by the Ministry, FFB's goal is to strengthen fjord fisheries management in the three counties where Sami fisher communities are most prevalent.[6] More specifically, the FFB shall:

- Evaluate the situation and development of particularly (but not exclusively) Sami resource use, and recommend locally adapted measures, including those of the Sami, in order to support coastal communities. The measures should aim at securing new recruitment to the industry and building necessary infrastructure such as fish landing facilities. The FFB shall exhaust scientific and local/traditional knowledge about important spawning and nursery areas in the fjords, and consider regulatory methods (including gear use) in order to secure sustainable yield and local resource exploitation.
- Evaluate the positioning of the fjord lines within which vessels over 15 m are not allowed, and if necessary, recommend acceptable exemptions from the regulations.
- Evaluate the need for an annual additional cod quota for the open access (small-scale) group, and decide how to distribute it to strengthen the local fjord fisheries, particularly those of the Sami.
- Evaluate which municipalities/Sami areas should be included in the specific regulations concerning rights, cod quotas, and quota additions, as well as in the fisher census.[7]

The FFBs mandate is not clear in terms of its role, namely whether it shall serve in an advisory capacity or autonomous capacity, i.e., decide and enforce its own rules and functions. The mandate only instructs the FFB to *assess* the situation in fjord fisheries with a particular focus on the Sami. The Marine Resources Act, in which the FFB is embedded, also says nothing definite in this regard, but mentions that the

[6] http://www.regjeringen.no/nb/dep/nfd/aktuelt/nyheter/nyheter-2013/fjordfiskenemnda-blir-oppretta/mandat-for-fjordfiskenemnda-.html?id=748027. Accessed 26 June 2015.

[7] Cf. Lov om forvaltning av viltlevande marine ressursar (havressurslova) § 11.

Ministry may establish tasks that are more specific for the FFB. This indicates that there is legal room for flexible interpretation of the FFB's status. However, the Norwegian Parliament, which is a legislative body, describes the FFB as an advisory organization (Prop. 70 L (2011–2012). In principle, the Ministry would therefore be of the same opinion. In a press release in 2012, the Ministry stated that the FFB "shall not have decision-making powers, but that it could become an important advisory body for the central authorities".[8] Although this would suggest that the FFB is not a co-management agency based on power sharing, it does not relegate the FFB to a merely reactive position. The mandate does not prohibit the FFB from being proactive in taking initiatives on matters that are within its mandate. Nonetheless, government consent is required for any of the FFB's suggestions.

The Sami Parliament has a somewhat different idea of the FFB's status; it has stressed that the FFB should be given decision-making power over certain issues, for instance with regard to gear restrictions, allocation of quotas, setting of fjord lines, and dispensation of rules that exclude bigger vessels fishing within these lines.[9] The Sami Parliament has repeatedly pointed out that local management arrangements that do not entail decision-making powers for the Sami are in conflict with international law on such matters, most notably with the ILO Convention 169, which Norway was the first to ratify.

Given this difference of opinion as to what role the FFB shall play, one may assume that the status of the FFB remains unsettled. The relationship between legal status and actual practice must be dealt with on a day-to-day basis within the FFB itself. One may imagine that practice eventually overrules principle rather than the other way around. The government might decide, however, to intervene if the gap between principle and practice becomes too wide. This has not happened thus far.

*First 2 Years of Operation* From the beginning, the fjord lines were the main issue. These lines decide how far into the fjord basin a vessel type can go and particular gear used (see Fig. 13.2). They are meant to avoid further resource degradation, help rebuild the coastal cod stocks by protecting their spawning grounds, and protect the small-scale fjord fishery from intrusion of large vessels that use trawl nets or long-lines.

The fjord lines were introduced in 2004, a decade before the FFB was established. Their calibration is now a task for the FFB. Even though the Ministry did not support the suggestion of a separate fishing act for Finnmark or acknowledge local

---

[8] Press release from the Ministry, March 16, 2012 (https://www.regjeringen.no/no/aktuelt/lov-forslag-om-oppfolging-av-kystfiskeutv/id675338/. Accessed 30 March 2014). In her speech on the 25th anniversary of the Sami Parliament in 2014, the fisheries minister Aspaker stated: The Fjord Fisheries Board shall be an important advisor to the fisheries authorities. The board shall help to ensure that Sami marine interests, local knowledge and practical concerns related to fjord fisheries are dealt with when regulations are determined. (Our translation) http://www.samedigge.no/Sammedikke-birra/Duogasj/Samediggi-25-jagi/Hilsningstale-av-fiskeriminister-Elisabeth-Aspaker. Accessed 19 January 2016).

[9] Cf. the Sami Parliament's resolution included as Appendix 5 to Proposition to the Parliament (Prop. 70 L 2011–2012).

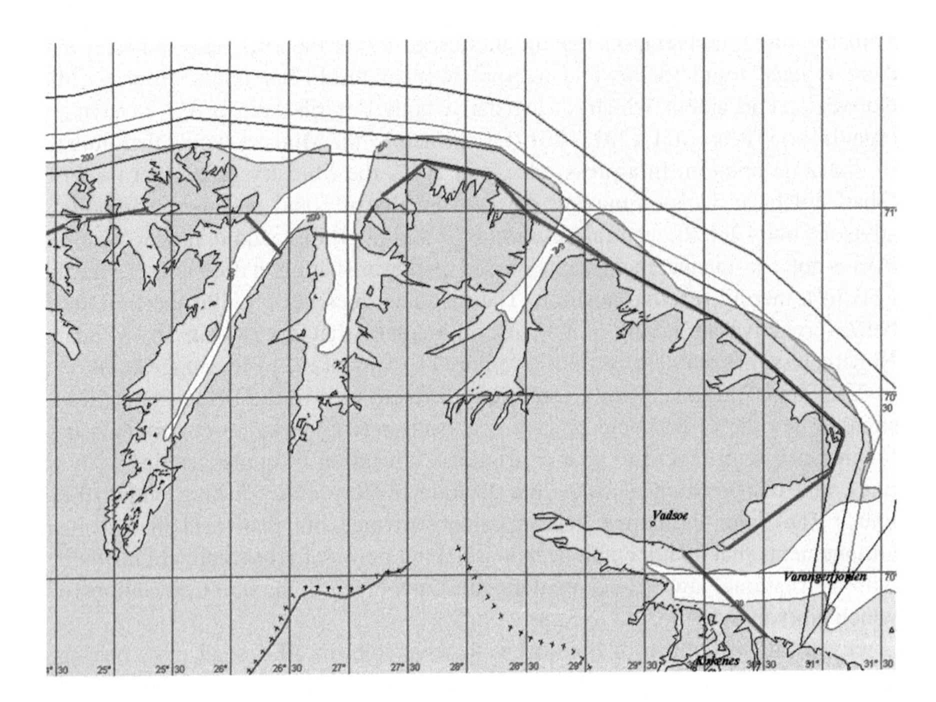


**Fig. 13.2** Fjord lines

fishing rights for fjord fishers, the Ministry deemed it necessary to ensure that the smallest vessels were able to harvest their quotas that were allocated within the annual quota regulation. Vessels over 15 m were banned from fishing inside the fjord lines, although exceptions to the rule could be made in particular circumstances (Prop. 70 L (2011–2012)). As of January 2013, this regulation is implemented in all Norwegian fjords.

As one could imagine, the positioning of the fjord lines has been a hot issue in fisheries circles. The NFA criticized the regulations and requested the Ministry to make them as lax as possible. The Ministry, in the first year of the FFB's existence itself, gave in and allowed for several exemptions vis-à-vis larger vessels fishing inside the fjord lines. This led to protests from fjord fishers as well as from the Sami Parliament.

The constitutional meeting of March 17th 2014 put the subject of the fjord lines on the agenda. A Ministry representative highlighted the challenges the fjord lines posed from the governments' point of view. At all the following meetings during the FFB's first 2 years of functioning, fjord lines were on the agenda (Table 13.1).

At their second meeting, the FFB agreed that until it has gathered sufficient information to consider adjusting the fjord lines, only vessels smaller than 15 m could harvest inside these lines. However, while the FFB accepted the exemptions made by the Ministry, it decided that it should not apply for vessels over 27 m (90 feet). The FFB's compromise met mixed reactions. Fjord and Sami fishers supported the compromise, even though some hoped for stricter recommendations

**Table 13.1** The Fjord Fisheries board's meeting agenda

| | FFB's meeting agenda |
|---|---|
| **1st meeting:** 17.3.2014 | Formal constitution of the FFB. |
| | Election of leader: Due to parity of votes, the leader sat for 1 year (2014–2015), and then switched roles with the deputy |
| | The fjord lines: proposal to prolong temporary exceptions until May 2014. |
| **2nd meeting:** 14.5.2014 | Internal work: Discussing the mandate, scheduling meetings in 2014, information channels/media, knowledge collection, and budget |
| | The fjord lines: In general, vessels less than 15 m allowed to harvest within the fjord lines. The FFB advices exceptions for saith seine/ herring seine/ mackerel fisheries vessels <=28 m |
| **3rd meeting:** 11–12.9.2014 | Internal work: Setting of action plan, meetings open for the public when resolutions on single issues are taken, and closed during working sessions. Economy/budget: Request more money for 2014 in order to be able to attend to its work |
| | The fjord lines: Starts in Finnmark and ends in Nordland with the work of setting permanent fjord lines, discussing what information is necessary for conducting this work (experience based- and scientific knowledge), mapping fishing grounds |
| | Additional quantity of cod accessible to Open group; discusses the effect of this regulation |
| | The spread of mackerel in the north; need for scientific input before making decisions |
| **4th meeting:** 6–7.11.2014 | Budget proposal 2015: 805.000 NOK (= ca.87500 EUR) |
| | The fjord lines: Request the secretariat to prepare a progress plan for the work on the fjord lines, including the work methodology |
| | Decision on advising the Ministry that the FFB become an observer at national regulation meetings |
| | Resolutions concerning mackerel fisheries: Concern expressed regarding the northward spreading of mackerel and the effect on the fjord ecosystems |
| | Resolutions: Requests a review of whether the fisheries reform proposal from the Ministry will affect coastal and fjord fisheries |
| **5th meeting:** 26–27.1.2015 | The fjord lines: Agree on the progress of gathering information concerning the fjord lines. Decision on arranging a public meeting in East-Finnmark |
| | Resolution concerning additional quantity of cod to open group: Advice to increase the quota from 12 to 18 tons |
| | Resolution concerning fisheries task force (NOU 2014: 16). Recommends that consequences for Sami fisheries, including Sami/local communities are investigated |

(continued)

**Table 13.1** (continued)

| | FFB's meeting agenda |
|---|---|
| **6th meeting:** 5–6.03.2015 | Changing leader: According to the rules, the new leader sits for the next year |
| | The budget for 2015: Not set as the Ministry requests a detailed budget plan. Resolution: Decides to prepare a revised budget |
| | The fjord lines: Discusses the progress of making decisions for placing permanent fjord lines in Varangerfjord. Emphasizes need for experience-based data from fishers. For the next meeting, members requested to evaluate what data are needed to decide on the fjord line. Considers arranging public meetings on the fjord line issue |
| | Resolution: the FFB invites tender for a report on the consequences of the NOU 2014: 16 proposals received, under presumption that the FFB is allocated the necessary economic resources |
| | Discusses a concrete infrastructure case in Troms that may have negative consequences for local fisheries. Resolution: The FFB sends a statement to the Troms county authority that particularly stressed the effects on local Sami conditions |
| **7th meeting:** 29.05.2015 | The fjord lines: The Varangerfjord case. Wants to avoid user conflicts between vessels using mobile and stationary gears, in which the drawing of the fjord lines is taken into consideration. Preliminary resolution: The FFB advices drawing a new fjord line in Varangerfjord |
| | Resolution on Red King Crab fisheries |
| **8th meeting:** 26–27.08.2015 | The fjord lines: Preliminary resolution concerning Finnmark based on data collection |
| | Additional quantity of cod to open group: Resolution on the recruitment of fishers younger than 30 years |
| | Budget: Resolution to apply for increased funding to carry out the FFB through 2015 |
| | Resolution to prepare a note about the development of auto line fisheries within the fjord lines for vessels >15 m, and to increase tourist fisheries in Northern Norway |
| **9th meeting:** 4–5.11.2015 | The fjord lines: Preliminary resolution that the fjord lines in Troms County remain unaltered |
| | Tourist fisheries: Resolution to request the Ministry to regulate tourist fisheries. Point to the difficulties related to lack of knowledge about tourist fisheries harvests |
| | Resolution to postpone the setting of a fjord line in Finnmark as the collection of local information is still in progress |
| | Requests a note about the salmon marine fisheries. |
| **10th meeting:** 20–21. 01.2016 | The fjord lines: What are the FFB and fisheries authorities' responsibilities? |
| | Discussions about the FFB's role; internal-external perception |
| | Discuss its mandate; focus on securing Sami fisheries, or small-scale fisheries independent of ethnicity |
| | The budget-situation; need for more money to fulfill its mandate |

Source: http://www.fiskeridir.no/Yrkesfiske/Regelverk-og-reguleringer/Lokale-reguleringer/Fjordfiskenemnda/Protokoller regarding restrictions on bigger vessels. The NFA was, and still is, negative about the exclusion of vessels over 90 feet, as it feels this will result in the neglect of other fisheries such as mackerel and saithe-seine fisheries. In protest, some within the NFA also said that members of the FFB who held membership in the NFA should resign from the NFA. The Ministry was slow to follow up on the FFB's advice to ban vessels over 90 feet from fishing inside the fjord line. For this, the Ministry received criticism from both the FFB leader and the Sami Parliament. After some hesitation, however, the Ministry decided to side with the FFB.

As would be expected of a new organization, the FFB has deliberated about how it should work, what kind of knowledge it needs to generate to fulfil the mandate, and how data ought to be collected. The division of labor between the FFB and the FFB's secretariat (The Directorate of Fisheries, Region Finnmark) appears to be blurred. The FFB's members complained that the data that the secretariat provided on the fjord lines delayed their work during the first year. The FFB planned to undertake investigations about the fjord areas throughout 2015, and decided to start with public hearings in four municipalities in the Varangerfjord (east Finnmark – adjacent to Russia). The position of the line proposed for this fjord was based on interviews with fishers who harvest both within and outside the prevailing fjord lines. As of January 2016, the evaluation of fjord lines as part of the FFB's mandate continues.

After spending time and effort on collecting relevant information on the fjord lines, the FFB's members soon realized that they needed to broaden their focus. A new task force (in Norway known as the 'Tveterås committee', after its leader), established in 2014, submitted its recommendations to the Fisheries Minister in 2014. In the report, it argued for restructuring the Norwegian seafood industry (NOU 2014: 16). The FFB focused much of its time in 2015 on the report and its findings. The report suggested a rather drastic liberalization of the overall management system, with privatization of quota rights into individual transferable quotas (ITQs) as a central component. One issue of concern was its silence on issues pertaining to the Sami. At the task force's 5th meeting, FFB members demanded that attention be given to how the reports' proposals would affect the situation and the general terms for fjord and coastal fisheries in northern Norway, with an emphasis on Sami use, including consequences for coastal communities and Sami communities (ref. minutes). The Ministry never responded. While the FFB could have initiated its own investigation because, as its chairperson commented in the press ('Ságat' 28 February 2015), it has a right to initiate examination of "all the issues we want within our mandate", the fact that the FFB only has been provided half of the needed budget, made it impossible for it to do so. The broad responsibility of the FFB and the challenges it faces to fulfil those responsibilities given restricted resources, amongst other things, has been a topic of internal discussion among the FFB members during the first 2 years of its existence.

## Discussion

*Indigenous Peoples in the SSF Guidelines* The SSF Guidelines emphasize the need to address the concerns of indigenous peoples by focusing on their impoverished situation and vulnerability. The Guiding Principles "emphasize "respect for cultures", which involves recognizing existing forms or organization, traditional and local knowledge and practices of small-scale fishing communities, including indigenous peoples…"'. Section 3.6 talks about consultation and participation and the need to take into account the UN Declaration on the Rights of Indigenous Peoples in the whole decision-making process related to fisheries resources and areas where small-scale fisheries operate, and taking existing power imbalances between different practices into consideration". Paragraph 5.4 says that states should protect all forms of legitimate tenure rights…, including those of indigenous peoples. States should recognize the role of small-scale fisheries communities and indigenous peoples to restore, conserve, protect and co-manage local aquatic and coastal ecosystems." In paragraph 6.2, the need for preferential treatment of indigenous peoples is emphasized to ensure equitable benefits". Paragraphs 11.6 says that "All parties should ensure that the knowledge, culture, traditions and practices of small-scale fishing communities, including indigenous peoples, are recognized and, as appropriate, supported and that they inform responsible local governance and sustainable development processes". These are among the paragraphs that talk specifically about indigenous peoples. The SSF Guidelines as a whole are also relevant to the Sami.

The Sami fishing population, being part of a wealthy country and a functioning welfare state, do not experience the poverty and food insecurity that their indigenous counterparts in many other countries face, particularly in the south. Still, the same forces of globalization and political marginalization that make their livelihoods and culture insecure confront them. Their communities and customary institutions are not sufficiently robust to withstand the current trends of neoliberalization, where their traditional rights of land and water are up for grabs. These concerns, as the SSF Guidelines stress, are not just about fisheries management but about human rights. From such a perspective, the Sami are also struggling for recognition of their historical rights to fish in Norway. They are broadening the discourse about fisheries rights in the manner that the SFF Guidelines have done. Their struggle has been evolving over time, and was well underway when the SSF Guidelines were endorsed, but it is the SSF Guidelines that might bring a new momentum as they speak to issues that have been on the Sami political agenda for decades.

*Post-colonialism* The situation of the Sami in Norway, Sweden, Finland, and Russia is not very different from that which exists in post-colonial settings. The need for legal reform that empowers indigenous fishing people through institution building and other measures is essential to decolonization (Woodman 1995; Zips 2003). This requires that indigenous legal norms and principles somehow be integrated in existing law, as the SSF Guidelines say should happen when appropriate. These legal norms should, however, not be compromised. The Norwegian Sami land and fisheries rights investigations initiated by the Norwegian government aimed to correct the problem of legal colonization, as opposed to the more pragmatic approach of the Norwegian authorities. In other words, the Ministry now thinks that the existing order goes far enough in accommodating Sami claims. However, the Sami Parliament seems united in thinking that much still remains to be done to recognize Sami historical rights and self-governance and meet Norway's obligations with regard to international law. There is reason to believe, therefore, that the issue will linger on into the future.

The Sami Parliament had put forth a subsidiarity principle similar to those of the SSF Guidelines before the latter came into existence when it proposed that those living in an area and who are dependent on local resources to maintain culture, industry and language should have the first right to use the [natural] resources (Sametinget 2004, 47). This principle takes issue with the existing Norwegian fisheries governance order – which works from the premise that fish in the ocean are a common pool resource (Davis and Jentoft 2001; Jentoft 2013). It is important to point out that this principle may also find support amongst Norwegian small-scale fishers in general, as it does not apply exclusively to Sami fisheries (Søreng 2008, 2013a). Moreover, it would fundamentally change the working conditions of the FFB.

The FFB is exposed to competing political positions, having to maneuver between Sami and Norwegian expectations and legal perceptions that do not always correspond and easily mix, for instance, on the question of access and preferential treatment of resident fjord fishers. Given the mandate to strengthen Sami fisheries dependent communities, Sami customary practices and Sami rights perceptions should be recognized not just in principle but also in management practice. However, any preference or privilege to Sami fisheries to correct for previous neglect, as the SSF Guidelines indicate may be needed, can easily trigger opposition within the ethnic Norwegian fishing community, which does not always distinguish between equality and equity, i.e., between equal and fair treatment. Likewise, for the Sami, any compromise may be seen as surrender. This is, however, 'realpolitik' since the FFB is now an integral part of the national fisheries governing system and hence must be able to balance between Norwegian and Sami political and institutional demands. At the 11th meeting of the FFB, the former FFB leader, Mr. Pedersen, summed up the first 2 years of the institution's existence: "We have had a process, which has not been easy, to locate ourselves within the system."

The FFB must know how to address normative diversity and complexity. It must also be able to deal with dynamism and with 'living law', where legal perceptions are changing as the problems living law tries to address are also changing (Svensson 2005). In a natural resource-based industry like fisheries, nothing is stable. Legal perceptions are no exception to this rule. The crisis in the cod fishery that triggered the quota system in 1990 and the political process among the Sami that subsequently led to the FFB is now largely gone, while new challenges have appeared such as the growth of salmon aquaculture in the fjords of northern Norway and the migration north of mackerel due to rising sea temperatures. Thus, the FFB finds itself in a difficult situation. On the one hand, it must adhere to the rules and regulations of the Norwegian fisheries governance system. On the other hand, the

FFB must deliver on the expectations of the Sami community, including those of the Sami Parliament. Unless it performs well in this balancing act, the FFB's legitimacy is likely to be challenged.

*Positive Discrimination* The FFB is criticized, often in hostile ways, particularly by the NFA and parts of the fisheries political establishment who would like to see it removed. The NFA was also very negative about the Coastal Fisheries Committee (CFC) report in 2008 and strongly opposed its recommendations. With reference to the FFB, the leader of the NFA has stated that Norway is one kingdom, and all areas ought to have the same rules. "No one should be given special treatment" (NRK Sami, 21 March 2014). There is reason to assume that the same opinion will be voiced against the SSF Guidelines once they become better known.[10] (Now that the FFB is a reality, the NFA demands representation in the board).[11] The opposition to special treatment – or positive discrimination – has also been articulated by parts of the political right in Norway, who currently control government. Being against 'special treatment' as a governance principle translates into opposing the very existence of the FFB as an institution and consequently its practical management proposals regardless of the content (Søreng 2007, 2008, 2013b; Jentoft and Mikalsen 2014).

It is probably helpful that the FFB's mandate arose from a series of consultations between the Sami Parliament and the Ministry. The mandate accommodates the different normative perceptions of both parties. Despite its shortcomings, the legal reform that resulted in the FFB is, as Josefsen (2014) notes, an opportunity for the Sami community more generally and the Sami Parliament in particular to gain formal access to the country's fisheries governance decision-making in a way that it was previously not possible. With a foot in the door, it might be able to advance its agenda.

The Ministry does not share the idea that the FFB has broken in to the decision-making process. More likely, it takes the view that FFB has been co-opted (Selle and Falch 2015). Whatever the case might be, the government may want to bring the FFB on board and thereby ensure a less boisterous Sami Parliament. From a governability perspective (Bavinck et al. 2013), this may well be a win-win situation, but only to the extent that the FFB can maintain its support base amongst Sami fishing communities and the fishing industry as a whole. The contradictions are numerous and the challenges complicated, which for the FFB means a slippery slope. For instance, Sami-Norwegian ethnic fisheries conflicts overlap with other conflicts, such as those regarding gear use and space (Søreng 2013a). Likewise, fisheries communities often have a mixed population of Sami and non-Sami, which makes 'positive discrimination' complicated. This is why the FFB's mandate relates to geographical communi-

---

[10] Similarly, in a letter dated December 12, 2015, to the Ministry and copied to the FFB, the FFA and FFA Nordland, Ministry, the NFA and the Nordland Herring Fisher's Association expressed the view that it is 'highly discriminatory that ethnicity should decide which fisheries regulations should apply'. The association also disagreed with the idea that residence should have any significance, as 'it is also discriminatory to fishers who do not live in the municipalities which are covered by this arrangement'. (Our translation).

[11] http://www.fiskarlaget.no/~fiskabhe/index.php/tariffseksjoner/bateierseksjonen/bateierseksjonen-virksomheten/details/104/731-vil-med-i-fjordfiskenemnd

ties in northern Norway rather than to the Sami as an ethnic group distinct from fishing residents of Norwegian heritage. The FFB's mandate also extends to areas where Sami presence is scant or non-existent, most notably in Nordland, which is the southernmost of the three counties that the FFB is covering.

The advisory role of the FFB, and the fact it is integrated into the overall Norwegian fisheries governance system, means that confusion may arise as to which side of the table the FFB occupies. Such confusion may emerge when advice involves conflicting viewpoints and dialogue takes the form of negotiation. The FFB can perhaps play the ethnic card with the government, but this is not possible to the same degree with local fisher communities in Finnmark, where there will be a pressure on the FFB to be ethnically neutral. Even if its mandate says that it should especially consider Sami interests, doing so in practice on a daily basis is likely to meet criticism among non-Sami fisheries stakeholders who may feel that they are being discriminated against, like for instance when gear groups have mixed ethnic representation. One could easily imagine that the FFB ends up being criticized in the same way as The Finnmark Estate was (see above), namely by being accused on the one hand of favoring Sami interests and on the other for being too passive in representing Sami interests (Broderstad et al. 2015).

The FFB, like the Finnmark Estate, carries the history of Sami discrimination, marginalization, and assimilation. This adds symbolic value to the FFB as it highlights the indigenous rights base of Sami small-scale fishers. The FFB represents a normative order of its own, which is not only circumscribed by the fisheries governance system of Norway but is also embedded in a legal system that is beyond Norway, expressed in the UN Declaration of the Rights of Indigenous Peoples and codified in international human rights law (Jentoft et al. 2007). The FFB is therefore a manifestation of 'legal pluralism' (von Benda-Beckmann 2002; Jentoft 2011).

*Governing Orders* 'Interactive governance' theory (Kooiman 2003; Kooiman et al. 2005; Kooiman and Jentoft 2009) perceives governance as happening at three distinct but related 'orders'. At the 'meta-order' basic values, norms, and principles are the focus. The SSF Guidelines belong to this order, whereas its implementation would occur at lower orders. 'Second order governing' is about the formation of institutions, including law, whereas 'first order governing' is about the day-to-day process of implementation, stakeholder participation, and decision-making.

The exact relationship between Sami and Norwegian fisheries governance is still being negotiated at all 'orders'. There are on-going controversies regarding the basic normative principles at the 'meta-order'. There are those who would dispute that indigenous Sami rights have any role to play in fisheries management. The idea of positive discrimination for Sami small-scale fishers is controversial. The government's rejection of CFC regarding Sami historical rights to fishing may end up in court, as its chairperson (Carsten Smith) has suggested. It would be up to the Sami Parliament or perhaps some other Sami organization to push it through the justice system.

The FFB is not yet consolidated at the 'second order', i.e., as an institution. The mandate is formalized, and the rules of representation are settled, but opinions differ as to what should be its status, especially in terms of its autonomy. In its fourth meeting, the FFB expressed interest in obtaining observer status at the national regulatory meeting. This would be unusual for an agency that assumes public functions and is an integral part of Norwegian fisheries governance. Observers are usually affected outsiders. At the 11th meeting, FFB members expressed frustration over the fact that hostile stakeholders think that the FFB has more power than it actually has according to their mandate as an advisory body. The Sami Parliament is of the view that it should not play only an advisory role, as the Norwegian government wants and current legislation defines. Rather the Sami Parliament believes it should have the opportunity to make its own rules pertaining to fishing practice: quotas, gear, fjord lines etc. This, undoubtedly, challenges the principles of the current Norwegian fisheries governance system, which is based on 'centralized consultation', with the NFA in a privileged position and in a rather cozy relationship with the Ministry. Both parties therefore have a history of opposing governance reforms that involve decentralization and delegation of fisheries decision-making powers, regardless of whether indigenous Sami interests and rights are implicated in it or not (Jentoft and Mikalsen 2014). It should therefore be of no surprise that the FFB is at best getting reluctant support and at worst is facing opposition from other fisheries stakeholders.

The developments occurring now are at the "first order" of governance, i.e., with regard to the operation of the FFB, how it goes about delivering its mandate, what happens when the board meets to discuss and vote on concrete issues, and how the Ministry addresses the FFB's concerns. Of course, the question remains as to what happens if there is a disjuncture between meta-order norms and values reflected in the advice given and those of government and non-Sami stakeholders. Over time, the pragmatism of FFB decision-making may have a bearing on the institution. Eventually meta-order principles may even change.

The NFA has more power in fisheries governance affairs than its formal advisory status would suggest. This is a result of a long process of mutual adaptation and trust building. In many instances, the Ministry just confirms what the NFA says and does (Hernes et al. 2005; Jentoft and Mikalsen 2014). The same may also occur with the FFB. Informal power is still power. In its first 2 years of existence, much of the FFB's effort has been concentrated on the fjord lines, which is no doubt important from a Sami small-scale fisheries perspective. It is also important from a non-Sami perspective in areas where Sami and non-Sami challenges overlap. The better the FFB does its job of investigating issues and the more balanced its advice, the more likely this advice will become the rule. If this is the case, we may see a situation of 'interlegality' (Svensson 2005) arise where Norwegian statutory and Sami customary law undergo mutual adaptation as opposed to them existing side by side in a case of legal pluralism. This may also help legitimize the institution in the eyes of critical stakeholders. The recent Norwegian fisheries law reforms that accommodate Sami interests is possibly an example of interlegality that also falls in line with the SSF Guidelines which talk a lot about recognition of customary and indigenous law.

The FFB may not get the space it wants even within the framework of its mandate. The positioning of the fjord lines continues to be contested as they involve the exclusion of fishing interests that are usually more powerful than those they are there to protect. The FFB's advice on this and other matters can also be challenged in the hearing process following the giving of advice and before it eventually becomes a formal regulation. On several occasions, criticism of the FFB's advice has been highlighted in the news media. However, there is reason to believe that once decisions are settled, for instance concerning fjord lines, over time they will be less controversial and taken as given, as often happen with institutions and regulations (Jentoft 2004). The fjords were not open access and free of spatial demarcation to begin with, and although rules were routinely negotiated they eventually became an 'objective reality' that fishers learned to live with.

*Legal Pluralism* Vanderlinden (1989, 151) points out that the individual "is the converging point of the multiple regulatory orders which each social network necessarily includes …". Although this is undeniably true for Sami small-scale fishers (as individuals) in the fjords of northern Norway, it is also true for the FFB as an organization, as it must maneuver between Norwegian and Sami legal perceptions. The conflicts it must handle are, therefore, not just economic but also normative, i.e., about orders (Bavinck 2005). This adds complexity to Sami small-scale fisheries governance, which means that most decisions have multiple contexts and dimensions. By asserting local indigenous norms, values, and principles, which is what the Sami community expect of it, the FFB is operating in a political and institutional landscape where Sami values and principles are not generally shared, regardless of existing international and domestic law. The implementation of the SSF Guidelines in multi-ethnic communities would therefore require a balancing act.

To succeed, the FFB must be a link between the conflicting normative orders. The FFB has a vertical function in terms of integrating national and local normative systems. The FFB also has a horizontal function, which means bridging between conflicting interests and governance principles that exist at the local level. The FFB's ability to function as a horizontal integrating body depends on its vertical function, i.e. how it finds its place in the overall and long established fisheries governing system that it is a part of, and vice versa. While allowing local variation in rules and regulation, the FFB must ensure that what is allowed in one fjord cannot be forbidden in another fjord unless there are very good reasons for it. There are also limitations to how many exceptions to the rule can occur, like those pertaining to fjord lines. It would also be impossible to discriminate between small-scale fjord fisheries on ethnic grounds, as the CFC has also advised against (NOU 2008: 5). Still each fjord system has its own particularities. The governability of small-scale fisheries in northern fjords requires dexterity and attention to detail and context. This calls for the use of local knowledge (Eythórsson 1993), which is also something the SSF Guidelines talk about in Chapter 3.1. The FFB has a big geographical area to cover and it remains to be seen how it will manage to integrate such knowledge.

Whether the FFB succeeds in realizing Sami fishing rights and fulfils the expectations that rest upon it, even more so with the SSF Guidelines now in place, depends partly on how the FFB organizes its internal work and how its members operate as a team, given that they are divided ethnically. With an expectant and critical stakeholder audience, the FFB can hardly afford internal conflicts which, given the equal representation of Sami and non-Sami stakeholders, is a risk. Success also depends on how perceptions over local fishing rights are translated into practical regulations, as in the case of fjord lines. Sami small-scale fishers must deem these regulations legitimate. In the end, these regulations must also be accepted within Norwegian fishing communities and authorities, the latter having the final word.

## Conclusion

By endorsing the SSF Guidelines, FAO member states like Norway committed themselves to recognize and support the role of small-scale fishing communities and indigenous peoples, and help them become more sustainable and secure. Whether these marginal legislative reforms and the establishment of the FFB are sufficient to accommodate this commitment is questionable. There will obviously be different views about whether these mechanisms go far enough. It is too early to know but it is clearly not sufficient to show only good intent. The Norwegian government must also prove that things work to safeguard Sami livelihoods and culture.

This case study provides an illustration of what implementing the SSF Guidelines may involve as far as indigenous peoples are concerned, what solutions one might think of, and what challenges and obstacles one should be prepared for. It has, therefore, value beyond Norway and the Sami. There is also reason to assume that the SSF Guidelines when implemented may encounter similar challenges elsewhere. If modest, inadequate steps are contested, a more comprehensive, progressive reform, as envisioned in the SSF Guidelines, will be contested even more. The FFB has met opposition at all 'governing orders'; the basic principles that pertain to the rights of indigenous peoples in Norway with the Sami are still disputed both among the public and Norwegian authorities, particularly with regard to fishing rights. In Norway there are even voices questioning the legitimacy of the Sami Parliament. Those who disagree that the Sami are an indigenous people that deserve secure rights and justice through positive discrimination also dispute the need for the FFB (and thereby the SSF Guidelines).

If a case is lost at one governing order, it can be brought to another order. One may also be able to live with a disconnect between orders. One may, for instance, be willing to accept governing principles as long as they remain symbolic only. If one has to unwillingly accept that human rights principles are relevant and that the FFB is here to stay, it is still possible to oppose the existence (second order) of the FFB and raise concerns regarding the functioning (first order) of the FFB. It is likely that negotiations over conflict will involve movement along the governing ladder, from the 'meta' to the 'second' to the 'first governing order,' or in the opposite direction. Thus, the implementation of the SSF Guidelines is likely to be challenged in practice by those who oppose it in principle. Now that Norway has endorsed the SSF Guidelines, there is still a possibility to obstruct their operationalization and implementation in Sami small-scale fisheries. Implementation involves a process of contextualization at lower governing orders. In the process of doing so, it is possible to be pragmatic at first order and flexible at second order, while being firm at meta-order without compromising the principles.

The future will tell how this new governance system will work. Although Sami small-scale fjord fisheries have diminished in number, they have also transformed, and in local communities even grown, thereby proving resilient through adaptability (Brattland 2014; Broderstad and Eythórsson 2014). One hypothesis is that once the FFB has been in operation for a while, controversies will fade away or be isolated at first order. When things at the first order of governing calm down, it is possible that higher order conflicts also will subside. Thus, it is safe to conclude that even if the FFB is inadequate to secure a sustainable future for Sami fisheries, it is still a valuable contribution to the fisheries governing system and a useful instrument for implementing the SSF Guidelines.

## References

Bavinck, M. (2005). Understanding fisheries conflicts in the South – A legal pluralist perspective. *Society and Natural Resources, 18*, 805–820.

Bavinck, M., Chuenpagdee, R., Jentoft, S., & Kooiman, J. (2013). *Governability of fisheries and aquaculture: Theory and applications*. Dordrecth: Springer Publication.

Benda-Beckmann, F. von (2002). Who's afraid of legal pluralism? *Journal of Legal Pluralism, 47*, 37–83.

Brattland, C. (2010). Mapping rights in coastal Sami seascapes. *Arctic Review of Law and Politics, 1*(1), 28–53.

Brattland, C. (2014). A cybernetic future for small-scale fisheries. *Maritime Studies, 13*(18). doi:10.1186/s40152-014-0018-1.

Broderstad, E. G., & Eythórsson, E. (2014). Resilient communities? Collapse and recovery of a social-ecological system in Arctic Norway. *Ecology & Society, 19*(3). doi:10.5751/ES-06533-190301.

Broderstad, E. G., Josefsen, E., & Søreng, S. U. (2014). *Finnmarklandskap i endring. Omgivelsenes tillit til FeFo som forvalter, eier og næringsaktør*. Norut Alta Rapport 2015(1).

Broderstad, E. G., Josefsen, E., & Søreng S. U. (2015). *Finnmarkslandskap i endring – Omgivelsenes tillit til FeFo som forvalter, eier og næringsaktør*. Alta: NORUT (Northern Research Institute).

Davis, A., & Jentoft, S. (2001). The challenge and promise of indigenous peoples' fishing rights: From dependency to agency. *Marine Policy, 25*(3), 223–237.

Eythórsson, E. (1993). Sami fjord fishermen and the state; traditional knowledge and resource management in northern Norway. In J. Inglis (Ed.), *Traditional ecological knowledge: Concepts and cases* (pp. 132–142). Ottawa: International Development Research Centre.

Eythórsson, E. (2008). *Sjøsamene og kampen om fiskeressursene*. Karasjok: ČálliidLágádus.

Hernes, H. K., Jentoft, S., & Mikalsen, K. H. (2005). Fisheries governance, social justice and participatory decision-making. In T. Gray (Ed.), *Participation in fisheries governance* (pp. 103–118). Dordrecht: Kluwer.

Jentoft, S. (1989). Fisheries co-management: Delegating government responsibility to fishermen's organizations. *Marine Policy, 13*(2), 137–154.
Jentoft, S. (2004). Institutions in fisheries: What they are, what they do, and how they change. *Marine Policy, 28*, 137–149.
Jentoft, S. (2011). Legal pluralism and the governability of fisheries and coastal systems. *Journal of Legal Pluralism, 64*, 149–172.
Jentoft, S. (2013). Governing tenure in Norwegian and Sami small-scale fisheries: From common pool to common property? *Land Tenure, 1*, 91–115.
Jentoft, S., & Mikalsen, K. H. (1994). Regulating fjord fisheries. Folk management or interest group politics? In C. L. Dyer & J. R. McGoodwin (Eds.), *Folk management in the world fisheries, Lessons for modern fisheries management* (pp. 287–316). Niwot: University Press of Colorado.
Jentoft, S., & Mikalsen, K. H. (2014). Do national fisheries resources have to be centrally managed? Vested interests and institutional reform in Norwegian fisheries governance. *Maritime Studies, 15*, 5. http://www.maritimestudiesjournal.com.content/13/1/5.
Jentoft, S., Bavinck, M., Johnson, D., & Thomson, K. T. (2007). Fisheries co-management and legal pluralism: How an analytical problem becomes an institutional one. *Human Organization, 68*(1), 27–38.
Josefsen, E. (2014). *Selvbestemmelse og samstyring: En studie av Sametingets plass ipolitiske prosesser i Norge*. Doctoral dissertation. UiT Norges arktiskeuniversitet, Fakultet for humaniora, samfunnsvitenskap og lærerutdanning, Tromsø.
Kooiman, J. (2003). *Governing as governance*. London: Sage Publications.
Kooiman, J., & Jentoft, S. (2009). Meta-governance: Values, norms and principles, and the making of hard choices. *Public Administration, 87*(4), 818–836.
Kooiman, J., Bavinck, M., Jentoft, S., & Pullin, R. (2005). *Fish for life: Interactive governance for fisheries*. Amsterdam: Amsterdam University Press.
Minde, H. (2005). Assimilation of the Sami – Implementation and consequences. *Gáldu Čála. Journal of indigenous peoples rights, 3*(2005), 1–34.
Nilsen, R. (2003). From Norwegianization to coastal Sami uprising. In S. Jentoft, H. Minde, & R. Nilsen (Eds.), *Indigenous peoples: Resource management and global rights* (pp. 163–184). Delft: Eburon Academic Publishers.
Norwegian Official Report. (2008). *NOU 2008:5 Retten til å fiske i havet utenfor Finnmark*. Oslo: Departementenes servicesenter. Informasjonsforvaltning.
Norwegian Official Report. (2014). NOU 2014:16. *Sjømatindustrien. Utredning avsjømatindustriens rammevilkår*. Oslo: Departementenes sikkerhets- og Serviceorganisasjon, Informasjonsforvaltning.
Paine, R. (1965). *Coast lapp society* (Vol. 2). Oslo: Universitetsforlaget.
Proposition to the Parliament. (2012). *Prop. 70L 2011–2012* Endringar i deltakarloven, havressurslova og finnmarksloven (kystfiskeutvalet). http://www.regjeringen.no/nb/dep/fkd/dok/regpubl/prop/2011-2012/prop-70-l20112012.html?id=675139. Accessed 11 Dec 2016.
Sametinget. (2004). *Sametingets melding om fiske som næring og kultur i kyst og fjordområdene*. https://www.sametinget.no/nor/Dokumenter. Accessed 11 Dec 2016.
Selle, P., & Falch, T. (2015). Staten og Sametinget. In B. Bjerkli & P. Selle (Eds.), *Samepolitikkens utvikling* (pp. 63–88). Oslo: Gyldendal.
Smith, C. (1990). Om samenes rett til naturressurser – særlig fiskerireguleringer. *Lov og rett*, 507–534.
Søreng, S. U. (2006). Moral discourse in fisheries co-management: A case study of the Senja fishery, northern Norway. *Ocean & Coastal Management, 49*, 147–163.
Søreng, S. U. (2007). Fishing rights struggles in Norway: Political or legal strategies? *Journal of Legal Pluralism, 55*, 187–211.
Søreng, S. U. (2008). Fishing rights discourses in Norway: Indigenous versus non-indigenous voices. *Maritime Studies, 6*(2), 77–99.

Søreng, S. U. (2013a). Legal pluralism in Norwegian inshore fisheries: Perceptions on Sami fishing rights. *Maritime Studies, 12*(9). doi:10.1186/2212-9790-12-9.

Søreng, S. U. (2013b). *Lokale rettighetsforståelser i fiskeriforvaltningen: med særlig fokus på samiskefiskerettigheter*. Doctoral dissertation. UiT Norges arktiske universitet, Fakultet for biovitenskap, fiskeri og økonomi, Tromsø.

Svensson, T. G. (2002). Indigenous rights and customary law discourse: Comparing the Nisg'a and the Sámi. *Journal of Legal Pluralism, 47*, 1–34.

Svensson, T. G. (2005). Interlegality, a process for strengthening indigenous peoples'autonomy: The case of the Sámi in Norway. *Journal of Legal Pluralism, 51*, 51–78.

Vanderlinden, J. (1989). Return to legal pluralism: Twenty years later. *Journal of Legal Pluralism, 28*, 149–157.

Ween, G. (2006). Sedvaner og sedvanerett: Oversettelsesproblemer i møte mellom rettsvesen, samer og antropologi. *Tidsskrift for menneskerettigheter, 24*(1), 15–30.

Woodman, G. R. (1995). The common law as folk law of the lawyers. In H. Finkler (Ed.), *Proceedings from the Xth International Congress, Legon, Ghana* (pp. 329–356). Dordrecht: Commission on Folk Law and Legal Pluralism.

Zips, W. (2003). The double–bladed sword: A comparative analysis of the bond between rule of law and the concept of justice in Akan legal thought. In R. Pradhan (Ed.), *Legal pluralism and unofficial law in social, economic and political development* (pp. 181–212). Papers of the XIII international congress, 7–10 April 2002.

# Chapter 14
# Protections for Small-Scale Fisheries in India: A Study of India's Monsoon Fishing Ban

**Surathkal Gunakar, Adam Jadhav, and Ramachandra Bhatta**

**Abstract** In India, fisheries governance suffers from weak regulation and poor compliance, with a primary exception – a collection of coastal seasonal fishing bans or closures. Much other fisheries policy (e.g., fuel subsidies or incentives for deep-sea fishing) promotes increasing production over conservation. The benefits of such measures have generally accrued to owners of industrial and semi-industrial operations, often at the expense of the small-scale fisheries sector. Viewed critically, Indian fisheries governance can be described as out of compliance with the FAO's new Voluntary Guidelines for Securing Sustainable Small-Scale Fisheries (the SSF Guidelines). In this chapter, we analyze the coastal seasonal fishing bans in light of the SSF Guidelines and, in particular, the provisions for sustainable resource management (Section 5b of the Guidelines). Details of the monsoon bans have varied by time and place, but a diverse group of stakeholders have generally accepted the principle of a seasonal ban. However, there remains a complicated history of policy, legal, and social contestations – in short, politics – around the particulars of the bans, which we review. We also consider the specific case of Karnataka state. We find that weak scientific arguments generate a contested ecological justification and reduced support for seasonal closures. We suggest the monsoon bans are better justified when framed as safeguards for the small-scale fisheries sector. The SSF Guidelines provide a normative foundation for strengthening the monsoon fishing bans as part of dynamic fisheries management to privilege and protect India's small-scale fisher communities.

S. Gunakar
Pompei College, Mangalore, India
e-mail: gunakarsurathkal72@gmail.com

A. Jadhav (✉)
University of California at Berkeley, Berkeley, CA, USA

Panchabhuta Conservation Foundation, Kagal, India

Dakshin Foundation, Bangalore, India
e-mail: ajadhav@berkeley.edu

R. Bhatta
Indian Council of Agricultural Research Emeritus Scientist, College of Fisheries, Mangalore, India
e-mail: rcbhat@gmail.com



S. Jentoft et al. (eds.), *The Small-Scale Fisheries Guidelines*, MARE Publication Series 14, DOI 10.1007/978-3-319-55074-9_14

**Keywords** SSF Guidelines • Fisheries governance • Monsoon fishing ban • Traditional fishers • Conflict • Management • India

## Introduction

The fishing harbor at Mangalore, on the southwest coast of Karnataka, India, stretches for nearly a kilometer. During most of the year, the flat, narrow dock grounds are packed with gear to be loaded, baskets of catch, raucous auctions, the come-and-go of cargo trucks and a steady flow of workers. In the background, small fishing supply stores, tea stalls and cheap eateries buzz, especially when boats are heading out or coming back. Low-rise buildings house fisher business offices, trading firms, government departments, and fisher unions – the political and economic elite of the fishery. Nearly 1200 trawl and purse seine boats (the workhorses of India's industrial fisheries) call this harbor home.

Yet this harbor – one of the largest in India – practically shutters for weeks each year. During those times, hundreds of boats are moored to docks and one another, packed so densely that none can move (see Fig. 14.1). The harbor can feel all but abandoned.

This period of idling of the industrial/semi-industrial fleet happens at fishing docks across mainland India. It is a consequence of years of fisher politics, debate, and maneuver resulting in so-called monsoon fishing bans. Annually, for parts of

**Fig. 14.1** Trawl and Purse Seine Boats sit idle at Mangalore Bunder during the Monsoon Ban in 2012; these "Mechanized" Boats represent India's Industrial/Semi-industrial Sector (Photo by Adam Jadhav)

April, May and June, large boats are banned from east coast seas. Boats on the west coast are subsequently banned during June and July.

Nearly three decades ago, individual states began adopting such seasonal bans – from several weeks to a few months each year – on fishing in their own territorial waters (the water up to 12 nautical miles from shore that each state governs). Eventually the central government did likewise in national waters beyond 12 nautical miles. The bans developed as part of larger negotiations between conflicting sections of the fishery, and each state's own politics produced its own slightly different ban.

Annual newspaper stories on the bans offer a brief window into the politics of India's diverse fisheries. In coverage, the owner of a large boat often complains about hard times during the non-fishing weeks. A government regulator offers a justification about protecting stock. Journalists also sometimes refer to the bans as fishing holidays – seemingly apolitical periods of rest when fishers tend to social matters such as festivals or the marriages of children.

The smallest scale of fishers – those using oars, sails or tiny trolling motors – are typically exempt from the bans. This means the bans operate as a *de facto* protection by giving space to the most marginalized. Yet these same beneficiaries of the bans are usually left out of media coverage.

India's monsoon bans entered a new chapter in 2015, when the central government set out to implement an extended, 61-day uniform ban on fishing in the national waters off each coast – from 15th April through 14th June on the east and 1st June through 31st July on the west. The government justified the extended ban as necessary 'for conservation and effective management of fishery resources and also for sea safety reasons' (GOI 2015a). The central government also negotiated with state governments, and most adopted identical ban periods in their own waters.

Collectively, the bans arguably represent India's single most successful marine fisheries[1] regulation in terms of compliance. Yet rancor remains, and camps of fishers, government officials, scientists, and civil society invoke various justifications and anti-justifications for lengthening, shortening, or even ending the fishing season ban.

In this chapter we examine the evolution of the fishing ban periods. We do so because the bans represent a generally successful regulation that nonetheless remains a continual site of fisher politics. Our analysis is motivated by a new internationally agreed instrument – the Voluntary Guidelines for Securing Sustainable Small-Scale Fisheries in the Context of Food Security and Poverty Eradication (known as the SSF Guidelines) – which call for securing and sustaining small-scale fisheries. We pay particular attention to Section 5b of the guidelines, which mandates sustainable resource management specifically to protect small-scale fisheries.

This chapter relies on a combination of document, policy, and literature review, mixed with observations and field interviews in multiple sites around India. This field research comes out of our collective and individual research engagements with

[1] We focus exclusively on marine fishing (and estuarine fishing, in so far as those fishing communities, markets, and governance overlap the marine sector).

Indian fisheries; where appropriate, we note a year and month when observations come from specific interviews. We also wish to explicitly acknowledge our position of scholars who are also advocates in Indian fisheries and marine conservation policy at various levels.

To frame this chapter, we first present an overview of the SSF Guidelines, which apply to (albeit voluntarily) India. We follow with some basic information of fisheries and management in India. This leads to a discussion of the development of and debates over the monsoon bans. We then examine the case of fishing ban in the state of Karnataka, as a place that exemplifies ban politics up to the present but was not an outlier or leading edge of fisheries management or development. We ultimately find that the exclusively ecological justification that officials most frequently make for the ban may actually weaken or confuse support, and hence conclude with an argument for reimagining the seasonal fishing ban explicitly as a protection for small-scale fishers, in line with India's commitments to the SSF Guidelines.

## The SSF Guidelines

Decades of international fisher activism and negotiation by stakeholders led the Committee on Fisheries of the Food and Agriculture Organization of the United Nations (FAO) in 2014 to endorse a normative international instrument calling for protecting and privileging of small-scale fisheries. Thousands of stakeholder representatives worldwide were involved in creating the SSF Guidelines. Notably, many Indian activists and international advocates based in India played key roles in the institutional processes and underlying sociopolitical foment that culminated in the guidelines.

Though voluntary, the SSF Guidelines call on a range of fishery actors – international bodies, national governments, civil society, academia, fishers themselves, and other stakeholders – to recognize and address small-scale fisheries challenges (FAO 2015). The Guidelines are ambitious as they offer overarching principles of governance such as recognition of human rights, attention to equality, and the need for accountable and participatory action. The Guidelines also include specific provisions relating to sea tenure, resource management, development, employment, trade, gender, disasters, climate change, and policy.

For this chapter, the most relevant section of the SSF Guidelines is 5b, which addresses the duties and responsibilities of "states and all those engaged in fisheries management" to attend to sustainability and conservation (FAO 2015, 6). This section emphasizes that small-scale fishers themselves have a role to play in conserving resources – especially those to which they could claim tenure – and that states should enable participation, training, monitoring, co- or community-based management, equity in decision-making about fisheries management, and development. States are also encouraged to avoid policies that contribute to "fishing overcapacity and, hence, overexploitation of resources that have an adverse impact on small-scale fisheries (FAO 2015, 7–8)." Readers of this paper should hold the Guidelines in mind while encountering subsequent sections on the evolution and implementation of monsoon fishing bans.

An introduction to the Guidelines notes that small-scale fisheries are often juxtaposed with large- (or at least larger-) scale outfits and firms. The Guidelines, however, intentionally do not prescribe a specific definition of small-scale fisheries. This is not surprising; many scholars have argued that an abstract or international definition would be impossible; for example, Charles writes, 'the many categorizations of SSF (subsistence, artisanal, etc.) and their diversity of forms imply that any broad discussion of these fisheries cannot deal with all the nuances of specific situations – a small-scale fishery in one location will not necessarily look similar to one elsewhere' (Charles 2011, 85–86). In India's hyper-diverse fisheries, this holds true at scales far below the national level; what might be called small-scale in one Indian state would be more akin to large-scale in another. Additionally, technical, political, and economic forces continually remake Indian fisheries; increasingly "old dichotomies of non-mechanized and mechanized, small and big, artisanal and modern are irrelevant" (Kurien 2016, 30).

The lack of a definition "creates flexibility as to (the Guidelines') interpretation and, hence, implementation. Countries may therefore decide themselves who the SSF Guidelines are relevant for – or, indeed, if they are relevant at all" (Jentoft 2014, 4). This subjectivity and the Guidelines' voluntary nature mean that implementation is inherently a political project that could change or reorient fisheries. The Guidelines "are meant to intervene in situations where different interests are in conflict and where small-scale fisheries are the weaker party. They will inevitably interfere with power" (Jentoft 2014, 7). This is the key intersection between the Guidelines as a political and normative instrument and our analysis here of fisher politics in the origins and debates about monsoon fishing bans in India. As a site of varying fisher politics, the bans arose as actors wielded different kinds and levels of power. We certainly do not ignore ecological discussions, but these are raised largely in the context of political discourse.

## *Fisheries in India*

Indian fisheries present intractable problems: institutional failures, ecological uncertainty, political and socioeconomic considerations, pernicious market forces, and declines in traditional fisheries governance. India has attempted to govern and 'develop' fisheries for food security, livelihoods, and foreign exchange often with mixed results. More than 60% of Indian fishing households are officially categorized as poor (CMFRI 2012) and officials often view fishers, particularly the small-scale fisheries sector, as 'backward' in Indian parlance. Traditional governing institutions in communities have largely been replaced by weak command-and-control regimes, which often fail to make or enforce policies in the face of politics, technological change, economic realities, and capacity constraints. Fishers particularly small-scale fishers – have felt threatened by multiple forces (international boats, industrial pollution, larger-scale fishing capital, and even state bureaucracy) for decades (Kocherry and Achary 1989).

The breadth of India's fisheries complicates the fisheries governance. The 10-year average annual marine mainland catch from 2005 through 2014 is estimated at 3.27 million metric tons (CMFRI 2016), making India one of the largest fishing nations by catch volume. Catch is also diverse as nearly 70 taxa (some only at the level of order or superorder) contribute enough to merit government reporting. Yet understanding and management are hampered by limited or incomplete stock assessments (for example, see Ghosh et al. 2015) and weak capacity.

India's fisheries are also complex and diverse from economic and technical perspectives. Government statistics count three meta-classes of fishing boats based largely on power:

- **Mechanized craft**, typically up to 20 meters long, with engines (sometimes up to 500 horsepower) specifically to shoot/haul fishing gear in addition to propel the boat.[2] This category covers trawl, purse seine, line, and some larger gillnet boats.
- **Motorized craft**, an intermediate gear class using an engine only for propulsion. Some boats approach the size of a mechanized craft (e.g., ring-seiners in Kerala) but the category includes much smaller outboard motor boats, both traditional and contemporary.
- **Non-motorized craft**, the mostly traditional boats that dot India's coastline without a motor. They are often employed by fishers with the least economic or social power or when larger craft are banned. These range from oversized canoes with an outrigger to small rafts poled around a calm estuary.

These categories are neither discrete, nor do they align neatly with distinct social classes or economic models of fishing. Some large boats operate as capitalist firms owned by investors; others are household owned and operated. Motorized and non-motorized boats often serve as the primary fishing vessel of a household but sometimes work in a fleet. Fishing labor moves between sectors and geographies; deckhands in one season in one place become self-employed subsistence fishers at another time of year in a different region. Some social and religious classes/castes have ties to particular fisheries, but demographic, political, and economic changes have altered the profile of the traditional fisher. (For a detailed look at sociopolitical shifts in fisher communities rooted a particular geography, see Subramanian 2009). India's fishing economy is also heavily stratified in economic terms; while most fishing households live in poverty (CMFRI 2012) there is also considerable wealth in fishing (Bavinck 2008). Even some estuarine small-scale fisher communities in Karnataka – often presumed to be destitute by policy makers – report that they catch enough for household consumption and basic income (field interviews March 2015).

[2] Larger fishing boats comparable to international fleets are nearly nonexistent in India.

## *Indian Fisheries Governance*

Some brief context is needed to understand how Indian fisheries are managed and governed; here, we take a broad view of governance as state controls, restrictions, and policies as well as non-state stakeholder activity, politics, and maneuver that cumulatively set directions and agendas (in India's fisheries).

With no official barrier to entry, Indian fisheries in general can be described as open access in terms of property rights, where the theoretical race for fish ensues among rational economic actors (e.g., Gordon 1954; Scott 1955). However, this purely economistic description of fishers ignores social norms and institutions (such as class, religion, and caste); traditional tenure to the marine and estuarine commons; political mobilization of fishers; and the role of the state and private capital in directing or incentivizing fishing behavior.

Individual Indian states regulate fisheries in the waters up to 12 nautical miles from their respective shorelines. Each state's fishing bureaucracy[3] regulates (often weakly) and licenses fishing boats. This regime creates an identity fishing boats as belonging to the location of their license (e.g., a Karnataka fishing boat) and generates a sense of quasi ownership of a state's fish or waters (e.g., Karnataka fish caught in Karnataka's waters 'belong' to Karnataka fishers).[4] More onerous central government rules apply for all vessels, fishing or not, beyond state waters and up to the boundary of India's Exclusive Economic Zone (EEZ). In practice, however, state-licensed boats that are 20 meters or less in length are generally exempted from central regulation, so nearly all owners build only to that length; this creates a *de facto* technical cap on boat size even if those fishing vessels go deep within the EEZ in pursuit of fish.

Many formal regulations suffer from middling compliance. Much fisheries policy is intended to bolster production, which largely benefits successful large capital owners through fuel subsidies, loan discounts or incentives for deep-sea fishing. On paper, states sometimes demarcate fishing zones for different gear classes or restrict small mesh sizes. Additionally, managers may attempt to discourage entry by not issuing licenses, documentation, or paperwork to operate a new boat or secure financing. But spatial regulations are often ignored, unenforced, or irrelevant, while new boats receive licenses and approvals via loopholes or exceptions (field interviews July, August 2012). Spatial reservations are further undermined as fish populations in Indian waters vary seasonally and spatially. Dozens of fish populations overlap, such that gear classes which theoretically should not compete end up fishing for the same stocks.

As this chapter makes clear, fisher organizations – of gear-specific, social, socio-economic, or geographic constituencies – are sometimes politically influential and

[3] The particulars of the marine fishing regulation acts and rules of each individual state are beyond the scope of this chapter; however, we note that they were typically patterned after a central government 'model' legislation.

[4] This bears resemblance to the international system of national 'ownership' of the Exclusive Economic Zone.

can alter, resist, or generate regulation. Other government institutions also play roles in governance (or, related, development). The National Fisheries Development Board has a mission to expand fishery and aquaculture production. The Marine Products Export Development Authority, another central bureaucracy, promotes production for trade generating foreign exchange. The Central Marine Fisheries Research Institute (CMFRI) and other government science organs advocate policy, often with a top-down managerial bent. Simply put, fisheries governance is subject to multiple, often countervailing pressures and interests (particularly between the small-scale fisheries and larger-scale sectors).

Scientists and activists have argued that India's long-term fishery health is in question (e.g., Bhathal and Pauly 2008; Fernandes and Gopal 2012). Indian waters overall are widely regarded as overcapitalized (field interviews July, August 2012; February, March, April 2014; September 2015) and the costs of unsustainability fall most heavily on the small-scale fisheries sector, which has suffered in recent decades. Complex management systems suggested by classical theory, such as Individual Transferable Quotas (ITQs), are unavailable or wildly impractical for fishery managers.[5] Furthermore, strict imposition of fishery access limits raise equity concerns while still not addressing political and other external incentives for unsustainable catch and capital levels (c.f. Mansfield 2001). In some cases, officials attempt more basic regulations such as setting a minimum net mesh size.

Meanwhile, traditional fishery governance embedded within small-scale fisher communities has mostly declined since the 1950s in the face of capital intensification, commercial expansion, and state control. Present-day fisheries governance lacks strong participatory approaches (field interviews July, August 2012) such as prescribed by the SSF Guidelines' Section 5b, though in a few geographies traditional communities continue to govern aspects of fishing, fishing space, and coastal life with tacit or explicit government support (Bavinck 2001a, b; field interviews January 2016). Bavinck and Karunaharan (2006) review cases of gear restrictions created by small-scale fisher community institutions (as opposed to state bureaucracy) in Tamil Nadu. Elsewhere, larger fisher political movements have occasionally allowed small-scale fisher groups to redefine themselves and wield clout with (or against) state rule (Subramanian 2009).

In 2012, one of the authors witnessed parts of the fisheries governance problematic in a microcosm in Karnataka. During the monsoon fishing ban, with hundreds of mechanized boats idle, the state fisheries directorate abruptly and without fisher consultation banned engines above 250 horsepower ostensibly to fight overcapacity. The move was justified primarily by an assessment from the Central Institute of Fisheries Technology (field interviews July, August 2012). Despite the widespread use of much larger engines, boat owners were given only months to retrofit. As the notice was still circulating, associations and unions of mechanized boat owners went to Bangalore to lobby political patrons, who in turn pressured fishery bureaucrats to back down. Within 2 weeks, and with almost no public debate at any point,

[5] A fishery management official actually laughed at one of the authors when he once asked if an ITQ could be possible.

the state had unceremoniously circulated and rescinded its order. Afterward, managers blamed self-interested, short-sighted political fishers. Fishers decried a non-participatory, heavy-handed, illogical regime.

India fisheries are also increasingly subject to market forces beyond the reach of the state. Increasingly marine harvest – from grouper to prawns to illegally harvested sea cucumbers – are auctioned at global prices and destined for export. Additionally, the rise of a fishmeal and oil industry in many parts of India has created a new demand for previously unmarketable catch; while this may lower bycatch (as more caught fish can be sold) and reduce 'wasted' fishing effort, it also incentivizes more indiscriminate harvesting (field interviews 2012). Furthermore, the income reaped from such economic restructuring accrues largely to the commercial sector as small-scale fisheries are less likely to be linked to such markets. At the same time, sustainability concerns about Indian fisheries – brought in part by excessive harvests – affect all fishers. These market developments affect material fishery outcomes with little or no recourse to government.

In sum, the fisheries management regime struggles with diverse and increasingly capitalized and commercialized fisheries (Jayasankar 2008; Narayanakumar 2008; Vivekanandan 2008; field interviews June 2012), while varied political interests pull governance in different, sometimes oppositional directions. Policy recommendations exist (Vivekanandan et al. 2010; Mohamed et al. 2014) but capacity to design and implement rules is lacking; political will often also favors growth in production rather than conservation or restriction. Our field interviews (2012, 2014) found some support within the fishing community for additional rules, but fishers also view state managers skeptically. Viewed critically, this scenario can be understood as out of compliance with India's commitments under the new SSF Guidelines regarding sustainable resource management.

## *India's Fishing Ban(s)*

The standout exception to India's weak fisheries governance is the collective seasonal ban. Sometimes called time zoning, such temporal restrictions have a long-standing place in the fishery management and conservation toolkit (FAO 2011). Research shows seasonal closures can bolster fish populations if the timings are biologically appropriate (Arendse et al. 2007). Common reasons for seasonal closures are to protect particular spawning aggregations (Sadovy et al. 2005), life stages or habitats. In some cases, seasonal closures may not be biologically justified as a spawning protection, but they still constitute an accepted bulwark against overfishing (Halliday 1988).

In 2015, the central government and states implemented simultaneous coast-wide bans prohibiting most if not all fishing for 61 days (GOI 2015a). Off the east coast, the ban now lasts from 15th April through 14th June, which is the pre-monsoon period. Off the west coast, fishers are prohibited from 1st June through 31st July, which covers approximately the first half to two thirds of the monsoon season. Though the central government only has authority beyond 12 nautical miles from shore, state officials agreed to create bans in their own waters, with some variation on the type of craft restricted. Some states allow fishers claiming traditional status as well as small engines (for example, less than 25 horsepower) or no engine at all to fish during the ban; others such as Gujarat and Maharashtra do not.

Notably, Kerala observes a shorter ban period and exempts 'traditional' fishers, including those that might be large-scale by Indian standards. The reduced ban time as well as exemption for larger but still 'traditional' boats is seen as an act of defiance (Philip 2015) and can be read as a political win by that state's mobilized traditional community that often uses semi-industrial gears.

Prior to 2015, however, bans varied from state to state and sometimes year-to-year (Table 14.1). All coastal states and territories have had some kind of a ban for many years, but details were subject to political maneuver and contestation.

## *Fishing Ban History*

The earliest government seasonal fishing ban (specifically on trawlers) arose in Kerala in the 1980s after years of political maneuver. Kurien (1991) provides a history of maneuvering in Kerala by small-scale fisheries activists, trawl boat owners, the government, and even the Catholic Church. A small-scale fisheries federation first launched the call for the monsoon ban in 1981, Kurien writes, in the name of fighting "the anarchic and destructive fishing of trawlers in coastal waters" (Kurien 1991, 19).

A government committee half-heartedly studied the issue in 1982, but concluded that a monsoon fishing ban was not needed (Kurien 1991; Vivekanandan et al. 2010). Fishers answered with hunger fasts, roadblocks, and public marches. The traditional fishing sector also strengthened itself, pursuing intermediate technology and motorizing boats (with government support). We note how pressure for the monsoon fishing ban originated from traditional fishers seeking protection from a capitalist fisher class. Though also couched in conservation terms, the underlying politics, as chronicled by Kurien in 1991, advocated security for small-scale fishers, which remains a goal of today's SSF Guidelines.

However, Kerala's formal 1988 monsoon season trawling ban – a national first – was not particularly successful initially (Kurien 1991). One harbor remained open and many trawl boats relocated there. In subsequent years, a kind of fishing class struggle ensued over the details and impacts of the ban. This included court challenges, expert reviews, deal making, introduction of wider fishery regulations, and even a boycott (of artisanal fishers by prawn exporters). Kurien's analysis of that early period of politicking for the ban ended in a warning: "the fish economy of Kerala is in the throes of a crisis… it is also clear that, in the long run, it is the coastal commons and the working fishermen, rather than the capitalists, that will be most affected" (Kurien 1991, 29).

**Table 14.1** Variations of state seasonal Fishing Bans (Subject to yearly change)

| State/Territory | Year of official introduction[a] | Period | Days | Exceptions to the ban |
|---|---|---|---|---|
| Gujarat | 1998–1999 | 10th June to 15th Aug | 67 | None |
| Daman and Diu | Unknown | 1st June to 15th Aug | 76 | 'Traditional' and motorized craft |
| Maharashtra | 1990 | 10th June to 15th Aug | 67 | None |
| Goa | 1989 | 10th June to 15th Aug | 67 | None |
| Northern Karnataka (Uttara Kannada District) | 1989 | 15th June to 29th July | 45 | Non-motorized and motorized craft with engines up to 25 horsepower |
| Southern Karnataka (Udupi and Dakshin Kannada Districts) | 1989 | 15th June to 10th August | 57 | Non-motorized and motorized craft with engines up to 25 horsepower |
| Kerala | 1988 | 15th June to 31st July | 47[b] | 'Traditional' or motorized craft with engines up to 10 horsepower |
| Western Tamil Nadu (Kanyakumari District) | 2001[a] | 15th June to 29 July | 45 | Non-motorized and motorized craft with engines up to 25 horsepower |
| Eastern Tamil Nadu (multiple districts) | 2001 | 15th April to 31st May | 47 | Non-motorized and motorized craft with engines up to 25 horsepower |
| Puducherry | 2001 | 15th April to 31st May | 47 | Non-motorized and motorized craft with engines up to 25 horsepower |
| Andhra Pradesh | 2000 | 15th April to 31st May | 47 | Non-motorized and motorized craft with engines up to 25 horsepower |
| Odisha | 2000 | 15th April to 15th June | 62 | Non-motorized and motorized craft with engines up to 25 horsepower |
| West Bengal | 1995 | 15th April to 31st May | 47 | Not available |

Combined and reproduced from sources: Jayasankar (2008) and Vivekanandan et al. (2010)

[a]In some cases, bans were introduced earlier, sometimes through local orders and agreements. For example, Bavinck (2003) reports the monsoon fishing ban existed in western Tamil Nadu as early as 1993

[b]61 days in 1988 and 67 days by 2006

The ban in Kerala eventually became permanent (if still contested) and also traveled to other states, often with similar politics and arguments. In Tamil Nadu, the ban was woven into contests between perceived winners and losers in the fishery. This is a point to stress: “As the closed season was imposed only on mechanized boat fishers, small-scale fishers were automatically given additional breathing space – an event of considerable political importance” (Bavinck et al. 2008, 371).

The evolution of each state’s ban is beyond the scope of this chapter, but we note how struggle brought to light tensions between capitalist and small-scale fisher classes. Conflicts also arose between in-state and out-of-state fishers, as the lack of ban uniformity meant that the larger mobile fishing boats could more easily find an open fishing season in one geography or another.

For example, Kerala’s trawlers in the early 1990s – rather than stay idle during their ban period – shifted to unrestricted waters off Tamil Nadu’s Kanyakumari District to the south, provoking the local, mostly small-scale fleet. Under pressure, Kanyakumari officials in 1993 instituted a ban concurrent to Kerala’s (Bavinck 2008). Once again, the ban in Kanyakumari arose to protect the small-scale fisheries community from an influx of capital. The entire state of Tamil Nadu would implement a seasonal ban in 2001.

Civil society and courts have also joined fishers and governments in the fray, a reflection of the breadth of fishery stakeholders. In 2000, after Goan officials reduced a ban there from 90 to 54 days, a public interest lawsuit asked the High Court of Bombay to intervene; the court ordered a ban of approximately 65 days alongside numerous other restrictions on the large-scale sector. The court also asked the central government to intercede in inter-state conflicts. In December 2002, the central Ministry of Agriculture banned monsoon fishing in the EEZ beyond 12 nautical miles of the west coast and directed states to agree on ban dates for their own waters (GOI 2003). State ministers gathered in 2003 in New Delhi though a uniform period never materialized.

In 2005, in response to a petition from the Goa Environment Federation, the Supreme Court of India issued an interim order for a uniform 67-day ban in all west coast states from 10th June to 15th August, ‘keeping in view the prime need to preserve the natural fishing resources as also to protect the traditional fishermen.’ This order exempted boats with engines only up to 10 horsepower (Supreme Court of India 2005). Despite the court order, subsequent revisions to rules at the state and central government levels ultimately meant that neither the ban period nor exemptions became uniform. The door remained open to conflicts, particularly on the west coast (Chari 2014), until the uniform ban began in 2015.

## *Scientific Justifications for the Ban*

Vivekanandan et al. (2010) survey reviews by various expert committees – led by Babu Paul in 1982; A.G. Kalawar in 1985; N.B Nair in 1989; E.G. Silas in 1994; Nair, again, in 2000; Mohan Joseph Modayil in 2005; D.K. Singh in 2007 – and note that committees have increasingly deemed the bans justifiable on general conservation grounds as a brake on fishing effort. Vivekanandan et al. also analyze catch and effort data and generally find that without monsoon fishing bans, given increasing fishing effort, average catch and fish sizes would have been lower. Ammini (1999) and Sreedevi and Kurup (2001) report the seasonal Kerala trawl ban did indeed boost stock regeneration. More recently, a government committee found that Kerala's ban boosted yields and economic values for a decade or more, until 2000 when benefits began to decline (Mohamed et al. 2014). Another still more recent central government review (GOI 2014) confirmed post-ban increase in fish harvest for up to two months after the resumption of fishing.

Today, a common official justification for a monsoon ban is to protect peak fish spawning periods that purportedly coincide with the season. This is a common refrain from fishery managers. For example, in justifying the seizure of ban-violating boats in Odisha, an official said, "the authorities imposed fishing ban to conserve fish during its breeding period [sic], so that fishermen can get maximum benefits later. Otherwise, fish productivity will fall drastically" (Times of India 2016). The official may have been reading from a 2015 central government commentary which stated the ban's goal is to "to ensure conservation of fish during its breeding-period, so that fishermen can get the maximum benefit... Otherwise, the fish productivity will be decrease, and the damage will be caused mostly to the traditional fishermen" (GOI 2015b, 1).

However, fisheries regulations often rely on approximate knowledge, rather than specific data; detailed historical catch data or reliable ecological information on India's fish stocks – including peak breeding times or spawning grounds – remain elusive. Many stock assessments are incomplete or available only at a coarse resolution (see Ghosh et al. 2015), and studies suggest many species actually spawn outside the Indian monsoon period as well. James (1992) found that important species caught on the southwest coast – mackerels, sardines, and prawns – breed not only during the monsoon but also during much of the rest of the year. Reviews of published sources (Vivekanandan et al. 2010; GOI 2014) suggest that more than 40 species along the West coast spawn many months every year during the monsoon and not:

> It is very common to find species that spawn for six months or for much longer duration in a year... Moreover, the same species spawns during different seasons in different localities. Hence, spawning season could not be considered as the sole criterion for deciding the season of fishing ban. (Vivekanandan et al. 2010, 25)

Fisher traditional knowledge – recognized by the SSF Guidelines – also suggests that the bans may not be justified based on conservation of spawning fish alone. Karnataka fishers tell stories about traditional Mangalorean fish roe curry harvested from brooders caught April or May, outside the monsoon ban (field interviews July 2012). Some trawl fishers in eastern Tamil Nadu argue that their ban period should be timed to November and December when they say important crustaceans and shrimp are breeding (Vivekanandan et al. 2010). Schaap and Haastrecht (2003) found that Tuticorin fishers supported a trawl ban but they, too, disagreed on timing; the authors concluded that this lack of consensus on breeding periods weakens ecological justifications.

At least some large-scale fishers want to scrap the ban entirely (Vivekanandan et al. 2010; field interviews July, August 2012). Others meld equity claims with strategic arguments: If there is indeed an ecological justification for the monsoon ban, they say, then small-scale fishers should also be prohibited (field interviews July, August 2012).

Some officials (e.g., GOI 2015a; Times of India 2016) offer a secondary justification, namely that restricting boats from rough monsoon seas removes perverse incentives to risk lives and gear in competition for fish. In other words, the government claims it helps fishers overcome a barrier to collective action. But this paternalist justification likely misunderstands fisher ability and knowledge. Some older fishers report that caste-based community councils would collectively ban fishing during bouts of inclement weather, irrespective of season. These fishers contrast their traditional practices to the state's blanket ban for an entire season (field interviews July, August 2012; March 2014). Meanwhile, on becalmed days during the monsoon, fishers would go to the sea, relying on their own reads of weather. 'There used to be some stretches where the sea used be very calm to enable the highly skilled fishermen to venture into sea. These stretches were known as *palke* or *madi*. The fishermen attributed this phenomenon to be the divine intervention to help the poor' (Hosbet unpublished, 1).

These debates continue, even with the onset of the 2015 (mostly) uniform ban, yet the prohibitions have been largely successful as the vast majority of the restricted fishing fleet is idled in compliance. That the basic premise of a banned fishing season is accepted largely reflects an uneasy consensus among fishers and multiple layers of government (Bavinck et al. 2008), through a combination of politics and science.

## Fishing in Karnataka

We now turn to the case of Karnataka. The state's 300-km coastline is home to a variety of fishing communities and sectors. As recently as the 1950s, many fishers used small craft and large, communal beach seines called *rampani* nets. Annual catch was perhaps only 50,000 metric tons (Bhathal 2005), and caste-based councils often provided governance for their own villages.

Like elsewhere in India, fishery development and state intervention for so-called modernization upended traditional governance and reoriented fishing away from small-scale fisheries gears. Some small-scale fishers became industrialists with state help (e.g., subsidized loans) and additional outside capital incentivized bigger boats. In the late 1960s, ice plants and cold storage appeared, and the Dakshina Kannada District Co-operative Fish Marketing Federation began to construct and distribute trawlers. By 1975, catching mackerel and sardines in Karnataka was an industrial business (Haywood and Curr 1987).

This capital influx shook traditional fisher communities that resisted mechanization. In later years, the state also promoted motorization to 'develop' traditional and small-scale fisher communities. But the playing field never leveled as large-scale fishers employed technologies such as sonar, radios, GPS, and mobile phones to great effect (Hosbet unpublished; field interviews July, August, 2012; March, April 2014). Night fishing and multi-day fishing trips by mechanized boats became commonplace. Coastal aquaculture development during the 1990s increased demand for fishmeal and further incentivized industrial fishing.

In 1980, as mechanization had just taken hold, Karnataka still had fewer than 1100 mechanized boats (CMFRI 1981). Three decades later, the state had more than 3600 mechanized boats, the majority being trawlers (CMFRI 2012). Engines larger than 300 horsepower proliferated and in early 2014, fishers reported the arrival of 600 horsepower engines in the fishery. Meanwhile, another 7500 motorized boats (often traditional craft with outboard motors) remain as an intermediate fishery.

Karnataka fisheries continue to provide employment for many coastal households. In 1980, roughly 25,000 people worked as active fishers. By 2010 that figure had risen to more than 40,000, nearly tracking with statewide population growth (CMFRI 1981, 2012; Census of India 2011). Secondary fisheries employment has accompanied industrialization; as of 2010, another 34,000 men and women worked in secondary and ancillary fishery activities such as marketing, dock labor, and processing. Additional labor migrates from other states (field interviews September 2015), though how much is unknown because migrant populations have not been counted in fisher censuses.

This increase in fishing capital and capacity has boosted catches.[6] Statewide marine fish production landings in 1977 – as mechanized craft had begun to spread – were still only 62,000 metric tons annually; by 2013, landings had increased almost six-fold to 357,000 metric tons (GOK 1997, 2015). This level is likely unsustainable as fishing remains concentrated above the continental shelf, where the state's fishery potential is estimated at just 225,000 metric tons (GOK 2015).[7]

The explosion of mechanical fishing power shifted the distribution of the returns to fishing; in 1977, just 40% of the total catch went to the mechanized and motorized sectors of the fishery. Less than a decade later, in 1985, more than 90% of fish catch was brought in by mechanized or motorized boats (GOK 1997). This skew has become permanent; from 2000 to 2013, the trawl and purse seine sectors have averaged 79.7%of all harvests; the motorized sector (including large gill net vessels) has taken 15.6%. The remainder falls to communal *rampani* shore seines and the non-motorized sector, which has averaged just 4.2% of the catch (GOK 2015).

The state government openly calls the *rampani* net 'almost obsolete' (GOK 2015), implying that small-scale fisheries are vestigial or fading. Certainly many traditional boats have been abandoned. Yet the percentage of state catch by *rampani* seines and non-motorized boats has remained relatively steady for more than decade. More than 2800 non-motorized craft still officially exist (CMFRI 2012), and many are put to use when labor is not engaged in the mechanized sector, during calmer parts of the monsoon season or in sheltered estuaries that dot the coast. Traditional canoe building with mango and jackfruit wood, stained and sealed with

[6] Though state statistics on catch differ some from CMFRI's data, we use Karnataka's figures here because the state counts catch by gear type.

[7] Fish catch data in India only track landings, offering no insight into exactly where the fish came from.

cashew oil – all local materials – remains an active small-scale fisher practice in parts of coastal Karnataka (field interviews March, May 2015). Hence, we argue small-scale fisheries are more appropriately understood as persistent.

This redistribution caused conflict between a new capitalist fisher class and the remainder. Tensions have at times erupted in violence – such as when traditional fishers have burned mechanized boats (Bhatta et al. 2000). In response, the state has generally adopted two strategies – the spatial separation of fishing sectors and capital-intensive 'development' of the small-scale fisheries sector so they are functionally less like small-scale fisheries.

The first state fishery regulations in 1978 created spatial demarcations (GOK 1978). For example, up to 5 km from the shore was reserved for *rampani* nets and traditional, non-mechanized boats. Small trawlers were allowed in this area only during September specifically to catch shrimp. Meanwhile, purse seine boats were not allowed within 8 km from shore.

By 1989, the Karnataka Marine Fisheries Regulation Act created a top-down regime that continues today. With the act, the state government imposed its hallmark rule – the monsoon ban on mechanized fishing, initially for June, July, and August. As a protection for small-scale fisheries, traditional boats fitted with outboard engines remained exempt (GOK 1989). This failed to alleviate inter-sector tensions, and in 1994 the state government extended the reserved zone for traditional fishers (including the motorized class) to 10 km from shore (GOK 1994). However, such spatial reservations are routinely ignored (or irrelevant) today. This lack of compliance contrasts with the relative success of the monsoon ban.

As elsewhere, the monsoon ban has been a product of contestations and politics. Facing lobbying from mechanized fishers, the state government in 2000 reduced the ban from 90 days to 65 days (GOK 2000). The next year the *Mogaveera Mahajana Sangha,* an institution representing more than 140 fishing villages in southern Karnataka, would largely accede to the shortened period, ordering trawl and purse seine boats (at least those of its members) to not fish from June 6 to August 9. This perhaps signaled that many fishers belonging to the traditional *mogaveera* caste had effectively accepted capitalist identities (either as owners or laborers).

Meanwhile, a new fracture appeared within the small-scale fisheries sector when four traditional fishers from northern Karnataka petitioned the state High Court in 2003 to prevent monsoon fishing by boats with any size of engine. In an interim response, the court agreed and the state followed suit, but given widespread motorization of traditional craft, many other traditional fishers lobbied again for an exception for small engines. The government conceded, but the state High Court, in its final judgment in 2004, again ordered a complete ban. Many traditional fishers then formed the Karnataka Coastal Traditional Fishers Association (KCTFA) mostly to represent the intermediate, motorized small-scale fisheries sector. The KCTFA successfully lobbied the government to yet again allow fishing by boats with engines up to 10 horsepower.

Meanwhile, influential mechanized fishers convinced Karnataka officials to reduce the ban period from 65 days to 45 days (GOK 2005a). The KCTFA went to battle again, this time against the mechanized lobby. After protests and politicking, both sides reached a compromise, in 2005, of a 57-day fishing ban for all boats with engines above 25 horsepower. The state government (GOK 2005b) fixed this agreed-to ban from 15th June to 10th August.

Various sectors continue to lobby officials and argue – publicly and privately – over the ban (field interviews July, August 2012, March 2014). Some in the small-scale fisheries sector wish to see horsepower exceptions curtailed; the intermediate class of small-scale fishers want to maintain horsepower exceptions while extending the ban period for mechanized fishers. Mechanized boat fishers generally want to shorten the ban period. And many fishers representing all gear classes raise complaints about trespass of boats from neighboring states into Karnataka waters. Given that the SSF Guidelines call for participatory management interventions, Karnataka's complex fishing politics demonstrate the range of stakeholders who may want a seat at the table.

## A Concluding Argument

Our chapter has examined India's evolving monsoon fishing bans. We note that seasonal closures have a longstanding place in fisheries management. We also have highlighted how the monsoon fishing bans have generally risen out of the politics of small-scale fisher communities, facing increased power of larger-scale sectors. The political issues raised by small-scale fishers in support of the monsoon ban receive normative backing from calls for equity and justice in the new SSF Guidelines.

In this context, we have argued that the ban is best seen as a specific privilege for the small-scale fisher community to protect livelihoods and lifestyles. These fishers, who remain at least partly in the subsistence economy, persistently rely on fish during the monsoon for even basic subsistence and income (Sehara et al. 1992; field interviews July, August 2012; February, March, April, November 2015; January 2016). While these politics often split between small-scale fishers and the interests of larger boat owners, these are not completely discreet groups. The monsoon fishing ban, as a protection for small-scale fishers, also creates a safety net for fishery laborers, who can return to their traditional craft. By protecting resources specifically for small-scale fishers, the monsoon fishing ban is in line with India's commitments under the Guidelines and, in particular, Section 5b, which prescribes sustainable fisheries management for small-scale fishers' wellbeing and livelihoods.[8]

Yet the most popular justifications for a fishing ban specifically during the monsoon remain debatable. Evidence suggests different species do breed throughout much of the year. We argue that the breeding season claim may be a canard that undermines and confuses support for the ban. The secondary, questionable argument regarding sea safety may further weaken consensus around the ban's value.

[8] Before the new SSF Guidelines, such consideration was also called for by the 1995 FAO Code of Conduct for Responsible Fisheries.

To be clear, we do not mean to completely dismiss ecological or conservation reasoning for the bans. While the specific ecological justification regarding a monsoon breeding season may be weak, a general brake on fishing effort by the overcapitalized fleet may allow for stock recovery to the benefit of all. More robust ecological data – currently nascent in Indian fisheries – may provide new arguments or counter arguments. Relying on fishers' traditional ecological knowledge may also provide a better ecological basis for the current or a different ban regulation.

But if India's fisheries policy is to be socially just and pro-poor – a normative motivation for the SSF Guidelines – then we argue the monsoon ban is justified at present as a specific benefit for small-scale fisheries. The ban's explicit exemption for small boats acts as a management measure, supported by Section 5b, to bolster viability of those stocks most relied upon by India's marginalized but persistent small-scale fisheries sector.[9] Considering the analysis in this chapter, we would even support lengthening the ban period coupled with adequate – if politically difficult – policies to compensate those who may actually suffer.

Critics of the ban (and developmentalist policy adherents) often point out that many of the smallest-scale fishers report not actually wanting to fish. Fears about unsustainability of fishing loom large, and we, like other scholars and activists, suspect that the monsoon ban alone is unlikely to be sufficient to balance ecological, economic, and social inequity. Indian fisheries remain in crisis, as Kurien identified in 1991; the larger issues of weak fisheries governance and a development-at-all-costs agenda needs critical consideration.

Ultimately, we do not argue that the monsoon fishing ban is the *best* way to manage fisheries, but we do call attention to its political and normative underpinnings. Furthermore, we suggest that paying attention to the politics that have generated the ban can highlight other ways that officials, activists and fishers may work toward more sustainable fisheries governance. Recognizing the manifold political stakeholders may open doors for new forms of participation in management (in line with Section 5b of the Guidelines which calls for participatory decision making or co-management). Empowered and engaged fishers will be necessary for a more responsive and locally appropriate governance system. Top-down, command-and-control regimes are prone to failure (e.g., Scott 1999; Kompas and Gooday 2007), but decentralization of management and more actively including fishers in self-governance (traditional or not) may support the long-term health of natural resources. For example, training and equipping small-scalefishers themselves for monitoring and enforcement could increase compliance with rules (from spatial reservations to minimum net mesh sizes). More participation by a wide array of stakeholders (especially fishers themselves) might even reverse the trend toward uniformity of ban periods, as fishers develop, monitor, and enforce locally appropriate measures. Stronger fisheries self-governance and political representation might also better address the range of external pressures on the marine commons, from

[9] We agree with Johnson (2006) that small-scale fisheries are not categorically sustainable even if they are often valorized as such. Unsustainable small-scale fisheries practices should also be addressed, as called for by the SSF Guidelines.

pollution to minerals extraction to tourism. We admit that a revolution in Indian fisheries governance sounds utopian; yet we see precedent in the politics and participation that created monsoon fishing bans to privilege marginalized small-scale fishers in the first place.

## References


Ammini, P. L. (1999). Status of marine fisheries in Kerala with reference to ban of monsoon trawling. *Marine Fisheries Information Service T & E Series, 160*, 24–36.

Arendse, C., Govender, A., & Branch, G. (2007). Are closed fishing seasons an effective means of increasing reproductive output? A per-recruit simulation using the limpet Cymbula granatina as a case history. *Fisheries Research, 85*, 93–100.

Bavinck, M. (2001a). Caste panchayats and the regulation of fisheries along Tamil Nadu's coromandel coast. *Economic and Political Weekly, 36*(13), 1088–1094.

Bavinck, M. (2001b). *Marine resource management: Conflict and regulation in the fisheries of the Coromandel coast*. New Delhi: Sage Publications.

Bavinck, M. (2003). The spatially splintered state: Myths and realities in the regulation of marine fisheries in Tamil Nadu, India. *Development and Change, 34*(4), 633–657.

Bavinck, M. (2008). Investigating poverty through the lens of riches – Immigration and segregation in Indian capture fisheries. *Development Policy Review, 32*(1), 33–52.

Bavinck, M., & Karunaharan, K. (2006). A history of nets and bans: Restrictions on technical innovation along the Coromandel coast of India. *Maritime Studies, 5*(1), 45–59.

Bavinck, M., de Klerk, L., van Dijk, D., Rothuizen, J., Blok, A., Bokhorst, J., van Haastrecht, E. K., van de Loo, T. J. C., Quaedvlieg, J. G. J., & Scholtens, J. (2008). Time-zoning for the safe-guarding of capture fisheries: A closed season in Tamil Nadu, India. *Marine Policy, 32*, 369–378.

Bhathal, B. (2005). Historical reconstruction of Indian marine fisheries catches, 1950–2000, as a basis for testing the 'marine trophic index'. *Fisheries Centre Research Reports, 13*(5), 1–122.

Bhathal, B., & Pauly, D. (2008). Fishing down marine food webs' and spatial expansion of coastal fisheries in India, 1950–2000. *Fisheries Research, 91*(1), 26–34.

Bhatta, R., Rao, K., & Bhat, M. (2000). *Alternative fishery regulations and its impact on generation and demographic transition in the maritime state of India*. New Delhi: Report Submitted to the Indian Council of Agricultural Research.

Census of India. (2011). *Provisional population totals Karnataka* (Series 30). *Directorate of Census Operations, Bangalore*. http://censusindia.gov.in/2011-prov-results/prov_data_products_karnatka.html. Accessed 1 Dec 2016.

Chari, M. (2014). Why Goa's fishermen are angry this monsoon. *Scroll.in* July 7. http://scroll.in/article/668991/why-goas-fishermen-are-angry-this-monsoon. Accessed 6 Dec 2016.

Charles, A. (2011). Small-scale fisheries: On rights, trade and subsidies. *Maritime Studies, 10*(2), 85–94.

CMFRI. (1981). All Indian census of marine fishermen, craft and gear 1980. *Marine Fisheries Information Service, 30*, 2–32.

CMFRI. (2012). *Marine fisheries census 2010*. New Delhi: Ministry of Agriculture.

CMFRI. (2016). *Fish catch estimates*. http://www.cmfri.org.in/fish-catch-estimates.html. Accessed 1 Dec 2016.

FAO. (2011). *Fisheries management 4, marine protected areas and fisheries* (FAO technical guidelines for responsible fisheries, Vol. 4, pp. 1–198). Rome: Food and Agriculture Organization of the United Nations.

FAO. (2015). *Voluntary guidelines for securing sustainable small-scale fisheries in the context of food security and poverty eradication*. Rome: Food and Agriculture Organization of the United Nations.

Fernandes, A., & Gopal, S. (2012). *Safeguard or squander: Deciding the future of India's fisheries*. Bangalore: Greenpeace India.

Ghosh, S., Muktha, M., Hanumantha Rao, M., & Behera, P. (2015). Assessment of stock status of the exploited fishery resources in northern bay of Bengal using landed catch data. *Indian Journal of Fisheries, 62*(4), 23–30.

GOI. (2003). *Order on uniform fishing ban in West Coast including the Lakshadweep Islands*. Order No. 59811/3/2001. New Delhi: Government of India, Ministry of Agriculture.

GOI. (2014). *Report of the echnical committee to review the duration of the ban period and to suggest further measures to strengthen the conservation and management aspects*. New Delhi: Government of India, Ministry of Agriculture.

GOI. (2015a). *Order NO. 30035/15/97-Fy(t-1) Vol. IV.* New Delhi: Government of India, Ministry of Agriculture.

GOI. (2015b). *Clarification from agriculture ministry in relation to rights of fishermen and fishing ban*. Press Release 27 May 2015. New Delhi: Government of India, Press Information Bureau Press.

GOK. (1978). *Order on demarcation of operational areas for fishing vessels of different types*. No. SWL 316. Government of Karnataka, Social Welfare and Labour Department, Bangalore.

GOK. (1989). *Notification on prohibiting operation of mechanized boats in the coast of Karnataka*. No. AHFF266 SFM 88. Bangalore: Government of Karnataka.

GOK. (1994). *Order on delimitations of area for exclusive use of traditional fishermen*. No. AHF365, SFM 93. Government of Karnataka, Animal Husbandry and Fisheries Department, Bangalore.

GOK. (1997). *Statistical bulletin of fisheries*. Bangalore: Government of Karnataka, Directorate of Fisheries.

GOK. (2000). *Notification on prohibition of fishing operation by mechanized boats in the coast of Karnataka*. No. AFH/107/SFM/98. Bangalore: Government of Karnataka.

GOK. (2005a). *Notification on prohibition of fishing operation by mechanized boats in the coast of Karnataka*. No. HF148 SFM 2004. Bangalore: Government of Karnataka, Directorate of Fisheries.

GOK. (2005b). *Notification on prohibition of fishing operation by mechanized boats in the coast of Karnataka*. No. AHF 126 SFM 2005. Bangalore: Government of Karnataka, Directorate of Fisheries.

GOK. (2015). *Statistical bulletin of fisheries*. Bangalore : Government of Karnataka, Directorate of Fisheries.

Gordon, H. S. (1954). The economic theory of common property resource: The fishery. *Journal of Political Economy, 62*(2), 124–142.

Halliday, R. G. (1988). Use of seasonal spawning area closures in the management of haddock fisheries in the Northwest Atlantic. *NAFO Scientific Council Studies, 12*, 27–35.

Haywood, K., & Curr, C. (1987). *Project identification in small pelagic fisheries: Indonesia, India, Thailand, Morocco*. FAO Fisheries technical paper. No 289. Rome: The Food and Agriculture Organization of the United Nations.

James, P. (1992). Impact of fishing along the west coast of India during southwest monsoon period on the finfish and shellfish resources and the associated management considerations. In P. Rao, V. Murthy, & K. Rengarajan (Eds.), *Monsoon fisheries of west coast of India- prospects, problems and management*. Kochi: Central Marine Fisheries Research Institute.

Jayasankar, J. (2008). Fishing regulation and fishing ban. In P. Thomas (Ed.), *Proceedings of training programme on sustainable fishing and fisheries conservation for NETFISH*. Kochi: Central Marine Fisheries Research Institute.

Jentoft, S. (2014). Walking the talk: Implementing the international voluntary guidelines for securing sustainable small-scale fisheries. *Maritime Studies, 13*(16), 1–15.

Johnson, D. (2006). Category, narrative, and value in the governance of small-scale fisheries. *Marine Policy, 30*(6), 747–756.
Kocherry, T., & Achary, T. (1989). Fishing for resources: Indian fisheries in danger. *Cultural Survival Quarterly, 13*(2), 31–33.
Kompas, T., & Gooday, P. (2007). The failure of 'command and control' approaches to fisheries management: Lessons from Australia. *International Journal of Global Environmental Issues, 7*(2/3), 174–190.
Kurien, J. (1991). *Ruining the commons and responses of the commoners: Coastal overfishing and fishermen's actions in Kerala state, India.* Discussion paper No. 23. Geneva: United Nations Research Institute for Social Development.
Kurien, J. (2016). SSF guidelines: The beauty of the small. *Samudra Report, 72*, 30–36.
Mansfield, B. (2001). Property regime or development policy? Explaining growth in the U.S. Pacific groundfish fishery. *The Professional Geographer, 53*(3), 384–397.
Mohamed, K., Puthra, P., Sathianandan, T., Baiju, M., Sairabanu, K., Lethy, K., Sahadevan, P., Nair, C., Lailabeevi, M., & Sivaprasad, P. S. (2014). *Report of the committee to evaluate fish wealth and impact of trawl ban along Kerala coast.* Kochi: Government of Kerala, Department of Fisheries.
Narayanakumar, R. (2008). Sustainable fishing and fisheries regulations. In P. Thomas (Ed.), *Proceedings of training programme on sustainable fishing and fisheries conservation for NETFISH.* Kochi: Central Marine Fisheries Research Institute.
Philip, S. (2015). Why Kerala won't stop fishing. *The Indian express,* 4 June 2015.
Sadovy, Y., Colin, P., & Domeier, M. (2005). Monitoring and managing spawning aggregations: Methods and challenge. *SPC Live Reef Fish Information Bulletin, 14*, 25–29.
Schaap, M., & Haastrecht, E. (2003). *A critical look at fisheries management practices: The 45-day ban in Tuticorin district, Tamil Nadu, India (Master's thesis).* Amsterdam: University of Amsterdam.
Scott, A. (1955). The fishery: The objective of sole ownership. *Journal of Political Economy, 63*, 116–124.
Scott, J. C. (1999). *Seeing like a state: How certain schemes to improve the human condition have failed.* New Haven: Yale University Press.
Sehara, D., Pannikar, K., & Karbhari, J. (1992). Socio-economic aspects of the monsoon fisheries of the west coast of India. In P. Rao, V. Murthy, & K. Rengarajan (Eds.), *Monsoon fisheries of west coast of India - prospects, problems and management.* Kochi: Central Marine Fisheries Research Institute.
Sreedevi, C., & Kurup, B. M. (2001). Temporal variations in polychaete population in 0–50 meters along coastal Kerala, S. India. In AFS (Ed.), *Sixth Asian fisheries, forum book of abstracts.* Taiwan: Asian Fisheries Society.
Subramanian, A. (2009). *Shorelines: Space and rights in South India.* Palo Alto: Stanford University Press.
Supreme Court of India. (2005). Writ petition (Civil) 393, Goa environment federation v. union of India, State of Karnataka; State of Gujarat; State of Maharashtra; State of Kerala; State of Goa.
Times of India. (2016, May 26th). Paradip officials seize 10 boats for violating fishing ban. *Times of India.*
Vivekanandan, E. (2008). Conservation of marine fisheries resources. In P. Thomas (Ed.), *Proceedings of training programme on sustainable fishing and fisheries conservation for NETFISH.* Kochi: Central Marine Fisheries Research Institute.
Vivekanandan, E., Narayanakumar, R., Najmudeen, T., Jayasankar, J., & Ramachandran, C. (2010). *Marine fisheries policy brief – 2: Seasonal fishing ban.* Kochi: CMFRI Special Publication 103.

# Chapter 15
# Implementing the Small-Scale Fisheries Guidelines: Lessons from Japan

**Alyne Delaney and Nobuyuki Yagi**

**Abstract** The Food and Agriculture Organization of the United Nations (FAO) recently began the implementation phase of its Voluntary Guidelines for Securing Sustainable Small-Scale Fisheries in the Context of Food Security and Poverty Eradication (the SSF Guidelines). The SSF Guidelines emphasize food security, poverty reduction, and ecological sustainability. Japan is an industrialized nation with a relatively low poverty rate and good food security. Thus, what utility, if any, could the SSF Guidelines hold for Japan? And what lessons can the Japanese case provide for other nations around the world? Outsiders to fisheries may assume that developing countries are characterized by small-scale fisheries while industrialized nations have large-scale fleets and a minority of small-scale fishers. Yet fisheries in Japan are overwhelmingly small-scale and based in local communities with historic links to nearby coastal resources and characterized by strong local community culture, values, and identities, representing a way of life for these practitioners. With this reality in mind, this chapter focuses on three of the SSF Guidelines objectives for which Japan presents a positive case: sustainable fisheries management; equitable development in coastal communities; and the contribution of small-scale fisheries to an economically, socially, and environmentally sustainable future. This chapter provides a brief overview of the history of Japanese coastal fisheries, with a special emphasis on the community-based management styles and how these sustain both cultural and environmental resources. The current challenges of Japanese coastal fisheries are also discussed, highlighting both lessons learned and potential challenges ahead for other nations as they work toward implementation of the SSF Guidelines.

**Keywords** UN SSF Guidelines • Small-scale fisheries • Japan • Social sustainability • Marine Protected Areas (MPAs)

---

A. Delaney (✉)
Innovative Fisheries Management, Aalborg University, Aalborg, Denmark
e-mail: ad@ifm.aau.dk

N. Yagi
Global Agricultural Sciences, The University of Tokyo, Tokyo, Japan
e-mail: yagi@fs.a.u-tokyo.ac.jp



S. Jentoft et al. (eds.), *The Small-Scale Fisheries Guidelines*, MARE Publication Series 14, DOI 10.1007/978-3-319-55074-9_15

## Introduction

In 2011, the United Nations Food and Agriculture Organization (FAO) Committee on Fisheries recommended that an international instrument on small-scale fisheries be developed as a means to strengthen world-wide efforts to alleviate poverty and improve food security. This resulted in the Voluntary Guidelines for Securing Sustainable Small-Scale Fisheries in the Context of Food Security and Poverty Eradication (the SSF Guidelines) (FAO 2015a). It has been demonstrated that small-scale fisheries provide food security, livelihoods, and human well-being. Though the SSF Guidelines are global in scope, they focus particularly on the needs of developing countries and, among other goals, support initiatives for equitable social and economic development and to advocate to secure small-scale fisheries and related livelihoods (FAO 2015a, b). There is often an assumption that small-scale fisheries are not particularly important in industrialized nations today. Yet, they make up a complex mix of artisanal, subsistence, and even commercial sectors; using a wide variety of gear types and harvesting numerous different species (Berkes 2001; Pauly et al. 2005; Chuenpagdee 2011) and remain important for a number of industrialized nations such as Spain, Portugal, and Japan (Camiñas et al. 2004; Pita et al. 2017). Small-scale fisheries also are significant for certain regions within other industrialized countries such as in Europe and North America. This importance is derived from both the economic contributions of these fisheries as well as cultural values held by fisheries-dependent communities and the broader society (e.g., Berkes 2001; Cinner 2005; Poepoe et al. 2007). Therefore, governance frameworks which support and protect small-scale fisheries, rather than discourage them, are critical for ensuring social and ecological sustainability.

In this chapter, we argue that governance frameworks which support and protect small-scale fisheries, rather than discourage them, are critical for ensuring social and ecological sustainability. Japan provides a clear case of supportive governmental policies which enabled small-scale fisheries through equitable development in coastal communities, sustainable fisheries management, and the contribution of small-scale fisheries to an economically, socially, and environmentally sustainable future. We contend that social and environmental sustainability came about in large part due to Japan's fisheries co-management and associated institutions, such as cooperatives. These cooperatives, which have a long history in Japan, hold the legally recognized fishing rights which are included with membership. Even so, as Uchida and Makino (2008) point out:

> That history is not the end of the story, nor is that history determinative of Japanese success today. Rather, Japan faces fisheries management challenges that are similar to those of contemporary fisheries elsewhere. However, the breadth of experience with co-management in Japan can yield valuable lessons for other fishery management systems. (Uchida and Makino 2008, 222)

## *Japanese Small-Scale Fisheries*

Japan has historically had one of the world's highest per capita rates of consumption of fish and fisheries products (MAFF 2010, 2011a). Japan's fisheries product imports are second only to that of the United States, with recent declines linked in part to imports being more expensive for Japan due to a weaker currency (FAO 2016). Japanese fisheries are locally and nationally oriented, with most harvests going to domestic consumption. There is no legal definition of small-scale fishing in Japan, but for the purpose of fisheries production statistics, fishing boats smaller than 10 gross tons are recognized as coastal fishing vessels and also as small-scale fishers in Japan. Overwhelmingly, Japanese fisheries comprise small-scale fleets and fisheries: according to 2013 government statistics, Japanese fishing boats numbered 94,507 and, of these, 89,107 entities (94% of the total) were small coastal fishers with fishing boats smaller than 10 gross tons (MAFF 2016a). This gives small-scale fisheries a unique position within Japan compared to many other industrialized nations.

Japanese small-scale fisheries are also based in local communities with historic links to nearby coastal resources and with strong local community culture, values, and identities; fisheries represent a way of life for these practitioners (Delaney 2003; Yagi et al. 2010, 2012; Makino 2011). Given Japan's position as an industrialized nation which has risen from poverty in only a few generations, the chapter focuses upon sections of the SSF Guidelines in which equitable development in coastal communities, sustainable fisheries management, and the contribution of small-scale fisheries to economically, socially, and environmentally sustainable futures are highlighted.

Despite historical and current success in meeting most of these objectives, Japanese fisheries are facing difficulties due to on-going demographic and societal changes. The elderly population makes up an average 23% of Japanese communities, but this figure increases to 32% in coastal communities (MAFF 2011b). In addition to a greyer population, many coastal regions remain remote from large urban centers, which provide better employment, education, and social opportunities for younger residents. The Pacific coast of Tōhoku, one of Japan's most important fisheries regions, also continues to struggle with a slow recovery following the tsunami generated by the 2011 Great East Japan Earthquake.

These on-going challenges notwithstanding, small-scale fisheries remain important for Japan and Japanese coastal society. While it is true that some sources show there has been a worldwide decline in marine capture fisheries landings (Pauly and Zeller 2016), and it is true for the case of Japanese long-distance fishing operating in high seas and exclusive economic zones (EEZ) of other countries as well as Japanese off-shore fishing operating in Japan's EEZ, the Japanese coastal fisheries have had relatively stable landings since the 1960s, with relatively slow decline over the last 20 years (Table 15.1, MAFF 2016a). With the decline in off-shore fishing

**Table 15.1** Japanese fisheries production statistics

| Year | Total | Inland fisheries | Marine aquaculture | Coastal | Offshore | Long-distance |
|---|---|---|---|---|---|---|
| 1988 | 12,785 | 198 | 1327 | 2115 | 6897 | 2247 |
| 1989 | 11,914 | 202 | 1272 | 2123 | 6340 | 1976 |
| 1990 | 11,052 | 209 | 1273 | 1992 | 6081 | 1496 |
| 1991 | 9978 | 205 | 1262 | 1894 | 5438 | 1179 |
| 1992 | 9257 | 188 | 1306 | 1968 | 4534 | 1270 |
| 1993 | 8707 | 177 | 1274 | 1861 | 4256 | 1139 |
| 1994 | 8103 | 169 | 1344 | 1807 | 3720 | 1063 |
| 1995 | 7489 | 167 | 1315 | 1831 | 3260 | 917 |
| 1996 | 7417 | 167 | 1276 | 1901 | 3256 | 817 |
| 1997 | 7411 | 153 | 1273 | 1779 | 3343 | 863 |
| 1998 | 6684 | 143 | 1227 | 1582 | 2924 | 809 |
| 1999 | 6626 | 134 | 1253 | 1605 | 2800 | 834 |
| 2000 | 6384 | 132 | 1231 | 1577 | 2591 | 855 |
| 2001 | 6126 | 117 | 1256 | 1545 | 2459 | 749 |
| 2002 | 5880 | 113 | 1333 | 1489 | 2258 | 686 |
| 2003 | 6083 | 110 | 1251 | 1577 | 2543 | 602 |
| 2004 | 5776 | 106 | 1215 | 1514 | 2406 | 535 |
| 2005 | 5765 | 96 | 1212 | 1465 | 2444 | 548 |
| 2006 | 5735 | 83 | 1183 | 1451 | 2500 | 518 |
| 2007 | 5720 | 81 | 1242 | 1287 | 2604 | 506 |
| 2008 | 5592 | 73 | 1146 | 1319 | 2581 | 474 |
| 2009 | 5432 | 83 | 1202 | 1293 | 2411 | 443 |
| 2010 | 5312 | 79 | 1111 | 1286 | 2356 | 480 |
| 2011 | 4733 | 73 | 869 | 1129 | 2264 | 431 |
| 2012 | 4841 | 67 | 1040 | 1090 | 2198 | 458 |
| 2013 | 4791 | 61 | 997 | 1150 | 2168 | 396 |
| 2014 | 4793 | 64 | 988 | 1098 | 2274 | 369 |
| 2015 | 4600 | 69 | 1067 | 1071 | 2115 | 347 |

Source: Japanese Ministry of Agriculture, Forestry and Fisheries (2016)

following the introduction of EEZs, the coastal fisheries became the most important sector, value-wise, in the 1980s, a position retained today (Makino 2011). Marine aquaculture, which developed greatly in the 1980s and which primarily consists of small-scale household operations, is the second largest sector in terms of production value (Makino 2011).

Focusing on the case study of small-scale fisheries in Japan, we pose two questions: what utility, if any, could the SSF Guidelines hold for a country like Japan; and, what can the continued importance of small-scale fisheries in Japan teach the rest of the world in terms of ways to implement the SSF Guidelines?

In order to answer these questions and understand what lessons can be drawn from the Japanese case for small-scale fisheries worldwide and the SSF Guidelines, this chapter provides a brief overview of the history of Japanese coastal fisheries, with a special emphasis on the community-based management styles and how these sustain both cultural and environmental resources. The current societal challenges faced by Japanese coastal fisheries and on-going debates on how to best meet these difficulties are also discussed.

The data in this chapter are primarily drawn from secondary sources. However, the authors each have a lengthy history of work and research in Japan. Consequently, their work is also influenced by their experiences, fieldwork, and the gathering of empirical data in Japanese coastal communities.

## History and Description of Japanese Small-Scale Fisheries

Currently, approximately 94% of Japanese fishing entities operate boats less than ten tons, roughly equivalent to the boat size less than ten meters with 87% of fishers working in small-scale coastal fisheries (Makino 2011, 2016a). Small-scale fisheries tends to be defined differently around the world with varying definitions based on boat length, artisanal versus commercial, gear type, or other factors. In the Japanese case, small-scale fishers use small boats and practice commercial, coastal fishing with landings at local ports. Japanese small-scale fisheries are also broadly characterized by the following points: (1) many boats and landing ports, making it difficult for the central government to monitor and control; (2) varied catch of more than 400 species, making the setting of a Total Allowable Catch costly for the government; (3) emphasis on sales to local and domestic consumption, rather than export driven business; and (4) relatively high priority placed on social issues such as fair and equitable distribution of the catch, rather than maximizing economic efficiency. These points not only characterize the current fisheries, but also impacted

**Fig. 15.1** Shellfish fishers sort their catch. Yogasakihama, Shichighama, Miyagi, Japan, 2013. (Source: Alyne Delaney)

the formation of the SSF Guidelines and have the potential to impact their future development and implementation (Fig. 15.1).

## History of Japanese Small-Scale Fisheries Management[1]

Japanese ancient fisheries have a history of the collective use of common property resources, open to all, and managed by local user groups, as first described in the Taiho Code in the year of 704 A.D. (Makino 2011). Once the population increased after the seventeeth century, additional measures were made. For example, the Tokugawa Dynasty passed the Urahō, or 'beach law,' which stipulated that coastal fishing grounds in coastal, near-shore waters should be accessed only by residents of local fishing communities under the control of village headmen (Makino 2011). Under these regulations, coastal waters were regarded as extensions of the land and were thus a part of the domain of feudal lords (Akimichi and Ruddle 1984; Makino 2011). In general, communities controlled adjacent coastal areas, and were responsible for establishing appropriate rules for use of the area. This community management, essentially an autonomous management body of local fishers, formed the basis for subsequent formal management institutions such as Fisheries Societies and Fishing Cooperative Associations (FCAs) (Makino 2011).

During the modernization period following the opening of Japan in 1868, the Japanese government briefly experimented with a Western, centralized top-down model by issuing licenses (Weinstein 2000; Makino 2011). With licenses, individuals no longer needed to reside in a local fishing community so there was a sudden and extremely large increase in fishers and landings; governmental statistics show that annual fisheries production tripled in only 7 years (Makino 2011). Such an increase was unsustainable. Thus, a trial of (1870s) Western-style fisheries management was proven to be a failure (Weinstein 2000; Makino 2011).

In the 1880s, fishers were encouraged to form fisheries unions, bringing back management by local groups (Weinstein 2000). Fishing Cooperatives became official in 1891 as a modern version of guilds which required residence and an apprenticeship period (Makino 2011). The 1901 Fisheries Law awarded exclusive rights to inshore waters to these local groups (Makino 2011). A characteristic of these laws was the possibility to transfer fishing rights, which meant by the 1930s many money-lenders actually owned such rights since poor fishers often transferred their rights to gain capital for purchasing fishing gear (Weinstein 2000). Credit federations began in the 1930s to provide safe and secure credit to fishers, and by the 1948 Fisheries Cooperative Association Law (passed 1949), safeguards were put in place to maintain rights and credit into the modern version of cooperatives.

Today, all Japanese fishers in the inshore areas are members of Japanese Fishing Cooperative Associations (FCAs). FCAs are locally, regionally, and nationally

[1] For more in depth discussion on the history of Japanese FCAs and small-scale fisheries, please see, for example, Makino 2011, Makino and Matsuda 2005, and Weinstein 2000.

linked organizations which market products, provide gear and insurance, and work as credit unions (Delaney 2015a). FCA membership entitles fishers to use (usufruct) rights to resources found within the territory of their local FCA; the FCA hold the right to exclusively use the resource in their tenure area. Rights for different resources and technologies must be applied for separately by the FCA to the prefectural government and may include small-scale net and trap fisheries, aquaculture, and large-scale set-net fisheries (Ruddle and Akimichi 1984; Delaney 2003). "The fishers, as committee members of their local and prefectural FCAs, are the primary managers of each local resource, working together with government fisheries regulatory commissions and scientific staff at the prefectural and national levels" (Delaney 2003, 71). This management tool includes maximum catch quota amounts decided by FCAs as local rules for species with no national TAC or local no-take zones/seasons.

Fishing Cooperative Associations thus are the membership and management group for small-scale fisheries in Japan. These groups also are often the focus of social activity. In addition to formal management boards, FCAs have women's groups, who often take part in activities such as beach clean-ups, environmental campaigns, and leadership positions in local activities such as local festivals, and young men's groups – which often organize study groups and workshops with fisheries scientists (Delaney 2003).

## Voluntary SSF Guidelines

Japanese small-scale fisheries are based in local communities with historic links to nearby coastal resources with strong local community culture, values, and identities; fisheries represent a way of life for these practitioners (e.g., Norbeck 1954; Kalland 1980; Cordell 1989; Delaney 2003; Martinez 2004). With these social and cultural considerations in mind, this chapter focuses on three objectives of the SSF Guidelines: (1) sustainable fisheries management; (2) equitable development in coastal communities; and (3), the contribution of small-scale fisheries to an economically, socially, and environmentally sustainable future.

According to the SSF Guidelines, reaching equitable development and eradicating poverty requires working "within the context of sustainable fisheries management" (FAO 2015a, b, 1). What makes fisheries sustainable is not defined; key requirements for sustainable fisheries and resource management are highlighted, however. Sustainable fisheries management relies upon responsible governance (Objective 5a.) such as community and co-management (Objective 5.17), culturally appropriate and secure tenure rights (Objective 5.3), legislation which recognizes the rights of the communities (Objective 5.4), and monitoring, surveillance, and control systems should be promoted (Objective 5.16).

Equitable development is another primary objective of the SSF Guidelines (Objective 1). This is in response to the fact that small-scale fishers throughout the world – and the communities where they reside – are currently, as well as historically, often marginalized, vulnerable, and poor. These fishers and community members are often highly dependent on having and maintaining access to fisheries and marine resources. An implicit understanding of the importance of access to fisheries resources is the understanding that these must be sustainably managed. In a mutually-reinforcing way, social sustainability in natural resource–based communities relies on environmental sustainability and vice-versa. A key aspect for social and environmental sustainability is the organization of communities, management, and resource rights. If these are organized in an equitable manner, small-scale fisheries can then contribute to an economically, socially, and environmentally sustainable future. Indeed, if properly managed, they can, as the SSF Guidelines state, 'contribute to the equitable development of small-scale fishing communities and poverty eradication and to improve the socio-economic situation of fishers and fish workers within the context of sustainable fisheries management' (FAO 2015a, b, 1). Japan offers a case where the legal protections of fishing rights and the protection and organization of fishing cooperative associations (FCAs) and membership in FCAs helped improve the livelihoods of small-scale fishing families.

An additional objective of the SSF Guidelines is the promotion of the contribution of small-scale fisheries to an economically, socially, and environmentally sustainable future (Objective 1.1d).

A representative of Japan's Fisheries Agency recently highlighted some of the aspects of the Japanese system which they feel help meet the goals of social and environmental sustainability, such as protected resource rights which sustain and protect community-based fisheries management. They also believe the governmental support provided to their form of community-based fisheries management (FCAs) has been key. This support includes providing Prefectural Fisheries support officers who "cooperate with local research institutes to respond to FCAs needs in terms of resource management, research, and technology development" (FAO 2015b, 31).

## *Ecologically Sustainable Fisheries Management*

Japan governs its coastal areas with collective fisheries management regimes. In these areas, stakeholders have been given the opportunity to regulate their own activities without the need for stringent top-down management (Yagi et al. 2012). The stakeholder initiative is legally supported by the Japanese national and prefectural governments as a form of co-management. In this system, a limited entry system for fishing operations is regulated though a strict government fishing licensing scheme and other legally binding effort control mechanisms which limit the boat sizes, gears, seasons, areas, and other aspects of the fishery. Area-based management is also an important tool for Japanese coastal fisheries management. Such traditional effort controls, in addition to area-based management, are regarded as the primary elements of the domestic fishery governing scheme in Japan; even after a national total allowable catch (TAC) system was introduced in 1997, effort controls are viewed as the main pillar of management because there are only seven species covered by the national TAC. In fact, in interviews with many officials in Fisheries Agency of Japan, the interviewer sensed that the main motivation of introducing TAC was not to fulfil the management needs of domestic fisheries but rather to fulfil the international requirements of UNCLOS (United Nations Convention on the Law of the Sea), which Japan ratified in 1996.

After the TAC system was introduced, annual catch quotas for seven fishery species (mackerel, Jack mackerel, Japanese sardine, Pacific saury, Alaskan pollock, common squid, and snow crab) were decided based on scientific advice and enforced by the government. The state of stock conditions had been assessed regularly even before the introduction of TAC. According to the most recent assessment by the Fisheries Agency of Japan in March 2016 – which includes 20 stocks consisting of the seven TAC species, four stocks had high stock levels, nine stocks had middle levels, and seven stocks had low levels (Fisheries Agency of Japan 2016). TAC limits are, in general, set for pelagic species which are not as important for the coastal fisheries.

It is not easy to evaluate the effectiveness of the TAC system in Japan because traditional effort controls exist together with the TAC system and the individual contributions to resource conservation cannot be separated. In addition, due to the fact that a large fluctuation on resource recruitment occurs under the influence of ocean temperature or other environmental factors, non-human factors could contribute to the increase and decrease of the fish stocks. For now, the use of TACs is limited, with traditional systems remaining strong.

Traditional systems also include traditional Japanese area-based management for coastal fisheries, a system which shows striking parallels with contemporary Marine Protected Areas (MPAs) and management methods found with Territorial Use Rights for Fishing (TURF) programs. Currently, 1161 MPAs are spread more or less evenly throughout Japanese waters (Yagi et al. 2010). Of these MPAs, 1055 locations are implemented in conjunction with fishery regulations and they take the form of no-take fishing zones (Yagi et al. 2010). Most of them are situated near the coastal residential areas where peer-monitoring activities can be implemented at a relatively low cost (Yagi et al. 2010). A number of these MPAs are believed to have originated from self-imposed community fishery rules established in or before the nineteenth century (Yagi et al. 2010).

The SSF Guidelines call for sustainable fisheries management. Though conflicts among different groups will always exist and some stocks face pressures and overharvesting, in general, Japanese small-scale coastal fisheries are, as evidenced by the relatively stable catch record for decades (MAFF 2016a), sustainably managed. In the nineteenth century, Japan experimented by making fisheries essentially open access, with licenses available to all – catches tripled in only 7 years; making the experiment a failure as fisheries managed this way were unsustainable. This led Japan to fall back and build upon traditional management strategies. Such management includes controls such as area-based management, access restrictions (e.g., available through membership in FCAs), with limitations including local residence and inheritance, and co-management with control and responsibilities shared with prefectural and national governments. The co-management process also includes consensus-based decision-making and ensures equitable access to marine resources.

In Japanese co-management, collaboration takes place among the three key stakeholder groups: management, in the form of fishers (the resource users), regulators (the authorities), and scientists. Fishers have their own fisheries ecological knowledge and experience which is an advantage in this system; scientific knowledge is also indispensable for ecologically sound management; and regulators also contribute through the coordination and facilitation of multi-jurisdiction management arrangements (Uchida and Makino 2008). The relationships between regulators and fishers are close in Japan. For example, one of the roles of FCAs is to inform their members of new and changing national fisheries policies.

## *Equitable Development in Coastal Communities*

Equity does not equal equality. Notions of fairness are always based on local understandings from within one's own society and culture (Delaney 2015a) as well as one's own position within it. Japanese society includes both vertical and horizontal social ties and developments in coastal communities and fisheries management reflects both aspects.

As a hierarchical society (Nakane 1970), vertical relations are important in Japan with social institutions such as the household, schools, workplaces, and clubs organized vertically internally. Within such, individuals are socialized to grant respect to those of higher status – people who are older, better educated, and often wealthier (though this often coincides with higher education), or those who have seniority. Respect towards these individuals can often be seen through the use of language and customs (Lebra 1992).

Yet despite the important of hierarchy in Japanese culture and society, horizontal relationships also remain important. Horizontal ties can be seen among same age mates (e.g., Traphagan 1998) as well as among groups such as *kumi* and *han* which operate without hierarchy (Benjamin 1997; Delaney 2003). In the old village structure, for example, individual households theoretically stood equal to other households as members of the village (Nakane 1967) and it is in the sub-group (not dependent on a ranking hierarchy) where this ideal of equality plays out. In groups, members usually cooperate with one another on communal tasks on an equal basis (Beardsley et al. 1959) such as with the clean-up of irrigation ditches or in the case of port villages, installing the bamboo into communal seaweed seeding beds. "Within the villages, there may be different levels of groups of mutual assistance … but all are characterized by equal level relationships – all are cooperative work and mutual assistance units" (Delaney 2003, 248). Fishing cooperative associations are an extension of these groups with many formed along *kumi* (group) lines for the purpose of mutual assistance and the spread of information (Fukutake 1967).

Yet, historically in Japan, fishers and fishing communities were poor and considered of low status. Smethurst (1986) points out that the poorest of groups have neither the agency nor the ability to work for change, but instead are simply struggling to survive. "It is poor, but not destitute groups, along with outsiders … who are actually able to work for change" (Delaney 2003, 11). This is why the on-going collective action seen in many coastal regions in the world today (e.g., India in Agrawal and Ostrom 2001) is vital for improving fisheries-based livelihoods. Similar collective action took place in Japan in the 1930s with political activists and social-minded government servants protesting and working along with the farmers and fisheries to increase prices and conditions in order to improve social welfare (Smethurst 1986; Ando 1995).

These efforts coincided with the Japanese government's passage of its Farms and Fisheries Revitalization Policy (1933), a policy which was criticized and termed the 'Self-Rehabilitation Movement' by many due to the very little it actually did for the fishers: low interest loans were made available to fishers, at least until the financing ran out, but nothing else was enacted (Ando 1995).

The 1930s were a time of great poverty, with emigration to other countries being very high. For example, in the first half of the decade, more than 70,000 Japanese immigrated to Brazil (IBGE n.d.). At this time, many fishers were indebted to money-lenders and middlemen for the loans they needed not only to buy gear, but also to pay for social obligations such as wedding and funerals. During this period, the Japanese government, through the efforts of individuals such as Takatoshi Ando, a public servant in Hokkaido, worked to break the fishers free from the hold merchants and middlemen had in controlling the capital and marketing of their catches. In this period, credit federations were established and fishers began to market their own catch (Ando 1995). During the Occupation in the immediate post-World War II period, legislation was finally passed that prevented fishing rights from being transferable, thereby preventing the rights from being turned over to money-lenders again (Ruddle and Akimichi 1984; Sato 1992; Weinstein 2000).

An important aspect of fishing right holder-ship in Japan is the fact that rights are protected by law. These legal protections ensure that if coastal development is planned, fishing right-holders have the opportunity to protest or advocate, vote and, if accepted, receive monetary compensation from the sale of their rights. Especially since a disproportionate amount of industrialization and development has taken place along coasts, this is an important protection (McKean 1981; Delaney 2003).

Fishing rights are also conferred upon the household head. In what is perhaps a serendipitous result of the Japanese kinship system and its focus on households, the system thus also innately protects and supports small-scale fisheries since it is the household head upon which the right is conferred and it is the household unit which operates the small-scale fisheries operations. National and local governments in Japan also emphasize keeping employment in local communities. Consequently, small-scale fisheries and agriculture are often treated as key stakeholders in decision-making processes. The combination of legal regulations and policies surrounding fishing rights and the cultural tradition of enterprise households combined to protect Japanese small-scale fisheries as Japan was industrializing.

In the post-war era, FCAs played a strong role in the equitable development of coastal communities. Furthermore, the modern FCA system was organized around "community territoriality and livelihood rights" (Ruddle and Akimichi 1984, 333). Thus, cooperatives were formed first and foremost for community development, to provide livelihoods and eradicate the extreme poverty of the pre-War era.

The FCAs, as cooperatives, are egalitarian in nature. As shown by Barrett and Okudaira, "cooperatives are … likely to place a strong emphasis on leveling individual differences and competition between their members" (1995, 209). Yet, this is still set within local understandings of equity. In one community, full members (*sei kumiaiin*) and junior members (*jun kumiaiin*) had differences in rights. Also, the length of a household's membership impacted resource distribution. For example, in one Pacific coast town (e.g., Delaney 2015a), with the expansion of membership in one of the newer FCAs formed after WWII, the original members were allocated more fishing ground space than the newcomers.

However, society and culture is fluid and ever-changing. In the 1980s in this FCA, pressure for equality could no longer be resisted as active membership continued to decline while the extras space in aquaculture (zones) allocated to families with seniority were no longer deemed fair and equitable. Thus, at this time, access was equalized regardless of one's history in the FCA (Delaney 2003). There are other methods and means to ensure equity among FCA members. For example, an annual lottery is held prior to each new season to allocate seeding and cultivation space. This is done as the seabed and water quality can vary significantly from place to place. Thus, a lottery provides a way to vary the space one receives ensuring fairness and parity. Similar examples of notions of equity and fairness can also be seen in how fish auction stalls are allocated (Bestor 2004) and even in the allocation of temporary housing following the 2011 Great East Japan earthquake and tsunami (Delaney 2015a).

The SSF Guidelines call for equitable development in coastal communities. The Japanese case shows that notions of equity may vary among societies and cultures, and that such notions can change over time and be influenced by other developments. Japanese FCAs played an important role in not only developing communities through helping members economically, but also in fostering and strengthening horizontal ties and emphasizing equality.

## *Contributions to an Economically, Socially, and Environmentally Sustainable Future*

Small-scale fisheries can contribute to an economically, socially, and environmentally sustainable future. Yet, fisheries and maritime resources around the world appear to be under great pressure; more than 28% of assessed stocks are fished at biologically unstustainable levels (FAO 2016). Despite governments' best efforts to manage their resources sustainably, stocks remain at risk of overfishing. Recently, it has been estimated that 85% of global fish stocks are fully and over-exploited (FAO 2010).

**Fig. 15.2** Japanese landing market, Kagoshima, Japan, 2014. (Source: Nobuyuki Yagi)

An economically stable future for Japanese fishers could be in doubt. Recently, the economic competitiveness of Japanese fisheries in the international market is being questioned. According to the Ministry of Agriculture, Forestry and Fisheries, approximately 45% of fish and fish products consumed in Japan were imported in recent years (MAFF 2016b) (Fig. 15.2).

Imported fisheries products became more price competitive in the 1980s when the Japanese Yen strengthened considerably. In other words, the value of foreign products has gotten cheaper due to the strong Yen. During this period (after 1992), retailers started to gain market power over other players in the fishery products value chain (i.e. fishers) (Nakajima et al. 2014). With the emergence of large supermarkets, buyers in wholesale markets (i.e., supermarkets) gained negotiation power, while the sellers (i.e. domestic fishers) have lost power (Demura 2002). Japanese fishers began to lose market power in the 1990s, despite a concurrent decline in domestic fisheries production. This result can be partly attributable to the increased import of price competitive fish from foreign countries.

It is argued that an ITQ (Individual Transferable Quota) system can increase economic efficiency. Japan has not yet officially introduced a governmental ITQ system although informal private arrangements to set individual quotas exist in some fisheries such as Bluefin tuna in international waters. There has been a push for the privatization of fishing rights, especially in Miyagi Prefecture following the 2011 tsunami (Delaney 2015a; Hishida and Shaw 2015). However, taking into account that the current economic difficulty of domestic fishers is caused by lower prices of fish which are primarily due to increased imports and increased market power of supermarkets and little to do with fishing production methods (such as ITQ or traditional management), it is the authors' view that effectiveness of the introduction on new production related measures (such as ITQs) remain questionable.

Rather than using ITQs as an instrument to increase income and wealth, small-scale fishers in Japan focus on alternative strategies such as pooling their harvests for increased market power, branding of their products (e.g., Matsushima oysters; Kessennuma saury, etc.), and using media and technology for direct sales (e.g., Delaney 2015a, b). Many of fisheries branding and marketing activities are run by FMOs and research has shown that "FMOs actively engaged in marketing activities tend to earn higher revenue per member (Uchida 2007)" (Uchida and Makino 2008, 226).

The SSF Guidelines argue that small-scale fisheries can contribute to an economically, socially, and environmentally sustainable future. One vital means for doing so is to maintain human and tenure rights of fisheries-based communities and to evaluate the use of different fisheries management tools vis-a-vis their impacts on these rights.

## Challenges to Small-Scale Fisheries in the Future

Despite historical and current success in meeting much of these three objectives, Japan is facing difficulties due to on-going demographic (and societal) changes as well as general low income in small-scale coastal fisheries. The elderly population is almost 10% higher in coastal communities than in the rest of Japan. The implications of such an aging population are related to not only the workforce and community social life and economics, but also the use of resources. The population change has resulted in changes in internal and external practices. Among some Miyagi FCA subgroups, for example, the allocation structure of fishing grounds has changed and others are working with private groups to encourage younger people to join the fisheries. Meanwhile, externally, the Miyagi Prefectural government proposed increasing the allocation of fishing rights to private firms over FCA members.

### *Communities and Society*

As mentioned above, Japan is currently faced with a greying society. The push for the consolidation of fishing cooperatives in the late 1990s and early to mid-2000s, for example, was directly related to concerns regarding falling membership (Delaney 2015a).

Fishers and their families make up the backbone and the heart of coastal communities. These are people who organize community festivals and local commemorations (e.g., Hama O-bon), who do volunteer work as FCA sub-group members, who run beach clean-ups and tree planting activities, who volunteer with the fire brigade, and who are physically around, chatting with neighbors and available in case of unexpected trouble.

Today, with fewer working-age small-scale fishing households, the dynamics of life in port communities has changed. When one author (Delaney) first began fieldwork in Miyagi in 1991, in one port, for example, there were 21 active small-scale fisheries households, four small shops, and traveling salesmen who came and went throughout the day. The local fishing cooperative had 91 households as full members with three full-time staff working to arrange auctions, supply gear, help with marketing activities, provided credit union services, organize educational workshops, and liaison with the prefectural and national level cooperative associations. In many ways, the cooperative was the heart of the community, with daily activities and shops surrounding it.

By 2011, the local situation had changed completely with the port becoming a quiet and still place. The decline was gradual: in 1999, the local, port-level cooperative merged with others in the town with many of the activities becoming centralized. However, the local staff remained, with gradually decreasing hours and services until this larger FCA merged with all other FCA in the prefecture in 2007. At this point, most of the port FCAs (now termed 'branch' FCAs) closed with services no longer available locally, including banking. A retired FCA staff member lamented the change. "It was easier in the old days, when one could ask directly" for services and help (August 10, 2013). By 2011, staffing was already quite limited, until the tsunami generated by the Great East Japan Earthquake struck, with the damage to the building closing it for good.

The SSF Guidelines call for collective and community-based management. The declining fishing population and resulting decline in population made a strong argument for consolidating small FCAs into larger units. Miyagi Prefecture, for example, now only has one, official FCA for the entire prefecture (Delaney 2015a). The consolidation of FCAs decreases the management decisions being undertaken at the community level. Such consolidations also included the closing of local FCA offices, impacting the activity taking place in the local port, thereby also impacting the quality of life in the local community. If the SSF Guidelines – which call for local, community-based management – are followed instead of the singular path of economic efficiency (which was a major rationale for consolidating cooperatives), community and daily life at the local level would be improved.

## *Resource Usage*

As laid out above, the current situation in Japan, marked by an aging workforce and negative recruitment, impacts the community negatively. Such a greying population also impacts the rational use of resources. For example, with a generally declining strength and stamina, fishers on small trawlers may not hoist the nets into their boats as often, leaving the nets in the water for hours. The consequence is a decrease in the commercial value of the fish given how the fish would be compressed into the cod end (Yagi 2007). The survival rates of undersized fish and non-targeted species will also be lower given the length of time they remain in the water before released (Yagi 2007).

The SSF Guidelines call for sustainable and ecologically sound food production. The decrease in quality increases the likelihood of fish being discarded. Such discards, along with the decrease in survival rates of those stocks in nets, decreases the ecological sustainability of such fisheries. There are number of ways to improve the situation. One possible way is to encourage younger generations to join the small-scale fisheries sector to reduce the adverse effect caused by the greying workforce. The SSF Guidelines underline the important contribution of the small-scale fisheries to the world and could provide moral support for younger Japanese generations to join the small-scale fisheries sector.

## Conclusion

To re-examine our primary research questions, what utility, if any, could the SSF Guidelines hold for a country like Japan; and, what can the continued importance of these fisheries in Japan teach the rest of the world in terms of ways to implement the SSF Guidelines? Japan serves as a case study highlighting the criteria (e.g., protected resource rights, local management) needed to reach the objectives of sustainable fisheries management and equitable development in coastal communities in developing societies through its historical and current practices. Japan also presents a case of the challenges faced by small-scale fisheries in a post-industrialized world: they can continue to contribute to an economically, socially, and environmentally sustainable future, but the management, social and economic challenges faced are varied.

In thinking of small-scale fisheries from a social and economic point of view, collective management by FCAs in remote coastal ports can strengthen local social cohesion. Participatory approaches of decision making for fishing regulations and peer monitoring systems also enhance social integrity of these areas. The current Japanese system of fishery management can be regarded as traditional knowledge to best utilize social capital of the local community. However, ageing workforces and declining rural populations are serious challenges.

Japanese small-scale fisheries have also suffered from increased competition with imported products as well as high prices for fuel and other supplies. To reverse this situation, efforts should be made on the market side as influencing changes to the production system may be limited in certain circumstances. Although, a need for a change of traditional tenure management system in Japanese coastal fisheries may not be high, non-fishery factors such as currency exchange rates or fuel prices can bring adverse impacts to the economic aspects in fisheries.

When looking at small-scale fisheries from an environmental perspective, Japanese rights-based management with area allocation can provide incentives for fishers to conserve their fishing grounds (rather than simply fish). For example, more than 1000 MPAs are located along the coast of Japan thanks primarily to the efforts of fishers. Area-based management and quota-based management both can achieve proper environmental conservation if the rights are attached to the resource users. In a number of areas, watershed management has also been taken on by FCA members, as planting trees and is believed to increase the health and productivity of the coasts. Given the overall, sustained continuation of the rights-based management system with stable production in the coastal fishery, a need to change the traditional tenure management system in coastal fisheries may not be high. However, coastal development such as constructing huge sea-walls or land reclamation can be a threat to the environment. Such cases call for better representation in land-use planning for integration coastal zone management and especially for true stakeholder involvement in coastal planning, especially in the case of rebuilding seawalls in the post-tsunami.

The current success of Japanese small-scale fisheries management in social, economic and environmental terms stems in large part to the Japanese government choosing to go back to using traditional area and rights-based management systems after flirting with an unlimited licensed-based, effectively open access fisheries in the 1870s. By the 1930s, as more and more fishers lost their access to fisheries resources due to poverty and became heavily indebted to moneylenders, formal legislation was put in place to protect their rights and ensure they had access to credit and marketing resources. Through FCAs, the local fishers not only strengthened their position in the fisheries, they also often served as the center of coastal communities, with FCA groups and members taking charge of community activities such as serving on festival organizing boards and working as volunteer firemen. Japan's small-scale fisheries success in social and ecological sustainability stems from relying on traditional systems which were given formal, legal protections, and by adapting to the local conditions.

In terms of the future of Japanese small-scale fisheries in social, economic, and biological terms, an incredible amount of diversity exists in Japanese fisheries, and there is no single approach that will fit every fishery (Yagi et al. 2012). In general, however, many of the problems facing small-scale fisheries today come from external pressures such as social and demographic change and pressure to change to ITQ and quota systems. Consequently, increased observations and vigilance against the non-fishing factors of future challenges is needed.

Many of the objectives of the SSF Guidelines for sustainable fisheries management are already in place in Japanese small-scale fisheries. The goals of eradicating poverty and reaching equitable development, for example, were secured and protected in national-level legislation by providing for legally-protected fishing rights and for enabling co-management of resources. The system also fostered equitable community development. The Japanese system of co-management includes responsible governance (Objective 5a.) through community co-management (Objective 5.17), culturally appropriate and secure tenure rights (Objective 5.3), and monitoring, surveillance, and control systems (Objective 5.16). The system is based on traditional customs and culture, but it is also one based on trial and error and the evolution of institutions. The Japanese governmental experimented with different licensing and fishing rights systems, for example, but abandoned these with they appeared unsustainable. The current fishing cooperative associations (FCAs) are based upon earlier, community practices with area-based management, which also evolved from fishery guilds.

Japan's recent history thus provides a roadmap on the path towards poverty reduction and equitable community development, marking both some of the pitfalls to be found along the way, as well as successful pit-stops necessary to achieve the goals of economically, socially, and environmentally sustainable futures.

## References


Agrawal, A., & Ostrom, E. (2001). Collective action, property rights, and decentralization in resource use in India and Nepal. *Politics & society, 29*(4), 485–514.

Akimichi, T., & Ruddle, K. (1984). The historical development of territorial rights and fishery regulations in the western Pacific. In K. Ruddle & T. Akimichi (Eds.), *Maritime institutions in the western Pacific* (pp. 37–88). Osaka: National Museum of Ethnology.

Ando, T. (1995). *The Hokkaido Fishermen's liberation movement: The philosophy and works of Takatoshi Ando, the pioneer of the fishery cooperative movement in Hokkaido, Japan.* New York: Vantage Press.

Barrett, G., & Okudaira, T. (1995). The limits of fishery cooperatives? Community development and rural depopulation in Hokkaido, Japan. *Economic and Industrial Democracy, 16*, 201–232.

Beardsley, R. K., Hall, J. W., & Ward, R. E. (1959). *Village Japan.* Chicago: University of Chicago Press.

Benjamin, G. (1997). *Japanese lessons: A year in a Japanese school through the eyes of an American anthropologist and her children.* New York: New York University Press.

Berkes, F. (2001). *Managing small-scale fisheries: Alternative directions and methods.* Ottawa: International Development Research Center (IDRC).

Bestor, T. C. (2004). *Tsukiji: The fish market at the center of the world* (Vol. 11). Berkeley: University of California Press.

Camiñas, J. A., Cerdán, R. A., & Domínguez, J. B. (2004). *La pesca en el Mediterráneo andaluz.* Málaga: Servicio de publicaciones de la Fundación Unicaja.

Chuenpagdee, R. (2011). *World small-scale fisheries: Contemporary visions.* Delft: Eburon Academic Publishers.

Cinner, J. (2005). Socioeconomic factors influencing customary marine tenure in the Indo-Pacific. *Ecology and Society, 10*(1), 36.

Cordell, J. (1989). *A sea of small boats. Cultural survival report no. 26.* Cambridge, MA: Cultural Survival.

Delaney, A. E. (2003). *Setting nets on troubled waters: Environment, economics, and autonomy among nori cultivating households in a Japanese fishing cooperative* (Doctoral dissertation). University of Pittsburgh, USA.

Delaney, A. E. (2015a). Japanese fishing cooperative associations: Governance in an era consolidation. In S. Jentoft & R. Chuenpagdee (Eds.), *Interactive governance for small-scale fisheries: Global reflections, MARE Publication Series* (Vol. 13, pp. 263–280). New York: Springer.

Delaney, A. E. (2015b). Social sustainability in post-3.11 coastal Japan. In S. Assmann (Ed.), *Sustainability in contemporary rural Japan: Challenges and opportunities* (pp. 3–17). New York: Routledge.

Demura, M. (2002). Change in distribution system of fisheries products and correspondence in production sites: Toward construction of new wholesale markets at production areas (in Japanese). *Norin-Kinyu, 2*, 38–51.

FAO. (2010). *The state of world fisheries and aquaculture. FAO Fisheries and Aquaculture Department*. Rome: Food and Agriculture Organization of the United Nations.

FAO. (2015a). *Voluntary guidelines for securing small-scale fisheries in the context of food security and poverty eradication*. FAO, ISBN 978-92-5-108704-6. Rome: Food and Agriculture Organization of the United Nations.

FAO. (2015b). *Towards the implementation of the SSF guidelines in the southeast Asia region: Proceedings of the southeast Asia regional consultation workshop on the implementation of the voluntary guidelines for securing sustainable small-scale fisheries in the context of food security and poverty eradication,* 24–27 August 2015 Bali. N. Franz, S. Funge-Smith, S. Siar, & L. Westlund (Eds.) FAO Fisheries and Aquaculture Proceedings 24. ISBN 978-92-5-109027-5.

FAO. (2016). *The state of world fisheries and aquaculture. FAO fisheries and aquaculture department*. Rome: Food and Agriculture Organization of the United Nations.

Fisheries Agency of Japan. (2016). 2015 *classification of annual fish species resources*. http://abchan.fra.go.jp/digests27/index.html. Accessed 11 Nov 2016.

Fukutake, T. (1967). *Japanese rural society*. Ithaca: Cornell University Press.

Hishida, N., & Shaw, R. (2015). The situation and challenges of recovery of fishing industries in Tohoku region. In R. Shaw (Ed.), *Tohoku recovery* (pp. 27–36). Tokyo: Springer Japan.

IBGE. (n.d.). *Instituto brasileiro de geografia e estatística* [Brazilian institute of geography and statistics]. http://www.ibge.gov.br/home/. Accessed 9 Sep 2016.

Kalland, A. (1980). *Shingu: A Japanese fishing community*. London: Curzon Press.

Lebra, T. S. (1992). *Japanese social organization*. Honolulu: University of Hawaii Press.

MAFF. (2010). *Fisheries white paper* (in Japanese). Ministry of agriculture, forestry, and fisheries, fisheries division. Tokyo: Government of Japan.http://www.jfa.maff.go.jp/j/kikaku/wpaper/h22_h/trend/1/t1_2_1_1.html. Accessed 11 July 2014.

MAFF. (2011a). *Fisheries white paper*. Ministry of agriculture, forestry, and fisheries, fisheries division. Tokyo: Government of Japan. http://www.jfa.maff.go.jp/j/kikaku/wpaper/pdf/2011_jfa_wp.pdf. Accessed 11 July 2014.

MAFF. (2011b). *Statistics of FCAs*. Ministry of agriculture, forestry, and fisheries, fisheries division. Tokyo: Government of Japan. Accessed 11 July 2014.

MAFF. (2016a). *Fisheries white paper*. Ministry of agriculture, forestry, and fisheries, fisheries division. Tokyo: Government of Japan. http://www.jfa.maff.go.jp/j/kikaku/wpaper/H27/pdf/27suisan-sankou.pdf. Accessed 11 Nov 2016.

MAFF. (2016b). *Outline of food supply and demand table*. Ministry of agriculture, forestry, and fisheries, fisheries division. Tokyo: Government of Japan. http://www.jfa.maff.go.jp/j/kikaku/other/pdf/270807oshirase.pdf. Accessed 11 Nov 2016.

Makino, M. (2011). *Fisheries management in Japan: Its institutional features and case studies, Fish and fisheries series* (Vol. 34). London: Springer.

Makino, M., & Matsuda, H. (2005). Co-management in Japanese coastal fisheries: Institutional features and transaction costs. *Marine Policy, 29*, 441–450.

Martinez, D. P. (2004). *Identity and ritual in a Japanese diving village: The making and becoming of person and place*. Honolulu: University of Hawaii Press.

McKean, M. A. (1981). *Environmental protest and citizen politics in Japan*. Berkeley: University of California Press.

Nakajima, T., Matsui, T., Sakai, Y., & Yagi, N. (2014). Structural changes and imperfect competition in the supply chain of Japanese fisheries product markets. *Fisheries Science, 80*, 1337–1345.

Nakane, C. (1967). *Kinship and economic organization in rural Japan*. London: Athlone Press.

Nakane, C. (1970). *Japanese society*. Berkeley: University of California Press.

Norbeck, E. (1954). *Takashima: A Japanese fishing community*. Salt Lake City: University of Utah Press.

Pauly, D., & Zeller, D. (2016). Catch reconstructions reveal that global marine fisheries catches are higher than reported and declining. *Nature Communications, 7*, 10244.

Pauly, D., Watson, R., & Alder, J. (2005). Global trends in world fisheries: Impacts on marine ecosystems and food security. *Philosophical transactions of the royal society B: Biological sciences, 360*(1453), 5–12.

Pita, C., Pereira, J., Lourenco, S., Sonderblohm, C., & Pierce, G.J. (2017). The traditional small-scale octopus fishery in Portugal: Framing its governability. In S. Jentoft & R. Chuenpagdee (Eds.), *Governing the Governance of small-scale fisheries: A global scan*. Manuscript submitted for publication.

Poepoe, K. K., Bartram, P. K., & Friedlander, A. M. (2007). The use of traditional Hawaiian knowledge in the contemporary management of marine resources. In N. Haggen, B. Neis, & I. G. Baird (Eds.), *Fishers' knowledge in fisheries science and management* (pp. 117–141). Paris: UNESCO Publishing.

Ruddle, K., & Akimichi, T. (1984). Maritime institutions in the western Pacific. *Senri ethnological studies, 17*, 1–9.

Sato, M. (1992). Fisheries cooperatives in Japan as fisheries management organization. In T. Yamamoto & K. Short (Eds.), *International perspectives on fisheries management with special emphasis on community-based management systems developed in Japan*. Tokyo: National Federation of Fisheries Cooperative Associations (ZENGYOREN).

Smethurst, R. (1986). *Agricultural development and tenancy disputes in Japan, 1870–1940*. Princeton: Princeton University Press.

Traphagan, J. W. (1998). Emic weeds or etic wildflowers? Structuring 'the environment' in a Japanese town. In K. Aoyagi & P. Nas (Eds.), *Toward sustainable cities: Readings in the anthropology of urban environments* (pp. 37–52). Leiden: Leiden Development Studies, Institute of Cultural and Social Studies, University of Leiden.

Uchida, H., & Makino, M. (2008). Japanese coastal fishery co-management: An overview. In R. E. Townsend, R. Shotton, & H. Uchida (Eds.), *Case studies in fisheries self-governance, FAO fisheries technical paper no. 504* (pp. 221–229). Rome: FAO.

Weinstein, M. S. (2000). Pieces of the puzzle: Solutions for community-based fisheries management from native Canadians, Japanese cooperatives, and common property researchers. *The Georgetown international environmental law review, 12*(2), 375–412.

Yagi, N. (2007). Implications of an ageing fisheries labour force in Japan. In E. A. Chouicha (Ed.), *Structural changes in fisheries dealing with the human dimensions* (pp. 121–130). Paris: OECD Publishing.

Yagi, N., Takagi, A., Takada, Y., & Kurokura, H. (2010). Marine protected areas in Japan: Institutional background and management framework. *Marine Policy, 34*, 1300–1306.

Yagi, N., Clark, M., Anderson, L. G., Arnason, R., & Metzner, R. (2012). Applicability of ITQs in Japanese fisheries: A comparison of rights based fisheries management in Iceland, Japan, and United States. *Marine Policy, 36*, 241–245.

# Part V
# Empowerment and Collective Action

Empowerment through collective action and organization is stressed in the SSF Guidelines as a way out of poverty and marginalization. Small-scale fishing people need a voice in the policy making process, but they would also benefit from more control of the conditions under which they work. The case study from Madagascar by Charlie Gardner, Steve Rocliffe, Charlotte Gough, Adrian Levrel, Rebecca Singleton, Xavier Vincke, and Alasdair Harris (Chap. 16) illustrates this need. As one of the world's poorest countries, Madagascar is sustained significantly by small-scale fisheries, which provide food and income for local people. Unequal power relations with post-harvest actors, however, hinder small-scale fishers and fish workers from moving out of poverty. The authors argue that fishers and fish workers need better involvement in the management process and that more data are needed to provide a more solid foundation for decision-making. Gabriela Sabau provides ideas of what this involvement might possibly entail in Chap. 17. Her case study is situated in Costa Rica, a country that has pioneered the SSF Guidelines' implementation. The chapter describes how the government and fishers' organizations are in alignment with respect to conservation and participative management policies. She notes that involving small-scale fishers in dialogue and design of their own future is key to the implementation of the SSF Guidelines. Oscar Amarasinghe and Maarten Bavinck offer further evidence of the merits of collective action and organization in Chap. 18. Situating their case study in Sri Lanka, they hold that fisheries cooperatives can be an important instrument for the implementation of the SSF Guidelines, given their long history of working for the betterment of small-scale fishing people throughout the entire value chain. They conclude with a set of recommendations for improving the performance of cooperatives. Collective action, also in terms of co-management, is the topic that Sílvia Gómez Mestres and Joseph Lloret explore in Chap. 19, with a case study from a marine protected area in Mediterranean Spain. They argue for the need to understand small-scale fisheries in their particular social and cultural context, where they make up an important part of community heritage.

# Chapter 16
# Value Chain Challenges in Two Community-Managed Fisheries in Western Madagascar: Insights for the Small-Scale Fisheries Guidelines

**Charlie J. Gardner, Steve Rocliffe, Charlotte Gough, Adrian Levrel, Rebecca L. Singleton, Xavier Vincke, and Alasdair Harris**

**Abstract** Madagascar, among the world's poorest countries, depends heavily on small-scale fisheries for food security and income. Many of its fisheries have transitioned from subsistence- to market-oriented in recent decades, driven by the emergence of new export markets. In this chapter, we consider the *Voluntary Guidelines for Securing Sustainable Small-Scale Fisheries in the Context of Food Security and Poverty Eradication* ('SSF Guidelines') in light of experiences from two small-scale fisheries in Madagascar: octopus (*Octopus cyanea*) and mud crab (*Scylla serrata*). We focus on articles related to value chains, post-harvest, and trade. The dispersed nature of these fisheries means fishers rely on private sector collectors to access markets. Post-harvest actors hold disproportionate negotiating power, with benefits from management initiatives accruing mainly to actors high in the value chain rather than the fishers who implement them. To address these imbalances and increase the contribution of these fisheries to poverty reduction and food security, it is critical to empower fishers and improve their representation in management processes. Data deficiencies must also be tackled, to enhance transparency and provide an evidence base for decision-making.

**Keywords** Community-based natural resource management • Locally Managed Marine Area (LMMA) • Market-based approaches • Mud crab • Octopus

C.J. Gardner (✉)
Blue Ventures Conservation, Omnibus Business Centre,
39-41 North Road, London N7 9DP, UK

Durrell Institute of Conservation and Ecology (DICE), University of Kent,
Canterbury CT2 7NR, UK
e-mail: cg399@kent.ac.uk

S. Rocliffe • C. Gough • A. Levrel • R.L. Singleton • X. Vincke • A. Harris
Blue Ventures Conservation, Omnibus Business Centre,
39-41 North Road, London N7 9DP, UK
e-mail: steve@blueventures.org





S. Jentoft et al. (eds.), *The Small-Scale Fisheries Guidelines*, MARE Publication Series 14, DOI 10.1007/978-3-319-55074-9_16

# Introduction

Small-scale fisheries directly or indirectly support the livelihoods of over 500 million people worldwide (Béné et al. 2007; FAO 2016). Thus, ensuring that they are managed sustainably is critical to food security and poverty reduction efforts (FAO 2005; Bell et al. 2009; Garcia and Rosenberg 2010; Smith et al. 2010). They are particularly important in tropical developing countries where the majority of the world's fishers live (Pomeroy and Andrew 2011), but where productivity and sustainability are often threatened by a suite of factors including competition with industrial fleets (Pauly 1997, 2006), climate change (Allison et al. 2009; Hoegh-Guldberg and Bruno 2010), inadequate environmental governance (Garcia and Rosenberg 2010; Allison et al. 2012), and the marginalisation of small-scale fisheries in policy and planning (Andrew et al. 2007; Mills et al. 2011). As a result, many such fisheries are "failing to fulfil their potential as engines of social and economic development" (Andrew et al. 2007, 228).

Recognizing that a lack of policy guidance was hindering the sustainable development of the sector, in 2015 the Food and Agriculture Organisation of the United Nations (FAO) published the *Voluntary Guidelines for Securing Sustainable Small-Scale Fisheries in the Context of Food Security and Poverty Eradication* (FAO 2015), the first internationally agreed instrument focusing on small-scale fisheries. Developed through a bottom-up, participatory process involving over 4000 fishers, academics, government, and civil society stakeholders from over 120 countries (da Silva 2015), the SSF Guidelines are viewed as a potential turning point for small-scale fishers worldwide (Jentoft 2014). However, the impact of the SSF Guidelines will depend entirely on their implementation, and this should be a "cyclical, interactive, and iterative process, where original objectives are subject to repeated questioning, debate, evaluation and reformulation" (Jentoft 2014, 6).

In this chapter, we contribute to this refinement process by considering the SSF Guidelines in light of experiences from two small-scale fisheries in Madagascar, an island state with extremely high poverty (World Bank 2015), and high dependence on small-scale fisheries for both food security and income (Le Manach et al. 2012; Barnes-Mauthe et al. 2013). The two fisheries target reef octopus (*Octopus cyanea*) and mangrove mud crab (*Scylla serrata*). We focus specifically on Chap. 7 of the SSF Guidelines 'value chains, post-harvest and trade', since the isolated, dispersed nature of the fisheries, combined with their export-oriented markets, mean that interventions centred on reducing value-chain inefficiencies have great potential to improve the economic returns received by small-scale fishers.

## Study System

The fourth largest island on Earth, Madagascar spans 14 degrees of latitude and has more than 5500 km of coastline, along which over half the population is concentrated (WRI 2003). The third least food secure nation in the world (GFSI 2015), Madagascar suffers absolute poverty rates of around 90% (World Bank 2015). Its largely rural population depends heavily on renewable natural resources for subsistence and household income, particularly in western coastal regions, where poor soils and climatic constraints impede agriculture, and where small-scale fisheries are concentrated (Razanaka et al. 2001; Horning 2008; Harris 2011). This area is home to the Vezo, a semi-nomadic ethnic group of traditional fishers (Astuti 1995). Traditionally dominated by subsistence, the Vezo economy has become increasingly trade-oriented since the 1970s, and has focused on export markets for octopus, sea cucumbers and shark fin since the turn of the twenty-first century (Iida 2005; Muttenzer 2015; Cripps and Gardner 2016). However, several stressors threaten the safety net provided by small-scale fisheries in this region, including reef sedimentation (Maina et al. 2012; Sheridan et al. 2015), coral bleaching (McClanahan et al. 2009), destructive capture techniques (Andréfouët et al. 2013), and overfishing driven by increases in fisher populations, improved technologies and new or growing markets (Bruggemann et al. 2012; Grenier 2013; Muttenzer 2015). In addition, the state continues to promote access to its fisheries by foreign fleets, despite evidence that this may damage the sector's productivity, and lacks the capacity to regulate the small-scale fisheries sector effectively (Harris 2011; Le Manach et al. 2012, 2013).

Given the lack of central fisheries management capacity, management initiatives in the country's small-scale fisheries sector are largely driven by non-governmental organizations (NGOs), often with assistance from international development agencies. Blue Ventures (BV), a British environmental conservation NGO working with communities to rebuild tropical fisheries, has been supporting local fisheries management and conservation initiatives in Madagascar since 2003. From its initial experience with octopus fishery management and the development of community-managed protected areas, the organization now implements diverse programs in community health, education, aquaculture, and mangrove-based carbon emissions reductions, and contributes to the management of several fisheries (Fig. 16.1).

### *Octopus Fishery*

The octopus fishery of southwest Madagascar targets three species, although *Octopus cyanea* makes up 95% of landings (Raberinary and Benbow 2012). Octopus is harvested by all sectors of Vezo society and is either gleaned from coral reef flats during low tides, or caught by free divers in deeper waters (Westerman and Benbow 2013). Traditionally of minor local importance, the fishery expanded at the turn of the twenty-first century, when private sector export companies started to send trucks

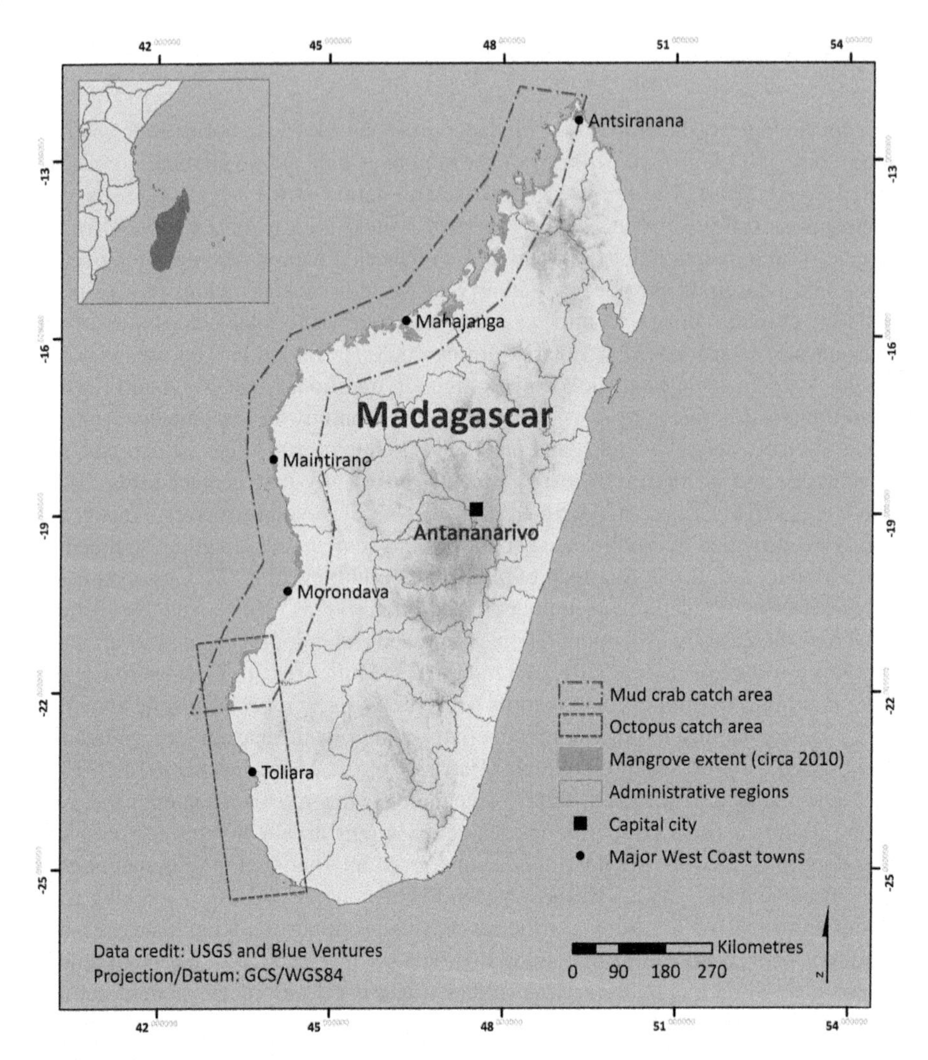


**Fig. 16.1** Map of Madagascar showing major towns of western regions and approximate extent of octopus and mud crab fisheries

carrying ice buckets to remote villages along the coast, buying octopus from intermediaries and transferring them to processing facilities and the export cold chain (L'Haridon 2006).

In order to improve the productivity of the fishery, in 2004 a trial closure was implemented over 150 hectares of reef flat near the village of Andavadoaka by the local community with support from BV and several private sector, NGO, and government partners (Harris 2007). The success of this approach stimulated the adoption of the periodic closure system by neighboring villages and, in the decade since, more than 200 hundred closures have been implemented along 500 km of coastline in southwest Madagascar, as well as others in Northern Madagascar, Pemba

(Tanzania), Cabo Delgado (Mozambique) and Bahia de los Angeles (Mexico). Recent evaluations have demonstrated that these closures, when well-managed, can create net economic benefits for fishers (Oliver et al. 2015). Oliver et al. (2015) analyzed 8 years of data from more than 30 sites and found that octopus landings increased by more than 700% in the month following the lifting of a closure, boosting the catch per fisher per day by almost 90% over the same period. They suggested that the success of these closures was principally underpinned by three factors: (i) provisions within Madagascar's legal code to allow local marine management; (ii) backing from seafood exporters, who supported the closures (a considerable interruption to revenues, followed by a sudden surge in production) and facilitated access to export markets; and (iii) the growth rates of the targeted species, which are so rapid that stocks can respond favourably to protection periods of 2–3 months (Oliver et al. 2015). The periodic closure system has also influenced national fisheries policy, with the introduction of a minimum catch size of 350 grams and an annual national fishery closure for 1.5 months in 2005. National closures have also been adopted by Mauritius and Rodrigues in the Western Indian Ocean as a result of this program.

At present (2016), periodic octopus fishery closures form the bedrock of broader marine resource management initiatives centred on Andavadoaka. 25 villages are grouped into a local marine environmental management association named Velondriake, which is governed by an elected committee. The association is legally empowered to set and enforce resource use rules using a form of by-law known as *dina* (see Article 7.4; Andriamalala and Gardner 2010), and manages over 600 km$^2$ of shallow coastal seas as Madagascar's first Locally Managed Marine Area (LMMA). In addition to periodic closure sites, the LMMA incorporates permanent reserves in reef and mangrove areas and in 2015 was officially gazetted as an IUCN category V protected area, co-managed by the Velondriake Association and Blue Ventures.

## *Mud Crab Fishery*

Madagascar's mud crab (*Scylla serrata*) fishery is based on traditional methods using simple gears, with fishers operating on foot or from small dugout canoes in the mangrove forests of the west and northwest coasts. Small-scale collectors and sellers operate locally, often within informal markets, and sell their produce to seafood export companies. Market demand has grown significantly in recent years, particularly for live crabs which are exported to China, leading to price increases of 500% since 2011 and subsequent pressure on wild stocks. In 2014, national production reached 3087 tons, of which 75% was exported to China (Fisheries Hygiene Authority Statistics, unpublished data).

Since 2011, local communities and the Ministry of Fisheries and Marine Resources have implemented management measures, including 3–5 month periodic closures. National legislative changes in 2014 set a minimum catch size of 11 cm, an annual quota of 5000 tons (including 4250 tons for export), and an annual nationwide fishery closure from July to October. Management priorities include improving stock management through further periodic closures, decreasing post-harvest losses, developing crab aquaculture and/or fattening to reduce pressure on wild stocks, and improving monitoring, transparency, and regulation along the value chain.

## Application of SSF Guidelines Chap. 7: Value Chains, Post-harvest and Trade

The bulk of BV's management interventions in Madagascar's small-scale fisheries were implemented prior to 2015, making it impossible to link their level of success with the SSF Guidelines. Nevertheless, experiences in Madagascar's small-scale fisheries sector over the last decade offer valuable insights that can contribute to their iterative improvement, particularly due to the public-private partnership-based approaches to fisheries management and post-harvest value chain improvements that have been developed in these fisheries. The lessons learned from these experiences may prove instructive to other parties – be they state, private or civil society – seeking to implement the SSF Guidelines in similar socio-ecological contexts. Here, we discuss our experiences of value chain interventions in Madagascar's small-scale octopus and mud crab fisheries using the framework of the SSF Guidelines Chap. 7 on 'value chains, post-harvest and trade'.

*Article 7.1* "All parties should recognize the central role that the small-scale fisheries post-harvest subsector and its actors play in the value chain. All parties should ensure that post-harvest actors are part of relevant decision-making processes, recognizing that there are sometimes unequal power relationships between value chain actors and that vulnerable and marginalized groups may require special support."

Both the octopus and mud crab fisheries occur largely in isolated locations lacking transport infrastructure or cold chains, but are primarily export-oriented. As such, post-harvest actors are particularly important for economic productivity, providing collection, processing, and access to markets (Oliver et al. 2015; Smartfish 2015). BV's actions in both fisheries integrate post-harvest actors in all management activities, but focus predominantly on traditional small-scale fishers, the most vulnerable and marginalized actors in the supply chain (see Article 7.4).

The seafood export company Copefrito, one of two main collectors operating out of the city of Toliara in southwest Madagascar, has been a key partner in the development of periodic fishery closures for octopus since initial trials in 2004 (Harris 2007; Oliver et al. 2015). By coordinating collections on closure openings and offering a price premium of 15–20% for octopus at openings, as well as a premium for larger octopus, the partnership provides economic incentives for good closure management. This move by a key commercial actor has since been followed by the region's other principal buyer, Murex, as well as by some smaller collectors.

All octopus fishery stakeholders (including fisher associations, collectors/export companies, the Ministry of Fisheries and Marine Resources, the marine research institute (IHSM) and supporting NGOs) convene quarterly within the octopus fishery management platform (*Comité de Gestion de la Pêche aux Poulpes* (CGP)). An informal platform with no official fisheries management mandate, the CGP nonetheless serves as an effective forum for engaging stakeholders. Decisions made by the body mostly relate to the management of periodic closures, the negotiation of prices, and the implementation of the southwest Octopus Fishery Improvement Project (FIP) – an action plan intended to move the fishery towards certification by the Marine Stewardship Council's (MSC) ecolabel.

In the mud crab fishery, buyers are, whenever possible, involved in decisions over the timing of periodic closures, or notified of opening dates, in order to ensure prompt collection following openings. There is some evidence from the Smartfish project (see Article 7.3) that buyers catalysed the adoption of improved storage and transport technologies, since they valued healthier crabs (Smartfish 2015). To identify the key actors in the value chain to be integrated into management activities, a stakeholder analysis for this fishery will be completed in 2016.

*Challenges* Within the octopus fishery, the development of effective partnerships between private sector buyers, fishers, and NGOs is threatened by rivalry and mistrust, particularly between those operating without the necessary authorizations. While illegal, such informal trade is practically unavoidable, given the country's vast coastline, poor transport and communications infrastructure, and limited central fisheries management capacity. While the CGP helps to engage community representatives in an open forum, it lacks the legal authority to bring about necessary but unpopular actions and, as such, currently operates based on cross-party consensus. Even within the CGP, fishers have limited negotiating power over prices as they remain heavily dependent on post-harvest actors for access to markets.

Madagascar's mud crab fishery lacks a coordination platform and thus the negotiating power of fishers is even more limited. There is no fixed price for mud crab, and buyers are in a strong position to dictate prices since fishers have little capacity for post-harvest conservation of their produce, and remain unorganized. Coordination efforts are more complex than in the octopus fishery, since there are more buyers, collaboration is lower, and the market is much more dynamic.

*Article 7.2* "All parties should recognize the role women often play in the post-harvest subsector and support improvements to facilitate women's participation in such work. States should ensure that amenities and services appropriate for women are available as required in order to enable women to retain and enhance their livelihoods in the post-harvest subsector."

Women play a key role in both fisheries, as fishers and in the immediate post-harvest sub-sector. Between 2004 and 2011, 58% of recorded octopus fishing outings in 14 villages were by women (Westerman and Benbow 2013). Women comprise the majority of local intermediaries employed by export companies to buy octopus from fishers at the village level. Nevertheless, men tend to dominate discussions and decision-making related to octopus closures (Westerman and Benbow 2013), as well as the CGP.

In Velondriake an awareness-raising campaign called *Ampela Tsy Magnavake* ('women not segregated') was launched in 2014, aimed at encouraging the participation of women in octopus fishery management and culminating in the establishment of women's fora in 19 villages. Each forum has two focal points participating in key meetings and relating news back to their group. The initiative has helped to increase the participation of women in meetings.

*Challenges* The transition of octopus from subsistence food to commercially valuable commodity over the last decade has increased the participation of men in the fishery, reducing the proportion of landings from women, and marginalizing them from decision-making processes. While efforts are underway to address the latter issue, traditional Malagasy societies are strongly patriarchal, and women traditionally play only a limited role in collective decision-making (Gezon 2002). Although the *Ampela Tsy Magnavake* initiative has stimulated increased female participation in fishery management, an exclusive focus on women risks the disempowerment and disenchantment of male stakeholders. Further, while women's attendance has increased, many remain reluctant to actively participate and speak out in meetings.

*Article 7.3* "States should foster, provide and enable investments in appropriate infrastructures, organizational structures and capacity development to support the small-scale fisheries post-harvest subsector in producing good quality and safe fish and fishery products, for both export and domestic markets, in a responsible and sustainable manner."

State investment in post-harvest infrastructure was virtually non-existent until the launch of the African Development Bank-funded *Projet d'Appui aux Communautés de Pêcheurs* (PACP) in 2006. Project activities included the construction of fisheries landing and pre-processing stations ('*débarcadères*') in 20 octopus fishing villages in the southwest costing 1.5 million UC. The aim of the *débarcadères* is to ensure landings meet the hygiene standards expected of Western markets, and their planning included provisions for solar power and ice production facilities. Unfortunately, the project failed to consult local fishers, and whilst facilities were completed in 2013, most remain locked and unopened (ICAI 2015). The poor planning and execution of this project has been highly controversial, making headlines in the British press as an example of wasted overseas aid money (Martin 2015; Parfitt 2015).

Although there has been little state involvement in the mud crab fishery, the European-Union funded Smartfish programme has focused on reducing post-harvest mortality in the mud crab fishery since 2013. Initiatives have included the promotion of low-tech crab storage and transport equipment such as improved ox-carts and fishing vessels, fixed cages, tidal pens and storage sheds, as well as a multimedia campaign on mortality reduction. The project, which has been implemented in 33 villages across four regions has decreased post-capture mortality by about 15%, benefitting all actors in the value chain (Smartfish 2015; Smartfish unpublished data).

*Challenges* State agencies are virtually absent through much of rural Madagascar, and major-donor-agency funded development interventions are not always well conceived, implemented or received. Consequently, there is widespread apathy amongst fishers and buyers regarding the state's capacity to mobilize support for infrastructure improvements in support of small-scale fisheries. The persistence of pre-existing collection stations for octopus, in which fishers sell catches to intermediaries in unsanitary conditions under a makeshift wooden shelter – often against a backdrop of a sophisticated yet securely locked and vacant landing station – provides a compelling illustration of the origin of such sentiments.

*Article 7.4* "States and development partners should recognize the traditional forms of associations of fishers and fish workers and promote their adequate organizational and capacity development in all stages of the value chain in order to enhance their income and livelihood security in accordance with national legislation. Accordingly, there should be support for the setting up and the development of cooperatives, professional organizations of the small-scale fisheries sector and other organizational structures, as well as marketing mechanisms, e.g., auctions, as appropriate."

As part of a broader initiative to devolve natural resource management rights and responsibilities to rural communities (Ferguson et al. 2014), in 1996 the Malagasy state permitted local social norms known as *dina* to be formalized and ratified as by-laws in the context of community forestry contracts (Antona et al. 2004). *Dina* have since been used to formalize resource use regulations in a range of community-based conservation and natural resource management initiatives, including community forestry contracts (Pollini et al. 2014), LMMAs (Andriamalala and Gardner 2010), and new protected areas (Virah-Sawmy et al. 2014).

Periodic octopus fishery closures in the Velondriake LMMA are regulated by the Velondriake *dina*, which specifies rules for each management zone, as well as penalties and enforcement mechanisms. Developed in a participatory process, it can be applied against offenders locally, but serious cases can be taken to a magistrate's court if required (Andriamalala and Gardner 2010). Similar *dina* have been developed with local communities in several LMMAs along the southwest coast to support octopus fisheries management efforts.

In the mud crab fishery BV has supported the creation of five fisher associations in the Menabe region, enabling a limited number of fishers to receive licenses to sell their products in urban markets, thus bypassing middlemen and gaining higher prices. It is anticipated that the associations will also empower fishers to negotiate better prices with collectors. These associations use *dina* to regulate their fisheries (including management measures such as periodic mangrove closures), although the *dina* have yet to be ratified in a District court and thus can only be applied at the local level without recourse to formal legal procedures.

*Challenges* While *dina* have allowed the legal recognition of community-developed (or in many cases NGO-suggested) rules over local resource use, in practice the Velondriake *dina* is rarely fully applied within the community because of strong social cohesion, and is difficult or impossible to apply against outsiders such as migrants, motorized small-scale fishers or illegal industrial fleets (Andriamalala and Gardner 2010; Pollini et al. 2014; Cripps and Gardner 2016). In addition, although communities are legally empowered to manage the fishery, they have little capacity to influence the post-harvest value chain, since they are unable to transfer their product to market and thus remain entirely dependent on buyers to support their efforts. While the CGP provides an informal platform for price negotiations, buyers retain a position of disproportionate strength in discussions. This is compounded by a lack of awareness amongst fishers of their rights and ability to dictate terms, as well as by the impoverished conditions of many fishers which compel them to sell their produce daily. Further, all forms of formal social organization within small-scale fisheries in Madagascar are hampered by very low literacy rates, high isolation, and a lack of infrastructure, as well as by the costs and legal/administrative complexity of existing mechanisms. Training in collective bargaining is currently being trialled within BV's aquaculture programme and may be extended to fisheries following evaluation. Fisheries management efforts for mud crab are further constrained by jurisdictional complexities regarding the mangrove habitats in which they live, since mangroves are legally considered terrestrial forests under national legislation and therefore do not fall under the remit of the Ministry of Fisheries and Marine Resources.

*Article 7.5* "All parties should avoid post-harvest losses and waste and seek ways to create value addition, building also on existing traditional and local cost-efficient technologies, local innovations and culturally appropriate technology transfers. Environmentally sustainable practices within an ecosystem approach should be promoted, deterring, for example, waste of inputs (water, fuelwood, etc.) in small-scale fish handling and processing." Post-harvest losses were identified as the key inefficiency in the mud crab value chain, with losses on average 23% (rising to 50% in the rainy season) due to low investment in storage and transport materials by collectors (Smartfish 2015). The Smartfish project aims to reduce such losses through the spread of simple innovations made using readily-available local materials (see Article 7.3). These techniques reduce waste and ensure that crabs are kept in good condition, facilitating entry into more lucrative export markets for live, rather than frozen, animals. Independently of Smartfish, BV has launched a feasibility study into 'crab fattening' aquaculture, since the value can double from an 'empty' crab to a 'full' crab of the same size. Undersized and badly damaged crabs are consumed by fishers or sold locally, thus waste during immediate post-harvest stages is low, though losses during onward transportation may be significant.

Waste in the octopus fishery is minimal. Undersized octopus and those not processed to export standards are dried for sale in the domestic market, thus contributing to domestic food security, while ink sacs are used as bait to fish rabbitfish (*Siganus* spp.) (Gough et al. 2009). Because current harvesting methods (spearing) damage the product and reduce its value, the possibility of introducing alternative methods (e.g., traps) is being investigated. However, these may trigger overfishing, for example by allowing the harvesting of octopus in deeper waters than where they are currently fished. Post-harvest inputs are low in both fisheries, and we do not believe that they generate serious environmental impacts.

*Challenges* The octopus fishery suffers from overwhelming dependence on a small number of buyers with exclusive access to the cold chain for preservation. Although some commercial operators are sensitive to fishers' needs and interests, fishers remain entirely distinct from collection businesses, and have little negotiating power.

*Article 7.6* "States should facilitate access to local, national, regional and international markets and promote equitable and non-discriminatory trade for small-scale fisheries products. States should work together to introduce trade regulations and procedures that in particular support regional trade in products from small-scale fisheries and taking into account the agreements under the World Trade Organization (WTO), bearing in mind the rights and obligations of WTO members where appropriate."

There is no regional cooperation between Western Indian Ocean coastal states with regards to the octopus or mud crab trade. Existing trade regulations relating to the mud crab fishery work against the interests of small-scale fishers since quotas are based on vague and obsolete mangrove productivity studies and are likely to have overestimated productivity of stocks, potentially stimulating overfishing. Further, by requiring all collector-exporters to have an aquaculture pond of at least 1500 square meters, new export regulations discriminate against small collectors seeking to export live crabs (Fig. 16.2).

*Article 7.7* 'States should give due consideration to the impact of international trade in fish and fishery products and of vertical integration on local small-scale fishers, fish workers and their communities. States should ensure that promotion of international fish trade and export production do not adversely affect the nutritional needs of people for whom fish is critical to a nutritious diet, their health and well-being and for whom other comparable sources of food are not readily available or affordable.'

The impacts of octopus and mud crab export markets on the food security of small-scale fisher communities, as well as of inland communities with which they may previously have traded, have not been investigated. Many Vezo fishers now prioritize fishing for trade (i.e. sea cucumbers, shark fin) over fishing for subsistence (Muttenzer 2015; Cripps and Gardner 2016). There is also some anecdotal evidence that Vezo communities consume more octopus during the national closure (when buyers are not operating) than during other periods, suggesting that octopus which is usually traded may otherwise have been consumed. Trade is additionally likely to reduce the amount of protein locally available, with children the most liable to suffer as a result. Conversely, by providing a ready source of cash in isolated communities that largely rely on subsistence fishing, export markets offer a rare opportunity for fishers to earn income to purchase foodstuffs, as well as invest in fishing equipment. The question of whether these markets serve to alleviate or reduce poverty and food security in participating fisher communities requires additional research.

**Fig. 16.2** Images illustrating the fisheries and value chains of small scale octopus (*Octopus cyanea*) and mud crab (*Scylla serrata*) fisheries in Madagascar. (**a**) octopus are collected from fishers by intermediaries in coastal villages; (**b**) seafood export companies then transport the octopus to export processing centres in ice buckets loaded on trucks; (**c**) fishers collecting mud crab from mangroves in northwest Madagascar; (**d**) live mud crabs stored in rice sacks by intermediaries, prior to transport to an export processing centre (Image credits: A, Xavier Vincke; B, Alejandro Castillo Lopez; C and D, Adrian Levrel)

Blue Ventures supports MIHARI, a national peer-to-peer learning network of more than 150 community associations working on LMMAs. The network gives small-scale fishers increased capacity to reach and influence policy makers and is currently raising the profile of the small-scale fishing sector in discussions around Madagascar's plan to triple the total coverage of its marine protected areas, with a focus on LMMA-based approaches (Mayol 2013; Rajaonarimampianina 2014).

*Challenges* International trade remains very lightly regulated but can have rapid and profound impacts on both small-scale fisher communities and the resources/ ecosystems they exploit. For example, fishing intensity in the mud crab fishery strictly follows international demand and is now leading to overexploitation in the most accessible areas (Blue Ventures, unpublished data). While there is chronic food insecurity in south and southwest Madagascar, there is little understanding of the impacts of international trade on such issues and the resilience of small-scale fisher communities. Beyond the local point of collection, small-scale fishers have no representation in trade or related decision-making processes, and because trade monitoring systems are very poor, existing data may not be reliable. Small-scale fishers need to be better represented in national-level decision-making, and an improved understanding of the importance of small-scale fisheries to food security and local resilience is an urgent research priority (Le Manach et al. 2012).

*Article 7.8* "States, small-scale fisheries actors and other value chain actors should recognize that benefits from international trade should be fairly distributed. States should ensure that effective fisheries management systems are in place to prevent overexploitation driven by market demand that can threaten the sustainability of fisheries resources, food security and nutrition. Such fisheries management systems should include responsible post-harvest practices, policies and actions to enable export income to benefit small-scale fishers and others in an equitable manner throughout the value chain."

The impact of new and growing export markets on the sustainability of the octopus and mud crab fisheries remains poorly understood, in part due to a lack of monitoring systems and reliable data. However, effective fisheries management systems (i.e. periodic closures) have been developed and are spreading rapidly amongst small-scale fisher populations (Mayol 2013). LMMAs focused on ensuring resource productivity and sustainability are likely to play a key role in Madagascar's plans to triple marine protected area coverage, greatly increasing the proportion of small-scale fisheries managed through community-based mechanisms. While many initiatives that reduce fishing effort may favour certain institutions or segments of society over others, octopus closures do not impose differential access restrictions, theoretically avoiding the issue of elite capture: all community members may access closure areas following their opening, whether or not they are members of the management association. That said, there is some evidence that arrangements like LMMAs may favour wealthier resource users (Cinner et al. 2012). Notions of equity have received little research attention in this context, and could benefit from further study.

*Challenges* The dispersed nature of the fisheries and existence of a strong informal trade sector hinders the development of effective stock and trade monitoring systems. As a result, the data required for adaptive management of the fisheries are lacking. At the national level, state policies to ensure that maximal additional value from export trade filters down to small-scale fishers are required to provide strong management incentives and promote poverty reduction.

*Article 7.9* "States should adopt policies and procedures, including environmental, social and other relevant assessments, to ensure that adverse impacts by international trade on the environment, small-scale fisheries culture, livelihoods and special needs related to food security are equitably addressed. Consultation with concerned stakeholders should be part of these policies and procedures."

All fisheries management measures supported by BV over the last 14 years have been developed through bottom-up, participatory processes that integrate marginalized members of small-scale fisher communities (e.g., women, migrants) as well as actors in the post-harvest value chain. The spread of LMMAs and their anticipated recognition within Madagascar's protected area system should legally empower small-scale fisher communities to implement management initiatives and thereby minimize the adverse impacts of trade. The development of the MIHARI network of LMMA managers will also provide small-scale fishers with increased opportunities and power to reach and influence policy-makers (see article 7.7).

*Challenges* There are currently no mechanisms through which concerned small-scale fisheries stakeholders can reach policy-makers effectively, although the MIHARI network has been designed to address this deficiency. Existing fisheries legislation (i.e. minimum catch size and national closure period) for mud crab is inappropriate, in that it is not based on current information about the species' biology and life cycle (Le Vay 2001; Leoville 2013). As a result, the national closure does not correspond to the peak of reproduction, and minimum catch sizes are set too low, permitting the capture of sexually immature females (Blue Ventures, unpublished data). In addition, the economic and social impacts of national and periodic fishery closures remain poorly understood.

*Article 7.10* "States should enable access to all relevant market and trade information for stakeholders in the small-scale fisheries value chain. Small-scale fisheries stakeholders must be able to access timely and accurate market information to help them adjust to changing market conditions. Capacity development is also required so that all small-scale fisheries stakeholders and especially women and vulnerable and marginalized groups can adapt to, and benefit equitably from, opportunities of global market trends and local situations while minimizing any potential negative impacts."

Participatory monitoring systems using locally trained data collectors and mobile technology to monitor landings are being developed for each fishery, though these covers only a small proportion of fishers. The CGP has implemented a dashboarding system to facilitate the communication of these landings data, but it remains insufficiently understood by stakeholders. Further, information on market conditions is not available to fishers, limiting their ability to adjust to market conditions or negotiate prices.

*Challenges* Small-scale fishers have no access to market and price information other than the prices offered by local buyers. NGOs and partner fisher communities struggle to keep pace with the rapid expansion of fisheries improvement initiatives, to implement effective participatory monitoring systems at scale, and to find ways of effectively disseminating data back to communities and into local decision-making processes. Efforts to build management capacity within fisher communities have been launched in order to address these issues.

Lack of access to broader market information prevents the identification and design of fisheries management interventions that may improve the value of the products at a local level or in some cases may reduce post-harvest losses. For example, if trap-based fishing for octopus were viable, it could result in a product with the potential to meet quality standards of new higher value markets. There is thus an urgent need for a centralized and transparent monitoring and support system for small-scale fisheries landings and trade, one that includes broader market data.

Knowledge of international market-based initiatives to encourage sustainable fishing through ecolabel schemes is low among fisheries authorities and seafood exporters, and none of Madagascar's fisheries have yet been certified through the Marine Stewardship Council (MSC) standard. Given the lack of central management of the small-scale fisheries sector, as well as the high data deficiency and uncertainty around stock status, accessing such schemes – and thus potentially higher value export markets which reward sustainable fishing practices – remains out of reach for this sector. Notwithstanding these limitations, southwest Madagascar's octopus fishery underwent pre-assessment for the MSC standard in 2010, and an ambitious FIP is being implemented by BV and the CGP with a goal of potentially entering full assessment for the fishery. Notably, this FIP is not being led by the collection and export companies that would have the most to gain from certification, largely due to scepticism of the likely future benefits of MSC certification within these businesses.

## Discussion

The post-harvest subsector is critical to efforts to improve the capacity of Madagascar's small-scale fisheries to contribute to poverty reduction and food security. This issue is especially important since these fisheries are widely dispersed around the country's 5500 km of coastline and fishers rely on post-harvest actors such as seafood export companies to reach markets and derive maximum revenues from their produce. However, to date most fisheries improvement initiatives have focused on building local capacity for fisheries management and governance, and initiatives in the post-harvest subsector remain in their infancy.

At present, the post-harvest value chains of both fisheries are opaque, poorly understood, and managed exclusively by commercial actors, notably private sector seafood collection and export companies. Beyond direct transactional relationships, these companies have little engagement with fishers, whose interests are not necessarily considered at higher levels of the supply chain. Moreover, commercial actors have little incentive to address imbalances in post-harvest power relationships or to maximize the value received by fishers for their fisheries management efforts. Given widespread poverty and low literacy among small-scale fishing communities, severe transport and communications challenges, and a general disregard of the small-scale sector by central fishing authorities, fishers remain particularly vulnerable to exploitation and unfair distribution of benefits by actors higher in the supply chain. Although communities have developed notable experience in local fisheries management efforts over the past decade, there is no history of formal fishers' organisations, cooperatives or trade associations representing the interests of these marginalized groups.

To improve the capacity of fishers to engage with, influence, and benefit from post-harvest processes, greater fisher representation in national fisheries policy decision-making will be needed, for example through the establishment and formalization of multi-stakeholder fisheries management platforms. It will also require the fundamental imbalances in bargaining power between actors to be addressed through capacity building of fisher communities and external pressure from civil society organisations. The vulnerability of small-scale fishers through these unequal power relationships notwithstanding, seafood collection and export businesses have a strong interest in ensuring a sustainable supply of high quality produce and have generally proved willing to collaborate on fisheries management initiatives. Indeed, it is for this reason that their importance in providing access to markets that their partnership is considered indispensable.

While much of the SSF Guidelines focus on state parties, experience from the octopus and mud crab fisheries in Madagascar indicates that partnerships between the state, small-scale fisher communities, civil society organisations, academic institutions, and private sector businesses provide the most realistic hope for rapid and far-reaching advances in the management of the Madagascar's small-scale fisheries. Like many tropical developing countries, Madagascar's state lacks the capacity to regulate its fisheries sector effectively and has only recently recognized the importance of small-scale fisheries in meeting its domestic food security and poverty reduction goals (Harris 2011). NGOs and fishers have played an important role in bringing small-scale fisheries to the attention of policy-makers, and are likely to remain at the forefront of fisheries improvement efforts, notably through the innovation and adoption of appropriate, participatory management initiatives. However, the state, international development actors, and the private sector all have a crucial role in popularizing and scaling these initiatives, and enshrining them in policy.

Several key challenges remain for states and other small-scale fisheries stakeholders to reduce value chain inefficiencies and ensure maximum returns to fisher communities. Most importantly, data deficiencies and the lack of transparency regarding market, catch, and price information throughout the supply chain should be urgently addressed. Knowledge of what and how much fishers are catching, where they are catching it, how much they receive and how much is consumed or traded locally, domestically, and internationally is fundamental for understanding the importance of the small-scale fisheries sector for poverty alleviation and reduction. Such understanding will help raise awareness of the importance of these 'not so small-scale' fisheries in the eyes of decision-makers. Low-cost mobile technology-based approaches offer the potential to collect the required data at scale. Preliminary results from Blue Ventures trials between 2014 and 2016 show encouraging signs that such approaches can help to overcome geographic and infrastructure barriers, increasing both the speed of data entry and extent of monitoring coverage, while delivering cost savings and secondary benefits through use as a communication tool (Blue Ventures, in prep).

Further regulation and professionalization is required of all steps in the supply chain, including the registration of fishers, centralisation and sanitation of landing stations and the supervision of middle-men and collectors, particularly those operating in the informal sector. Such steps would promote improved quality standards and could thus increase the revenues accruing to fishers, while facilitating the collection of data required to inform decision-making.

Ensuring that the export of seafood products does not exacerbate food insecurity amongst small-scale fisher populations or the populations they previously traded with will not only require a better comprehension of the social impacts of decisions to fish for trade rather than subsistence, but also the implementation of robust, science-based stock management to ensure that existing fisheries can sustainably meet domestic fish protein needs. The allocation of export quotas should be informed by the best available data and include the promotion of food security as a guiding principle, while simultaneous mechanisms to promote sustainability, such as the introduction or revision of minimum catch sizes and other harvest control rules should be included in legislation and enforced.

All proposed measures require substantial investment, particularly in building the capacity of central fisheries agencies. However, such investment would be modest relative to the importance of the small-scale fisheries sector from both a poverty and food security perspective. Alongside the implementation of improved fisheries management initiatives, structural investment in the post-harvest subsector will be key to maximizing – and sustaining – the value that fishers can recover from these resources in the long term. As such, the publication of the SSF Guidelines is particularly timely, providing a valuable starting point to raise awareness of small-scale fisheries among policy-makers and guidance to enhance their contribution to food security and poverty reduction. To maximize the impact of the SSF Guidelines, and avoid the risk of them losing relevance over time, we need to continuously refer to and consult with small-scale fisher communities throughout implementation. Improving the value of fisheries products such as octopus and mud crab is only one element; it is also critical to ensure that the added value is passed on to the fishers themselves, providing incentives to adhere to best practice in management, and develop socio-economic monitoring systems to enable the detection of any negative impacts arising from international trade. Independent organizations such as BV that have a permanent field presence and established, trusting relationships with fishing communities are in an excellent position to facilitate dialogue between fishers, other stakeholders, and the authors of the SSF Guidelines for the iterative, bottom-up improvement of the SSF Guidelines, and thus help ensure their contribution to global food security and poverty eradication.

**Acknowledgements** BV's efforts to support management of mud crab and octopus fisheries in Madagascar are supported by MacArthur Foundation, Helmsley Charitable Trust, DFID, and Smartfish. We thank Leah Glass for the preparation of Fig. 16.1, the editors for the opportunity to contribute to this volume, and the reviewers for comments which greatly improved the manuscript.

## References


Allison, E. H., Perry, A. L., Badjeck, M.-C., Adger, W. N., Brown, K., Conway, D., Halls, A. S., Pilling, G. M., Reynolds, J. D., Andrew, N. L., & Dulvyet, N. K. (2009). Vulnerability of national economies to the impacts of climate change on fisheries. *Fish and Fisheries, 10*(2), 173–196.

Allison, E. H., Ratner, B. D., Åsgård, B., Willmann, R., Pomeroy, R., & Kurien, J. (2012). Rights-based fisheries governance: From fishing rights to human rights. *Fish and Fisheries, 13*, 14–29.

Andréfouët, S., Guillaume, M. M. M., Delval, A., Rasoamanendrika, F. M. A., Blanchot, J., & Bruggemann, J. H. (2013). Fifty years of changes in reef flat habitats of the Grand Récif of Toliara (SW Madagascar) and the impact of gleaning. *Coral Reefs, 32*(3), 757–768.

Andrew, N. L., Béné, C., Hall, S. J., Allison, E. H., Heck, S., & Ratner, B. D. (2007). Diagnosis and management of small-scale fisheries in developing countries. *Fish and Fisheries, 8*(3), 227–240.

Andriamalala, G., & Gardner, C. J. (2010). L'utilisation du *dina* comme outil de gouvernance des ressources naturelles: Leçons tirés de Velondriake, sud-ouest de Madagascar. *Tropical conservation science, 3*, 447–472.

Antona, M., Biénabe, E. M., Salles, J.-M., Péchard, G., Aubert, S., & Ratsimbarison, R. (2004). Rights transfers in Madagascar biodiversity policies: Achievements and significance. *Environment and Development Economics, 9*, 825–847.

Astuti, R. (1995). *People of the sea: Identity and descent among the Vezo of Madagascar*. Cambridge: Cambridge University Press.

Barnes-Mauthe, M., Olesen, K. L. L., & Zafindrasilivonona, B. (2013). The total economic value of small-scale fisheries with a characterization of post-landing trends: An application in Madagascar with global relevance. *Fisheries Research, 147*, 175–185.

Bell, J. D., Kronen, M., Vunisea, A., Nash, W. J., Keeble, G., Demmke, A., Pontifex, S., & Andréfouët, S. (2009). Planning the use of fish for food security in the Pacific. *Marine Policy, 33*, 64–76.

Béné, C., Macfadyen, G., & Allison, E. H. (2007). *Increasing the contribution of small-scale fisheries to poverty alleviation and food security*. Rome: FAO.

Bruggemann, J. H., Rodier, M., Guillaume, M. M. M., Andréfouët, S., Arfi, R., Cinner, J. E., Pichon, M., Ramahatratra, F., Rasoamanendrika, F., Zinke, J., & McClanahan, T. R. (2012). Wicked social-ecological problems forcing unprecedented change on the latitudinal margins of coral reefs: The case of Southwest Madagascar. *Ecology and Society, 17*, 47. doi:10.5751/ES-05300-170447.

Cinner, J. E., McClanahan, T. R., MacNeil, M. A., Graham, N. A. J., Daw, T. M., Mukminin, A., Feary, D. A., Rabearisoa, A. L., Wamukota, A., Jiddawi, N., Campbell, S. J., Baird, A. H., Januchowski-Hartley, F. A., Hamed, S., Lahari, R., Morove, T., & Kuangel, J. (2012). Comanagement of coral reef social-ecological systems. *Proceedings of the National Academy of Sciences, USA, 109*, 5219–5222.

Cripps, G., & Gardner, C. J. (2016). Human migration and marine protected areas: Insights from Vezo fishers in Madagascar. *Geoforum, 74*, 49–62.

Da Silva, J. G. (2015). Foreword. In FAO (Ed.), *Voluntary guidelines for securing sustainable small-scale fisheries in the context of food security and poverty eradication*. Rome: FAO.

FAO. (2005). *Increasing the contribution of small-scale fisheries to poverty alleviation and food security*. Rome: FAO.

FAO. (2015). *Voluntary guidelines for securing sustainable small-scale fisheries in the context of food security and poverty eradication*. Rome: FAO.

FAO. (2016). *Fishing people*. FAO Fisheries and Aquaculture Department, FAO. http://www.fao.org/fishery/topic/13827/en. Accessed 9 Feb 2016.

Ferguson, B., Gardner, C. J., Andriamarovololona, M. M., Healy, T., Muttenzer, F., Smith, S., Hockley, N., & Gingembre, M. (2014). Governing ancestral land in Madagascar: Have policy reforms contributed to social justice? In M. Sowman & R. Wynberg (Eds.), *Governance for justice and environmental sustainability: Lessons across natural resource sectors in sub-Saharan Africa* (pp. 63–93). London: Routledge.

Garcia, S. M., & Rosenberg, A. A. (2010). Food security and marine capture fisheries: Characteristics, trends, drivers and future perspectives. *Philosophical Transactions of the Royal Society B, Biological Sciences, 365*, 2869–2880.

Gezon, L. L. (2002). Marriage, kin, and compensation: A socio-political ecology of gender in Ankarana, Madagascar. *Anthropological Quarterly, 75*, 675–706.

Gough, C., Thomas, T., Humber, F., Harris, A., Cripps, G., & Peabody, S. (2009). *Vezo fishing: An introduction to the methods used by fishers in Andavadoaka, southwest Madagascar*. London: Blue Ventures Conservation.

GFSI. (2015). *Global food security index 2015*. London: The economist intelligence unit.

Grenier, C. (2013). Genre de vie vezo, pêche "traditionnelle" et mondialisation sur le littoral sud-ouest de Madagascar. *Annales de géographie, 693*, 549–571.

Harris, A. (2007). "To live with the sea": Development of the Velondriake community managed protected area network, southwest Madagascar. *Madagascar Conservation & Development, 2*, 43–49.

Harris, A. (2011). Out of sight but no longer out of mind: A climate of change for marine conservation in Madagascar. *Madagascar Conservation & Development, 6*, 7–14.

Hoegh-Guldberg, O., & Bruno, J. F. (2010). The impact of climate change on the world's marine ecosystems. *Science, 328*, 1523–1528.

Horning, N. R. (2008). Strong support for weak performance: Donor competition in Madagascar. *African Affairs, 107*, 405–431.

ICAI (Independent Commission for Aid Impact). (2015). *How DFID works with multilateral agencies to achieve impact*. London: ICAI.

Iida, T. (2005). The past and present of the coral reef fishing economy in Madagascar: Implications for self-determination in resource use. *Senri Ethnological Studies, 67*, 237–258.

Jentoft, S. (2014). Walking the talk: Implementing the international voluntary guidelines for securing sustainable small-scale fisheries. *Maritime Studies, 13*, 16.

Le Manach, F., Gough, C., Harris, A., Humber, F., Harper, S., & Zeller, D. (2012). Unreported fishing, hungry people and political turmoil: The recipe for a food security crisis in Madagascar? *Marine Policy, 36*, 218–225.

Le Manach, F., Andriamahefazafy, M., Harper, S., Harris, A., Hosch, G., Lange, G. M., Zeller, D., & Sumaila, U. R. (2013). Who gets what? Developing a more equitable framework for EU fishing agreements. *Marine Policy, 38*, 257–266.

Leoville, A. (2013). *Diagnostic de l'état de la mangrove et de la pêche du crabe de mangrove (S. serrata) dans la région de Belo-sur-Mer, Menabe, Sud-Ouest de Madagascar*. MSc thesis, Université du Littoral Côte d'Opale, Dunkerque.

Le Vay, L. (2001). Ecology and management of mud crab *Scylla* spp. *Asian Fisheries Science, 14*, 101–111.

L'Haridon, L. (2006). *Evolution de la collecte de poulpe sur la côte Sud Ouest de Madagascar: éléments de réflexion pour une meilleure gestion des ressources*. London: Blue ventures conservation.

Maina, J., de Moel, H., Vermaaqt, J. E., Bruggemann, J. H., Guillaume, M. M. M., Grove, C. A., Madin, J. S., Mertz-Kraus, R., & Zinke, J. (2012). Linking coral river runoff proxies with climate variability, variability, hydrology and land-use in Madagascar catchments. *Marine Pollution Bulletin, 64*, 2047–2059.

Martin, D. (2015, June 11). Millions of UK aid 'being squandered': International agencies failing to spend British money in ways that effectively help the poor. *Daily Mail*. http://www.dailymail.co.uk/news/article-3119170/Millions-UK-aid-squandered-International-agencies-failing-spend-British-money-ways-effectively-help-poor.html. Accessed 2 Nov 2016.

Mayol, T. L. (2013). Madagascar's nascent locally managed marine area network. *Madagascar Conservation & Development, 8*, 91–95.

McClanahan, T. R., Ateweberhan, M., Omukoto, J., & Pearson, L. (2009). Recent seawater temperature histories, status, and predictions for Madagascar's coral reefs. *Marine Ecology Progress Series, 380*, 117–128.

Mills, D. J., Westlund, L., De Graaf, G., Kura, Y., Willman, R., & Kelleher, K. (2011). Under-reported and undervalued: Small-scale fisheries in the developing world. In R. S. Pomeroy & N. L. Andrew (Eds.), *Small-scale fisheries management: Frameworks and approaches for the developing world* (pp. 1–15). Cambridge: CABI International.

Muttenzer, F. (2015). The social life of sea cucumbers in Madagascar: Migrant fishers' household objects and display of a marine ethos. *Etnofoor, 27*, 101–121.

Oliver, T. A., Olesen, K. L. L., Ratsimbazafy, H., Raberinary, D., Benbow, S., & Harris, A. (2015). Positive catch and economic benefits of periodic octopus fishery closures: Do effective, narrowly targeted actions 'catalyze' broader management? *PloS One, 10*, e0129075.

Pauly, D. (1997). Small-scale fisheries in the tropics: Marginality, marginalization, and some implications for fisheries management. In E. K. Pikitch, D. D. Huppert, & M. P. Sissenwine (Eds.), *Global trends: Fisheries management* (pp. 40–49). Bethesda: American Fisheries Society Symposium.

Pauly, D. (2006). Major trends in small-scale marine fisheries, with emphasis on developing countries, and some implications for the social sciences. *Maritime studies, 4*, 7–22.

Parfitt, T. (2015, June 11). Revealed: Britain 'wasted MILLIONS on foreign aid project that was never used'. *Daily Express*. http://www.express.co.uk/news/uk/583652/foreign-aid-government-African-Development-Bank-Madagascar-fishing. Accessed 2 Nov 2016.

Pollini, J., Hockley, N., Muttenzer, F. D., & Ramamonjisoa, B. S. (2014). The transfer of natural resource management rights to local communities. In I. R. Scales (Ed.), *Conservation and environmental management in Madagascar* (pp. 172–192). Abingdon: Routledge.

Pomeroy, R. S., & Andrew, N. L. (2011). *Small-scale fisheries management: Frameworks and approaches for the developing world*. Cambridge: CABI International.

Raberinary, D., & Benbow, S. (2012). The reproductive cycle of Octopus cyanea in Southwest Madagascar and implications for fisheries management. *Fisheries Research, 125–126*, 190–197.

Rajaonarimampianina, H. (2014, November 12–19). 'Sydney Vision' declaration. *Speech presented at VIth World Parks Congress*, Sydney.

Razanaka, S., Grouzis, P., Milleville, P., Moizo, B., & Aubry, C. (Eds.). (2001). *Sociétés paysannes, transitions agraires et dynamiques écologiques dans le sud-ouest de Madagascar*. Antananarivo: CNRE-IRD.

Sheridan, C., Baele, J. M., Kushmaro, A., Frejaville, Y., & Eeckhaut, I. (2015). Terrestrial runoff influences white syndrome prevalence in SW Madagascar. *Marine Environmental Research, 101*, 44–51.

Smartfish. (2015). *Enhancing value-chain performance for mud crab in Madagascar*. http://commissionoceanindien.org/fileadmin/projets/smartfish/Fiche/FICHE3ENGLISH.pdf. Accessed 2 Nov 2016.

Smith, M. D., Roheim, C. A., Crowder, L. B., Halpern, B. S., Turnipseed, M., Anderson, J. L., Asche, F., Bourillón, L., Guttormsen, A. G., Khan, A., Liguori, L. A., McNevin, A., O'Connor, M. I., Squires, D., Tyedmers, P., Brownstein, C., Carden, K., Klinger, D. H., Sagarin, R., & Selkoe, K. A. (2010). Sustainability and global seafood. *Science, 327*, 784–786.

Virah-Sawmy, M., Gardner, C. J., & Ratsifandrihamanana, A. N. (2014). The Durban vision in practice: Experiences in participatory governance of Madagascar's new protected areas. In I. R. Scales (Ed.), *Conservation and environmental management in Madagascar* (pp. 216–252). Abingdon: Routledge.

Westerman, K., & Benbow, S. (2013). The role of women in community-based small-scale fisheries management: The case of the south west Madagascar octopus fishery. *Western Indian Ocean Journal of Marine Science, 12*, 119–132.

World Bank. (2015). *Madagascar systematic country diagnostic*. Washington, DC: The World Bank Group.

WRI (World Resources Institute). (2003). *Earth trends: Coastal and marine ecosystems, Madagascar*. Washington, DC: World Resources Institute.

# Chapter 17
# Costa Rica: A Champion of the Small-Scale Fisheries Guidelines

**Gabriela Sabau**

**Abstract** This chapter uses a case study approach and a transdisciplinary perspective to shed light on Costa Rica's implementation of the Voluntary Guidelines for Securing Sustainable Small-Scale Fisheries in the Context of Food Security and Poverty Eradication (SSF Guidelines), and assess their potential to make Costa Rican small-scale fisheries ecologically and socially sustainable. The chapter identifies how the government, together with fishers' organizations, is aligning its conservation and participative management policies to the provisions of the SSF Guidelines. It also discusses how a small-scale fishery cooperative's participative management initiatives have led to the establishment of an institutional arrangement which has the potential to successfully promote the implementation of the SSF Guidelines not only in Costa Rica but also in the Central American region as a whole. Involving small-scale fishers in dialogue and design of their own sustainable future is key to how Costa Rica has become a champion of the SSF Guidelines.

**Keywords** Small-scale fisheries • Marine areas of responsible fishing • Responsible co-management • Dialogue • Participation

## Introduction

Small-scale (artisanal) coastal fisheries make up 80% of Costa Rica's fisheries. They are an important source of fishery products for local, national, and international markets and provide livelihoods for 16,411 small-scale artisanal fishers living in 86 coastal communities in Costa Rica (OSPESCA 2012). Yet, Costa Rican small-scale fisheries face multiple challenges. One challenge comes from large-scale fisheries, developed in the 1950s and expanded after 1975, when Costa Rica acquired,

G. Sabau (✉)
School of Science and the Environment, Memorial University of Newfoundland, Grenfell Campus, Corner Brook, NL, Canada
e-mail: gsabau@grenfell.mun.ca



S. Jentoft et al. (eds.), *The Small-Scale Fisheries Guidelines*, MARE Publication Series 14, DOI 10.1007/978-3-319-55074-9_17

with its Exclusive Economic Zone (EEZ), a huge marine area of over 500,000 km$^2$. The second challenge is Costa Rica's network of conservation and protected areas, which covers 26% of the national terrestrial area and 3% of the country's marine area (SINAC 2014). Many of the protected areas have been established without coastal communities being consulted, and so far no legal framework has been developed to govern coastal and marine areas in ways that would "include conservation and sustainable use perspectives rather than just a resource production perspective" (Alvarado et al. 2012). In 2007, the government adopted a 'National Strategy for Integrated Management of Coastal Marine Resources' aimed at reforming its top-down marine management system by effectively involving fishers' organizations in small-scale fisheries management and allowing low-impact (small-scale) fishing in some of the marine protected areas.

The Voluntary Guidelines for Securing Sustainable Small-Scale Fisheries in the Context of Food Security and Poverty Eradication (SSF Guidelines) (FAO 2015) were adopted in June 2014. Costa Rica hailed them as "an important international tool of enormous value" (Meneses Castro 2014) which could help the country develop public policies for promoting "the development of decent living conditions and the human well-being of the coastal and seafaring communities" (ibid). In November 2014, the government included the SSF Guidelines in the country's National Development Plan (2015–2018) (MIDEPLAN 2014). With financial support from the FAO, the plan's goal is to increase the number of fishing communities applying the Guidelines from one to eight by 2018. Moreover, in September 2015, the official implementation of the SSF Guidelines was decreed; it was made mandatory for government institutions to include the Guidelines in their operational plans and allocate the budgetary and economic resources necessary for their implementation (Decreto 2015). Through these landmark actions, Costa Rica became the first country in the world to develop a national policy for implementing the SSF Guidelines. A few questions follow: Why Costa Rica, and what factors led to such a decision being taken? How are the SSF Guidelines being implemented – are they contributing to a meaningful transformation of small-scale fisheries, towards both ecological and social sustainability? These are difficult questions that require complex answers. A simple answer is that Costa Rica became a champion of the SSF Guidelines due to the vision of some national champions who understood that it was wrong to keep Costa Rican small-scale fishers marginalized and in poverty in spite of their important contribution to providing livelihoods in the coastal areas and food for local, national, and regional markets. Some of these leaders were Vivienne Solis Rivera, Patricia Madrigal, David Chacón, Fernando Mora, and Gustavo Meneses Castro (M.M. Chavarría, personal communication, March, 2015). All of them have been actively engaged in making small-scale fishers more visible and supporting their legitimate claim to have a say in fisheries management. These individuals value the human rights approach of the SSF Guidelines because it helps promote both economic and human development of small-scale fisher communities (G. Meneses

Castro, personal communication, March, 2015). However, such a simple answer does not explain the complexity and uncertainty involved in the process of implementing the SSF Guidelines both from a legal and regulatory perspective, nor their intent. Their implementation implies not only legislative reform but also real change in Costa Rican small-scale fisheries and local fishing communities. This chapter assumes that effective implementation of the SSF Guidelines was facilitated by Costa Rica's national policy of biodiversity conservation and started when the government recognized small-scale fishers as partners in dialogue vis-a-vis the development of an original co-governance tool, namely marine areas for responsible fishing (AMPRs). The AMPRs guarantee small-scale fishers the right to use a designated marine and coastal area, provided they do so responsibly with the aim of achieving both long-term ecosystem and human wellbeing. Before the SSF Guidelines were introduced in Costa Rica, the government had already started a process of transitioning from a top-down approach to fisheries management to more participative approaches. The process was prompted by rampant poverty in the coastal communities at the beginning of the 2000s, due to depletion of fish stocks, an inadequate development model, and inequity in wealth distribution (V. Solis Rivera, personal communication, August 2016). The reform was also required to balance the sustainable use of coastal resources with the conservation requirements imposed by the country's network of protected areas. An intensive dialogue between small-scale fishers, industrial fishers, and INCOPESCA (the National Institute for Fisheries and Aquaculture) led to a slow alignment of the national (institutional) goals with the coastal and marine areas users' interests. The reform also helped clarify small-scale fishers' social identity, giving them a tool and a voice in the sustainable co-governance of their resources through their involvement in AMPRs. Several AMPRs have been created with the goal of contributing to marine conservation and sustainable and responsible use of fishery resources. This new type of fisheries management includes an element of a human rights approach to protecting small-scale fisheries. In this way, the AMPRs seem to be the most important tool for implementation of the SSF Guidelines in Costa Rica.

This chapter uses a case study approach and a transdisciplinary perspective to shed light on Costa Rica's implementation of the SSF Guidelines, and assesses their potential to produce a sustainable change in Costa Rican small-scale fisheries by balancing ecological and social sustainability concerns. The chapter aims to discuss how the government, fishers' organizations, and NGO leaders are aligning the country's conservation and participative management policies with the provisions of the SSF Guidelines by identifying patterns of transdisciplinary thinking and using dialogue heuristics in the process of transitioning from a top-down approach to a more participatory approach to fisheries' management in Costa Rica. The chapter further aims to answer the question whether this transition can impact and drive small-scale fisheries towards long-term sustainable development by implementing the SSF Guidelines.

## Research Methods

### *Theoretical Background*

A case study approach (Baxter and Jack 2008; Yin 2003) is employed for this research, given its potential to provide an in-depth analysis of a complex environmental problem that is complicated by social challenges and shaped by context. The research question asked is whether Costa Rica's implementation of the SSF Guidelines is a viable way to secure sustainable development of its small-scale fisheries. Sustainability, in this case, is defined as the 'potential for long-term maintenance of wellbeing' with reference to "the interaction between the dynamics of nature and dynamics of society" (Kates et al. 2001). Costa Rica's small-scale fisheries are complex socio-ecological units, embedded in particular ecological and social environments. Their sustainable development involves not only economic prosperity and environmental protection but also social inclusion and good governance (Sachs 2015). These multiple goals are amenable to multiple solutions due to the numerous and sometimes conflicting aspects that need to be addressed, and to the wide range of meanings, perceptions, and interpretations by various stakeholders involved in the decision-making process (Paloniemi and Vainio 2014). A transdisciplinary approach is employed as it allows a systemic view on real-life complexities and a more realistic integration of academic and non-academic perspectives (including fishers' abilities to engage with complexity). Such an approach will hopefully address the social, political and normative priorities of small-scale fisheries without ignoring their ecological context. The integration of various perspectives has the potential not only to produce shared transdisciplinary knowledge that can inform decision-making, but also build stakeholders' accountability, as it "diverts attention from the epistemological basis of knowledge claims towards a more democratic and socially robust culture of knowledge production" (Huutoniemi 2014). Such an approach seeks consensus rather than certified knowledge. A transdisciplinary approach encourages participation in public dialogue, which often aims not only to reconcile opposing points of view but also conceptualize indeterminate situations in purposeful ways (ibid), which can challenge the status quo and envision better futures. The toolkit of transdisciplinary research includes problem-solving heuristics, an 'art' or type of 'logic' that is more fluid and ad hoc than linear and rote deduction of an answer or a solution from a prescribed set of rules (Klein 2010). In fisheries, overfishing can be analyzed quantitatively in terms of stock size and species or unemployment rates, but a heuristic approach can define overfishing as a historical decline in sea habitat or as a social consequence of fishers' lives, for example lack of hope or the dissolution of community social bonds.

Focusing on heuristics can shed light on the particular way Costa Rica is implementing the SSF Guidelines, highlighting especially the collaborative efforts to design implementation policies or the specific strategies used by stakeholders to implement them. Dismantling the top-down approach in fisheries management has involved intense stakeholder communication and the building-up of a willingness to collaborate. Yet these social actions are seldom discussed in the relevant literature. They are, however, important, as "fisheries co-management is not so much about the rules *per se* as it is about the communicative and collaborative process through which these rules are formed" (Jentoft et al. 1998). One heuristic tool relevant for this research is dialogue.

Dialogue was at the outset considered one of the most ethical forms of communication (Kent and Taylor 2002). It was defined as "an effort to recognize the value of the other – to see him/her as an end and not merely as a means to achieving a desired goal" (Buber 1970). At the level of organizations, dialogue was considered essential for ethical public relations that should involve dialogic procedures rather than monologic 'policies' (Pearson 1989). The concept has evolved from a communication tool into a public relations strategy for "building, nurturing and maintaining relationships" (Ledingham and Brunning 2000). As such, dialogue has been defined not as a process but as 'a product of ongoing communication and relationships' (Kent and Taylor 2002). According to Kent and Taylor (2002), any dialogue has five tenets: mutuality (recognition of organization-public relationships); propinquity (temporality and spontaneity of interactions with the public); empathy (supportiveness and confirmation of public goals and interests); risk (willingness to interact with individuals and the public on their own terms); and commitment (the extent to which an organization gives itself over to dialogue, interpretation, and understanding in its interactions with the public). Using Kent and Taylor's tenets, a 2008 study empirically tested dialogue's capacity to build sustainable relationships between a state organization and the public. The results demonstrated that a careful, deliberate, and mutually beneficial dialogue can be a "system-changing heuristics" (Paloniemi and Vainio 2014) by helping "align organizational goals and public interest" (Bruning et al. 2008). Involving stakeholders in dialogical relationships is important in promoting environmental sustainability goals (Brugnach and Ingram 2012). Investigating whether a dialogical interaction between small-scale fishers and the government and other organizations has implications for SSF Guidelines implementation in Costa Rica can be a useful tool for assessing the chance of the SSF Guidelines to bring sustainable social change in artisanal fishers' lives.

Data collection for this chapter includes a thorough review of the literature as well as primary field research conducted for a 5 week period (February–March 2015) in Costa Rica, where the researcher held both formal and informal interviews with key informants, and engaged in personal observation.

## Study Site

### *The General Context*

Costa Rica is a small Central American country (51,100 km$^2$) dubbed as an 'environmental paradise' (Davis 2014). The country has mountains, active volcanoes, tropical rain forests, beaches on both the Pacific Ocean and Atlantic Ocean coasts, and a rich biodiversity. It is estimated that approximately 94,753 known species exist in Costa Rica, and that the country owns 5% of the world's biodiversity (SINAC 2014), being one of the 20 most biodiverse countries in the world (Alvarado et al. 2012). The biodiversity extends to marine species also; the country has approximately 7000 marine species (SINAC 2014) living in the country's huge territorial waters (589,000 km$^2$).

Costa Rica is a middle-income country with very high literacy rates (96.3%), and a growing population (4.8 million) (UNICEF 2012). It is a stable democracy which in 1948 also decided not to have an army. However, in recent years, the State of the Nation Report (2013) indicated that it has experienced a "slow and uncertain progress in human development." During the first half of the twentieth century, the economy focused on agriculture, in particular the export of coffee beans and bananas. Fishing at that time was mostly a subsistence activity practiced with small-scale artisanal boats with little or no preserving (freezing) capacity (Trujillo et al. 2015). Since the mid-1990s, Costa Rica has specialized in services such as tourism, including ecotourism and medical tourism, finance, and pharmaceutical research. Tourism is one of the fastest growing activities: in 2014 the revenue from tourism accounted for 5.3% of the country's GDP (ICT Statistics 2014).

The country, since the early twentieth century, has a long tradition of protecting nature. The first nature reserve, Cabo Blanco in the Nicoya Peninsula, was established in 1955 – a private initiative led by Olof Wessberg, a Swede (Davis 2014). Though this area had a coastal zone, a marine protected area (MPA) was not established here until 1982 (Alvarado et al. 2012). Today, Costa Rica has a network of conservation areas and biological corridors called Sistema Nacional de Áreas de Conservación (SINAC). These areas constitute 26% of the total land area of the country. However, only 3.2% of the marine territory is estimated to be under some kind of protection (Solis Rivera et al. 2013). The discrepancy between conservation efforts in terrestrial and marine areas may be due to three main reasons: (1) the history of country's economic development based on agriculture and less on fisheries (Quesada Alpizar 2006); (2) the fact that most MPAs were established as extensions of terrestrial protected areas (Alvarado et al. 2012) since they happened to be within or adjacent to a terrestrial protected area; or (3) a lack of legislative and institutional coordination and poor management by government agencies (Quesada Alpízar 2006). One state agency, the Ministry of Environment, Energy, Waters, and Seas (MINAEM), has jurisdiction over marine resources within MPAs; resources outside MPAs are under the jurisdiction of INCOPESCA and the Institute of Tourism (ICT) monitors the maritime-terrestrial zone (MTZ), the strip of 200 m width along the entire Atlantic and Pacific coasts, encompassing approximately 45% of Costa Rica's 1466 km – long coastline.

## Costa Rica Fisheries

The establishment of fishing as an economic activity in Costa Rica has been slow, and was preceded by the adoption of a series of fishing regulations. In 1949, Costa Rica became a founding member of the Inter-American Tropical Tuna Commission, and in the 1970s, based on the Law of the Sea Treaty (UNCLOS), it acquired an EEZ almost 11 times larger than the country's terrestrial area.

In the 1970s the fishing sector expanded rapidly as a result of financial incentives given by the state, such as tax exemptions and tax credits, and fuel subventions. Expansion of fishing activity in the 1980s was driven by an increase in demand from the North American seafood market, and a cultural change in Costa Rican seafood perception and consumption that increased internal demand for seafood (Trujillo et al. 2015). However, fishing was still an 'incipient sector' of fringe importance to national economic development. In 1983, the contribution of commercial fisheries to the country's GDP was a mere 0.25% (FECOP 2013) and in 2002 still only 0.32% (FAO 2004). By 2008, the fishing industry's contribution had increased, with commercial fisheries contributing 1.88% to GDP and recreational fisheries 2.13% (IICE 2010).

Small-scale fisheries' contribution to landings was small. In 2002, small-scale fisheries accounted for only 6% of all catches reported by Costa Rica to the FAO (FAO 2004). If the composition of the fishing fleet is considered, a pattern arises: the national fishing fleet consists of two distinct sectors, a large small-scale fishery (~75% of the fleet) fishing mostly in coastal waters, and a small industrial fleet fishing mostly in offshore waters (Trujillo et al. 2015). It is assessed that this relatively small longline (industrial) fishing fleet accounts for approximately 80% of all catches, while the large small-scale fleet fishing coastal resources accounts for only 20% of landings (Quesada Alpizar 2006).

Costa Rica's small-scale fishery has grown rapidly in the past 20 years as a result of structural changes in the agricultural sector, which have led to large-scale unemployment in the countryside and human migration to the coast. Because of this, and due to the lack of effective fisheries legislation (and consequently a lack of deterrents for those breaking fishing rules), fisheries resources have come under strain. This has resulted in a substantive fall in fish stocks, in smaller catches of the most valuable species per unit effort, a reduction in the sizes of fish and shrimp caught (FAO 2004), and a deterioration of marine ecosystems through indiscriminate use of fishing gear (SINAC 2014).

Overfishing by trawling in the shrimp fishery has caused a severe economic recession in the sector; shrimp production has decreased from a record high of 5000 metric tons per year during the 1990s to 1000 metric tons per year in 2013 (SINAC 2014). The granting of new shrimp trawling licences and renewal of existing ones was prohibited in 2013, through a Constitutional Court order, "until INCOPESCA, through technical and scientific studies, reports that more environmentally responsible practices are used and sustainability of trawling practices can be proven. If there is no proof of sustainability studies, then in 2019 the last trawling license will be terminated" (V. Solis Rivera, personal communication, August 2016). Even more of a problem are the 24–35 foreign vessels targeting tuna through the use of purse seines in Costa Rica's EEZ under the INCOPESCA licences system. While the international tuna industry caught an annual average of 26,163 metric tons of tuna in Costa Rican waters between 2002 and 2009 (IATTC 2013), the national longline industry caught an annual average of 1484 metric tons of tuna.

Overfishing in Costa Rican waters is also due to illegal and unreported fishing and the use of fish aggregation devices (FADs), especially in fishing for tuna. In July 1999, INCOPESCA banned illegal tuna fishing and use of FADs (INCOPESCA 1999). The problem is that a substantial portion of fisheries' catches in Costa Rica are not reported and ignored. This catch comes either in the form of bycatch or from the small-scale, subsistence, or recreational fishing sectors (Trujillo et al. 2015). With the exception of tuna, Costa Rica's post-catch industry is fairly small, as almost all products are exported fresh or frozen, with little value added (FAO 2004).

## *Small-Scale Fisheries*

The small-scale fishing sector in Costa Rica has established itself in the last 30 years as an 'occupation of last resort' (Fargier et al. 2014) for surplus rural labor and for immigrants (FAO 2004). The sector has received little support from the government (Jiménez 2013), despite its considerable socio-economic importance for coastal communities and potential to conserve the environment by using low impact fishing gear (Solis Rivera et al. 2015). The number of small-scale (artisanal) fishers in Costa Rica is significant and growing: in 2009, there were 13,850 small-scale fishers living and working in 75 communities on the Pacific coast and 950 small-scale fishers living and working in 11 communities on the Caribbean coast (OSPESCA 2010). In 2010 the number of small-scale fishers grew to 16,411 (Table 17.1).

For small-scale fishers the preferred fishing gear is the line or hand rope, followed by the gillnet. Fishers minimally process their catches, 60% conserve them on the boats in ice, and only 0.7% have infrastructure on boats to freeze them (OSPESCA 2012).

A social profile of the Costa Rican small-scale fisher (OSPESCA 2012) illustrates that the sector has considerable socio-economic importance, but it has been marginalized (Elizondo Mora 2005). Almost 75% of the fishers live in large families and have economic responsibilities to members of their household, most of the time as sole income providers in the household (Table 17.2).

Most of the fishers (84%) own their house and have been fishers for more than 20 years. They most often lack higher education, but 73% have finished primary school (Tables 17.2 and 17.3).

The fishers seem very dependent on their profession, as 80% have never been involved in other economic activities. However, about 65% have taken training courses in small-scale fishing, organized either by the government or by international

**Table 17.1** Costa Rican artisanal fishery profile (2009–2011)

| Fishers | Numbers | Fleet length (m) | Numbers | Harvest field (in 2010) | Harvest metric tons (MT) | Species harvested | Significance in harvest (from 1 to 5) |
|---|---|---|---|---|---|---|---|
| Artisanal | 14,800 | .60–7 | 1854 | Pacific | 21,000 | Tuna | 2 |
| Aquaculture | 1611 | .50–2.5 | 4246 | Caribbean Sea | 2100 | Bass, croaker | 3 |
| Private investors | 1603 | Privately owned | 5961 | Continental waters | 0 | White marlin | 4 |
| Private and public investors | 8 | Equipped for trawling | 547 | | | Shrimp, prawn | 5 |
| Total | 16,411 | | 6100 | | 23,200 | | |

Source: OSPESCA (2012)

organizations, on topics such as fish conservation and processing, use of fishing gear, or boat maintenance and repair. Access to the profession is not easy. It is estimated that an initial investment of US$ 13,000 is needed and 78% of fishers do not have access to credit. Community infrastructure is poorly developed to support fishing; most communities do not have processing plants, shipyards, or refrigeration units. The only infrastructure that might be available in some communities is a receiving and distribution center, an ice producing plant, and a gas station. Only 46 of the communities have public roads and 11 communities have dock infrastructure. The network of community public services is not well developed, except for basic health care, education, communications, and domestic services which are present in 74% of communities. Only four of the 86 communities have sewage and drainage systems, and only three have a market (OSPESCA 2012). Implementation of the SSF Guidelines, therefore, could end marginalization of small-scale fishers and bring sustainable wellbeing to their communities.

## *Fisheries Management: From Top-Down Regulations to Responsible Co-management*

Management of fisheries in Costa Rica has been slowly evolving in the last decade from a centralized top-down approach to a more participative management which involves multiple stakeholders such as municipalities, NGOs, universities, and community groups (Fargier et al. 2014). Up to 1994, as a relatively unimportant economic sector in an economy dominated by agriculture, fisheries were managed by the Ministry of Agriculture as an 'exploitable resource' (Quesada Alpizar 2006). The main management tools were command and control regulations, such as species size limits, gear restrictions, licence limitations, and time and area closures. Since 1994, fisheries were given more prominence in Costa Rica's economy with

**Table 17.2** Costa Rica artisanal fisher's social profile (in numbers)

| Fishers' gender | | Fishers' age | | Marital status | | Persons in household | | Wage – earners / household | | House ownership | |
|---|---|---|---|---|---|---|---|---|---|---|---|
| Male | 13,860 | Under 20 | 340 | Single | 3197 | 1 | 903 | 1 | 9886 | Yes | 12,432 |
| Female | 940 | 21–40 | 5832 | Married | 10,848 | 2–3 | 3878 | 2–3 | 4248 | House rented | 1406 |
| | | 41–50 | 4144 | Widowed | 222 | 4–7 | 8969 | More than 4 | 666 | House sharing | 962 |
| | | Over 51 | 4484 | Separated/divorced | 533 | More than 8 | 1050 | | | | |
| Total | 14,800 | | 14,800 | | 14,800 | | 14,800 | | 14,800 | | 14,800 |

Source: OSPESCA (2012)

**Table 17.3** Costa Rica artisanal fisher's social profile 2 (in numbers)

| Fishers level of education | | Number of years fishing | | Fields of income-earning activity | | Time dedicated to other economic activities | |
|---|---|---|---|---|---|---|---|
| No studies | 1154 | Less than 10 | 2456 | Fishing only | 11,810 | 1–3 months/year | 962 |
| Primary school | 10,849 | 11–20 | 4840 | Agriculture | 400 | 4–6 months/year | 932 |
| Secondary school | 2560 | More than 20 years | 7504 | Animal husbandry | 74 | 7–11 months/year | 237 |
| University | 237 | | | Commerce | 636 | Different every year | 829 |
| | | | | House work | 104 | | |
| | | | | Other | 1776 | | |
| Total | 14,800 | | 14,800 | | 14,800 | | 14,800 |

Source: OSPESCA (2012)

the opening of INCOPESCA. The institute was established with the goals of promoting and regulating the development of fisheries, maritime hunting, aquaculture and research, and the conservation, exploitation, and sustainable use of the ocean and of aquaculture-based biological resources (FAO 2004). INCOPESCA is also the executive body of the new Law on Fisheries and Aquaculture (2005) and in charge of the National Development Plan for Fisheries and Aquaculture. This plan establishes as a national priority the sustainable development of fisheries and aquaculture activities (Decree No.35260/MAG- 2009). Another national priority is the promotion of responsible fishing, initiated in1999 when Costa Rica officially approved the application of the FAO Code of Conduct for Responsible Fisheries (Decree No.27.919/MAG-1999). Since its inception, INCOPESCA has faced challenges in implementing its mandate because of the existing complex system of protected areas under the control of diverse institutions for management and monitoring, such as ICT, MINAEM, and municipal governments.

The lack of institutional coordination in the management of coastal and marine resources has led to poor control of the fishing and aquaculture sectors. The consequences of this have been overexploitation of marine resources, contamination of coastal zones, inequity in distribution of benefits derived from tourism and conservation, illegal fishing and use of fishing gear (SINAC 2014), and privatization of public spaces in the MTZ. For instance, in Guanacaste and Puntarenas, 20% of the MTZ is now privately-owned (State of the Nation 2013). First attempts at institutional coordination between relevant authorities and stakeholders were made in 1995 with the introduction of a legal tool called multiple use marine areas (AMUM). In 2004, the Inter-Institutional Commission of the Costa Rican EEZ was created with the mission to elaborate a national marine strategy for integrated management of coastal marine resources. A national marine strategy was adopted in 2007. It signaled an important policy change, as it regulated the integrated management of coastal marine resources 'led by the government and promoting involvement of the rest of society' (CIZEE 2008). No tradition of civil society participation in environmental management in Costa Rica existed, due to the country's strong tradition of centralized management of natural resources. Some forms of civil society participation in environmental management have existed since ratification of the Convention on Biological Diversity in 1994. However, they have been limited to the involvement of local and regional conservation bodies created under SINAC in conservation areas, bodies that are considered poor representatives of the communities (Solis Rivera et al. 2013; Fargier et al. 2014).

Small-scale fishers have become involved in some forms of participative management of their coastal and marine resources mainly as a result of the intensified conflict between the government and fishing communities who had lost access rights to their resources after the establishment of a protected area. For instance, in 1997, a co-management agreement (mostly a conflict resolution tool) was reached between the community of Cahuita and MINAE, following the designation of the Cahuita National Park in 1978 as a protected area (Quesada Alpizar 2006). Outside conservation areas, small-scale fishers have initiated some forms of collective management of their coastal resources, mostly in response to their cultural and original tenure rights and declining small-scale catches. For instance, the hook-and-line fishers of Palito in Isla Chira have decided to protect a coastal reef area in their village by banning the use of any harmful fishing gear in a particular area of their fishing territory. The community was to monitor enforcement of this voluntary rule by patrolling. At the request of the fishers, in 1995, the government declared this zone "exclusive for hand line fishing" but did not provide institutional support for its implementation (Fargier et al. 2014). The government's vision for involving fishers in management of their resources implied a "co-management regime in which the fishing industry has been empowered to manage marine fishery resources in partnership with the State" (Quesada Alpizar 2006). Small-scale fishers, being the least empowered of all fishers, have been encouraged by the government to organize themselves in cooperatives or in local committees of artisanal fishers (COLOPES). These attempts by the government to organize small-scale fishers through top-down initiatives failed and very few cooperatives and COLOPES still exist today. However, they have paved the way for small-scale fishers to organize themselves in fishers' associations and become more pro-active in seeking participatory solutions to the management of their resources. In 2011 about 30% of small-scale fishers were members of small-scale artisanal fisheries associations (OSPESCA 2012).

In 2004, the members of the small-scale fishers' cooperative CoopeTárcoles R.L, with support from CoopeSoliDar R.L, a cooperative oriented to give professional services in human rights and environmental conservation, initiated a process which led, in 2008, to government approval of a new model of participatory management of small-scale fisheries in Costa Rica, namely marine areas for responsible fishing (AMPRs) (Spanish acronym). An AMPR is 'an area with significant socio-cultural, fishery or biological characteristics in which fishing is especially regulated to ensure long-term use of fishery resources and in which INCOPESCA can count on the support of coastal communities and/or other institutions for its conservation, use and management' (Decree 35502-MAG 2009). The AMPR is a form of community-based resource management, which does not create exclusive fishing rights for a community. Anyone with a fishing licence and following the fishing rules established in the area's fishery management plan can fish in the area. For this reason, the AMPR has been legally approved as a co-governance model of a coastal and marine area, negotiated between small-scale and commercial fishers, government agencies and other stakeholders. It prioritizes the protection of marine resources from non-selective fishing practices at both the commercial and small-scale scale levels (Garcia Lozano and Heinen 2015). Eight AMPRs have been legally established in Costa Rica so far, and others are in various stages of approval (D. Chacón, personal communication, March, 2015).

## Case Study: CoopeTárcoles R.L.

CoopeTárcoles R.L. is the only small-scale fishing cooperative in Costa Rica still functional as such (D. Chacón, personal communication, March, 2015). The cooperative has been instrumental in the establishment of the AMPR model and is the only fishing organization already implementing the SSF Guidelines. In what follows, we indicate in parentheses the SSF Guidelines' articles that are being implemented by CoopeTárcoles R.L. (Fig. 17.1).

The SSF Guidelines are well known to fishers in Tárcoles. David Chacón, the president of CoopeTárcoles R.L, participated in their development, contributing first-hand insights from small-scale fishers, their experience, needs, and aspirations. After their adoption, the Guidelines have been disseminated in a simplified and translated version to the residents of Tárcoles (13.3). The cooperative is located in Tárcoles, a small coastal community (population 4300) in the central Pacific. The area is rich in valuable fish species, such as shrimp, and in mangrove forests. This abundance has shaped the economic and cultural identity of Tárcoles in the last 50 years, when most of its inhabitants became fishers (Solis Rivera et al. 2015). The cooperative was established in 1985 with the main goals of improving fishers' working conditions and finding better marketing opportunities for their fishery products. These efforts have paid off, as today CoopeTárcoles is the only Costa Rican fishing group involved in the entire value chain from production to marketing to selling (6.5, 7.1) (Fargier et al. 2014). Fishers eliminated intermediaries (7.5, 7.7) and sell approximately 70% of their catches directly to an exporter (Martec) and local businesses for domestic consumption (D. Chacón, personal communication, March, 2015). This way fishers have managed to keep fish prices sufficiently high, so that they can operate their boats and also make a living, while the cooperative can cover its operation costs and sometimes make an end-of-year profit, which is fairly distributed to all the members of the cooperative (7.8).

**Fig. 17.1** Tárcoles AMPR, March 20, 2015 (Credit: G. Sabau)

The cooperative has also signed an agreement with the Costa Rican social security department, securing pensions to all members when they retire (6.3, 7.4) (D. Chacón, personal communication, March, 2015). The cooperative has 45 members, six of whom are women (5.18, 8.1, 8.2), but it provides jobs to approximately 250 community members (many of them women or youth) (6.5, 7.1, 7.2, 7.5). Community members participate in fishing or in pre- and post-harvest fishing operations, like baiting and untangling fishing lines, or processing and selling fishing products (UNDP 2012). At the beginning of the 2000s, various fish catches started to diminish due to overfishing and high levels of pollution coming through Rio Tárcoles. Also, fishers' access to fishing grounds was threatened by trawling of large-scale fishing boats and the rapid development of tourist amenities. The cooperative, therefore, included among its priorities the promotion of sustainable management of its natural and cultural resources (5.14). At the same time, it initiated a 'relationship of mutual fortification', grounded in social and environmental responsibility, with the cooperative CoopeSoliDar R.L. (7.4, 10.6). (CoopeSoliDar 2005; Solis Rivera et al. 2015). The relationship, characterized as "a process of continual informed consent, wherein cooperative members agree and decide upon the role of the external organization" (Garcia Lozano 2014), has contributed to training and capacity building for collective action in Tárcoles (11.4, 11.8, 12.1, and 12.3).

One of the first results of this fruitful cooperation was CoopeTárcoles R.L adopting its Code of Conduct for Responsible Fisheries in 2004. Inspired by the FAO document (FAO 1995), the Tárcoles Code of Conduct contains rules for sustainable fishing through the use of low impact fishing gear (longline, small boats, and fishing nets with mesh size larger than three inches to reduce bycatch) and through voluntarily giving up fishing of species that showed signs of distress (5.14). The two cooperatives also started a series of participatory research projects, which included building a database recording daily fish catches, gear used, and site locations (10.6, 11.4).

Another research project involved a process of participatory zoning, with fishers drawing maps of their fishing spots and keeping inventories of their ecological assets. The information collected was used to monitor the fishing grounds and identify vulnerable species, as well as to plan for the future. By disseminating this data to government institutions, fishers provided useful local information for decision-making purposes (11.2, 11.8, and 11.9). For instance, based on their research, fishers have recommended that the government should monitor the fishing of manta ray species. They have also identified and communicated the need to carry out participatory research on lobster stocks along the Pacific coast (UNDP 2012). Through these collective research efforts, fishers started building social capital and learned how to sustain cooperation, "not only within the cooperative but also to facilitate networking and lobbying efforts" (12.1, 12.3) (Garcia Lozano and Heinen 2015). The Tárcoles Code of Conduct was presented to government institutions in 2005, together with the fishers' request for the creation of a community-managed marine area. This proposition was accepted by INCOPESCA in January 2009, after a long process of political advocacy (UNDP 2012) and cooperative work within a mixed government-cooperative-NGO commission (12.4). After the regulation for establishing AMPRs at the national level was published, the two cooperatives worked together to develop a comprehensive fisheries management plan required for approval of the Tárcoles AMPR. The fisheries governance plan (Fig. 17.2) has several thematic axes, including areas with total or partial fishing bans, biology and fishery management, and permissible fishing methods and practices. It also included training and outreach, community outreach and knowledge-building, strengthening of local organizational structures, strategic partnerships and marketing, enforcement and compliance with current laws, and research and monitoring programs (CoopeSoliDar R.L 2010).

The Tárcoles Marine Area for Responsible Fishing was officially recognized on August 19, 2011 (Decree A.J.D.I.P./193–2011) (5.1, 5.5, 5.6) after a voluntary mutual agreement was reached by the Tárcoles small-scale fishers and semi-industrial trawlers. The agreement established a 1 year no-fishing zone for shrimp, extending from the Tárcoles coast to the 15 m-depth bathymetry line, which aimed to restore the shrimp populations negatively impacted by trawling. The agreement was the result of more than 2 years of negotiations between the small-scale fishers who initiated the negotiations, with support from CoopeSoliDar, R.L, and trawler fishers. The ban on shrimp capture had several important benefits. It helped shrimp populations recover both in quantity and body weight. In 2014 fishers caught nearly 5000 kilos of shrimp in their AMPR with substantial increases in weight (25–26 shrimps/kg). Moreover, some species of anchovies recovered, and some species of molluscs – which had disappeared 20 years prior – started to reappear (D. Chacón, personal communication, March, 2015). Fishers also learned to understand the importance of shrimp ecology and the value of an AMPR. They felt 'empowered by this knowledge and by the fact that they could make informed management decisions' (11.4) (D. Chacón, personal communication, March, 2015).

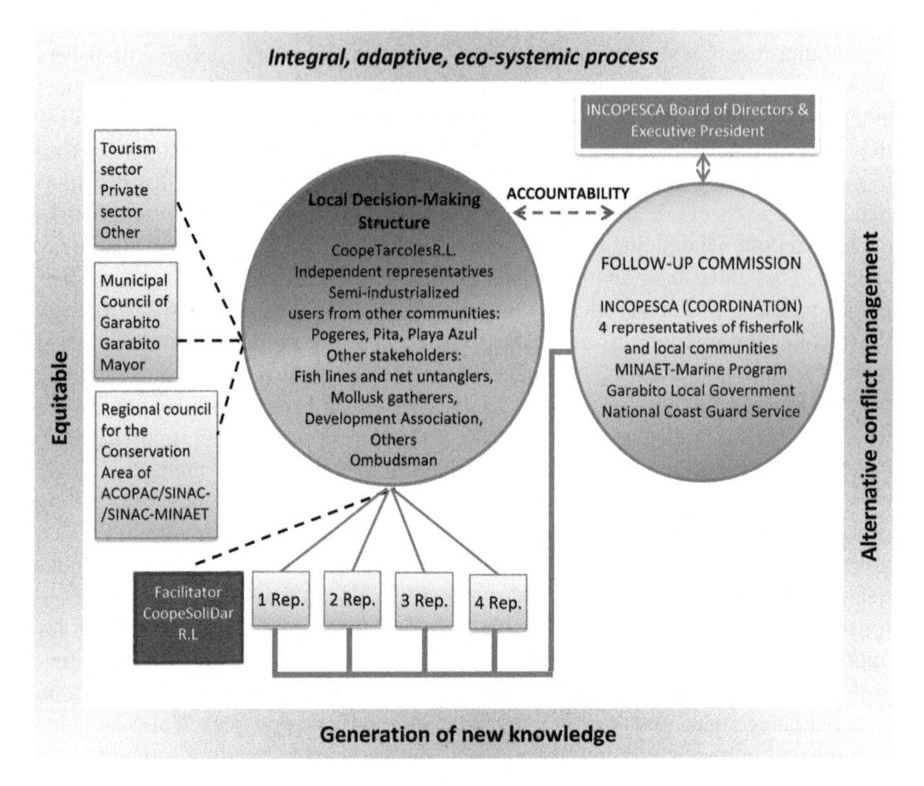


**Fig. 17.2** Proposal for a governance model for the Tárcoles AMPR (Source: CoopeSoliDar R.L 2010)

The Tárcoles AMPR covers approximately 108.8 km$^2$ (Salas et al. 2012). Sustainable fishing is regulated and practiced within six different zones there, each with specific fishing and gear restrictions (Garcia Lozano 2014). In 2007, CoopeTárcoles R.L and CoopeSoliDar R.L created a sustainable ecotourism venture, Consorcio Por La Mar R.L. The goals of this venture are to develop a model for rural community-based tourism and provide a complementary and alternative source of income for the fishing cooperative (6.8). The main product on offer with the Por La Mar consortium are guided tours that give individual or organized visitors a glimpse into the daily life of a small-scale Costa Rican fishing community. For visitors who want to spend more days in Tárcoles, there are 10 small-scale fisher families who are ready to provide accommodation and share their lives with visitors. (D. Chacón, personal communication, March, 2015).

More recently, the Tárcoles cooperative, alongside CoopeSoliDar R.L, initiated a national network of AMPRs (10.6). The group is now in the process of developing a regional network for implementation of the SSF Guidelines. In December 2015, the Costa Rican AMPR network organized a regional workshop aimed at establishing an action plan for the implementation of the SSF Guidelines both nationally and at the regional (Central American) level. The workshop produced a common declaration and a draft of an action plan, which were both discussed at the OSPESCA meeting in El Salvador (15–16 December 2015). Costa Rica's plan of action for the implementation of SSF Guidelines between 2016 and 2018 identified several priority actions aimed at promoting the implementation of articles 7, 5a, 13, and 8 of the Guidelines. Some of these actions are: training of fishers and fishers' organizations about the Guidelines; designing a national certificate for responsible small-scale fishing and for the responsible exploitation of molluscs; developing a marketing strategy for marketing national products; and working together with state institutions to ensure the quality, traceability, and safety of fishery products.

## Discussion

The human rights approach of the SSF Guidelines appealed to Costa Rica's government because these Guidelines had the potential for "eradicating poverty holistically and restoring the dignity of fishers" (G. Meneses Castro, personal communication, March, 2015), while promoting the sustainable use of the country's rich natural resources. These are the lofty goals of sustainable development (Sachs 2015). Their implementation calls for collective action based on clarifying values. It also requires collective decision-making based on knowledge sharing and creation of a common understanding concerning sustainable solutions for both environmental and social challenges.

This chapter has identified dialogue (Brugnach and Ingram 2012), based on fair procedures, as a heuristic tool to facilitate social interaction for decision-making which can encourage societal change towards sustainability. There are two levels of dialogue that have facilitated the adoption and implementation of the SSF Guidelines in Costa Rica. The first level is a nation-wide consensus about the need to restore and promote values that have traditionally shaped Costa Rican identity; the second level is a transdisciplinary dialogue aimed to design and implement sustainable solutions for small-scale fisheries' co-governance. A third level dialogue is a regional dialogue that has just been initiated.

### *Democracy and Biodiversity Conservation*

Two of the very important values that have shaped Costa Rican's identity are their rich democratic traditions and concern for biodiversity conservation. In Costa Rica there is a "deep-seated commitment to democracy which goes beyond issues of inter-personal trust" (Seligson 2001). Democracy is defined as "liberty, respect for the rule of law, and willingness to hold government accountable for its actions" (ibid). Democracy is protected by the country's Constitution of 1949 which provides for a unicameral legislature, a fair judicial system, and an independent electoral body (Elbow et al. 2016). Human rights are protected in Costa Rica both by the Constitution and the international treaties on human rights which the government has been 'eager' to sign (Quesada-Alpizar 2015). However, in 2000 the Constitutional Chamber of the Supreme Court decided to expand the scope of enforcement of those international treaties domestically when those treaties (even those not ratified by the state) bestowed greater guarantees to citizens than those contained in Costa Rica's Constitution. This shows awareness that the country's democratic traditions have been eroded and that at least the judicial system is trying to restore them.

Biodiversity preservation is embedded in the everyday life of Costa Rica. The following sign was displayed on the inner door of a five-star hotel room at the foothill of Arenal volcano: "Dear guests, in this room you might find insects. Please do not kill them, call someone from reception to deal gently with them. They belong to our rich biodiversity, we love them and we protect them" (personal observation 2015). This philosophy was embraced, since 1998, as Costa Rica's policy of biodiversity conservation. It is based on the principle of 'respect for all forms of life'. This means that "all the living things have the right to live, independently of actual or potential economic value" (Biodiversity Law 1998). This philosophy was later refined to show that biodiversity protection makes economic sense, as "protected areas and the biodiversity which exists within and outside them represent economic uses which can generate multiple benefits" (SINAC 2014). In 2009, national parks and biological reserves contributed $1357 million to the socio-economic development of the country, 5% of the Costa Rican GDP, and "provided livelihoods for indigenous people and for people in other geographic areas" (SINAC 2014).

While there is a wide national consensus about the value of biodiversity conservation, a consensus is lacking on the best governance model to achieve it, and the most equitable way of sharing its benefits. Despite numerous attempts at integrative governance, Costa Rica is still stuck with the country's tradition of centralized decision-making vis-a-vis the management of natural resources. The country's Constitution does not recognize the concept of co-management of protected areas as the "identification, development and implementation of strategies, plans and budgets concerning conservation areas are considered exclusive powers of the State" (DFOE-AM 2005). However, new regulations based on the recognition of local and indigenous communities' contribution to conservation and sustainable use of natural resources are slowly breaking the government's exclusive monopoly on management of natural resources.

In 2009, two new management categories for MPAs were established: marine reserves and marine management areas, both aimed at conservation of ecosystems and habitats and the sustainable use of coastal and marine resources to "satisfy the needs of human populations and their quality of life" (MINAET 2009). A recent regulation addresses the problem of natural resource governance by recognizing four models of governance in protected areas: state governance, co-governance, private governance, and governance by indigenous people and local communities. In the co-governance model, "various actors, from public administration or outside it, formally and informally, share responsibilities, decision-making and the benefits" of protected areas (Decreto 2016). The document is important as it establishes, among the principles of governance for protected areas, 'the legitimacy and voice' of social actors in a 'transparent dialogue based on mutual respect and efforts to reach solutions through consensus' and 'respect for justice and rights', including the human rights of local communities and indigenous populations.

## *Fisheries Co-governance*

Our study shows that the 'transparent dialogue based on mutual respect' between government agencies and small-scale fishers did not come easily. Small-scale fishers needed to learn how to act as dialogue partners and regain trust in government institutions. They tended to be distrustful of government agencies as these agencies tended to represent the interests of large-scale fishers. Indeed, INCOPESCA has shown a "lack of responsiveness to fishers' and 'poorly enforced and implemented the regulations" (Garcia Lozano and Heinen 2015). This changed in 2014 when Gustavo Meneses Castro became Chief-Executive of INCOPESCA. He made a commitment to 'bring some justice and equity to the fisheries' by initiating a "public policy [that] promotes the development of decent living conditions and the human well-being of the coastal and seafaring communities" (Meneses Castro 2014). He understood that the government would have to work with small-scale fishers to solve the difficult problems of fish stock sustainability and community resilience. A small-scale fishery representative was invited to sit on the board of INCOPESCA (D. Chacón, personal communication, March, 2015).

Small-scale fishers started acquiring 'legitimacy and voice' as social dialogue partners when they decided to use and share their fishing knowledge not only with government agencies but with civil society organizations as well. The Tárcoles case study shows the significance of the small-scale fishers' cooperative strategic alliance with a like-minded civil society organization, CoopeSoliDar R.L. This cooperative, "a voluntary association of persons not of capitals" (CoopeSoliDar R.L. 2016), was initiated in 2000 by a group of professionals from different disciplines, interested in promoting both development and environmental conservation, mostly for local communities. By working with the fishers' cooperative, both institutions built social capital (trust) and produced transdisciplinary knowledge which is more socially useful than academic knowledge for solving sustainability problems (Gibbons et al. 1994). This knowledge has been used by small-scale fishers in their negotiations with semi-industrial fishers regarding the cessation of shrimp trawling. It has also been used in designing the AMPR model, an original and innovative co-governance model, which has been a breakthrough in participative management of small-scale fisheries. The model was conceived and built by the two cooperatives which carefully followed the FAO discussions on the Guidelines and the discussions in ICSF, the international NGO in support of fish workers (V. Solis Rivera, personal communication, August 2016). CoopeSoliDar R.L has been part of the FAO discussions related to the Guidelines' development and implementation since 2010 (FAO Workshop 2014). The AMPR model has been negotiated and finalized by a joint working committee that included government representatives, representatives of CoopeTárcoles R.L, and CoopeSoliDar R.L, and representatives of other NGOs involved in marine conservation. Considering the way it was conceived and developed, i.e. by many stakeholders, and the features it includes, the AMPR model is the perfect tool for implementation of the SSF Guidelines in Costa Rica.

An important finding of this research is that Tárcoles' small-scale fishers have participated in the dialogue not only in terms of articulating their 'voice', but also as initiators of the dialogue who were able to exercise some degree of control over the decision-making process. This might have been the result of the fact that fishers had the experience of being organized as a small-scale fishing cooperative, interested not only in the wellbeing of its members but also in the long-term sustainability of its coastal and marine resources. The AMPR model legitimizes the role of small-scale fishers in decision-making processes concerning management and monitoring of the use of the ocean, and recognizes their rights to have jobs, and to enjoy the bounty of their responsibly managed coastal and marine areas, either in fishing or in tourism activities. With these characteristics, the AMPRs have the potential to implement various aspects of the SSF Guidelines (5.3, 5.6, 5.15, 5.16, 5.17, 12.4, 13.5) in small-scale fishing communities throughout Costa Rica.

A significant step in the process of the implementation of the SSF Guidelines has been the establishment of a network of AMPRs in 2014. The network provides 'an organizational space' where small-scale fishers can consolidate the practices of responsible fishing that can contribute to improving the quality of life for their families (Declaracion 2014).

Costa Rica's model of SSF Guidelines implementation seems to be working. There is appropriate legislation in place, and enthusiasm among small-scale fishers about the SSF Guidelines. One fisher said: "We are already implementing the Guidelines in my community and at the national level. We started with a 3 day workshop with a dynamic team, which included eight fishermen and fisherwomen from the whole country. It was a very important workshop; the first thing we did was to believe that fishermen could train other fishermen. Also understand that the Guidelines are ours, the artisanal fishers." (Matarrita 2015). But there is also frustration at the difficulty of developing management plans required for an AMPR's approval and lack of support from the government (R. Rojas, personal communication, February, 2015). As the process of implementing the SSF Guidelines matures, real social inclusion that goes beyond dialogue to embrace deeper changes in power relations through transformative social policies (UNRISD 2014) will need to happen for small-scale fishers in Costa Rica. Effectively protecting the human rights of small-scale fishers is an important investment in making their ecological and economic future sustainable. It will also boost Costa Rica's democracy.

## Conclusion

Costa Rica has become a champion for implementing the SSF Guidelines. The government has listened to small-scale fishers and taken note of their desire to use sustainably their coasts and marine areas for the benefit of their generation and future generations. It was a right decision, whose value can be seen with every step of the implementation process. Its value will become more obvious as food insecurity and climate change present greater challenges to countries in the future. Marine areas for responsible fishing, the manner in which the Guidelines have been implemented, are an example of an original, innovative co-governance model that gives small-scale fishers access to decision-making and monitoring of the sustainable use of their AMPR. The establishment of a network of AMPRs in Costa Rica as well as the incipient regional dialogue are encouraging signs that the model can proliferate in Central America. However, not all countries in the region have Costa Rica's rich democratic tradition and concern for biodiversity conservation. Moreover, to remain a champion, Costa Rica needs to do more for effectively protecting the human rights of its small-scale fishers.

## References


Alvarado, J. J., Cortés, J., Esquivel, M. F., & Salas, E. (2012). Costa Rica's marine protected areas: Status and perspectives. Revista de Biología Tropical. (Int. J. Trop. Biol. ISSN-0034-7744), *60*(1), 129–142.

Baxter, P., & Jack, S. (2008). Qualitative case study methodology: Study design and implementation for novice researchers. *The Qualitative Report, 13*(4), 544–559.

Biodiversity Law. (1998). *Costa Rica Ley 7788*. http://theredddesk.org/countries/laws/biodiversity-law-law-7788-1998-costa-rica. Accessed 12 Feb 2016.

Brugnach, M., & Ingram, H. (2012). Ambiguity: The challenge of knowing and deciding together. *Environmental Science and Policy, 15*(1), 60–71.

Bruning, S., Dials, M., & Shirka, A. (2008). Using dialogue to build organization-public relationships, engage publics, and positively affect organizational outcomes. *Public Relations Review, 34*(1), 25–31.

Buber, M. (1970). *I and thou* (W. Kaufmann, Trans.). New York: Charles Scribner's Sons.

CIZEE (Comisión Interinstitucional de la Zona Económica Exclusiva). (2008). *Estrategia nacional para la gestión integral de los recursos marinos y costeros de Costa Rica*. San José, Costa Rica.

CoopeSoliDar R.L. (2005). *Sueños de mar: una fantástica historia de la vida real, ilustrada por los niños de la Escuela de Tárcoles*. Recop. de Ana Coralia Fernández; il. por Carla Amador. – 1 ed. – San José, Costa Rica.

CoopeSoliDar R.L. (2010). *Plan de ordenamiento de la pequeña pesquería. Area marina de pesca responsable de Tárcoles*. CoopeSoliDar R.L., CoopeTárcoles R.L., Incopesca,1 ed. – San José, C.R.

CoopeSoliDar R.L. (2016). http://www.coopesolidar.org/index.php?option=com_content&view=article&id=22&Itemid=29. Accessed 21 Oct 2016.

Davis, D. H. (2014). *Comparing environmental policies in 16 countries*. Boca Raton: Taylor and Francis Group/CRC Press.

Declaracion. (2014). *Constitucion de la Red de Areas Marinas de Pesca Responsable de Costa Rica*, San José.

*Decree No.27919/MAG-1999 adopting the Code of Conduct for Responsible Fisheries*.

*Decree No.35260/MAG Declaratoria de política nacional de pesquera* (2009).

*Decree No. 35502/MAG* of August 3 2009, allows fishery products caught inside responsible fishing marine areas be commercialized.

*Decree A.J.D.I.P./193–2011* published in La Gaceta No. 159 on 19 August, 2011.

*Decreto*. (2015). *N° 39195 MAG-MINAE-MTSS*, La Gaceta N° 184, 22 Setiembre 2015.

Decreto. (2016). *Decreto ejecutivo No. 39519-MINAE, DAJ-D-010-2016.*

DFOE-AM. (2005). *Informe sobre los resultados del estudio especial efectuado en el Ministerio del Ambiente y Energia sobre el "comanejo" del Parque Nacional Marino Ballena*. Reporte DFOE-AM -38-2005 San José, Costa Rica.

Elbow, G. S., Karnes, T. L., Parker, F. D., & Stansifer, C. L. (2016). *Costa Rica in Encyclopedia Britanica*. https://www.britannica.com/place/Costa-Rica. Accessed 21 Oct 2016.

Elizondo Mora, S. (2005). *Pesca y procesos de trabajo. El caso de los pescadores de isla Cabbalo, Golfo de Nicoya, Costa Rica*. Disertación, Universidad de Costa Rica, San José.

FAO. (1995). *Code of conduct for responsible fisheries*. Food and Agriculture Organization of the United Nations Rome, FAO. http://www.fao.org/docrep/005/v9878e/v9878e00.htm. Accessed 6 Dec 2016.

FAO. (2004). *Fisheries country profile Costa Rica*. http://www.fao.org/fi/oldsite/FCP/en/cri/profile.htm. Accessed 2 Mar 2016.

FAO. (2014). FAO fisheries and aquaculture department. *Technical consultation on the international guidelines for securing sustainable small-scale fisheries*. Italy, Rome, 3–7 February 2014. ftp://ftp.fao.org/FI/DOCUMENT/SSF/SSF_guidelines/TC/2014/default.htm Accessed 6 Dec 2016.

FAO. (2015). *Voluntary guidelines for securing sustainable small-scale fisheries in the context of food security and poverty eradication*. Rome: Food and Agriculture Organization of the United Nations.

Fargier, L., Hartmann, H. J., & Molina-Ureña, H. (2014). Marine areas of responsible fishing: A path toward small-scale fisheries co-management in Costa Rica? Perspectives from Golfo Dulce. In F. Amezcua & B. Bellgraph (Eds.), *Fisheries management of Mexican and central American estuaries, estuaries of the world*. Dordrecht: Springer Science + Business Media.

FECOP. (2013, August). *Tuna catches and fisheries management in Costa Rica's exclusive economic zone*. A proposal for alternative development.

Garcia Lozano, A. J. (2014). *An Institutional, socio-economic, and legal analysis of fisheries co-management and regulation in the Gulf of Nicoya, Costa Rica*. Master's thesis. http://digitalcommons.fiu.edu/cgi/viewcontent.cgi?article=2645&context=etd. Accessed 2 Mar 2016.

Garcia Lozano, A. J., & Heinen, J. T. (2015). Identifying drivers of collective action for the co-management of coastal marine fisheries in the gulf of Nicoya, Costa Rica. *Environmental Management*. doi:10.1007/s00267-015-0646-2.

Gibbons, M., Limoges, C., Nowotny, H., Schwartzman, S., Scott, P., & Trow, M. (1994). *The new production of knowledge: The dynamics of science and research in contemporary societies*. London: Sage.

Huutoniemi, K. (2014). Introduction: Sustainability, transdisciplinarity and the complexity of knowing. In K. Huutoniemi & P. Tapio (Eds.), *Transdisciplinary sustainability studies. A heuristic approach*. London: Routledge.

IATTC. (2013). *Fishery status report No. 11: Tunas and billfishes in the eastern Pacific Ocean in 2012*. Inter-American Tropical Tuna Commission, California: La Jolla.

ICT Statistics (Instituto Costarricense de Turismo). (2014). http://www.ict.go.cr/es/estadisticas/informes-estadisticos.html#2014. Accessed 5 Mar 2016.

IICE (Instituto de Investigaciones en Ciencias Económicas). (2010). Universidad de Costa Rica. *Análisis de la contribución económica de la pesca deportiva y comercial a la economía de Costa Rica*. IICE-UCR.

INCOPESCA. (1999). AJDIP/241-99.

Jentoft, S., McCay, B. J., & Wilson, D. C. (1998). Social theory and fisheries co-management. *Marine Policy, 22*, 423–436.

Jiménez, J. (2013). Crisis de ecosistemas marinos y costeros en Costa Rica. *Ambientico, 230–231*, 4–8.

Kates, R. W., Clark, W. C., Corell, R., Hall, J. M., Jaeger, C. C., Lowe, I., McCarthy, J. J., Schellnhuber, H. J., Bolin, B., Dickson, N. M., Faucheux, S., Gallopin, G. C., Grübler, A., Huntley, B., Jäger, J., Jodha, N. S., Kasperson, R. E., Mabogunje, A., Matson, P., Mooney, H., Moore, B., O'Riordan, T., & Svedin, U. (2001). Sustainability science. *Science, 292*, 641–642.

Kent, M. L., & Taylor, M. (2002). Toward a dialogical theory of public relations. *Public Relations Review, 28*(1), 21–37.

Klein, J. T. (2010). A taxonomy of interdisciplinarity. In J. T. Klein & C. Mitcham (Eds.), *The oxford handbook of interdisciplinarity*. Oxford: Oxford University Press.

Ledingham, J. A., & Brunning, S. D. (2000). *Public relations as relationship management: A relational approach to the study and practice of public relations*. Hillsdale: Lawrence Erlbaum.

Matarrita, S. M. (2015). *Sonia María Matarrita is Costa Rica representative of the network of responsible fisheries areas*. Puntarenas.

Meneses Castro, G. (2014). *A future commitment. Costa Rica is working towards a national policy for implementing the SSF Guidelines recently adopted by the Food and Agriculture Organization of the United Nations* (SAMUDRA Report 69).

MIDEPLAN (Ministerio de Planificación Nacional y Política Económica). (2014). *Plan Nacional de Desarrollo 2015–2018* "Alberto Cañas Escalante"/Ministerio de Planificación Nacional y Política Económica. San José.

MINAET. (2009, July 20). *Executive Decree No 35369*, in *La Gaceta* No. 139.

OSPESCA. (2010). *Encuesta Estructural de la Pesca Artesanal y la Acuicultura en Centroamérica, 2009*, La Libertad, El Salvador.

OSPESCA. (2012). *Encuesta Estructural de la Pesca Artesanal y la Acuicultura en Centroamérica, 2009–2011*.

Paloniemi, R., & Vainio, A. (2014). Understanding environmental euristics: Trust and dialogue. In K. Huutoniemi & P. Tapio (Eds.), *Transdisciplinary sustainability studies. A Heuristic approach* (pp. 158–174). London: Routledge.

Pearson, R. (1989). *A theory of public relations ethics*. Unpublished doctoral dissertation, Ohio University. Quoted from Kent and Taylor (2002).

Quesada Alpizar, M. A. (2006). Participation and fisheries management in Costa Rica: From theory to practice. *Marine Policy*, 641–650. doi:1016/j.marpol.2005.09.001.

Quesada-Alpízar, T. (2015, October 19). Rights in vitro: What Costa Rica's IVF struggle means for democracy. *The Tico Times, news*. http://www.ticotimes.net/2015/10/19/rights-in-vitro-what-costa-ricas-ivf-struggle-means-for-democracy. Accessed 21 Oct 2016.

Sachs, J. D. (2015). *The age of sustainable development*. New York: Columbia University Press.

Salas, E., Ross, E., & Arias, A. (2012). *Diagnóstico de áreas marinas protegidas y áreas marinas para la pesca responsable en el Pacífico costarricense*. San José: Fundación MarViva.

Seligson, M. A. (2001). *Costa Rican exceptionalism. Why the Ticos are different, in citizen views of democracy in Latin America*. In: R. I. Camp (Ed.), Pittsburgh: University of Pittsburgh Press, pp. 90–106.

SINAC. (2014). *Informe nacional al convenio sobre la diversidad biológica, Costa Rica*. GEF-PNUD, San José, Costa Rica.

Solis Rivera, V., Borrás, M. F., Gallardo, D. B., CoopeSoliDar, R. L., Ochoa, M., Castañeda, E., & Castillo, G. (2013). *Regional study on social dimensions of MPA practice in central America: Cases studies from Honduras, Nicaragua, Costa Rica and Panamá*. Monograph. Edited by CoopeSoliDar, R. L, Chennai, India: International Collective in Support of Fishworkers. www.icsf.net

Solis Rivera, V., Muñoz Rivera, A., & Fonseca Borrás, M. (2015). *Integrating traditional and scientific knowledge for the management of small scale fisheries: An example from Costa Rica, in Fishers' knowledge and the ecosystem approach to fisheries. Applications, experiences and lessons in Latin America* (FAO Fisheries and Aquaculture Technical Paper 591, Rome).

State of the Nation Report. (2013). *Costa Rica*. http://www.centralamericadata.com/en/article/home/Costa_Rica_State_of_the_Nation_Report_2013. Accessed 21 Oct 2016.

Trujillo, P., Cisneros-Montemayor, A. M., Harper, S., Zylich, K., & Zeller, D. (2015). *Reconstruction of Costa Rica's marine fisheries catches, 1950–2010*. Fisheries Centre, University of British Columbia, Working Paper #2015–31.

UNDP. (2012). *United Nations Development Programme. CoopeTárcoles, Costa Rica. Equator Initiative Case Study Series*. New York.

UNICEF. (2012). *At a glance: Costa Rica*. http://www.unicef.org/infobycountry/costarica_statistics.html. Accessed 12 Mar 2016.

UNRISD. (2014). *Social inclusion and the Post-2015 sustainable development agenda*.

Yin, R. K. (2003). *Case study research: Design and methods* (3rd ed.). Thousand Oaks: Sage.

# Chapter 18
# Furthering the Implementation of the Small-Scale Fisheries Guidelines: Strengthening Fisheries Cooperatives in Sri Lanka

**Oscar Amarasinghe and Maarten Bavinck**

**Abstract** This chapter proposes that fisheries cooperatives can play an important role in furthering the implementation of the Voluntary Guidelines for Securing Sustainable Small-Scale Fisheries (SSF Guidelines) in Sri Lanka. These organizations have a long history of supporting the fisheries sector, both in northern and southern Sri Lanka, with strong contributions to fisher wellbeing and the functioning of the value chain. Their involvement in resource management, however, is still relatively minor. The authors evaluate the performance of cooperatives against the outcomes of the South Asia consultation on the SSF Guidelines (November 2015). They argue that while fisheries cooperatives in Sri Lanka have many weaknesses, they are uniquely positioned to aid the small-scale fisheries sector. The chapter concludes with a set of recommendations for improving the performance of cooperatives.

**Keywords** Cooperatives • Governance • Rights • Management • Value chain

## Introduction

Cooperatives have a long and chequered history in Sri Lanka's fisheries. They have functioned as a prime governmental policy channel for the dissemination of material and immaterial benefits to the coastal population since Independence (1948). They have also become important agents of self-governance. Although the contribution of cooperatives has not always been judged positively, and their roles have varied from one region and one time period to another, we argue that they are an important party to the implementation of the SSF Guidelines in Sri Lanka.

O. Amarasinghe (✉)
University of Ruhuna, Matara, Sri Lanka
e-mail: oamarasinghe@yahoo.com

M. Bavinck
University of Amsterdam, Amsterdam, The Netherlands

University of Tromsø, Tromsø, Norway
e-mail: j.m.bavinck@uva.nl



S. Jentoft et al. (eds.), *The Small-Scale Fisheries Guidelines*, MARE Publication Series 14, DOI 10.1007/978-3-319-55074-9_18

This chapter commences with a brief summary of the South Asia consultation on the implementation of the SSF Guidelines that took place in the latter months of 2015 in Colombo. Sections "Small-scale fishing in Sri Lanka" and "Fisheries cooperatives in Sri Lanka" describe the small-scale fisheries and the cooperative system as it functions currently in Sri Lanka, taking note of the differences that existed between northern and southern Sri Lanka as a result of the civil war (1983–2009). Section "Role of cooperatives in implementing SSF Guidelines" then examines the contribution cooperatives could make to the implementation of the SSF Guidelines, referring back to the South Asia consultation. The final section summarizes the argument and points out necessary action required for cooperatives to play a more effective role.

## The South Asia Consultation

The *South Asia FAO-BOBLME Regional Consultation on the Implementation of the Voluntary Guidelines for Securing Sustainable Small-Scale Fisheries in the Context of Food Security* and Poverty Eradication (henceforth the SSF Guidelines) was held in Colombo, Sri Lanka, between the 23rd and 26th November 2015. The objective of the workshop was to raise awareness about and support the implementation of the SSF Guidelines in the region. An array of public, private, and civil society actors highlighted the status of fisheries in Pakistan, India, Bangladesh, Maldives, Indonesia (observer), and Sri Lanka. The major issues highlighted by all participants included the need to promote sustainable use of fisheries resources, promote participatory decision making and management, empower small-scale fishers, provide them with market access, enforce laws, and protect aquatic resources. Moreover, emphasis was laid on gender concerns, especially the need to empower women. All participants stressed the importance of identifying and recognizing the rights of fishers. Some of the important considerations that emerged during discussions included the importance of the ecosystem approach to fisheries management, the need for engagement of fishing communities in decision-making, integration of research outputs into policy, and capacity development of all parties concerned in the implementation of the SSF Guidelines. Several voids in fisheries research, especially in the area of small-scale fisheries development, were also identified, such as the need to find out the most appropriate mechanisms for intervention, and addressing issues of legal pluralism.

## Small-Scale Fishing in Sri Lanka

Being surrounded by sea, it is no surprise that fisheries play an important role in Sri Lankan society. Sri Lanka's population consume large amounts of fish (an average of 10.8 kg/year per person), while 6.3% of the country's labour force finds direct or indirect employment in the fishing industry. The marine fishing population consists of 190,000 households and 221,000 active fishers spread out along the coastline, while another 199,000 fishing households and 51,000 active fishers fish in lagoons and inland water bodies (Ministry of Fisheries 2015) All of these fishers, it could be argued, engage in small-scale fishing, as the size of vessels is invariably small. In this chapter, however, we choose to limit ourselves to fishers who employ beach-landingcraft and whose fishing operations normally last no longer than 1 day.[1] This excludes the multi-day, harbor-based fishing fleet,[2] but includes the beach seining industry that operates along the coast. Small-scale fishers in Sri Lanka are of many kinds. While some are beach-based, others use a variety of craft – *orus*, *kattumarams*, or fiber-glass boats – to ply nearshore, offshore, lagoon, or inland fishing grounds. Gears include a variety of gillnets, cast nets, long lines, fixed nets, and traps. While men generally dominate the harvesting process, women play an important part in processing, marketing, and in support activities.

Fishers inevitably belong to communities, sometimes segregated physically from other professional groups, such as farmers, and sometimes intermingled with them. Caste plays a role, both in Sinhala and in Tamil society, with the *Karava* dominating in the south, and the *Karaiyar* in the north. Muslims play an important role in fishing in the west and east. But only a fraction of the *Karava* presently engages in fishing, and in the north there are other fishing castes too (Scholtens 2016). Finally, while all small-scale fishers have permanent abodes, they frequently also engage in camp-based, seasonal migration to other coasts (Fig. 18.1).

**Fig. 18.1** Small-scale fishing crafts (Vallum) beached at Jaffna Lagoon in the North of Sri Lanka

[1] Representing nearly 87% of the fisher population operating 90% of the fishing fleet.

[2] For more information on harbour-based, multi-day fishing, see Amarasinghe et al. (2005a).

There is substantial evidence of fisher engagement in resource and market management. Alexander (1982) documents the complex system of beach-seine rotation in southern Sri Lanka and patron-client relations that prevail (also see Amarasinghe 1989). Stirrat (1988) analyses fish marketing systems in migratory and non-migratory settings of western Sri Lanka. Wickramasinghe and Bavinck (2015) highlight the informal, beach-based institutions that boat fishers in southern Sri Lanka devise to regulate their operations, and Bavinck (2015) describes the fisheries tenure system that has re-emerged from the ruins of war in Jaffna District. All of these customary institutions are predicated on the control of marine space, and include territorial rights.

Unlike in south India, where customary organizations, such as caste councils (*or panchayats*) or sea courts (*kadakkodis*) still play a prominent role in fisheries (Kurien 2000; Bavinck 2001), the rules and regulations found in Sri Lanka do not have a firm, customary anchor. On the one hand, rules and regulations were established with the involvement of the state and, on the other their activities are also regulated to some extent by the state. To a large extent, state laws are embedded in fisheries cooperatives, as we will show in greater detail in the following section. Nevertheless, cooperatives constitute the only fisheries-related organization at the local level.

Before moving to the topic of cooperatives, we must point out the influence of the civil war that plagued the country from 1983 to 2009 on developments within the fisheries sector. Civil war erupted in Sri Lanka when the guerilla forces of the Liberation Tigers of Tamil Eelam (LTTE), started to fight the Sri Lankan government demanding a separate Tamil State in the Northern and Eastern provinces of Sri Lanka claiming that these areas formed the traditional homeland of the minority Tamil population. During the civil war, when the Northern and Eastern Provinces were largely controlled by the LTTE and thus located in the war zone, fishing was virtually impossible (Scholtens et al. 2012). The fishing population of this region went into internal or external exile, or led a hand to mouth existence. Although the government maintained an administrative presence even in guerilla-held territories, its control was limited. It was the LTTE that largely called the shots. A difference thus emerged between fisheries patterns in the southern districts and those in the north, where the fishing industry has been rebuilding itself only since 2009. This is most evident in the performance of fishing cooperatives.

## Fisheries Cooperatives in Sri Lanka

Sri Lanka has a strong post-Independence history of cooperatives. Cooperatives are defined by the International Cooperative Alliance as: "An autonomous association of persons united voluntarily to meet their common economic, social, and cultural needs and aspirations through a jointly-owned and democratically-controlled enterprise (ICA, n.d.)." This definition emphasizes what could be seen as the 'bonding capital function' (see Woolcock 2001) of cooperative organizations. But others point out that: "Cooperatives can also be isolated, and may, like other local groups, be high in 'bonding social capital', but not be able to find the "bridging social capital" that will link them to others" (Birchall 2004, 47). Many of the cooperatives found in developing countries have actually been instigated by government. This is often perceived of as a crucial weakness, and even contrary to the essence of the cooperative movement (Jentoft 1986). The Overseas Cooperative Development Council (OCDC 2007, 4) thus concludes that: "Government-controlled parastatals are not true cooperatives."

Indeed fisheries cooperatives in Sri Lanka do not have an unblemished history. While published information on the functioning of cooperatives is quite scanty, Amarasinghe (1995) demonstrated that cooperatives were used in the early days (1960s and 1970s) by politicians to give favors to their political clientage by channeling public goods to them. Generally, when there was a change of government, some existing cooperatives were dissolved and new office bearers who had links to the political party in power elected. This also meant that cooperatives with a 'colour' different from that of the party in power did not receive any government assistance. The system was corrupt and many cooperatives collapsed due to lack of faith and trust expressed by the membership. The present cooperatives evolved through this process. Today, they are more self-sustaining.

Fisheries cooperatives in Sri Lanka are organised through the collaborative intervention of the Department of Cooperative Development and the Department of Fisheries and Aquatic Resources, which make them 'formal organisations.' Many of the fishing cooperatives in Sri Lanka can be characterized as multipurpose, combining functions such as the provision of credit, technology, and insurance; and occasionally, the organization of marketing. The Blue Revolution initiated in the post-WWII period constituted an important stimulus to the formation of fisheries cooperatives, as the Sri Lankan government strove to channel subsidies and credit to enable asset-poor fishers to adopt new capital-intensive technology (Amarasinghe 2005). What is important to note is the fact that membership in cooperatives is in principle voluntary and that individual cooperatives enjoy a great freedom in planning, organizing, and implementing activities aimed at meeting the diverse needs of the community. Today, the activities of fisheries cooperatives are guided by the Cooperative Societies Act No. 5 of 1972 and the Fisheries Cooperative Constitution.

Fisheries cooperatives throughout the country are therefore basically alike in structure and activities. However, it is to be noted that fishing activities in Sri Lanka have a distinctive geographical flavor to them and are shaped by ethnicity and religon to a significant extent (Tamil fishers in the north, Muslim fishers in the east, Sinhala Catholic fishers in the west, Sinhala Buddhist fishers in the south, etc.). Therefore, within a certain cooperative, the membership is highly homogeneous with respect to ethnicity and religion, leaving little room for internal conflicts. Moreover, members in all fisheries cooperatives are small scale and artisanal fishers while those engaged in offshore fishing, especially multiday fishing, have their own boat owner associations. We must note that because of the 25 years of civil war cooperatives in the Northern and Eastern Provinces have undergone a radically different trajectory than those in other parts of the country. In the Northern Province, at least, the civil war has meant that cooperatives are now still more tightly organized, and nested in regional organizations, than elsewhere (Scholtens 2016). In northern Sri Lanka, cooperative membership is still considered a 'must' for being involved in fisheries at all, and membership rates are concomitantly high.

By 2010, 579 fisheries cooperatives were functioning in Sri Lanka (Ministry of Fisheries and Aquatic Resources, personal communication, 20th May, 2012). The total membership of fisheries cooperatives (active and defunct) was 87,895 persons, which is roughly equivalent to a fifth of the total fishing population. Some reasons for the less than optimal rate of membership are poor management, lack of interest among office bearers, inadequate training of personnel in business management, lack of awareness among members of principles of cooperation, and poor loan recovery rates (Amarasinghe and Bavinck 2012). Some of the best functioning fisheries cooperatives, with high rates of savings and lending, have been restructured as Fisheries Cooperative Banks (*Idiwara* banks). There were about 107 such banks in operation in 2010.

A notable change in the history of fisheries cooperatives in Sri Lanka was the large-scale withdrawal of state assistance to fisheries cooperatives in 1994 because of the prioritization of defence expenditure. This would have made some cooperatives defunct or dormant, though data is not available to prove this point. Moreover, repeated changes in government policy have had serious impacts on fisheries cooperatives. Such changes have occurred twice since the new millennium, first in 2004 and then in 2010. In aiming for effective management of fisheries resources, the Ministry of Fisheries introduced, in 2004, a new type of community organization called the 'Landing Site Management Committees.' However, this program was improperly implemented, resulting in the disintegration of some of the existing cooperative societies. The second threat to fisheries cooperatives came in 2010 when the Ministry of Fisheries, largely for political reasons, established a parallel, multi-layered system of Rural Fisheries Organisations (RFOs).

Interestingly, a number of cooperatives studied by the first author in southern Sri Lanka were functioning well prior to 1994, and continued to do so without government assistance thereafter. Strong levels of social capital are one possible reason for self-sustenance (Amarasinghe and Bavinck 2012). Strong interpersonal relationships among members (bonding social capital) ensured continuity of operations with satisfactory levels of savings, which allowed them to effectively deal with risks inherent in fisheries, inadequately developed markets, and other vulnerabilities. Informational problems associated with insurance have been minimized by the high degree of personal -knowledge that exists among members. Further, collateral problems in the 'credit market' have been resolved through 'group guarantees'. High unpredictability of fish catches, and risks and damage to fishing equipment causing income shortfalls have been managed through 'instant loans.' It has also been pointed out that various loan schemes performed both a credit and an insurance function, hedging fishers against a host of risks and uncertainties. However, success of cooperatives has also been related to their ability to horizontally link up with other societies (bridging social capital) and closely associate with supra-local agencies such as government, NGOs, and development agencies (linking social capital). Apart from their engagement in fisheries related activities and services, they have also been involved in providing the membership with a range of livelihood capitals (provision of credit facilities to take up other self-employment activities and training and capacity building by linking with diverse state and non-state organisations). This has helped them meet cultural and religious aspirations through the organization of festivals and promotion of women's involvement in the sector. This is the reason why people have faith in cooperatives and tend to save with and promote them. Amarasinghe and Bavinck (2012) argued that the utility of cooperatives for coastal populations is because they have successfully addressed core vulnerabilities of fishers. Institutions allow social actors to accomplish things that they otherwise could not (Jentoft 2005). In contrast, when organizations are uni-task, such as RFOs, which only channel limited state fisheries assistance to their membership, it is easier for interested parties to hijack them for their own advantage, especially to meet their political ends.

Probably the greatest shortcomings of fisheries cooperatives in Sri Lanka are found in fish marketing and resource management. Fisheries cooperatives have rarely got involved in marketing and resource management. Generally, fish marketing is in the hands of fish merchants who are specialized marketing middlemen wielding both economic and political power. The marketing networks include local fish assemblers, transport agents, wholesalers, and retailers. This is a closely knit network controlled by powerful middlemen. Some of them have even integrated all the above activities so as to benefit from economies of scale and also control the market share. Although some cooperatives have tried to engage in wholesale fish marketing by forming cooperative unions (or federations), they failed due to various forms of power exerted by merchants, limiting entry into fish marketing Cooperatives have also failed to engage effectively in resource management. In fact, the need for resource management was felt quite late, i.e. after the Blue Revolution in the 1960s and 1970s. Signs of increasing fishing pressure, use of destructive techniques, and declining resource health began to appear by the early 1990s, which finally led to the formulation of the Fisheries and Aquatic Resources Act of 1996. Here we witness a situation where the old community laws were under pressure from the forces of modernisation. The new market opportunities have prompted people to extract more and more returns from common property resources causing degradation of these resources. Amarasinghe (2009a) and Amarasinghe and Bavinck (2012) have shown that fisheries cooperatives have actually induced entry into fisheries by providing both physical and financial capital, resulting in high rates of exploitation. Thus, state intervention in fisheries could be seen as a means of protecting resources from further degradation, or a means of achieving sustainable use of fisheries resources. In contrast to community norms and laws, which were based on the principles of equality, wellbeing, and harmony, state laws are mainly based on the principle of resource conservation and resource health. One can also witness a time dimension in the formation of different rules and norms. Community rules in fisheries were more relevant in the pre-Blue Revolution era, while state laws in a period of modern fisheries that has witnessed population growth, market expansion, technological development, and state intervention. While in the Northern Province, evidence of fisheries cooperatives protesting vehemently against Indian trawlers invading their fishing grounds (Scholtens et al. 2012), and playing a role in the regulation of harmful technology in inshore waters are found (Bavinck 2015), these activities do not appear to be replicated in other parts of the country. In a situation of increasing fishing effort and declining resources, we suggest that the success of cooperatives in the long run depends on the will and ability to introduce more stringent management measures.

## Role of Cooperatives in Implementing SSF Guidelines

The SSF Guidelines argue for the governance of tenure in small-scale fisheries, resource management (Article 5), and improving the effectiveness of the fish value chain (Article 7). Emphasis is placed on the role that fisher associations such as cooperatives can play in addressing these concerns (Article 7 and 10).

In the sections that follow, we highlight instances where fisheries cooperatives could play an important role in implementing the SSF Guidelines: (i) the recognition and protection of customary rights to aquatic resources (Article 5.4); (ii) the adoption of measures for sustainable use of fisheries resources (Article 5.13); (iii) the promotion of participatory management systems, such as co-management (Article 5.15); (iv) the need for integrated and holistic approaches, including cross-sectoral collaboration (Article 6.1); (v) the acknowledgement of the small-scale fisheries post-harvest subsector and the role its actors play in the value chain (Article 7.3); (vi) the inclusion of women and marginalised groups (found in Articles 5.15, 7.2, 8.3); ); and (vii) risk management (Article 9.3).

### *Recognition of Customary Rights*

The Sri Lankan government and the constitution recognize only one legal system in fisheries: that of the state. Customary law is not acknowledged, except where it has been incorporated into state law. The consequence of this is, as in many other post-colonial countries where colonial law was imposed on existing bodies of law, that there exists a body of norms and regulations outside the state, which is relevant to fisheries yet not formally acknowledged, protected, and respected. Thus, Wickramasinghe (2010) demonstrates convincingly the existence of an extensive body of customary norms and regulations applied to fisheries along the south coast, which include norms and regulations pertaining to jurisdiction over marine and coastal territory. These often conflict with state rules. Bavinck (2015) provides a briefer, yet similar account, for the north coast where fishers claim authority over who is allowed to fish and by which means. Fisher rulings tend to prevail on the ground as government enforcement is often weak.

Instances of conflicts among community and government actors often follow from the failure of the state governors to understand the existence of other legal systems, especially those at the community level. Thus, clashes may emerge between the state and communities which at times could lead to intense conflicts and violence (Peramunagama and Amarasinghe forthcoming). The non-acceptance of state law is evident from incidents such as the use of dynamite all around Sri Lanka, use of wing nets mounted on galvanized pipes in Mannar, use of purse seines in near shore waters in the southern coast, expansion of brush pile fisheries in Jaffna and catching lobsters during the breeding season in Hambantota. Although these conflicts are not always related to the availability of fisher law, they do suggest the limitations of the governmental regulatory framework.

In addition, the perceptions fishers have of government officials in the field of fisheries are poor at best. Field studies carried out in the south of Sri Lanka to find out what fishers considered as the most important relationships in fishing revealed that, although fishers considered relationships with the Department of Fisheries as very important (ranking 'very high'), their satisfaction with the relationship was ranked 'very low'. The main reason for this lack of satisfaction appeared to be the perceived unwillingness of government officials to listen to fisher opinions.

Fishing cooperatives sometimes align with customary law by protecting the rights of access to resources and controlling entry of outsiders. No person outside a fishing village is allowed to anchor his craft on their beach (community barriers to entry). Moreover, cooperatives also preserve the rights to beach seine sites and state actors do not intervene. Some cooperatives, for example in Godawaya, South Sri Lanka, even control access to the beach by tourists to prevent any nefarious activity from taking place which are deemed adverse to the culture of the community.[3]

## *Sustainable Management of Resources*

The SSF Guidelines argue for measures that realize long-term conservation and sustainable use of fish resources (Article 5.13). Customary fisher law in Sri Lanka generally aims to regulate the distribution of access to and income from marine fisheries between different categories of users, also taking future use into consideration. While government has been most interested in the development potential of fisheries, it is now cognizant of the need for sustainable resource management.

Lagoon fisheries provide an indication of the possible resource management potential of Sri Lanka's marine fisheries cooperatives. These fisheries were earlier managed solely by fisheries cooperatives until the government started forming co-management platforms such as the 'Lagoon Fisheries Management Authorities' in 1996 (CCD 1996). Prior to this the cooperatives completely controlled access to resources and ensured that they are used in a sustainable manner. Amarasinghe et al. (1997) reported on a centuries old stake-net community-based fisheries management system in the Negombo lagoon, where three communities of fishers shared fishing days and members of each community shared fishing sites using a lottery system. The system not only prevented further entry into the fishery but also ensured better management through the invocation of traditional rights to the fishing sites of the three communities.

However, sustainable management of fisheries is not something that fisheries cooperatives have explicitly addressed in the past. While use of environmentally

[3] Field studies of a research project titled "Tangled in their own (safety) nets- Resilience, adaptability and transformability in small scale fisheries in a world fisheries crisis", a project initiated by the Institute of Development Studies (IDS), University of Sussex, UK. (2012–2015) 2006 (Resilience Study).

unfriendly gear by members has been discussed and users warned at monthly meetings, there is hardly any evidence of cooperatives adopting effective measures against resource degradation. Probably, fisheries cooperatives, as community institutions, do both operate in a vacuum and are embedded in social networks, which themselves are institutions (with institutions being both the rules of the game and the organizations implementing them, see Jentoft 2005). It has been shown that such institutions rest on the foundations of principles of the peasantry; equality, harmony, reciprocity, etc. which might prevent the cooperatives from taking action against some rule breakers (Amarasinghe 2009b). Key informant discussions revealed that fisheries cooperatives often welcome the idea of co-management, where government actors intervene in managing resources that cooperatives are unable to manage effectively (Ranjith, personal communication, 8th July, 2014).

## Participatory Management

The SSF Guidelines call upon states to facilitate, train, and support small-scale fishing communities to participate in and take responsibility for the management of resources (Article 5.15). The recent efforts made by the Ministry of Fisheries and Aquatic Resources to establish Fisheries Co-Management Committees, especially in lagoons, under provisions made in the Fisheries and Aquatic Resources Act of 1996, with the participation of a wide variety of stakeholders including fishers (men/women) can be considered as an important step towards establishing effective interactive platforms. However, it is premature to make an assessment of the effectiveness of these committees. Bavinck et al. (2013) investigate the potential for co-management in various South Asian fisheries characterized and suggest that the bridging of governance efforts by state authorities and fisher organizations is more than a technical matter. The success of this process depends on proper interaction between community, state, and civil society actors where the participating actors recognize each other's rights and responsibilities, as well as their legal systems.

## Holistic and Integrated Approaches to SSF Development

The SSF Guidelines emphasize the need to adopt a holistic and integrated approach to development (Article 6.1). It is now recognized that the major development goal today is the improvement of human well-being,[4] which could come from various

[4] Today, the term wellbeing is often used in policy statements and in theories of development. In simple terms, wellbeing as used by us borrowing from Gough and McGregor (2008) is a combination of "what a person has" "what a person does with what he has" and "how he feels about what

sources (McGregor et al. 2009). Fishing alone may not generate sufficient incomes, as in the case of lagoon fishing (Silva et al. 2012), and may not meet al wellbeing aspirations. Low incomes coupled with seasonality in fisheries often make fishers vulnerable to poverty, threatening their livelihoods and causing 'illbeing'. In coping with vulnerability (improving their resilience), fishers and their household members often engage in a set of income generation activities (alternative livelihoods such fish marketing, agriculture, animal husbandry, etc.

Marine, inland, and lagoon ecosystems generate a number of values. Models pertaining to the total economic value (TEV) of wetlands (Munasinghe 1992; Freeman 1993; Costanza 2000; FAO 2002, 2010; Anthony et al. 2009; Boateng 2010) include various social and economic values both in the present and in future.[5] These values could be exploited by resource users for wellbeing improvements through the adoption of appropriate livelihood strategies. Table 18.1 provides an example of the non-extractive values that exist in Sri Lanka's lagoon systems (see Silva et al. 2012) (Fig. 18.2).[6]

Non-fisheries values, such as nature tourism in mangrove forests, bird watching, boat tours, and associated cultural and religious values are often exploited by other stakeholders such as hoteliers, tour operators, and wealthy individuals. In general,

**Table 18.1** Potential for development of ecotourism, recreational facilities, scientific research, and aquaculture in lagoons, by coastal district

| | Potential that exists in lagoons for the development of | | | | |
|---|---|---|---|---|---|
| Coastal District | Ecotourism (Rank) | Recreation (Rank) | Research (Rank) | Aquaculture (Rank) | Average (Rank) |
| Mannar | 4 | 4 | 4.5 | 4 | 4.1 |
| Colombo | 4 | 5 | 2 | 4 | 3.8 |
| Hambantota | 3.6 | 4.3 | 3.7 | 3.1 | 3.7 |
| Mullativu | 3.3 | 3.8 | 3.5 | – | 3.5 |
| Batticaloa | 1.7 | 2 | 4 | 4 | 2.9 |
| Gampaha | 3 | 3 | 3 | 2 | 2.8 |
| Galle | 3.3 | 3.3 | 2 | 2.3 | 2.7 |
| Ampara | 2.4 | 3 | 2.1 | 1.5 | 2.7 |
| Chilaw | 1.5 | 3.5 | 2 | 3.5 | 2.6 |
| Jaffna | 1.5 | 2 | 2 | 2 | 1.9 |

Key to ranking: Very high = 5: High = 4: Moderate = 3: Low = 2: Very low = 1
Source: Silva et al. (2012)

he has and do". "Wellbeing is a state of being with others, which arises where human needs are met, where one can act meaningfully to pursue one's goals, and where one can enjoy a satisfactory quality of life."

[5] Any good or service which people use at present or in the future has a value because people are 'willing to pay' for it.

[6] Identification of diverse values that exist in lagoons and their ranking was done using a five point Likert scale by officers of the Department of Fisheries, based on their knowledge and experience.

**Fig. 18.2** Non-fisheries values at Kokilai Lagoon in the North East of Sri Lanka

tourism is a costly activity demanding large investment funds, training, and skill development in a diversity of fields: gastronomy, culinary arts, linguistics, ITC, etc., which rural financiers find difficult to provide. Thus, those who invest in tourism are urban-based investors who have funds to invest and the trained staff to cater to diverse demands. This means that incomes produced by village resources give way to those generated by outsiders, and suggests the need for fisheries cooperatives to broaden their scope to include development. An integrated sustainable tourism initiative launched in Rekawa (Amarasinghe et al. 2010) was meant to build capacities of coastal populations in tourism related activities (guiding, bird watching, mangrove studies, etc.) and form networks of diverse stakeholders, including fisheries cooperatives, bird clubs, local ecotourism operators, and local guest houses. Although this gives the impression that these new tasks, completely outside the sphere of fisheries, would overburden the cooperatives, Silva et al. (2012) present examples of fishers undertaking tourism-related activities in Rekawa (south of Sri Lanka) and Pothuvil (east of Sri Lanka) lagoons. Moreover, the Lagoon Fisheries Management Authority of Rekawa (the former fisheries cooperative) once requested state assistance in building/modifying fishing crafts to transport tourists (Amarasinghe 2010).

The other concern is the need to adopt integrated approaches. The Special Area Management Planning (SAMP) initiative, established and implemented by the Coast Conservation Department (CCD) of the Sri Lankan government, is an integrated approach to coastal zone management. The Revised Coastal Zone Management Plan, Sri Lanka (CCD 1996, 99) points out that "…a comprehensive strategy was needed to cope with the impacts of these individual resource use decisions and conflicts over an area that might include resources not in the legally designated coastal zone," the result of which was SAMP. The most important characteristic of SAMP is that it is community-based and collaborative. In the process, both marine and lagoon fisheries cooperatives joined hands with all other stakeholders to form Special Area Management Coordinating Committees. CCD argues that SAMP has contributed positively towards both the development and management of environmentally sensitive areas by dealing effectively with multi-stakeholder conflicts and ensuring sustainable use of resources. Rekawa Lagoon and Negombo Lagoon have often been cited as good examples of successful implementation of SAMP (ibid.)

## Improving and Strengthening Fish Value Chain

The SSF Guidelines lay a strong emphasis on the fish value chain, especially in terms of developing the necessary infrastructure, institutions, and capacity of small-scale fishers including vulnerable and marginalized groups, to effectively deal with more powerful actors, such as fish merchants. We will show in this section that fisheries cooperatives have played an important role in the market, and also engage in the provision of diverse livelihood capitals and training and capacity building for their members.

Small-scale fishers are involved in chains that run from production to value addition to marketing to export of fish and fish products. Collective interventions in these chains can significantly improve their wellbeing. Studies carried out in the southern province of Sri Lanka (RUEDA 2010) revealed that fisheries cooperatives, if well organized, are in a position to provide fishers with access to important livelihood assets (Amarasinghe and Bavinck 2012). Table 18.2 provides information on the proivision of physical capital by two cooperative societies in southern Sri Lanka: Bata Atha (which is a well functioning cooperative) and Rekawa (where the cooperative is now dormant).

As apparent in Table 18.2, 60–5% of current craft owners in Bata Atha have procured their fishing equipment under diverse assistance schemes operated by the Fisheries Cooperative.[7] Although a measure of assistance was provided by the Rekawa cooperative, which was less well organized, the proportion of fishers benefiting there has been small. This shows the important role played by well-functioning cooperatives in assisting small-scale fishers to acquire physical capital.

Training is another area where cooperatives have an advantage over individuals. Most of the training programs designed for fisheries, especially in respect of post-harvest handling of fish (such as fish drying and preparation of other fisheries products), are offered by institutions such as Sri Lanka's National Aquatic Resources and Development Agency (NARA) and the Industrial Development Board (IDB). The National Institute of Fisheries and Nautical Engineering (NIFNE) is also involved in offering on-site training on fishing and net-mending techniques. Many

[7] Fishing cooperatives were able to borrow from state banks under various loan schemes and disbursed such funds through their own programs, earning income in interest.

**Table 18.2** Fishers' access to craft and gear through sample cooperatives (2010)

| Access to crafts and gear | Fishers reporting procurement of Physical Capital through assistance provided by cooperative (%) | | | |
|---|---|---|---|---|
| | Bata Atha | | Rekawa | |
| | NTRB | OFRP | NTRB | OFRP |
| Crafts | 60 | 65 | – | – |
| Gear | 75 | 70 | 28 | – |
| Engines | – | – | – | 65 |
| Three wheel vehicles | 25 | 15 | – | – |
| Motor cycles | – | – | 2 | – |

Source: Amarasinghe and Bavinck (2012)
*NTRB* Nan-motorized traditional Boat, *OFRP* Fibre glass boat with outboard engine

of these training programs are offered at a very moderate cost or free of charge. Fisheries cooperatives are in an advantageous position to request such training facilities because of their official recognition and broad membership. Many cooperatives in the south of Sri Lanka (such as the Bata Atha fisheries cooperative and Godawaya fisheries cooperative) have engaged in training women members in post-harvest handling of fish. In fact, the monthly meetings of both cooperatives are generally being attended by more women members than men. This can be attributed to two reasons. First, women members too benefit directly from the cooperative (training, capacity building, and loans for self employment) and therefore have an interest in managing the cooperative well. Second, this leaves more space for men to concentrate on fishing. These are clear examples of how fisherfolk, by grouping themselves, have collectively developed social and other forms of capital, contributing towards increased incomes for fishing households, empowering women, and making fishing livelihoods more sustainable. It is important to note that in all of these cases, cooperatives constitute channels for outside resources. Their success is thus closely associated with access to supra-local agencies: government, NGOs, and development agencies. Evidence from post-tsunami recovery of fisheries shows that NGOs that rushed to Sri Lanka to help the tsunami affected coastal populations always looked for well-functioning cooperatives with a good record of effectively addressing community issues to channel such assistance. Where such organizations existed, they have often taken the leadership in identifying affected people, assessing damages, distributing relief assistance among the affected persons and families, etc., by working closely with NGOs and government officials (Amarasinghe 2006).[8] On another front, some well-functioning fisheries cooperatives have also been able to approach diverse state and non-state institutions to obtain help in training and capacity building of their membership (Amarasinghe and Bavinck 2012). The linking capital which fishing cooperatives in Sri Lanka have thus created is maintained

[8] Field studies carried out in 2006 in the southern province to assess work associated with post tsunami re-building of the fisheries sector revealed that, proper assessment of damages, beneficiary selection and livelihood assistance had been effectively carried out in areas where there existed well-functioning fisheries cooperatives.

through efforts of local leadership and through reputation. Evidently, institutions allow social actors to accomplish things and without them these actors would be handicapped. Some things would simply be beyond their reach (Jentoft 2005).

One of the serious problems in fisheries that has been well documented in studies of Asian fishing communities is the oligopsonistic powers enjoyed by fish merchants and the low prices of landed resource imposed on fisher-producers (Platteau and Abraham 1987; Amarasinghe 1989; Amarasinghe and Wanasinghe 1993). Even the benefits of international trade in fish and fish products have mostly been reaped by these middlemen (Kurien 2005). One way of increasing the bargaining power of fishers is to strengthen their fisheries cooperatives and organize them into larger cooperative unions. Such a change would increase their bargaining power and also provide them with the capacity to engage in fish marketing, enjoying economies of scale,[9] By establishing links with large urban centers, super markets, and with fish processing and exporting firms, the unions could sell the fish at a fair price, bypassing middlemen. This would lead to a fair price for fish, for both producer and consumer.

However, experience with cooperative unions trying to undertake fish selling is not always very encouraging.[10] Bata Atha cooperative once made an effort to assemble fish from their members and others in the district and sell the consignment by themselves (personal communication with President, Bata-Atha South Fisheries Cooperative Society on 10th August 2008). The fish merchants immediately got themselves organized and sent a message to all wholesale markets asking the buyers to reject the fish. Another example is the formation of the Fisheries Cooperative Union of Echchelampattu. In the year 2005, as a means of breaking monopolitic fish buying by merchants, the Trincomallee District Development Association (TDDA) and the Danish Refugee Council (DRC) took the iniative to organize nine fisheries cooperatives in the Echchelampattu area into a fisheries cooperative union, which was provided with a Rs. 3 million revolving fund to provide required funds to the individual cooperatives (TDDA 2008). The power wielded by fish merchants can be attributed to three factors. First, there is limited competition in fish trade (oligopsony). Second, there are economic barriers to entry because of heavy demand for investment funds for insulated fish vans, cold rooms, etc. Third, there are physical barriers of entry, where merchants resort to physical harm, using their army of musclemen (Amarasinghe et al. 2005b).

Amarasinghe (2003a) suggests that cooperative unions can develop into strong regional organizations, even representing fishers at the national level in organizations such as fisheries trade councils. Such representation would ensure trickeling down of benefits of foreign trade to fishing communities, while ensuring that trade in fish and fishery products promotes food security and does not result in environmental degradation or adversely impact the nutritional rights and needs of the people who depend on fish for their health and well-being (see Article 11.2.15 of the Code of Conduct for Responsible Fisheries).

[9] With an increase in the scale of operation, unit cost of marketing would fall.

[10] Cooperative Unions in the Northern Province have apparently been more successful in marketing local, high-value catches on the Colombo market than cooperative unions in the south.

## Inclusion of Women and Marginalised Groups

Women play an increasingly important role in the fisheries sector. Household responsibilities of women are growing along with the development of the ultra-modern deep sea fisheries sub-sector in which men are away on longer duration fishing trips. Not only do women have to feed, educate, and protect children as well as manage the household, but they are also supposed to confront and resolve all health and other household problems and fulfil social obligations. Field observations suggest that the involvement of women in fisheries organizations, especially cooperatives, is now perhaps higher than that of men (Amarasinghe 2003b, 2009a, b).

Since cooperatives are community institutions that respect principles and laws of the community, all community members[11] have equal rights of joining them, irrespective of caste, class, religion, and creed (Government Gazette 1974). By joining cooperatives, which are considered as strong forms of social capital, those vulnerable and marginalized groups can in principle cope with stressors by securing diverse livelihood capitals (Amarasinghe and Bavinck 2012). Cooperatives also provide avenues for poor and vulnerable groups to 'voice' their issues and grievances, rather than doing the same as 'individuals.'

## Safety Issues and Fisheries Insurance

Safety at sea and designing proper insurance schemes are other concerns expressed in The SSF Guidelines (Article 9.3), as well as at the SSF Guidelines regional consultation meeting. Fisheries is an area where there is less chance for the emergence of formal insurance companies providing insurance cover for the loss or damage to fishing equipment due to well-known moral hazard and adverse selection problems, which arise from informational asymmetries between insurance agents and those who seek insurance (Amarasinghe et al. 2005a). Moral hazard problems arise due to incentives provided to the insuree to take less care of contingencies that give rise to claims. Adverse selection problems arise when the insurer is unable to distinguish between honest and dishonest agents demanding insurance and is, therefore, forced to offer all the same insurance. Of course, the insurance agents can reduce informational asymmetries by collecting information. But then the insurance policy is unlikely to attract fishers because the insurance rates will be set too high.

Due to the above imperfections in the insurance market, the Sri Lankan government designed subsidised fisheries insurance schemes a few decades ago in order to help fishermen manage the risks associated with modern technology. Yet, the state insurance agencies too were confronted with incentive problems mentioned above and failed to provide effective insurance to meet the demand. Studies carried out in

[11] Since many cooperatives today have become 'savings societies' and banks (*Idiwara* Banks), even-non fishers are members.

Kudawella, a fishing village in the south of Sri Lanka in 1997 (Amarasinghe et al. 2005a), revealed that fishers have withdrawn from participating in state insurance schemes because of the state's long delays in making indemnity payments due to informational problems and high insurance rates, which reflect the high cost of information collection.

One way of overcoming the information problem is to establish insurance schemes jointly with fisheries cooperatives. Information about craft operations (fishing times, fishing locations, fishing techniques) are known to all because fishing is carried out during particular time periods by groups of fishers operating similar craft-gear combinations. Therefore, if a fisher is confronted with a shock, such as an accident, damaging his craft and/or gear, this is known to others. Thus, any insurance scheme operated with the involvement of a cooperative is unlikely to be severely constrained by informational asymmetries. This would lower the cost of information collection and the insurance rates, and minimize delays in indemnity payments.

Generally cooperatives offer 'instant loans' (loans lent 'over the counter'), immediately on request. This mechanism, devised by cooperatives, is intended to assist fishers to cope with short term risks of catch fluctuations and idiosyncratic shocks such as craft and gear damage (Amarasinghe and Bavinck 2012). Interestingly, instant loans or 'instant credit' perform an important insurance function, in addition to a credit function.

## Conclusions

This chapter has argued that fisheries cooperatives in Sri Lanka are in a position to address many of the important issues outlined in the SSF Guidelines. For these to be realized, however, cooperatives should be reorganized as effective co-management platforms.

With regard to governance and management, it is evident that a considerable distance between the state and community organizations exist, which has often led to governance failures. On the one hand, this situation has emerged due to the failure of the government to understand the existence of legal systems other than their own. On the other hand, the fisheries cooperatives have promoted entry into fisheries (leading to increased pressure on the resources) without much regard for resource governance. Evidently, community principles based on equality, harmony, and well-being impede cooperatives from adopting strict resource management measures, while government efforts to manage resources are constrained by informational problems and high transaction costs. It is now understood that both the state and fisher organizations have to play equally important roles in resource governance. Yet, whether the present co-management platforms, where cooperatives have joined hands with state actors in managing resources, are able to design appropriate resource management measures is questionable on the grounds of power imbalances. State dominance is quite evident in the functioning of these platforms, with little space for the 'voice' of the people at the bottom. Another important issue highlighted is the need to deal with broader ecosystems rather than fisheries resources alone, which requires an integrated and holistic approach to management. Cooperatives in Sri Lanka have already been engaged by government in processes such as SAMP and Lagoon Co-Management Committees. A broader approach will have to take into account both fisheries and non-fisheries values of the ecosystem, as well as the interests of all stakeholders. While representing fisher interests, we argue that cooperatives can jointly make decisions along with other stakeholders in managing the ecosystem in which they are a part.

The contribution of cooperatives to restructuring the lower reaches of the value chain, which is the production and landing of fish, remains quite high. One of the principal functions of cooperatives has been the provision of membership with access to physical resources. However, the role of cooperatives in post-harvest activities is relatively weak. Despite their early focus on marketing, cooperatives have not been very successful in breaking the oligopsonistic behavior of fish merchants and improving the market position of their members substantially. But here further advances can be made through the more effective bundling of individual cooperatives into cooperative unions.

We suggest that as community organizations, based on trust and reciprocity, cooperatives have the potential of realizing many of the wellbeing aspirations of small-scale fishers in Sri Lanka. Cooperatives could therefore also play a vital role in taking the SSF Guidelines forward. However, there are two important concerns. First, cooperatives will have to be recognized as representatives of fishing communities (and not as arms of the state) – their structure and functions will have to be appropriately revised, changed, restructured, and revitalized to take up the new challenges, as outlined in the SSF Guidelines. In doing so we wish to make the following suggestions:

- Today fisheries cooperatives in Sri Lanka face competition from a parallel set of institutions – Rural Fisheries Organisations (RFO) – that have recently been established by the Minister of Fisheries for reasons that seem to have little to do with fisheries. We think this creates unnecessary confusion, and that the Sri Lankan government would do well by reverting to the original situation wherein fisheries cooperatives constitute the main organizational body at the local level. We suggest that fisheries cooperatives should be transformed into interactive co-management platforms, where fishers' rights to resources are clearly defined and recognized, and appropriate responsibilities for resource management, safety issues, and insurance assigned. One of the pre-requisites for successful implementation of such a strategy is that state actors recognize that customary legal systems are available and have a role to play.
- Deviating from their traditional roles, cooperatives should function as co-management bodies, incorporating both the participation of community representatives and government officials, and the participation of women, vulnerable groups and other relevant stakeholders. Although co-management platforms are now being established under the Fisheries and Aquatic Resources Act of 1996, the government must ensure that they are true interactive platforms, providing space for all stakeholders to voice their interests. This means the state needs to revisit the concept of co-management.

- As participatory management platforms, cooperatives shall adopt, among other things, a more holistic and integrated approach to managing systems other than individual resources. Cooperatives would be strengthened by training and capacity building of both community representatives and government officials and through the establishment of appropriate legal frameworks.
- Governments may help cooperative join hands with similar organizations to form strong 'bridging social capital' (cooperative unions) and link them up with state marketing institutions and fish exporting firms to challenge trade oligopsonies. The government can also help form multi-tier district, provincial, and national level fisheries consultative committees.
- Such a change requires the incorporation of the relevant SSF Guidelines into the National Plan of Action.

## References

Alexander, P. (1982). *Sri Lankan fishermen – Rural capitalism and peasant society.* Canberra: Australian National University.

Amarasinghe, O. (1989). Technological change, transformation of risks and patronage relations in a fishing community of South Sri Lanka. *Development and Change, 25*(4), 684–734.

Amarasinghe, O. (1995). Political Clientelism and underdevelopment. *Rohana, 5*, 112–134.

Amarasinghe, O. (2003a). *International trade in fish and fish products and food security in Sri Lanka*. Report submitted to FAO, Rome.

Amarasinghe, O. (2003b). *Changing role of women in the deep sea fisheries sub-sector of Sri Lanka*. Report submitted to ICSF (International Collective in Support of Fishworkers), Chennai.

Amarasinghe, O. (2005). Technological change in Sri Lankan fisheries. In O. Amarasinghe (Ed.), *Modernization and change in marine small-scale fisheries of southern Sri Lanka* (pp. 52–102). Colombo: Navamaga Printers.

Amarasinghe, O. (2006). *Post-tsunami recovery of the fisheries sector of Sri Lanka.* Proceedings of the seminar on Post-tsunami Recovery of the Fisheries Sectors of India, Indonesia, Thailand and Sri Lanka (12–14). International Collective in Support of Fishworkers (ICSF), Chennai India, 16–18 January 2006.

Amarasinghe, O. (2009a). *Social capital to alleviate poverty: Fisheries cooperatives in Southern Sri Lanka. Study 15, Newsletter 51.* Amsterdam: International Institute of Social Studies, Hague, the Netherlands.

Amarasinghe, O. (2009b). *Rule breaking and conflicts in marine small-scale fisheries of south Sri Lanka- the need for interactive governance.* Preparatory Paper presented at workshop in July 2009 at University of Ruhuna, ESPA project on Building Capacity for Sustainable Governance in South Asian Fisheries: Poverty, Wellbeing and Deliberative Policy Networks, Institute of Development Studies, University of Sussex, 20 p.

Amarasinghe, O. (2010). *Building governance for sustainable small-scale fisheries: Bridging human and ecosystem wellbeing.* A Policy Brief, ESPA project on Building Capacity for Sustainable Governance in South Asian Fisheries: Poverty, Wellbeing and Deliberative Policy Networks, Institute of Development Studies, University of Sussex, 4 p.

Amarasinghe, O., & Bavinck, M. (2012). Social capital and the reduction of vulnerability: The role of fisheries cooperatives in southern Sri Lanka. In S. Jentoft & A. Eide (Eds.), *Poverty mosaics: Realities and prospects in small-scale fisheries.* Dordrecht: Springer.

Amarasinghe, O., & Wanasinghe, W. A. G. (1993). Risk management in modern fisheries: A case study of a marine fishing community of southern Sri Lanka. *Journal of the Sri Lankan Agricultural Economics Association., 1*(1), 46–58.
Amarasinghe, U. S., Chandrasekara, W. U., & Kithsiri, H. M. P. (1997). Traditional practices for resource sharing in an artisanal fishery of a Sri Lankan estuary. *Asian Fisheries Science, 9*, 311–323.
Amarasinghe, O., Wanasinghe, W. A. G., & Jayantha, S. P. M. (2005a). Risks and uncertainties in fisheries and their management: Fisheries insurance. In O. Amarasinghe (Ed.), *Modernization and change in marine small-scale fisheries of southern Sri Lanka* (pp. 103–180). Colombo: Navamaga Printers.
Amarasinghe, O., Wanasinghe, W. A. G., Jayantha, S. P. M., Wishantha Malraj, A. S., Nishani, D., & Somasiri, A. (2005b). Market for fisheries credit: The functioning of the market for fixed capital in fishing communities in southern Sri Lanka. In O. Amarasinghe (Ed.), *Modernization and change in marine small-scale fisheries of southern Sri Lanka* (pp. 181–252). Colombo: Navamaga Printers.
Amarasinghe, O., Wickramathilake, S., Champika, J., Ktishanthi, J., & Premachandra, P. (2010). *Potential for Integrated Sustainable Tourism (IST) in Rekawa of Southern Sri Lanka. An Inception Paper*. University of Ruhuna/UNDP-ARTGOLD. Ruhuna Economic Development Agency (RUEDA), Sri Lanka.
Anthony, A., Atwood, J., August, P., Byron, C., Cobb, S., Foster, C., Fry, C., Gold, A., Hagos, K., Heffner, L., Kellogg, D. Q., Lellis-Dibble, K., Opaluch, J. J., Oviatt, C., Pfeiffer-Herbert, A., Rohr, N., Smith, L., Smythe, T., Swift, J., & Vinhateiro, N. (2009). Coastal lagoons and climate change: Ecological and social ramifications in U.S. Atlantic and Gulf Coast ecosystems. *Ecology and Society, 7*(1), 8.
Bavinck, M. (2001). *Marine resource management: Conflict and regulation in the fisheries in the Coromandel coast*. New Delhi/Thousand Oaks: Sage Publications.
Bavinck, M. (2015). Fishing rights in post-war Sri Lanka: Results of a longitudinal village enquiry in the Jaffna lagoon. *Maritime Studies, 14*(1), 1–15.
Bavinck, M., Johnson, D., Amarasinghe, O., Rubinoff, J., Southwold, S., & Thomson, K. T. (2013). From indifference to mutual support – A comparative analysis of legal pluralism in the governing of south Asian fisheries. *European Journal of Development Research, 25*(4), 621–640.
Birchall, J. (2004). *Cooperatives and the millennium development goals*. Geneva: ILO.
Boateng, I. (Ed.) (2010). *Spatial planning in coastal regions – Facing the impact of climate change*. Copenhagen: FIG Publication No. 55.
CCD (Coast Conservation Department). (1996). *Revised Coastal zone management plan*, Sri Lanka, CRMP & CCD. 128 p.
Costanza, R. (2000). Social goals and the valuation of ecosystem services. *Ecosystems, 3*, 4–10.
FAO. (2002). Corporate document repository, Forestry department. http:/www.fao.org/docrep/007/ae212e/ae212e04.htm. Accessed 11 Dec 2011.
FAO. (2010). *Payments for environmental services (PES)*. Rome: Food and Agriculture Organization of the United Nations.
Freeman, M. (1993). *The measurement of environmental and resource values: Heory and methods*. Washington, DC: Resources for the future.
Gough, I., & McGregor, J. A. (Eds.). (2007). *Wellbeing in developing countries: From theory to research*. Cambridge: Cambridge University Press.
Government Gazette. (1974, January 10). *The cooperative societies law no. 5 of 1972*. Gazette of the Democratic Socialist Republic of Sri Lanka. 93/5.
ICA International Cooperative Alliance. (n.d.). International Cooperative Alliance. http://www.ica.coop/coop/index.html. Accessed 18 Aug 2010.
Jentoft, S. (1986). Fisheries co-operatives: Lessons drawn from international experiences. *Canadian Journal of Development Studies, 7*(2), 197–209.

Jentoft, S. (2005). Institutions for fisheries governance; Introduction. In J. Kooiman, M. Bavinck, S. Jentoft, & R. Pullin (Eds.), *Fish for life: Interactive governance for fisheries*. Amsterdam: Amsterdam University Press.

Kurien, J. (2000). The Kerala model: Its central tendency and the "outlier". In G. Parayil (Ed.), *Kerala: The development experience*. London: Zed Books.

Kurien, J. (2005). *Responsible fish trade and food security* (FAO fisheries technical paper 456). Rome: Food and Agriculture Organization of the United Nations.

McGregor, J. A., Camfield, L., & Woodcock, A. (2009). Needs, wants and goals: Wellbeing, quality of life and public policy. *Applied Research Quality Life, 4*, 135–154.

Ministry of Fisheries. (2015). *Fisheries statistics 2015*. http://www.fisheries.gov.lk/content.php?cnid=ststc. Accessed 3 Mar 2016.

Munasinghe, M. (1992). *Environmental economics and sustainable development*. Rio Earth Summit. Washington, DC: World Bank.

OCDC. (2007). *Cooperatives: Pathways to economic, democratic and social development in the global economy*. Arlington: US Overseas Cooperative Development Council.

Peramunegama, S. S. M. & Amarasinghe, O. (forthcoming). *Implications of technological change in a multi-stakeholder fishery for fisheries development and livelihoods; the case of small scale fisheries in the north of Sri Lanka*. M.Phil, thesis. Sri Lanka: Faculty of Agriculture, University of Ruhuna.

Platteau, J. P., & Abraham, A. (1987). An inquiry into quasi-credit contracts: The role of reciprocal credit and interlinked deals in small-scale fishing communities. *Journal of Development Studies., 23*(4), 461–490.

RUEDA (Ruhunu Economic Development Agency). (2010). *Fisheries and fish processing in the Rakawa Sam Area. An inception paper*. University of Ruhuna/UNDP-ARTGOLD/Ruhuna Economic Development Agency (RUEDA). Sri Lanka.

Scholtens, J. (2016). Fishing in the margins; North Sri Lankan fishers' struggle for access in transboundary waters. Ph.D. thesis. Amsterdam: University of Amsterdam.

Scholtens, J., Bavinck, M., & Soosai, A. S. (2012). Fishing in dire straits: Transboundary incursions in the Palk Bay. *Economic Political Weekly, Vol XLVII, 25*, 87–96.

Silva, E. I. L., Katupotha, J., Amarasinghe, O., Manthrithilake, H. Y., & Ariyaratna, R. (2012). *Lagoons of Sri Lanka: From the origins to the present*. Colombo: International Water Management Institute.

Stirrat, R. L. (1988). *On the beach. Fishermen, fishwives and fishtraders in post-colonial Lanka*. Delhi: Hindustan Publishing Corporation.

TDDA. (2008). *Strengthening fisheries cooperative societies to ensure sustainable livelihoods to fishers in the Echchilampattu area of the Trincomalee District of Sri Lanka*. A Project Proposal for funding By Cordaid, the Netherlands.

Wickramasinghe, W. A. R. (2010). *Livelihoods, institutions and fish resource use: The performance of small-scale marine fisheries in Hambantota district, Sri Lanka*. Amsterdam: University of Amsterdam.

Wickramasinghe, W. A. R., & Bavinck, M. (2015). Institutional landscapes affecting small-scale fishing in southern Sri Lanka – Legal pluralism and its socio-economic effects. *Maritime Studies, 14*, 18.

Woolcock, M. (2001). *The place of social capital in understanding social and economic outcomes*. Development Research Group. The World Bank and Kennedy School of Government, Harvard University.

# Chapter 19
# The Small-Scale Fisheries Guidelines as a Tool for Marine Stewardship: The Case of *Cap de Creus* Marine Protected Area, Spain

**Sílvia Gómez Mestres and Josep Lloret**

**Abstract** This case study examines how the *Voluntary Guidelines for Securing Sustainable Small-Scale Fisheries in the Context of Food Security and Poverty Eradication* (SSF Guidelines) can be used, as socially-legitimized basic principles, to establish regulations in Marine Protected Areas (MPAs) that take into account the spatial, ecological, and cultural specificities of small-scale fisheries which are not reflected in national and European legislations. Based on a case study carried out in the *Cap de Creus* MPA in Catalonia, Spain, this chapter argues that the implementation of the SSF Guidelines in Mediterranean MPAs is a slow process, and highlights the difficulties involved in it in the context of small-scale fisheries. Through an empirically informed ethnography, it is argued that the idea of small-scale fisheries in *Cap de Creus* embodies a particular definition of fishing that sees the economic practice of fishing as one of *socio-ecological embeddedness*. Fishing is embedded in local culture and in a set of social relations with nature. In this chapter, we suggest ways in which the idea of socio-ecological embeddedness can contribute to the institutionalization of sustainability in fisheries. This can be done by seeing fishing as 'heritage value' and considering biological studies in a co-management plan. The SSF Guidelines, by recognizing small-scale fisheries' tenure rights as a way to promote small-scale fishers' stewardship over resources, can help meet the objectives of sustainable resource management.

**Keywords** Cap de Creus • MPAs • Heritage value • Stewardship • Sustainability

S. Gómez Mestres (✉)
Faculty of Humanities, Department of Social and Cultural Anthropology, Autonomous University of Barcelona, Catalonia, Spain
e-mail: silvi.gomezmestres@gmail.com

J. Lloret
Institute of Aquatic Ecology, University of Girona, Catalonia, Spain
e-mail: Josep.lloret@udg.edu



S. Jentoft et al. (eds.), *The Small-Scale Fisheries Guidelines*, MARE Publication Series 14, DOI 10.1007/978-3-319-55074-9_19

## Introduction

Small-scale fisheries are often considered to have less ecological impact and be more ecologically sustainable than large-scale, industrial fisheries (Leleu et al. 2014). Those who highlight their potential as environmentally, economically, and socially sustainable alternatives to large-scale fisheries do so by illustrating that small-scale fisheries accounts for 24 times more fishers than large-scale fisheries at an equivalent level of total annual catch (Jacquet and Pauly 2008). Moreover, small-scale fisheries provide direct employment for about 100,000 people in the European Union. There European Union fleet has approximately 70,000 small-scale fishing vessels (84% of the European Union fishing fleet) (Guyader et al. 2013).

Raising awareness of the importance of small-scale fisheries is particularly relevant not only because the livelihoods of many people depend on the sustainable use of marine resources, but also because such fisheries provide vital nutritious food and basic income for over 90% of small-scale fisheries households (about 116 million around the world) that are living in low-income countries (World Bank 2010; Naji 2013), as well as in countries affected by war and economic crises. In contrast, in high-income countries (e.g. European Union member states and the US), fisheries as a whole play a much smaller role because these countries are more reliant on terrestrial animal protein as opposed to seafood protein (Westhoek et al. 2011; EEA 2016). According to some studies (Westhoek et al. 2011; EEA 2016), recent annual consumption of meat in the EU (about 52 kg per capita) is double that of seafood consumption (about 23 kg per capita). In countries like Portugal, which consume high amounts of meat and fish, the annual consumption of meat in 2007 was about 90 kg per capita whereas the consumption of seafood (including fish) in 2011 was about 57 kg per capita. Even in Mediterranean countries with a diet that traditionally contained less meat, the consumption of meat has increased sharply and is currently higher than the consumption of seafood. In Spain and Cyprus, for example, the consumption of meat in 2007 was around 65 kg per capita whereas in Greece it was around 81 kg per capita (Westhoek et al. 2011). This compares with an annual consumption of seafood (including fish; per capita) of 42 kg in Spain, 23 kg in Cyprus, and 20 kg in Greece (EEA 2016). For that reason, in high-income countries much attention has been given to the contribution of seafood to a healthy diet because of the health benefits provided by the long-chain omega-3 (or n-3) fatty acids contained in seafood (Lloret et al. 2016).

Perhaps a main reason that not enough attention is given to small-scale fisheries is because, as compared to large-scale industrial fisheries, the volume of catches and the economic importance of small-scale fisheries are relatively low. The total value of landings of the fisheries sector in EU-27 is estimated at almost 8 billion euros, of which only 2.1 billion euros (23%) is from small-scale fisheries (Macfayden et al. 2011). However, small-scale fisheries are especially important when socio-cultural and local economy considerations are taken into account. This is the case in the Mediterranean coastal zone as well (Di Franco et al. 2014). Whereas in high-income countries such as Spain, the primary sector (including fisheries) is not very important for Gross Domestic Product (GDP) (only 25% of total GDP), it is often significantly more in regional and local economies. This is especially notable in Galicia. Much of fisheries contribution in regional and local economies comes from small-scale fisheries because small-scale fisheries' catches have a high unit value and are destined for these regional and local (tourist) markets (Guyader et al. 2013). Equally important is that small-scale fisheries embody rich cultures with long histories (Gómez, Lloret, Riera, Demestre 2006; Gómez and Lloret 2016).

Also worth noting is that though small-scale fisheries have historically been a major source of food, employment, and economic benefit to the inhabitants of Mediterranean coastal villages (Bellido et al. 2014), currently they are declining in many parts of the Mediterranean (Guillou and Crespi 1999; Colloca et al. 2004; Gómez, Lloret, Riera, Demestre 2006). The small-scale fleet (vessel numbers) has declined in many European Union Mediterranean countries during the the period 2000–2010, with a reduction of up to 40% in Spain, leading to a reduction in the catch from artisanal fisheries in many areas (Gómez et al. 2006; Lloret et al. Forthcoming).

In addition to changes in consumer patterns, which has resulted in a decline of seafood consumption in Mediterranean countries, also of importance is the fact that fish stocks are facing increased levels of disturbance from human activities. For instance, commercial and recreational fisheries have had adverse impacts on both target and non-target species and their habitats, and have resulted in fewer and more expensive fish (Milazzo et al. 2002; Lloret et al. Forthcoming). Hence, the use of fishing gears that actively select certain species, sizes and sexes, the deployment of fishing gears on certain fragile habitats, the loss of fishing gears and the use of non-native species as bait are examples of how SSFs can threaten the sustainability of vulnerable coastal species and habitats (Lloret et al. Forthcoming)

Several management options are available to address resource degradation and promote sustainable fisheries such as temporal closures, limits to catch and effort, and minimum landing sizes. Marine protected areas (MPAs), which have for many years been used to meet conservation objectives and promote sustainable development have used such management options (Di Franco et al. 2014). Some Mediterranean MPAs have been able to enhance fish stock abundance and improve fisheries through specific agreements with fishermen in order to decrease fishing effort (Font and Lloret 2015). For example, in Torre Guaceto (Italy) and Port-Cros (France), these agreements have allowed them to increase both the quantity and quality of fish yields and related revenues (Claudet and Guidetti 2010; Guidetti and Claudet 2010; Guidetti et al. 2010).

Not all MPAs have been successful, however, in avoiding a decline in small-scale fisheries. Take the case, for example, of the rocky headland of the *Cap de Creus Natural Park* (henceforth CCNP) on the northeast coast of Catalonia in the northwest of the Mediterranean (Gómez, Lloret, Riera, Demestre 2006). In this region small-scale fisheries have for centuries provided livelihoods for entire local communities (e.g. Roses, Cadaqués, El Port de la Selva, and Llançà). However, in the second half of the twentieth century semi-industrial fishing fleets (trawlers and purse seiners) and tourism have grown in importance to the detriment of traditional small-scale fisheries in the area (Pi Sunyer 1977). The establishment of the *Cap de Creus* MPA in 1998 has not improved the situation for small-scale fisheries. While a small-scale fisheries fleet still persists, the number of vessels in the fleet has declined from 32 in 2006 to 19 in 2015 (Gómez, Lloret, Riera, Demestre 2006; Gómez 2015; Gómez and Lloret 2016). As a result of the current economic crisis that began in 2008 in southern Europe in some regions such as Catalonia the number of small-scale fishers has stabilized, with the fisheries attracting some young men who cannot find other jobs or have lost their jobs on land (such as in construction or real estate). This has been the case in *Cap de Creus*, where small-scale fisheries still provide employment opportunities, despite uncertainty, for people from fishing families and others who have some fishing history behind them (Gómez 2015).

Like many others elsewhere, the small-scale fisheries in *Cap de Creus* face a number of difficulties in adapting to the changing economic, societal, and environmental context in which they are immersed. As competition with other fisheries increase and recreational activities too, small-scale fishers claim their rights to fishing grounds. Small-scale fisheries distinguish themselves from large-scale fisheries, not only legally but also in terms of their local cultural heritage that entails a special relationship with the environment. For this reason, not all small-scale fisheries are perceived as artisanal fisheries by small-scale fishing communities (according the *emic* perspective).

A co-management plan is seen to be the way to implement the SSF Guidelines and address questions of stewardship. Such a plan will help secure the livelihoods of small-scale fishers in *Cap de Creus* and their claims to sea tenure, something that at present is only protected by MPA regulations. Such a plan cannot be realized, however, without thorough understanding about small-scale fisheries, including how they are defined and how fishers themselves identify with their fishing livelihoods. Therefore, in this chapter we present the results of our research aimed at understanding these specific concerns in relation to small-scale fisheries and fishers in *Cap de Creus*. Doing so will facilitate the proper implementation of the SSF Guidelines and consequently stewardship measures.

Our field research highlights that not only do the definitions of small-scale fisheries vary across EU Member States, but also that national definitions of small-scale fisheries do not encompass the social conceptions of small-scale fisheries, which account for cultural heritage and help distinguish them from large-scale fisheries.

The chapter is divided into five sections. Following this introduction, we present the theoretical perspectives that shape our analysis in section "Theoretical perspectives", with a focus in particular on the socio-ecological embeddedness perspective of small-scale fisheries. This is done keeping in mind the SSF Guidelines and small-scale fisheries in *Cape de Creus*. The socio-ecological embeddness perspective, as we shall see, refers to how fishing practices are embedded in local culture and in social relationships with nature that are in keeping with the local concept of small-scale fisheries in the CCNP. In section "Methods: the ethnographic fieldwork", we outline the methods used and the data collected during the research. In section "What are small-scale fisheries?", we describe the findings and data in relation to a contextualized understanding of small-scale fisheries, rooted in culture and adapted to the local environment. Section "Discussion: how to implement the SSF guidelines as marine stewardship?" discusses the results of the research vis-à-vis implementing the SSF Guidelines so as to regulate and secure locally-identified small-scale fisheries in the specific case of the CCNP. In section "Conclusions", we present the key lessons learned as far as the implementation of the SSF Guidelines is concerned.

## Theoretical Perspectives

In Catalonia small-scale fisheries is called 'artisanal fisheries' by fishermen.[1] The definition of artisanal fisheries has not only to do with a question of "scale" but also with a local culture that underlines the difference between –artisanal- small-scale fisheries (considering the existing broad variety of fishing fleets and gears, size of vessels, fishermen profiles, working hours, etc) and large-scale fisheries. The SSF Guidelines recognize the great variability in existing small-scale fisheries definitions, and therefore it is necessary to understand what small-scale fisheries are on a case by case basis. As stated in the Section 2.4 of the SSF Guidelines (FAO 2015, 1–2):

> These Guidelines recognize the great diversity of small-scale fisheries and that there is no single, agreed definition of the subsector. Accordingly, the Guidelines do not prescribe a standard definition of small-scale fisheries nor do they prescribe how the Guidelines should be applied in a national context. These Guidelines are especially relevant to subsistence small-scale fisheries and vulnerable fisheries people. To ensure transparency and accountability in the application of the Guidelines, *it is important to ascertain which activities and operators are considered small-scale, and to identify vulnerable and marginalized groups needing greater attention. This should be undertaken at a regional, subregional or national level and according to the particular context in which they are to be applied.* [...] (our emphasis).

Giovanni et al. (1996), referring to the case of Catalonia, highlighted that under the restructuration and modernization process, it was not enough to differentiate industrial fisheries from artisanal small-scale ones by the types of fishing gear used. Other aspects also needed to be taken into account to establish the boundaries of small-scale fisheries. That is, aspects such as the level of capitalization, type of property, size of the vessels, and the technology being used also needed to be considered. Going a step further, we would emphasize that small-scale fisheries include a specific, culturally-inherited conception of fishing reflected, for example, in fishing techniques, fishing knowledge, and a locally-established relationship with the environment, among other aspects.

The SSF Guidelines (which on page ix of the preface refer to small-scale fisheries as 'small-scale and artisanal fisheries', although this is the only use of the term 'artisanal' in the document) consider small-scale fisheries to be a fishing sector

---

[1] National legislation currently regulating small-scale fisheries merely state that they are notably artisanal in character without specifying what artisanal fisheries actually refers to. (Statutory Order AAA/2794/2012. BOE (Official State Gazette) from the 28th of December, 2012, n° 312, Sec.I. Pp. 88675)

firmly rooted in local communities, traditions, and values. They don't, however, specify whether there is a difference between small-scale and artisanal fisheries or whether we should use one term or the other or both. A FAO report from 2007, however, distinguished between artisanal and small-scale: 'artisanal' indicates low levels of technology adopted during fishing operations without any reference to vessel size, while 'small-scale' indicates small vessel size regardless of the level of technology used (Griffiths et al. 2007). Although, as Griffiths et al. (2007) state, small-scale fishing is a difficult term to define because the criteria vary over space and time, also often depending on the socio-cultural and historical context.

The idea of small-scale fisheries can also be expressed through the perspective of socio-ecological embeddedness proposed by Kirk and Memon (2010). The socio-ecological embeddedness approach theorized by Kirk and Memon (2010) has its roots in the social embeddedness theory of Polanyi (1944), later revised by Grannovetter (Granovetter 1985). In the words of Paavola and Ropke (2008, 15 as cited in Kirk-Memon 2010), whereas the social embeddedness approach emphasizes embeddedness of the economy in social and cultural institutions, the ecological embeddedness approach emphasizes embeddedness of the economy in the bio-physical environment. Embeddedness refers to systematic interdependencies of socio-economic and bio-physical factors (Von-Benda-Beckmann et al. 2006), which are paramount to the success of sustainable resource management (McCay 1996, 2001). Socio-ecological embeddedness can contribute to institutionalizing sustainability through collective action (Jentoft 2005). Collective action, in our case, involves small-scale fishing communities, the environmental managers of CCNP, and fishers' own organizations (known as Confraries).[2]

Small-scale fisheries in CCNP, as viewed from the socio-ecological embeddedness perspective do not necessarily match the definition of small-scale fisheries as defined in law. Whereas small-scale fisheries as defined in law include certain legally-authorized fishing techniques, specific boat lengths, a limited number of nets, and fishery rights within a particular area,[3] small-scale fisheries in our case study comprise various *social representations* that involve an expertise bonded to a common cultural substrate inherited from fishers' ancestors. However, just as there are different interpretations of small-scale fisheries by EU Member States within their own national contexts ("there is no single definition of small-scale fisheries in European legislation that can then be applied across all EU Member States [Macfadyen et al. 2011, 19])," nor is there a definition of small-scale fisheries in any national legislation that matches the manner in which fishers define themselves.

---

[2] Confraria is the name given to ancient and traditional fishers' organizations. Although these associations function much like cooperatives and, historically, have had an important role in the co-management of marine resources in the Cap de Creus (Pi-Sunyer 1977), and elsewhere in the Mediterranean (see, e.g., Dufour 1988; Alegret 1995; Weber 1992), small-scale fishers have never been sufficiently represented in these associations because they are much fewer in number compared to purse seiners and trawlers.

[3] Statutory Order AAA/2794/2012. BOE (Official State Gazette), 28th of December, 2012, n° 312, Sec.I. p. 88675)

The legal standards regulating small-scale fisheries at national or European level's do not consider the spatial, ecological, and cultural specificities of small-scale fisheries. This results in a certain lag between legal norms and social practices that are adapted to specific environmental realities cultural knowledge. This legal pluralism[4] in fishing regulations in the context of co-existing access rights to the same sea also results in 'complex normative orders' (Von Benda-Beckmann 2002) compounded by state law and common law, by whether one or the other takes precedence, and by whether or not common law is formally recognized (Techera 2010).

To sum up, there are a number of important nuances in the various definitions of small-scale fisheries as conceived in legal provisions, although these nuances are not captured adequately in legislations within European countries and European legislation as a whole. That is, law does not capture the social and cultural specificities of Mediterranean small-scale fisheries (which account for these different cases of socio-ecological embeddedness). Reyes et al. (2015) have already pointed out these paradoxes for the small-scale fisheries of the Languedoc-Roussillon lagoons of France. It is precisely this difference between what the law says and what the social conception of fisheries is among small-scale fishing communities that produce tensions among fishers in general about what small-scale fisheries *can* be and *should* be.

## Methods: The Ethnographic Fieldwork

In order to get an idea of what fishers themselves understand to be small-scale fisheries, we carried out empirically informed socio-cultural research. We undertook an extended ethnography based on interviews which shed light on the differences between small-scale and large-scale fisheries, as they are understood culturally. Maritime socio-anthropology is emerging as a valuable research field with regard to the assessment and management of fisheries activities, especially in MPAs (Carter 2003; Himes 2003; Gómez, Lloret, Riera, Demestre 2006). Maritime socio-anthropology methods were used to analyze the most important social attributes of small-scale fisheries in the *Cap de Creus*. Maritime socio-anthropology is concerned with the way coastal communities and fishers perceive, mentally represent, and express their own perceptions of the transformations, regulation, and management of their fisheries (Acheson 1981; Cadoret 1984; Geistdoerfer 1984; Galvan-Tudela 1988).

During our ethnographic fieldwork, conducted over 6 months in 2015 (from April to December), 19 interviews were carried out with small-scale fishers from

[4] Following Von Benda-Beckmann, we define legal pluralism as the 'multiplicity of normative orders in a single social space and of not privileging state law over other normative orders Von Benda-Beckmann (2002).'

different villages within the CCNP (Roses, Cadaqués, Port de la Selva, and Llançà) (Gómez 2015). We undertook semi-structured, face-to-face interviews that involved extended conversations with fishers following a set of open-ended questions. Five fishers were interviewed in Roses, nine in Port de la Selva, five in Cadaqués, i.e. a total of 19 of the 21 enrolled fishers working in the MPA area. Furthermore, we had informal conversations with the directors (*Patró Major*) of each of the *Confraries* from the different villages to complement and enhance the overall information we collected.

With the semi-structured interview format, the interviewer had the flexibility to change the order and wording of questions to suit the respondent and situation. Individuals were asked to answer several questions concerning their knowledge and beliefs about their activities in the *Cap de Creus*, their conception of small-scale fisheries, and their main concerns regarding the immediate future. More specifically, the questions in the Table 19.1. were planned to assess:

In addition, we used data provided by the fishing guilds and *Confraries* to quantitatively evaluate the current situation of small-scale fisheries in CCNP. This data supported our fieldwork analysis.

**Table 19.1** Interview questions used in the fieldwork

| |
|---|
| Who goes fishing? (crew) |
| What role the family plays in different small-scale fisheries' tasks? |
| Where the fishers fish in*Cap de Creus*? (Show it on the map) |
| How fishers decide where to fish and for what kind of fish species? |
| How often do fishers go fishing? |
| Are there new species that they fish? (explain the appearance/disappearance of species) |
| Where do fishers sell the catches? |
| Are fishers well paid? (Their own understanding)id? |
| What are the differences between the catch of small-scale fishers and this of large-scale fishers? |
| Have there been changes in terms of the consumption of seafood? |
| What are the indicators of conflicts and/or difficulties in the small-scale fisheries sector (with recreational boats, trolling boats, the administration etc.)? |
| Do fishers define themselves as small-scale fishers? Why? How do they characterize small-scale fishers? |
| How do fishers feel about CCNP and it has improved the sector? Or do they think it has not improved the sector? |
| How canCCNP improve the small-scale fisheries sector? |
| Who and how should decisions be taken that affect the small-scale fisheries sector?And how should regulations apply? |
| Do you think that promoting local marine products (KM0) would be a viable option to promote the sector? |
| Would this be (e.g initiatives promoting local marine products such as through short food supply chains) a solution that can be combined with eco-tourism initiatives? |

## What Are Small-Scale Fisheries?

According to fieldwork data analyzed, small-scale fisheries can be defined in general by (1) their limited fishing capacity; (2) the traditional ecological knowledge (TEK) needed to decide what, when, and how to fish according to the season (otherwise known as 'selective fisheries'); (3) specific relationships (of care and reciprocity) between people and environment; (4) small units of production; and (5) socio-ecological embbededness, what actually summarizes all the before parameters.

Our focus will now be on how small-scale fisheries are more broadly defined in the *Cap de Creus* MPA (Fig. 19.1).

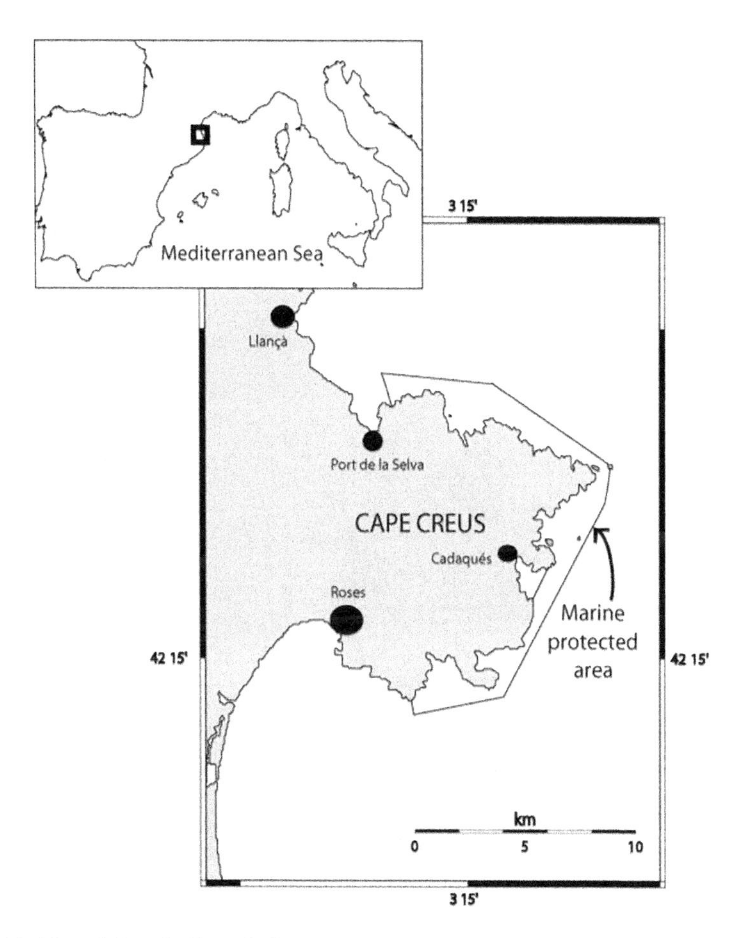


**Fig. 19.1** Map of Cape de Creus, indicating the protected area zone

## *Fishing Capacity: Small-Scale Fisheries as a 'Limited Class'*

Small-scale fishers are culturally defined as being a limited class or as fisher's say:' small-scale fisheries have a limited framework (interview 4).' A limited class means fisheries are limited by their own characteristics that are specific to local environmental conditions. Overall, boat lengths are less than 9 m, the numbers of nets are between 12 and 60 (units) at most,[5] and there are maximum two crew.

As one fisher put it: "For me small-scale fishing is that carried out with a boat of less than 9 m in length with just a one person crew, and respectful of weather conditions. Small-scale fishing is the "tradition" that our grandparents followed: when the weather was bad, they did not go fishing! They repaired the nets, they worked in the vineyard or in horticulture, or in the olive groves ...This is what has been done lifelong. We are not about to build bigger boats to carry more crew or set long lines, traps, and I do not know what else ... (interview 5) (Fig. 19.2)."

In this part of the Mediterranean, the small length of the boats makes fishers vulnerable to the strong northerly winds, known locally as the *Tramuntana*. The small boat size of fishers in *Cap de Creus* imposes constraints upon fishers that make them respect the adverse meteorological and environmental conditions.

Given small boat sizes, loading capacity is also restricted. Likewise, the space on boats to carry crew members is also limited which places further constraints on

**Fig. 19.2** Artisanal fishermen working at sea (Photo credit: Josep Lloret)

[5] We have calculated, based on the data gathered that small-scale fishers use an average of 30.8 nets per year.

fishing effort. That is, fishing effort is not only regulated by legislation pertaining to the exploitation of resources and by permits limiting on-board capacity to two, but also by the traditional design of the boats.

This limited class of vessels also means restrictions on the type of gears used. Given all these constraints, we can say that small-scale fishers have limitations imposed on them that consequently helps them limit the overexploitation of resources. Nevertheless, there are fishers who may not heed to the constraints that define this class of fishers. But in general small-scale fisheries is an attitude and an ideology about how to relate to the environment.

## *Traditional Ecological Knowledge and Nature-Culture Relationships*

Despite some small technological improvements, fishers define small-scale fisheries as a 'manual technique.' Although this does not imply the rejection of technological progress, what it does mean is that small-scale fishers continue to make use of first-hand knowledge (so-called traditional ecological knowledge) of meteorology and marine currents, their fishing grounds, the seasonality of target species, and their understanding of what fish to catch which season. That is, they embrace knowledge that will help them in their decisions about where, how, and which nets (trammel nets or gillnets) to use for fishing and what to fish at particular points of time.

Maintaining the manual character of small-scale fishing requires using traditional knowledge. Doing so is understood as part of a process of constant interaction with the environment, whose resources are carefully selected and used. It implies a constant dialogue between fishers and their environment. Therefore, fishers try to maintain a symbiotic relationship between human culture and nature. Selective methods of fishing and seasonal practices are all products of this symbiotic human culture-nature relationship. What can be viewed as a relationship of care makes fishers responsible for quantifying their daily fishing effort and calculating its long term impact so that their continual harvesting of the in the long term if they want to continue harvesting the sea's resources does not result in depleting these same resources.

Working hours are another important dimension to fishers' relationship with nature. Working hours means, for example, how long nets remain at the bottom of the sea. The less time fish spend in the nets, the less the fish suffers and the less it can get damaged, resulting in improved quality and appearance for the market. Thus, fishers place restrictions on the time allowed between setting and hauling certain trammel nets. Trammel nets used to catch striped red mullet (*Mullus surmuletus*) cannot remain in the water for more than 4 h, approximately.[6] Nets are set in the early morning and hauled in at dawn just as the large-scale trawlers return from fishing.

[6] Information gathered from the fieldwork (Interview 10).

### *Small Units of Production and Social Reproduction in Small-Scale Fisheries*

Household production systems within the small-scale fisheries of *Cap de Creus* were based primarily on family labor (both on land and at sea). They were underpinned by intergenerational divisions of labor that combined fishing with horticulture. Traditionally, many fishers owned smallholdings (olive groves, vineyards, or vegetable gardens) and divided their time between working the land and fishing. A typical Mediterranean coastal household,[7] engaged in fishing as part a wider diversified livelihood strategy that also included the cultivation of horticultural crops and to some degree wage employment (Pi-Sunyer 1977).

However, in the 1960s and 1970s, the Spanish and Catalan economies switched from being a primary sector driven economy to a service sector driven economy. This greatly affected coastal villages as many members of households were enticed by new job opportunities. Women, in particular, found employment in the tourism industry (such as hotels for tourist accommodation and restaurants). The families of these women would gradually invest part of their savings to open new businesses (related to real estate and catering among other things) which, it was hoped, would provide a more stable basic family income. Meanwhile, small-scale fishing gradually became more of a subsidiary activity.

Consequently, the production unit of small-scale fisheries is no longer the family. In the words of one fisher (interview 04), diversification of livelihoods has transformed 'the world of small-scale fisheries.' In the past, kinship was embedded in all aspects of fishing community and, therefore, in all production activities as well. The reduced number of family members available for fishing has meant that workers are employed from outside the family. The additional costs of are hard to sustain in a fishing system that does not produce enough profit. On balance, all these changes that have taken place reduce the profitability of fisheries and increase the dependency on market forces, which consequently hampers the social reproduction of small-scale fisheries.

Small-scale fisheries cannot absorb the costs of salaried workers unless fishing effort is intensified. Traditionally, small-scale fishing involved piecework and the means of production expenses were assumed by both skipper and crew. Rubio-Ardanaz (1994) called it the 'ideology of common costs,' the aim being to reduce uncertainty and randomization to which the fisheries were subjected. Faced with these difficulties, and in order to obtain fixed, stable, and constant incomes, fishers preferred to go fishing alone or, at most, with a family relative. Only in the odd case did two or more fishers associate with each other.

As our study reveals, the progressive commoditization of production factors (vessels, maintenance, and net repairs, as well as the labor force) has meant that

[7]According to Braudel (1972, 144 in Pi-Sunyer 1977) "this is the traditional wisdom of the old mediterrancan way of life where the meagre resources of the land are added to the meagre resources of the sea".

production and social reproduction of small-scale fisheries is no longer in the control of families and depends more and more on market forces. At the same time, fisheries are unable to guarantee basic family incomes that other sectors do. Consequently, pluriactivity and complementary sources of income play a much more significant role for fisher families than they ever did before.

This situation has been abetted by the fact that since the 2008 global economic crisis, which particularly affected southern European countries, a large number of jobs have been lost in the *Cap de Creus* area. The economic crisis has also adversely affected fishing households by rendering unemployed the sons of traditional fishers who already had shore-based jobs. This has resulted in a greater dependency on parental income. There is a feeling, therefore, that sustaining the fisheries sector requires ensuring fishery rights to those who fish. Notwithstanding, fishers regret that ensuring their rights would not be enough so that they can meet the requirements of market competition and make a living according the modus vivendi of western contemporary society.

### *Socio-ecological Embeddedness*

Small-scale fishing in CCNP involves an economy embedded in techniques and traditional ecological knowledge (TEK), and in a wider set of social relations with nature (of reciprocity and caretaking).

Fishing rights have been established upon culturally and environmentally defined relationships between humans and nature, over many centuries. This is what drives the claims of fisher communities for sea tenure rights in CCNP. Fishers must compete against extractive industries for these rights but they do so also because they believe it will prevent environmental degradation. That is, fishers say that their age-old rights of access to resources respond to precise human-nature interactions and that they involve a certain 'attitude to fishing' (interview 13) that comprises a culturally-inherited legacy that has survived to the present day. Quite literally, they say it is "fisheries of our ancestors" (interview 5) and "fishing as it has always been done here" (interview 5). It is a legacy they have inherited from their history, their identity, and local fishing culture.

## Discussion: How to Implement the SSF Guidelines as Marine Stewardship?

The case study of the *Cap de Creus* MPA sheds light on the complexity of regulating small-scale fisheries in the Mediterranean and adopting a management strategy that fits the interests of all the diverse stakeholders that are commonly found in Mediterranean coastal zones. Despite the fact that, since 2004, various social and biological studies on small-scale fisheries (Gómez, Lloret, Riera, Demestre 2006; Gómez 2015; Lloret 2015) have been oriented towards implementing a co-management plan[8] called *Pla Rector d'Ús i Gestió* (PRUG) which aims to establish a local regulation of all marine activities adapted to the environmentally-specific characteristics and protection needs of CCNP, this plan has not yet been implemented. In fact, it is now under juridical review by the Catalan Government and open to suggestions and objections of citizens (thus addressing the participatory dimension recommended in Section 2.5 of the SSF Guidelines).

Tourist businesses, e.g. scuba diving clubs, boat cruise operators, and recreational fisheries complain about the rigidity of environmental protection measures which impose restrictions on tourism. Tourist operators run by local native people (which are not fishers) are asserting their rights with the assumption that the Natural Park is part of an age-old 'domain' inherited from their ancestors, legitimizing their claims over the territory. Regulation plans are perceived as a form of territorial dispossession and intrusion and are denounced by civil platforms such as 'We Defend Catalonia' (Plataforma Defensem Catalunya, n.d.). Accompanying these rights claims by tourist operators, somewhat paradoxically, are also demands for 'responsible fisheries' (with reference to recreational fisheries and diving) which are seen as rational and counterposed with rigidness of protection measures. Once the consultation process has finished, the claims of the recreational fisheries sector will be considered in the hope of agreeing to a set of regulations that take account of their points of view but which also meet the objectives of sustainable management of resources which will be decided by the governing board of CCNP.

In contrast, small-scale fishers have pointed out that PRUG focuses only on small-scale fisheries – understood in its standard sense. They say current rules leave tourism unregulated which consequently hinders fishing and depletes resources. The coastal marine environment of the Mediterranean Sea, and particularly that of MPAs, is facing increased levels of disturbance by recreational activities, including recreational fisheries and scuba diving (see for example Badalamenti et al. 2000; Lloret and Riera 2008) which often compete with small-scale fisheries for limited coastal space and resources (Lloret et al. Forthcoming). This competition is all the more intense because both small-scale fishers and recreational fishers target mostly a few species despite abundant diversity in the waters (Lloret and Font 2013; Lloret et al. Forthcoming). Competition between sectors also involves competition over human and socio-cultural resources. Small-scale fisheries have been the losers as they have progressively declined in Cape Creus (Gómez, Lloret, Riera, Demestre 2006) whereas recreational fishing has continued to expand (Lloret 2010). Other leisure activities too are expanding and impinging on the remaining marine resources and habitats available small-scale fisheries (Lloret and Riera 2008; Lloret 2010). The disappearance of small-scale fishing is also threatening the area's cultural heritage which we have illustrated contributes to sustainable resource management (Gómez 2006). As we have said before, the preservation of this cultural heritage

[8] Involving the managers of CCNP, the small-scale fishers and the fisheries community organizations (the *Confraries*).

needs to go hand in hand with the preservation of the natural heritage in Mediterranean coastal areas.

CCNP is also part of the MPA network called MedPAN North (Mediterranean Protected Areas Network; www.medpan.org), a program headed by WWF-France with the participation of various countries from the northern Mediterranean that have MPAs in their territories. Some research studies on CCNP have served as a basis for developing the MedPAN North project, with financial assistance from the EU INTERREG program. The objective of this initiative is to build a network of MPAs in the Mediterranean in order to coordinate a number of management actions, following the recommendations of the conference on small-scale fisheries organized by FAO-GFCM-MedPAN, held in Algiers (Algeria) between 7th and 9th March 2016. That conference established the need to put into practice the principles of the SSF Guidelines so as to promote and secure small-scale fisheries in the Mediterranean and Black Sea.

The conclusions of this conference, based on an in-depth ethnography and various biological studies conducted in *Cap de Creus* in 2015, led to plans to implement diverse stewardship measures using the SSF Guidelines as a benchmark. That is to say, in a reciprocal manner, the SSF Guidelines lend support as socially-legitimate basic principles to prevailing claims by small-scale fishers for sea tenure rights while at the same time enforcing small-scale fishers' obligations regarding sustainable resource management (in line with Sections 5.13, 5.14, and 5.15 of the SSF Guidelines). CCNP, as the main stewardship entity, will assist fishers in meeting this commitment while at the same time ensuring fishers' tenure rights are upheld (in line with Section 5.2 of the SSF Guidelines).

Sections 5.13, 5.14, and 5.15 of the SSF Guidelines stipulate that small-scale fishers' tenure rights must be balanced by duties and co-responsibility by including fishers in a participatory process of sustainable resource management. By implementing stewardship measures that ensure small-scale tenure rights, it is expected that small-scale fishers will embrace the idea of responsible fishing (in line with the FAO's Code of Conduct for Responsible Fisheries) by means of voluntary commitments within the framework of national and European regulations.

This will ensure continuity with the participatory process involved in the implementation of CCNP's own co-management plan (PRUG). Regulation guidelines are to be laid down that value small-scale fisheries not only as sustainable extractive fisheries but also as fisheries with an unique 'attitude to fishing' as described in Section 4.4, namely an attitude that is shaped by a set of principles inherited from their ancestors and which is socio-ecologically embedded. The value of small-scale fishing is that it fortifies a respectful relationship between fishers and the environment; a relationship that is economically and socially beneficial. These stewardship measures will ensure that small-scale fishers engage voluntarily in sustainable resource management as part of their livelihood.

The situation of small-scale fisheries throughout the Mediterranean is similar. The few remaining are underrepresented in the *Confraries* and in the fishing sector as a whole. Consequently, it has been considered appropriate to establish alliances with other small-scale fisher communities in order to achieve more visibility and weight in the political decision-making process at the European level, particularly within the new Common Fisheries Policy (CFP). The need for greater representation is reflected in Section 5.17 of the SSF Guidelines which highlight the importance of involving community organizations in small-scale fisheries.

Complementing the stewardship initiatives, the managers of CCNP will seek the support of the *Confraries* in developing practical initiatives to promote the exceptionally healthy produce of local fisheries as part of the Mediterranean diet, which is traditionally rich in seafood, and broaden the appeal of this produce to a greater range of consumers by making it more affordable. These initiatives aim to directly link the fishers' commitment to resource protection with livelihoods so as to demonstrate the positive economic outcomes derived from protecting resources as part of a feedback process involving the environmental, social, and economic changes at the 'interface' of production, distribution, and consumption.

To meet this goal, namely to balance respect for sustainability measures and livelihoods in small-scale fisheries, CCNP being in charge of stewardship will try to ensure support for small-scale fishers in the face of market competition. Market competition is one of the challenges of the globalization of fish markets, especially considering that one of the defining characteristics of small-scale fisheries is the limited size of catch, which puts small-scale fishers at a disadvantage vis-à-vis others.

Whereas landing prices for fish have fallen, consumers complain that the final price they pay has risen. On the other hand, consumption of marine products has changed; what used to be an everyday food is now more sporadically available and limited species dominate the market, reducing local food biodiversity at these markets. Moreover, seafood from outside the EU and frozen fish products from trawling or purse-seining are usually cheaper than the higher quality fish caught by small-scale fishers and are therefore a more affordable option for low-income families and for low income consumers. In order to address the disparity between landing price and final price, a number of initiatives by the *Confraries* of Port de la Selva and Cadaqués are already underway. One initiative aims to establish *short food supply circuits* (SFSC) that avoid the costs of intermediaries while another seeks to promote healthy local marine products and broaden food biodiversity to include once well-known species, by providing traditional recipes aimed at recovering culinary traditions (see *Slow fish initiatives in Catalonia*: http://slowfood.com/slowfish/pagine/esp/pagina.lasso?-id_pg=242). Although such initiatives are, for the moment, aimed at promoting products from all kinds of fisheries, it is expected that priority will soon be given to small-scale fisheries' products in particular.

Promoting a 'fish culture' by recovering local marine food biodiversity is another goal of the stewardship initiatives. People are being encouraged to buy fish caught using sustainable small-scale fishing techniques that respect what the seas offer. This is also known as the 'stewardship market' where consumers can purchase 'stewardship fish,' something which is already in place at other Catalan MPAs (Stewardship fish of Submon company n.d.). Similarly, various initiatives introducing a closed, on-line platform for buying and selling fish are being developed in CCNP fishing communities. At the moment these initiatives too are aimed at all kind of fisheries products without distinguishing between small-scale and industrial.

Therefore, one part of the stewardship program in small-scale fisheries is to involve SFSC in promoting sustainable fisheries' products and making the sustainable food supply chain more efficient, which would be healthy for humans and the seas and ensure the livelihoods of socio-ecologically embedded small-scale fishers. All these actions are aimed at supporting sustainable small-scale fisheries (including natural resources and fishers' livelihoods) in CCNP, which is at present not fully guaranteed. One small-scale fisher from Port de la Selva said that he hopes "that at least in tomorrow's future, my children can make a living from this if they want to" (interview 07).

## Conclusions

In view of the crisis facing small-scale fisheries and the conflicts we identified regarding the implementation of a management plan for fisheries and the park in general, we conclude that using the voluntary SSF Guidelines as stewardship measures constitutes a unique opportunity to establish effective regulations for MPAs such as *Cap de Creus*. A set of regulations based on the SSF Guidelines that takes into account small-scale fisheries as they are understood in relation to their particular cultural and environmental context can go a long way in addressing the needs of small-scale fishers. It is important to understand small-scale fisheries as they are defined locally as this helps identify vulnerable groups locally. National legislation at present does not do this and is often at odds with social perceptions as to who is vulnerable. In the context of Mediterranean MPAs, the SSF Guidelines have emerged as a useful framework to highlight the value of the embedded nature of small-scale fisheries in local culture and environment - a value that needs preserving and strengthening in all contexts. These Guidelines can be used to support the great diversity of small-scale fisheries which are collectively recognized on paper through national legislation. However, this legislation does not leave room for local differences within the fisheries, often resulting in small-scale fisheries being confused with large-scale fisheries or not being clearly established the differences.

Furthermore, the SSF Guidelines may enable the establishment of regulations at the local level that are adjusted to the reality of small-scale fisheries in practice and that take into account real social needs.

Finally, it should be remembered that the fisheries sector has always been traditionally regulated by their own regulatory systems established through their own autonomous organizations (the *Confraries*) that function in much the same way as cooperatives. In this context, the SSF Guidelines, which consist of a non-binding declaration of good intentions under the auspices of the FAO, will give greater impetus to co-management and inspire much greater respect from stakeholders.

**Acknowledgements** We thank all artisanal fishermen who have participated in this study, as well as the director of the Natural Park of Cap de Creus, Victoria Riera, for their support and contribution to this research. We thank also the anonymous reviewers of this manuscript for their comments and suggestions than have helped to improve it.

## References

Acheson, J. M. (1981). Anthropology of fishing. *Annual Review of Anthropology, 10*, 275–316.

Alegret, J. L. (1995). *Co-management of resources and conflict management: The case of the fishermen's confreries in Catalonia*. Hojbjerg: MARE.

Badalamenti, F., Ramos, A., Voultsiadou, E., Sanchez-Lizaso, J. L., D'Anna, G., Pipitone, C., Mas, J., Ruiz Fernandez, J. A., Whitmarsh, D., & Riggio, S. (2000). Cultural and socioeconomic impacts of Mediterranean protected areas. *Environmental Conservation, 27*, 1–16.

Bellido, J. M., Quetglas, A. C., Garcia, M., Garcia, T. & González, M. (2014). *The obligation to land all catches – Consequences for the Mediterranean*. European Parliament, Directorate-General for Internal Policies, Policy Department B: Structural and Cohesion Policies, Fisheries. http://www.europarl.europa.eu/studies. Accessed 2 Feb 2016.

Braudel, F. (1972). *The Mediterranean and the Mediterranean world in the age of Philip II*. New York: University of California Press.

Cadoret, B. (1984). Sources pour l'ethnographie maritime. *Terrain, 2*, 33–44.

Carter, D. W. (2003). Protected areas in marine resource management: Another look at the economics and research issues. *Ocean and Coastal Management, 46*, 439–456.

Claudet, J., & Guidetti, P. (2010). Fishermen contribute to protection of marine reserves. *Nature, 464*, –673.

Colloca, F., Crespi, V., Cerasi, S., & Coppola, S. R. (2004). Structure and evolution of the artisanal fishery in a southern Italian coastal area. *Fisheries Research, 69*, 359–369.

Di Franco, A., Bodilis, P., Piante, C., Di Carlo, G., Thiriet, P., Francour, P., & Guidetti, P. (2014). *Fisherfolk engagement, a key element to the success of artisanal fisheries management in Mediterranean marine protected areas, MedPAN north project*. Port-Cros: WWF France.

Dufour, A. H. (1988). Pêcheurs et prud'hommes. A propos des Salins d'Hyères. *Terrain*, 66–84.

EEA (2016). *Seafood in Europe. A food system approach for sustainability* (European Environmental Agency Report No 25/2016). http://www.eea.europa.eu/publications/seafood-in-europe-a-food. Accessed 2 Feb 2016.

FAO. (2015). *Voluntary guidelines for securing sustainable small-scale fisheries in the context of food security and poverty eradication*. Rome: Food and Agriculture Organization of the United Nations.

Font, T., & Lloret, J. (2015). Improving the efficiency of MPAs as fisheries management tools and benefits from involving the small-scale fisheries sector. *MedPAN Background Report for Panel 3. FAO/GFCM Regional conference for building a future for small-scale fisheries in the Mediterranean and Black Seas*. Algers: Algeria.

Galvan-Tudela, A. (1988). *La Antropología de la pesca: Problemas, teorías y conceptos*. La Laguna: Universidad de la Laguna.

Geistdoerfer, A. (1984). Ethnologie des activités halieutiques. *Anthropologie Maritime, 1*, 5–10.

Giovanni, V. et al. (1996). *Anthropologie et droit comparé des pêches en Mediterranée nord-occidentale. Les propriétés de résistance des systèmes de gestion*. Contrat TR/MED 92/017, commission des communautés européennes DG XIV.

Gómez, S., & Lloret, J. (2016). La pesca artesanal a Cap de Creus: una mirada al futur. In *Parc Natural de Cap de Creus*. Girona: Consell Social Universitat de. ISBN 978-84-393-9511-9.

Gómez, S. (2015). *Estudi social de la pesca artesanal a Cap de Creus*. Parc Natural de Cap de Creus: Departament de Medi Ambient de la Generalitat de Catalunya.

Gómez, S., Lloret, J., Riera, V., & Demestre, M. (2006). The decline of the artisanal fisheries in Mediterranean coastal areas: The case of cap de Creus. *Coastal Management, 34*, 217–232.

Granovetter, M. (1985). Economic action and social structure: The problem of embeddedness. *American Journal of Sociology, 91*, 481–510.

Griffiths, R. C., Robles, R., Coppola, S. R., & Camiñas, J. A. (2007). *Is there a future for artisanal fisheries in the western Mediterranean?* Rome: Food and Agriculture Organisation of the United Nations.

Guillou, A., & Crespi, V. (1999). *Enquête-cadre concernant la répartition, la composition et l'activité des petits métiers dans le Golfe du Lion*. Sète: IFREMER.

Guidetti, P., & Claudet, J. (2010). Co-management practices enhance fisheries in marine protected areas. *Conservation Biology, 24*, 312–318.

Guidetti, P., Bussotti, S., Pizzolante, F., & Ciccolella, A. (2010). Assessing the potential of an artisanal fishing co-management in the marine protected area of Torre Guaceto (southern Adriatic Sea, SE Italy). *Fisheries Research, 101*, 180–187.

Guyader, O., Berthou, P., Koutsikopoulos, C., Alban, F., Demanèche, S., & Gaspar, M. B. (2013). Small scale fisheries in Europe: A comparative analysis based on a selection of case studies. *Fisheries Research, 140*, 1–13.

Himes, A. (2003). Small-scale Sicilian fisheries: Opinions of artisanal fisheries and sociocultural effects in two MPA case studies. *Coastal Management, 31*(4), 389–408.

Jacquet, J., & Pauly, D. (2008). Funding priorities: Big barriers to small-scale fisheries. *Conservation Biology, 22*, 832–835.

Jentoft, S. (2005). Beyond fisheries management: The *Phronetic* dimension. *Marine Policy, 30*, 671–680.

Kirk, N., & Memon, A. (2010). Sustainable governance of marine fisheries: A socio-ecological embeddedness perspective. In *Proceedings of the 4th international conference on sustainability engineering and science*. Auckland: Society for Sustainability Engineering and Science.

Leleu, K., Pelletier, D., Charbonnel, E., Letourneur, Y., Alban, F., Bachet, F., & Boudouresque, C. (2014). Métiers, effort and catches of a Mediterranean small-scale coastal fishery: The case of the Côte Bleue Marine Park. *Fisheries Research, 154*, 93–101.

Lloret, J. (2010). Environmental impacts of recreational activities on the Mediterranean coastal environment: The urgent need to implement marine sustainable practices and ecotourism. In A. Krause & E. Wier (Eds.), *Ecotourism: Management, development and impact* (pp. 135–157). New York: Nova Science Publishers Inc.

Lloret, J. (2015). *Seguiment de la pesca artesanal al parc natural de Cap de Creus*. Parc Natural de Cap de Creus: Departament de Medi Ambient de la Generalitat de Catalunya.

Lloret, J., & Font, T. (2013). A comparative analysis between recreational and artisanal fisheries in a Mediterranean coastal area. *Fisheries Management and Ecology, 20*, 148–160.

Lloret, J., & Riera, V. (2008). Evolution of a Mediterranean coastal zone: Human impacts on the marine environment of Cap de Creus. *Environmental Management, 42*, 977–988.

Lloret, J., Cowx, I. Cabral, H., Castro, M. Font, T., & Gonçalves, J.M.S., et al.. in press. Small-scale coastal fisheries in European Seas are not what they were: Ecological, social and economic changes. Marine Policy (in press). http://dx.doi.org/10.1016/j.marpol.2016.11.007

Lloret, J., Ratz, H. J., Lleonart, J., & Demestre, M. (2016). Challenging the links between seafood and human health in the context of global change. *Journal of the Marine Biological Association of the United Kingdom, 96*(1), 29–42.

Macfadyen, G., Salz, P., & Cappell, R. (2011). *Characteristics of small-scale coastal fisheries in Europe: Study*. Directorate general for internal policies, Policy department B: Structural and cohesion policies: Fisheries.

McCay, B. (1996). Common and private concerns. In S. Hanna, C. Folke, & K. G. Maler (Eds.), *Rights to nature: Ecological, economic, cultural and political principles of institutions for the environment* (pp. 111–126). Washington, DC: Island Press.

McCay, B. (2001). Environmental anthropology at sea. In C. Crumley (Ed.), *New directions in anthropology and the environment* (pp. 254–272). Walnut Creek: Altamira Press.

Milazzo, M., Chemello, R., Badalamenti, F., Camarda, R., & Riggio, S. (2002). The impact of human recreational activities in marine protected areas: What lessons should be learnt in the Mediterranean Sea. *Marine Ecology, 23*, 280–290.
Naji, M. (2013). *Enhancing small-scale fisheries value chains in the Mediterranean and Black Sea*. Rome: General Commission for the Mediterranean.
Paavola, J., & Ropke, I. (2008). Sustainability and environment. In J. Davis & W. Dolfsma (Eds.), *Elgar companion to social economics* (pp. 11–27). London: Edward Elgar.
Pi-Sunyer, O. (1977). Two states of technological change in a Catalan fishing community. In M. E. Smith (Ed.), *Those who live from the sea* (pp. 41–56). Saint Paul: West Publishing Co..
Plataforma Defensem Catalunya. (n.d.). Homepage: 'We defend catalonia'. http://www.defensem-catalunya.org/. Accessed 11 Dec 2016.
Polanyi, K. (1944). *The great transformation: The political and economic origins of our time*. Boston: Beacon Press.
Reyes, N., Bahuchet, S., & Wahiche, J.D. (2015). Quelle définition des petits métiers de la pêche ? *Revue d'ethnoécologie, 7*.
Rubio-Ardanaz, J. A. (1994). *La Antropología marítima subdisciplina de la Antropología Sociocultural*. Bilbao: Universidad de Deusto.
Stewardship fish of Submon company. (n.d). *Stewardship mission statement*. http://www.canyons-delmaresme.cat/peix-de-custodia-2/peix-de-custodia/. Accessed 14 Dec 2016.
Techera, E. J. (2010). *Legal pluralism, customary law and environmental management: The role of international law for the South Pacific* (Macquarie University Law School Legal Studies Working Paper No. 2010–01).
Von Benda-Beckmann, F. (2002). Who's afraid of legal pluralism? *The Journal of Legal Pluralism and Unofficial Law, 34*(47), 37–82.
Von Benda-Beckman, F., Von Benda-Beckman, K., & Wiber, M. (2006). *Changing properties of property*. New York: Berghahn.
Weber, J. (1992). *Le rôle des organisations professionnelles dans la gestion des pêches en Méditerranée (Espagne, France, Italie, Grèce)*. ASCA Report.
Westhoek, H., Rood, T., Van den Berg, M., Janse, J., Nijdam, D., Reudink, M., & Stehfest, E. (2011). *The protein puzzle: The consumption and production of meat, dairy and fish in the European Union*. The Hague: PBL Netherlands Environmental Assessment Agency.
World Bank. (2010). *The hidden harvests: The global contribution of capture fisheries*. Washington, DC: World Bank.

# Part VI
# Broadening Participation

The SSF Guidelines do not only speak to states but also to civil society organizations, recognizing their important contribution to overall fisheries governance. In Chap. 20, María José Espinosa-Romero, Jorge Torre, José Alberto Zepeda, Francisco Javier Vergara Solana, and Stuart Fulton, who are themselves civil society organization representatives, take us to Mexico, which has a huge small-scale fisheries sector. Their chapter provides insights into, and recommendations on, how civil society organizations and other stakeholders can help the implementation of the SSF Guidelines. In Chap. 21, Patrick McConney, Terrence Phillips, Nadine Nembhard and Mitchell Lay carry us across the Atlantic and into the Caribbean, where fishing industry workers and their organizations became engaged early on in the process of developing the SSF Guidelines. Here, civil society groups have become the champions of the Guidelines, but are facing limited collaboration from governments. However, fishing people are showing increasing capacity for self-organization, advocacy and policy influence. Sérgio Macedo Gomes de Mattos, Matias John Wojciechowski, Alison Elisabeth Macnaughton, Gustavo Henrique G. da Silva, Allyssandra Maria Lima R. Maia, and Joachim Carolsfeld bring us to Brazil in Chap. 22. They describe a project that builds institutional and community capacity and linkages between government, university researchers, and local fishing associations, which aims to improve the livelihoods and well-being of 'marisqueiras', women and families. Much like the SSF Guidelines, the project promoted an integrated approach to equitable development of sustainable fisheries that include various components such as participatory research, co-management, and the empowerment of women. Involving stakeholders at various levels as a way to build community capacity and empowerment is also the theme of Chap. 23 by Ratana Chuenpagdee, Kim Olson, David Bishop, Meike Brauer, Vesna Kereži, Jonas Plaan, Sarah Pötter, Victoria Rogers, and Gabriela Sabau. Their chapter is an illustration of how the SSF Guidelines have relevance also for the Global North. Their case study is situated in Newfoundland, Canada, where small-scale fisheries remain active, despite the cod fishery moratorium in 1992, which had devastating effects on coastal communities. Although poverty is not a main issue, rural small-scale livelihoods are still at risk and need support by civil society participants and local communities.

# Chapter 20
# Civil Society Contributions to the Implementation of the Small-Scale Fisheries Guidelines in Mexico

**María José Espinosa-Romero, Jorge Torre, José Alberto Zepeda, Francisco Javier Vergara Solana, and Stuart Fulton**

**Abstract** Small-scale fisheries contribute about half of global catches whilst employing approximately 90% of the people directly dependent on capture fisheries. Taking into account the importance of this sector in the global economy, and its contribution to nutrition and livelihoods, in 2015 the Food and Agriculture Organization of the United Nations published the *Voluntary Guidelines for Securing Sustainable Small-Scale Fisheries in the Context of Food Security and Poverty Eradication (SSF Guidelines)*. This chapter describes the contributions, challenges, and lessons learned from implementing the SSF Guidelines, from the perspective of a marine conservation civil society organization (CSO) that works on providing effective solutions for small-scale fisheries management in Mexico in direct collaboration with stakeholders. Mexico is a developing country, with a small-scale fishing force of over 74,000 registered boats, in which diverse fisheries face many challenges to secure livelihoods whilst simultaneously ensuring sustainability and adapting to changing environmental conditions. The SSF Guidelines represents a landmark document that highlights the importance of the small-scale fisheries sector and provides significant guidance to states and stakeholders for ensuring the long-term sustainability of small-scale fisheries. Finally, the chapter provides

M.J. Espinosa-Romero (✉) • J. Torre • S. Fulton
Comunidad y Biodiversidad, A.C, Guaymas, SO, Mexico
e-mail: mespinosa@cobi.org.mx

J.A. Zepeda
Centro Interdisciplinario de Ciencias Marinas, Instituto Politécnico Nacional, La Paz, B.C.S, Mexico

Independent Consultant, Prolongación Baja California s/n, ampliación Centenario, La Paz, B.C.S, Mexico
e-mail: joalzedo@hotmail.com

F.J. Vergara Solana
Centro Interdisciplinario de Ciencias Marinas, Instituto Politécnico Nacional, La Paz, B.C.S, Mexico
e-mail: vs_fj@yahoo.com



S. Jentoft et al. (eds.), *The Small-Scale Fisheries Guidelines*, MARE Publication Series 14, DOI 10.1007/978-3-319-55074-9_20

insights into and recommendations on how CSOs and other interested stakeholders can foster the implementation of the Guidelines.

**Keywords** Civil society organizations • Small-Scale fisheries • Governance of tenure • Decent work • Gender equality • Climate change • Market incentives • FAO guidelines

## Introduction

Small-scale fisheries contribute about half of global catches whilst employing about 90% of the people directly dependent on capture fisheries (FAO 2014). Strengthening small-scale fisheries has been recognized as an important strategy not only for employment, but also for addressing food security and poverty issues. Mismanagement of small-scale fisheries has ecological, socioeconomic, and governance implications (FAO 2014).

It has been recognized that a proper fisheries management system not only involves the proper use of fish stocks, but a balance between environmental, social, and economic objectives (Charles 1988; Ahmed 1991; Hanna 1994). In 1995, the Food and Agriculture Organization (FAO) published the *Code of Conduct for Responsible Fisheries* to provide guidance on how to balance these objectives (see FAO 1995). Furthermore, in 2015, FAO published the first international instrument focused on small-scale fisheries in the context of food security and poverty eradication, the *Voluntary Guidelines to Securing Sustainable Small-Scale Fisheries* (SSF Guidelines). These guidelines provide significant advice to states and stakeholders for ensuring the long-term sustainability of small-scale fisheries, whilst sustaining people's livelihoods.

### *Small-Scale Fisheries in Mexico*

For the past 20 years Mexico has been one of the top 20 fish producing nations in the world. During this period, annual catches have fluctuated around 1.2 million tons (FAO 2014). Small-scale fisheries account for 40% of total catch, including 74,055 registered boats (CONAPESCA 2013) and an estimated 222,165 fishers (Moreno-Báez et al. 2010). These numbers do not include the illegal, unreported, and unregulated (IUU) sector, which is estimated to be equal to, or even greater than, the legal fishing effort and catch (Cisneros-Montemayor et al. 2013). At present, 17% of Mexican fisheries are overexploited, 70% at the maximum sustainable yield (MSY), and only 13% underexploited (DOF 2012a). With 41% of the Mexican population living in coastal municipalities (CIMARES 2010) and 11,500 coastal communities with less than 15,000 habitants (Gabriel-Morales and Perez-Damian

2006) relying heavily on marine resources and ecosystems, sustainable fisheries are a necessity for securing employment, income, and food for many people.

Examples of small-scale fishers and communities taking leadership in managing their resources through sophisticated and sustainable approaches for fisheries management exist (Espinosa-Romero et al. 2014a). To create and enforce local rules for sustainable resource use, groups of fishers have used their traditional ecological knowledge, have established key alliances with other stakeholders, and have learned how to successfully implement ecosystem-based approaches in their region (see Espinosa-Romero et al. 2014a). There are examples of small-scale fisheries fulfilling international standards for sustainable fishing set by certifying bodies such as the Marine Stewardship Council (MSC) and the Seafood Watch Program of Monterey Bay Aquarium (MBA). Small-scale fishers are already implementing practices for maintaining the health of stocks and ecosystems (e.g., Senko et al. 2014) and for building robust, participatory, and transparent governance systems (Espinosa-Romero et al. 2014b).

However, Mexican small-scale fisheries still face several challenges. Existing top-down policies in Mexico have disenfranchised small-scale fishers (Finkbeiner and Basurto 2015). The lack of socioeconomic information at the local level has limited the development of integrated public policies for inland (Mendoza-Carranza et al. 2013), coastal, and marine fisheries (Ramírez-Rodríguez and Ojeda-Ruíz 2012; Robles-Zavala 2014; Méndez-Medina et al. 2015; Marín-Monroy and Ojeda-Ruíz 2016; Zepeda-Domínguez et al. 2016). The conflict between conservation and fisheries management objectives continues, and is particularly conspicuous in the Upper Gulf of California where environmental policies conflict with those who are directly dependent on fisheries (Bobadilla et al. 2011; Senko et al. 2014; López-Torre et al. 2016; Zepeda-Domínguez et al. Forthcoming). Public participation in fisheries management faces operational challenges (Ramírez-Rodríguez and Ojeda-Ruíz 2012; Zepeda-Domínguez et al. In press) as, according to the General Law for Sustainable Fisheries and Aquaculture, only 'legal and well-organized' fishers can participate in decision-making, excluding a significant percentage of traditional fishers who operate without legal permits or access rights (Robles-Zavala 2014; Finkbeiner and Basurto 2015) and stakeholders (e.g., civil society organizations or CSOs) who have played a significant role in fisheries management (see Espinosa-Romero et al. 2014b).

## *Role of CSOs in Mexican Fisheries Management*

Worldwide, CSOs are playing a key role in fisheries management. They usually represent public interests in decision-making forums (Ratner and Allison 2012), have demonstrated an ability to move from problem identification towards problem solving (Dunn 2005), and have acted as bridging organizations between fishers, managers, academics, and other stakeholders (Berkes 2010). This phenomenon has been observed in Latin America (McConney et al. 2014) and continues to spread due to international tendencies towards more participatory approaches to fisheries management (Hernández and Kempton 2003).

In Mexico, the Federal Law to Promote Activities of Civil Society Organization Law (CSOs Law) states that CSOs can enhance public participation in addressing the public interest, namely; promoting the inclusion of women, the defense of human rights, the development of indigenous communities, the sustainable use of natural resources, and the creation of social capital. CSOs have the right to participate in forums and structures created by the federal government for stakeholder participation; support the government in the mentioned activities; and take part in the planning, implementation, and follow-up on policies, programs, projects, and processes led by the federal government (DOF 2012b).

In practice, national, and international CSOs operating in Mexico have moved from being environmental conservation advocates to multi-scale governance operators (Espinosa-Romero et al. 2014b; Espinoza-Tenorio et al. 2015), and have found a niche in balancing conservation with livelihood and development objectives (McConney et al. 2014). In addition, they have supported capacity building programs, brought technical expertise to the fishing sector (Espinosa-Romero et al. 2014b; Méndez-Medina et al. 2015), and facilitated the participation of small-scale fishers in decision-making forums (Cinti et al. 2014). Also noteworthy is the increasing number of collaborations and alliances between CSOs, governments at different levels, and academia to work together towards the sustainability of small-scale fisheries management (Espinosa-Romero et al. 2014a, b; Moreno et al. 2016).

Comunidad y Biodiversidad (COBI) is a Mexican marine conservation CSO founded in 1999. Its mission is to develop effective participatory approaches for fisheries management and marine biodiversity conservation (Espinosa-Romero et al. 2014b). A multidisciplinary team operates along four national strategies: (1) capacity building to strengthen skills of local leaders and fishing organizations for achieving sustainable fisheries; (2) implementation of international standards for sustainable fishing; (3) implementation of no-take zones for fisheries and ecosystem restoration; and (4) development of formal institutional arrangements for sustainable fishing and marine conservation (Table 20.1). COBI's programs operate in four of the 17 Mexican coastal states (Fig. 20.1), where 70% of the country's total industrial and small-scale fisheries production is concentrated. COBI has actively participated in the eco-certification of four Mexican fisheries (Caribbean Spiny Lobster, Pacific Red Rock Lobster, Pacific Yellowtail, and Monterey Sardine in the Gulf of California), and promoted fisheries improvement plans in five other small-scale fisheries (abalone, clams, pen-shell, swimming crab and the snapper-grouper complex). To date, 80,000 hectares have been fully protected by communities, with COBI's support, under the three existing management instruments in Mexico (core zones in marine protected areas, fish refuges, and voluntary fully-protected marine reserves). COBI has distinguished itself from other CSOs by producing applied science papers on sustainable fisheries and marine conservation each year (an average of three academic papers) that make use of traditional knowledge and which have been utilized as a reference for fisheries management in data-poor areas. Finally,

**Table 20.1** COBI's national strategic lines

| **Effective participation of all sectors** | | | | | | |
|---|---|---|---|---|---|---|
| **Strategy** | **Capacity building** | | | **Sustainable fisheries** | **Marine reserves** | **Public policy** |
| *Goal* | *Human development and collective action* | | | *Adoption of international standards for sustainable fishing* | *Fishery and ecosystem restoration* | *National impact through the escalation of local models* |
| | Community leaders | Cooperatives | Committees | | | |
| Main components | Human dimensions | Cooperativism values and legal framework | Creation with key representatives of the fishery | Healthy stock | Design | Collaborative |
| | Technical knowledge | Administration and financial support | Training program | Healthy ecosystem | Implementation | Work with the Executive |
| | Implementation of sustainable fishing projects | Investment in sustainable fisheries | Evaluation and adaptation of training program | Participatory and transparent governance system | Monitoring | Collaborative work with the Legislative |
| | | | | Social fairness | Evaluation and adaptation | Collaborative work with the |
| **Citizen science** | | | | | | |

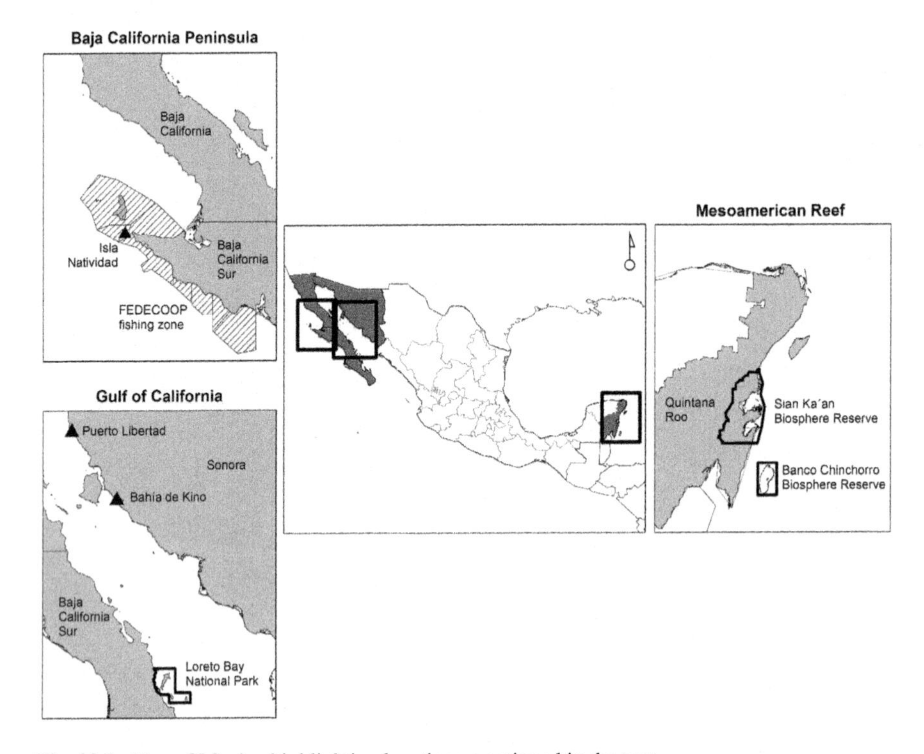


**Fig. 20.1** Map of Mexico highlighting locations mentioned in the text

COBI's efforts have informed two initiatives at the public policy level for restoring marine ecosystems and improving public participation in fisheries management.

This chapter provides insights and lessons learned based on 17 years' of experience working with small-scale fisheries in Mexico and implementing diverse principles and elements of the SSF Guidelines. It also illustrates ways in which CSOs and other interested stakeholders can foster the implementation of the SSF Guidelines.

## Contributions to the SSF Guidelines

Progress was made on the implementation of the principles and elements of the SSF Guidelines prior to their publication. To analyze COBI's contributions to the implementation of the SSF Guidelines, we focus on part two: *Responsible fisheries and sustainable development* and part three: *Ensuring an enabling environment and supporting implementation* of the document.

Two external reviewers analyzed 65 documents (peer review papers, internal reports, conference proceedings, and book chapters) developed and published by COBI to identify the contributions to the main components of the SSF Guidelines as well as to identify recommendations for implementation. Following this, one external reviewer undertook 11 interviews with current and former staff members to delve in to more detail on COBI's contributions, to assure the relevant information was included, and to reflect on lessons learned.

In the following section we first present the title of each SSF Guideline, followed by a brief summary of the main components of the guideline; the Mexican context with respect to formal institutional arrangements in Mexico, specifically with regard to the General Law for Sustainable Fisheries and Aquaculture (Fisheries and Aquaculture Law), General Law of Cooperatives (Cooperatives Law), Marine Transportation and Trade Law, and General Law for Climate Change (Climate Change Law); a summary of the main contributions; and a summary of the lessons learned. Finally, conclusions and recommendations are drawn from the contributions and learning process of implementing the SSF Guidelines in a developing country.

## *Guideline: Governance of Tenure in Small-Scale Fisheries and Resource Management*

### Tenure in Fisheries

*Components of the SSF Guideline* Responsible governance considers equitable tenure as central for human rights, food security, poverty eradication, and sustainable livelihoods. Tenure of existing communities and cultures should be respected and protected. Tenure can play a key role in restoration, conservation, and protection of co-managed marine ecosystems. Disputes over tenure rights should be resolved in a timely, affordable, and effective manner with equitable outcomes (FAO 2015).

*Mexican Context* According to the Fisheries and Aquaculture Law, access rights to Mexican fisheries are through permits and concessions. Both instruments focus on individual species and may contain a spatial component. Permits can be issued from 2 to 5 years, while concessions range from 5 to 20 years. The process for issuing permits and concessions has to be transparent, based on the best available information, and subject to the availability of natural resources. The process is meant to seek social equity and give preference to local communities, as long as the communities are committed to sustainability (DOF 2007).

*Contributions* In the mid-1990s, a founding member of COBI was the first person to study the territorial use rights for fishing (TURFs) in Mexico using the Comcaác (Seri, an indigenous group from Northwest Mexico) and the swimming crab fisheries as case studies (Bourillón 2002). Further research was completed in the same area to examine the effect of territorial rights (which was the prevalent scheme in the 1970s) on the pen-shell fishery (Basurto 2005) and compare the fishery performance with that of nearby fishing towns that lacked such rights (Basurto et al.

2012). The recognition of the importance of fishing rights for ensuring the sustainable use of resources led to the creation of a program for assisting traditional fishing organizations obtain legal access to fisheries (e.g., Sanchez-Bajo and Roelants 2011), as well as make sustainability commitments and fulfill the obligations of the Fisheries and Aquaculture Law and other applicable legal instruments.

Examples of success include a cooperative led by women in the Gulf of California (Loreto Bay National Park). The cooperative obtained fishing permits to extract aquarium species and implemented a sophisticated management scheme that includes individual quotas and no-take zones and is based on citizen science – the community and CSOs have participated in stock evaluations (Germain et al. 2015). The federal government adopted this model to create a framework to assign permits of this kind at the national level (SEMARNAT 2012). In the same region (Bahia de Kino and Puerto Libertad), four cooperatives have obtained permits to fish (clam, penshell, and octopus) and participated in the definition and enforcement of management rules (quotas, no-take zones, closed seasons, minimum sizes, gear limitations, and spatial boundaries) (Fernández-Rivera Melo et al. 2015). In addition, two of the cooperatives have recently obtained permits to repopulate penshell under a mariculture scheme, combined with quotas and no-take zones.

*Lessons Learned* All support to help fishers access fishing rights should be linked to sustainability. During the aforementioned processes, official access to the fisheries was not the only benefit obtained by fishers. The learning process, the capacities developed, and the changes in behavior also contributed to improving fishers' livelihoods and the sustainability of each fishery. It is important to acknowledge that many fishers have spent years (in some cases decades) dealing with corruption (internal or external to the cooperative), long delays or lack of response to permit applications, and limited access to information. Through COBI's experience, it has been observed that by having access to basic information on the process of how to access fishing rights (application and resolution), fishers are empowered to apply for permits and follow the rules (obligations under national laws). Once fishers obtain the permits, they are unlikely to continue their participation in illegitimate practices (e.g., corruption, illegal fishing) due to the risk of losing their permits. Instead, fishers are more willing to design and implement sophisticated management schemes; they want to prove that they are good candidates to possess permits and, given the increased responsibility, they will work hard to maintain them. In addition, fishers know that if they follow rules, they will also have access to other benefits such as federal subsidy programs or new markets. Finally, with permits, the catches of these fishers enter the system and are reported to, and by, the government, reducing the proportion of IUU catches and allowing science-based management decisions to be made. To ensure compliance and commitment, COBI encourages fishers to sign formal sustainability agreements as part of the requirements to receive technical support with the permit applications.

### Sustainable Resource Management

*Components of the SSF Guidelines* Long-term conservation and sustainable use of fisheries resources should be assured. Management systems should be consistent with existing obligations under national and international laws and voluntary commitments. Small scale fishing communities should participate in monitoring and provide the information required for fishery management. Small-scale fishers should be represented in relevant associations and fisheries bodies with special attention to the equitable participation of women, vulnerable, and marginalized groups (FAO 2015).

*Mexican Context* The Fisheries and Aquaculture Law indicates that fisheries should be sustainable and management decentralized (DOF 2007). Although the law does not include the principles of sustainability, one can find diverse elements that can be used as reference. For example: *fishing should be compatible with the capacity of the resources and ecosystems to recover*; *only selective gear types should be used to reduce the impact on the environment and fish populations*; *the fishing sector should be developed with a sustainability approach that balances economic, social, and environmental aspects*; *the precautionary principle should be adopted*; *impact evaluations should be undertaken to ensure sustainability*. The law also mentions that permits and concessions can be revoked if fishing activity is putting marine ecosystems at risk. There is also a penalty system for infringements. In terms of the decentralization of fisheries management, the law devolves authority to regional and local governmental agencies and defines structures for stakeholder participation at the national and subnational levels (DOF 2007). The Cooperatives Law also states that cooperatives should ensure ecological considerations are embraced by members (DOF 2009).

*Contributions* COBI's approach to the inclusion of sustainability in fisheries management has evolved over time. At first, the principal strategy was the use of no-take zones for fishery and ecosystem restoration. This strategy seemed to be the most suitable, particularly in small, remote, data-poor sites (Espinosa-Romero et al. 2014b) and represents a mechanism to protect a resource for future needs (e.g., Isla Natividad, Micheli et al. 2012). With time, fishers have proposed diverse management schemes depending on the fishery. These include individual quotas, rotation of fishing grounds, and restoration through seed and larval collection, amongst others. Technical expertise was acquired by COBI to support the implementation of diverse management tools whilst ensuring that the focus on ecosystems was not lost. To account for the different dimensions of sustainability, COBI has integrated a model (Table 20.1) based on international standards for sustainable fishing (MSC, Seafood Watch Program of the MBA, and Fair Trade [FT]) that integrates four objectives: (1) stock health, (2) ecosystem health, (3) social justice, and (4) robust governance systems. The fisheries model is complemented by COBI's capacity building program for cooperatives and multi-stakeholder committees for the sustainable and transparent management of resources. The latter applies in situations where management involves diverse fisheries or multiple stakeholders.

COBI has promoted the pairing of sustainable fishing efforts with market incentives (Fujita et al. 2013). Fifteen fishing cooperatives, six in the Mesoamerican Reef (that together form a collective 'Integradora de Pescadores de Quintana Roo') and nine in the Pacific (grouped in the Regional Baja California Cooperatives Federation, 'FEDECOOP') decided to go through the MSC certification process. In 2004 the lobster fishery in the Baja California Peninsula (*Panulirus interruptus*) became the first small-scale fishery in Latin America, and in a developing country, to obtain the MSC certification (Bourillón 2009). After a decade, this fishery continues to be managed by sustainable practices (Smith et al. 2010). Eight years after 2004, in 2012, the Caribbean Spiny Lobster (*P. argus*) fishery of Sian Ka'an and Banco Chinchorro (both Federal Biosphere Reserves) obtained the MSC certification. The six cooperatives involved in the process have rigorously followed the certification requirements. One of the nine MSC certified cooperatives (Buzos y Pescadores de la Baja California of Isla Natividad) in Baja California also operates a Yellowtail (*Seriola lalandi*) fishery and decided to go through the Seafood Watch Program. In 2014, MBA scored the hook and line yellowtail fishery in Isla Natividad as the 'best choice' (green).

To increase fisher participation in monitoring and management, COBI created a citizen science program directed at fishers interested in becoming surveyors of fisheries and ecosystems. This program has been successfully implemented for over a decade, initially having a focus on the evaluation of the effects of no-take zones. Now, it has been expanded to include the evaluation of fisheries. Fishers are trained and certified in SCUBA diving and biological monitoring techniques. Fishers start fishing for data. Over the last 17 years, this program has involved 222 fishers, including 28 fisherwomen. Fishers have produced data for over 300 species by undertaking 12,000 transects in three ecosystems: coral reefs, rocky reefs, and kelp forest (Fulton et al. Forthcoming). In the most successful cases, fishers have organized themselves into formalized groups and have begun hiring out their services to those in need of data. Government agencies have hired the certified fishers to conduct the monitoring of marine protected areas, fish populations, and to calculate quotas.

*Lessons Learned* Applying sustainability standards to small-scale fisheries requires collective action, strong cooperatives to invest in sustainability, and a clear understanding that it is no panacea to the problem of overfishing. From COBI's experience, it is clear that sustainability can neither rely nor depend on a single stakeholder nor on a single management tool. As transitioning to sustainability always implies a cost for the fishers, it is important to anticipate how expensive the process will be and how fishers will cover the cost of this transition in the future. To address this, we included two main elements in COBI's capacity building program: (1) administrative and financial stability, and (2) investment in sustainable fishing.

When fishers decide to be certified, the biggest challenges are covering the additional costs involved in the process (audits) and opening access to premium markets that could potentially offset some of the upkeep costs. In 2016, the lobster fishing cooperatives of Sian Ka'an and Banco Chinchorro decided to withdraw from the

MSC certification because the costs of maintaining the certification were too high and they had not received additional economic benefits. When eco-certification is included as a component for sustainable fishing, the cost must be quantified in advance and included in financial plans, with the knowledge that fishers must assume responsibility for the maintenance costs in the long-term.

When managing a fishery involves diverse stakeholders, different groups face the challenges related to differing scales of governance (Espinosa-Romero et al. 2014a, b). Work needs to be conducted to align scales (local, regional, and national), build shared visions, and build trust between stakeholders. From COBI's experience with management committees, after completing the learning process, committees have built capacity to craft and enforce appropriate management rules and decision-making has moved from being self-interested (a race for fish) or sector-interested to being collective action oriented.

Finally, empowering community members to collect scientific data creates responsibility, pride, and a deeper understanding of the ecosystem in which they live and work. This in turn provides social, economic, and ecological benefits to the community and marine ecosystems ensuring the long-term sustainability of the fishery (Fulton et al. In press) (Fig. 20.2).

## *Guideline: Social Development, Employment, and Decent Work*

*Components of the SSF Guidelines* Holistic, inclusive approaches that consider human rights and the complexity of livelihoods should be considered. Investments should be made in human resource development. Social security protection for the entire value chain should be promoted as well as access to services such as savings, credit, and insurance schemes. Improved safety at sea and reductions in the multiple causes behind deficient safety should be made following appropriate national laws and regulations that are consistent with FAO guidelines and the International Maritime Organization (IMO) (FAO 2015).

*Mexican Context* According to the Fisheries and Aquaculture Law, the federal government should enhance the sustainable and equitable development of fishing communities and indigenous groups, and provide incentives, resources, and technologies to these groups to increase their productive capacities (DOF 2007). The Cooperatives Law establishes that rights and obligations will be the same for men and women. It also indicates that cooperatives should create three types of saving funds: (1) a saving fund to cover periods of low income, (2) a social security fund to cover sickness, pensions, retirement packages, medical insurance, life insurance, leave of absence, education for children of cooperative members, child care, cultural and sporting activities, and other social security activities, (3) and a fund for cooperative training (DOF 2009). The Marine Transportation and Trade Law establishes the basic requirements and procedures for safety at sea (DOF 2014a).

**Fig. 20.2** Fiorenza Micheli (researcher at Stanford) and Miguel Castillo (fisher) comparing data after monitoring no-take zones in Natividad Island. Credits: Comunidad y Biodiversidad, A. C.

*Contributions* COBI has included a component of social justice in its sustainable fishery model. This component includes promoting access rights to fish and ensuring the presence of safety equipment during fishing trips (GPS, radio, and SOS signaling device), social security benefits (including health insurance), and access to fair markets.

COBI's capacity building program has work at individual (for community leaders) and cooperative levels. The work at the individual level focuses on human development. It includes modules on leadership, common wellbeing, communication, conflict resolution, and negotiation. A final module on the technical aspects of marine conservation and sustainable fisheries to orientate fishers towards the development of future projects is given at the end. Coaching sessions and personality assessments are also given to leaders throughout the training to better understand their personality type, common behavior, and leadership style according to their level of risk, patience, capacity to follow norms, and decision-making. The program was designed by fishers, governments, and CSOs, and piloted in the Gulf of California in 2013. After 3 years, 20 leaders (including four women) have graduated and started marine conservation and sustainable fisheries projects in their communities. The projects range from a fisherman teaching nature photography to children to a group of women establishing a sustainable fishing cooperative (Meza-Monge et al. 2015).

Training at the cooperative level focuses on developing social enterprises. It includes modules on legal fishing, administration, competiveness, and financial mechanisms to invest in marine conservation and sustainable fisheries. This program has been implemented in 26 cooperatives in Mexico. The modules are tailored to the cooperative's needs. Coaching sessions for the cooperative's leaders are also one of the elements of this program, and have been seen as vital for creating the conditions under which the collaborative work plans can be completed.

The SSF Guidelines mention the importance of digital inclusion to add value to fishing resources and raise awareness. In 2015, 81 interviews were undertaken with fishers on the use of digital social network (DSN). Results showed that 90% knew about DSN and 58% used it daily, mostly to maintain communication with their family members and friends, as well as to look for information and commercialize fishing products. Fishers are using Facebook (47%), YouTube (29%) and WhatsApp (17%) mainly through smartphones (57%) (Gastelum et al. 2015). These platforms and project specific websites have been successfully used to communicate fisheries information in Mexico. Examples include Facebook pages for the community leader program and the websites for the swimming crab fishery management plan (INAPESCA 2016), the MSC certification process of the Monterey Sardine fishery (CANAINPES 2016), and the multi-sectoral Kanan Kay Alliance (Kanan Kay Alliance 2016).

*Lessons Learned* The fulfillment of obligations and legal requirements for safety at sea and social security are generally lacking within traditional small-scale fisheries communities. In COBI's experience these are key elements to dignify the fishing activity. This problem cannot be addressed in isolation as fishers operating without permits or on low margins are unlikely to invest time and money in acquiring or maintaining the necessary safety equipment. Promoting safe working conditions at sea, and security nets on land, enables conditions for sustainable fishing.

Adding a human dimension approach to the leadership program has been strategic in ensuring personal development and growth of local leaders. Our program modernized capacity building in Mexico. For decades, capacity building usually focused on marine conservation and best practices for fishing. There are always coastal community leaders working towards a common wellbeing that, with a little help and skill development, can make significant changes in their communities. Through our capacity building program we have been able to create a network of change makers. This network continues to grow through exchanges and social networks such as Facebook.

Training for fishing cooperatives needs to have a business focus, beginning with a legal constitution, having robust finances and governance systems, and finally having access to financial support mechanisms (e.g., access to loans). Without any of these elements, it is unlikely that cooperatives will be able to fully engage in sustainable fisheries.

The importance of DSN for communicating information with and amongst fishers must be recognized. Fishers are highly active and informed, and build strong communities through DSN

## *Guideline: Value Chains, Post-harvest and Trade*

*Components of the SSF Guidelines* Parties should recognize the role of small-scale fisheries post-harvest subsectors in the value chain and assure participation in decision-making, with a particular emphasis on the role of women. Traditional forms of association and capacity development in all stages of the value chain should be promoted, including support for the development of cooperatives and professional organizations. The end result should be quality and safe fishery products for both export and domestic markets, as well as robust marketing mechanisms. Effective fisheries management systems and policies should be in place to prevent over-exploitation driven by market demand (FAO 2015).

*Mexican Context* The importance of setting the basis for the implementation of measures to ensure high-quality, safe fish products during extraction, transportation, storing, and distribution is established in the Fisheries and Aquaculture Law. The law also establishes that the federal government should have a designated budget to strengthen the value chain for fisheries and promote production, industrialization, commercialization, quality improvement, and export of fishery products. In addition, the federal government should promote national consumption of seafood, added value to fishing products, access to premium and international markets, capacity building to link producers and strengthen value chains, and loans and financial mechanisms for sustainable fishing, research, technology and for improving competitiveness in the value chain. Finally, the federal government will establish the basis for the implementation of traceability systems for fishing resources (DOF 2007). The Cooperatives Law establishes that financial mechanisms should be available for investment projects proposed by cooperatives that demonstrate feasibility, economic benefits, as well as financial and operational plans. It provides the basis for cooperatives to be organized through federations and confederations (DOF 2009).

*Contributions* COBI has focused on two main strategies: the use of market incentives such as eco-labeling and consumer guides (Fujita et al. 2013) and connecting small-scale fishers with new markets. These strategies are accompanied with a capacity development program for fishers to know which options they have for eco-labels and to give them creative communication channels to connect to better markets.

The case of the Baja California lobster fishery is particularly interesting. Fishers have stated that they have not received a better price for their lobster since certification by the MSC. They export 90% of the catch (total catch in 2014 was 1446 tons) to markets in France, USA, and Asia (Smith et al. 2010). These markets were already paying a good price before certification and continue to do so. However, since certification, the cooperatives have become a model for sustainable fishing in Mexico, giving them access to subsidies and providing intangible benefits such as a feeling of pride for members for having certification and being leaders in the field.

Only one of the six cooperatives in Quintana Roo that were certified by the MSC perceived higher economic benefits, mostly due to the fact that the cooperative improved their negotiation skills and invested heavily in searching for markets.

Currently, a small portion (three out of 90 tons) of Yellowtail production (scored as best choice by MBA) is sold through SmartFish, an organization dedicated to improving onboard management of catch and commercialization of environmentally-friendly seafood. This label has been affordable to small-scale fishing cooperatives; however, to really maximize potential benefits, more consumer awareness is required. Whilst this label is recognized by consumers in the USA, it is not well known in Mexico.

COBI has linked fishers with main retailers in Mexico (e.g., Walmart, Chedraui) through participation in seafood fairs. The most important national event is organized by the federal government through the National Commission of Fisheries and Aquaculture (CONAPESCA). In this annual event, fishers are invited to exhibit their products and attend one-on-one meetings with the main Mexican retailers. Unfortunately, this type of business transaction is unattractive to many small-scale fishers as they often sell the product on the beach, get cash payment at the point of sale, and do not have access to processing, packaging, or transportation facilities. COBI, after taking note of these obstacles, began to change course. The current approach is to identify small-scale buyers that are willing to incentivize sustainable fishing practices. Meetings are being organized and digital platforms being created for fishers and buyers to connect. Chefs and restaurants have shown particular interest in buying small-scale fisheries products, considering the volume and season. There is a growing trend, particularly in big cities such as Mexico City to sell and consume high quality gourmet, local, and organic food.

Finally, COBI is starting to involve buyers in the fishery improvement plans to make sure buyers are aware of fisher efforts towards the sustainability of the small-scale fisheries.

*Lessons Learned* Three main obstacles for concretizing negotiations between retailers and small-scale fishers have been identified. First is that of volume: retailers generally expect high and consistent volumes throughout the year, not usually possible due to resource seasonality and closed seasons. Second is the issue of payments: retailers have payment policies that imply that the seller will receive payment weeks or months after they deliver the product. Third is the logistical concern: retailers usually expect fishers to bring products to the stores, and in particular presentations.

Small-scale fisheries require links to markets that not only incentivize sustainable fishing, but also understand the characteristics of small-scale fisheries (seasons, volumes etc.). Small-scale buyers represent a good opportunity. They are willing to pay more under the characteristics of small-scale fisheries. But, they expect high quality seafood, formality, and professionalism from the fishers. Thus, fishers need to be willing to improve harvest and post-harvest practices

Although there is support from the Mexican government for the commercialization of seafood products, it is important to recognize that small-scale fisheries have to be well organized and legal before looking for access to better market opportunities. Additionally, fish buyers should be included in marine conservation and sustainable fishery efforts.

The experience with the three certification processes described above shows that cooperatives are strongly aligned to sustainability standards. The challenge has been to cover the costs of the certification process and access preferential markets that provide incentives such as higher prices for sustainably caught products. COBI's experience of working with MSC certifications for small-scale fisheries has highlighted that whilst certification may not provide economic benefits, it may bring intangible benefits that will be reflected in the governance structures, including more inclusive, transparent, and collaborative processes, as well as access to other incentives such as subsidy programs.

## *Guideline: Gender Equality*

*Components of the SSF Guidelines* Gender mainstreaming should be an integral part of all small-scale fisheries development strategies including compliance with obligations under international human rights law. Measures should be put in place to address discrimination against women and improve organizational development and equal participation in decision-making processes. There should be equal access to extension and technical services, fishery relevant legal support and evaluation systems to improve women's status, develop better technologies, and generate appropriate work in small-scale fisheries (FAO 2015).

*Mexican Context* The Fisheries and Aquaculture Law has no emphasis on gender equality (see DOF 2007). The Cooperatives Law only establishes that there will be equity in rights and obligations for the members of cooperatives, including women (see DOF 1994). The Climate Change Law includes a wide array of mechanisms to promote gender equality. For example, it has a national program for gender equality that aims at no discrimination against women and promotes women's access to decent work and resources. It also aims for greater, inclusion of women in environmental decision-making, productive projects, and evaluation processes, offers support to indigenous women, and incorporates a gender equality approach to environmental and sustainability policies, aligned to international agreements (DOF 2012c).

*Contributions* Fisherwomen have been invited to, and participated in, COBI's programs. Two fisherwomen cooperatives, each consisting of eight people, have self-organized and become legal fishers; four women from the Bahia de Kino community are part of the first generation of community leaders in the Gulf of California; 28 women have been trained in subtidal monitoring to evaluate aquarium species and the effect of no-take zones; two of these 28 women collect data registered by oceanographic sensors (pH, salinity, temperature, and dissolved oxygen) as a part of a long-term project to monitor the effects of hypoxia on benthic resources reported in their community (Micheli et al. 2012) and along the California Current caused by climate change (Chan et al. 2008). In addition, in 2013 the first workshop to exchange experiences among women involved in fisheries was held in the small town (5000 people) of Bahia de Kino. Early in 2016, one of the community leaders participated in a panel organized by the Mexican Senate and CSOs about the role of fisherwomen in fisheries management and research. Mrs. Delfina said, "*It is our [women's] time to show others how to fish and protect our seas for our grandchildren, the time of our husbands is over.*"

*Lessons Learned* Women have decided to participate in COBI's programs because of personal challenges. However, once they have become part of projects, women have shown themselves to be great leaders, entrepreneurs, citizen scientists, and speakers on what they do and why they do it. In addition, they have demonstrated a strong commitment to marine conservation and the long term sustainability of small-scale fisheries. Including women in fisheries management and research brings new perspectives, visions, solutions, and knowledge to the table (Fig. 20.3).

## *Guideline: Disasters Risks and Climate Change*

*Components of the SSF Guidelines* Combating climate change in the context of small-scale fisheries requires urgent and ambitious action. Actions and support should be in accordance with the United Nations Framework Convention on Climate Change (UNFCCC) and The Future We Want from the United Nations Conference on Sustainable Development (Rio + 20). Strategies for adaptation, mitigation, and building resilience should be developed, including special support for small islands (where climate change may have particular implications for food security, nutrition, housing, and livelihoods), indigenous peoples, and vulnerable and marginalized groups (FAO 2015).

*Mexican Context* The Fisheries and Aquaculture Law establishes the importance of promoting, regulating, and implementing mitigation and adaptation actions to address the impacts of climate change. In addition, it suggests that national policies for fisheries and aquaculture need to be designed with a climate change adaptation and mitigation approach (DOF 2007). The Climate Change Law establishes the basis to address the adverse impacts of climate change. It states that policies need to be designed to take these changes into account, and highlights the importance of reducing the vulnerability of human populations, the relevance of building national capacity to respond to environmental changes, and of education, research, technology, and innovation in this field. According to this law a fund will be created to gather public, private, national, and international financial resources to be used to implement actions to address climate change. The federal government will invite Mexican society to participate in the design and implementation of a National Policy on Climate Change Adaptation and Mitigation, and will provide incentives for the best efforts in this field (DOF 2012c). The Strategic Program on Climate

**Fig. 20.3** Fisherwomen in Estero Santa Cruz, Sonora. The cooperative "Mujeres del Mar de Cortés" is self-organizing to have rights to fish under sustainability standards. Credits: Eunice Adorno/Comunidad y Biodiversidad, A. C.

Change has been created. The program includes the creation of a National Oceans Policy to address the effects of climate change, the implementation of the Code of Conduct for Responsible Fishing, and the implementation of additional safety at sea measures (DOF 2014b).

*Contributions* COBI, along with local fishers, has begun designing and implementing regional networks of reserves in the Mexican Pacific Ocean, Gulf of California, and Mesoamerican Reef principally as a strategy to restore fisheries and marine ecosystems, but also to adapt to environmental changes. The process in the Gulf of California is the first to include ecological connectivity and climate change into the regional design of a network of marine reserves (Suarez et al. 2014).

Fishers have started to see and understand the potential benefits of marine reserves for adapting to climate and environmental changes. In 2006, the fishing cooperative Buzos y Pescadores de Baja California established two marine reserves within their fishing concession. In 2008 and 2009, two climate-driven hypoxia events produced mass mortality of benthic invertebrates leading to 75% and 50% mortality of Pink Abalone, (*Haliotis corrugate),* in the fishing grounds and marine reserves, respectively (Micheli et al. 2012). These hypoxia events have been well documented along the California Current (Chan et al. 2008), with significant drops in oxygen concentrations in shallow waters resulting from changes in upwelling intensity and patterns caused by abnormally strong winds; events that have become more common in the last two decades. After these mortality events, monitoring results showed that pink abalone inside the reserves were more abundant, larger (45% above the legal fishing size), and more mature (92%), producing 40% more eggs than in fishing areas and hence resulting in higher recruitment inside the reserves and in adjacent areas (Micheli et al. 2012). In the case of the Caribbean, no-take zones established in 2012 in the Sian Ka'an Biosphere Reserve have shown significant increases in spiny lobster abundance (Fulton et al. 2015). Additionally, despite heavy rains through the 2013–2014 lobster season, which reduced state-wide catches by half due to freshwater influx in fishing grounds, lobster abundance in the no-take zones remained the same and in some cases continued to increase (Fulton et al. 2015).

*Lessons Learned* The inclusion of oceanographic monitoring equipment is vital for adequately monitoring components of climate change in the marine reserves and fisheries. Using the example from Isla Natividad (Micheli et al. 2012), if there had not been access to physical data generated in the marine reserves during the hypoxia events, mass mortality could have been attributed to some other factor, limiting the management options available. In some areas, implementation of oceanographic monitoring has been slower than desired. Unfortunately, it is often difficult to justify the installation of expensive oceanographic equipment *before* unpredictable climate variations occur. However, not having this equipment means the most critical data is lost before monitoring begins. Luckily, fishers are much attuned to their environment and act as early warning systems for adverse events. In most areas, their openness, willingness to share information, and find a solution with groups with whom they have long collaborated creates favorable conditions in which to come up with proposals aimed at adapting to climate change.

## *Guideline: Ensuring an Enabling Environment and Supporting Implementation of the SSF Guidelines*

*Components of the SSF Guidelines* All parties are encouraged to support, communicate, and monitor the implementation of these Guidelines, including the facilitation of national platforms in collaboration with CSOs to oversee implementation. States should recognize the need for policy coherence with regard to national and international laws and agreements, institutional coordination (ecosystem approaches, marine spatial planning, and integrated coastal zone management), and collaboration (international, regional, sub regional and with CSOs, fisheries cooperatives, etc.). Fisheries information systems should be used to promote effective decision-making, transparency, sustainability, research, and communication. Capacity development should be in place to improve participatory decision-making, benefit from market opportunities, develop co-management arrangements, and improve knowledge transfer (FAO 2015).

*Mexican Context* The SSF Guidelines have not been explicitly incorporated into national policies. However, as mentioned above, some of the principles were implemented before the SSF Guidelines were published. To be systematic, the implementation and monitoring of the SSF Guidelines should be incorporated into national policies.

*Contributions* Efforts have been made to create enabling environments along with communities of change makers who can help continue the implementation and enforcement of the SSF Guidelines and improve policy coherence as well as transparency and participation in decision-making and management schemes.

For example, COBI has implemented a marine conservation course directed at decision-makers (mainly legislators from the Senate Chamber) and graduate students to increase the national capacity for addressing Mexican marine conservation and fisheries issues. In the course, participants analyze the current situation of marine fisheries and ecosystems and develop creative and feasible solutions. The course has been run four times and has facilitated the development of new relationships between decision-makers, experts (including fishers), and students.

COBI also organizes a fisher exchange called 'Pescador a Pescador' (Fisher to Fisher). Four exchanges have happened thus far, each time bringing together up to 100 fishers, mostly from Mexico, but also at times from USA, Guatemala, Argentina, Belize, Chile, Philippines, Jamaica, Honduras, Colombia, and Brazil. These events create dialogue spaces for small-scale fishers and allow discussion of common problems. Each meeting has a theme: 2003 – *Improving fishing through the use of marine reserves;* 2006 – *The responsibility of the fisher in creating an ordered and sustainable fishery*; 2011 – *Fishers' role in fisheries management*; and 2015 – *Organization is key*. These interchanges build long lasting connections between fishers that the confidence of fishers, and allow the exchange of ideas aimed at coming up with common solutions for sustainable fisheries.

In collaboration with other CSOs, governments, and researchers, COBI is part of a citizen's initiative to improve public participation in fisheries management and research in the Fisheries and Aquaculture Law. This initiative has been presented to the Senate and was well received.

*Lessons Learned* Most of the recommendations of the SSF Guidelines rely heavily on state intervention; however, a diverse set of other stakeholders also play an important role in building policy coherence, institutional coordination, collaboration, capacity development, and monitoring the implementation of the SSF Guidelines. Creating enabling environments and alliances with diverse sectors and disciplines are key for the design, implementation and monitoring of the SSF Guidelines.

Much of the work in which COBI has been involved has consisted of multi-sectoral collaboration in which researchers, users, governments, CSOs, and other stakeholders work towards a common goal and provide important perspectives on the work being completed. Alliances and exchanges bring diversity of knowledge, resources, and expertise to the management of a resource, co-production of knowledge, and social learning. This helps make more informed and collective decisions, and improve transparency and trust between the different sectors. And as mentioned above, the use of platforms to report progress on national and international goals and projects is essential for increasing participation, collaboration, commitment, as well as transparency.

## Main Lessons and Challenges

The SSF Guidelines provide significant impetus for the long-term sustainability of small-scale fisheries. They can be used as a framework to document experiences and guide local and national initiatives. From the Mexican experience of implementing the principles on small-scale fisheries, the following can be ascertained:

*Property and Access Rights Are Fundamental for the Sustainability of Small-Scale Fisheries* In Mexico the Fisheries and Aquaculture Law allows exclusive access to marine resources in the form of permits and concessions, but does not allow for property rights *per se*. Access rights are sufficient to promote long-term sustainability in many fisheries, however in many parts of Mexico, small-scale fishers do not operate under these schemes. In many cases this is simply because they have not requested them, given the lack of information and misconceptions over what the process entails. Many fishers end up getting trapped in a downward cycle of disorganization or corruption and continue to fish illegally. It was observed in some cooperatives and organized groups that sharing even the most basic information on how to apply for a permit can result in radical changes in fisher behavior. Fishers are willing to follow this process and commit to sustainability in order to get and maintain permits. Preference for receiving fishing rights should be given to traditional small-scale fishers to ensure that the rights are not concentrated in the hands of a few powerful players, a situation that is common in some areas and one that can lead to unregistered fishers and illegal fishing.

*Capacity Building with Fishers, Leaders, and Fishing Organizations Is of Utmost Importance* Coordinated, collective action is a key component of movements towards sustainability, but it is recognized as particularly difficult in the remit of fisheries (Feeny et al. 1990). In the first half of the twentieth century, Mexican public policy favored the creation of fishing cooperatives and organizations (Cruz-Ayala and Igartúa-Calderón 2006); however, in the last 30 years policies and privatization in the fishery sector has removed many of the incentives for fishers to organize. COBI's experience has been that working with organized fishers is more likely to lead to successful implementation of sustainable fishing projects. Thus, investments should be made in creating leaders, transforming cooperatives into social enterprises, and creating multi-stakeholder management committees with common goals and objectives.

*International Standards for Sustainable Fishing Can Be Used as a Framework for Small-Scale Fisheries* Sustainability has been extensively defined. International standards can provide the framework and main elements for sustainable fishing. In COBI's experience, the fulfilment of the sustainability requirement of eco-certifications requires collective action, strong cooperatives that are prepared to cover the transition to sustainability, and the understanding that there is no panacea: multiple tools may need to be applied depending on the resources and governance system. The costs of maintaining certifications are often too high for small producers. It is important to explore the diversity of financial mechanisms available to fishers to ensure that they can absorb the costs of certification in the mid to long term.

*Market Incentives Are Yet to Guide Small-Scale Fisheries Towards Sustainability* Whilst market incentives do help, access to preferential markets is not necessarily easily obtained (at least in Mexico where only a few exceptional sustainable fishery cases have had access to premium markets), and hence cannot be relied upon as the sole means for fishers to commit to sustainability. In addition, the idea that major retailers can be linked to sustainable small-scale fishers has not come to pass in Mexico as of yet, due to the small catch volume, variable fishing seasons, and logistical hurdles. COBI is currently exploring the opportunity of connecting small-scale fishers directly with small-scale buyers, financial institutions, and novel options for making these connections (digital platforms and events) more viable. It is important to recognize that small-scale buyers focus on the quality of the product. For small-scale fishers to access these markets, they have to improve the way they catch and manage products onboard and during processing.

*Gender Equality Is Essential for Sustainable Fisheries* The role of women has been widely recognized but poorly quantified. In Mexico, the role of women is not explicitly addressed in the Fisheries and Aquaculture Law or in decision-making processes more generally. By including women in such processes richer discussions, more representative visions, and more creative solutions to fisheries matters can be obtained. In Mexico, fisherwomen have proven to be great leaders, entrepreneurs, and fisheries scientists.

*Environmental Change Is Happening* Fisheries management tools have to consider changing environmental conditions as part of design and implementation. No-take zones have proven to be effective tools to cope with these changes but more work is needed. Examples on how to incorporate environmental change into marine reserve designs are being developed. Fishers in the meantime are adapting and obtaining experience of how to adapt. Resource distribution and seasonality are shifting fishing patterns and frameworks based on the SSF Guidelines need to be in place to ensure that small-scale fisheries continue to provide food security and livelihoods to the people that depend on them.

*Alliances Create Enabling Environments and Support Implementation* Multi-sectoral collaborations are vital for the successful implementation of the SSF Guidelines. Trust must be built between sectors, based on transparency, openness, and mutually-beneficial objectives. Having fishers participate in public policy and speak directly top-level government officials helps transmit the message that organized fishers and organized citizens (the CSOs) together are pushing the move towards implementation of sustainable fishing practices.

## Conclusions

The SSF Guidelines are a landmark document that highlights the importance of the small-scale fishery sector and provides the framework on which FAO member organizations should base their efforts to not only ensure the survival of the sector, but allow it to thrive, particularly in developing countries. As mentioned in the SSF Guidelines, an enabling environment should be created in each state that promotes collaboration between the state, fishing organizations and CSOs. Whilst the guidelines highlight the importance of the role of the state in the implementation of the SSF Guidelines, other stakeholders can use these principals to assist the transition towards sustainability. CSOs should not be limited to filling the gaps left by states because the latter do not have the necessary resources to adequately manage an extensive coastline and diverse fisheries. Using examples from Mexico, we have described how a CSO, working in direct collaboration with fishers and fishing communities, can use the framework and guiding principles of the SSF Guidelines to create replicable models of social development, gender equality, and sustainable resource use in the coastal communities of a developing country. The long-term success of these models will be measured by the number of successful replicas. We intend that these replicable models result in a paradigm shift in sustainable fisheries management at the national level.

There continue to be many challenges on the path towards sustainable small-scale fisheries that balance conservation and livelihood objectives; however, CSOs are bridging the gap between the state and fishers, and the SSF Guidelines provide vital guidance and direction.

**Acknowledgements** Thanks to L. Bourillon, F. Fernandez, A. Gómez, A. Hernández, A. Lejbowicz, A. Moreno, E. Rolón, M. Rojo, N. Zamora García and the reviewers for their contribution to this work. Thanks are also due to Alianza World Wildlife Fund–Fundación Carlos Slim, David and Lucile Packard Foundation, Fondo Mexicano para la Conservación de la Naturaleza, International Community Foundation, Marisla Foundation, Oak Foundation, Sandler Supporting Foundation, Summit Foundation, and Walton Family Foundation for their support over the years. Finally, thanks are due to our community partners, and fisherwomen and fishermen, from whom we have learned so much.

## References

Ahmed, M. (1991). *A model to determine benefits obtainable from the management of riverine fisheries of Bangladesh* (ICLARM Technical Report, 28). 133 p.

Basurto, X. (2005). How locally designed access and use controls can prevent the tragedy of the commons in a Mexican small-scale fishing community. *Society and Natural Resources, 18*(7), 643–659.
Basurto, X., Cinti, A., Bourillón, L., Rojo, M., Torre, J., & Weaver, A. H. (2012). The emergence of access controls in small-scale fishing commons: A comparative analysis of individual licenses and common property-rights in two Mexican communities. *Human Ecology, 40*(4), 597–609.
Berkes, F. (2010). Linkages and multilevel systems for matching governance and ecology: Lessons from roving bandits. *Bulletin of Marine Science, 86*(2), 235–250.
Bobadilla, M., Alvarez-Borrego, S., Avila-Foucat, S., Lara-Valencia, F., & Espejel, I. (2011). Evolution of environmental policy instruments implemented for the protection of totoaba and the vaquita porpoise in the upper gulf of California. *Environmental Science & Policy, 14*(8), 998–1007.
Bourillón, L. (2002) *Exclusive fishing zone as a strategy for managing fishery resources by the Seri Indians, gulf of California*. Doctoral dissertation, University of Arizona.
Bourillón, L. (2009). Eco-certificación de la pesca de langosta roja en Baja California. *Biodiversitas, 86*, 7–11.
CANAINPES. (2016). *Pesquería de pelágicos menores en el Golfo de California*. www.sardina-golfodecalifornia.org. Accessed 12 Dec 2016.
Chan, F., Barth, J. A., Lubchenco, J., Kirincich, A., Weeks, H., Peterson, W. T., & Menge, B. A. (2008). Emergence of anoxia in the California current large marine ecosystem. *Science, 319*(5865), 920–920.
Charles, A. T. (1988). Fishery socioeconomics: A survey. *Land Economics, 64*(3), 276–295.
CIMARES. (2010). *Política Nacional de Mares y Costas de México* (65p). Ciudad del Mexico: Secretaría de Medio Ambiente y Recursos Naturales.
Cinti, A., Duberstein, J. N., Torreblanca, E., & Moreno-Báez, M. (2014). Overfishing drivers and opportunities for recovery in small-scale fisheries of the Midriff Islands region, gulf of California, Mexico: The roles of land and sea institutions in fisheries sustainability. *Ecology and Society, 19*(1), 15.
Cisneros-Montemayor, A. M., Cisneros-Mata, M. A., Harper, S., & Pauly, D. (2013). Extent and implications of IUU catch in Mexico's marine fisheries. *Marine Policy, 39*, 283–288.
CONAPESCA. (2013). *Anuario estadístico de pesca y acuacultura 2013*. México: SAGARPA.
Cruz-Ayala, M. B., & Igartúa-Calderón, L. E. (2006). La transformación de la legislación pesquera en México: un acercamiento en el contexto político-económico (1925–1992). In P. Guazmán & D. F. Fuentes (Eds.), *Pesca, acuacultura e investigación en México* (p. 400). México: Centro de Estudios para el Desarrollo Rural Sustentable y la Soberanía Alimentaria, Comisión de Pesca, Cámara de Diputados.
DOF. (2007). *Ley General de Pesca y Acuacultura* 67p Ciudad de Mexico: Diario Oficial de la Federación.
DOF. (2009). *Ley General de Sociedades Cooperativas* (34p). Ciudad de Mexico: Diario Oficial de la Federación.
DOF. (2012a). *Carta Nacional Pesquera* (236p). Ciudad de Mexico: Diario Oficial de la Federación.
DOF. (2012b). *Ley Federal de Fomento a las Actividades Realizadas por Organizaciones de la Sociedad Civil* (16p). Ciudad de Mexico: Diario Oficial de la Federación.
DOF. (2012c). *Ley General de Cambio Climático* (44p). Ciudad de Mexico: Diario Oficial de la Federación.
DOF. (2014a). *Ley de Navegación y Comercio Marítimos* (77p). Ciudad de Mexico: Diario Oficial de la Federación.
DOF. (2014b). *Programa Especial de Cambio Climático* (96p). Ciudad de Mexico: Diario Oficial de la Federación.
Dunn, E. (2005). The role of environmental NGOs in fisheries governance. In T. Gray (Ed.), *Participation in fisheries governance* (pp. 209–218). Dordrecht: Springer.

Espinosa-Romero, M. J., Cisneros-Mata, M. Á., McDaniels, T., & Torre, J. (2014a). Aplicación del enfoque ecosistémico al manejo de pesquerías artesanales. Caso de estudio: Puerto Libertad. *Sonora. Ciencia Pesquera, 22*(2), 65–77.

Espinosa-Romero, M. J., Rodriguez, L. F., Weaver, A. H., Villanueva-Aznar, C., & Torre, J. (2014b). The changing role of NGOs in Mexican small-scale fisheries: From environmental conservation to multi-scale governance. *Marine Policy, 50*, 290–299.

Espinoza-Tenorio, A., Espejel, I., & Wolff, M. (2015). From adoption to implementation? An academic perspective on sustainable fisheries management in a developing country. *Marine Policy, 62*, 252–260.

FAO. (1995). *The code of conduct for responsible fisheries*. Rome: Food and Agriculture Organization of the United Nations.

FAO. (2014). *The state of world fisheries and aquaculture 2014*. Rome: Food and Agriculture Organization of the United Nations.

FAO. (2015). *Voluntary guidelines for securing sustainable small-scale fisheries in the context of food security and poverty eradication*. Rome: Food and Agriculture Organization of the United Nations.

Feeny, D., Berkes, F., McCay, B. J., & Acheson, J. M. (1990). The tragedy of the commons: Twenty-two years later. *Human Ecology, 18*(1), 1–19.

Fernández-Rivera Melo, F. J., Espinosa-Romero, M. J., Rojo, M., Soria, G., & Torre, J. (2015). Innovating management bivalve fisheries in Mexico. In *Abstracts of the Florida State University Mote ymposium, Propagating TURFs into the 21st Century: The value constraints and limitations to territorial user rights in fisheries*. Mote, Florida, 11–12 October 2015.

Finkbeiner, E. M., & Basurto, X. (2015). Re-defining co-management to facilitate small-scale fisheries reform: An illustration from Northwest Mexico. *Marine Policy, 51*, 433–441.

Fujita, R., Lynham, J., Micheli, F., Feinberg, P. G., Bourillón, L., Sáenz-Arroyo, A., & Markham, A. C. (2013). Ecomarkets for conservation and sustainable development in the coastal zone. *Biological Reviews, 88*(2), 273–286.

Fulton, S., Caamal, J., & Marcos, S. (2015). *Resultados del monitoreo biológico de los refugios pesqueros de Quintana Roo 2012–2015*. Sonora: Comunidad y Biodiversidad A.C.

Fulton, S., Hernández-Velasco, A., Suarez-Castillo, A., Fernández-Rivera Melo, F. J., Rojo, M., Sáenz-Arroyo, A., Weaver, A. H., Cudney-Bueno, R., Micheli, F., & Torre, J. (Forthcoming). From fishing fish to fishing data: The role of artisanal fishers in conservation and resource management. In Salas et al. (Eds.), *Navigating small-scale fisheries towards viability and sustainability: Experiences from Latin America and the Caribbean*, MARES Series, Springer.

Gabriel-Morales, J., & Perez-Damian, J. L. (2006). Crecimiento poblacional e instrumentos para la regulación ambiental de los asentamientos humanos en los municipios costeros de México. *Gaceta ecológica, 79*, 53–77.

Gastelum, E., Torre, J., Espinosa-Romero, M. J., Fernandez-Rivera Melo, F. J., Zepeda, J. A., & Salas, S. (2015). The use of social digital networks by artisanal fishers in the Gulf of California, México. In *Abstracts of the 68th conference of the Gulf and Caribbean Fisheries Institute Two oceans…same coastal issues*. Panama City, Panama, 9–13 November 2015.

Germain, N., Hartmann, H. J., Melo, F. J. F. R., & Reyes-Bonilla, H. (2015). Ornamental reef fish fisheries: New indicators of sustainability and human development at a coastal community level. *Ocean & Coastal Management, 104*, 136–149.

Hanna, S. (1994). Property rights and performances of natural resource systems. *The Common Property Resource Digest, 29*, 1–3.

Hernández, A., & Kempton, W. (2003). Changes in fisheries management in Mexico: Effects of increasing scientific input and public participation. *Ocean & Coastal Management, 46*, 507–526.

INAPESCA. (2016). *Proceso de elaboración del Plan de Manejo de Jaiba*. http://plandemanejo-jaiba.blogspot.mx. Accessed 12 Dec 2016.

Kanan Kay Alliance. (2016). The Kanan Kay Alliance. www.alianzakanankay.org. Accessed 12 Dec 2016.

López-Torres, V. G., Moreno-Moreno, L. R., & Marín-Vargas, M. E. (2016). Un acercamiento a los actores ribereños en la pesca de camarón en San Felipe, Baja California. *Región y Sociedad, 23*(67), 5–44.
Marín-Monroy, E. A., & Ojeda-Ruíz de la Peña, M. A. (2016). The role of socioeconomic disaggregated indicators for fisheries management decisions: The case of Magdalena-Almejas Bay, BCS. Mexico. *Fisheries Research, 177*, 116–123.
McConney, C. P., Pomeroy, R., & Khan, Z. (2014). ENGOs and SIDS: Environmental interventions in small island developing states. In S. M. Garcia, J. Rice, & A. Charles (Eds.), *Governance of marine fisheries and biodiversity conservation. Interaction and coevolution*. West Sussex: Wiley-Blackwell.
Méndez-Medina, C., Schmook, B., & McCandless, S. R. (2015). The Punta Allen cooperative as an emblematic example of a sustainable small-scale fishery in the Mexican Caribbean. *Maritime Studies, 14*(1), 1.
Mendoza-Carranza, M., Arévalo-Frías, W., & Inda-Díaz, E. (2013). Common pool resources dilemmas in tropical inland small-scale fisheries. *Ocean & Coastal Management, 82*, 119–126.
Meza-Monge, A., Figueroa, A. L., Torre, J., Espinosa-Romero, M. J., Rojo, M., & Hernández, G. (2015). Programa de líderes comunitarios marinos en el Golfo de California, México. In *X Convención internacional sobre medio ambiente y desarrollo*. La Habana, Cuba, 6–10 July 2015.
Micheli, F., Saenz-Arroyo, A., Greenley, A., Vazquez, L., Montes, J. A. E., Rossetto, M., & De Leo, G. A. (2012). Evidence that marine reserves enhance resilience to climatic impacts. *PloS One, 7*(7), e40832.
Moreno, A., Bourillón, L., Flores, E., & Fulton, S. (2016). Fostering fisheries management efficiency through collaboration networks: The case of the Kanan Kay alliance in the Mexican Caribbean. *Bulletin of Marine Science, 92*, 0.
Moreno-Báez, M., Orr, B. J., Cudney-Bueno, R., & Shaw, W. W. (2010). Using fishers' local knowledge to aid management at regional scales: Spatial distribution of small-scale fisheries in the northern gulf of California, Mexico. *Bulletin of Marine Science, 86*(2), 339–353.
Ramírez-Rodríguez, M., & Ojeda-Ruíz, M. Á. (2012). Spatial management of small-scale fisheries on the west coast of Baja California Sur, Mexico. *Marine Policy, 36*(1), 108–112.
Ratner, B. D., & Allison, E. H. (2012). Wealth, rights, and resilience: An agenda for governance reform in small-scale fisheries. *Development Policy Review, 30*(4), 371–398.
Robles-Zavala, E. (2014). Coastal livelihoods, poverty and well-being in Mexico. A case study of institutional and social constraints. *Journal of Coastal Conservation, 18*(4), 431–448.
Sanchez-Bajo, C. B., & Roelants, B. (2011). *Capital and the debt trap: Learning from cooperatives in the global crisis*. Houndmills/New York: Palgrave Macmillan.
SEMARNAT. (2012). *Plan de Manejo Tipo para Peces Marinos de Ornato. Secretaria de Medio Ambiente y Recursos Naturales, Dirección General de Vida Silvestre, Mexico*. http://www.semarnat.gob.mx/archivosanteriores/temas/gestionambiental/vidasilvestre/Documents/Planes%20de%20Manejo/PM%20Peces%20Ornato%2031%20octubre%202012.pdf. Accessed 10 May 2016.
Senko, J., Mancini, A., Seminoff, J. A., & Koch, V. (2014). Bycatch and directed harvest drive high green turtle mortality at Baja California Sur, Mexico. *Biological Conservation, 169*, 24–30.
Smith, M., Roheim, C. A., Crowder, L. B., Halpern, B. S., Turnipseed, M., Anderson, J. L., Asche, F., Bourillon, L., Guttormsen, A. G., Khan, A., Liguori, L. A., McNevin, A., O'Connor, M. I., Squires, D., Tyedmers, P., Brownstein, C., Carden, K., Klinger, D. H., Sagarin, R., & Selkoe, K. (2010). Sustainability and global seafood. *Science, 327*(5967), 784–786.
Suarez, A., Torre, J., & Alvarez-Romero, J. (2014). Diseño de una red de reservas marinas para los arrecifes costeros en la Región de las Grandes Islas, Golfo de California. *Quinto Informe Nacional de México ante el Convenio sobre la Diversidad Biológica Comisión Nacional para el Conocimiento y Uso de la Biodiversidad*. México: Comisión Nacional para el Conocimiento y Uso de la Biodiversidad

Zepeda-Domínguez, J. A., Zetina-Rejón, M., Espinoza-Tenorio, A., Ponce-Díaz, G., Lluch-Belda, D., Espinosa-Romero, M. J., Torre-Cosío, J., & Cisneros-Mata, M. A. (2016). Mapeo topológico de los actores involucrados en la pesquería de jaiba café *Callinectes bellicosus* del estado de Sonora. *Ciencia Pesquera. N° especial, 23*, 81–90.

# Chapter 21
# Caribbean Fisherfolk Engage the Small-Scale Fisheries Guidelines

**Patrick McConney, Terrence Phillips, Nadine Nembhard, and Mitchell Lay**

**Abstract** Small-scale fisheries are prominent features of the small island developing states (SIDS) of the Caribbean Community (CARICOM). Small-scale fisheries contribute to foreign exchange earnings, income, food security, employment and culture in most CARICOM SIDS. Fishing industry workers (fisherfolk) and their organisations became engaged in the process of developing the Voluntary Guidelines for Securing Sustainable Small-Scale fisheries in the Context of Food Security and Poverty Alleviation (SSF Guidelines) in 2012. Given the subdued responses of most national fisheries authorities to the SSF Guidelines, these civil society formal and informal groups have become the champions of the Guidelines into the current implementation phase. At the same time, they are struggling to engage with the Caribbean Community Common Fisheries Policy (CCCFP) in an environment of limited policy coherence and collaboration. This case study, conceptually grounded in social-ecological system and resilience thinking, examined the engagement of fisherfolk with the SSF Guidelines through the lens of institutional analysis. Through their activities in communication, advocacy, policy influence and capacity development, we examined patterns of interaction and outcomes. Fisherfolk are demonstrating increasing capacities for self-organisation, advocacy and policy influence, but face a rather passive policy domain in which active engagement with state agencies can be challenging. The SSF Guidelines process has helped to

P. McConney (✉)
Centre for Resource Management and Environmental Studies, The University of the West Indies Cave Hill Campus, St. Michael, Barbados
e-mail: patrick.mcconney@gmail.com

T. Phillips
Coastal and Marine Livelihoods and Governance Programme, Caribbean Natural Resources Institute, Laventille, Trinidad and Tobago
e-mail: terrence@canari.org

N. Nembhard
Caribbean Network of Fisherfolk Organisations, Belize City, Belize
e-mail: nembhardnadine@gmail.com

M. Lay
Caribbean Network of Fisherfolk Organisations, St John's, Antigua and Barbuda
e-mail: mitchlay@yahoo.co.uk



S. Jentoft et al. (eds.), *The Small-Scale Fisheries Guidelines*, MARE Publication Series 14, DOI 10.1007/978-3-319-55074-9_21

empower fisherfolk, and if they maintain their trajectory they should realise their potential as change agents in Caribbean policy despite the challenges.

**Keywords** Caribbean • Fisherfolk • Organisation • Policy

## Introduction

Small-scale fisheries for inshore and offshore living marine resources contribute to national food security, livelihoods, foreign exchange earnings, personal income, social relations, culture and well-being in the Caribbean (Fanning et al. 2011). Small-scale fisheries are prominent features of the 17 small island developing states (SIDS) that comprise the Caribbean Regional Fisheries Mechanism (CRFM).[1] The CRFM is an advisory regional fisheries body, established by treaty to address (CRFM 2002):

(a) *the efficient management and sustainable development of marine and other aquatic resources within the jurisdictions of Member States;*
(b) *the promotion and establishment of co-operative arrangements among interested States for the efficient management of shared, straddling or highly migratory marine and other aquatic resources;*
(c) *the provision of technical advisory and consultative services to fisheries divisions of Member States in the development, management and conservation of their marine and other aquatic resources.*

In 2003 the heads of government of the member states decided that a fisheries policy and regime were necessary to implement the CRFM treaty. Almost a decade of technical and legal talks ensued, resulting in the Caribbean Community Common Fisheries Policy (CCCFP). Although not yet signed by the countries, the CCCFP was recognized in 2014 as a treaty that was legally binding on the members of the CRFM (2015). Also in 2014, the FAO Committee on Fisheries adopted the Voluntary Guidelines for Securing Sustainable Small-scale fisheries in the Context of Food Security and Poverty Alleviation (SSF Guidelines) following a global civil society consultation and official technical negotiation process that started in 2008 (FAO 2015).

Despite different start dates and development processes, the CCCFP and SSF Guidelines have converged to have generally compatible aims and content. For example, while the CRFM was leading the formation of the Caribbean Network of Fisherfolk Organisations (CNFO), fishing industry workers (fisherfolk) were minimally engaged in the formulation of the CCCFP. Fisherfolk, through the CNFO as

[1] Anguilla, Antigua and Barbuda, Bahamas, Barbados, Belize, Dominica, Grenada, Guyana, Haiti, Jamaica, Montserrat, St. Kitts and Nevis, St. Lucia, St. Vincent and the Grenadines, Suriname, Trinidad and Tobago, Turks and Caicos.

a regional network of national fisherfolk formal and informal bodies, were more engaged in the process of developing the SSF Guidelines. While both fisheries instruments have received limited support from Caribbean fisheries authorities, the SSF Guidelines are receiving more attention from fisherfolk than the CCCFP largely due to the engaging consultative process. The formal and informal groups (fisheries cooperatives and associations respectively) have also paid more attention to the SSF Guidelines in their current implementation phase than have the national fisheries authorities. Fisherfolk promote the links between the SSF Guidelines and CCCFP more than fisheries authorities. Thus non-state actors are attempting to influence fisheries policy. This reversal of roles in promoting new directions for Caribbean fisheries policy deserves close examination. It may signal changes in the role of fisherfolk in the policy environment.

This chapter, conceptually grounded in social-ecological system and resilience thinking, examines how fisherfolk engage with fisheries policy, it seeks to learn lessons from recent fisheries governance. Viewed through the lens of institutional analysis, it examines Caribbean fisherfolk engagement with the SSF Guidelines process globally and regionally. It focuses upon a partnership of academic, NGO and fisherfolk organisation stakeholders that coalesced to interact with national and regional fisheries bodies and policy processes in order to promote and implement the SSF Guidelines, and to link the Guidelines to the CCCFP along with various fisheries projects, plans and programmes. The focus is mainly on the SSF Guidelines but, like the fisherfolk, we take note of links with the CCCFP.

In reviewing fisheries activities in communication, policy influence and capacity development we look for patterns of interaction, outcomes, and learning in relation to Part 3 of the SSF Guidelines aimed at *Ensuring an enabling environment and supporting implementation*. The authors also surveyed fisherfolk leaders for their perspectives on achievements and future priorities relevant to the SSF Guidelines. The chapter addresses how the Guidelines are providing a strong rallying point and focus for fisherfolk collective action and engagement with fisheries policy at multiple levels through many different means.

The next section describes the conceptual approach used in the chapter. The following two sections respectively examine the early engagement of fisherfolk in the SSF Guidelines process up to 2014, and then the initiation of implementation from 2014 to mid-2016. There is diversity and complexity in the Caribbean situation. We can learn from the engagement whether there is an enabling policy environment, and the extent to which the environment supports implementation of the SSF Guidelines or other fisheries instruments. The final sections draw out the main lessons.

## Conceptual Approach

Our analysis is rooted in the notion that small-scale fisheries are complex and adaptive social-ecological systems (Mahon et al. 2008). People-centred marine ecosystem-based management that includes small-scale fisheries is necessary in the Caribbean for sustainable living marine resources (Fanning et al. 2011). Governance of the living marine resources of the Caribbean Large Marine Ecosystem (CLME) and North Brazil Shelf LME is currently receiving much attention to improve formal transboundary governance arrangements and their performance through linked multi-level policy cycles (Mahon et al. 2014). Attention is also being paid to the multi-level engagement of fisherfolk with policy in CRFM countries (McConney and Phillips 2011).

Several complementary analytical frameworks can be used to examine fisherfolk and other stakeholder engagement in the SSF Guidelines' development and implementation. Jentoft (2014) argues that implementation can be framed as interaction. In this light, prominent among the frameworks is the interactive governance approach (Jentoft and Chuenpagdee 2015). Regional governance framework assessment (Mahon et al. 2014) and collaborative planning (McConney and Phillips 2011) have been used with Caribbean fisheries stakeholders. The institutional analysis and development framework (McGinnis and Ostrom 2014) has been used for Caribbean coastal resources, especially to investigate collaborative management interactions (Pomeroy et al. 2004). We use institutional analysis to draw out lessons from limited qualitative data on CRFM fisheries events and interactions related primarily to the SSF Guidelines (Fig. 21.1).

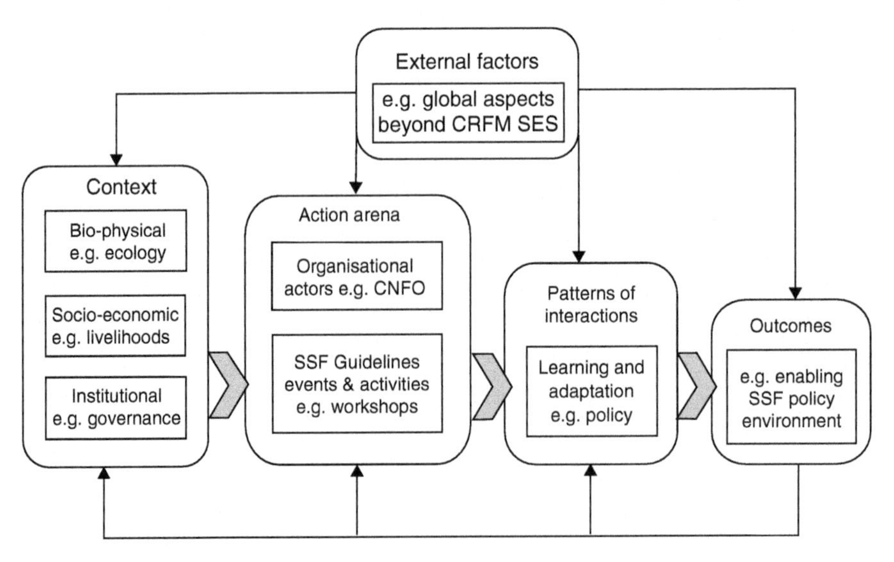


**Fig. 21.1** Institutional analysis and development framework applied to the Caribbean situation (Based on McGinnis and Ostrom 2014)

FAO documents chronicle global development of the SSF Guidelines, but most Caribbean engagement in the global SSF Guidelines process, and the entire CCCFP process, are either undocumented or in grey literature. In addition, there are stakeholders, such as the authors, who were participant observers in both processes. We draw heavily on these observations to fill in the documentation gaps and furnish interpretation in the analysis.

The analysis in this chapter starts by examining the engagement of Caribbean fisherfolk in events and initiatives at national, regional and global levels tied to the SSF Guidelines, and to fisheries policy and projects at these multiple levels of governance. We reviewed capacity building and fisheries advisory workshops, calls for a SSF Guidelines protocol to the CCCFP, links to 'blue economy' initiatives, and inclusion in transboundary governance in the CLME. Multi-faceted projects with connections to the SSF Guidelines are in progress for improving food security, organisational leadership and advocacy for policy influence. Despite the long list of events and initiatives, it is worth asking whether the many interactions and their outcomes have assisted in forging an enabling environment for the SSF Guidelines that supports their implementation in combination with regional fisheries policy such as the CCCFP. Like Jentoft (2014), we ask whether the countries and stakeholders are able to walk the talk. Perhaps it is too soon to say for sure, but we can share the signs of progress.

In the next sections we use annotated timelines to present and discuss interactions from the perspectives of the authors who participated in the global and regional SSF Guidelines and regional CCCFP processes. The results and discussion include views from some national fisherfolk leaders, via a survey, on what they have done and hope to achieve in relation to the main operational provisions of the SSF Guidelines.

## External Factors and Context

Our institutional analysis starts with an appreciation of external factors and the context for interactions and outcomes in the fisheries social-ecological system of the CRFM. World events and SSF Guidelines processes occur outside of the CRFM system at higher external institutional levels and in other geographic locations. However, most of these are connected to the CRFM regional and national levels. Defining the CRFM by maritime jurisdictions alone does not define the SES within the CLME since fishery resources and human operations extend beyond the political boundaries. Defining the CRFM SES as fishery resources plus member countries, observers, partners and other functional or interactive actors within a multi-level governance network is more useful for determining the most relevant bio-physical, socio-economic and institutional contexts (Fanning et al. 2011; Mahon et al. 2014).

Within the CRFM SES the predominant bio-physical or ecological contextual feature is that the most valuable fishery resources accessible to the countries (e.g. shrimp, lobster, conch, reef fish and large pelagic fishes) are characterised by either increasingly restrictive conservation measures or by over-exploitation and depletion. Either way, fishery resources are scarce. The CLME transboundary diagnostic analysis also identified habitat degradation with ecosystem modification and pollution as additional ecological threats with governance implications (Mahon et al. 2011). In this context, CRFM countries are little different from others that recognise resource sustainability as an issue for small-scale fisheries requiring immediate action on several fronts, often complicated by the impacts of climate change and variability.

In the socio-economic context, the fronts include livelihoods, markets, food security, poverty and other human dimensions. Fisheries livelihoods tend to be seasonal, either alternating among fisheries or switching between fisheries and other forms of work such as in tourism or construction. Many fisherfolk only work part-time in fisheries, and the official statistics cannot capture, either in real time or accurately, the numbers involved. These range from hundreds to tens of thousands depending on the size of the country. Similarly, the contribution to Gross Domestic Product from fisheries is difficult to calculate accurately, but official statistics range from 0.5% to 3.0% in many reports (CRFM 2012). This pattern of uncertainty surrounding a variety of socio-economic metrics extends to market information such as prices and income, the contribution to food security and levels of poverty, despite regional studies that attempt estimation (Mahon et al. 2011; CRFM 2012). The CRFM is similar to many developing regions in that small-scale fisheries are often marginalised by other economic sectors (Jentoft 2014; Nayak et al. 2014). The quantitative comparisons within and across sectors demanded by policy analysts to develop scenarios for policy choices are challenging undertakings.

Regarding the governance institutions for decision-making among the above choices, it was previously noted that many fisheries processes do not reflect the policy coherence called for in the SSF Guidelines (Mahon et al. 2011, 2014). This is especially the case for shared transboundary living marine resources and improving the situation is the rationale behind the CLME series of governance projects. This is where instruments such as the SSF Guidelines and CCCFP are particularly relevant along with collaboration between fisheries authorities and fisherfolk organisations to implement full linked policy cycles for the issues facing the region (Mahon et al. 2014). The next sections elaborate upon these points.

## Actors, Interactions, and Outcomes

### *Early Engagement*

The CRFM spearheaded the establishment of the CNFO from 2004 in a project that was separate from the country-level technical and legal discussions on the Common Fisheries Policy and Regime (as it was called before becoming the CCCFP) getting underway at the same time (McConney and Phillips 2011). The SSF Guidelines

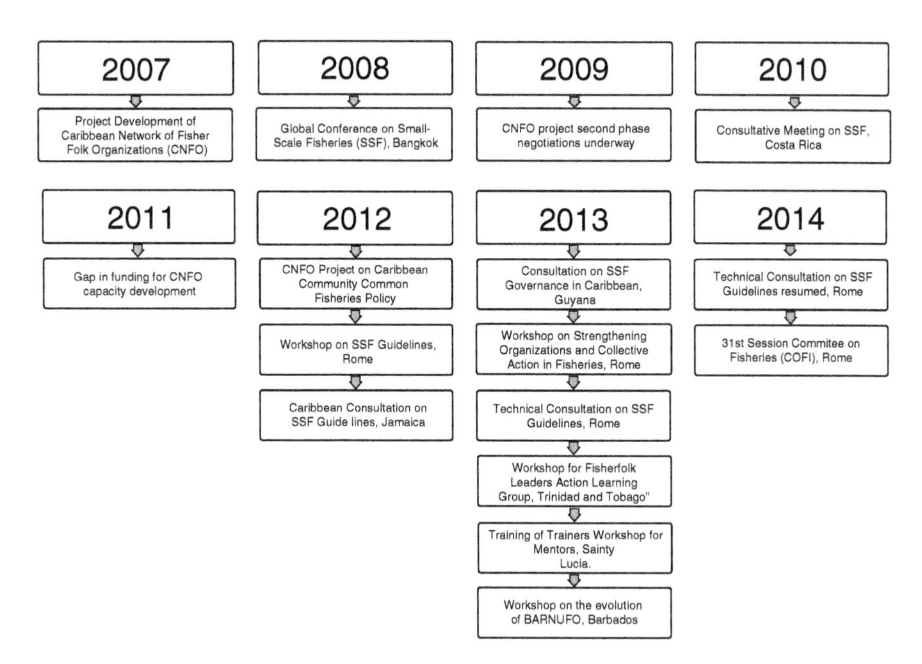


**Fig. 21.2** Timeline of Caribbean fisherfolk engagement related to the SSF Guidelines up to 2014

process started in 2008 with the Global Conference on Small-Scale Fisheries in Bangkok (FAO 2009). Formal approval of the CCCFP and adoption of the SSF Guidelines (FAO 2014a, b) both took place in 2014 although there was little interaction between the two processes along the way. Figure 21.2 sets out a timeline of important SSF Guidelines activities and events from a fisherfolk perspective, with relevance to interactions and outcomes, including those germane to the CCCFP.

Caribbean fisherfolk leaders were networked to form the CNFO and also drawn into the SSF Guidelines process by the CRFM Secretariat and its NGO project partners rather than by the national fisheries authorities. Tangible incentives for fisheries authorities to become involved were either absent or weak. When there was a break in the supply of external funding for the CNFO establishment from 2009 to 2011 the national fisheries authorities did not fill the gap either in cash or in kind through national and local initiatives. The fisherfolk also took no action to address the situation. By the time funding resumed, the priorities of the Technical Centre for Agricultural and Rural Cooperation (CTA), as donor, had changed, with the emphasis being on policy formulation and execution in agriculture (including fisheries) and rural development and promoting the use of ICT in these areas, rather than on organisational development. The successor project was to focus on policy development and implementation, but still within the context of sustainable resource use. At the 2010 FAO Latin America and Caribbean Regional Consultative Meeting on Securing Sustainable Small-scale Fisheries in Costa Rica the CRFM participants were fisheries officials from Belize, Dominica, Grenada and the CRFM Secretariat

(FAO 2011). Caribbean fisherfolk representatives were not present. The need to engage the Caribbean SIDS was pointed out to the FAO (2012).

As a consequence, the regional workshop at the end of 2012 was organised as the major Caribbean event to draw diverse stakeholders together for a few days to examine the SSF Guidelines (FAO 2013). Together fisheries officers, fisherfolk, NGO staff, academics and others crafted a Caribbean perspective. This workshop also marked the beginning of the international networking in which the CNFO would become engaged until approval of the Guidelines. Unlike the CNFO, most of the CRFM national fisheries authorities did not take part in the SSF Guidelines process at the FAO, for reasons that remain unclear, despite encouragement from the CRFM Secretariat and the CNFO. National fisheries authorities also did not strongly engage fisherfolk in the final stages of the CCCFP negotiations that took place in 2011.

The following year, 2013, saw a flurry of activity on several fronts. FAO, CRFM Secretariat, Caribbean Natural Resources Institute (CANARI) and University of the West Indies-Centre for Resource Management and Environmental Studies (UWI-CERMES) all engaged the CNFO in various projects and events related to the SSF Guidelines. These organisations became the Caribbean partners of the FAO in promoting the SSF Guidelines regionally. Some of the interactions among the Caribbean partners directly tackled the approach of the fisherfolk towards self-organisation to develop capacity and influence policy, sometimes via mentors (CANARI 2013a). Leadership at the levels of the individual, the national fisherfolk organisation and CNFO was of particular interest. An example is offered (Table 21.1) through the lens of a Barbados National Union of Fisherfolk Organisations (BARNUFO) workshop on the SSF Guidelines in 2013.

It is evident from Table 21.1 that the fisherfolk were acutely aware of both the leadership role that BARNUFO could potentially play as well as the capacity development required to realise that potential. The policy environment did not appear to be enabling especially with regard to knowledge mobilisation and information flows. Much relied upon developing human capital within the fishing industry as a prerequisite for taking advantage of the SSF Guidelines that the FAO COFI approved for worldwide implementation in June 2014. Importantly, the CNFO attended the rounds of technical consultations that took place in Rome. Perhaps even more critical was that CNFO participation was part of the wider global civil society organisation engagement within which the CNFO was becoming embedded on its own merit, without the direct assistance of state actors in the Caribbean. Among CRFM member states, only Suriname stood out as exceptional by both actively engaging in the SSF Guidelines technical consultations in Rome and also including a fisherfolk representative in the national delegation to give them a voice.

**Table 21.1** Fisherfolk perspectives on organisation leadership roles and the capacity required to lead implementation of the SSF Guidelines

| Sections of SSF Guidelines | BARNUFO leadership role | Capacity development required |
|---|---|---|
| *Part 1: Introduction* | | |
| Objectives | Link BARNUFO's objective to SSF Guidelines | Ensure objective is understood within the fishing industry |
| Nature and scope | Adapt SSF Guidelines to national context | Workshops, video etc. to promote SSF Guidelines |
| Guiding principles | Insert into national fisheries management plans | Fisherfolk meetings on specific fisheries plans and projects |
| Relationship with other international instruments | See above | See above |
| *Part 2: Responsible fisheries and sustainable development* | | |
| Governance of tenure and resource management | Help fisheries to become less marginalised | Knowledge on tenure rights; public awareness for industry |
| Social development, employment, decent work | Advocate more funds to be spent on fisheries issues | Data availability and access; awareness of fisheries value |
| Value chains, post-harvest and trade | Adapt policies, procedures outlined in the Guidelines | Advocate for continued training, storage facilities, marketing |
| Gender equality | Gender mainstreaming | Know about gender mainstreaming |
| Disaster risks and climate change | Integrate disaster and climate into fisheries plans | Workshops and resources to inform of climate, disaster risks |
| *Part 3: Ensuring an enabling environment and supporting implementation* | | |
| Policy coherence, institutional coordination, collaboration | Strengthen communication among stakeholders | Improve organisation website; learn and use more ICT tools |
| Information, research and communication | Strengthen communication among stakeholders | Get information out to more fisherfolk to get them engaged |
| Capacity development | Build capacity for effective NGO management overall | Workshop on effectively managing boards of NGOs |
| Implementation support, monitoring, evaluation | Conduct all above with sustainable financing | Sustainable financing that can combine all above activities |

Adapted from Blackman et al. (2013)

## *Implementation Phase*

The period from mid-2014 to mid-2016 is the start of the SSF Guidelines implementation phase globally. In Fig. 21.3 are some of the major activities, events, interactions and outcomes for that period in the CRFM member states.

The approval of the SSF Guidelines by COFI did not immediately release an abundance of international resources for implementation. FAO worked on formulating the required Global Assistance Programme (GAP) and Caribbean fisheries non-state actors participated in GAP-related events in Rome and the Caribbean throughout the period. The FAO provided small grants to implement Component 3 of the GAP for *Empowering stakeholders: capacity development and institutional*

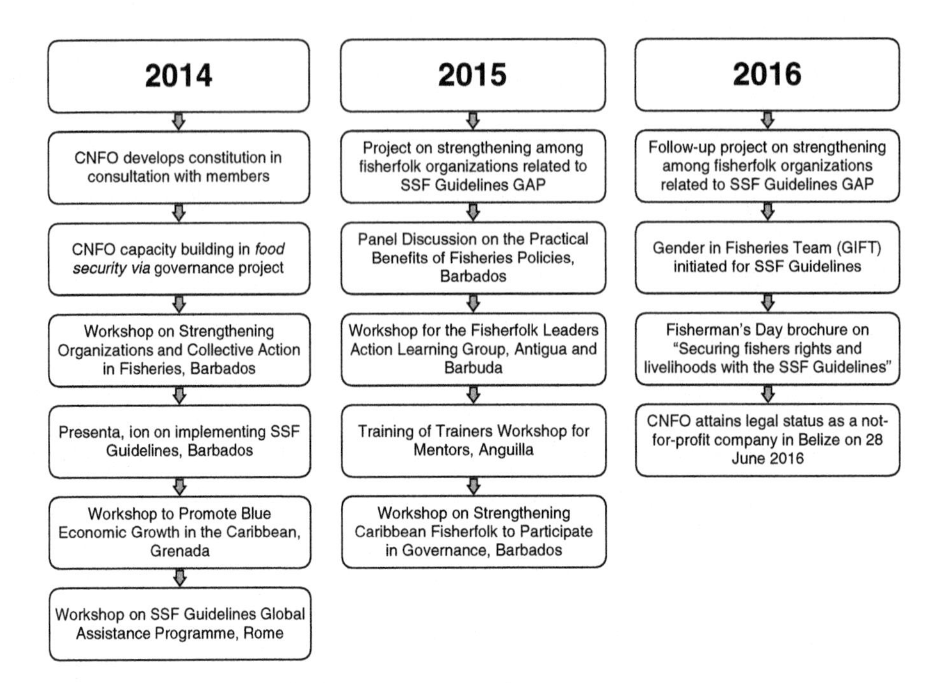


**Fig. 21.3** Timeline of SSF Guidelines implementation phase from mid-2014 to mid-2016

*strengthening* which was focused on forming a more formal partnership by agreement among the main Caribbean actors previously mentioned and on the further strengthening of the CNFO. On 28 June 2016, the CNFO reached the milestone of gaining legal status in Belize as a not-for-profit company. This status is expected to open new doors for the CNFO to achieve greater autonomy and regional policy influence. For example, big international NGOs such as The Nature Conservancy have been attracted to the CNFO as project partners in conservation. Non-state partners assisted the CNFO formalisation process and the CRFM Secretariat has provided office space. The Caribbean partners identified earlier also linked sustainable resource use, managed marine areas, food security, leadership, gender and other aspects of fisheries to the provisions of the SSF Guidelines in project activities. While not opposing either of the instruments or any of the actors, national fisheries authorities have done little to advance either the SSF Guidelines or the CCCFP to date. The potential for synergies between the two policy instruments was pointed out in several activities (CANARI 2015a). Their compatibility is shown in Fig. 21.4 by comparing their content. The absence of attention to gender in the CCCFP stands out in the comparison, but much content is similar or compatible. For example, both the CCCFP and SSF Guidelines embrace responsible fisheries and the ecosystem approach to fisheries as fundamental for sustainable management and livelihoods. Both instruments also recognise the importance of the postharvest sector, data management and communication. While the CCCFP does not employ a human rights based approach and civil society organisation to the extent that the SSF Guidelines

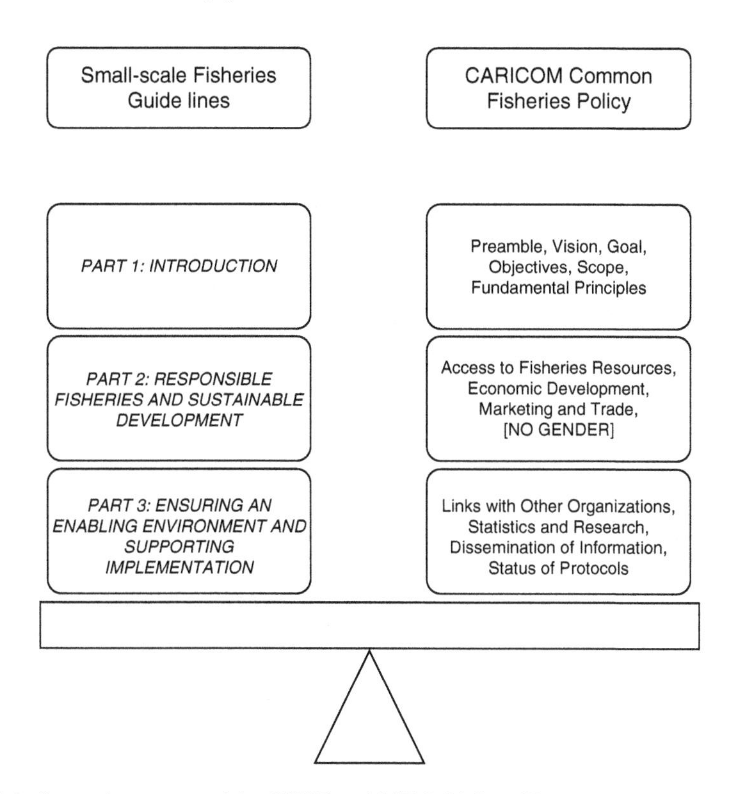


**Fig. 21.4** Comparing content of the CCCFP and SSF Guidelines illustrates compatibility

does, it addresses fisherfolk participation, livelihoods and empowerment in a few articles. Given that a global instrument would not develop the same level of specificity to norms, values, culture, society, economy and ecology as a regional instrument would, the degree of compatibility is high considering their separate processes.

In this early implementation phase, the CNFO has been actively encouraging its fisherfolk leaders to contemplate what they wanted to focus on as priority under the GAP to create the enabling environment for SSF Guidelines implementation. Table 21.2 shows the outputs of a fisherfolk working group that considered GAP implementation in the context of promoting the 'blue economy' (CRFM 2014). Due mainly to CNFO limited capacity the recommendations remain much more aspirational than achievable.

The recommendations in the first three components of Table 21.3 demonstrate that fisherfolk were keenly aware of practical steps to be taken. The fourth component suggests that it is necessary but not sufficient for them to be aware, since the deficiencies in collaboration with state authorities were serious constraints to implementing the recommendations. This was one of the events in which the CNFO recommended the development of a protocol for the CCCFP based on the SSF Guidelines, and to incorporate the Guidelines into national fisheries policies (CRFM 2014). The latter strategy is similar to that taken to insert the guiding principles of

**Table 21.2** Outputs of the CNFO SSF Guidelines working group at the blue growth workshop

| Components of SSF Guidelines global assistance programme | CNFO SSF Guidelines working group recommendations |
|---|---|
| Raising awareness and providing policy support: knowledge products and outreach | Strengthen fisherfolk capacity to communicate effectively |
| | Prepare CNFO and member fisherfolk organizations to take the lead in awareness raising in the fishing industry – peer learning |
| | Take the information into schools to prepare youth stewardship |
| | Develop a marketing programme for the SSF Guidelines |
| | Obtain a champion or mascot, and develop a catch phrase |
| | Initiate partnership with the news media for reliable coverage |
| | Use billboards, TV and radio (interviews, morning programmes) |
| | Utilize more personal ICT: social media e.g. Facebook, Twitter |
| | Use major events to disseminate information; graphics and print |
| | Outreach focused on sustainability of livelihoods not only of fish |
| | Use case studies that mainly demonstrate practical successes |
| | Link awareness to issues, so that information is demand driven |
| Strengthening the science-policy interface: sharing of knowledge and supporting policy reform | Research agendas must be driven by the needs of the industry |
| | Develop scientific models appropriate to small scale fisheries |
| | Holistic approach: stock assessment, ecosystem services, socio-economic and cultural aspects, occupational health and safety |
| | Mainstreaming of gender issues and pursuit of gender equality |
| | Fisherfolk need to be involved in the process at each stage and make sure the science gets into policy and gets implemented |
| | Collaboration of fisherfolk with recognized scientists in the field should be facilitated so that research done is better recognized |
| | National policy should include reference to the SSF Guidelines and other global guidelines that support policy implementation |
| | Insurance supports more risk-based science to ensure protection of fisherfolk against natural hazards and other livelihood threats |

(continued)

**Table 21.2** (continued)

| Components of SSF Guidelines global assistance programme | CNFO SSF Guidelines working group recommendations |
|---|---|
| Empowering stakeholders: capacity building and institutional strengthening | Build capacity of national fisherfolk organisations to engage in policy discussion and be well represented in policy arenas |
| | Develop organisational culture of accountability and transparency |
| | Improved oversight of fisherfolk organizations by authorities such as Cooperative Departments |
| | Intensify training in administration, leadership, succession plans |
| | Use study attachments to learn about success stories first-hand |
| | Make funding available for fisherfolk organisations to participate at various meetings and other key events |
| | Recognise informal cultural institutions for social protection |
| | Enabling environment: policy, infrastructure, market access |
| | Fisherfolk representation on fisheries committees at national, regional and global levels (e.g. CNFO attends Forum meetings) |
| Supporting implementation of monitoring, evaluation, adaptive management, and learning | Include fisherfolk at all levels, including implementation process |
| | Informal arrangements for civil society and state to collaborate |
| | NGOs need to sit more at the table with state representatives |
| | Use simple score card system to evaluate implementation of the SSF Guidelines in three main areas: socio-economic, governance and ecology |

Source: CRFM (2014)

the Code of Conduct for Responsible Fisheries (CCRF) into national fisheries management plans. The strategy elevated the visibility and awareness of the Code throughout the CRFM region.

In order to obtain a fisherfolk perspective on the implementation of the SSF Guidelines in the context of fisheries management, a small purposive sample of fisherfolk leaders who had been to SSF Guidelines events indicated via a self-administered open-ended survey what fisherfolk were already doing that was consistent with the Guidelines, and what additional measures they would recommend to further implement the Guidelines in their countries. The CNFO's aim was to also use the information to develop a demand-led programme of work to further engage members as well as to interact with fisheries authorities. Responses were received from fisherfolk leaders in Barbados, Guyana, and St. Lucia. Table 21.3 summarises combined responses.

The table shows a clear recognition by fisherfolk leaders that they are already on the path to implementing the SSF Guidelines and that their accomplishments are substantial. It also shows a well thought out set of priorities for the future. The recommendations are generally compatible with the CCCFP as well. Many of the future priorities, however, require state agency or other external assistance to implement due to capital cost, need for a prior legal-institutional framework, scale of implementation or level of technology. These may prove to be constraints given the restrained interest of the fisheries authorities in the SSF Guidelines and the scarcity of state resources to allocate to the fishery sector in most cases.

## Discussion

The institutional analysis and development framework acknowledges the influence of external factors such as donor funding, the FAO global consultative process, attracting big international NGOs; and these contributed significantly to the interactions within the social-ecological system. In particular, the FAO played a major role in shaping the patterns of interactions and outcomes of all of the Caribbean actors at critical points in the timeline. In many respects, these global to lower level interactions contributed to policy coherence by maintaining a focus that seems to have wavered less than in the CCCFP process. The CNFO's global networking and their entry into the international fisheries policy domain was greatly facilitated by FAO events and various forms of support. On the other hand, external actors had limited influence on national fisheries authorities and their interactions.

The context for the Caribbean experience with the SSF Guidelines process was firmly rooted in institutional arrangements and governance interactions. Resource sustainability remained in the background as there was little dialogue on the status of fisheries stocks and habitat. Due in part to dependency on external resources, the CNFO capacity development and engagement shifted from socioeconomics and fisherfolk livelihoods towards policy influence. In the institutional context the CNFO broke new ground, becoming through advocacy an actor in the evolving policy domain of the CRFM. The CNFO engaged in advocacy with the support of other non-state actors and the CRFM Secretariat (McConney and Phillips 2011) despite the conspicuous absence of fisherfolk from the centre stage of the CCCFP process (CANARI 2013b). The institutional arrangements towards the end of the period saw the CNFO active in both the Guidelines and CCCFP arenas, but still much more involved in the former. Advocacy has gained the CNFO a foothold in regional fisheries policy domains (e.g. CLME+ Project, Caribbean Fisheries Forum). However, the CNFO still needs to work on developing the capacity, collective action and resource mobilization skills to influence policy significantly. An example would be to marshal resources to formulate the protocol for the CCCFP to integrate the SSF Guidelines into regional fisheries policy as recommended.

**Table 21.3** Results of an open-ended survey of fisherfolk leaders on implementing the SSF Guidelines

| SSF Guidelines sub-headings, examples | What fisherfolk have done, or are doing in policy now | What fisherfolk want to do in the future mainly by policy influence |
|---|---|---|
| Governance of tenure and resource management | Helped develop policy, management practices, institutional collaboration | Influence fisheries governance and more resource ownership |
| Reasonable rights to where and how you fish responsibly | Access resources guided by fisheries authority | Involve fisherfolk in decisions impacting fisheries livelihoods |
| Manage fisheries to protect ecosystems and livelihoods | Adhere to management practices e.g. closed seasons, gear regulations | Co-ownership of government fish marketing infrastructure |
| Responsible fishing, co-management, monitoring, control and surveillance | Co-management of fishery resources and MPAs | Promote ecosystem based fisheries management |
| | Fishers often reminded by peers to fish responsibly | Artisanal vessels fitted with vessel monitoring systems |
| | Larger boats fitted with vessel monitoring systems | Train fishers in monitoring, control and surveillance; and data collection and use |
| | | Researchers and fisherfolk co-investigate traditional fishing combined with community life |
| Social development, employment and decent work | Benefit from capacity building, fiscal incentives and public infrastructure | Improve social services and productivity incentives |
| Reducing poverty, improving food security, quality of life | Training and education increased productivity and reduced postharvest loss | Support fisherfolk organization human capacity development |
| Safe working environment | Participation in safety at sea training, compliance | Subsidize fishers safety equipment and supplies |
| Social security, insurance | Fuel rebates provided to fisherfolk in cooperatives | Improve facilities at landing sites e.g. ramps, lockers |
| Financial services and credit | Better fish landing sites | Create awareness of the need to save and live a better life |
| Good public infrastructure | Social security to insure artisanal vessels, workers | Stricter law enforcement for compliance with safety laws |
| | Minor self-help repairs to maintain infrastructure | Gain access to more financial services and credit facilities |
| | | Collaborate on support for trade and entrepreneurship |
| Value chains, post-harvest and trade | Institutional collaboration, market access and trade | Build capacity of secondary cooperatives to increase fish production, reduce postharvest losses and improve safety |
| Include postharvest sector in fisheries decisions | Use technology to sustain fish quality at sea via ice | Improved marketing standards |

(continued)

**Table 21.3** (continued)

| SSF Guidelines sub-headings, examples | What fisherfolk have done, or are doing in policy now | What fisherfolk want to do in the future mainly by policy influence |
|---|---|---|
| Reduce wastage, loss of fish | Training for fisherfolk in fish handling, processing | Transfer ownership of markets to fisherfolk organizations |
| Support fisherfolk organizing | Availability of marketing facilities for storage, processing and retailing | Create incentives to generate more investment in processing |
| Market access for seafood to be traded locally and export | Created comprehensive fisheries incentives regime for fisherfolk islandwide | Involve fish processing plant representatives in policy, etc. |
| Fair distribution of fisheries benefits among fisherfolk | Fish vendors are included | Involve more fisherfolk in the sector decision-making |
| | Fishers learning to reduce product wastage, loss | Develop value added products and alternative livelihoods |
| | | Fisherfolk know how to access regional, international markets |
| Gender equality | Gender sensitivity in fisheries legislation, management practices | Women in fisheries leadership, ownership and governance |
| Equal opportunities for the participation of men and women in decision-making | Involved women at all levels of fisheries policy, planning, management | Greater participation of women in fisheries decision-making processes and organizations |
| Policies and laws support livelihoods and rights of women as well as men | Appreciate the critical role women play in sustaining many fisherfolk livelihoods | Highlight contribution (social and economic) of women in fisheries, rural development |
| Women are equally assisted in capacity development and new technologies or services | Recruitment of women into fisheries management and co-operatives | Change mind set of women to get them to participate more in process of decision making |
| | No restrictions on women | Clear policy addresses the marginalization of fisherfolk |
| | No barriers preventing women from being trained | Develop a women in fisheries committee to focus on the issues of women and work with authorities on specific tasks |
| | Equality via Fisheries Advisory Committee | |
| Disaster risks and climate change | Policies and coordinated action among fisherfolk and other stakeholders | Improved coordination between state and fisherfolk on action |
| Taken into account in fishery strategies, policies and plans | Fisherfolk involved in consultation, conferences, climate change committee | Consensus on best practices to reduce the impacts of climate change on fisheries, fisherfolk |

(continued)

**Table 21.3** (continued)

| SSF Guidelines sub-headings, examples | What fisherfolk have done, or are doing in policy now | What fisherfolk want to do in the future mainly by policy influence |
|---|---|---|
| Emergency response and good disaster preparedness | Coordinated responses to emergencies; developed resilient adaptive strategy | Compensation for fisherfolk after extreme weather systems |
| Introducing technology and practices for climate change adaptation and mitigation | Modify gear and fishing methods including FADs | Assistance to replace gear and repair boats, livelihood assets |
| Improving energy efficiency | Small vessels assist emergency response | Need to put in place risk-based fisheries policies and plans |
| | Use of sails in assisting motorized vessels during long trips to conserve fuel | Develop a disaster mitigation, resilience and response plan |
| | Concessions, incentives for fisherfolk to handle climate change impacts | Develop a more harmonized relationship between fisherfolk and emergency organizations |
| | | Introduce more energy efficient systems on board fishing vessels e.g. solar panels |

Source: Fisherfolk Leader 2016 Survey

The action arenas changed little over the period regarding actors, events and activities. The period was characterised by workshops and project activities of several types, but mainly at the regional level (CANARI 2015a). Although the projects concerned with strengthening governance for food security and for implementing the SSF Guidelines GAP included national and local arenas, these were seldom sustained beyond specific events and initiatives. A major reason for this was the low level of national fisheries authority activity in either policy process, perhaps due to scarce or weak tangible incentives. Caribbean fisheries authorities tend to set the national policy agendas. Most fisherfolk organisations have too low capacity, or willingness, to engage in policy entrepreneurship. Thus, unless there is greater leadership from the organisations, these policies will remain neglected and isolated from new thrusts such as the 'blue economy' (CRFM 2014). For the CNFO to have greater impact upon the work of fisheries authorities and bring the SSF Guidelines to centre stage there must be more constant engagement among the stakeholders with the fisherfolk persuasively demonstrating how implementing the Guidelines will be in the interest of the authorities. Showing, for example, how the Guidelines' content on climate change adaptation and disaster risk management can be practically implemented to reduce these major threats to fisheries would be likely to elevate levels of state agency interest and policy support.

The patterns of governance interaction and their outcomes, taken together, reveal how the CNFO is acquiring adaptive capacity through networking and developing an informed set of fisherfolk leaders that facilitate self-organisation (CANARI 2015a). These features confer resilience, but questions remain as to whether the extent and rate of change in the CNFO are sufficient to adapt to the rapidly changing and increasingly unpredictable ecological, socio-economic and institutional contexts with faster variables in the social-ecological system. If the CNFO and partners effect change too slowly, then implementing the SSF Guidelines will always lag behind the fisheries' circumstances to be addressed. However, if fisherfolk try to change quickly and too often, the chaotic lack of institutionalised adaptive capacity could cause system resilience to be impaired or not to emerge. There needs to be a balanced approach through adaptive governance that allows for monitoring, evaluation and learning in order to improve institutions based on changing circumstances and knowledge. Neither the CNFO nor fisheries authorities are currently invested sufficiently in participatory monitoring and evaluation within fisheries policy cycles. This is one of the shortcomings that the CLME+ Project is intended to address through national intersectoral consultative mechanisms, for example.

Through fisherfolk interaction with policy actors an enabling policy environment appears to be emerging (McConney and Phillips 2011). The main deficiency, remarked upon above, is the inability to attract national fisheries authorities to actively champion the SSF Guidelines and CCCFP implementation. Whereas CRFM countries incorporated the guiding principles of the CCRF into national and regional fisheries statements, management plans and policy instruments, the uptake of the SSF Guidelines has, so far, been much slower and less enthusiastic, As previously mentioned, we speculate that inadequate incentives explain the lack of engagement although the countries typically do not reveal their reasons. Yet, these are early days, and the pace may quicken as global external factors and regional pressures exert more influence on national decision-making in favour of paying attention to the SSF Guidelines. These factors are expected to enhance fisherfolk policy influence whether or not they become better self-organised and empowered through capacity building and advocacy.

Compared to the CCRF, the SSF Guidelines are less technical and emphasise a human rights approach that has not been a part of the CRFM fisheries policy discourse as reflected in the development of the CCCFP. The latter omits gender, for example, and has little on decent work, social protection, climate change adaptation, disaster risk management and more human dimensions of fisheries. Fisheries authorities are generally still ill-equipped to deal with social science and human dimensions (Mahon and McConney 2004). In addition, the inter-agency (e.g. with gender and poverty units) and inter-sectoral (e.g. with tourism) interaction is typically well below the ideal for an ecosystem approach to fisheries (Mahon et al. 2011). The regional initiative to integrate climate and disasters into fisheries and aquaculture (McConney et al. 2015) remained disconnected from the final stages of the CCCFP process. In many respects, the SSF Guidelines are also less congruent with the fisheries' economic development trajectory of these countries than was the CCRF.

The above observations do not, however, entirely explain the low interest by most fisheries authorities in the SSF Guidelines. Explanations are uncertain, but the low level of active participation in global fisheries and marine science policy-level events coupled with minimal engagement with national stakeholders both before and after such events (Mahon et al. 2010) could be contributing factors besides inadequate tangible incentives. Other possible contributing factors include the absence of actively implemented fisheries policies and management plans in most countries, out-dated fisheries legislation in some, low status of fisheries authorities in the public service, and low levels of staffing and functional capacity in many fisheries authorities. These factors should be taken into account regarding the action arena, patterns of interactions and outcomes. While the CCCFP currently does not fully provide an enabling environment for the SSF Guidelines, there is sufficient similarity and complementarity for improvement to occur. The CRFM context is conducive to advancing both documents to address small-scale fisheries. Although CRFM countries show low interest in the SSF Guidelines, there is no indication that they are opposed to them. This situation beckons fisherfolk organisations to strengthen their roles in advocacy as a means of policy influence to strengthen the enabling environment. To some extent the CNFO has responded to this call by, for example, its interventions at the annual CRFM Caribbean Fisheries Forum that is convened to advise the Ministerial Council on fisheries policy (McConney and Phillips 2011).

The CNFO's response through engagement in the SSF Guidelines process from global to national level has largely been through interactions with Caribbean non-state actors as partners (CANARI 2015a) and the FAO (2015). Within the CNFO, a few national fisherfolk organisations have been strong advocates of the SSF Guidelines, but one cannot say that they pervade the regional network yet. The interactions and outcomes show that fisherfolk capacity development workshops have been the primary means of engagement. The CNFO has not yet led major advocacy of its own aimed at fisheries policy decision-makers or managers. Since 2015, through a FAO project with the Caribbean partners, there has been more emphasis on linking the SSF Guidelines to developing capable fisherfolk leadership and succession planning. These strategies mainly target improving fisheries sustainability awareness and action among fisherfolk.

To date, CNFO engagement has been shaped or executed largely to fit the projects and programmes of its partners in promoting the SSF Guidelines. Recently, the CNFO and its members have been attempting to formulate their own activities, with assistance as needed, to pursue a more immediate and practical agenda of demonstrating how the SSF Guidelines can be applied to fisherfolk issues. For example, fisherfolk promotion of social protection was described in connection with decent work and gender in the SSF Guidelines (CANARI 2015b). This change from passive partner to action leader offers evidence of the CNFO's potential for self-organisation that the SSF Guidelines is catalyzing and CNFO partners are encouraging. Some of this potential can be demonstrated by the CNFO's involvement in an increasing number of regional projects of the TNC, and the Caribbean partners in which it has considerable voice.

The CRFM Secretariat has engaged with the SSF Guidelines process and the CNFO initiatives much more than the national fisheries authorities. However, as demonstrated in a brochure (CRFM 2015) developed to promote the Guidelines, creating synergy with the CCCFP is not a priority. If the CRFM Secretariat were to make the SSF Guidelines more central to its operations such as the technical-scientific working groups, its strategic plan, and Caribbean Fisheries Forum, then the opportunities for engaging more fisherfolk would expand. The current governmental separation of the engagement with, and the implementation of, the two fisheries policy instruments, and hence fisheries management based upon them, remains a critical constraint to fisherfolk engagement in the SSF Guidelines spreading to the CCCFP and ultimately linking them whether in protocols or practice.

In an interactive multi-stakeholder policy environment Jentoft (2014, 10) suggests that "over time, partly due to power struggle, goal displacement is to be expected. This may cause disappointment among those who initially had high expectations of the Guidelines and for whom they were primarily intended". While it appears that the Caribbean fisherfolk, and other non-state actors, have higher expectations for the SSF Guidelines making a difference than the CCCFP, there appears to be no power struggle. Indeed, the parties are passive to the extent that the actor who takes the initiative is most likely to set the policy agenda and achieve a desired outcome. To date, it is primarily the engaged fisherfolk who are driving this process. Lack of engagement by fisheries authorities rather than exercise of power is most likely to be the reason for the SSF Guidelines not achieving the required traction in Caribbean fisheries. It is, however, still early in the process and there is every reason to be optimistic that the SSF Guidelines process has ushered in a change in the fisheries policy arena that may encourage fisherfolk to self-organise and become change agents for their own benefit.

## Conclusion

The institutional analysis of the Caribbean cases of fisherfolk engagement with the SSF Guidelines has illustrated that although CRFM countries were late to fully engage in the global civil society process of developing the SSF Guidelines, the year and a half of involvement was meaningful. This was especially so for the concurrent establishment of the CNFO. While the CCCFP formulation took place in parallel with the previously mentioned two processes, it did so in isolation, not benefitting from the many synergies achievable through linkages. Fisheries authorities remained detached from the SSF Guidelines process despite urging from the CRFM Secretariat. Consequently, while an enabling environment for regional fisheries policy that is supportive of the SSF Guidelines is slowly emerging, the rate of emergence and the extent of enabling are cause for concern. The scarce resources of Caribbean small-scale fisheries stakeholders are in danger of being squandered unless the pace of governance adaptation can be accelerated while allowing for institutionalisation of best practices that confer resilience. While national fisherfolk groups, and especially the CNFO, have proven to be change agents in fisheries policy, there is still much dependence on fisheries authorities. The non-state SSF Guidelines project partners have also proven themselves to be useful in advocating, promoting and helping to develop capacity for policy influence in interactive governance to implement the SSF Guidelines. External actors and factors clearly provided direction to the engagement and widened the governance network.

The SSF Guidelines are still in the early stages of implementation in the Caribbean as elsewhere. Engagement with them in the region needs to be interpreted in various contexts and within an influential external environment, bearing in mind that the action arena, patterns of interaction and outcomes also have an element of path dependency. Considering the latter, and the previous positive experience with the CCRF, there is scope for cautious optimism that fisherfolk along with other state and non-state stakeholders will become more engaged with the SSF Guidelines. Similarly, a measure of realism suggests that the SSF Guidelines are unlikely to become a centrepiece of marine policy engagement given the several constraints on the evolving enabling environment for fisheries policy unless fisherfolk organisations and fisheries authorities collaborate through already established and new institutional arrangements.

## References


Blackman, K., Selliah, N., & Simmons, B. (2013). *Report of the CERMES/BARNUFO workshop on the evolution of BARNUFO and its future in fisheries governance*, Bridgetown, Barbados, 12 December 2013. Centre for resource management and environmental studies, University of the West Indies.

CANARI. (2013a). *Report of the regional training of trainers workshop for mentors*, Saint Luciac, 19–22 November 2013. Held as part of the strengthening Caribbean fisherfolk to participate in governance project. Laventille: CANARI.

CANARI. (2013b). *Report of the CRFM/CNFO/CTA consultation on the implementation and mainstreaming of regional fisheries policies into small-scale fisheries governance arrangements in the Caribbean*. CRFM Technical & Advisory Document No. 2013/2.

CANARI. (2015a). *Report of the third regional workshop for the fisherfolk leaders action learning group*, Saint John's, Antigua and Barbuda, 5–8 October 2015. Held as part of the strengthening Caribbean fisherfolk to participate in governance project. Laventille: CANARI.

CANARI. (2015b) *Report of the national fisherfolk workshop,* Barbados, 13 October 2015. Held as part of the strengthening Caribbean fisherfolk to participate in governance project. Laventille: CANARI.

CRFM. (2002). *Agreement establishing the Caribbean regional fisheries mechanism*. Belize City: CRFM Secretariat.

CRFM. (2012). *Diagnostic study to determine poverty levels in CARICOM fishing communities – Policy document*. CRFM Technical & Advisory Document No. 2012/3, Volume II. Belize City: CRFM Secretariat.

CRFM. (2014). *Report of the CRFM/CNFO/CTA regional fisheries workshop: Investing in blue growth,* St. George's, Grenada, 20–21 November 2014. CRFM Technical & Advisory Document No. 2014/5. Belize City: CRFM Secretariat.

CRFM. (2015). *Caribbean fisheries – Our treasure, our life*. CCCFP pamphlet. Belize City: CRFM Secretariat.

Fanning, L., Mahon, R., & McConney, P. (Eds.). (2011). *Towards marine ecosystem-based management in the wider Caribbean*. Netherlands: Amsterdam University Press.

FAO. (2009). *Report of the global conference on small-scale fisheries – Securing sustainable small-scale fisheries: Bringing together responsible fisheries and social development*. Rome: Food and Agriculture Organization of the United Nations.

FAO. (2011). *Report of the Latin America and Caribbean regional consultative meeting on securing sustainable small- scale fisheries: Bringing together responsible fisheries and social development*. FAO Fisheries and Aquaculture Report No. 964. Rome: Food and Agriculture Organization of the United Nations.

FAO. (2012). *Report of the Workshop on international guidelines for securing sustainable small-scale fisheries*. FAO Fisheries and Aquaculture Report No. 1004. Rome: Food and Agriculture Organization of the United Nations.

FAO. (2013). *Report of the FAO/CRFM/WECAFC Caribbean regional consultation on the development of international guidelines for securing sustainable small-scale fisheries*. Fisheries and Aquaculture Report No. 1033. Rome: Food and Agriculture Organization of the United Nations.

FAO. (2014a). *Securing sustainable small-scale fisheries: Update on the development of the voluntary guidelines for securing sustainable small-scale fisheries in the context of food security and poverty eradication*. (COFI/2014/3). Rome: Food and Agriculture Organization of the United Nations.

FAO. (2014b). *Voluntary guidelines for securing sustainable small-scale fisheries in the context of food security and poverty eradication*. Rome: Food and Agriculture Organization of the United Nations.

FAO. (2015). Towards the implementation of the SSF Guidelines. *Proceedings of the Workshop on the development of a global assistance programme in support of the implementation of the voluntary guidelines for securing sustainable small-scale fisheries in the context of food security and poverty eradication*. FAO Fisheries and Aquaculture Proceedings No. 40. Rome: Food and Agriculture Organization of the United Nations.

Jentoft, S. (2014). Walking the talk: Implementing the international voluntary guidelines for securing sustainable small-scale fisheries. *Maritime Studies, 13*(1), 1–15.

Jentoft, S., & Chuenpagdee, R. (2015). *Interactive governance for small-scale fisheries: Global reflections*. Dordrecht: Springer.

Mahon, R., & McConney, P. (2004). Managing the managers: Improving the structure and operation of fisheries departments in SIDS. *Ocean and Coastal Management, 47*, 529–535.

Mahon, R., McConney, P., & Roy, R. (2008). Governing fisheries as complex adaptive systems. *Marine Policy, 32*, 104–112.

Mahon, R., McConney, P., Parsram, K., Simmons, B., Didier, M., Fanning, L., Goff, P., Haywood, B., & Shaw, T. M. (2010). *Ocean governance in the wider Caribbean region: Communication and coordination mechanisms by which states interact with regional organisations and projects*. CERMES Technical Report No. 40. Barbados: Centre for Resource Management and Environmental Studies, University of the West Indies.

Mahon, R., Fanning, L., & McConney, P. (2011). *CLME TDA update for fisheries ecosystems: Governance issues. The Caribbean large marine ecosystem and adjacent areas (CLME) project*. Cartagena, Colombia: CLME Project.

Mahon, R., Fanning, L., & McConney, P. (2014). Assessing and facilitating emerging regional ocean governance arrangements in the wider Caribbean region. *Ocean Yearbook, 28*, 631–671.

McConney, P., & Phillips, T. (2011). Collaborative planning to create a network of fisherfolk organisations in the Caribbean. In B. Goldstein (Ed.), *Collaborative resilience: Moving through crisis to opportunity* (pp. 207–229). Cambridge: MIT Press.

McConney, P., Charlery, J., Pena, M., Philips T., Anrooy, R. V., Poulainet F., & Bahri T. (2015). *Disaster risk management and climate change adaptation in the CARICOM and wider Caribbean region – Formulating a strategy, action plan and programme for fisheries and aquaculture*. FAO Fisheries and Aquaculture Proceedings No. 35. Rome: Food and Agriculture Organization of the United Nations.

McGinnis, M. D., & Ostrom, E. (2014). Social-ecological system framework: Initial changes and continuing challenges. *Ecology and Society, 19*(2), 30. doi:10.5751/ES-06387-190230.

Nayak, P. K., Oliveira, L. E., & Berkes, F. (2014). Resource degradation, marginalization, and poverty in small-scale fisheries: Threats to social-ecological resilience in India and Brazil. *Ecology and Society, 19*(2), 73. doi:10.5751/ES-06656-190273.

Pomeroy, R., McConney, P., & Mahon, R. (2004). Comparative analysis of coastal resource co-management in the Caribbean. *Ocean and Coastal Management, 47*, 429–447.

# Chapter 22
# Implementing the Small-Scale Fisheries Guidelines: Lessons from Brazilian Clam Fisheries

**Sérgio Macedo G. de Mattos, Matias John Wojciechowski, Alison Elisabeth Macnaughton, Gustavo Henrique G. da Silva, Allyssandra Maria Lima R. Maia, and Joachim Carolsfeld**

**Abstract** From 2008 to 2011 the Brazilian Ministry of Fisheries and Aquaculture and the Canadian charity World Fisheries Trust implemented a project known as *Gente da Maré* (GDM), or 'People of the Tides'. GDM worked strategically to build institutional and community capacity and linkages between government, university researchers, and local fishing associations involved in projects to improve the livelihoods and well-being of 'marisqueiras,' women and families that depend on clam and oyster extraction, mainly the Venerid clam *Anomalocardia brasiliana,* in the Northeast Region of Brazil where the country's highest number of coastal and estuarine small-scale fishers are concentrated. Consistent with many of the principles and guidelines in FAO's Voluntary Guidelines for Small-Scale Fisheries in the Context of Food Security and Poverty Eradication (SSF Guidelines), GDM promoted an integrated approach to equitable development of sustainable fisheries that included: co-management including participatory research and stronger research-policy interface; empowerment of women in fisheries occupations and improved opportunities for women; and value chain upgrading and democratization focusing on the decent work agenda. In this chapter, we analyze the clam fisheries component of GDM as an example of steps towards the implementation of the SSF Guidelines in Brazil. We examine the context in which the project was carried out, the results that were achieved, lessons learned, and indications on how a regional government could act to implement the new SSF Guidelines to the benefit of the clam fisheries.

S.M.G. de Mattos (✉)
Ministry of Planning, Brasilia, Brazil
e-mail: smgmattos@outlook.com

M.J. Wojciechowski • A.E. Macnaughton • J. Carolsfeld
World Fisheries Trust – WFT, Victoria, BC, Canada
e-mail: matias.john.w@gmail.com; alison@worldfish.org; yogi@worldfish.org

G.H.G. da Silva • A.M.L.R. Maia
Universidade Federal do Semiárido – UFERSA, Mossoró, Brazil
e-mail: gustavo@ufersa.edu.br; allyssandramr@hotmail.com



S. Jentoft et al. (eds.), *The Small-Scale Fisheries Guidelines*, MARE Publication Series 14, DOI 10.1007/978-3-319-55074-9_22

**Keywords** Clam fishery • Value chain • Decent work • Small-Scale Fisheries Guidelines • Brazil

## Introduction

Small-scale fisheries are complex, plural, and individual activities that variably perform cultural, economic, food security, poverty alleviation, and livelihood functions within the constraints of socio-environmental sustainability (da Silva et al. 2014a). One of the key features of a small-scale fishery is the multiplicity of fishers' decision-making relationships along the whole value chain that affect production, income, and participation (Maldonado 1986), along with the involvement of an array of other stakeholders and practitioners. It is important to understand which factors influence decision-making by fishers and managers, not only in relation to a *technical-scientific* concern for conservation and sustainability of the resource, but also to attend *empirical-traditional* pressures to achieve social, cultural, and economic expectations (Mattos 2011).

Supporting the visibility, recognition and enhancement of small-scale fisheries, fishers, fishworkers and fisheries-related activities, through a human rights-based approach, is a central element of the Food and Agriculture Organization's (FAO) *Voluntary Guidelines for Securing Sustainable Small-Scale Fisheries in the Context of Food Security and Poverty Eradication* (FAO 2015), hereafter referred to as the SSF Guidelines. Recognition of unequal power relationships between value chain stakeholders, specifically the marginalization of vulnerable groups involved at the production and pre- and post-harvest levels of fisheries is critical. In this context, vulnerable and marginalized groups may benefit from special support to enhance their participation in decision-making processes.

In this chapter, we examine the steps needed to implement the SSF Guidelines in Brazil. We analyze the specific case of the clam fishery in the Northeast region of Brazil, where significant international cooperation was invested between 2008 and 2011, through a project known as *Gente da Maré* (GDM), or 'People of the Tides', to build institutional and community capacity for improving the livelihoods of fisherwomen and families in coastal communities.[1] In our analysis, we try to answer the question, "In what ways does the work of the GDM project, and the features of the clam fishery itself, support the implementation of the SSF Guidelines?"

Implementing supportive policies and monitoring systems may provide a way forward for the future development of small-scale fisheries and the implementation of the objectives and recommendations in the SSF Guidelines. We focus on a subset of specific principles of the SSF Guidelines that are most relevant to the case of the

[1] The Gente da Maré (GDM) project was co-coordinated by the Brazilian Ministry of Fisheries and Aquaculture and the Canadian charity organization World Fisheries Trust (WFT) and funded by the Canadian International Developmental Agency (CIDA) through a cooperation agreement with the Brazilian Cooperation Agency (ABC) of the Ministry of Foreign Affairs for the "*Development of Coastal Communities in the Northeast of Brazil*".

clam fishery and GDM, namely: (1) governance of tenure in small-scale fisheries and resource management; (2) social development, employment and decent work; (3) value chains, post-harvest and trade; (4) gender equality; (5) policy coherence, coordination and collaboration; (6) information, research and communication; and (7) capacity development.

We first examine the pre-existing context of small-scale fisheries and specifically the clam fishery in northeast Brazil. Following this, for each of the identified principles, we discuss the results achieved by GDM. Based on lessons learned from the project and our analysis, we provide recommendations with regard to what is still needed to advance public policies for regional implementation of the SSF Guidelines to the benefit of the clam fisheries.

## The Brazilian Small-Scale Fishing Sector – A Brief

The last available official statistics (MPA/Brasil 2010) placed Brazil as the 18th largest fish producer in the world, with around 65% of production coming from marine fisheries, and approximately half of this from the country's Northeastern Region, the leading regional producer. The available numbers suggest that over 60% of the total estimated catch comes from small-scale fisheries. Fishing is one of the most traditional and important activities for coastal communities in Brazil, in many cases providing the main source of food and income (Isaac et al. 2006), generating direct jobs and income for an estimated one million fishers and fishworkers, as well as indirect employment for another three million. Small-scale fisheries are marked by local and regional diversity, resulting from differences in habitats, ecosystems, and target species of fish, as well as in the availability of fishing resources, and the technology and practices of fishers.

Brazilian national policy has historically promoted natural resource extraction, including fisheries development, as an important contributor to economic growth and development, at both national and local levels. However, Dias-Neto and Dornelles (1996) estimated that over 80% of Brazil's main fisheries were already fully exploited, overfished, depleted, or recovering; a situation that does not appear to have improved since, though data for many of these fisheries is limited or unavailable. Given the existing mismatch between policy and reality, poverty in coastal fishing communities is a wicked problem (Jentoft and Chuenpagdee 2009), a continuing and complex issue that cannot be resolved by policies promoting a simple increase in fishery production. Fifty-nine percent (9.6 million) of the 16.3 million Brazilians who live below the poverty level are located in the country's Northeast Region (IBGE 2010), many in small coastal communities where fishing is a main livelihood (Fig. 22.1).

Following Brazil's independence in 1822, decades of civil war, the declaration of the Brazilian Republic, and the official end of slavery in 1888, poor landless peasants and former slaves migrated within the country in search of land and work, many settling in riverine and coastal areas to fish. Subsequently, in the early 1900s,

**Fig. 22.1** Although proud, clam fishing communities in Igarassu, Pernambuco State, Notrheast Region of Brazil, live in very poor conditions, lacking basic health conditions and citizenship, surrounded by disordely growths of urban centers (Photo credit: Sergio Mattos 2008)

male fishers were individually registered and organized into 'capatazias' (a regional unit within the context of a naval reserve), largely to facilitate the supervision of the activity and people involved, and to promote surveillance of the coast and waterways which were strategically important to national security and for moving goods and people (Silva 1988). During the twentieth century, with continuing economic and political upheaval, the movement of people into urban centers and then out to smaller communities continued and more poor people entered the fisheries, often as an occupation of last resort.

Since the 1960s, institutional crises have marked discussions on fisheries management in Brazil, and the institutions responsible for governance of fisheries have gone through cycles of interventions, from emptying and re-starting institutional structures, to expansion through specific planning authorities and public policies (Mattos 2011). Such policies for the fisheries sector, from the1960s to the mid-1980s, led to great increases in fish harvests, but without appropriate consideration for the long-term sustainability of the marine resource, leading to the decline in fish catch in the following years (Abdallah et al. 2007). Public policies at the start of the twenty-first century have not helped, and possibly will not help, in reducing overexploitation, because these policies were too optimistic about the abundance of fish in

Brazil's EEZ, and were not accompanied by a fisheries management plan that is likely to work.

The perception that fish stocks were inexhaustible led to the development of the small-scale fishing sector by Superintendência do Desenvolvimento da Pesca – SUDEPE (Superintendency for the Development of Fishery), linked to the Ministry of Agriculture. SUDEPE's aim was to industrialize small-scale fisheries to enhance productivity. It promoted this strategy through tax incentives. The result was environmental degradation, decline of numerous fish stocks, breakdown of many fishing communities, and impoverishment of traditional fishing families.

Following the Rio Summit in 1992, Brazilian environmental policy shifted to a more conservationist stance under the governance of the newly created Instituto Nacional do Meio Ambiente e dos Recursos Naturais Renováveis – IBAMA (National Institute for the Environment and Renewable Natural Resources). This was accompanied by a growing public awareness of the importance of ecosystem conservation, and more recently, the importance of managing this ecosystem for the sustainable use of fisheries resources by local communities. Despite this, the Ministry of Agriculture and Supply in the 1990s continued to focus on developing fishing as a production-oriented industry, formulating and implementing policies to increase production and international competitiveness in various segments of the value chain.

In 2003, with the creation of the Secretaria Especial de Aquicultura e Pesca da Presidência da República – SEAP/PR (Special Secretariat of Aquaculture and Fisheries of Brazilian Presidency), the government directed its efforts to structure an integrated national policy for fishing and aquaculture activities. The main goal was to increase production and revenue, including through the promotion of fish consumption nationally for enhanced food security. SEAP was upgraded to the status of Ministry of Fisheries and Aquaculture in 2009, at the same time as new Fisheries and Aquaculture Law that established a National Plan on the Sustainable Development of Fisheries and Aquaculture (Law #11,959/2009) came into being. The overarching aim of this law was to promote sustainable development in harmony with environmental and biodiversity protection, representing the most significant step forward in Brazilian fisheries policy in the last 50 years. The guiding principles of the Brazilian Fisheries and Aquaculture Policy (Brasil 2009), reinforced by the Strategic Plan of Action, include: (1) social, economic and environmental sustainability; (2) transparency; (3) innovation; (4) guaranteed rights; (5) equity and social participation; (6) recognition of local cultures; (7) respect for regional diversity; (8) efficiency, efficacy and effectiveness; (9) commitment; and (10) development and growth with a focus on value chains.

The new institutional and legal framework of the Ministry established guidelines for the planning, promotion, and supervision of fisheries and aquatic resources as well as their preservation, conservation, and recovery. Moreover, it sought to promote socio-economic, cultural, and institutional capacity building. The promotion of sustainable development, shared and participatory management of resources with fishing communities, and research and development of new technologies and value chain development were central components. Development was recognized as not purely an economic goal, but as something that should also consider well-being, citizenship, and democracy outcomes (SEAP/Brasil 2008).

While these reforms in the structure and mandate of public sector agencies responsible for fisheries represent a significant advance, policy implementation and local engagement remain a challenge. Despite its broad mandate, the Ministry of Fisheries and Aquaculture lacked an adequate budget and the number of personnel necessary for outreach to the small-scale fisheries sector and implementation of these reforms, relying heavily on regional superintendents with limited staff and budgets to attend to the large numbers of fishers, who in many cases are socio-economically marginalized with very low effective access to other public agencies and programs for public health, education, and other basic services. In this context, and in an effort to support and strengthen civil society, it was important to work closely with local organizations representing fishers. What were once *capatazias*, now known as *colônias* (fishing guilds) and in some cases registered as Producer Associations, continue to be the main institutional mechanism through which the public sector reaches fishers. Producer Associations help with the registration of professional fishers and facilitate fisher access to social benefits. They are also the platform from which fishers organize and lobby for fisheries and social rights. However, a legacy of cronyism, nepotism, clientelism, and corruption continue to plague many of such groups, in some cases challenging their ability to effectively and fairly meet the needs of their membership. In parallel, a number of civil society groups, supported by associated social movements, for example the pastoral non-governmental organization 'Conselho Pastoral dos Pescadores,' (CPP) have developed alternative local organizations. Generally the goal of these organizations is to strengthen fishers' rights through advocacy for public recognition of the sector's economic and social contribution, and associated policy to support its development. As stated by Mattos (2014), these groups are also plagued by issues of legitimacy, representativeness, and lack of balance between political rhetoric and the achievement of real social justice.

Overall, despite significant advances in public policy, increased visibility and political voice of the sector, especially in the past decade or so, the small-scale fishing sector continues to lack adequate institutional and political support at all levels for sustainability and social balance. Fishers still face poor working conditions, lack of infrastructure, and low levels of education (MPA/Brasil 2013), all aggravated by the demise of the Ministry of Fisheries and Aquaculture in 2015. Even though government policies after SUDEPE brought major changes in the fisheries sector, required modernization became a fallacy and a 'myth,' generating very heterogeneous production structures (Mattos 2007). Modernization, according to Diegues (1983), did nothing more than hasten the irrational exploitation of fishing resources, and the gradual impoverishment of thousands of small-scale fishers. In fact, SUDEPE warned that the low level of technology at the time and the non-adoption of a research incentive policy to propitiate their improvement, creation and/or adaptation, became limiting factors of the process of development.

The construction and implementation of public policies for the sector requires dialogue and a close relationship with representative fishers associations, not only to induce intervention processes, but also to meet the challenges of a bottom-up implementation of public policies that take notice of empirical and traditional knowledge (Mattos 2011). There is still a lack of access to basic rights, such as adequate health care (including primary care), recognition, prevention, and treatment of occupational diseases that affect fishworkers, documentation (many fishers do not have basic documents such as birth certificates or social insurance numbers that are necessary to register for most social support programs), and basic information about individual rights and how to access social security programs.

## *Clam Fisheries and Women*

Fishing continues to be considered primarily a male activity in Brazil, with a lack of visibility and recognition of fisherwomen and their contribution to household food security, income, and regional economies. Almost half of registered professional fishers (46.3%), both men and women, live in the Northeast Region, working mainly in coastal and estuarine areas (MPA/Brasil 2012). Across Brazil's more than 8000 km of coastline, women carry out the bulk of harvesting in estuarine fisheries, through the collection of clams (mainly the Venerid clam *Anomalocardia brasiliana*, hereafter referred to as the 'tiny venus clam') (Fig. 22.2), oysters, crabs, other types of shellfish, and other aquatic organisms.

**Fig. 22.2** Tiny venus clam (*Anomalocardia brasiliana*) collected during low tide, mainly by fisherwomen, in Grossos, Rio Grande do Norte State, Northeast Region of Brazil. Clams are placed in bucket for further processing (Photo credit: Gustavo Henrique G. da Silva 2012)

Bivalve mollusks are a particularly common target species (Dias et al. 2007; Rios 2009) which are extensively distributed and harvested year-round. Tiny venus and other similar beach clams are easily extracted and do not require boats or specialized fishing gear to harvest (Rodrigues et al. 2013), providing an accessible source of income for many small-scale fishing families (Oliveira et al. 2014), as well as constituting an important source of protein, contributing to food security (Nishida et al. 2004). Exploitation levels of the tiny venus clams are considered high throughout their range, with some evidence of reduced abundance attributed to fishing pressure and coastal degradation (Nishida et al. 2004; Rocha 2013; Rodrigues et al. 2013). However, historical studies on the impacts of fishing on the resource are scarce, and existing records largely inaccurate (Chiba et al. 2012), making resource management and the decision-making process of collecting and processing shellfish difficult (Rocha and Lopes 2014).

## People of the Tides – The GDM Project – Capacity Building for Institutions and People Together

The GDM project was an agreement of cooperation between the Brazilian Cooperation Agency (ABC) of the Ministry of Foreign Affairs and the Canadian International Developmental Agency (CIDA). The Agreement was also co-coordinated by the Brazilian Ministry of Fisheries and Aquaculture and the Canadian charity organization World Fisheries Trust (WFT), and implemented from 2008 to 2011. The aim was to build institutional and community capacity for improving the livelihoods of women and families that depend on clam and oyster extraction in coastal communities in the Northeast of Brazil, through a partnership initiative involving national and local governments, researchers (universities) and local fishing associations (guilds). The project proposed to mainstream social equity, reduce poverty, improve access to citizenship rights and duties, and develop technology for fisheries management, culture, processing and commercialization of bivalve mollusks. It worked in collaboration with local institutions already engaged in research and extension partnerships with traditional coastal clam harvesting communities in four Brazilian northeastern states: Pernambuco, Paraiba, Rio Grande do Norte and Bahia. The GDM project actively pursued the participation of fishing community representatives at all levels from the inception stage and promoted an affirmative approach to enhancing the participation of women in particular. It was possible to create strong, locally supported, and doable activities which fostered some exciting opportunities for collaboration and exchange among Brazilian and Canadian partners, from fishing community representatives to private sector and federal government representatives. By emphasizing participatory processes, GDM initiatives built multilateral partnerships and strengthened local resource users' capacity, in particular fisherwomen and vulnerable and marginalized groups, to monitor their own progress in support of adaptive management of local development and sustainable shellfish fisheries management (Macnaughton et al. 2010).

Prior to this work, many clam fisherwomen self-identified themselves as housewives rather than professional fishers, because of the negative social stigma associated with the activity. The project created a national public profile for these fishers, through a variety of activities including a national exposition of a clam-fishers photo-voice project, and support for their active participation at state and national-level meetings, congresses, and consultations on the 2009 Fisheries and Aquaculture Law. Further, it helped promote networking with and support of the work of researchers and civil society organizations already working in partnership with strong female leaders in fishing associations and *Colônias* (Macnaughton et al. 2010). A variety of university research groups involved with GDM developed new projects on clams and oysters and associated fisheries (da Silva et al. 2014a), some of which continue to be pursued. This has greatly improved knowledge on *Anomalocardia brasiliana* biology and this fishery.

## Advances in the Implementation of Key SSF Guidelines – The Case of the Clam Fishery and GDM Experience in Brazil

### *Guideline 5 – Governance of Tenure in Small-Scale Fisheries and Resource Management*

The right to fish for subsistence by traditional fishing communities is guaranteed in the Brazilian Constitution. However, specific tenured access to fishing areas and resources continues to be a contentious issue due to unclear and overlapping jurisdictions in shared commons, resulting in many conflicts between different users of both the space and the resource. Access to traditional fishing grounds and coastal lands by small-scale fishing communities is often very limited in practice. On the other hand, there are often significant systems of informally regulated tenured use of resources or fishing spaces within fishing community environments, including some rules and practices that contravene federal fishing regulations related to minimum catch size and fishing non-take zones.

Government-led fisheries management in Brazil generally focuses on controlling effort and fleet capacity through seasonal and spatial closures and gear restrictions. Legislation does not generally allow for exclusive access rights for either large or small-scale fisheries. However, extractive reserves (RESEXs[2]), marine protected areas (MPAs) defined in Brazilian legislation as conservation units as per the National Protected Areas System[3] (Brasil 2000), provide an interesting anomaly. Under this system, the request to establish a RESEX must come from traditional resource users, based on a concern for conserving biodiversity and resources for

[2] RESEX – Reserva Extrativista (Extractive Reserve); Sistema Nacional de Unidades de Conservação – SNUC: Law # 9,985/2000.

[3] Sistema Nacional de Unidades de Conservação – SNUC. Law # 9.985/2000.

sustainable use. Following the designation of RESEX areas, resources within them should be allocated to creating local resource management plans and monitoring systems in the protected areas. Simply put, traditional fishing communities are allowed to extract resources because their 'tenure rights to the resources that form the basis for their social and cultural well-being, their livelihoods and their sustainable development' are recognized (FAO 2015, 5.15). Some extensions of this principle are being trialed by clam fisherwomen, primarily within the envelope of community co-management that may implement gear or temporal restrictions that make fishing by 'outside' users difficult (Kalikoski et al. 2006; Almeida et al. 2009).

The high degree of manual labour and low economic returns associated with clam fisheries, as well as the wide distribution of tiny venus clam, means there is relatively little competition or conflict among harvesters over acceptable fishing grounds (e.g., beaches). However, there are considerable conflicts associated with restricted access to beaches through irregular occupation of public lands by private interests, especially in the vicinity of tourist resorts. This goes against what SSF Guidelines (FAO 2015, 5.3) propose to ensure, namely, that small-scale fishers, fish workers, and their communities have secure, equitable, and socially and culturally appropriate tenure rights to fishery resources and small-scale fishing areas and adjacent land. Special attention, moreover, is given to women in the Guidelines. Through the GDM project, a group of female leaders of local fishing associations raised concerns regarding user-conflicts and tenure and access rights. Making use of the network of GDM partners they were able to articulate their demand for support to the appropriate public actors involved. Accountability was improved with some inroads made with regard to consolidating legal access to fishing beaches and promoting a better understanding of the need for a comprehensive restriction on the number of users in the fishery and the prioritization of zoning to avoid contaminated areas.

Sustainable resource management in an open-access situation such as that of coastal beaches is a daunting task, particularly in the general absence of fisheries and stock data and the lack of effective policing. The value of traditional knowledge among small-scale fishers has long been promoted by both researchers and social advocates (Begossi et al. 2006; Silvano 1997 cited by Clauzet et al. 2005), resulting in a significant institutional focus on community co-management and the importance of community participation in decision-making for local resource management. However, Kalikoski et al. (2006) and Rocha and Pinkerton (2015) stress that while many state-proposed co-management arrangements express a willingness to assign rights and responsibilities to communities, few actually delegate decision making powers. Mattos (2014) emphasizes the need to minimize, even halt, continuing predominance of centralized control, the demand for education at all levels, and the need for greater recognition that community-based organization can build *de facto* effective policy for participatory and equitable resource management.

During the course of the GDM Project, there were examples in smaller communities such as Grossos, Rio Grande do Norte State, of fisherwomen collectively controlling resource access effectively, and demonstrating great interest for sustainable use and a willingness to adjust fishing pressures as needed. In areas of the project closer to large urban centres in Pernambuco State, interests were more focused on gaining adequate access to the beaches so as to increase economic returns. In both cases, platforms suitable for discussions on sustainable use of fishing resources were provided.

This information recorded by GDM indicates that an important starting point for sustainable fisheries lies in improving the knowledge base necessary for management, while appreciating the broader integrated community context, including the need for community development and poverty reduction. Following the recommendations of the SSF Guidelines (FAO 2015, 5.15), it is necessary to involve small-scale fishing communities in the design and planning of, and, as appropriate, in the implementation of management measures.

## *Guideline 6 – Social Development, Employment and Decent Work*

Although Brazilian fishing policies commonly engage *social development* and social justice approaches, the concepts of equality and effective *employment* or *decent work* are not necessarily guaranteed through this discourse. Due attention to ensuring that small-scale fishing communities are empowered and can enjoy their human rights are a must (FAO 2015, 6.1). Also, fisheries activities in both the formal and informal sectors must be taken into account in order to ensure the sustainability of small-scale fisheries (FAO 2015, 6.6).

Prior to the creation of the Ministry of Fisheries and Aquaculture, responsibility for the implementation of fishing policies resided in the Ministry of Agriculture. This resulted in the adoption primarily of an agribusiness model, generating little, if any, opportunity for small-scale fishing, and consequently little social inclusion. Unsustainable development models were the order of the day. In the push to create a distinct ministry and institutional framework, specifically to govern fisheries and promote sustainable policies that brought actors and actions together, the lobbying by civil society was instrumental.

One of the main accomplishments of GDM was to improve the public visibility of female clam diggers and their concerns, and increase their recognition as professional fishers, improving their access to associated rights. However, actual implementation of such rights has remained problematic. A key priority identified by several female leaders was a concern for occupational health and safety – the heavy manual labour and extreme exposure and difficult working conditions associated with digging and processing of the clams contribute to a series of health issues that are being largely ignored by the medical profession. Recognition of these distinct health concerns of the small-scale fishing sector (e.g. users or groups of users) may support "fishers, particularly women, to be able to earn a fair return from their labour, capital and management, and encourage conservation and sustainable management of natural resources" (FAO 2015, 6.7).

## *Guideline 7: Value Chains, Post-Harvest, and Trade*

Value chain optimization for equality has been a key target of fisheries-based community development goals for quite a long time. Potential latent income through value-added processing, losses due to inadequate handling practices, and considerable inequities and potential short-cuts in fish trading and marketing have been recently documented. Timely and accurate market information has been shown to be a key element in affording fishers equitable return for their products, in order to enhance their income and livelihood security (FAO 2015, 7.4). Unequal power relationships sometimes observed in the Brazilian context mean that vulnerable and marginalized groups may require special support (FAO 2015, 7.1). For example, women are often involved in the post-harvest processing, with a need for appropriate working conditions and processes for good quality and safe fish products (FAO 2015, 7.2 & 7.3) and reduced waste (FAO 2015, 7.5). The Brazilian government has recognized these needs, and has built them into their institutional strategies, providing the enabling legislation that allows for effective fisheries management systems to prevent overexploitation driven by market demand threatening the sustainability of fisheries resources, food security, and nutrition (FAO 2015, 7.8).

A participatory value chain analysis was carried out through GDM for clam fisheries (Wojciechowski et al. 2014), and identified limitations in fishing and processing practices, as well as unpreparedness of local actors for marketing and low organizational collaboration in the critical early stages of the value chain. Supportive public policies to aid development have not been implemented, undermining isolated efforts to improve production and shellfish quality, especially with regard to distribution and marketing. Advances were made in processing and value-added production in Rio Grande do Norte State, with the help of the Universidade Federal do Semi-Árido – UFERSA (Semi-Arid Federal University) team. Interventions were also made in Pernambuco State with the introduction of more efficient stoves that produced less smoke when cooking clams (as part of processing). These new stoves had less of an adverse impact on fisherwomen's lungs and visions and reduced the use of mangrove trees for combustion up to 80%. However, many needs still remain to be addressed with respect to marketing, fishing, and processing in manners that are equitable and adequately address cultural and social concerns. Creation and adoption of significant innovations of this kind is particularly challenging, where poverty is high and resilience to impacts from change low. Efficiency enhancement along the value chain, improving post-harvest technology, and helping create better environmental conditions, all aimed at adding value and trading safe and healthy fishing products, may contribute in engaging fishing communities in issues regarding their own development, in line with many assumptions raised by the SSF Guidelines.

## *Guideline 8 – Gender Equality*

Brazil has made substantial investment in improved rights and opportunities for women. Policies that have aimed to do so recognize the need for gender mainstreaming, with strategies and approaches adapted to different cultural contexts (FAO 2015, 8.1). Nevertheless, gender inequality, including violence against women, continues to be significant in most rural communities. According to Nishida et al. (2004), the clam fishery provides the main source of income or is a complementary source of income for a significant number of women. They indicate that the clam fishery is important from an economic, social, and cultural point of view, as well as being a critical source of protein and food security for families.

The clam fishery component of GDM was specifically designed to work with women, and promote opportunities for them. The project worked to catalyze the formation of a gender and fisheries working group within the Ministry of Fisheries and Aquaculture and supported a variety of successful associated lobbies that aimed to improve the language in the new fisheries legislation in 2009, recognizing clam fishers and others in the post-harvest sector (in large part women) as professional fishers and promoting more equitable rights. These women also became some of the most effective political lobbyists at the federal level for clam fisherwomen diggers, being instrumental in gaining recognition for the group as professional fishers.

## *Guideline 9: Disaster Risks and Climate Change*

Fishing resources are quite sensitive to changes in weather patterns or other environmental changes. Such changes acutely affect small-scale fishers who are limited in how far they can move to find new resources. Mechanisms to mitigate such impacts are evolving, generally as a patchwork of locally adapted solutions, with different levels of effectiveness. These mechanism recognize the need, following the SSF Guidelines (FAO 2015, 9.3), for integrated and holistic approaches, including cross-sectoral collaboration, in order to address disaster risks and climate change in small-scale fisheries.

Clam resources could be seriously affected by changing weather patterns – particularly in areas with seasonal freshwater inputs or changing coastal erosion. Changes of this kind were documented in the project area (da Silva et al. 2014b), with fishers shifting to more remote clam beds where possible, which required a participatory management strategy. The project also developed protocols for clam aquaculture, in preparation for eventual supplementation of natural beds and/or specific beach culture. However, adequate time for field implementation was not available, and wild resources appeared to be adequate to maintain fisheries at the time.

## *Guideline 10: Policy Coherence, Coordination, Collaboration*

While both the small- and large-scale fisheries sectors have shaped policy and development strategies, as Chuenpagdee (2011) warns, their needs and priorities are not necessarily similar. In Brazil, both sectors lobbied for relief from the 'conservationist' regulatory system imposed by the state, albeit with substantially different agendas (Acselrad 2004). The large-scale fishery sought deregulation and weakening of the institutional framework to overcome restrictions to fishing. The small-scale fishery sector, on the other hand, felt persecuted and repressed by the state, and lobbied for similar relief from regulation. But in the name of 'environmental justice', its aim was to overcome inequalities, promote rights to the benefits of environmental goods, and gain entry to the decision-making process regarding use and access of local natural resources.

GDM raised concerns that clam fisherwomen should be part of discussion forums and councils under the existing institutional framework, where stakeholders, practitioners and government representatives participate in the implementation of initiatives for participatory fisheries management; socio-educational issues focusing on gender mainstreaming and promoting empowerment of women in fisheries occupations; and value chain upgrading and democratization focusing on the decent work agenda. This would provide an opportunity to the Government, through the Ministry of Fisheries and Aquaculture, to ascertain the strengths of the various interventions made in terms of the direction of public policies. It was also stated that this should be done "through consultation, participation and publicizing, gender-sensitive policies and laws on regulated spatial planning" (FAO 2015, 10.2). Substantial gains were achieved with the capacity development and value chain upgrading components of the proposal. The fact that the process was semi-autonomous from the Ministry of Fisheries and Aquaculture allowed for some supplementary synergies, i.e. co-management, that may not otherwise have been possible.

Fisheries co-management implementation for many species faces great challenges, though clam fisheries are *de facto* largely unmanaged and of such low profile that their management is not a high institutional priority. This highlights the difficulties of adopting specific measures that ensure the harmonization of policies affecting the health of coastal ecosystems to fishing communities' livelihoods and well-being. Overcoming these challenges is necessary for achieving policy coherence, cross-sectoral collaboration, and the implementation of holistic and inclusive ecosystem approaches in the fisheries sector (FAO 2015, 10.5). It will also facilitate fishing community involvement in policy- and decision-making processes relevant to small-scale fisheries (FAO 2015, 10.6).

## *Guideline 11: Information, Research, Communication*

Most Brazilian fisheries have limited information about their stock and catches. Data from small-scale fisheries is particularly difficult to obtain and keep up-to-date. The existing institutional and legal framework does not support detailed reporting from fishers. To date, production statistics have long delays and are primarily for total production. They do not have information with regard to fishing activity and catch composition, and hence small-scale fishing communities are not effectively recognized as "holders, providers and receivers of knowledge" (FAO 2015, 11.4).

Despite this paucity of fishing data and information, the main difficulties faced by small-scale fishers are known. For example, low educational levels in fishing communities is related to the difficulty of reconciling 'working time' with 'study time', thus resulting in a high illiteracy rate. This also interferes with efforts to organize fishers and to work in a cooperative and/ or associated manner, making the search for solutions to community problems more complicated, as well as reducing access to benefits of public policies and communication with governmental agencies and civil society organizations at national, state and local levels.

Outcomes from the project recognized that an improvement in continuous and systematic monitoring of coastal and estuarine small-scale fisheries is needed, as well as more information about environmental conservation for the recovery of shellfish stocks. It is also necessary for development that ensures ecosystem sustainability, "for an improved understanding and visibility of the importance of small-scale fisheries and its different components, including socioeconomic aspects" (FAO 2015, 11.1). And, perhaps even more applicable, in line with the SSF Guidelines (FAO 2015, 11.6) is the need "to investigate and document traditional fisheries knowledge and technologies in order to assess their application to sustainable fisheries conservation, management and development". Equally important is that "research priorities should be agreed upon through a consultative process focusing on the role of small-scale fisheries in sustainable resource utilization, food security and nutrition, poverty eradication, and equitable development" (FAO 2015, 11.9).

## *Guideline 12: Capacity Development*

The Brazilian government launched several educational initiatives for fishers to increase literacy. Lessons were given in the fishing communities' space and time and to improve access to technical training by recognizing practical experience, through a civil society institutional framework. Despite these substantial efforts and good will to enhance the capacity of small-scale fishing communities in order to enable them to participate in decision-making processes (FAO 2015, 12.1), substantial challenges remain in reaching remote communities and the most vulnerable populations, many of whom may depend on local small-scale fisheries for part of their food security and livelihoods.

Capacity development in the GDM included technical training for processing and value-added product development, but also training in participatory value chain evaluation, leadership, gender equality, and peer-peer networks. The most striking results were empowerment and increased visibility of fisherwomen, a "possible way to develop knowledge and skills to support sustainable small-scale fisheries development and successful co-management arrangements" (FAO 2015, 12.4).

## A Way Moving Forward: Small-Scale Fisheries Proposed Model and Roadmap for Brazilian Small-Scale Fishing Sector

Based on our experiences with the GDM project, and analysis of the evolution and main challenges facing small-scale fishing policies for Brazilian coastal communities, we have highlighted some critical areas that need improvement. Although the challenges facing small-scale fisheries development in Brazil may seem straightforward in general, they are not yet broadly acknowledged among fisheries scientists, government representatives, and other stakeholders. This is particularly the case for clam fisheries where information about the sector is still scattered.

In Brazil, variable political support and associated difficulty in providing continuity to national initiatives has been a paradox. Empowerment, supportive networking, appropriate visibility, and civil society support are thus essential in creating the necessary lobbies for program and policy continuity (e.g. Oliveira 2013). Of course, programs at all levels may be favourable or unfavourable for the small-scale fisheries sector, depending on lobbies of different fishing groups, creating barriers for implementation and continuity of some initiatives. Long-term monitoring depends on both governmental support (through political will and institutional steadiness), and more localized research projects, and both have historically been patchy. Ongoing support programs and monitoring for sustainable small-scale clam fisheries with equitable socio-economic benefits seem unlikely without ongoing financial and technical support from the outside. Nevertheless, the patchwork of projects that have reached communities over the years do provide benefits that are of great local significance to the people that carry out this activity within a complex reality of diverse livelihoods and social environments "that allow a better understanding and documentation of the true contribution of small-scale fisheries to sustainable resource management for food security and poverty eradication" (FAO 2015, 13.4).

Finally, we present the following road map, constructed with stakeholders during the final workshops of the GDM Project in 2010, of priority action fronts for the sector:

1. *Co-management of Small-Scale Fisheries*

   Goal: Effective and timely monitoring and regulation of small-scale fishing resources in estuarine and coastal areas.

   Deliverable 1.1: Instruments and mechanisms for monitoring and regulation elaborated for clam fisheries.

   Activity 1.1.1: Conceptual and methodological studies on instruments and mechanisms for monitoring.

   Activity 1.1.2: Conceptual and methodological studies on processes, instruments, and mechanisms for regulation.

   Activity 1.1.3: Implementation of mechanisms for monitoring and regulation.

Deliverable 1.2: Processes, tools, and mechanisms for small-scale fisheries monitoring and regulation through ecosystem approaches and ecosystem-based management, considering human dimensions and creating appropriate space to support empowered participation of clam fishers.

Activity 1.2.1: Carry out pilot case studies of clam fisheries co-management.

Activity 1.2.2: Publications of case studies allowing for their replication and evolution.

Activity 1.2.3: Evaluation of the effectiveness of processes, tools, and mechanisms.

2. *Gender Mainstreaming and the Empowerment of Women in Fisheries Occupations.*

Goal: On-going education and technical training, including monitoring and effectiveness, of small-scale fishers and fishworkers, in particular clam fisherwomen.

Deliverable 2.1: Integration of existing databases of the clam fishery

Activity 2.1.1: Studies on the integration of existing databases.

Activity 2.1.2: Gender mainstreaming and promoting empowerment of women in fisheries occupations.

Deliverable 2.2: Programs of education and capacity building developed and implemented.

Activity 2.2.1: Carry out pilot studies of education and professional training for clam fisheries value chains.

Activity 2.2.2: Assist stakeholder involvement to implement, refine, and monitor pilot studies.

Activity 2.2.3: Develop training documents in accessible language based on pilot experiences.

3. *Value Chain Upgrading and Democratization Focusing on the Decent Work Agenda.*

Goal: Development and monitoring of small-scale fisheries value chains with a focus on decent work for clam fishers in estuarine and coastal areas.

Deliverable 3.1: Instruments and mechanisms for the monitoring and socio-economic optimization of the clam fisheries value chains.

Activity 3.1.1: Conceptual, methodological, and technical studies for monitoring and developing technical assistance.

Activity 3.1.2: Develop and test processes, instruments, and monitoring for regulatory mechanisms.

Activity 3.1.3: Create parameters and criteria to evaluate the effectiveness of instruments and mechanisms for monitoring and control.

Activity 3.1.4: Carry out surveys and systematic mapping of needs for technical assistance.

Deliverable 3.2: Technologies and innovations for the clam fisheries value chain developed and tested.

Activity 3.2.1: Pilot studies to improve technologies and innovations.

Activity 3.2.2: Intersectoral dialogues for the analysis and proposal of sustainable development.
Activity 3.2.3: Develop training documents in accessible languages for the different audiences based on the pilot experiences.
Deliverable 3.3: Methods and instruments develop for the evaluation and monitoring of working conditions in the clam fisheries value chain.
Activity 3.3.1: Conceptual, methodological, and technical multi-stakeholder studies on decent work to develop monitoring instruments and mechanisms.
Activity 3.3.2: Conduct a diagnostic study of working conditions.
Activity 3.3.3: Develop and test a monitoring plan for promoting decent work conditions.
Activity 3.3.4: Perform conceptual and technical studies to support knowledge of causal diseases related to clam fisheries working conditions.

Each strategy should include communications and dialogues of the ascertained goals, considering a combination of: (a) deliverables and activities with *economic* issues that require a diversification and consolidation of production flows for improved economic and social returns; (b) *social* issues to promote societal equality in access rights and reduction of social gaps – particularly for women – including social security, conditions for decent work, and access to appropriate health services; and (c) *environmental* issues for proper resource management, reduction of waste, and general minimization of the environmental impact during the production, processing, distribution, and marketing processes.

## Final Remarks

National richness is the human development of its citizens. For this to be possible, it is necessary to create a situation in which all people enjoy a long, good quality with sound health and avenues for creativity. In this sense, economic income should not be seen as an end in itself but as a means for welfare. That is why the role of small-scale fisheries in local economies and the links of the subsector to the wider economy need to be recognized and benefit from sustainable resource utilization and livelihood diversification (FAO 2015, 6.8).

Initially our belief was that the key challenge for improved small-scale fisheries in Brazil rested in the implementation processes rather than in the policy itself. Mattos (2011, 2014) notes that implementation should benefit the most vulnerable and marginalized groups of fishers, such as clam harvesters, through special support on a long-term basis, e.g. through gender mainstreaming and decent work agendas.

Even if a consensus can be achieved in participatory decision making processes, with legitimate, democratic and representative structures, challenges remain in providing access to market opportunities and increased transparency and information-sharing in the small-scale fisheries value chains. While the current Brazilian government incorporates many of these principles in its development strategies, at least on paper, implementation is an ongoing challenge, particularly if combating social inequality is to become ineffective due to a decline in economic growth that hampers wealth creation and equitable distribution.

Most important among the findings and recommendations of the GDM project is improving the knowledge base necessary to support both sustainable fishery management and more broadly community development and poverty reduction programs and policy. Clam fisheries are vitally important to food security and as a source of income and should be protected and supported as such.

It is important to state that there has been significant progress in implementing the SSF Guidelines for Brazilian fishers, through civil society organizations' activities, and supported by national and international non-governmental organizations. In particular, a national workshop on capacity building for the implementation of the SSF Guidelines was held in Brazil, in June 2016, which paved the road ahead for implementation taking a co-design approach to identifying critical areas and priorities.

The aim of the workshop was to promote awareness about the SSF Guidelines and mobilize support for their implementation across several countries of the Global South through a methodological, analytical, and descriptive approach. An additional objective was to build capacity, in particular, among fishers' organizations to position them as the key actors in the implementation process. The workshop was also an opportunity to document existing governance practices of tenure and resource management to enhance small-scale fishers' rights to resources and territories, and guarantee respect of human rights. Values, norms, and principles embedded in governance and resource management practices, which are essential for facilitating the implementation of the SSF Guidelines in Brazil, were adopted keeping in mind the need for autonomy. This, in turn, stimulated discussion on suitable training and exchanges of experiences to disseminate local knowledge, using a broad participatory and communicative process.[4]

Considering the characteristics of Brazilian small-scale fisheries, in particular the expressed features of clam fishery, and the recognition of similarities with states' existing obligations under national and international law and voluntary commitments, including the Code of Conduct for Responsible Fisheries (FAO 1995), that give due recognition to the requirements and opportunities of small-scale fisheries (FAO 2015, 5.13), we believe the discussion and analysis provided about the outcomes and lessons learned from the GDM project can help implement the SSF Guidelines, assisting the Brazilian state in the process to promote and implement appropriate management systems.

[4] Ouvidoria do Mar. https://sites.google.com/site/ssfguidelines/brazil/resources

## References

Abdallah, R., Sumaila, U. R., & Abdallah, P. (2007). An historical account of Brazilian public policy on fisheries subsidies. *Marine Policy, 31*(4), 444–450.

Acselrad, H. (2004). As práticas espaciais e o campo dos conflitos ambientais. In H. Acselrad (Ed.), *Conflitos ambientais no Brasil* (pp. 13–35). Rio de Janeiro: Relume Dumará.

Almeida, O. T., Lorenzen, K., & McGrath, D. G. (2009). Fishing agreements in the lower Amazon: For gain and restraint. *Fisheries Management and Ecology, 16*(1), 61–67.

Begossi, A., Hanazaki, N., Peroni, N., & Silvano, R. A. M. (2006). Estudos de ecologia humana e etnobiologia: uma revisão sobre usos e conservação. In C. F. D. Rocha, H. G. Bergallo, M. A. S. Alves, & M. Van Sluys (Eds.), *Biologia da Conservação: Essências* (pp. 320–331). Rio de Janeiro: Editora da UERJ.

Brasil. (2000). *Lei n° 9.985, de 18 de Julho de 2000*. Regulamenta o art. 225, § 1o, incisos I, II, III e VII da Constituição Federal, institui o Sistema Nacional de Unidades de Conservação da Natureza e dá outras providências. Brasília: Diário Oficial da República Federativa do Brasil.

Brasil. (2009). *Lei n° 11.959, de 29 de Junho de 2009*. Dispõe sobre a Política Nacional de Desenvolvimento Sustentável da Aquicultura e da Pesca, regula as atividades pesqueiras, revoga a Lei no 7.679, de 23 de novembro de 1988, e dispositivos do Decreto-Lei n° 221, de 28 de fevereiro de 1967, e dá outras providências. Brasília: Diário Oficial da República Federativa do Brasil.

Chiba, W. A. C., Assunção, A. W. A., Takao, L. K., Rocha, G. S., Janke, H., Valsko, J., Ebert, L. A., Figueroa, M. E., & Cunha, S. (2012). Caracterização da produção pesqueira ao longo do tempo, no município de Cananéia, litoral Sul de São Paulo. *Boletim do Instituto de Pesca, 38*(3), 265–273.

Chuenpagdee, R. (2011). A matter of scale: Prospects in small-scale fisheries. In R. Chuenpagdee (Ed.), *World small-scale fisheries contemporary visions* (pp. 21–36). Delft: Eburon Academic Publishers.

Clauzet, M., Ramires, M., & Barella, W. (2005). Pesca artesanal e conhecimento local de duas populações caiçaras (Enseada do Mar Virado e Barra do Una) no litoral de São Paulo, Brasil. *MultiCiência,* p. 4.

da Silva, G. G., Costa, R. S., Belem, T. P., Rodrigues, A. M. L., Moura, R. S. T., Martins, L. P. C., et al. (2014a). Ecologia populacional e manejo pesqueiro de *Anomalocardia brasiliana*. In G. G. da Silva, J. Carolsfeld, & A. Olivera Gálvez (Eds.), *GENTE DA MARÉ: Aspectos Ecológicos e Socioeconômicos da Mariscagem no Nordeste Brasileiro* (pp. 117–156). Mossoró: EdUFERSA.

da Silva, G. G., Macnaughton, A., & Carolsfeld, J. (2014b). Projeto Gente da Maré: aspectos sociais, ambientais e econômicos. In G. G. da Silva, J. Carolsfeld, & A. Olivera Gálvez (Eds.), *GENTE DA MARÉ: Aspectos Ecológicos e Socioeconômicos da Mariscagem no Nordeste Brasileiro* (pp. 25–34). Mossoró: EdUFERSA.

Dias, T. L. P., Rosa, R. S., & Damasceno, L. C. P. (2007). Aspectos socioeconômicos, percepção ambiental e perspectivas das mulheres marisqueiras da reserva de desenvolvimento sustentável Ponta do Tubarão (Rio Grande do Norte, Brasil). *Gaia Scientia, 1*(1), 25–35.

Dias-Neto, J., & Dornelles, L. C. C. (1996). *Diagnóstico da pesca marítima no Brasil*. Brasília: Ibama.

Diegues, A. C. S. (1983). *Pescadores, Camponeses e Trabalhadores do Mar*. São Paulo: Ática.

FAO. (1995). *Code of conduct for responsible fisheries*. Rome: Food and Agriculture Organization of the United Nations.

FAO. (2015). *Voluntary guidelines for securing sustainable small-scale fisheries in the context of food security and poverty eradication*. Rome: Food and Agriculture Organization of the United Nations.

IBGE. (2010). *Censo 2010*. Brasília: Instituto Brasileiro de Geografia e Estatística.

Isaac, V. J., Martins, A. S., Haimovici, M., Castello, P., & Andriguetto, J. M. (2006). Síntese do estado de conhecimento sobre a pesca marinha e estuarina do Brasil. In V. J. Isaac, A. S.

Martins, M. Haimovici, & J. M. Andriguetto (Eds.), *A pesca marinha e estuarina do Brasil no início do século XXI: recursos, tecnologias, aspectos socioeconômicos e institucionais* (pp. 181–186). Belém: Editora Universitaria UFPA.

Jentoft, S., & Chuenpagdee, R. (2009). Fisheries and coastal governance as a wicked problem. *Marine Policy, 33*, 553–560.

Kalikoski, C. D., Almudi, T., & Seixas, C. S. (2006). O Estado da Arte da Gestão Compartilhada e Gestão Comunitária da Pesca no Brasil. In Informativo do Projeto Manejo dos Recursos Naturais da Várzea. *Revista Jirau*, p. 15.

Macnaughton, A., Rocha, L. M., Wojciechowsk, J. M., & Carolsfeld, J. (2010). Tools for understanding the complexities of small-scale coastal fisheries economies in Northeastern Brazil: Participatory value-chain mapping and economic feasibility studies. In A. L. Shriver (Ed.), *Proceedings of the IIFET- International Institute of Fisheries Economic and Trade*. Corvallis: International Institute of Fisheries Economics & Trade.

Maldonado, S. C. (1986). *Pescadores do Mar*. São Paulo: Editora Ática.

Mattos, S. M. G. (2007). Contribuição dos modelos bio-econômicos para a gestão participativa e o ordenamento da pesca artesanal e de pequena escala. *Revista Brasileira de Engenharia de Pesca, 2*(2), 52–68.

Mattos, S. M. G. (2011). Desafios à implementação de políticas públicas e à gestão participativa da pesca artisanal. In S. M. G. Mattos, R. T. Moura, & W. M. Maia Jr. (Eds.), *Gestão de Pescarias Costeiras e da Maricultura. Anais da II Oficina de Trabalho de Aquicultura e Pesca do Nordeste* (pp. 79–96). Brasília: Ministério da Pesca e Aquicultura; Superintendência do Desenvolvimento do Nordeste.

Mattos, S. M. G. (2014). Políticas Públicas e Gestão Participativa da Pesca Artisanal: O Extrativismo Pesqueiro em Ecossistemas Estuarinos. In G. H. G. da Silva, J. Carolsfeld, & A. Olivera Gálvez (Eds.), *GENTE DA MARÉ: Aspectos Ecológicos e Socioeconômicos da Mariscagem no Nordeste Brasileiro* (pp. 315–352). Mossoró: EdUFERSA.

MPA/Brasil. (2010). *Boletim Estatístico da Pesca e Aquicultura 2008 e 2009*. Brasília: Ministério da Pesca e Aquicultura.

MPA/Brasil. (2012). *Registro Geral da Atividade Pesqueira e Aquícola (RGP)*. Brasília: Ministério da Pesca e Aquicultura.

MPA/Brasil. (2013). *Produção pesqueira e aquícola*. Brasília: Ministério Da Pesca e Aquicultura.

Nishida, A. K., Nordi, N., & Alves, R. R. N. (2004). Abordagem etnoecológica da coleta de moluscos no litoral paraibano. *Tropical Oceanography, 32*(1), 53–68.

Oliveira, V. R. (2013). O Processo de Participação Social nos Planos Plurianuais do Governo Federal. In E. M. Silva & L. B. Soares (Eds.), *Experiências de Participação Institucionalizada*. Belo Horizonte: UFMG.

Oliveira, I. B., Silva Neto, S. R., Lima Filho, J. V. M., Peixoto, S. R. M., & Galvez, A. O. (2014). Efeito do período chuvoso na extração do molusco bivalve *Anomalocardia brasiliana* (Gmelin, 1791). *Revista Brasileira de Ciências Agrárias, 9*(1), 139–145.

Rios, E. C. (2009). *Seashells of Brazil*. Rio Grande: Editora da Fundação Universidade do Rio Grande.

Rocha, L. M. (2013). *Ecologia humana e manejo participativo da pesca do búzio Anomalocardia brasiliana (Gmelin, 1791) bivalvia: veneridae) na Reserva de Desenvolvimento Sustentável Estadual Ponta do Tubarão (RN)* (Doctoral dissertation). Universidade Federal do Rio Grande do Norte, Natal.

Rocha, L. M., & Lopes, P. F. M. (2014). Ecologia humana e mariscagem de *Anomalocardia brasiliana* no Nordeste brasileiro. In G. H. G. da Silva, J. Carolsfeld, & A. Olivera Gálvez (Eds.), *GENTE DA MARÉ: Aspectos Ecológicos e Socioeconômicos da Mariscagem no Nordeste Brasileiro* (pp. 157–184). Mossoró: EdUFERSA.

Rocha, L. M., & Pinkerton, E. (2015). Co-management of clams in Brazil: A framework to advance comparison. *Ecology and Society, 20*(1), 7–16.

Rodrigues, A. M. L., Azevedo, C. M. S. B., Costa, R. S., & Henry-Silva, G. G. (2013). Population structure of bivalve *Anomalocardia brasiliana* (Gmelin, 1791) in semi-arid estuarine region of Northeast Brazil. *Brazilian Journal of Biology, 73*, 4.

SEAP/Brasil. (2008). *Mais Pesca e Aquicultura. Plano de Desenvolvimento Sustentável – Uma rede de ações para o fortalecimento do setor*. Secretaria Especial de Aquicultura e Pesca da Presidência da República.

Silva, L. G. (1988). *Os pescadores na história do Brasil*. Petrópolis: Vozes, Conselho Pastoral dos Pescadores.

Silvano, R. A. M. (1997). *Ecologia de Três Comunidades de Pescadores do Rio Piracicaba (SP)*. (Dissertação de Mestrado). Universidade Estadual de Campinas Instituto de Biologia, Campinas, SP.

Wojciechowski, M. J., Melo, K. S. G., & Nascimento, A. F. (2014). Caracterização da cadeia produtiva de moluscos bivalves nos Estados de Pernambuco e Rio Grande do Norte. In G. H. G. Da Silva, J. Carolsfeld, & A. Olivera Gálvez (Eds.), *GENTE DA MARÉ: Aspectos Ecológicos e Socioeconômicos da Mariscagem no Nordeste Brasileiro* (pp. 271–314). Mossoró: EdUFERSA.

# Chapter 23
# The Step Zero for Implementing the Small-Scale Fisheries Guidelines in Newfoundland and Labrador, Canada

**Ratana Chuenpagdee, Kim Olson, David Bishop, Meike Brauer, Vesna Kereži, Joonas Plaan, Sarah Pötter, Victoria Rogers, and Gabriela Sabau**

**Abstract** While global in scope, the Voluntary Guidelines for Securing Sustainable Small-Scale Fisheries (SSF Guidelines) have a strong orientation towards developing countries, especially with their focus on food security and poverty alleviation. Stakeholder consultations during the development phase, for instance, took place predominantly in developing countries, and none were held in Canada or the

R. Chuenpagdee (✉)
Department of Geography, Memorial University of Newfoundland, St. John's, Newfoundland and Labrador, Canada
e-mail: ratanac@mun.ca

D. Bishop • V. Kereži • J. Plaan
Too Big To Ignore Global Partnership for Small-Scale Fisheries Research, Memorial University of Newfoundland, St. John's, NL, Canada
e-mail: d.bishop@mun.ca; toobigtoignore@mun.ca; joonas.plaan@gmail.com

K. Olson
Government of Newfoundland and Labrador, St. John's, NL, Canada
e-mail: KimberlyOlson@gov.nl.ca

M. Brauer
Nordic Master Aquatic Food Production – Quality and Safety, Norwegian University of Life Science, Ås, Akershus, Norway
e-mail: meike.brauer@gmx.net

S. Pötter
West Nordic Studies, University of Akureyri, Akureyri, Iceland

University of the Faroe Islands, Thorshavn, Faroe Islands
e-mail: sarah.poetter@gmx.net

V. Rogers
Too Big To Ignore Global Partnership for Small-Scale Fisheries Research, Victoria, BC, Canada
e-mail: victoria.francis4@gmail.com

G. Sabau
School of Science and the Environment, Memorial University of Newfoundland, Grenfell Campus, Corner Brook, NL, Canada
e-mail: gsabau@grenfell.mun.ca



S. Jentoft et al. (eds.), *The Small-Scale Fisheries Guidelines*, MARE Publication Series 14, DOI 10.1007/978-3-319-55074-9_23

USA. Implementing the SSF Guidelines in developed countries may therefore be challenging, given also that the small-scale fisheries sector receives relatively little attention from governments compared to its large-scale counterpart. As in other cases, however, an understanding of what conditions and development may foster the implementation of the guidelines is imperative to gauge the feasibility and likelihood of success. Such knowledge can also serve as a starting point to engage fishers, governments and other stakeholders in a discussion about what they can draw from the SSF Guidelines to promote sustainability and viability of small-scale fisheries in their areas. We explore these questions in the context of small-scale fisheries in Newfoundland and Labrador (NL), Canada, where small-scale fisheries remain active, despite the cod fishery moratorium in 1992. Through a literature review, interviews, discussion sessions and community events, we gather information about various aspects of small-scale fisheries in NL and discuss the extent to which they may contribute to the successful implementation of the SSF Guidelines in this province. The chapter concludes with implementation challenges and ways forward.

**Keywords** Small-scale fisheries • SSF Guidelines • Enabling conditions • Developed countries • Newfoundland and Labrador

## Introduction

While small-scale fisheries exist around the world, they are generally not as visible in developed countries as those in developing and less developed ones. Pictures of an impressive amount of fishing people and fishing boats in a landing site often come from South or Southeast Asia, and Africa, not from Europe or North America. The focus on industrial development and export earnings in developed countries may explain some of this under-representation. Generally speaking, small-scale fisheries catches are lower in terms of trade values, when compared to those from large-scale, industrialized fisheries, because a good proportion of these catches are consumed within the fishing households. Among developed countries, small-scale fisheries in the USA are very prominent, contributing about 30% of the total catches, which is higher than the global average of about 26% (Sea Around Us n.d.). The contribution of small-scale fisheries catches in Europe and Canada are rather small, at 10% and 5% of total catches, respectively. These estimates may be useful in providing big pictures at the national level but they undermine the fact that the value of small-scale fisheries is wide reaching – beyond income, jobs, or Gross Domestic Product, which are often used to measure the contributions of sectors. A closer look at what goes on in fishing communities, in all areas of the world, would reveal other aspects such as the importance of small-scale fisheries to local food security, poverty alleviation, and heritage preservation, among other things.

This is certainly the case with small-scale fisheries in Newfoundland and Labrador (NL), the easternmost province of Canada. While it may be known for its infamous Atlantic cod (*Gadus morhua*) fishery collapse in 1992, the area is rich with fishing history and tradition dating back to nearly 500 years ago when fisheries were first explored by Europeans (Ryan 1990; Bryant and Martin 1996). The cod moratorium has transformed the fishing industry, which shifted the focus from groundfish to other species, particularly Northern prawn (*Pandalus borealis*) and snow crab (*Chionoecetes opilio*). Many processing plants closed down, and only a handful of them remain in operation today, after being refitted to process the new species. Social transformation also took place, with a massive job loss that resulted from the moratorium, and with it a dwindling social safety net due to high unemployment rates and associated out-migration (Palmer and Sinclair 1997; Harris 1998). Within a few months, the closure caused more than 40,000 people to lose their jobs. Over a span of 20 years, following the 1992 moratorium, employment in the fishery continued to decline, shrinking by about 60% during this time. Amidst these declines, small-scale fisheries[1] have managed to maintain their presence, dominating the sector by the number of vessels, accounting for approximately 12% of total landings in the province in 2015 and roughly 11% of total value (DFO 2016). Further, in the same year the small-scale fishery was responsible for approximately 56% (or about 6000 t) of the total cod production in the province from all fleets.

Almost 25 years after the moratorium, signs of cod recovery have been noted, bringing a lot of excitement and anticipation, but also some questions and concerns (Mather 2013). Small-scale fisheries of NL are well positioned to benefit from this opportunity, if different practices can be imagined when fisheries are re-opened. Some members of the fishing industry are apprehensive due to the current lack of processing facilities to handle large-scale commercial cod landings. Also, the price of cod is historically low, thus not motivating the young generation to enter cod fisheries. But perhaps this is not the only way forward. It may be possible, for instance, to consider allocating a small quota to small-scale fisheries allowing only certain low impact gears to be used, such as hand lines or cod pots. This can also be seen as a novel way to re-connect with cod, taking advantage of the new regulation that allows direct fish sales, the increasing use of locally caught fish in trendy restaurants and the increasing public interest in sustainable fisheries (Government of NL 2015).

Re-imagining the future of NL fisheries in this manner aligns well with the premise of the Voluntary Guidelines for Securing Sustainable Small-Scale Fisheries (SSF Guidelines) and their aim to support livelihoods and food security of small-scale fishing people, and promote overall fisheries sustainability (FAO 2015). Thus, in principle, the SSF Guidelines can be drawn upon to help the small-scale fisheries sector in NL reinforce their case with governments and other stakeholders as they navigate change in the fisheries. The utility of the SSF Guidelines for such a cause depends on several factors, however, starting from the extent to which the Guidelines are considered relevant to NL small-scale fisheries. Also, as argued by Chuenpagdee and Jentoft (2007), many conditions need to exist as part of the 'step zero' even before the implementation of the SSF Guidelines is contemplated.

This chapter examines the small-scale fisheries situation in NL, looking specifically at the likelihood that the SSF Guidelines could be implemented and how they

[1] Small-scale fisheries are referred to here as those using boats smaller than 35 feet in length. Locally, they are called 'small boat' fisheries, or officially known as 'inshore' sector.

may contribute to the revitalization of this sector, especially as new opportunities in fisheries unfold. The analysis is based mostly on information obtained during a series of community events in 2015, organized to gain insights about issues and concerns facing small-scale fisheries and to gauge public support of the sector. This is supplemented by interviews with fisheries stakeholders and a discussion session conducted to assess people's perception and awareness about the guidelines and to discuss their relevance to small-scale fisheries in NL.

The chapter begins with a brief account of small-scale fisheries in NL, followed by the description of research methods. Next, enabling conditions and recent development that could foster the implementation of the SSF Guidelines in NL are presented. Challenges in the implementation are discussed in the next section, along with some suggestions in the conclusion about what could be the first steps towards the implementation of SSF Guidelines in NL, and by extension, the rest of Canada.

## Small-Scale Fisheries of Newfoundland and Labrador: A Brief Account

NL has a long history of fishing tradition, the account of which can be found in numerous studies and archives. The Centre for Newfoundland Studies, a library based at Memorial University, for instance, has a vast collection of research articles, books, government documents, newspaper clippings, and historical maps about fisheries in the province. Entries about fisheries appear in official government sites, such as the Newfoundland and Labrador Heritage website, providing essential information about how the fishery was developed, its importance to cultural heritage, and the fundamental role it has played in NL society. Below is a brief history of NL fisheries, according to the Heritage Newfoundland and Labrador.[2]

The fishery began with the arrival of European fishers in the end of fifteenth century, when cod was fished seasonally. The first permanent settlements were established in the early 1800s. The cod fishery was predominately a family-based, small-scale cod fishery that was chiefly conducted in inshore waters using small boats and hand-lines with squid or capelin as bait. Since the beginning, the cod fishery was destined for export. Women and family members helped cure the fish, by removing the head, spine and gut, before salting and drying them on wooden flakes. Salted cod was integrated into a vast trade network, stretching from Europe to the Caribbean colonies. Not long after the fishery became lucrative, the price of cod dropped. As a result, fishing pressure increased, thus putting a strain on local stocks. As cod became scarce, particularly in areas of high fishing pressure, some fishers adopted more efficient gears, such as cod seines, trawls, gillnets, and cod traps. Some moved further from shore, using larger vessels and newer gears, which cost more but brought in a higher volume of catch. Competition and tension began between small-scale, traditional hand-line fishers and the new larger fishery. By the

[2] www.heritage.nf.ca

mid-1870s, the industrial cod fishery in the Grand Banks was fully developed and continued well into the 1970s. Competition between large-scale and international cod fisheries and small-scale fisheries was fierce, and added immense stress on the cod stocks. The intense fishing pressure on the resources could not be sustained, leading to the eventual collapse of the cod fisheries.

The economic downturn and community hardship that resulted from the 1992 cod moratorium has been well documented (see e.g., Milich 1999; Hamilton and Butler 2001). Stories about how some fishing communities have survived this difficult period are also known; among them are St. Anthony, Fogo Island, and Petty Harbour-Maddox Cove, acclaimed for their extraordinary efforts to deal with the fisheries crisis. In St. Anthony, St. Anthony Basin Resources Inc. (SABRI) was formed to manage the community shrimp quotas, allocated by the government as a replacement to cod, and acknowledged as the main driver of economic development and resource sustainability in the area (Khan and Chuenpagdee 2014; Foley et al. 2015). Fogo Island has a strong fisheries' cooperative, which is one of the oldest in the province and continues to play a vital role to support the local economy. While most communities faced huge outpourings of young people, the Fogo Island community stayed together largely due to the resistance to the quota systems (McCay 1999), and the resilience of the historical cooperative (McCay 2003). The community has also been able to capitalize on the recent development of the Fogo Island Inn, Fogo Island Arts, and other amenities, through the Shorefast Foundation which has attracted tourists who are interested in learning about fishing culture and experiencing local foods, including fish and berries. The work of SABRI and the efforts of the Shorefast Foundation are models for social enterprises and social development, which are critical elements to secure sustainable small-scale fisheries, as posited in the SSF Guidelines.

The story of Petty Harbour-Maddox Cove is rather unique and highly relevant to the discussion about SSF Guidelines in NL. Bryant and Martin (1996) conducted research and wrote about this community for the Protected Areas Association of Newfoundland and Labrador, as a tribute to the people of Petty Harbour-Maddox Cove for their 'wisdom and foresight to establish the Protected Fishing Area' (back cover text). The document is titled 'Ancient Rights' to reflect the collective effort by hand-line fishers to protect their traditional fishing grounds from trawling by provoking an Act that was passed in 1895 to reserve certain areas around the coast for hand-line fishers. According to Bryant and Martin (1996), because they were able to resist the use of more efficient gears and newer gear types, Petty Harbour-Maddox Cove remains one of the most prosperous fishing communities in the province. While considered successful, the fishers of Petty Harbour-Maddox Cove are also vulnerable to global change and other pressures. As will be further discussed, they may be able to draw upon the SSF Guidelines to strengthen their rights to protect the fishing grounds from destructive practices.

## Methods and Data

This paper draws from three main sources. First, under the Too Big To Ignore (TBTI) project,[3] a series of community events were organized in 2015 in four locations across the province (Fig. 23.1). This 'Great Fish for a Change' project had the objective of raising awareness about the important contribution of small-scale fisheries to the local economy and food security. These public events were also used to generate interest around the consumption of other fish species available in NL waters, particularly small pelagic species. As part of this, the Great Fish Recipe Challenge was launched, seeking recipes based on locally caught fish (also emphasizing pelagics). Three finalists from this recipe contest demonstrated their dishes for the general public to judge at the closing event on World Fisheries Day, November 21, 2015. While not the main goal of these events, issues and concerns affecting the viability of small-scale fisheries often emerged.

Each of the 'Great Fish for a Change' events involved a meal prepared by local people using locally sourced fish, as well as an informal discussions about the future of fisheries in the province with an emphasis on small-scale fisheries. Petty Harbour-Maddox Cove hosted the first Great Fish event on August 22, 2015, which was attended by approximately 50 people, mostly from the community and also neighbouring St. John's. The participants were treated with 'capelin on the stick,' (Fig. 23.2), which was an innovative way to prepare this small pelagic fish. In Monkstown, our research team partnered with the Big Fish Supper, an annual event organized by the town every Labour Day weekend. They sell tickets to the event to raise funds for renovations of the community centre. About 200 participants were treated with 20+ local seafood dishes, prepared by 20 households, and local entertainment.

The Port Union event was held on September 29th, 2015 in the historical 'Factory Building', restored by the Sir William F. Coaker Heritage Foundation. The Foundation hosted the event, in collaboration with the College of North Atlantic Culinary School who presented the 70 dinner guests with an exciting menu made from locally sourced and sustainable fish and seafood. Finally, in Stephenville, the event was hosted as part of a bi-monthly 'Community Café' organized by the Housing Stability Initiative, which provides support to people with housing-related issues. The event took place during their normal Thanksgiving dinner, but on that particular day (October 8, 2015), the approximately 100 guests were treated with fish instead of the traditional turkey, adding a new meaning to the phrase 'great fish for a change'.

---

[3] TBTI is a global partnership for small-scale fisheries research that aims to elevate the profile of small-scale fisheries, address their marginalization, and promote their viability and sustainability.

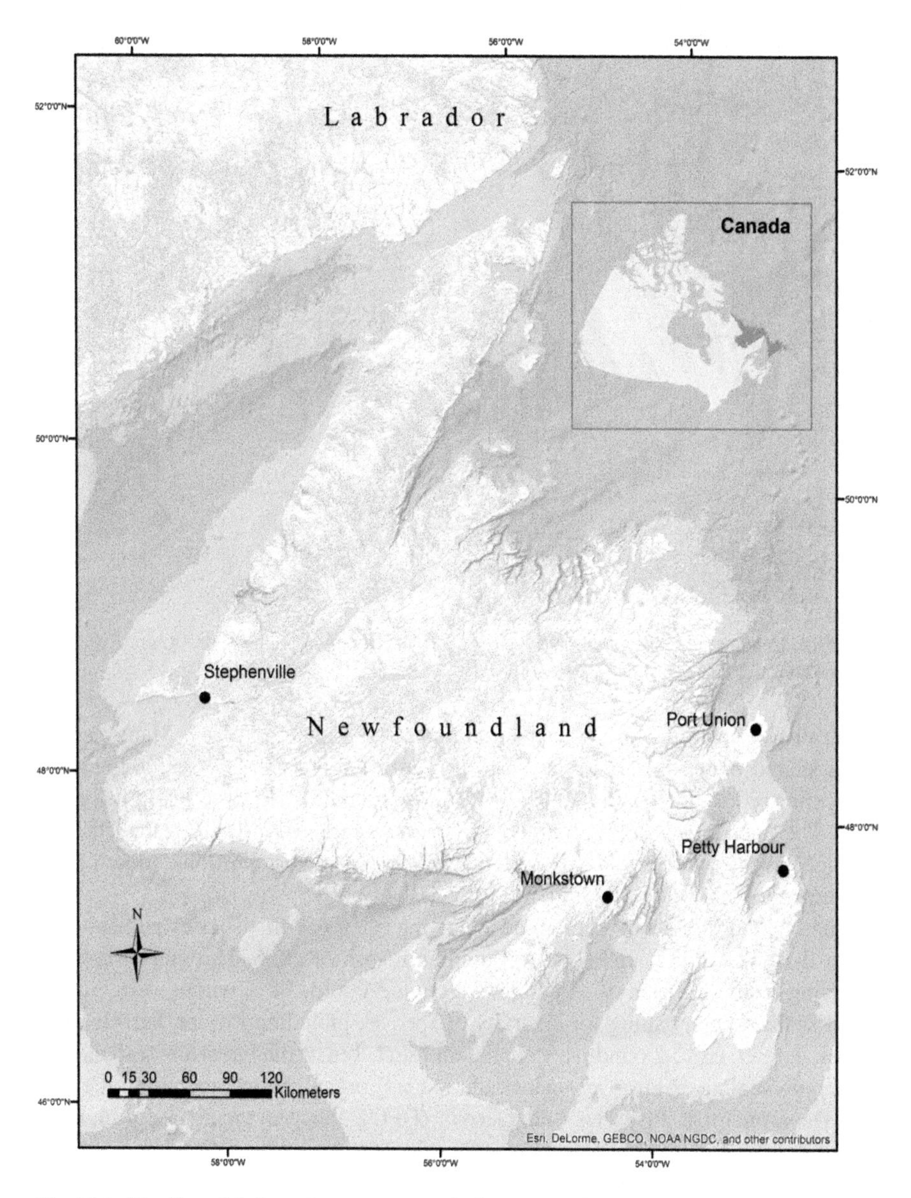


**Fig. 23.1** The Great fish for a change events took place across the province in the summer and fall of 2015

**Fig. 23.2** Capelin were skewered on the stick and pan-fried by a young Chef, Andre Aucoin, at Watershed Café, Petty Harbour

In all four events, good conversation about fisheries took place. Many participants shared lively fisheries-related stories, talked about fish and seafood recipes, and expressed their thoughts about the future of small-scale fisheries in the province. The anecdotal information and informal discussion at the 'Great Fish for a Change' events were analyzed to identify situations and conditions affecting the implementation of the SSF Guidelines in NL.

The second source of data came from interviews with fisheries stakeholders about the SSF Guidelines between March and April of 2016. Through a snowball sampling method, a total of 34 people were interviewed, 14 of whom were 'active' small-scale fishers, fishing for a variety of species, including lobster, herring, cod, turbot, halibut, crab, groundfish, capelin, and seal. Ten of the interview respondents were representatives of governments and fishers' organizations, including the federal Department of Fisheries and Oceans (DFO), the provincial Department of Fisheries and Aquaculture (DFA), municipal government, and the Fish Food and Allied Workers union (FFAW). The remaining 10 were academics and members of environmental and civil society organizations (referred to collectively here as non-governmental organizations, NGOs). The interviews took place in rural coastal communities throughout the province, including Port au Choix and Cow Head located on the western side of the island, and Petty Harbour on the eastern side. Interviews were also held in St. John's, the capital city, as it is the location of many government agencies. Interview questions focused on gauging the level of awareness about the SSF Guidelines, their applicability to NL small-scale fisheries, and the likelihood that they will be implemented in NL. The interviews were recorded, transcribed and coded for thematic analysis. The research was approved by the Interdisciplinary Committee on Ethics in Human Research of the host university (Memorial University of Newfoundland, Canada).

Finally, an open discussion session was held on April 22, 2016 at Memorial University, St. John's, as a collaboration with the Leslie Harris Centre of Regional Policy and Development and the Too Big To Ignore project, under the name 'Implementing the Small-Scale Fisheries Guidelines: What's in it for Newfoundland and Labrador?' The session aimed to share what was learned during the interviews, and generate focused discussion about key principles and relevant elements of the SSF Guidelines to NL small-scale fisheries and identify what may be required to implement them in the province. A total of 44 people participated in the two-hour session; 21 were on-site in St. John's, six participated as part of a satellite session at Memorial University's Grenfell Campus in Corner Brook, NL, and 17 joined virtually from their homes or offices via camera-assisted web conference. There was diverse stakeholder participation, including interested individuals, researchers, government officials, members of environmental and civil society organizations, members of fishing communities, as well as one FFAW representative. A prioritization exercise was conducted in small groups with the participants in St. John's and Corner Brook to capture what participants considered relevant and feasible concerning the SSF Guidelines' implementation. Discussion points and results from the exercise, supplemented by documents and media coverage (mainly newspapers, radio and TV programs and websites), were captured, reviewed and synthesized, as discussed below.

## Enabling Environment and Supporting Conditions

Key features and themes emerged out of the community events, stakeholder interviews, and the discussion session, suggesting the possibility for the successful implementation of the SSF Guidelines in NL. First and foremost, it was made evident that the practice of 'Responsible governance of tenure' was very relevant to fisheries in NL. Of particular relevance were Section 5.5 of the SSF Guidelines, which recognizes the role of small-scale fishing communities in protecting and co-managing fisheries resources, and corresponding Principle 10 (Economic, social and environmental sustainability), which speaks to the need to guard against undesirable outcomes such as overexploitation of resources or loss of livelihoods. Further, a recent change in provincial regulations related to the direct sale of fisheries products (DFA 2015) aligns well with Section 7 on 'Value chains, post-harvest and trade.' New market opportunities have emerged, as a result, to enhance values to small-scale fisheries (see below).

The next enabling condition is related to the existence of various organizations and community groups in the province that play a supporting role in small-scale fisheries. They work to preserve traditional fisheries and cultural heritage, promote local food security, educate children and the general public about fisheries and ocean ecosystems, raise environmental awareness and protect natural resources, and conduct research and training. This particular aspect speaks to the embedded nature of small-scale fisheries within local communities and culture, which is the foundation for social responsibility (Principle 12).

Finally, the emergence of a new organization representing small-scale fishers and their interests is likely to change the political landscape in the province (Randell 2016). While it may be too early to tell what prospect the new union, Federation of Independent Sea Harvesters of Newfoundland and Labrador (FISH-NL) will bring, the hope is that small-scale fishers will be better represented than through a union (i.e. FFAW) that serves a broad range of constituents. In addition to enhancing opportunities to participate in decision-making, which is promoted in the SSF Guidelines (Principle 6: Consultation and Participation), such political representation could also lead to improving working conditions and strengthening the economic viability of small-scale fishers through better employment options, as is suggested in Section 6 of the SSF Guidelines (Social Development, Employment and Decent Work). Together, these situations and development contribute to create an enabling environment for the implementation of the SSF Guidelines in the province, and are instrumental in promoting information sharing and communication about the guidelines. They also enhance the capacity of small-scale fishing communities to actively participate in decision-making processes, as stipulated in Section 12 of the guidelines.

## Protection Rights as Ancient Rights

> *Whereas the hand-line fishermen of Petty Harbour, who are all the fishermen in the place, have enjoyed protection from the use of trawls in their area for generations, which protection was confirmed by statute in 1895 and reconfirmed in 1943…*(Bryant and Martin 1996, front cover).

This opening sentence in the 1961 resolution petitioning the Canadian government to create the Petty Harbour-Maddox Cove Protected Fishing Area speaks volumes about the conservation values held by the community. It also asserts their historical rights to the resources for their livelihoods in accord with the adjacency principle, as well as the rights to protect the resource from harmful practices, in line with the precautionary principle. Importantly, the action by the fishers of Petty Harbour-Maddox Cove illustrates their active involvement in resource management as decisions about protected areas were, and still are, made at the local level through member voting. The level of organization in the community is remarkable. They formed a Fishermen's Committee, which has been in operation since 1923 and established the Petty Harbour Fishermen's Producer Co-operative Society in 1984. This is one of the only three fishery co-ops in the province. Among other things, the co-op owns a processing facility, which gives them control over the post-harvest activities, making them one of the most prosperous fishing communities in NL at that time (Bryant and Martin 1996).

While they were able to prevent trawling in their waters in 1961, their 'ancient rights' began to erode with a series of attempts to introduce trawls and gillnets; and with each introduction, their protection status was weakened, along with the community consensus. The protest that Petty Harbour fishers staged against the new law to allow trawling in their fishing grounds shows that the community was united in their opposition to the new law. The same opposition was shown in 1987 against lumpfish gillnets. In 1989, Petty Harbour-Maddox Cove fishers had again defended their fishing grounds against the amendment proposed by DFO to allow trawling in the southern portion of the Protected Fishing Area. Unfortunately, DFO went against the community's will by issuing special permits to trawlers from Bay Bulls, a neighboring town to the south. A couple more votes took place on the lumpfish gillnets issue and finally, in May 1996, the protected area status was significantly threatened when the majority of the Petty Harbour-Maddox Cove fishers voted, for the first time, to allow gillnets in the area.

The pressure from other fisheries on Petty Harbour-Maddox Cove small-scale fisheries initially resulted from the normal trend in fisheries development when lucrative resources enticed the introduction of gears considered more efficient. Later, in the early 1990s, gillnetters attempted to enter the area because of the cod collapse and decline in fisheries' resources, except those in the rich shoal-water fishing grounds of Petty Harbour. Despite all of these pressures, the small-scale fishers of Petty Harbour-Maddox Cove, who committed to the use of hand-lines as their sustainable fishing practice, continue to thrive. The observance of ancient fishing rights, the strong presence of the co-op, the existing leadership and other supporting organizations in the area (see below) make Petty Harbour-Maddox Cove one of the most promising candidates among NL communities to benefit from implementation of the SSF Guidelines.

## *Institutional Change Enhancing the Value Chain*

Until recently, small-scale inshore fishers in NL were not permitted to sell their catches directly to consumers. Instead, they were obliged to sell only to a licensed fish buyer or processor in the province, as part of the 'minimum processing requirements' clause legislated by the Fish Inspection Act (Song and Chuenpagdee 2015). This provincial regulation has restricted fishers from engaging in post-harvest trade that could result in better prices for their catch and thus also in higher income. This has not, however, affected local food security because of the 'personal use' provision made possible by the federal regulation, which allows a certain portion of the catch to be retained for household consumption or for sale by fish harvesters.[4] While not permitted under regulation, a report commissioned by the provincial fisheries department (DFA) estimated that at least half of the fishers were engaged in some kind of direct sale due to a policy loophole and to lack of enforcement (Eric Dunne Consulting Initiatives 2010). The problematic situation caused by this institutional mismatch and the controversy around it (Song and Chuenpagdee 2015), the different opinions about the 'minimum processing requirements' policy among fishers, processors and their union (The Navigator Magazine 2014), and the general support by consumers and restaurants in favour of direct fish sale (Murphy and Neis 2011) were among the key motivations for the regulatory change. In September 2015, the new 'Regulations for Direct Sales for Provincial Seafood' were announced, allowing individual consumers and restaurants to buy fish and seafood directly from the wharf or from the fishing establishments, with a maximum of 300 pounds per species per week as a limit (DFA 2015).

While it may be too early to assess the full impact of this regulatory change, new market opportunities have already surfaced, suggesting a positive outlook for enhancing the value chain for small-scale fishers. 'From the Wharf' is a 'virtual' market place that enables consumers to place an order for fish from harvesters through its website. Blaine Edwards, who founded the innovation in early 2016, differentiates his company from wholesale businesses or a fish store in that they neither carry inventory nor fill orders. Rather, their role is to facilitate an exchange between fishers and customers through the online site. While not exclusively for small-scale fishers, the policy of the company is to promote quality products coming from 'professional, hard working, seagoing, ocean loving local harvesters' and also to get the highest price possible for them (From the Wharf 2016). Thus, 'From the Wharf' is a good outlet for small-scale fishers.

Another initiative is Fogo Island Fish, established by Cobb and Thompson (Dickson 2015), as a way to offer a better product for consumers and better returns for small-scale fishers. The highlight of their venture is the responsible manner in which cod is harvested using hooks, small fishing boats, and a catch limit of 500 pounds a day for 3 weeks of the year. Equally important is the post-harvest handling of the fish, which involves bleeding of the fish as soon as it is brought to the boat, then placing fish in a bath of seawater and ice, before gutting and packing them on ice. The Fogo Island Fish initiative has attracted 33 harvesters this year (2016), including women, and is looking to increase that number in coming years. Fogo Island Fish does not only promote sustainable fisheries and premium quality seafood, but also facilitates social engagement and an inclusive local economy. While the model may not be readily applicable to other inshore fisheries in NL, it illustrates the importance of alternative approaches for achieving sustainable fisheries. In fact, some modifications and innovation should be considered in order to apply this model more broadly at the provincial level and in the rest of Canada.

Both initiatives are examples of the changing role of small-scale, independent fishers in NL, from harvesting alone to engaging in post-harvest activities. This new

[4] Term commonly used in NL to refer to anyone working in the harvesting sector, i.e. both large-scale and small-scale fishers.

direction aligns well with Section 7 of the SSF Guidelines, and more can be done following the stipulations in this section to enable fishers to benefit fully from participating in the enhanced value chain.

## Community Organizations Supporting Small-Scale Fisheries

Local communities and several groups and organizations have mandates and initiatives that support the sustainability and viability of small-scale fisheries in NL. These organizations span the entire fish chain in terms of their interests, starting from those promoting the conservation and protection of resources and habitats, to those working directly to protect the interests of fish harvesters and plant workers. Some of these organizations have specific mandates that are directly related to fisheries, for instance, by preserving fishery heritage, while others address broader issues like food security and poverty alleviation. Social cohesiveness, which is key to successful community development (Gutierrez et al. 2011), can also be observed with the existence of many voluntary organizations. Below are some examples of these organizations and the role they play in supporting small-scale fisheries in NL. They are chosen to suggest the diversity of organizational types and mandates.

*Island Rooms of Petty Harbour: Fishing for Success* Located in Petty Harbour, this non-profit organization was formed in 2014 to reconnect youth with their traditional fishing heritage. The founder, Kimberly Orren, and her team built traditional fishing premises reminiscent of those that once occupied the waterfront of Petty Harbour and coastal communities throughout the province. The Island Rooms forms a unique outdoor classroom designed to get kids outside, to gain practical fishing skills, to get in touch with their roots, and to cultivate a connection to the sea. Activities include trout and cod fishing, dory building, net mending, filleting demonstrations, camp cooking lessons, rinding sticks, and story sharing of Petty Harbour's unique history on a town 'walk about.' The organization puts a strong emphasis on re-connecting young people with nature, the fishery and their heritage through the development of fishing skills, improving self-sufficiency of the province's food system, and introducing young people to a possible career in the fishery (Fishing for Success n.d.; Poitevin 2015).

*Petty Harbour Mini Aquarium* Established in 2013, the Petty Harbour Mini Aquarium is located on the bottom floor of Petty Harbour's Fishermen's Co-op. The 'Mini' is among one of the few catch-and-release aquariums found within Canada. With a team of experienced divers, coastal animals are collected locally during the spring and are released in the fall. The aquarium offers to its guests a unique opportunity to gain insight into the rich biodiversity found in NL waters, including the rare gold and blue lobsters, bottom dwelling sculpin, the iconic codfish, as well as a variety of intertidal animals featured in six 'touch' tanks. The aquarium's motto, *foster curiosity to inspire change*, encapsulates their main mission: to get visitors of all ages excited about ocean life and to help them become more interested in conservation issues, and ultimately, to be motivated to take action within their own communities.

*Food First NL* Formerly known as the Food Security Network, Food First NL is a non-profit organization aiming to support community-based food security initiatives throughout the province. Food Fist NL provides education and awareness about food issues and advocates for change in food policy that can enhance production and delivery of food in the province. The organization recognizes the importance of having space for people to buy locally grown produce and homemade food, and has participated in the creation of the St. John's Farmer's Market and of a 'buy local' map for the Avalon Peninsula (Poitevin 2015). While their work is mainly land-based, such as community gardens, bulk-buying clubs, and food initiatives in schools, the organization is also interested in integrating fisheries into local food security, recognizing the high nutritional value of fish and seafood.

*World Wildlife Fund: Canada (WWF Canada)* As part of their activities in Atlantic Canada, WWF Canada is focused on finding solutions to rebuild Newfoundland's Grand Banks, an area once known for its prosperous cod fish industry, into a world-class model for ecosystem and economic recovery. WWF Canada hopes to achieve sustainable fishing and responsible ocean's management in the Grand Banks area by addressing some of the economic pressures, which are often the main driving forces threatening the sustainability of fishing resources, and by advocating for policies that will better protect this ocean ecosystem (WWF n.d.).

*Codfish Culinary Experience Project* Although not yet fully developed, this project is noteworthy because it links fisheries' sustainability with tourism, which could provide supplementary income to small-scale fishing communities. Located in Port Union, the project aims to leverage the history of the Trinity Bay inshore cod fishery through creating a sustainable model for harvesting, retailing, and devising a distribution system for cod products. With this project, Shelly Blackmore, the founder, hopes to develop a more direct approach of providing local fish to regional restaurants, and to serve both traditional and more innovative cod dishes on the spot. Visitors to the project will also be able to visit the nearby Factory Building, a museum and community centre, which showcases period woodworking machinery and the equipment used by the union newspaper – The Fishermen's Advocate. Port Union itself is unique as it is the only town in NL designed and built in the early twentieth century as a model town by Sir William Coaker and the members of his Fishermen's Union Trading Company and the Fishermen's Protective Union.

## *Fishers' Organizations for Small-Scale Fisheries*

The Fish, Food and Allied Workers Union (FFAW) is a fishers' union that was initiated in the early 1970s to represent all fishers in NL. The FFAW is the fishers' organization, representing workers from all sectors of the fisheries, from harvesters to plant workers, both small- and large-scale. In addition to serving as a link between fishers and the federal, provincial, and municipal governments, the FFAW acts as a significant advocacy group, playing a major role in negotiating and determining fishing policy (FFAW 2016). The most significant negotiation between the FFAW and the governments happened during the cod moratorium. When the cod stocks had collapsed and fishers were asked to haul up their nets for the last time, the FFAW fought for income support and training opportunities for those who were affected (Francis and Hart 2007). Since the moratorium, many have found overcapacity to be a significant issue in Newfoundland's inshore fishery, leading to lower income for harvesters (Song and Chuenpagdee 2015). The FFAW and the provincial government have worked toward resolving this issue by downsizing the industry to increase profitability, a decision that predominantly favours industrial, offshore fleets over small-scale, inshore harvesters (Song and Chuenpagdee 2015).

While overcapacity is seen as a major issue, the diverse representation held under the FFAW, which was a leading factor to the union's growth and success during its infancy, is regarded today as a cause for concern amongst NL's inshore fishers (Francis and Hart 2007). The conflicting issues between seafood processors, the offshore fleets, and the inshore fleets have led to the union's inability to keep in mind the best interests for all those involved. With the attention of the FFAW being spread amongst various sectors, poor communication between the union and its inshore fishers has ensued. This, and the FFAW's consistent lack of transparency with the inshore fishers regarding policy decisions made with DFO, and the use of money collected through union fees, is the underlying cause for a break away. According to Jason Sullivan, one of the co-founders of a new fishers' union 'FISH-NL', the aim of the new organization is to represent independent fishers, both under 40′ and over 40′, and offer equal opportunities and benefits for all of its members and executives (J. Sullivan, personal communication, 12 October 2016). It remains to be seen whether NL inshore fishers will choose to stay with FFAW or join FISH-NL when they cast their vote later this year (2016). As suggested in the SSF Guidelines, having their own organization could mean that small-scale fishers can exercise their rights to participate in the decision-making (Principle 6: Consultation and Participation) and maximize their interest, which is fundamental for improving their livelihoods and safe working conditions (Section 6), and for enhancing their resilience (Principle 13: Feasibility, and Social and Economic Viability).

## Implementation Challenges

Despite the existence of an enabling environment and of key conditions supporting the implementation of the SSF Guidelines, several challenges remain before implementation can take place. As suggested by Jentoft (2014), implementation of the Guidelines can follow a top-down process initiated by governments, a bottom-up process led by fishers at the community level, or can be a combination of both.

Either way, the basic ingredients need to be there, starting from whether these various stakeholders are aware of the guidelines, how they perceive them and the extent to which they consider them relevant and feasible to implement. Below are the results of the interviews and discussion session that offer insights into these questions.

## *Awareness and Perceptions About the SSF Guidelines*

As shown in Table 23.1, among 14 fishers interviewed for the study, only one indicated having heard about the SSF Guidelines. In the other two categories (governments and others), about 80% of the respondents reported that they were aware of the Guidelines. The low level of awareness about the SSF Guidelines in the small-scale fishers group is not surprising, given the general lack of effort to engage NL small-scale fishers in the process that was used to develop the Guidelines. None of the many stakeholder consultations organized around the world, which engaged over 4000 people in discussing the draft document, took place in Canada (or the USA). It is, however, encouraging that the majority of the respondents expressed the need for implementation of the SSF Guidelines in NL. It should be noted that some of the respondents who did not see the need to implement the guidelines in Canada explained that Canada was already doing well in managing the fisheries, including small-scale. Even if this were truly the case, there is always room for improvement, particularly in promoting the small-scale fisheries value chain, in accordance with Section 7 of the SSF Guidelines.

The respondents were generally positive about the future of small-scale fisheries in the province, mostly emphasizing the need for change in the fisheries systems and for improving conditions for harvesters. The proportion of fishers expressing doubts about the future was higher (about 42%) than the government and researchers/NGO representatives. With respect to the question about whether the current management considered the interests of small-scale fisheries, the overwhelming majority of fishers and the other stakeholders disagreed.

In addition to the above, other comments captured during the interviews indicate that the SSF Guidelines were considered beneficial to small-scale fisheries in NL since they could help address the ongoing recruitment problem, improve communication and participation of fisheries' stakeholders in the decision-making processes, and could result in enhanced economic viability.

## *Relevance of the Principles*

One of the interview questions was about the relevance of the principles in the SSF Guidelines for NL small-scale fisheries. Three principles were considered particularly relevant by all stakeholder groups, namely consultation and participation;

**Table 23.1** Stakeholders' awareness and perceptions about small-scale fisheries and the SSF Guidelines

| | Groups | | |
|---|---|---|---|
| | Small-scale fishers | Governments | Researchers/NGOs |
| #interview | 10 | 10 | 10 |
| Know the guidelines | 1 | 8 | 8 |
| Positive future | 8 | 10 | 7 |
| Management for SSF | 3 | 5 | 1 |
| Need to implement | 10 | 7 | 10 |

**Table 23.2** Relevance of the SSF Guidelines principles by stakeholder groups

| | Groups | | |
|---|---|---|---|
| | Small-scale fishers | Governments | Researchers/NGOs |
| Human right & dignity | – | – | 1 |
| Respect of culture | 1 | – | 1 |
| Non-discrimination | 2 | – | 2 |
| Gender equality & equity | – | – | – |
| Equity & equality | – | – | 2 |
| Consultation & participation | 10 | 3 | 6 |
| Rule of law | – | – | – |
| Transparency | 6 | 1 | 2 |
| Accountability | 2 | – | 1 |
| Econ, soc, env. Sustainability | 4 | 5 | 5 |
| Holistic & integrated approach | – | 2 | 2 |
| Social responsibility | 8 | 3 | 7 |
| Feasibility & soc/econ viability | – | – | – |

economic, social, and environmental sustainability; and social responsibility (Table 23.2). Principles not considered by any stakeholder group to be of high relevance to NL small-scale fisheries were gender equality and equity; rule of law; and feasibility and social and economic viability. Research and non-governmental organizations (environmental and civil society) indicated more principles to be relevant than other groups. This is contrary to the government group, which selected only a few principles to be relevant. Three of the respondents stated that all principles have already been implemented, while one mentioned that all principles are important and should be taken together.

Another interesting comparison is related to consultation and participation, which was considered by the majority of the fishers to be relevant, but not so by the government group. This is also to be expected since on one hand, it is the responsibility of the government to consult with fishers, according to the Fisheries Act. On the other hand, small-scale fishers may not feel sufficiently consulted, given their position about their representation within the FFAW. The general support of stakeholders for 'social responsibility' is an interesting result. The SSF Guidelines refer to this principle as 'promoting community solidarity and collective and corporate responsibility and the fostering of an environment that promotes collaboration among stakeholders should be encouraged' (FAO 2015, 3). As previously discussed, community groups and non-profit organizations in NL are already working towards sustainability and thus they may be able to encourage the small-scale fishing industry to embrace this principle.

## *Prioritization of Issues*

The main exercise at the discussion session was the consideration of the topics proposed in the SSF Guidelines in terms of their relevance to NL small-scale fisheries and according to the ease of implementation. The guidelines contain nine key topics (short names used in Fig. 23.3 are in brackets), i.e. (1) governance of tenure (Tenure); (2) sustainable resource management (SusDev); (3) social development, employment and decent work (SocDev); (4) value chain, post-harvest, and trade (VC); (5) gender equality (Gender); (6) disaster risks and climate change (Disaster); (7) policy coherence, institutional coordination and collaboration (Policy); (8) information, research and communication (Info); and (9) capacity development (Capacity). Participants at the discussion session in St. John's and Corner Brook worked in small groups to rate these topics based on two criteria – relevance and ease of implementation. Each group was given a chart, drawn with two axes as shown in Fig. 23.3, as well as a deck of nine index cards showing the topics. The cards were placed in the figure based on group consensus. Figure 23.3 shows the results of this exercise, compiled across all participant groups including those in Corner Brook.

All topics were considered relevant, with governance of tenure, sustainable resource management, value chain, disaster risks and climate change, and policy coherence among the top. The topic considered least relevant, but not irrelevant, to NL is gender equality. Participants explained that while the fishery in NL is male-dominated (and recruitment also targets men), many women work in fisheries, particularly in harvest and post-harvest activities, as well as in the resource management and decision-making sector, and that their work conditions are similar to men. It was mentioned that women are heavily engaged in fish processing, and that fish processing workers may not have a strong voice in fisheries' governance.

With respect to ease of implementation, participants at the discussion session considered the three moderately relevant aspects of the SSF Guidelines, i.e. capacity development, social development, employment and decent work, and information, research and communication to be relatively easy to implement. The other aspects were considered more difficult, especially disaster risks and climate change, policy coherence, institutional coordination and collaboration and value chain, post-harvest and trade. Among the aspects with high relevance, participants further elaborated that the difficulty with disaster risks and climate change was due to the high level of uncertainty and the general lack of information. Sustainable resource development,

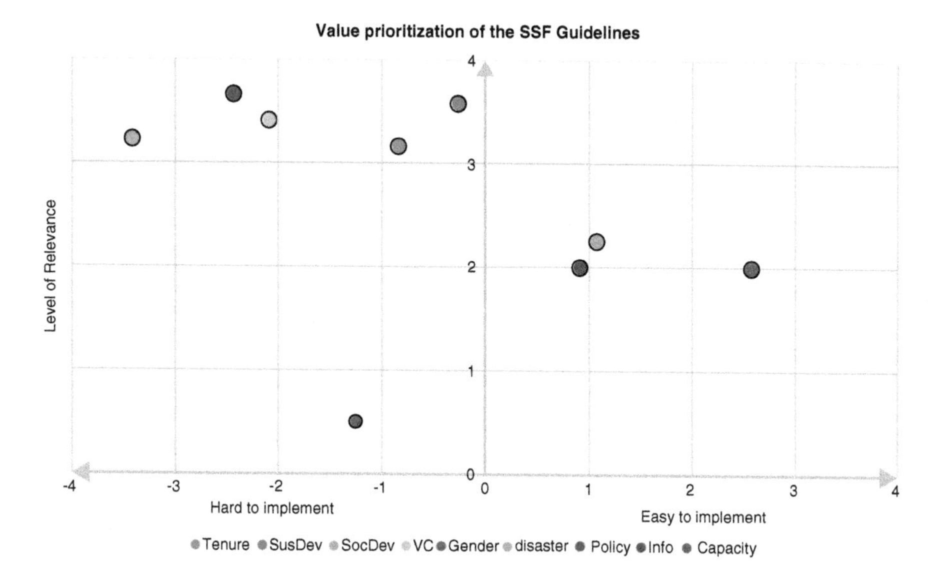


**Fig. 23.3** Distribution of key topics in the SSF guidelines according to relevance and ease of implementation

on the other hand, would be easier to implement because there are already several initiatives and good management plans that aim to protect and conserve fisheries' resources and habitats.

## Step Zero Recommendations

Despite the general lack of awareness about the SSF Guidelines, they were well received, and have been seen as beneficial to help address recruitment issues and to sustain fisheries in NL. Also, as expressed during the discussion session, the SSF Guidelines were considered highly relevant to NL, suggesting that they are applicable to small-scale fisheries in the North as much as they are in the South. This reinforces the purpose of the SSF Guidelines to be global in scope, even though they may be having a stronger focus on the needs of developing countries. Nevertheless, the implementation of the SSF Guidelines may not be a priority for policy makers and managers in NL, or the rest of Canada and other developed countries, for various reasons. In the case of NL, fisheries as a whole contribute only about 1% to the provincial GDP (Government of Newfoundland and Labrador 2015), which pales in comparison to the 28% contribution to GDP from the oil and gas industry. Further, within the fisheries sector, small-scale fisheries constitute a relatively small proportion of catches (12% of total landings, and 11% of total value). With the exception of the USA, this pattern is common in small-scale fisheries in developed countries, which could be the result of modernization, rationalization, and development in the industry. In this context, other considerations about the values and importance of small-scale fisheries must be brought to the surface beyond their number and economic contribution. The social and cultural aspects of fisheries, for instance, have been highlighted in many fishing communities in the developed economies (see examples about the UK in Brookfield et al. 2005; Acott and Urquhart 2014), and should be emphasized in the discussion about small-scale fisheries.

Although some of the existing governing structures and institutions seem appropriate for sustainable resource development, small-scale fishers are not as well represented in the decision-making processes as they could be, and their traditional rights are not always respected. The perception that the current management plans already follow some of the principles posited in the SSF Guidelines might lead to inaction by governments. In Canada, for instance, co-management of the oceans with various stakeholders is one of the top priorities of DFO, according to the Mandate Letter from the Prime Minister (DFO 2016). This implies, by default, the participation of all fishers, including small-scale fishers, in fisheries management. Small-scale fishers who were interviewed for the study and several participants at the discussion session indicated the lack of opportunity for small-scale fishers to engage in decision-making process, however. The SSF Guidelines, with their strong focus on the participatory principle, can be a good vehicle for small-scale fishing communities in NL, and Canada, to argue for stronger roles and meaningful involvement in fisheries' management and in policy development.

Finally, while direct fish sale is now possible, markets for high quality cod and other locally caught fish are still limited. Governments may have to consider different stakeholders' needs when allocating financial and human resources to support post-harvest activities. As demonstrated by initiatives like 'From the Wharf' and 'Fogo Island Fish,' local entrepreneurs can play an important role in enhancing the value chain for small-scale fisheries through social innovation. But small-scale fishers in NL are also self-motivated to engage in sustainable fishing practices, as seen in their conservation and stewardship practices, not only in Petty Harbour but also in other places like in the Eastport Peninsula with the protection of lobster fishing grounds (Blundon 1999). More support from the government in terms of market penetration and promotion could help them obtain higher prices for their catches, for instance, through the development of NL-owned label for locally caught fish and seafood.

Drawing from the step zero analysis (Chuenpagdee and Jentoft 2007), the implementation of the SSF Guidelines depends to some extent on knowing what could drive the process, what could inspire people to participate, and what could motivate or trigger policy change. As previously discussed, several conditions and infrastructure are already in place to enable implementation. These include, among other things, the historical rights of fishing communities and traditional and cultural ties to fisheries, new regulations allowing direct fish sales, and the possibility of a new fishers' organization for small-scale fisheries. Further, NL has the experience of many strong fishers' cooperatives and grassroots organizations working to promote the viability and sustainability of small-scale fisheries. In addition, the mandates of several environmental and civil society organizations, as well as the results of academic research projects, align very well with the principles and objectives of the SSF Guidelines. Civil society organizations are therefore key players in the implementation process and can help with various capacity developments. For instance, one of the TBTI research clusters is established to directly support the implementation of the SSF Guidelines around the world by raising awareness about the guidelines and building research and governance capacity among small-scale fishers and their organizations, as well as among government officials and practitioners.

More can be done on all of these topics. For instance, participatory and transdisciplinary research on small-scale fisheries would benefit the body of literature, offering an increased diversity of perspectives. Studies about what matters to small-scale fishers, what values they hold and what images they have of their future can prove useful in forming consensus around fisheries decisions, which are often contentious (see Song et al. 2013). Kooiman and Jentoft (2009) refer to this as meta-order governance, which is fundamental to influence change, at an individual level as well as collectively. From an institutional perspective (second-order governance), new platforms may need to be established in order to strengthen the fishers' ability to initiate and participate in the implementation process. Further, markets and infrastructure should be developed to enhance value chains for small-scale fisheries and improve their competitiveness with the large-scale, industrial sector. At the practical level (first-order governance), better coordination and concerted efforts among various groups and organizations should also be promoted, through strengthening of social networks and stakeholder partnerships, and through improving information sharing and communication, as suggested by Turner et al. (2014). Fundamentally, a stronger protection of traditional fishing rights is required.

While the current study involved only a small fraction of the population, the range of methods used to engage with stakeholders, from community events, interviews, to open discussion sessions, enabled a broad range of participation across the province. The interest shown by small-scale fishers in the discussions about the SSF Guidelines and the support from communities and various organizations throughout the study indicate the opportunity that small-scale fishers in NL have to initiate discussions about change in fisheries' policy in the province. Economic, social, and environmental sustainability, which is the principle shared by all stakeholders in terms of relevance to the province and a topic considered possible to implement, may be an initial goal for starting the process of implementation of the SSF Guidelines in NL. Similar studies could be conducted in other coastal provinces in Canada, with mechanisms established for sharing of lessons and best practices.

Unlike in developing countries, the situation in NL suggests that it may not be crisis that drives the implementation process, but opportunities, the first of which is the encouraging sign that the cod stocks may recover. As suggested by Jentoft (2014), however, small-scale fishers need to be listened to, inspired, and feel empowered to initiate the conversation about the SSF Guidelines, and through community support and collaboration with supporting organizations, they can call governments to task.

**Acknowledgements** We thank the Harris Centre of Regional Policy and Development of Memorial University, especially Rob Greenwood and Taylor Stocks, for co-hosting the synergy session. Funding for the research was provided by Social Sciences and Humanities Research Council of Canada (grant number 895-2011-1011).

## References

Acott, T. G., & Urquhart, J. (2014). Sense of place and socio-cultural values in fishing communities along the English channel. In J. Urquhart et al. (Eds.), *Social issues in sustainable fisheries management, MARE publication series* (Vol. 9). Dordrecht: Springer.

Blundon, J. (1999). *Co-management and the Eastport lobster fishery*. (Masters' thesis). National Library of Canada, Ottawa.

Brookfield, K., Gray, T., & Hatchard, J. (2005). The concept of fisheries-dependent communities – A comparative analysis of four UK case studies: Shetland, Peterhead, north shields and Lowestoft. *Fisheries Research, 72*, 55–69.

Bryant, S., & Martin, B. (1996). *Ancient rights: The protected fishing area of Petty Harbour-Maddox Cove*. St. John's: The Protected Areas Association of Newfoundland and Labrador.

Chuenpagdee, R., & Jentoft, S. (2007). Step zero for fisheries co-management: What precedes implementation. *Marine Policy, 31*(6), 657–668.

DFA. (2015). *Providing new opportunities for our oldest industry. New regulatory amendments permit direct sales for provincial seafood.* http://www.releases.gov.nl.ca/releases/2015/fishaq/0929n07.aspx. Accessed 2 Apr 2016.

DFO. (2016). *Fish landings and landed value reports*.http://www.nfl.dfo-mpo.gc.ca/NL/Landings-Values. Accessed 28 Nov 2016.

Dickson, S. (2015, November 17). Fogo Island Fish – Changing the cod industry one fish, one chef at a time. *Dine To*. http://www.dine.to/news/2015-11-17/fogo_island_fish_changing_the_cod_industry. Accessed 7 Oct 2016.

Eric Dunne Consulting Initiatives. (2010). *Report of the review of the regulations and policy for direct fish sales in Newfoundland and Labrador*. St. John's: Eric Dunne Consulting Initiatives. http://www.fishaq.gov.nl.ca/publications/Direct_Fish_Sales_NL_2010%20.pdf. Accessed 28 Nov 2016.

FAO. (2015). *Voluntary guidelines for securing sustainable small-scale fisheries in the context of food security and poverty eradication*. Rome: FAO.

FFAW-Unifor. (2016). History. http://ffaw.nf.ca/en/history#.WAjGKuArLIU. Accessed 7 Oct 2016.

Fishing for Success. (n.d.). Islands rooms of Petty Harbour – Fishing for success: About. https://www.islandrooms.org/about.html. Accessed 7 Oct 2016.

Foley, P., Mather, C., & Neis, B. (2015). Governing enclosure for coastal communities: Social embeddedness in a Canadian shrimp fishery. *Marine Policy, 61*, 390–400.

Francis, D. C., & Hart S. M. (2007). A unique pattern of collective bargaining: Fish, Food and Allied Workers Union and Fishery Products International Limited in Newfoundland and Labrador. In *Proceedings of the 37th Atlantic Schools of Business conference*, pp 76–84.

From the Wharf. (2016). Home. https://www.fromthewharf.com/. Accessed 28 Nov 2016.

Government of Newfoundland and Labrador. (2015). *The economic review Newfoundland and Labrador 2015*. St. John's: Economic Research and Analysis Division, Department of Finance.

Gutierrez, N. L., Hilborn, R., & Defeo, O. (2011). Leadership, social capital and incentives promote successful fisheries. *Nature, 470*, 386–389.

Hamilton, L. C., & Butler, M. J. (2001). Outport adaptations: Social indicators through Newfoundland's cod crisis. *Human Ecology Review, 8*(2), 1–11.

Harris, M. (1998). *Lament for an ocean: The collapse of the Atlantic cod fishery*. Toronto: McClelland and Stewart.
Jentoft, S. (2014). Walking the talk: Implementing the international voluntary guidelines for securing sustainable small-scale fisheries. *Maritime Studies, 13*(16), 1–15.
Khan, A. S., & Chuenpagdee, R. (2014). An interactive governance and fish chain approach to fisheries rebuilding: A case study of the Northern Gulf cod in eastern Canada. *Ambio, 43*, 600–613.
Kooiman, J., & Jentoft, S. (2009). Meta-governance: Values, norms and principles, and the making of hard choices. *Public Administration, 87*(4), 818–836.
Mather, C. (2013). From cod to shellfish and back again? The new resource geography and Newfoundland's fish economy. *Applied Geography, 45*, 402–409.
McCay, B. J. (1999). 'That's not right': Resistance to enclosure in a Newfoundland crab fishery. In D. Newell & R. E. Ommer (Eds.), *Fishing places, fishing people: Traditions and issues in Canadian small-scale fisheries* (pp. 301–320). Toronto: University of Toronto Press.
McCay, B. J. (2003). Women's rights, community survival, and the fisheries cooperative of Fogo Island. In R. Byron (Ed.), *Retrenchment and regeneration in rural Newfoundland* (pp. 158–176). Toronto: University of Toronto Press.
Milich, L. (1999). Resource management versus sustainable livelihoods: The collapse of the Newfoundland cod fishery. *Society and Natural Resources, 12*, 625–642.
Murphy, I., & Neis, B. (2011). *Navigating the legislative requirements for fisheries-tourism initiatives in Newfoundland and Labrador*. St. John's: Community University Research for Recovery Alliance (CURRA). http://www.curra.ca/documents/TCR_Fisheries_Tourism_Regulations_Report_Feb_14_Final_to_TCR_revised.pdf. Accessed 28 Nov 2016.
Palmer, C. T., & Sinclair, P. R. (1997). *When the fish are gone: Ecological disaster and fishers in Northwest Newfoundland*. Halifax: Fernwood Publishing.
Poitevin, C. M. (2015). *Fish as food: Examining a place for fish in Newfoundland's alternative food networks*. (Unpublished Master's thesis). Memorial University of Newfoundland, St. John's, Canada.
Randell, A. (2016, October 27). FISH-NL establishes foothold. *The Western Star*. http://www.thewesternstar.com/News/Local/2016-10-27/article-4673564/FISH-NL-establishes-foothold/1. Accessed 28 Nov 2016.
Ryan, S. (1990). *History of Newfoundland cod fishery. Archival material in the centre for distance learning and innovation*. Government of Newfoundland and Labrador. http://www.cdli.ca/cod/history5.htm. Accessed 13 Oct 2016.
Sea Around Us. (n.d.) Tools and data. http://www.seaaroundus.org/data/#/eez. Accessed 28 Nov 2016.
Song, A., & Chuenpagdee, R. (2015). A principle-based analysis of multilevel policy areas on inshore fisheries in Newfoundland and Labrador, Canada. In S. Jentoft & R. Chuenpagdee (Eds.), *Interactive governance for small-scale fisheries: Global reflections* (pp. 435–456). Dordrecht: Springer
Song, A., Chuenpagdee, R., & Jentoft, S. (2013). Values, images, and principles: What they represent and how they may improve fisheries governance. *Marine Policy, 40*, 167–175.
Turner, R. A., Polunin, N. V. C., & Stead, S. M. (2014). Social networks and fishers' behavior: Exploring the links between information flow and fishing success in the Northumberland lobster fishery. *Ecology and Society, 19*(2), 38.
Unknown Author. (2014). An argument for inter-provincial free trade and free enterprise. *The Navigator Magazine, 17*(9), 12–13.
WWF (World Wildlife Fund). (n.d.). *Atlantic Canada: Building a model for smarter fishing in the Grand Banks*. http://www.wwf.ca/conservation/oceans/atlantic_canada/. Accessed 7 Oct 2016.

# Part VII
# Managing Threats

Small-scale fisheries are often confronted with various forms of risk that expose their vulnerability. Among them are natural hazards, such as cyclones, that take the lives of fishers and ruin fishing communities, and climate change that threatens their viability. Mohammad Mahmudul Islam and Svein Jentoft illustrate this problem in Chap. 24. Their case study is located in Bangladesh, where coastal fishers are vulnerable to frequent extreme events and disasters. Their chapter argues that the government should incorporate fishers' adaptation strategies against disaster risk and climate change impacts in existing policies. Illegal, unreported, and unregulated (IUU) fishing is another problem affecting the wellbeing of fishing people. Two chapters in this part discuss different contexts related to the issue. Drawing from a case study of Lake Victoria, Tanzania, Joseph Luomba, Paul Onyango, and Ratana Chuenpagdee point to the problem of ineffective surveillance and control in Chap. 25, but also highlight the fact that government and small-scale fisheries communities tend to have a different perception about why IUU fishing occurs and how it should be addressed, making it a "wicked problem" without a clear solution. Moslem Daliri, Svein Jentoft, and Ehsan Kamrani focus on the same problem in their case study from the Hormuz strait of Iran (Chap. 26). While the problem of IUU fishing is also due to lack of capacity for surveillance, monitoring and control, they assert that focusing on these issues is just addressing the symptoms rather than the cause, which may be poverty and insecure food supply. Therefore, a broader approach would be needed in solving the problem. Chapter 27, by Lina María Saavedra-Díaz and Svein Jentoft, talks about another form of risk exposure – that of armed conflict, which is referred to in the SSF Guidelines, and drug-related violence. Their case study is from Colombia, where small-scale fishing people are often displaced from their communities and fishing grounds. However, another problem is a fragmented and unstable governing system, which is largely too dispersed to take responsibility for coordinating the implementation of international codes like the SSF Guidelines. With the peace process currently underway, there is now hope that the situation of small-scale fisheries communities will be improved, with the support of the SSF Guidelines.

# Chapter 24
# Addressing Disaster Risks and Climate Change in Coastal Bangladesh: Using the Small-Scale Fisheries Guidelines

**Mohammad Mahmudul Islam and Svein Jentoft**

**Abstract** This chapter examined the implementation of the Voluntary Guidelines for Securing Sustainable Small-Scale Fisheries (SSF Guidelines) which were endorsed by FAO member states in 2014, in the Bangladesh small-scale fisheries context with particular focus on disaster risks and climate change related guidelines (Para. 9). Given that small-scale coastal fisheries in Bangladesh are subjected to multifaceted vulnerabilities due to extreme events and disasters, they provide an important case to study the potential implementation process of disaster risks and climate change related rules of the SSF Guidelines. The Bangladesh government is yet to take any decision regarding implementation of the SSF Guidelines. Reference to small-scale fisheries is largely absent in current ideologies, perceptions, and policies targeting disaster risks and climate change discourse in Bangladesh, although small-scale fishers are among the most climate-vulnerable population. Responses from relevant government agencies, fishers' organizations, and NGOs are also inadequate in addressing the concerns of small-scale fisheries. This study identified priorities and potential entry points for implementation of the Guidelines in Bangladesh. It calls for the state to recognize that climate change induced disasters have intense but different impacts on small-scale fishing people than on other professional groups. Further, it is argued that effective and full consultation with fishing communities is needed, and that the government should incorporate fishers' adaptation strategies against disaster risk and climate change impacts in existing climate change adaptation policy.

**Keywords** Small-scale fisheries • Disaster risk • Climate change • Voluntary Guidelines • Bangladesh

M.M. Islam (✉)
Department of Coastal and Marine Fisheries, Sylhet Agricultural University, Sylhet 3100, Bangladesh
e-mail: mahmud.cmf@sau.ac.bd; mahmud2512@googlemail.com

S. Jentoft
Norwegian College of Fishery Science, UiT – The Arctic University of Norway, Tromsø, Norway
e-mail: svein.jentoft@uit.no



S. Jentoft et al. (eds.), *The Small-Scale Fisheries Guidelines*, MARE Publication Series 14, DOI 10.1007/978-3-319-55074-9_24

## Introduction

Situated in a low delta, Bangladesh is extremely vulnerable to a range of climate related extreme events like cyclones, to slow onset processes like sea level rise, and (almost) everything in between (Hossain et al. 2012). Since 1980, Bangladesh has experienced over 200 natural disasters leaving a total death toll of approximately 200,000 people and causing economic loss worth nearly $17 billion. Each year due to natural disasters the country incurs losses to the tune of 1.8% of GDP (CDMP II 2016). A severe cyclone strikes Bangladesh's coast every 3 years (GoB 2008). Though, cyclones that form in the Bay of Bengal constitute only 5–6% of annual cyclones worldwide, they are the deadliest of all cyclones (Chowdhury 2002). The severity of cyclones in Bangladesh is reflected in the fact that about 80–90% of global losses and 53% of global cyclone-related deaths occur in Bangladesh (GoB 2008; Paul 2009). For example, on 15 November 2007, cyclone Sidr struck the coastal region, the worst of its type in two decades, killing more than 3300 people and creating wide-scale damage and losses to the effect of $1.7 billion or 2.6% of GDP (GoB 2008).

Other climate-related events such as temperature extremes, erratic rainfalls, intensified floods and droughts, and rough weather in the bay have also increased. All these events lead to loss of lives and livelihoods with severe implications for the sustenance of poor and marginal populations, especially those who live on the coast. The coastal zone of Bangladesh is geo-physiologically and ecologically diverse, and environmentally vulnerable. The coastal region is particularly vulnerable to cyclonic storm surge floods due to it being in the path of tropical cyclones and having a wide and shallow continental shelf, as well as a funneling shape (Das 1972; Haque and Blair 1992). Due to the shallow continental shelf, the influence of large tidal water, and the vast amount of long and narrow shorelines or inlets from the land to the bay, the surge amplifies to a considerable extent as it approaches low-lying and poorly protected land. This causes disastrous floods along the coast (Murty 1984; Shamsuddoha and Chowdhury 2007; Karim and Mimura 2008).

Besides sea level rises and cyclones, the coastal zone of Bangladesh also faces several other challenges such as coastal and river bank erosion, saline water intrusion, water logging, vulnerable polders, conflicts over land use (e.g. shrimp vs. rice cultivation) etc. that affect the lives and livelihoods of coastal people. Coastal people comprise different occupational groups, viz. small agricultural farmers, agricultural labor, sharecroppers, shrimp farmers, salt farmers, honey collectors, small-scale fishers, shrimp fry collectors, boat builders, net makers etc. It is estimated that fishing is the predominant source of livelihood for 14% of households in 14 districts situated along the coastline of Bangladesh (GoB and FAO 2013). Additionally, thousands of other coastal people are involved in small-scale fishing as a part time or seasonal occupation, particularly during the peak fishing season of hilsa (Islam 2012).

While hazards are potentially damaging events or phenomena, they do not necessarily cause a disaster. Instead, disasters occur through a mix of physical exposure and socio economic pressure. Thus, hazards must be understood within the broader context of society, and vulnerability explained as a result of both biophysical dynamics and social, political, and economic processes (Blaikie et al. 2014). Moreover, vulnerability to disaster is not the same for all professional groups. Small-scale fisheries in Bangladesh are a recurrent victim of natural disasters. Their vulnerability to disasters relates to the nature of the occupation and the socio-economic position of the fishers involved. Small-scale fishers in Bangladesh are generally poor and considered to be part of a lower class profession. They live in hazardous landscapes and fish in turbulent coastal waters (Jentoft et al. 2010; Islam 2011). Such natural disasters create insurmountable pressure on the lives and livelihoods of fishers who already live on the margin of survival. Thus, climate change and extreme natural disasters are one of the major threats to the small-scale fisheries sector in Bangladesh. However, they often do not get sufficient consideration from academics and policy makers.

While over the past four decades, Bangladesh has been able to significantly reduce disaster mortality due to an early warning system, enabling large scale evacuation before any cyclone strike, post-disaster recovery programs often face criticisms for mismanagement and irregularities, and procrastination in rehabilitation and relief. Coastal fishing communities, moreover, continue to be exposed to water-borne disasters. Casualties during fishing in the rough bay are still rather common and fail to receive sufficient attention from policy makers. Recently the Bangladesh government adopted its 7th Five Year Plan, which emphasizes the need to build capacity in both disaster risk reduction and climate change adaptation. In the plan, the Bangladesh government envisages the reduction of risk to people, especially the poor and the disadvantaged, to the effects of natural, environmental, and human induced disasters/hazards to a manageable and acceptable humanitarian level and to have in place an efficient emergency response management system (Ministry of Planning 2015; CDMP II 2016). However, the plan does not explicitly mention small-scale fishers as a vulnerable group.

The present chapter argues that adopting the Voluntary Guidelines for Securing Sustainable Small-Scale Fisheries in the Context of Food Security and Poverty Eradication (hereafter the SSF Guidelines), promoted by the FAO, has particular relevance in the making of climate-resilient coastal small-scale fishing communities in Bangladesh. This study is primarily based on 50 individual fisher's interviews conducted in two phases in 2009 and 2013 in a remotely located char[1] known as Char Kukri Mukri in the Meghna estuary of Bangladesh. The study is supplemented by secondary data collected from a comprehensive literature review.

The following are the main objectives of the study: (i) to explore disaster risks of and the impact of climate change on coastal small-scale fishers that need to be addressed in implementing the SSF Guidelines in the Bangladesh context; (ii) to identify constraints in implementing the SSF Guidelines keeping these priorities in mind.

[1] Chars are silt islands, which are created in a delta by the swell of rivers, monsoon rains, and sand carpeting. More than five million people in Bangladesh, mostly the poor, live in this ever-changing land in rivers and inshore waters, which are regularly subjected to floods, massive and rapid erosion, accretion, and occasionally drought.

The next section describes the SSF Guidelines regarding disaster risk and climate change. In section "Assessment of existing emergency response and disaster preparedness" we evaluate the performance of disaster management in Bangladesh. Section "Disaster risk and climate change impacts on small-scale fisheries: a study from Char Kukri Mukri in the Meghna River Estuary of Bangladesh", depicts the vulnerabilities of coastal hilsa fishers to disaster risk and climate change. The final section presents some recommendations concerning disaster risk reduction and climate change adaptation.

## The SSF Guidelines: Addressing Disaster Risk and Climate Change

The SSF Guidelines is the first internationally agreed instrument dedicated entirely to the small-scale fisheries of the world (FAO 2015). About 108 million people (of which 54% are men and 46% women) in the developing world currently depend directly on small-scale fisheries and post-harvest activities for at least parts of their income (BNP 2009). Small-scale fisheries thus make important, though poorly quantified contributions to human society. Regardless of the geographical settings, small-scale fisheries are facing several challenges such as degradation, overfishing and overcapacity, conflicts over resources and space, and global changes, but yet they have not garnered adequate attention in national, regional, and global decision-making (Islam 2012; Jentoft 2014). In this context, FAO has been spearheading the initiative to develop international guidelines for small-scale fisheries. After years of planning, extensive consultation with civil society organizations (CSOs) and stakeholders, including the research community, and intense negotiation among member states, on June 9th, 2014, the Committee of Fisheries (COFI) of the FAO adopted the SSF Guidelines (Jentoft 2014). The SSF Guidelines have several sections about a wide range of issues. Section 9 particularly relates to natural disasters and climate change and how they affect the food-security and livelihoods of small-scale fishers.

Paragraph 9.2 reads as follows:

> All parties should recognize and take into account the differential impact of natural and human-induced disasters and climate change on small-scale fisheries. States should develop policies and plans to address climate change in fisheries, in particular strategies for adaptation and mitigation, where applicable, as well as for building resilience, in full and effective consultation with fishing communities including indigenous peoples, men and women, paying particular attention to vulnerable and marginalized groups.

The Guidelines request that states should recognize the vulnerability of small-scale fisheries (throughout the value chain) to the impacts of climate change and disasters. The Guidelines ask for urgent and ambitious action, in accordance with the objectives, principles, and provisions of the United Nations Framework Convention on Climate Change (UNFCCC). They also take (Para 9.1) into account the United Nations Conference on Sustainable Development's (Rio + 20) outcome document *The Future We Want.*

Natural and human-induced disasters and climate change have differential impacts on small-scale fisheries. Therefore, states need to develop specific policies and plans to address climate change adaptation and mitigation, and emergency response and disaster preparedness through consultation with vulnerable fishing communities. Hence, the SSF Guidelines call respective states to assist and support the fishers affected by climate change or natural and human-induced disasters through appropriate adaptation, mitigation, and aid plans (Para. 9.4). These strategies require integrated and holistic approaches, including cross-sector collaboration (Para. 9.2). If fishers are affected by human induced disasters, the responsible party for these disasters should be held accountable (Para. 9.5).

All parties, which include CSOs, should take into account the impact that climate change and disasters may have on the small-scale fisheries post-harvest and trade subsectors. Therefore, it is desirable that states and others provide support to small-scale fisheries stakeholders with regard to adjustment measures aimed at addressing changes so as to reduce their negative impacts (Para. 9.6). Broader emergency response and disaster preparedness of the state should be targeted at small-scale fisheries. In doing so, the concepts of relief-development continuum and 'building back better' should be applied in disaster response and rehabilitation (Para 9.7). All parties should promote the role of small-scale fisheries in efforts aimed at mitigating the effects of climate change and thus encourage and support energy efficiency in all levels of the fish chain. Finally, states should appropriately consider making available to small-scale fishing communities' transparent access to adaptation funds, facilities, and/or culturally appropriate technologies for climate change adaptation.

## Assessment of Existing Emergency Response and Disaster Preparedness

After the devastating Gorki cyclone in 1991, the Bangladesh government changed its response strategy to natural disasters (acting after the occurrence). It adopted an approach of total disaster management (which includes prevention, response, recovery, rehabilitation, mitigation, and preparedness) (Haque and Uddin 2013; Ministry of Planning 2015). The Bangladesh government's shift of strategy was well ahead of many other global initiatives. The overall shift in strategy comprised of a number of initiatives. Disaster risk reduction programs, extended volunteerism, emphasis on increased community resilience and extended social safety nets, adoption of legal and institutional frameworks including national and international frameworks for disaster risk reduction, and involvement of a vibrant NGO sector contributed to this success (Ministry of Planning 2015). Cyclone preparedness in Bangladesh has substantially improved over the last decade with the construction of additional public shelters in coastal districts, along with the expansion and reinforcement of embankments and coastal afforestation programs along the coast and estuary channels (Paul

2009). For example, in 1992, there were 512 cyclone shelters in coastal Bangladesh (Ikeda 1995) which increased to 3976 in 2007 (Shamsuddoha and Chowdhury 2007). Bangladesh has over 7500 km of embankments built along its rivers and coastal areas (UN 2007).

In 1985, the Bangladesh government introduced 'Standing Orders' (SOD) (updated in 2010) for natural disaster management (primarily for floods and cyclones) (GoB 2010). The SOD outlines specific directives, duties, and responsibilities regarding disaster risk reduction and emergency response management for public agencies at all levels, including regional and local administrations, relevant ministries, and defense forces. It is worth remembering that two structural cyclone mitigation measures had already been introduced in Bangladesh before independence in 1971. One was the coastal embankment project to protect villages from rising water and the other the coastal afforestation program to create bio-shields (Paul 2009). Both these projects are credited for their positive impact on disaster risk reduction.

The National Plan for Disaster Management (NPDM) was developed in 2010 to manage risk and the consequences of disasters through prevention, emergency response, and post-disaster recovery. The NPDM has identified seven strategic goals which are drawn from the SAARC Disaster Management Framework that include (1) professionalizing the disaster management system; (2) mainstreaming risk reduction; (3) strengthening institutional mechanisms; (4) empowering risk communities; (5) expanding risk reduction programming; (6) strengthening emergency response systems; and (7) developing and strengthening networks. All these initiatives have markedly increased the disaster management capacity of Bangladesh. This enhanced capacity reflects advancement at two levels. First, enhanced capacity has lowered the number of fatalities as well as the damage caused by cyclones. This was primarily because of the Bangladesh government's attempt to provide timely weather forecasting and advance warning systems. Second, enhanced capacity has contributed to the successful evacuation of people and to the construction of more cyclone shelters (Shamsuddoha and Chowdhury 2007). Heather Blackwell, head of Oxfam in Bangladesh, estimated that "Bangladesh's early warning system and preparation saved up to 100,000 lives' (Oxfam 2008)." To bolster early warning infrastructure, a coastal volunteer network has been established to disseminate warnings systems more widely (Murray et al. 2012).

In the aftermath of extreme events, the governmental and different CSOs deliver relief and support services to affected communities. Generally, these responses address only the immediate survival needs of distressed communities; they usually do not provide other services such as health or education, nor do they contribute to increasing resilience to command future extreme events. Thus, disaster recovery and 'building back better' has largely been overlooked. Although Bangladesh has made major efforts to prevent natural hazards from causing human disasters, the holistic approach the SSF Guidelines promote has been largely missing. As a result, many of the affected communities still suffer from the residual impacts of these devastating events and have not been able to fully restore their livelihoods. Consequently, overall resilience to disasters has been undermined (Ministry of

Planning 2015). Thus, despite efforts to reduce vulnerability, small-scale fishing communities continue to experience loss, often to an even larger extent than in the past (Shamsuddoha et al. 2013).

## Disaster Risk and Climate Change Impacts on Small-Scale Fisheries: A Study from Char Kukri Mukri in the Meghna River Estuary of Bangladesh

As mentioned above, this chapter undertakes a case study of coastal communities living in a char situated at the mouth of the Meghna River facing the Bay of Bengal (Fig. 24.1). Char Kukri Mukri is a very low-lying land area, situated only about 1.5–1.6 m above sea level. The char is part of the main island situated in Char Kukri Mukri union[2] under Char Fassaon upzilla of Bhola district. Due to its geographical location, Char Kukri Mukri is very vulnerable to disaster risks and climate change. To the south, no land or barrier exists that can cushion cyclones that boil over from the Bay of Bengal. When cyclone waters move in a funnel shape towards the Meghna River estuary, water surges and causes a disaster (cf. Shamsuddoha and Chowdhury 2007). Besides cyclones, community members say that they face frequent and severe risks from coastal erosion, tidal surges, water logging, and saline water intrusion.

According to popular belief, the char was formed four to five hundred years ago when silt carried by different rivers from the upstream Himalayan basin accumulated to form the island, which incidentally is also rich in biodiversity.[3] The landmass is still developing; thus, it is very prone to river bank erosion. About 40 years ago, people affected by river bank erosion in nearby localities migrated to the char, which is owned by the government, to take shelter. Easy access to fishing grounds, both in estuary and inshore waters also motivated fishers to settle on the char. The 12,000 inhabitants of the char are among the most disadvantaged population groups on the coast. About 38% of the population lives below the poverty line (national average below the poverty line is 17.6%). The inhabitants are overwhelmingly illiterate, with the literacy rate at only 28.55% (national rate is 57.91%). About 30% of houses are makeshift and 68% of houses thatched (*kutcha*) (national average for thatched houses is 8.11%). Only 2% of houses have concrete structures (national average is 25.12%). Less than 1% of the population has access to some sort of electricity (from solar panels) (national percentage is 55.26%), 6.28% has access to sanitary toilets (national percentage is 51.05%), and 14% have access to safe drink-

[2] Union is the lowest local administration unit of a three tiered local administration system in Bangladesh. The other two units are upzilla (sub-district), and zilla (district).

[3] Char Kukri Mukri is also rich with wildlife such as deer, monkeys, cougars, otters, foxes, wild hens, gray peacocks, migratory birds, and other wild species. Wildlife attracts an increasing number of tourists each year, particularly in the winter season. This char is considered as one of the most important wildlife sanctuaries (IUCN Category IV) of the forest department.

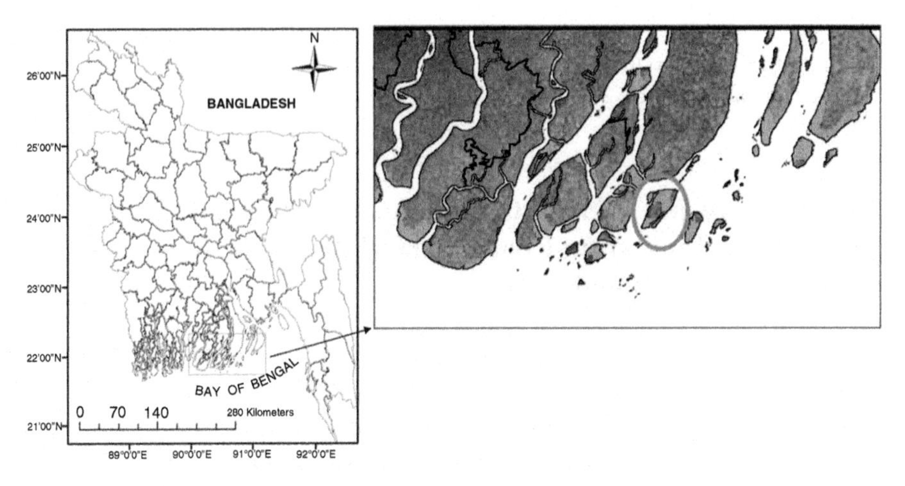


**Fig. 24.1** Location of the study area (Char Kukri Mukri)

ing water (national average is about 96%) (All data from BBS 2011). Boats are the only means of transportation to other nearby localities including Char Fasson upzilla. Though medical facility exists but is in very poor shape without appropriate manpower and doctor. There is no scheduled bank or post office, hence char dwellers mainly depend on money lenders and microcredit by NGOs for financial capital. Due to their physical isolation, char dwellers mainly depend on natural resources such as land and fisheries to make their living. Thus, agriculture and fishing are the two major professions. Most inhabitants are involved in both agriculture and fishing. Aquaculture is also practiced at a smaller scale. Some key informants estimated that 90% of local people are dependent on fisheries, particularly on hilsa (*Tenualosa ilisha*) fishing. The hilsa fishery is identified as the largest estuarine fishery in the world in terms of catch (Blaber 2000) and constitutes a long-standing economic activity in the Meghna basin where the study area is located. Hilsa shad contributes 11% of the total fisheries catch, provides employment for 0.5 million fishers directly and another 2 million people indirectly, and accounts for about 1% of GDP (Mohammed and Wahab 2013) totaling USD 1.3 billion per year (BOBLME 2012).

In Char Kukri Mukri men, women, and children are involved in fishing (Fig. 24.2). They use different fishing gears depending on economic capacity, skills, and target species. Women and children are mainly involved in shrimp fry collection using drag and/or pull nets. Crab collection is another important fishing occupation and is done by hand, bait and hook. Cast nets are also used for fishing in the floodplains of the char. Along with men, some women are also involved in estuarine set bag fisheries. The majority of fishermen are occupied in the hilsa fishery, which is an important economic activity in the area. Well-to-do fishers have their own boats and fishing gears (mainly gillnets) for hilsa fishing in the estuary and inshore waters. Other fishers work as crew and are either paid a share of the catch or wages as part of a fishing team. The duration of fishing ranges from one day to four to six days depending on the fishing zone (estuary or inshore/offshore waters). During the

**Fig. 24.2** Children catching fish in flood waters in Char Kukri Mukri

breeding season of hilsa in November, fishers are restricted from catching hilsa for 22 days. Fishers who were interviewed said that March to May was the slack period for hilsa fishing. In that period fishers repair their boats and gears for the upcoming hilsa season that starts when the monsoon comes. During the slack period demand for laborers goes down, and this pushes many fishers to migrate to nearby cities or the capital in search of alternative work. In the slack period, fishers who employed other fishers also face financial difficulties as repair and maintenance of fishing gears require financial capital. Therefore, they take advance loans (locally known as *dadon*) from fish traders from nearby fish landing centers in Char Fasson or other upzillas of Bhola district. The loan arrangement is such that the fisher has to sell his catch to the fish trader, usually at a discounted price; the trader also gets a commission, usually 10% of the total price. The majority of respondents we spoke to were indebted to fish traders for a long time *as* they face one crisis after another (such as illness, poor catch, ransom demands by criminal gangs etc.) and *dadon* is the only financial source they can avail. Thus, the hilsa fishery is the mainstay of livelihoods in the region and therefore any disruption to fishing operations or fish availability makes fishers vulnerable.

Hilsa species and fishers who engage in hilsa fishing are subjected to climate change and disaster risks. For instance, Fernandes et al. (2015) predicted that even if hilsa is fished sustainably, due to climate change hilsa shad catches could show a slight decline by 2030 and a significant (25%) decline by 2060. However, if overexploitation is allowed, catches are projected to fall much further, by almost 95% by 2060, compared to the 'business as usual' scenario for the start of the twenty-first century. High dependence on hilsa could lead to livelihood failures in case of stock collapse, due to either over-exploitation and/or climatic variability. As mentioned in the introduction, the Bay of Bengal is one of the most disaster-prone regions in the world. Increased intensity of cyclones and sea-borne depressions make fishing operations in the bay risky and end up limiting those (Jentoft et al. 2010). Most of the deaths and destruction attributable to cyclonic activity in Bangladesh are largely caused by abnormal storm surges (Chowdhury et al. 1993). Interviewed fishers reported that unfit weather conditions due to frequent cyclones and depressions (caused by lower atmospheric pressure) in the bay often force fishers to abandon their fishing trips and return to coast. Incomplete fishing trips incur a substantial financial loss estimated to be 76,400 BDT (955 USD) for a 10-day hilsa fishing trip. Fishers told us that such income loss is quite common nowadays in the hilsa fishery. They also estimated that depressions originating in the Bay of Bengal (Fig. 24.3) cause fishers to discontinue fishing for at least a week. To avoid economic loss in a situation of dire poverty, many fishers defy warnings and continue fishing, which results in fatalities and morbidities of fishers. The death of a household member capable of working can pull the whole family into extreme poverty and extended trauma.

Small-scale fishers of Char Kukri are also exposed to frequent cyclones originating in the Bay of Bengal. On 25th, May 2009, cyclone Aila struck the south-west coast of Bangladesh, killing 190, injuring 7103, and rendering more than half a million people homeless (UN 2007). On May 16th, 2013, cyclone Mahasen made landfall in Bangladesh and the associated torrential rainfall and strong winds destroyed hundreds of thatched houses in Char Kukri Mukri. Cyclone Mahasen was less severe than cyclone Aila, but yet it adversely affected many communities that were already in a stressed situation in the aftermath of cyclone Aila. Several studies have indicated that cyclone Aila caused devastation to the coastal landscape and people's livelihoods (Shamsuddoha et al. 2013; Islam et al. 2016), including in Char Kukri Mukri. The respondents of the present study detailed the dismal situation they faced after cyclone Aila. There were no coastal embankments in the char when it struck. Though the green belt of mangrove forest in the char helped reduce wind velocities, they were not able to shield communities from the swelling sea water. The tidal water washed away people's household assets, destroyed standing crops, flattened road infrastructure and thatched houses, placed a heavy toll on those who had livestock and poultry, and adversely affected those who had aquaculture farms (Fig. 24.4). Sanitation and safe drinking water were also severely impacted which resulted in the onset of water borne diseases. The long-term residual impacts were far-reaching and caused profound negative impacts on local food security, health, and the overall economic situation. For instance, salt water intrusion resulted in the soil being unfit for agricultural production; thus crop, vegetable and, fruit output reduced significantly. So too was the case with aquaculture production. Grazing areas for cattle were also damaged. Some respondents indicated that saline water

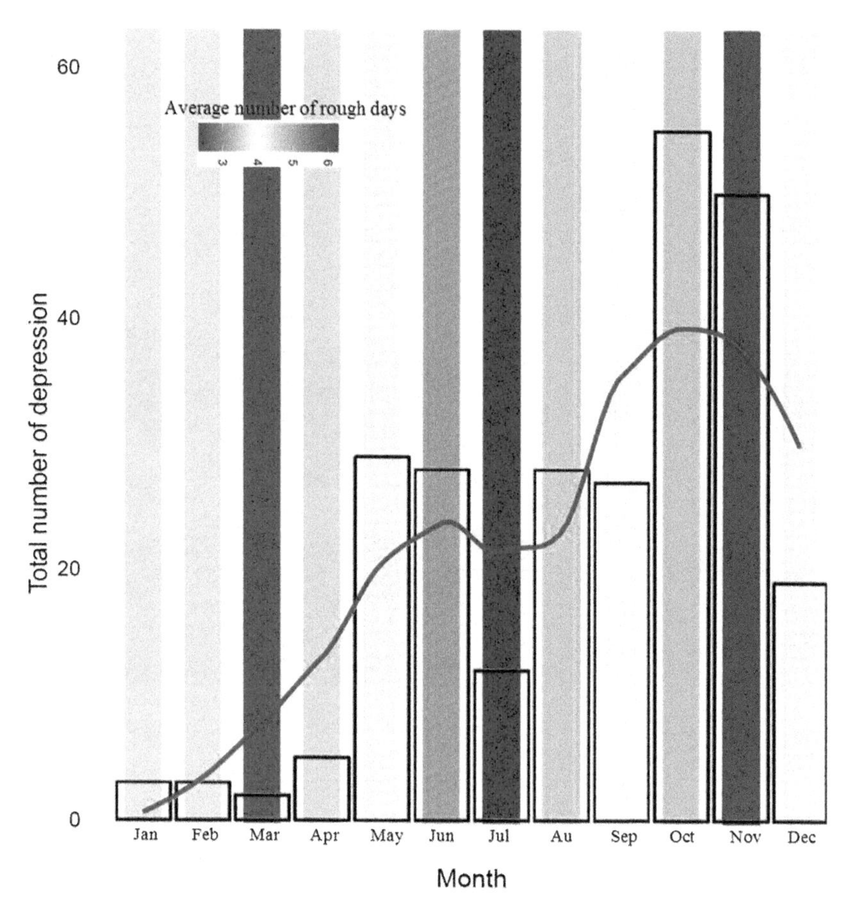


**Fig. 24.3** Long-term monthly depression scenario in the Bay of Bengal. Hollow bars indicate the total number of depression occurred in every month from 1975 to 2015. Blue smooth line indicates the long-term seasonal trend in total number of depression. Vertical solid lines with different colors (blue to white to red through a linear interpolation in between) indicate long-term monthly average number of rough days. (Source: Date collected from Bangladesh Meteorological Department)

intrusion and the unhygienic sanitary situation caused an increase in the morbidity of humans and livestock in the char.

Small-scale fishers in Char Kukri Mukri spoke about the economic burden created by repeated extreme events. The majority of fishers take *dadon* from fish-traders when productive assets are lost or damaged. To restore fishing operations they have to retake dadon, which entraps them into endless cycles of debt. A number of fishers spoke about extended trauma after they suffered at the hands of cyclones. When fishing gear are lost or damaged, marginal fishers in particular cannot afford to restore their fishing operations. One widowed fisherwoman described her dismal situation as a result of cyclone Aila in the following way:

**Fig. 24.4** After cyclone Aila, many households in Char Kukri Mukri had to start their livelihoods from scratch

> Our house was at the end of the village facing the river, so we were worst affected by cyclone Aila. All of my family members managed to survive by taking shelter in higher places. All my households' materials were washed away when we were struggling to survive against rising waters. Our fishing boat, the source of livelihood for the family, which was tied to a tree, broke away due to the rising tide. One *behundi jal* (estuarine set bag net), which was set up in the river was damaged another one was lost. Our thatch house collapsed, and we could not manage to rebuild it so we just put a roof structure on the plinth. This, however, makes it very damp and humid to live inside. Before Aila struck, our daily income was 300–500 BDT. After cyclone Aila, we had no boat, only a few torn nets. I also became ill. Now income from my two children is the financial mainstay of our family. My daughter goes for shrimp fry collection in the nearby river in the afternoon. My son goes hilsa fishing as waged labor from afternoon to late at night (1 am). He gets approximately 50 BDT for 8 hours work. After taking some rest, he then goes for shrimp fry collection early morning and gets back around 11 am. After he is back my daughter goes for fry collection. Thus they collectively make best use of one net for subsistence living. Despite our best efforts the family has became destitute.

Respondents spoke about a number of coping strategies that they had adopted in the aftermath of cyclone Aila. They took loans from money-lenders and microcredit from NGOs for immediate survival. Some small-scale fishers were able to restart the collection of shrimp fry and crab from tidal creeks and mangrove forests as these activities did not require expensive fishing gears. Some fishers took their children out of school and employed them in fishing. Other fishers migrated to nearby cities in search of work. Yet others got employment in 'food for work' or 'cash for work' programs run by different NGOs and local government for rebuilding physical infrastructure.

In terms of disaster responses, most of the interviewees reported that they receive early warnings of cyclones, through radio or from volunteers of the Red Crescent Society. Local people participated in disaster management committees in the area. However, they also mentioned that the cyclone center in the char can only accommodate half of the char dwellers. Many inhabitants do not want to abandon their assets (home, land, fishing gears, and boats) and leave for cyclone centers. Some respondents complained of nepotism and irregularities in distribution of relief and access to government support in the immediate aftermath of disasters. Repeated cyclone strikes have prompted the government and donor agencies to build the resilience of char dwellers. After cyclone Mahasen in 2013, the government built coastal embankments around the char to protect inhabitants from rising sea waters in the event of cyclones or floods. This hazard-prone char also received attention from different local NGOs that were in operation before and after cyclone Aila. These NGOs mainly work in microcredit, disaster management, health and sanitation, and physical infrastructure restoration so as to strengthen the long term resilience of char dwellers. The government of Bangladesh also provides relief to people and invests in the building of physical infrastructure such as roads and coastal embankments and afforestation programs to build green bio-shields. Fresh water reservoirs have been created for irrigation purposes that facilitate vegetable and fish production to meet local demands for food security. Installation of tube wells has helped improve the availability of safe drinking water. Char dwellers have noticed the positive changes in their well-being due to these interventions. Living in a sequestered island, they feel marginalized from mainstream economic opportunities. Therefore, they have now asked for support to diversify their livelihoods and strengthen the availability of good educational facilities.

## Discussion

Given that small-scale fishers are among the most vulnerable groups to disaster risks and climate change, the SSF Guidelines are an appropriate and timely instrument for initiating policy change aimed at making small-scale fisheries resilient. The Guidelines rightly call for states to recognize differential vulnerabilities of different sectors of small-scale fisheries and develop specific policies and plans to address climate change adaptation and mitigation, emergency response, and disaster preparedness through consultation with local fishing communities. These efforts could help in 'building back better' small-scale fisheries, particularly in developing countries such as Bangladesh where a combination of poverty, marginalization, and natural hazards have had devastating impacts on small-scale fishers in a way that has left them trapped in poverty. As the case study of Char Kukri Mukri illustrates, small-scale coastal fishers in Bangladesh are vulnerable to climate change and disaster risks in several ways. They largely rely on catching climate sensitive hilsa species. They face disaster risks while fishing in the bay; poor weather in the bay often forces them to abandon their fishing trips. Different extreme events also result in fishing communities losing household assets. While pre-disaster responses by the government have been successful in saving many lives, small-scale fishers remain vulnerable because of context specific factors.

In general, though small-scale fishers are more vulnerable than other professional groups on the coast, they do not get due political consideration when it comes to disaster risk reduction. While it is necessary to prioritize small-scale fishers in addressing disaster risks and climate change, doing so is a challenging task. Because fishing communities are located in remote regions and are spread sparsely across the landscape, there is also limited understanding of their differential vulnerabilities (Part 2, Chapter 9.2 of the SSF Guidelines) among policy makers, and even within the community itself. Implementation of the SSF Guidelines for disaster risk reduction could serve as a point of departure for the government in terms of 'what to do' to enhance the resilience of Bangladeshi small-scale fishers and fishing communities in ways that make them more self-reliant and engaged in the development process. Nevertheless, the Bangladesh government in the Seventh Five Year Plan emphasized the need to reduce the risk people, especially the poor and the disadvantaged, were exposed to vis-à-vis natural, environmental, and human induced disasters/hazards (cf. Dercon 2010). In order to do so, the government plans to develop an emergency response management system. Though small-scale fishers are not explicitly mentioned as a disadvantaged group, the SSF Guidelines may help make government aware of their special situation and needs.

Yet, as is made clear in the SSF Guidelines, full implementation of the Guidelines would require a major transformation in different aspects of governance in the form of institutional, social, political, and economic reforms. For instance, in Bangladesh, at present disaster management policy and practice across government institutions occur at a relatively small-scale and often on a pilot basis. Hence, a shift of perspectives requires a movement from theory to practice and linking policy with actions in the spirit of the SSF Guidelines. In our case study area, small-scale fishers participate in disaster management at the local level as part of a broader community and professional group involved in disaster management and climate change negotiations. Hence, it becomes critical and challenging to build strong and effective communities and/or fisher organizations empowered to participate in fisheries management and disaster management. Such transformation is essential in order to promote and deliver mainstream disaster management outcomes. Small-scale fisheries should be one of the target sectors of natural hazard reduction strategies. Particular attention must be given to small-scale fisheries because they have been considered as a 'lower' profession and thus remained marginalized (due to lack of education and lack of acceptance by others) politically and socially. Consequently, this has reduced their resilience (Islam 2011; Jentoft and Midré 2011). Further, small-scale fisheries are not just about fishing *per se*, as small-scale fishers and their dependents require support with regard to improving their overall wellbeing (health, education, livelihood diversification, and community development). Inclusion of the Department of Fisheries – the apex body of fisheries management – within the 'Disaster Management' team in the mandate of 'Standing Orders', will be a step in the right direction.

Disaster risk reduction is multi-sector affair, which requires strong collaboration among all relevant stakeholders such as government, civil society organizations, the private sector, media, research organizations etc., as is also emphasized in the SSF Guidelines (Part 2, Chapter 9.3). Building collaborative efforts is an interactive governance challenge that must be addressed if the SSF Guidelines are to be successfully implemented (Jentoft and Chuenpagdee 2015). While formulating the SSF Guidelines, NGOs played an important role (Jentoft 2014), which suggests that they can play a similar in the implementation of the Guidelines. Despite myriad numbers of non-governmental organizations working in Bangladesh, very few of them are working on issues directly related to small-scale fisheries. Thus, a lack of institutions targeting coastal and marine fisheries is clearly visible. NGOs could play an important role in building capacity of fisher organizations, particularly with regard to co-management, which will empower small-scale fishing communities and turn them into effective partners of government to reduce disaster risks.

Bangladesh has made remarkable progress towards addressing and managing disaster risk, but extreme events have also markedly increased. Moreover, Bangladesh's overall post-disaster recovery and reconstruction capacity is rather weak (Ministry of Planning 2015). Its response to natural disasters in the past has usually focused only on the reconstruction of physical assets. In our case study area, though rebuilding physical assets is important for livelihoods recovery, the provision of boats and nets to non-fishers as a means to relieve their economic constraints, often leads to over-capitalization of the fishery, and thus contributes to aggravating the consequences of disasters. While short-term loss and damage often draw the attention of policymakers and donor agencies, the long-term residual impacts of extreme events receive less attention. As the impacts of extreme events evolve with changing climatic conditions, any response must seek to instill the capacity and knowledge required to effectively anticipate and reduce future loss and damage.

In the process of implementing the SSF Guidelines, there are several others challenges that must be dealt with. Physical remoteness and the scattered location of fishing communities is a barrier for implementation. Yet even physical isolation can be mitigated through appropriate improvements to infrastructure, health, and education services, and improved access to information and markets. The need to identify and secure funding and build synergies between all stakeholders is another challenge. Interaction with relevant non-fisheries ministries and departments at all levels and the mainstreaming of the SSF Guidelines in relevant policies, strategies, and plans as well as public-private partnerships involving government, non-government, and fishers organizations in support of the SSF Guidelines could help in overcoming this challenge.

Another constraint is the poor implementation of the several plans policies regarding disaster risk reduction that Bangladesh has. Thus, even if the SSF Guidelines are translated into policy in Bangladesh, implementation at the local level will be a big challenge, as has been the case in the past. Community participation is a must. However, community participation is prone to elite capture by powerful people in society. Genuine accountability mechanisms for vulnerable communities that can make their voices heard is also missing. Thus, monitoring and accountability at all levels are essential to ensure implementation. Financing for disaster risk management of small-scale fisheries is another problem. As a developing country, Bangladesh has financed its investments in policies that address disaster risks largely through international cooperation. To implement the SSF Guidelines with the aim of making small-scale fishing people and communities more secure, will require funding from its own resources in addition to international funding (Part 2, Chapter 9.9 of the SSF Guidelines).

## Conclusion

Small-scale fisheries in Bangladesh are important for supplying animal protein, ensuring food security, and employing large numbers of people. It is estimated that one-tenth of the total population are somehow dependent on to fishing. Thus, small-scale fisheries in Bangladesh are definitely 'too big to ignore'. Still, not only are they poor and vulnerable to natural and other hazards, they are also marginalized, which is partly due to their often remote and scattered location, making them hard to reach. They are materially poor and they do not have organizations that can defend their interests. They therefore tend to be isolated from the political process, with no voice in issues that are of great concern to them. The sector is neglected and undervalued in terms of its food security and poverty alleviation potential. The fact that small-scale fishers live in low-lying coastal areas make them more vulnerable than any other professional groups with regard to climate change, sea-level rise, and cyclones. Their exposure to risk receives far less attention from policy makers than is necessary to keep them safe. They are in fact, largely 'ignored.' This is also why the SSF Guidelines are relevant in the context of Bangladesh

The situation of small-scale fisheries needs to be changed if the country wants to achieve not only the Sustainable Development Goals (SDGs), particularly goal 13 (taking urgent action to combat climate change and its impacts) and goal 14 (conserve and sustainably use the oceans, seas, and marine resources for sustainable development), but also the ambitions stated in the SSF Guidelines, as they are basically the same. The implementation of the SSF Guidelines could help create the impetus that is needed to ensure that the special situation that small-scale fisheries are in, especially in countries like Bangladesh where fishers and fishing communities are already weakened by poverty, can be addressed in the face of climate change and other natural hazards. In terms of disaster risk reduction, Bangladesh had already achieved commendable success. Yet, the SSF guidelines can help in specifically focusing on the vulnerability of small-scale fishers in ways that have not been done so far and that make them climate resilient. Given the strong disaster management mechanisms in place in Bangladesh, the importance given by the government to conserve hilsa, and the presence of CSOs at the grassroots level, there are clear opportunities to implement the SSF Guidelines in small-scale fisheries. However, there are also challenges. Reducing disaster risks and climate change vulnerability of small-scale fishers has largely failed so far to provide resilience and security to coastal communities. For that to happen, policies needed must be developed for the long term and aimed at empowering coastal communities through the building of more robust governance systems. The role of small-scale fisheries as a driver of economic growth and food security is still undervalued. Given this scenario, steps are required to minimize constraints and capitalize on the opportunity available with the SSF Guidelines.

The SSF Guidelines as a whole, and not only the section that deals with climate change, speak to the situation in Bangladesh's coastal areas and small-scale fisheries. For the government and other parties the SSF Guidelines represent a 'marching order' to do something about a general problem which is not just in the future, but which already exists. The SSF Guidelines are also relevant in their totality, as the vulnerability that small-scale fishers are experiencing with regard to natural hazards and climate change as well is due to their poverty and social marginalization. Thus, small-scale fishers' resilience to such events is dependent on the degree to which other parts of the SSF Guidelines are adopted as public policy and integrated in the planning process. This is what the 'holistic' approach in para 9.2 would imply in Bangladesh, and elsewhere. The vulnerability of small-scale fisheries to climate change and natural disasters may be extreme in Bangladesh, but they are not unique. Thus, we believe there are lessons to be learned from Bangladesh for other countries where small-scale fishers are poor and vulnerable.

## References


Blaber, S. J. (2000). *Tropical estuarine fishes: Ecology, exploration and conservation*. Oxford: Blackwell.

Blaikie, P., Cannon, T., Davis, I., & Wisner, B. (2014). *At risk: Natural hazards, people's vulnerability and disasters*. London: Routledge.

BNP. (2009). *Big number program. Intermediate report*. Rome: Food and Agriculture Organization of the United Nations and WorldFish Center.

BOBLME. (2012). *Management Advisory for the Bay of Bengal Hilsa Fishery*. Regional Fisheries Management Advisory Committee. http://www.boblme.org/documentRepository/BOBLME-2012-Brochure-02.pdf. Accessed 12 Dec 15.

CDMP II. (2016). *Comprehensive Disaster Management Programme, Phase 2*. Ministry of Disaster Management and Relief, Bangladesh Government.

Chowdhury, K. M. M. H. (2002). Cyclone preparedness and management in Bangladesh. In BPATC (Ed.), *Improvement of early warning system and responses in Bangladesh towards total disaster risk management approach* (pp. 115–119). Dhaka: BPATC.

Chowdhury, A. M. R., Bhuyia, A. U., Choudhury, A. Y., & Sen, R. (1993). The Bangladesh cyclone of 1991: Why so many people died. *Disasters, 17*(4), 291–304. doi:10.1111/j.1467-7717.1993.tb00503.x.

Das, P. K. (1972). A prediction model for storm surges in the Bay of Bengal. *Nature, 239*, 211–213.

Dercon, S. (2010). Risk, vulnerability and human development: What we know and what we need to know? In R. Fuentes-Nieva & P. Seck (Eds.), *Risk, shocks, and human development: On the brink* (pp. 13–35). Basingstoke: Palgrave MacMillan.

FAO. (2015). *Voluntary guidelines for securing sustainable small-scale fisheries in the context of food security and poverty eradication*. Rome: Food and Agriculture Organization of the United Nations.

Fernandes, J. A., Kay, S., Hossain, M. A., Ahmed, M., Cheung, W. W., Lazar, A. N., & Barange, M. (2015). Projecting marine fish production and catch in Bangladesh in the 21st century under long-term environmental change and management scenarios. *ICES Journal of Marine Science, 73*, 1357–1369. doi:10.1093/icesjms/fsv217.

GoB. (2008). *Cyclone sidr in Bangladesh: Damage, loss and needs assessment for disaster recovery and reconstruction*. Dhaka: Government of Bangladesh.

GoB. (2010). *Standing order on disasters*. Dhaka: Disaster Management Bureau, Government of Bangladesh.

GoB, & FAO. (2013). Master plan for agricultural development in the southern region of Bangladesh. In *The Government of the Peoples' Republic of Bangladesh & Food and Agricultural Organization of the United Nations*. Rome: Food and Agriculture Organization of the United Nations.

Haque, C. E., & Blair, D. (1992). Vulnerability to tropical cyclones: evidence from the April 1991 cyclone in coastal Bangladesh. *Disasters, 16*(3), 217–229.

Haque, C. E., & Uddin, M. S. (2013). *Disaster management discourse in Bangladesh: A shift from post-event response to the preparedness and mitigation approach through nstitutional partnerships* (Approaches to disaster management. Examining the implications of hazards, emergencies and disasters, p. 217).

Hossain, M. S., Reza, M. I., Rahman, S., & Kayes, I. (2012). Climate change and its impacts on the livelihoods of the vulnerable people in the Southwestern Coastal Zone in Bangladesh. In W. L. Filho (Ed.), *Climate change and the sustainable use of water resources*. Berlin: Springer.

Ikeda, K. (1995). Gender differences in human loss and vulnerability in natural disasters: A case study from Bangladesh. *Indian Journal of Gender Studies, 2*(2), 171–193.

Islam, M. M. (2011). Living on the margin: The poverty-vulnerability nexus in the small-scale fisheries of Bangladesh. In S. Jentoft & A. Eide (Eds.), *Poverty mosaics: Realities and prospects in small-scale fisheries* (pp. 71–95). Dordrecht: Springer.

Islam, M. M. (2012). *Poverty in small-scale fishing communities in Bangladesh: Contexts and responses*. Doctoral dissertation, Staats-und Universitätsbibliothek, Bremen.

Islam, M. M., Islam, N., Sunny, A. R., Jentoft, S., Ullah, M. H., & Sharifuzzaman, S. M. (2016). Fishers' perceptions of the performance of hilsa shad (*Tenualosa ilisha*) sanctuaries in Bangladesh. *Ocean and Coastal Management, 130*, 309–316.

Jentoft, S. (2014). Walking the talk: Implementing the international voluntary guidelines for securing sustainable small-scale fisheries. *Maritime Studies, 13*(1), 1–15.

Jentoft, S., & Chuenpagdee, R. (Eds.). (2015). *Interactive governance for small-scale fisheries*. Cham: Springer.

Jentoft, S., & Midré, G. (2011). The meaning of poverty: The conceptual issues in small-scale fisheries. In S. Jentoft & A. Eide (Eds.), *Poverty mosaics: Realities and prospects in small-scale fisheries* (pp. 43–70). Dordrecht: Springer Science.

Jentoft, S., Onyango, P., & Islam, M. M. (2010). Freedom and poverty in the fishery commons. *International Journal of the Commons, 4*(1), 345–366.

Karim, M. F., & Mimura, N. (2008). Impacts of climate change and sea-level rise on cyclonic storm surge floods in Bangladesh. *Global Environmental Change, 18*(30), 490–500.

Ministry of Planning. (2015). *7th five year planning of Bangladesh. Ministry of Planning*. Dhaka: Government of the Bangladesh.

Mohammed, E. Y., & Wahab, M. A. (2013). *Direct economic incentives for sustainable fisheries management: The case of Hilsa conservation in Bangladesh*. London: International Institute for Environment and Development.

Murray, V., McBean, G., Bhatt, M., Borsch, S., Cheong, T. S., Frian, W. F., Llosa, S., Nadim, F., Nunez, M., Oyun, R., & Suarez, A. G. (2012). Case studies. In C. B. Field, V. Barros, T. F. Stocker, D. Qin, D. J. Dokken, K. L. Ebi, M. D. Mastrandrea, K. J. Mach, G.-K. Plattner, S. K. Allen, M. Tignor, & P. M. Midgley (Eds.), *Managing the risks of extreme events and disasters to advance climate change adaptation. A special report of working groups I and II of the Intergovernmental Panel on Climate Change (IPCC)*. New York: Cambridge University Press.

Murty, T. S. (1984). Storm surges-meteorological ocean tides. *Canadian Bulletin of Fisheries and Aquatic Sciences, 212*, 876–897.

Oxfam. (2008). *Three months on from cyclone Sidr – 1.3 m Bengalis face monsoon rains in temporary shelter*. http://www.oxfam.org.uk/applications/blogs/pressoffice/2008/02/three_months_on_from_cyclone. Accessed 20 Oct 2016.

Paul, B. K. (2009). Why relatively fewer died? The case of Bangladesh's Cyclone Sidr. *Natural Hazards, 50*, 289–304.

Shamsuddoha, M., & Chowdhury, R. K. (2007). *Climate change impact and disaster vulnerabilities in the coastal areas of Bangladesh*. Dhaka: COAST Trust.

Shamsuddoha, M., Islam, M., Haque, M. A., Rahman, M. F., Roberts, E., Hasemann, A., & Roddick, S. (2013). *Local perspective on loss and damage in the context of extreme events: Insights from cyclone-affected communities in coastal Bangladesh*. Dhaka: Center for Participatory Research and Development (CRPD).

UN. (2007). *Cyclone Sidr: United Nations rapid initial assessment report with a focus on nine worst affected districts*. http://www.cdmp.org.bd/cdmp_old/publications/Cyclone_Sidr_UN_Rapid_Initial_Assessment_Report.pdf. Accessed 22 Jan 2013.

# Chapter 25
# Closing Loopholes with the Small-Scale Fisheries Guidelines: Addressing Illegal, Unreported and Unregulated Fishing in Lake Victoria, Tanzania

**Joseph Luomba, Paul Onyango, and Ratana Chuenpagdee**

**Abstract** Illegal, unreported, and unregulated (IUU) fishing poses a serious challenge to fisheries governance as it does not go away easily despite various efforts to combat it. The persistence of IUU fishing is a threat to securing sustainable small-scale fisheries, as it affects ecosystem health, food security, and viable livelihoods, and can worsen poverty conditions in the fishing communities. Thus, unless addressed, this problem may impede the implementation of the Voluntary Guidelines for Securing Sustainable Small-Scale Fisheries (SSF Guidelines). Knowing why IUU fishing occurs and persists is the first step to addressing the problem. Following interactive governance theory, this chapter draws on studies conducted to identify characteristics of the fisheries systems that give rise to IUU fishing in small-scale fisheries in Ijinga Island, Lake Victoria. The key findings indicate that the social system of Ijinga Island is highly diverse, complex, and dynamic, and is thus very difficult to govern. The governing system is also not as capable as it must be to combat IUU fishing due to the lack of interactions between responsible authorities at different levels, governing actors, and small-scale fishers. Studies also show that small-scale fishers of Ijinga Island differ significantly in their perception about IUU fishing problems when compared to government officials. This disparity contributes to making IUU fishing a wicked problem. Based on these findings, we conclude with some recommendations about how SSF Guidelines can be used to address the IUU fishing problems identified in the study, and thus fisheries governance challenges.

J. Luomba (✉)
Tanzania Fisheries Research Institute (TAFIRI), P. O. Box 475, Mwanza, Tanzania
e-mail: luomba@yahoo.com; josephluomba@tafiri.go.tz

P. Onyango
Department of Aquatic Sciences and Fisheries Technology, University of Dar-es Salaam, P. O. Box 35064, Dar-es Salaam, Tanzania
e-mail: onyango_paul@udsm.ac.tz

R. Chuenpagdee
Department of Geography, Memorial University of Newfoundland, St. John's, Newfoundland and Labrador, Canada
e-mail: ratanac@mun.ca



S. Jentoft et al. (eds.), *The Small-Scale Fisheries Guidelines*, MARE Publication Series 14, DOI 10.1007/978-3-319-55074-9_25

**Keywords** SSF Guidelines • IUU Fishing • Interactive Governance • Governability Assessment • Small-Scale Fisheries • Lake Victoria

## Introduction

Illegal, unreported, and unregulated (IUU) fishing is a worldwide problem, estimated to contribute to economic losses of between USD $10 and $23 billion annually (Agnew et al. 2009; EJF 2012). Concerns about IUU fishing are not related only to economics, but also to impacts on food security and the livelihoods of fisheries-dependent populations (FAO 2016), and on the health of fish stocks and the aquatic environment (MRAG 2005). While the problem is acute in high seas fisheries, IUU fishing also exists in nearshore and inland fishing areas where small-scale fisheries operate. Tackling IUU fishing in small-scale fisheries requires sound policy intervention because of the significant number of the people whose livelihoods depend on these fisheries (Andrew et al. 2007; Salas et al. 2007) Combatting IUU fishing in small-scale fisheries is necessary, but the interventions may need to be different, or have different foci from those applied to addressing IUU fishing in industrial fisheries.

Deterring IUU fishing has become a high priority agenda for policy makers, fisheries authorities, fishers, and other stakeholders (Agnew et al. 2009; Plagányi et al. 2011; FAO 2015, 2016). Two instruments are noteworthy in this endeavor. The Code of Conduct for Responsible Fisheries contains principles and standards for responsible fishing practices, with a view to ensuring the effective conservation, management, and development of fisheries resources (FAO 1995). More specifically, the International Plan of Action to prevent, deter, and eliminate IUU (IPOA-IUU), adopted in 2001, recognizes the increasing concern of IUU fishing and calls on member states, as well as Regional Fisheries Bodies, to put in place national legislation and action plans to prevent, deter, and eliminate the problem (FAO 2001). In Lake Victoria, a Regional Plan of Action on IUU fishing (RPOA-IUU) was adopted by the riparian states of Kenya, Tanzania, and Uganda in 2004. The plan focuses on the restriction of fishing gears and fishing methods, the protection of fish spawning and nursery areas, the prohibition of fish landing in undesignated landing sites, seasonal closures, and joint licensing mechanisms (LVFO 2004). Despite the efforts that have been made to eliminate IUU activities, the problem still exists. As shown by Agnew et al. (2009), there is an insignificant change in the trend of IUU activities globally between 1980 and 2003.

Why is the IUU fishing problem so persistent? Jentoft and Chuenpagdee (2015) argue that small-scale fisheries are very diverse, complex, and dynamic, and the problems they present are beyond simple governance solutions. Moreover, and similar to other sectors, small-scale fishing communities do not always welcome measures that they perceive to be against their interests (Onyango and Jentoft 2007). These considerations suggest that the persistence of IUU fishing can either be a result of a mismatch between alleviation strategies and the context in which these strategies are implemented, or dissent on the part of the fishing communities. In other words, IUU fishing is a good example of a wicked problem in fisheries governance, described by Jentoft and Chuenpagdee (2009), following Rittel and Webber (1973). In accord with interactive governance theory (Kooiman et al. 2005), addressing fisheries problems require careful examination of the natural and social systems that are being governed, the governing system, and their interaction. They further elaborate that problems can occur at all three orders of governance: routine decision-making (1st order), institutional arrangements (2nd order), and at the meta-level where values, principles, and images underlying the design of these institutions are of main concern. In the context of IUU fishing in Lake Victoria, this means that factors contributing to IUU fishing can be numerous and can be related to various aspects of the fisheries systems. For instance, they may be due to the biophysical features of the lake, the nature of fishing practices, the social structure of fishing people, economic conditions, the lack of capacity in the governing system, or a combination of these factors. It also implies that IUU fishing problems could be due to the mismatch between the values, perceptions, and expectations of fishing people and those of the governments.

For wicked problems such as IUU fishing, alternative approaches to address them are worth considering, starting from the way the problem is articulated, the choice of what instruments to use, and how to apply them effectively. The fact that many international agreements and conventions available to help combat IUU fishing have not resulted in desirable outcomes is an indication that a new way to approach IUU fishing problem is necessary. A new opportunity for this may be emerging with the recently released Voluntary Guidelines for Securing Sustainable Small-Scale Fisheries, or SSF Guidelines (FAO 2015). While it is obvious that IUU fishing problems can become an impediment to the implementation of these Guidelines, what is not evident is the extent to which the principles underlying the Guidelines can be used to help combat IUU fishing.

This chapter contributes to the discussion about the IUU fishing problem by drawing on a study carried out in Ijinga Island, a small island in Lake Victoria. Informed by interactive governance theory, the study examines various aspects of small-scale fisheries in this part of the lake and tries to elicit judgments and opinions from fisheries stakeholders about IUU fishing problems. It begins by describing the general situation of IUU fishing in Lake Victoria, followed by a brief description of the methods used to collect the information presented. Next, the chapter reports on the characteristics of Lake Victoria's system, including the natural and social system-to-be-governed, the governing system, and the governing interactions in terms of diversity, complexity, dynamics, and scales, and discusses whether these characteristics contribute to making IUU fishing problem occur and persist. We also present the results of a survey which aimed to assess the similarities and differences between the judgements of fishing people and those of government officers with respect to the severity of selected fishing practices, including those both considered legal and illegal by law. Finally, based on these results, we discuss options or opportunities provided in the SSF Guidelines to address IUU fishing in Lake Victoria more effectively. We note that in the case of Lake Victoria, the analysis is confined largely to illegal fishing activities (I) and practices that do not conform to the fisheries laws and regulations, and which are usually unreported (U). The second U (unregulated) is not applicable since Lake Victoria fisheries are significantly regulated. Thus, the term IUU is used in the chapter as a composite term to refer to the issues related to illegal and unreported problems in Lake Victoria fisheries.

## IUU Fishing in Lake Victoria

According to RPOA-IUU, IUU fishing practices include the use of illegal fishing gears and methods, use of unlicensed boats, fishing of undersized fish, landing them in undesignated landing sites, trading fish in undesignated areas, and fishing during closed seasons or in closed areas (LVFO 2004). IUU fishing in Lake Victoria is evidenced by the presence of illegal fishing gears such as gillnets with mesh sizes under 6 in., longlines and hooks under size 4 and above size 10, beach seines, and monofilament nets (see Table 25.1). The time series data in Table 25.1 show that IUU fishing remains widespread and is considered one of the major causes of fish decline (Ikwaput-Nyeko et al. 2009; Mkumbo and Marshall 2014; Njiru et al. 2009). IUU fishing is also known to have negative impacts on the livelihoods of fisheries-dependent communities (Onyango 2004; Ogwang et al. 2009). Although it received much attention during the Nile perch era from 1990s onwards, IUU fishing had been observed in Lake Victoria as early as 1900s with the decline in cichlid species due to the use of undersized gill nets (Graham 1929).

Given the negative impacts of IUU fishing on aquatic resources in Lake Victoria, and its consequences on the livelihoods of fisheries dependent communities, various efforts have been put in place to tackle IUU fishing. For example, Monitoring Control and Surveillance (MCS) activities have been carried out regularly, the Marine Police have been brought on board, and amendments to the fisheries regulations have been undertaken to tighten the loose ends in what the law states with regards to IUU fishing. In addition, the adopted RPOA-IUU complies with the FAO requirements that states and government should formulate mechanisms and strategies to address IUU fishing.

Despite efforts to address IUU fishing, Kayanda et al. (2009) and Ikwaput-Nyeko et al. (2009) have observed that illegal landings have persisted in the Lake Victoria fisheries. Njiru et al. (2007) and Johnson (2014), among others, have pointed out that IUU fishing persists due to the ineffectiveness of current management measures. According to these studies, these regulations need to be re-evaluated. In other words, they are critical of the technical regulations as a means of controlling IUU fishing. As a consequence of persistent IUU fishing and the observed decline in fish catches, the government introduced a co-management regime in the lake in 1999 (Hoza and Mahatane 2001). The situation has continued from bad to worse, judging from the number of illegal fishing practices that have been reported (see also Table 25.1). For instance, LVFO (2016) reports that, between 2012 and 2014, the use of undersized gillnets (mesh size <5 in.) increased by 16.8%, beach seines by

**Table 25.1** Number of illegal fishing gears in Lake Victoria, Tanzania

| Illegal fishing gears/year | 2000 | 2002 | 2004 | 2006 | 2008 | 2010 | 2012 | 2014 |
|---|---|---|---|---|---|---|---|---|
| Gill-nets <6 in. mesh size | 200,258 | 410,209 | 563,253 | 395,883 | 770,837 | 168,139 | 141,364 | 137,342 |
| Long lines hooks <4 | n/d | n/d | n/d | n/d | 10,280 | 5935 | 24,956 | 4.37,935 |
| Long line hooks >10 | n/d | n/d | n/d | n/d | 4,137,774 | 4,160,618 | 4,547,208 | 5,830,627 |
| Beach seine | 1996 | 1522 | 1592 | 1665 | 1776 | 1301 | 2079 | 1956 |
| Monofilament nets | 63 | n/d | n/d | n/d | 4801 | 2905 | 7944 | 22,064 |

Source: URT (2015)
n/d refers to no data

30.3%, and monofilament by 28.5%. These data suggest that new approaches in understanding and confronting IUU fishing beyond the existing measures are required.

The search for appropriate mechanisms to address IUU fishing has not stopped. The SSF Guidelines have also provided avenues for alleviating IUU fishing. They call for member states to ensure that there are effective monitoring and enforcement mechanisms to prevent, deter, and eliminate all forms of illegal and/or destructive fishing practices, which have a negative effect on marine and inland ecosystems. In addition, the SSF Guidelines seek to address the root causes of IUU fishing, while providing members with the flexibility to adopt the most suitable measures to their particular circumstances (FAO 2015). This requirement recognizes that, although small-scale fisheries share similar characteristics, the problems may differ from one area to the next. Thus, the methods used to approach these issues should be based on the local context. For this reason, understanding IUU fishing in Lake Victoria through an interactive governance and governability lens aligns well with the premise of the SSF Guidelines. First, the focus of the SSF Guidelines to support responsible fisheries and sustainable social and economic development for the benefit of current and future generations requires a shift from problem solving alone to exploring opportunities, which is the basis for interactive governance (Kooiman et al. 2005). Additionally, the SSF Guidelines recognize that small-scale fisheries are a diverse and dynamic sector, which is often characterized by competing interests. In this realm, any governability assessment is not only focused on finding technical measures or adjusting the existing governing structure alone, but also looks into the underlying values, images, and perspectives of stakeholders that make the problem occur and persist. Therefore, by taking into account system properties and the quality and type of interactions between the systems, governability assessment enables a contextualized understanding of the IUU fishing problem, which is critical in designing appropriate strategies to tackle it.

## Methods

The study included two main parts. The first involved a descriptive analysis of the fishery system. This was done using secondary information, which were gathered and synthesized from several published and grey literature, including government reports. We also drew from the experience gained by the authors from working in the area, and field observations in assessing the characteristics of the system-to-be-governed and the governing system in terms of their diversity, complexity, dynamics, and scale. In addition, we examined the formal and informal institutions used by the governing system in governing the lake, and the interactions between the system-to-be-governed and the governing system. This examination aided in understanding underlying governance issues required for sustainable fisheries conservation, management and development.

The second part of the study aimed at examining stakeholders' perceptions about IUU fishing. A survey questionnaire was developed, following a damage schedules method (Chuenpagdee 1998), and administered to 150 respondents. The respondents were sampled from the primary resource user groups (e.g. boat owners, fishing crews, processors, and traders), fisheries managers and scientists, and local residents. The questionnaire sought to examine the respondents' understanding about the damaging fishing activities and their respective conservation ethics. Eight fishing activities were presented in the paired comparison survey, four of which are considered IUU fishing by fisheries regulation while the other four are not. For each pair, the respondents were asked to choose a fishing activity that s/he considered more damaging. The percentage responses from the respondents were then normalized and ranked in order of the severity, with 1 denoting the most damaging while 8 the least. The data collection was carried out between August and October 2015, at Ijinga Island in Lake Victoria Tanzania.

## Lake Victoria's Fisheries System

### *Natural System-to-Be-Governed*

Lake Victoria is a diverse natural system, shared by three East African countries – Tanzania, Uganda and Kenya. Of about 68,000 km$^2$ of the lake surface, Tanzania occupies 51%, Uganda 43%, and Kenya 6%. Lake Victoria contains small and large islands that are inhabited by a significant number of people (Witte and van Densen 1995; World Bank 2009). Coastal development, agriculture, and other industries establishment around the lake cause impacts on the lake's biodiversity. For instance, effluents from farming and industrial activities have increased nitrogen and phosphorus levels. Together with the increased human population, these activities have contributed to eutrophication (Sitoki et al. 2010). In addition, water temperature has been altered, thus impacting ecosystem function and fish productivity (Ogutu-Ohwayo 1990).

Lake Victoria's fisheries used to be multi-species, comprising over 500 endemic species, mainly haplochromines and tilapiines belonging to the cichlidae family (Graham 1929). The lake also played host to several catfishes, e.g. *Bagrus docmak, Clarias gariepinus, syndontis, Schilbe intermedius,* Lungfish *Protopterus aethiopicus*, and carps *Labeo victorianus* (Kudhongonia and Cordone 1974). In addition, the lake contained the cyprinid *Rastrineobola argentea*, locally known as Dagaa. Between 1950 and 1960, Nile perch *Lates niloticus* and Nile tilapia *Oreochromis niloticus* were introduced in the lake (Ligtvoet et al. 1995). The predatory nature of the Nile perch had caused some of these indigenous species to disappear. Currently, Nile perch, tilapia, and Dagaa dominate the fisheries. The increased fishing pressure on Nile perch has seen the resurgence of haplochromines, which were once thought extinct (Balirwa et al. 2003; Mkumbo and Marshall 2014). At the same time, some catfish species, *Synodontis* and *Claria gariepinus*, have increased in number in recent years (Kayanda et al. 2009). The species diversity demands a management system that is cognizant of the presence of multiple species and their interactions. This is, however, a major challenge, not only in Lake Victoria but worldwide, due to lack of scientific knowledge, information, and research capacity. The lake fisheries have therefore been managed as if it was a single species system. For example, the recommended fishing gears for the Nile perch fish has been universally applied across other species, such as *Synodontis victoriae*, whose sizes do not reach Nile perch. Fishers targeting other fish species smaller than Nile perch and tilapia are therefore conditioned to use small-sized nets that are considered illegal by fisheries law. These situations illustrate the difficulties facing fisheries management and the consequences for IUU fishing problems.

## *Social System-to-Be-Governed*

Lake Victoria fisheries encompass a variety of stakeholders such as boat owners, fishing crews, fish traders, processors, fish agents, gear makers, and repairers. Adding to this diversity, there is a high dependency between these groups, making for complex relationships. For example, boat owners depend on fish agents to get loans to buy fishing gears while fishing crews depend on boat owners for employment. Further, fish traders and processors depend on people around the lake for fish trade. The majority of fishers come from about ten communities of diverse cultures and traditions that have lived riparian to the lake for long as their history is known. The type of fisheries activities they engage in defines the livelihoods of the people from these communities. For example, the Luo community, who live on the eastern side of the lake, derive their name from fishing. The word *Luo* came from the word *Luwo,* which has two meanings – one is following and another fishing. The *Luo* are believed to have originated from Southern Sudan and arrived by following fish in the Nile southwards until they settled in Uganda, Kenya, and Tanzania. Another example is the Wakerewe, found in the biggest island on the lake in Tanzania, whose identity is defined by their involvement in fishing activities. Although they practice agriculture, fishing has been a major activity, from which their culture and traditions have derived meaning. Times and periods for other activities in these communities are determined by fishing seasons, the type of gears used, and the method of fishing adopted, all of which reflect their culture and traditions. For instance, gill nets or beach seines are used because these gears enable them to fish as a group and thus remain together. When fishers bring fish after a fishing season, they hand over their catches to the clan leader, who distributes the whole catch to all the families in the clan. No one would go without a fish. For this reason, illegalizing some of the gears such as gill nets has been interpreted as dis-embedding the culture of many communities, something fishers vehemently resent.

The open access regime and the commercialization of Nile perch in 1990s led to increased fishing capacity and efforts (Onyango 2004; Ikwaput-Nyeko et al. 2009). As a result, some fishers resorted to the use of destructive fishing gears and methods (Mkumbo and Marshall 2014). This has negatively impacted ecosystem health, for instance, through a reduction of Nile perch stocks (Kayanda et al. 2009; Mkumbo and Marshall 2014). The decline in Nile perch has brought in new dynamics in the fisheries; a new fish product – gall bladder of Nile perch – has created another group of actors who focus their fishing activities on this product. Over the years, they have positioned themselves into a formidable group, competing even with the fish processing establishments, which have traditionally controlled the fisheries. The Nile perch gall bladder, which has found a lucrative market in China, is currently more profitable than the fish itself. Therefore, fishers sell their fish to fish processing establishments without the gall bladder. Such arrangements have created a conflict between the gall bladder traders with the owners of the fish processing establishments. The diversity of user groups, their social relationships, and the dynamics described above have not been considered when addressing IUU fishing problems. From a governability perspective, the social system-to-be-governed poses a major challenge to fisheries governance.

## *The Governing System*

The transboundary nature of the lake has contributed to multiple levels of governance, consisting of various institutions, management mechanisms, and instruments that guide planning and management processes. A system of co-management, where the government, fishing communities, and other stakeholders share management responsibilities and authorities, has been implemented since the 1990s (Onyango and Jentoft 2007; Ogwang et al. 2009). This shift from the centralized system was deemed necessary, not only because of the transboundary nature of the lake, but also to deal with management challenges such as inadequate staff and funding (Hoza and Mahatane 2001). At the regional level, LVFO is the main governing institution of the East African Community, responsible for fisheries management. In Tanzania, fisheries management falls under the Fisheries Division, which in turn belongs to the Ministry of Agriculture, Livestock, and Fisheries Development. Under the decentralized system, the Fisheries Division works in collaboration with the Ministry of Regional Administration and Local Government. The technical measures put in place to manage the fisheries are mainly derived from research advice provided by the Tanzania Fisheries Research Institute and academic institutions.

The National Fisheries Sector Policy, the Fisheries Act No. 22 of 2003, and the Fisheries Regulation of 2009 are the main instruments used to manage the fisheries resources. Consisting essentially of regulations restricting the use of certain fishing gears and fishing in certain areas, these laws and policies are linked with the international instruments developed by FAO and other development partners. Restrictions are also in place to eliminate fishing without a license and control the sizes of the fishing gears to be used in the lake. Besides these institutions and measures, there are also the Beach Management Units (BMUs) and civil society organizations. The BMUs are mainly responsible for collecting daily fisheries data at the landing site, participating in the vetting of fishers, and patrolling areas to weed out fishing illegalities (URT 2005). These responsibilities demand that they collaborate with fisheries staff at the landing site on a daily basis. They are also required to participate in village meetings to ensure that fisheries issues are discussed. Fisheries officers and the police carry out control and law enforcement. However, MCS in the lake is very limited due mostly to financial constraints and the lack of human resources to patrol the diverse areas of the lake (Ikwaput-Nyeko et al. 2009). This has given rise to the excessive use of destructive fishing gears and methods (Njiru et al. 2007). These wide arrays of governing actors, at various levels, display complex governance challenges for the lake. However, the system is slow to respond to changes and governance boundaries are relatively well defined. Thus, as shown in Table 25.2, the dynamics and the scale issues associated with the governing system are considered low and medium, respectively.

## *Governing Interactions*

The co-management system in practice in Lake Victoria requires that the multiple actors involved in management share information and communicate in a much more integrated and coordinated way than currently is done. In the case of Lake Victoria, these interactions should start at the beach level, from the BMUs to the Ministry. It

**Table 25.2** Summary of the assessed system properties

| System properties | System-to-be-governed | | Governing system | Governing interactions |
|---|---|---|---|---|
| | Natural | Social | | |
| Diversity | Moderate | High | High | Medium-high at regional level, but low at local level |
| Complexity | High | High | High | |
| Dynamics | High | High | Low | |
| Scale | High | High | Medium | |

is through the co-management system that the two main ministries responsible for fisheries – the Ministry of Agriculture, Livestock and Fisheries Development, and Ministry of Regional Administration and Local Government – liaise with local stakeholders. In order to foster these interactions, various co-management committees, such as district co-management committees, have been proposed at district levels, with defined roles and responsibilities. However, these committees have largely remained on paper. As such, the BMUs have maintained their status as the key institution facilitating formal interactions between fishers and the governments. Unfortunately, although the BMUs are required to hold periodic committee and assembly meetings to deliberate on various fisheries management issues before forwarding them to other relevant fisheries authorities for actions, they seldom do so. At the same time, interactions between fisheries officers and the fishers are only through *ad hoc* meetings, which are usually called when the former feel like doing so, or when they want to undertake an exercise at the landing site. As a result, the BMUs are ineffective at carrying out their functions as defined in the BMU guidelines, and their conservation knowledge has not been adequately tapped. Interactions at the ministerial level are much better, however, contributing to more effective governance of the lake. Onyango and Jentoft (2010) reason that although the lake's management has been decentralized, the government still retains power in making decisions and the implementation of management measures. Nevertheless, other challenges such as inadequate staff, the lack of reliable funding, and insufficient capability on the part of the BMUs have all influenced the interactions in the lake. Together, they contribute to making the existing governing interactions for Lake Victoria not profoundly conducive to facilitate effective governance.

## *Fisheries Stakeholders' Perceptions on IUU Fishing*

In addition to the inherent characteristics of the fisheries system, fisheries stakeholders' perceptions and opinions about IUU fishing are other elements which affect governability. As shown in Table 25.3, a certain level of agreement is found among the fisheries stakeholders group on damaging fishing activities. For example, use of non-selective fishing gears and fishing in breeding areas, the two activities classified by law as illegal, are considered the most damaging activities. However, the perceptions of other fishing activities varies between stakeholders groups. The most striking difference was found between managers/scientists and the primary resource users groups with respect to perceptions of other IUU fishing activities (e.g. fishing without a license and landing fish in non-gazetted sites). While boat owners, fishing crew, and traders viewed these activities in accordance with the law (i.e. they were harmful to the lake ecosystem), managers and scientists considered them to be less severe than activities not considered IUU. This finding points to a mismatch between local perceptions and universal understandings of what constitutes IUU fishing. Such disparities help explain why IUU fishing occurs and persists in Lake Victoria.

**Table 25.3** Ranking of damaging fishing activities by stakeholders groups in the fisheries

| Fishing activity | Boat owner | Fishing crew | Proc./ trader | Local resident | Managers/ scientist |
|---|---|---|---|---|---|
| Using non-selective fishing gears[a] | 1 | 1 | 1 | 1 | 1 |
| Fishing in breeding areas[a] | 2 | 2 | 2 | 2 | 2 |
| Increased number of fishers and gears | 3 | 5 | 6 | 3 | 3 |
| Fishing without a license[a] | 4 | 3 | 3 | 4 | 7 |
| Fishing around breeding areas | 5 | 6 | 7 | 8 | 4 |
| Landing fish in non-gazetted sites[a] | 6 | 4 | 4 | 7 | 8 |
| Fishing for longer hours | 7 | 7 | 5 | 5 | 6 |
| Many fishers targeting single species | 8 | 8 | 8 | 6 | 5 |

The ranking levels denote: 1 = most damaging while 8 = least damaging
[a]denotes activities considered IUU according to the regulations

## Discussion

The SSF Guidelines emphasize that, in tackling IUU fishing, member states should promote the application of appropriate measures that are reflective of the local context. In light of this, the governability assessment framework provides a good basis for identifying the reasons for persistent IUU fishing and context-based information required for addressing the IUU problem in Lake Victoria. The study reveals that the natural system-to-be-governed is very diverse, creating major challenges for the governing system as it aims to promote fisheries sustainability in Lake Victoria. At the same time, the social system-to-be-governed is also highly dynamic, making it difficult to maintain a good representation of various stakeholders and to encourage their participation in decision making, and by extension their involvement in the implementation of the SSF Guidelines. The governing system appears to be concentrated at the top (regional and national levels) to the extent that the participation of other actors in governance is minimized. Therefore, the government has remained the main actor in the management mechanism established through the co-management regime. In this way, fishing communities are not granted an enabling environment or forum for discussing issues that are most important to them. This has the potential of lowering the governability of the lake because any action taken by the governing system to manage the fisheries (in this case tackling IUU fishing) may be seen as infringement on fishers' rights, which goes against the very premise of the SSF Guidelines. The lack of representation and participation of all stakeholders in decision-making and fisheries policy-making processes is also problematic, as noted in paragraph 5.17 of the SSF Guidelines.

In addition, the weak interaction between the governing system and the social system-to-be-governed challenges the realization of desired goals. All of these factors contribute to governance and governability challenges in tackling IUU fishing. For example, the fisheries regulatory framework does not reflect the diversity and dynamics in the natural system-to-be-governed, and it does not match the values and expectations of the social system-to-be-governed. In other words, it falls short of the vision of an holistic and integrative approach which is promoted in the SSF Guidelines (Principle 11). These rules and regulations seem to target the three dominant fish species in the lake, thus failing to consider the livelihoods of fishing communities. Given that some of the fish species, such as haplochromines, Bagrus and Protopterus, which provide food sources to the majority of fishing communities, cannot be caught by the specified fishing gears, fishers are left with no option but to use fishing gears and methods that are considered illegal or destructive by regulation. The continued implementation of these regulations undermines the fishing communities' social and cultural tenure rights, a situation which is in direct contradiction to the main focus of the SSF Guidelines, as stipulated in Part 1, section 3.1. In the context of Lake Victoria, special attention is due since fishers consider fish to be a God-given right, and thus nobody should be denied access to fishing (Onyango and Jentoft 2010). The lack of appreciation of the social and cultural values of the lake's fisheries goes against one of the guiding principles in the SSF Guidelines, which emphasizes 'respect of cultures' (Principle 2). These social and cultural values come in the form of existing organizations, traditional and local knowledge, as well as the practices of fishing communities. From this perspective, the persistent IUU fishing in Lake Victoria cannot be attributed only to a mismatch between regulatory measures and the natural system-to-be-governed (Table 25.2), but also a dissent on the part of the social system-to-be-governed against laws they consider contrary to their beliefs. In addition, the laws are not explicitly focused on defending fishers' livelihood concerns and their access and use right of fisheries resources to achieve adequate food.

Importantly, the study also shows poor congruence between what fishers and governors/managers perceive about the level of severity of different fishing activities. Other studies have also reported differences in images between the governing system and the system-to-be-governed, which may therein result in lowering governability (Kooiman et al. 2008; Jentoft et al. 2010; Song and Chuenpagdee 2014; Voyer et al. 2015). Song et al. (2013), in particular, argue that stakeholders' values, images, and principles must be made explicit, understood, and articulated into decision-making process. Further, Chuenpagdee and Jentoft (2013) reason that the clear definition and formulation of images, instruments and actions can increase governability. In the case of Lake Victoria, this implies that actions taken to tackle IUU fishing should be consistent with the way the problem is perceived or understood by fisheries stakeholders. When this is not the case, as illustrated in Table 25.3, policies and measures that may have adverse impacts on small-scale fisheries should be avoided, as envisioned in paragraph 5.20 of the SSF Guidelines.

Effectively tackling IUU fishing requires multi-dimensional approaches to deal with the complex and unique concerns of the social and natural systems-to-be-governed. This is in line with paragraph 5.13 of the SSF Guidelines, which emphasizes that states should address IUU fishing based on the measures most suitable to their particular environment. In light of these, and as previously alluded to, we suggest that tackling IUU fishing should be done through interventions in all three orders of governance. At the first order, improving governing interactions and communication between actors is imperative. The fisheries division and the BMUs can work collaboratively in a platform provided by the existing co-management set-up to discuss emanating issues at the lower level of governance, which must be done in a more consultative way than the manner that is currently done. However, these consultations should be continous and not only be done at the whim of fisheries staff. More awareness on the part of the fisheries communities about some of the regulatory measures employed in tackling IUU fishing should result from these consultations. The SSF Guidelines (paragraph 5.18) emphasize that the participation of different actors in resource management is critical to providing guidance on sustainable fishing practices and fisheries management. Moreover, the governability challenges posed by the rules and regulations is a second-order matter. Here, there is a need to review fisheries regulatory measures with a view to aligning them to local people's mindsets and the diversity in the natural system-to-be-governed. Fishers should be allowed to target fish species other than the dominant ones, as long as this effort does not jeorpadize the health of the lake fisheries system and undermine conservation efforts. This approach is envisioned in Section 5a of the SSF Guidelines, which emphasizes that the fishing communities' cultures and traditions need to be respected. Additionally, Section 5b of the SSF Guidelines call for recognition of local conservation knowledge in formulating appropriate measures for responsible and sustainable fisheries. Finally, incorporate stakeholders' values and judgements in problem definition and goal setting is a task of third-order governance. This may lead to having shared and acceptable management measures which reach across stakeholder groups, which may then improve the legitimacy of the regulations. In addition, this may inspire and guide governing institutions on how to properly address IUU fishing problems. Some of these governing interventions could include awareness building and the use of alternative data for decision-making, which aligns with what the SSF Guidelines promote in the context of IUU fishing (Section 11.5).

## Conclusions

This chapter illustrates that the persistent IUU fishing in Lake Victoria is largely related to both the inherent and constructed characteristics of the governance system, especially those within the system-to-be-governed and the governing system. Efforts to eliminate IUU fishing need to begin with an appreciation that regulations and measures which are applicable to industrial fisheries may not be suitable to small-scale fisheries, and those which are appropriate to certain small-scale fisheries may not be pertinent in all cases. By examining what the issues are on the ground, in what contexts they occur, what the system's characteristics are, and how various stakeholders perceive them, proper institutions can be designed with suitable incentives put in place to guide behaviour.

The SSF Guidelines provide several principles and provisions that can be drawn upon to help address IUU fishing. IUU fishing, along with other damaging fishing activities, have negative consequences on fisheries ecosystems as well as socio-cultural/political systems. Long-term conservation and the sustainable use of fisheries resources (Section 5.13) need to be considered. In the case of Lake Victoria, this provision can be implemented through incorporating stakeholders' understandings about the levels of damages of different fishing activities (local conservation knowledge) to inform the design of regulatory measures. In eliminating IUU fishing, the policies adopted must ensure that other fisheries governance goals, such as community well-being, food security, and poverty alleviation, are not compromised. Further, in improving interactions between the governing actors and resource users, the existing co-management system in the lake offers a good starting point. The various stakeholders in the lake should be involved in fisheries decision-making and implementation, as stipulated in Section 5.15 of the SSF Guidelines and further emphasized in Section 5.17. These imply the need for government to recognize the fishing communities as important stakeholders in decision-making and implementation. In addition, it creates an environment for developing policies that may succeed because of potential support from stakeholders. All of these considerations constitute important issues and opportunities required to effectively improve the governability of the IUU problem in Lake Victoria.

**Acknowledgements** This chapter has been produced with the contribution of institutions and people worth mentioning. First, we thank the fishing communities in Ijinga island, Magu district for their willingness to provide information used here. Second, we also thank the fisheries staff at the landing site and district level for their logistical support. Lastly, we thank the Too Big To Ignore (TBTI), a global partnership for small scale fisheries research, and Social Sciences and Humanities Research Council of Canada (SSHRC) for financial support (Grant # 895-2011-1011) to conduct the study.

## References

Agnew, D. J., Pearce, J., Pramod, G., Peatman, T., Watson, R., Beddington, J. R., & Pitcher, T. J. (2009). Estimating the worldwide extent of illegal fishing. *PloS One*. doi:10.1371/journal.pone.0004570.

Andrew, N. L., Bene, C., Hall, S. J., Allison, E. H., Heck, S., & Ratner, B. D. (2007). Diagnosis and management of small-scale fisheries in developing countries. *Fish and Fisheries, 8*(3), 227–240. doi:10.1111/j.1467-2679.2007.00252.

Balirwa, J. S., Chapman, C. A., Chapman, L. J., Cowx, I. G., Geheb, K., Kaufman, L., & Witte, F. (2003). Biodiversity and fishery sustainability in the Lake Victoria basin: An unexpected marriage? *Bioscience, 53*(8), 703–715. doi:10.1641/0006-3568(2003)053-0703.

Chuenpagdee, R. (1998). *Scales of relative importance and damage schedules: A non-monetary valuation approach for natural resource management*. Doctoral dissertation, University of British Columbia, Vancouver, B.C., Canada.

Chuenpagdee, R., & Jentoft, S. (2013). Assessing governability–What's next. In M. R. Bavinck, R. Chuenpagdee, S. Jentoft, & J. Kooiman (Eds.), *Governability of fisheries and aquaculture: Theory and applications* (pp. 335–349). Dordrecht: Springer.

EJF. (2012). Pirate fishing exposed: The fight against illegal fishing in West Africa and the EU. A report of the environmental justice foundation. London. ISBN No. 978-1-904523-28-4.

FAO. (1995). *Code of conduct for responsible fisheries*. Rome: FAO.

FAO. (2001). *International plan of action to prevent, deter and eliminate illegal, unreported and unregulated fishing*. Rome: FAO.

FAO. (2015). *Voluntary guidelines for securing sustainable small-scale fisheries in the context of food security and poverty eradication*. Rome: FAO.

FAO. (2016). *The state of world fisheries and aquaculture 2015*. Rome: FAO.

Graham, M. (1929). *The Victoria Nyanza and its fisheries – A report on the fishing surveys of Lake Victoria (1927–1928)*. London: Crown Agents Colonies.

Hoza, R.B. & Mahatane, A.T. (2001). Establishment of collaborative fisheries management in the Tanzania part of Lake Victoria. In S. G. M. Ndaro & M. Kishimba (Eds.), *Proceedings of the LVEMP – Tanzania 2001 scientific conference*, 6–10 August 2001 (pp. 15–36). Mwanza: Tanzania Lake Victoria environmental management project Dar es Salaam.

Ikwaput-Nyeko, J., Kirema-Mukasa, C., Odende, T., & Mahatane, A. (2009). Management of fishing capacity in the Nile Perch fishery of Lake Victoria. *African Journal of Tropical Hydrobiology and Fisheries, 12*, 67–73.

Jentoft, S., & Chuenpagdee, R. (2009). Fisheries and coastal governance as a wicked problem. *Marine Policy, 33*, 553–560. doi.org/10.1016/j.marpol.2008.12.002.

Jentoft, S., & Chuenpagdee, R. (2015). *Interactive governance for small-scale fisheries: Global reflections*. Dordrecht: Springer.

Jentoft, S., Chuenpagdee, R., Bundy, A., & Mahon, R. (2010). Pyramids and roses: Alternative images for the governance of fisheries systems. *Marine Policy, 34*(6), 1315–1321. doi:10.1016/j.marpol.2010.06.004.

Johnson, J. L. (2014). *Fish work in Uganda: A multispecies ethno-history about fish, people, and ideas about fish and people*. Doctoral dissertation, University of Michigan, Michigan, USA.

Kayanda, R., Taabu, A., Tumwebaze, R., Muhoozi, L., Jembe, T., Mlaponi, E., & Nzungi, P. (2009). Status of the major commercial fish stocks and proposed species-specific management plans for Lake Victoria. *African Journal of Tropical Hydrobiology and Fisheries, 12*, 15–21.

Kooiman, J., Bavinck, M., Jentoft, S., & Pullin, R. S. V. (2005). *Fish for life: Interactive governance for fisheries* (3rd ed.). Amsterdam: Amsterdam University Press.

Kooiman, J., Bavinck, M., Chuenpagdee, R., Mahon, R., & Pullin, R. (2008). Interactive governance and governability: An introduction. *The Journal of Transdisciplinary Environmental Studies,* 7(1), 1–11. doi:http://hdl.handle.net/11245/1.293273.

Kudhongonia, A. W., & Cordone, A. J. (1974). Batho-spatial distribution pattern and biomass estimate of the major demersal fishes in Lake Victoria. *African Journal of Tropical Hydrobiology and Fisheries, 3*, 15–31.

Ligtvoet, W., Mouse, P. J., Mkumbo, O. C., Budeba, Y. L., Goudswaard, P. C., Katunzi, E. F. B., Temu, M. M., Wanink, J. H., & Witte, F. (1995). The Lake Victoria fish stocks and fisheries. In F. Witte & W. L. T. van Densen (Eds.), *Fish stocks and fisheries of Lake Victoria: A handbook for field observations* (pp. 11–53). Cardigan: Samara Publishing Limited.

LVFO. (2004). *Regional plan of action to prevent, deter, and eliminate illegal, unreported, and unregulated IUU fishing on Lake Victoria and its basin*. Jinja: Lake Victoria Fisheries Organization.

LVFO. (2016). *Regional Lake Victoria frame survey report 2015*. Jinja: Lake Victoria fisheries organization.

Mkumbo, O. C., & Marshall, B. E. (2014). The Nile perch fishery of Lake Victoria: Current status and management challenges. *Fisheries Management and Ecology, 22*(1), 56–63. doi:10.1111/fme.12084.

MRAG. (2005). *IUU fishing on the high seas: Impacts on ecosystems and future science needs*. London: MRAG Report.

Njiru, M., Nzungi, P., Getabu, A., Wakwabi, E., Othina, A., Jembe, T., & Wekesa, S. (2007). Are fisheries management measures in Lake Victoria successful? The case of Nile perch and Nile tilapia fishery. *African Journal of Ecology, 45*(3), 315–323.

Njiru, M., Getabu, A., Taabu, A. M., Mlaponi, E., Muhoozi, L., & Mkumbo, O. C. (2009). Managing Nile perch using slot size: Is it possible? *African Journal of Tropical Hydrobiology and Fisheries, 12*, 9–14.

Ogutu-Ohwayo, R. (1990). The decline of the native fishes of Lakes Victoria and Kyoga (East Africa) and the impact of introduced species, especially the Nile perch, Lates-niloticus, and the Nile tilapia, (Oreochromis-niloticus). *Environmental Biology of Fishes, 27*, 81–96. doi:10.1007/BF00001938.

Ogwang, V. O., Nyeko, J. I., & Mbilinyi, R. (2009). Implementing co-management of Lake Victoria's fisheries. *African Journal of Tropical Hydrobiology and Fisheries, 12*, 52–58.

Onyango, P. O. (2004). Refroming fisheries management: A case study of co-management in the Lake Victoria Tanzania. Master's thesis, University of Tromso, Tromso, Norway.

Onyango P. O. & Jentoft, S. (2007). Embedding co-management: Community-based fisheries regimes in Lake Victoria, Tanzania. In WorldFish Center (Ed.), *International conference on community based approaches to fisheries management, Dhaka* (Bangladesh), 6–7 March 2007 (pp. 1–21). Penang: WorldFish Center.

Onyango, P. O., & Jentoft, S. (2010). Assessing poverty in small-scale fisheries in Lake Victoria, Tanzania. *Fish and Fisheries, 11*(3), 250–263. doi:10.1111/j.1467-2979.2010. 00378.x.

Plagányi, É., Butterworth, D., & Burgener, M. (2011). Illegal and unreported fishing on abalone – Quantifying the extent using a fully integrated assessment model. *Fisheries Research, 107*(1), 221–232. doi:10.1016/j.fishres.2010.11.005.

Rittel, H. W. J., & Webber, M. M. (1973). Dilemmas in a general theory of planning. *Policy Sciences, 4*, 155–169.

Salas, S., Chuenpagdee, R., Seijo, J. C., & Charles, A. (2007). Challenges in the assessment and management of small-scale fisheries in Latin America and Caribbean (2007). *Fisheries Research, 87*(1), 5–16. doi:10.1016/j.fishres.2007.06.015.

Sitoki, L., Gichuki, J., Ezekiel, C., Wanda, F., Mkumbo, O. C., & Marshall, B. E. (2010). The environment of Lake Victoria (East Africa): Current status and historical changes. *International Review of Hydrobiology, 95*(3), 209–223. doi:10.1002/iroh.201011226.

Song, A. M., & Chuenpagdee, R. (2014). Exploring stakeholders' images of coastal fisheries: A case study from South Korea. *Ocean and Coastal Management, 100*, 10–19. doi:10.1016/j.ocecoaman.2014.07.002.

Song, A. M., Chuenpagdee, R., & Jentoft, S. (2013). Values, images, and principles: What they represent and how they may improve fisheries governance. *Marine Policy, 40*, 167–175. doi:10.1016/j.marpol.2013.01.018.

URT. (2005). *National guidelines for beach management units*. Dare-s Salaam: Ministry of Natural Resources and Tourism.

URT. (2015). *Lake Victoria fisheries frame survey results* (2014). Dare-s Salaam: Ministry of Livestock and Fisheries Development.

Voyer, M., Gollan, N., Barclay, K., & Gladstone, W. (2015). 'It's part of me': Understanding the values, images, and principles of coastal users and their influence on the social acceptability of MPAs. *Marine Policy, 52*, 93–102. doi:10.1016/j.marpol.2014.10.027.

Witte, F., & van Densen, W. L. T. (1995). *Fish stocks and fisheries of Lake Victoria: A handbook for field observations*. Cardigan: Samara Publishing Limited.

World Bank (2009). LVEMP II project appraisal document. Report.

# Chapter 26
# Illegal, Unreported, and Unregulated Fisheries in the Hormuz Strait of Iran: How the Small-Scale Fisheries Guidelines Can Help

**Moslem Daliri, Svein Jentoft, and Ehsan Kamrani**

**Abstract** According to a United Nations report, illegal, unreported, and unregulated (IUU) fishing is the main hindrance for sustainable fisheries in the open seas and inland waters. In this chapter, we explore how the Voluntary Guidelines for Securing Sustainable Small-Scale Fisheries (SSF Guidelines) can help address this problem and improve the sustainability of small-scale fisheries and local community livelihoods in Hormozgan Province of Iran and the Persian Gulf. At present, there is limited management planning for small-scale fisheries in Iranian waters, and information about IUU fishing is scarce. Nonetheless, IUU fishing is on the political agenda. In Daliri et al. (Ocean & Coastal Management 120:127–134, 2016), we explored what factors cause IUU fishing in the region, and highlighted culture, management, economic conditions, personal skills, and area features as important determinants of IUU fishing. We also concluded that co-management can help address this problem and promote more sustainable fisheries in this region, if implemented well. In this chapter, we argue that efforts to reduce IUU fishing must include multiple measures identified throughout the SSF Guidelines, measures that go beyond the paragraph that talks specifically about monitoring, control, and surveillance (MCS).

**Keywords** IUU fishing • Small-Scale fishing • Co-management • Hormuz Strait • Iran • SSF Guidelines

M. Daliri (✉) • E. Kamrani
Fisheries Department, Faculty of Marine and Atmospheric Sciences, Hormozgan University, Bandar Abbas, Iran
e-mail: moslem.daliri@yahoo.com; Ehsan.kamrani@hormozgan.ac.ir

S. Jentoft
Norwegian College of Fishery Science, UiT – The Arctic University of Norway, Tromsø, Norway
e-mail: svein.jentoft@uit.no


S. Jentoft et al. (eds.), *The Small-Scale Fisheries Guidelines*, MARE Publication Series 14, DOI 10.1007/978-3-319-55074-9_26

## Introduction

In 2014, the Committee on Fisheries (COFI) of the FAO adopted the Voluntary Guidelines on Securing Sustainable Small-Scale Fisheries (SSF Guidelines) for eradicating poverty and promoting food security in small-scale fishing communities globally. The SSF Guidelines contain not just a set of practical recommendations on a broad range of issues but also a set of normative meta-governance principles for small-scale fisheries, i.e. principles that would constitute good governance for this sector. The Guidelines stress the important contribution of small-scale fisheries in addressing societal concerns such as safe and nutritious food and employment (Jentoft 2014) and strongly advocate a human rights approach to fisheries management. They also complement the Code of Conduct for Responsible Fisheries, which, alongside the fishing provisions of the UN Convention on the Law of the Sea, is the most widely recognized and implemented international fisheries management instrument. Moreover, the SSF Guidelines are also closely linked to the Voluntary Guidelines on the Responsible Governance of Tenure of Land, Fisheries, and Forestry in the Context of National Food Security, the Voluntary Guidelines to Support the Progressive Realization of the Right to Adequate Food in the Context of National Food Security, and the Principles for Responsible Investment in Agriculture and Food Systems (Kurien 2015).

According to the United Nations report, illegal, unreported, and unregulated (IUU) fishing is the main hindrance for sustainable fisheries in the open seas and inland waters. IUU fishing has increased the concerns of fisheries administrators, traders, consumers, and fishers alike (Sodik 2007; Luomba et al. 2016). The international plan of action to prevent, deter, and eliminate this problem defines IUU fishing in an elaborate way (Baird 2006): *Illegal fishing* includes all fishery activities without license in coastal and inland waters or the area under the inspection of Regional Fisheries Management Organizations (RFMO), contravention of applicable laws and regulations noted in fishing licenses, invasion by vessels or harvester operators from neighboring countries, and fishing outside of designated areas or specified time periods. Other types of illegal fishing include the utilization of destructive fishing methods like fishing with explosive and poisonous materials and destructive gears like nets with small mesh size. *Unreported fishing* is fishing that has not been reported or has been misreported to the relevant national authority or RFMO, in contravention of applicable laws and regulations. *Unregulated fishing* generally refers to fishing activities by vessels without nationality or vessels flying the flag of a country not party to RFMO governance. It also refers to harvesting fish in unregulated areas. This is of importance in particular for straddling and highly migratory fish stocks such as tuna and tuna like fishes.

For developing countries such as the Islamic Republic of Iran, where small-scale fisher communities struggle with poverty, the implementation of the SSF Guidelines could be a solution to combat IUU fishing and thus achieve sustainability. There are two paragraphs in the SSF Guidelines (5.16 and 11.5) that especially relate to IUU fishing. The SSF Guidelines argue that states should ensure the establishment of monitoring, control, and surveillance (MCS) systems or promote the application of existing monitoring systems that are applicable for small-scale fisheries. States should also ensure that the information necessary for responsible small-scale fisheries and sustainable development is available, including information on IUU fishing.

As developing countries, Iran included, have poor monitoring capacities, small-scale fisheries are most at risk from IUU fishing (Erceg 2006). In fact, IUU fishing seems to be more common in the small-scale sector than in the industrial sector (Daliri et al. 2016). There is also limited management planning with regard to small-scale fisheries in Iranian waters, which the SSF Guidelines may help to change.

This chapter discusses how the implementation of the SSF Guidelines could contribute to addressing IUU fishing and consequently support more sustainable small-scale fisher communities in the Hormozgan Province of Iran. This province borders the northern part of the Persian Gulf and the Oman Sea, and is the major fisheries region of the country. In Daliri et al. (2016), we explored what factors cause IUU fishing in the region, and highlighted culture, management, economic conditions, personal skills, and area features as important determinants of IUU fishing. We also concluded that co-management can help address this problem and promote more sustainable fisheries in the region. This chapter draws on arguments made in that published article in the context of the SSF Guidelines. Additional interviews with people in fisheries administration and government were carried out. We posit that combatting IUU fishing requires a wide range of remedies beyond monitoring, surveillance, and control (MCS).

## IUU Fishing as a Global Concern

Most international fisheries regulations aim to curb IUU fishing. These regulations include the UN Convention on the Law of the Sea, the FAO Code of Conduct for Responsible Fisheries, the FAO UN Fish Stocks Agreement, the FAO Agreement to Promote Compliance with International Conservation and Management Measures by Fishing Vessels on the High Seas, and the International Plan of Action to Prevent, Deter, and Eliminate Illegal, Unreported, and Unregulated Fishing (IPOA-IUU).

Much academic literature has also been devoted to IUU fishing. For example, Pitcher et al. (2012) reviewed the implementation of the Code of Conduct for Responsible Fisheries by FAO member countries, and ranked countries in terms of their performance in managing and conserving resources in their EEZs. The authors argued that for most developing countries, finding and accessing relevant information about illegal fishing is highly problematic and challenging. Combatting IUU fishing would therefore have to start by actually detecting what is going on and how big the problem really is in specific fisheries. If IUU fishing is happening at a global scale, it does not necessarily mean that it occurs locally. If it is a problem that is noticeable, serious, and urgent, it is important to ask why it occurs, why fishers break rules, and how they justify it. Equally importantly, why do other fishers comply with the rules. Fishers may have good reasons to both break rules and comply with them (Stewart and Walshe 2008; Jagers et al. 2012). They may for instance have economic reasons to fish illegally, especially if the risk of arrest and/or penalties is low. Jensen and Vestergaard (2002), for example, highlighted that incentive schemes, such as taxes or subsidies, are potentially helpful in stopping illegal landings and discards. Cho (2012) observed that in South Korea the government, through a buy-back of vessels program, succeeded in eliminating illegal bottom trawl fishing in the coastal waters. What governments define as illegal, however, fishers may not necessarily consider as immoral, especially if rules exclude them from their traditional fishing grounds or take away their food security and income needs in other ways. In a study of IUU fishing in Lake Victoria, Luomba et al. (2016) found that scientists and managers do not share with fishers the same idea of what IUU fishing is and what consequences it has for fisheries resources. Fishers felt that legal fishing could in fact be more environmentally damaging than illegal fishing. Thus, if fishers do not believe in the rationality and legitimacy of certain regulations, they are not likely to abide to them, especially if they can get away with not doing so.

Like in other parts of the world, IUU fishing is also a major concern in the Persian Gulf. Al-Abdulrazzak and Pauly (2013) noted that catch data for intertidal fixed stake net traps are often unreliable and that in actual fact it is six times higher than the officially reported catch. This is in agreement with the findings of Daliri (2016) who reported that the annual catch for stake net traps is ten times higher in Iran than the officially reported catch.

Daliri et al. (2016) explored IUU small-scale fishing in Hormozgan Province by employing an integrated natural and social system perspective and developed a model with five categories, namely culture, management issues, economic conditions, personal skills, and area features. Daliri (2016) also estimated the annual IUU shrimp catch in Hormozgan waters to be between 461 and 520 tons, a total value of between 2.6 and 3 U.S. million dollars approximately. He also claimed that between 266 and 304 tons of Silver Pomfret (*Pampus argenteus*) is caught illegally in Hormozgan waters. According to a report of the Iran Fisheries Organization (IFO), known as Shilat in Persian, 3000 unlicensed outboard powered small boats are fishing in Hormozgan coastal waters (IFO 2014). IFO, which is a government organization under the Ministry of Agriculture-Jahad, was established to implement the Law of Protection and Exploitation of the Fisheries Resources of the Islamic Republic of Iran. In other words, the mandate of IFO includes policy-making, planning, and surveillance to promote sustainable exploitation of marine resources, the conservation and restocking of aquatic populations, fisheries and aquaculture management, and the maintenance of infrastructure, amongst other things.

Despite such reviews, information about IUU fishing in the Persian Gulf is very scarce. This is something that the SSF Guidelines encourages states to rectify. In Paragraph 11.5 it is stated that: "States should ensure that the information necessary for responsible small-scale fisheries and sustainable development is available, including on illegal, unreported, and unregulated (IUU) fishing." There is reason to believe that IUU is much more widespread than such studies have hitherto revealed. As there is limited management of small-scale fisheries in Iranian waters (Pitcher et al. 2012), IUU fishing in the small-scale sector is probably more common than in the large-scale sector, which is under more close surveillance.

Thus, introducing mechanisms to help prevent and reduce IUU fishing would make sense. The SSF Guidelines, therefore, call upon states to ensure that effective monitoring, control, and surveillance (MCS) are in place to "deter, prevent and eliminate" all forms of IUU fishing. Although such mechanisms are needed, we posit in this chapter that MCS will not be sufficient if it does not address the basic reasons why fishers do not comply. We believe it is important to recognize that small-scale fisheries are both victims and culprits of IUU fishing, and that the goals of combatting IUU fishing do not always justify the means. Small-scale fishers are sometimes exposed to poorly designed fishing regulations, which do not make much sense to them. We hold that the problem is complex, and that the answer does not only lie in those paragraphs that talk specifically about IUU fishing but also elsewhere in the SSF Guidelines.

## Small-Scale Fisheries in Iran

The Persian Gulf is within the mandate of the Regional Commission for Fisheries in the Persian Gulf and the Oman Sea (RECOFI). The aim of this commission is to promote the development, conservation, rational management, and best utilization of living marine resources, and promote sustainable development of aquaculture in the member countries (Iran, Bahrain, Iraq, Kuwait, Qatar, Oman, Saudi Arabia, and United Arab Emirates). The RECOFI continually reviews the state of these resources, including the level of fisheries exploitation. It also formulates and recommends appropriate measures for the conservation and rational management of living marine resources, and ensures the implementation of these recommendations, keeping in mind the economic and social aspects of the fishing industry and recommending any measures aimed at its development.

The region as a whole showed a steady increase in total catch from 350,000 tons in 1986 to about 700,000 tons in 2006. However, there has been a small decline in total catch over the last few years in the Persian Gulf area. The available catch data also indicate that most fish stocks are fully exploited in this region (FAO 2011). Lack of knowledge about the contribution of small-scale fisheries, which are frequently neglected and hence under-reported in terms of catch, and resultant illegal catches and discards (Hosseini et al. 2015; Daliri 2016), can lead to stock depletion that jeopardize food security, impair resource conservation, and undermine the livelihoods of small-scale fishing communities in the region.

Iran has the largest fishery in the RECOFI area. Fishing in Iranian waters of the Persian Gulf is both small-scale (using motorized dhows and sambuks, small wooden or fiberglass vessels, intertidal fixed stake net traps etc.) and large-scale (industrial-style trawlers). There are 51 industrial vessels and 10,574 small-scale vessels (3151 wooden vessels and 7423 outboard powered small boats) active in Iran's commercial fishing sector. They are mostly present in Iran's southern waters

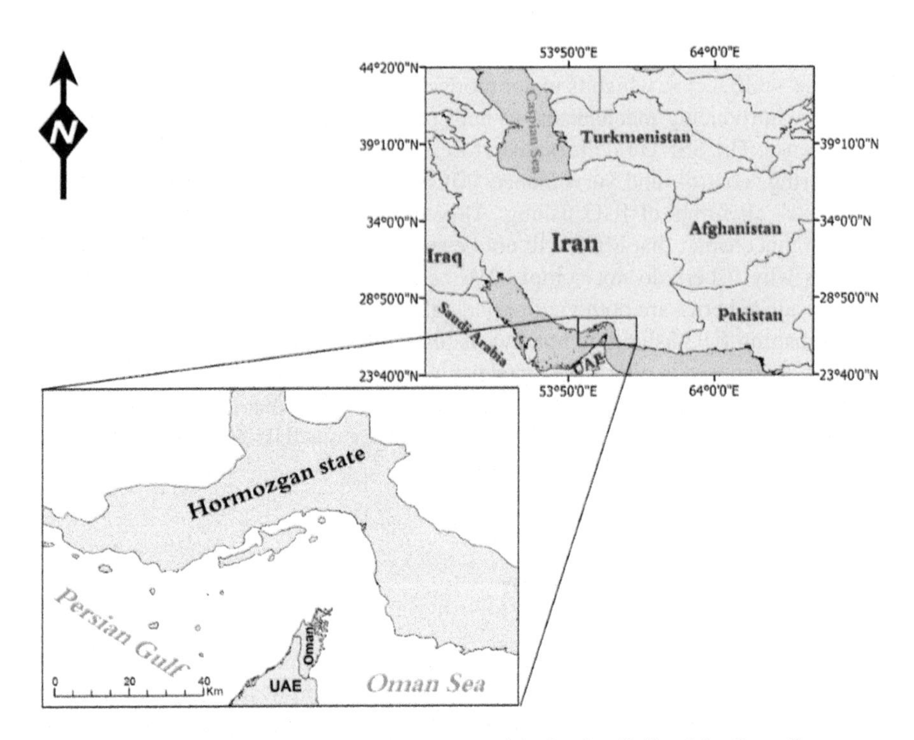


**Fig. 26.1** Location of Hormozgan in Iranian waters of the Persian Gulf and the Oman Sea

and open seas (IFO 2014). Iran has legislated fisheries management measures in the Persian Gulf, which include restrictions on net mesh size for commercial vessels, banning bottom trawl fishing (apart from shrimp trawling), designating closed areas (which are often marine protected areas), and defining closed fishing seasons for various fishing methods. Vessel monitoring systems (VMSs) are mandatory and used on all Iranian industrial-style trawlers.

Hormozgan Province has 14 islands and 1000 km (620 miles) of coastline (Fig. 26.1). Between 2003 and 2013, Hormozgan Province accounted for 60% of total landings in Iranian waters of the Persian Gulf and was home to the biggest fishery of the region. In 2012, total landings were 167,000 tons. Only 8.5% of the landings came from industrial vessels, which suggest that fishing in the area is predominantly small-scale and semi-industrial (trawlers) (Figs. 26.2 and 26.3). There are now approximately 22,500 and 1290 fishers employed respectively in the small-scale and large scale fishing sectors in Hormozgan (IFO 2014).

Small-scale fisheries communities in Hormozgan are generally poor, and fishing is considered as a low class profession. These communities have low living standards, for instance with regard to medical and welfare facilities, educational centers, and high schools. They do not get adequate consideration from academics and decision-makers for reasons because they are far from urban communities and lack visibility. Thus, relatively little is known about them. However, it is clear that small-scale fishers use different fishing gears such as drift or fixed gillnets, bottom trawls

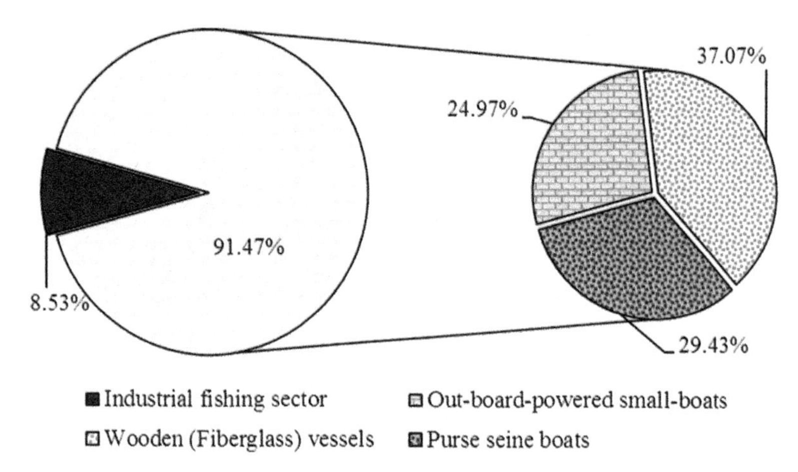


**Fig. 26.2** Total landings in Hormozgan in 2012

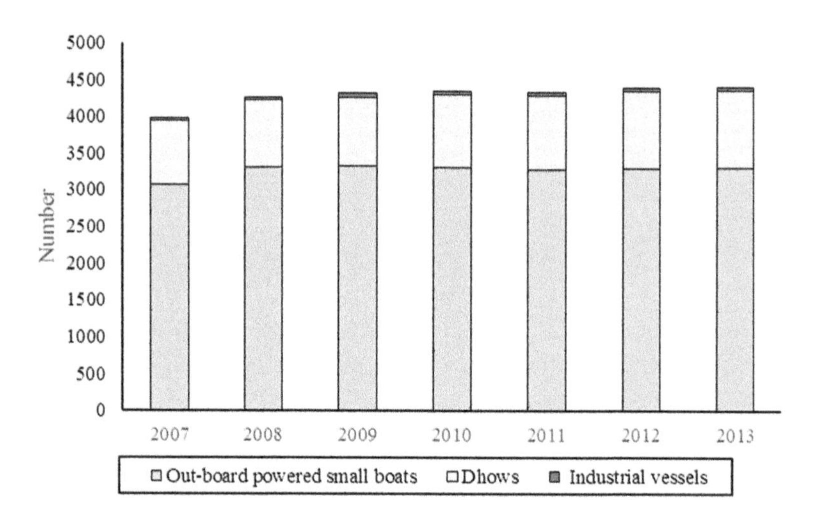


**Fig. 26.3** Number of active fishing vessels in Hormozgan between 2007 and 2013

for shrimp fishing, purse seines for small-pelagic fishes, wire traps (locally named: *gargoor*), fixed stake net traps (in Persian: *Moshtä*), bait and hooks, and some local fishing methods. Many boats use a combination of fishing nets depending on economic capacity, skills, and target species (Fig. 26.4). Small-scale fishers use a mix of traditional and modern methods. Some fishers (less than 10%), like those who fish along the beach, use fixed stake net traps, whereas others use new fishing equipment such as winches, GPS, etc. Fishers mostly target demersal fishes and have their own dhow or boat and fishing gears (gill nets) for use in the inshore waters and estuaries. In shrimp trawl fishing, fishers also work as crew in shared or wage based systems.

Small-scale fishers of Hormozgan are predominantly indigenous and reside in their birthplace. Through the traditional structure of these communities, there is a

**Fig. 26.4** Some small-scale fishing methods in Hormozgan waters (Source: Moslem Daliri et al. 2015)

strong tribal relationship among them and tribal elders have leading role to play. Fishing is by far the most important source of livelihood for households. Additionally, hundreds of other coastal people are involved in fishing as a part time or seasonal occupation combined with livestock, agriculture, and trade. Fishers view fishing as an ancestral tradition and have a sense of ownership over the sea. They also think that no one can treat the sea better than they do. Although most young fishers are dissatisfied with their occupation, they have few alternative employment opportunities available. Therefore, parents often insist that their children should continue their education and find jobs with more social standing.

Approximately half of small-scale fisher households in Hormozgan are living in poverty, and a minority, those who own a dhow, are wealthy. The household size in these fishing communities average between five and six people, which is higher than the average urban household size (3.8 on average) in this region (Hormozgan Governorship 2013). This could be due to reasons such as marriage at a young age (mostly 18–20 for men and 15–17 for women), low literacy levels of parents, and preference for sons leading to increased frequency of pregnancies. Divorce is an anomaly in Hormozgan small-scale fisher communities (Tavakol 1991).

Mostly men fish in this region. In Hengam Island, however, 30 women fish with hook and bait. Women and children are mainly involved in fish processing and

**Fig. 26.5** Local fish markets in Minab (Source: Moslem Daliri 2014)

selling (Fig. 26.5). For example; in anchovies fishing, which is done by purse seine boats and beach seines (locally named Jal-Sardin) in the regions of Qeshm Island, Bandar Lengeh, and Bandar Jask, family members are mostly involved in fishing and processing operations. After landing, part of the catch volume is transported to fish meal factories, while women and teenagers dry the rest for livestock and poultry feed (Fig. 26.6).

As highlighted by Daliri et al. (2015), illegal shrimp fishing is one of the most difficult problems facing fisheries management in Hormuz Strait. Industrial and small-scale fishers regularly undertake fishing for shrimps in the open fishing season (October and November). In addition to that, small out-board engine powered boats equipped with drift gill nets (locally named Semari), and in some cases also with bottom trawl nets, capture shrimp illegally in the closed season (late July to October). Silver Pomfret, due to its high commercial value in the Persian Gulf region (especially in the Arabian Peninsula), is also often caught illegally. This migratory fish species naturally reproduces between April to July in coastal areas and in the estuaries of Qeshm Island (Hormozgan Province). The latter is known as the hatchery for this species (Momeni et al. 2004). Although IFO declares these months as closed fishing seasons, local fishers equipped with fixed gill nets (locally named *Leeh*) catch the broodstocks illegally.

Recently, decreasing fish stocks and increasing industrial fishing activities in the region has led to lowering income and unemployment of small-scale fishers. This is a serious threat to the future of small-scale fishing communities in the region. In the Sixth Development Plan (IDP), which was submitted by the government to

**Fig. 26.6** Women drying anchovies in Qeshm Island (Source: Ali Salarpouri 2012)

Parliament in 2016, marine aquaculture development was prioritized for Hormozgan Province. According to this plan, contracts have been signed for marine fish cage farming in Qeshm Island and Gorzeh village (Bandar-Lengeh) amongst other places by the Hormozgan Fisheries Department (HFD), which is a subsection of IFO, and the private sector (IRNA 2016). The aim is to create job opportunities and improve the livelihoods of small-scale fishing communities.

## Monitoring, Control and Surveillance (MCS) in Small-Scale Fisheries

Iran, as detailed above, has committed itself to eliminate IUU fishing. However, to date, implementation and enforcement have not been very successful in combating IUU fishing, particularly by small-scale fishers where VMS is not in use as it is in the large-scale sector. In Paragraph 5.16 of the SSF Guidelines, it is mentioned that MCS is indispensable for sustainable fisheries and that there is a need to provide better information about IUU fishing: 'States should endeavor to improve registration of the fishing activity. Small-scale fishers should support the MCS systems and provide to the State fisheries authorities the information required for the management of the activity'. Below, we give a status overview of the use of MCS in the small-scale fisheries sector in Iran with emphasis on the Persian Gulf and Oman Sea.

## *MCS in Small-Scale Fisheries in Iran*

The Law of Protection and Exploitation of Fisheries Resources in Iran which includes 23 articles, is meant to ensure sustainable fisheries (Salari Shahr Babaki 1999). There was no major legislation for ensuring sustainable exploitation of fisheries resources in Iran until 1995, when the Parliament passed this law. In Articles 3 and 21, it is stated that the IFO is responsible for the management, development, and exploitation of fisheries resources alongside the aquatic resources protection unit of the coast guard. IFO is the authorized representative of the judiciary and is meant to carry out inspections in accordance with this law. The task of the aquatic resources protection unit is to ensure the proper implementation of regulations or legislations and arrest those who violate them. More specifically, Article 6 authorizes the IFO to combat IUU fishing in Iran's EEZ (Tghavi 1999). In addition, Articles 12, 16, and 22 of this law clearly explain what constitutes IUU fishing and what the penalties are. The law also stresses the necessity of recording statistics and data relevant to fishing vessels.

Although Daliri (2016) suggests that the MCS system is weak, he points to its gradual improvement. However, he also argues that financial resources and management plans are needed to strengthen the system. In terms of fisheries infrastructure, IFO recently established and improved ports and fishing docks in the southern part of Iran to increase the inspection of catches landed and active fishing vessels. Nonetheless, equipment for constant patrolling and monitoring of active fishing activities in the Persian Gulf and Oman seas are limited. In coastal waters, speed boats undertake surveillance and control, which is relatively effective.

The expansion of VMS technology will mean that all commercial fishing vessels are equipped with VMS systems to monitor the movement and activities of vessels. In addition, installation of VMS systems on small-scale fishing launches is currently being executed. According to the Hormozgan Fisheries Department (HFD), nearly 40% of fish landings in the province are subject to VMS surveillance.

Inspection and surveillance of vessels are adversely impacted by limited tools and finances. Therefore, inspection and surveillance of active small-scale fishing vessels is ineffective and takes place only once or twice a year and that too of only some commercial vessels and in fishing ports. In some fishing ports, vessels are inspected monthly before the extension of their license. However, inspections are random. Therefore, management plans for implementation of surveillance in coastal areas are limited, and sea patrolling limited to preventing illegal fishing (e.g. trawling) and use of nonstandard fishing gears during the open season.

Although there is a plan to record and collect fishing data from vessels, it is not comprehensive in nature. Industrial vessels are obliged to keep logbooks. In the case of small-scale vessels, inspectors obtain catch data and estimate catches based on surveys undertaken at landings centers. Not all catches are recorded and in some cases due to the lack of data, collected information from a few vessels are used to extrapolate about the whole fleet. However, some data including catch data, length frequency, and catch effort for major tuna species like *Thunnus albacores*, *Thunnus tonggol*, *Katsuwonus pelamis*, *Euthynnus affinis*, and *Auxis thazard* are gathered from landing sites and fishing ports of Hormozgan, and Sistan and Baluchestan.

Pramod and Pitcher (2006) found that plans regarding monitoring, control, and surveillance in IFO are not complied to properly. This is the case because in IFO, different units including the section of the fishing deputy, the special unit for protection of natural resources, the department of fishers (training or prevalence), and the department of fishing ports are all responsible for looking after catch and the exploitation of aquatic resources, new legalizations, data collection and fishing activities. Moreover, a special unit for protection of aquatic resources is responsible for preventing fishing activities in restricted seasons and places, enforcement of fishing laws and regulations, and disallowing illegal fishing activities. Construction and management of ports based on eco-biological laws regarding the berthing of fishing is the duty of the department of fishing ports. Therefore, given different responsibilities of different units, it is possible that the long-term management of catch and fishing activities in Iranian waters of the Persian Gulf and Oman Sea and the combating of IUU fishing are hindered.

Another weak point of IFO is its underfunding of the MCS system. During the last decade, international financial sanctions against Iran and consequently a shortage of government funds meant IFO did not receive an adequate budget. As a result, the economic situation in small-scale fishing communities deteriorated and triggered opportunistic behavior amongst fishers, including IUU fishing in the region.

## *IUU Small-Scale Fishing in Hormuz Strait*

Effective combating of IUU fishing requires an understanding of the factors that cause it and why fishers adhere to or break rules. Importantly, stricter rules and harsher penalties may not solve the problem if fishers have good reasons and ample opportunities to break them. According to Daliri et al. (2016), social and cultural factors are central to both causing and reducing IUU small-scale fishing in the Hormuz Strait. In general, non-compliance means breaking cultural norms, but on occasion it is the result of certain rules being considered irrelevant or inconsistent by fishers themselves. A regulatory system that provides incentives in accordance with cultural norms and fishers' own perceptions of how regulations should be designed is likely to be more effective than a system that is alien or nonsensical to them.

Daliri et al. (2016) noted that local people mostly have a sense of ownership over sea space where they fish and that they view fishing as an ancestral tradition. They also argued that fishers (like other people) act based on what they see, experience, and understand, and want to gain livelihood security within the prescribed rules that they must follow. Therefore, it would make sense to ask them why they (dis-) respect rules. Attitudes about following or breaking rules are transmitted to others over time. A person who breaks rules once is likely to do it again, until it becomes a routine and easy decision to make. Others are also likely to follow, especially if they see that one can get away with doing so unless of course there are strong norms that advise against breaking rules. Fishers, to sustain themselves, repeatedly break rules until it becomes accepted practice. It may in fact even become a cultural trait, driven on by poverty.

Daliri et al. (2016) also noted that personal traits of small-scale fishers influenced the extent of their IUU fishing in the region. In another study from Kerala, India, Kannan (1999) highlighted the importance of education, expertise, and literacy not only for human development but also for poverty alleviation in small-scale fishing communities. Lack of knowledge about existing rules also leads to IUU fishing. The ability to read is, in other words, an essential minimum to prevent IUU fishing. Literacy is relatively low among the small-scale fishers of Hormozgan. Daliri et al. (2016) posited that small-scale fishers of Hormozgan tend to behave opportunistically. In addition, they tend to mistrust managers and decision-makers and believe that they do not fully understand the situation of fishers. Lack of trust of officials is, therefore, another reason that fishers break rules. Economic concerns, as suggested above, were also important contributing factors to IUU fishing. Fishers complained mostly about unemployment and economic insecurity. Building trust and addressing poverty must therefore be part of the implementation of the SSF Guidelines in Iran. Doing so requires finding ways of involving fishing communities in decision-making pertaining to MSC.

The SSF Guidelines mention co-management as a means for handling IUU fishing and MCS (cf. paragraphs 5.15 and 5.18): States should provide support to such systems, involving small-scale fisheries actors as appropriate and promote participatory arrangements within the context of co-management. Co-management is also a way of building trust among fishers, and between fishing communities and government, because it is inclusive and participatory, which the SSF Guidelines mentions as a remedy for IUU. Co-management can provide incentives for fishers to view fish stocks as a long-term asset rather than discount future returns and thus risk overexploiting resources and ruining the marine ecosystem. There are also successful examples of co-management in Iranian fisheries. IFO has organized Iranian fishers in associations; there are about 200 such associations in the Caspian Sea (beach seine and *Kilka*) and 158 in the Persian Gulf and Oman Sea. These associations organize all licensed industrial and small-scale fishers and act as a bond between government managers and fishers. Furthermore, they play an important role in implementing fisheries regulations and legislations (IFO 2014). Beach seine fishers' associations participate in restocking of bony fish in the Caspian Sea. In the Persian Gulf, fishers' associations in cooperation with IFO manage most of the ten fishing ports in Hormozgan Province. The fisheries associations of Hormozgan also cooperated with the Statistical Center of Iran (SCI), for the compilation of the 2016 population census. IEPA (Iranian Environmental Protection Agency) currently supports a mangrove area co-management project in Hormozgan (Rood Shoor, Rood Shirin, and Minab reigns), in which stakeholders and local communities are involved. Hence, these associations may have an important role to play in the implementation of the SSF Guidelines by highlighting the importance of co-management and putting forward concerns regarding fisher wellbeing.

## Final Remarks

This chapter has focused on IUU fishing in the context of Hormozgan Province and the Persian Gulf of Iran. Here small-scale fishing communities, with a population of approximately 20,000 fishers, account for the overwhelming majority of fish landings. Thus, small-scale fisheries play a major role in providing food-security for the general population in the region. However, they themselves are ridden in poverty and are marginalized. The SSF Guidelines addresses the concerns of these small-scale fishing communities, namely low income, illiteracy, and poor health.

IUU fishing is common in this area and sector and threatens the sustainability of fisheries resources and the survival of fishing communities in the region. It is also a hindrance to bringing small-scale fishing communities out of poverty and on to a path of wellbeing. This is why fisheries management institutions in Iran are targeting IUU fishing, which the SSF Guidelines also suggest they should. However, this case study suggests that combatting IUU fishing requires a more comprehensive approach than just MCS. MCS is necessary, but it does not address the root causes of IUU fishing, only the practice. When fishers break rules, like when they fish with gear that is banned, in places that are closed to fishing, and at times where fishing is not supposed to happen, they do so not just because MCS is ineffective or to make more money, but because they feel compelled to do so because of their livelihood needs. Poor people cannot easily afford to limit their fishing effort as the regulatory system wants them to do, even if they realize that it is not sustainable and morally suspect. People also live in the short run; they need food on a daily basis and have expenses that cannot wait. This is also the situation for small-scale fishing families in the Hormozgan region. Therefore, if these needs are not satisfied, fishers will continue their IUU fishing, regardless of how effective the MCS system is and how harsh penalties are. This would especially be the case if MCS is established and implemented by authorities that fishing communities do not trust as fishers would justify their actions.

Therefore, combatting IUU fishing in small-scale fisheries would require efforts that are beyond MCS. Efforts should aim to address the livelihood concerns of fishers and their communities. Fishing communities should be empowered by involving them in fisheries management decision-making. Co-management, which is already being instituted in Iranian fisheries, is one option. This broad and holistic approach is exactly what the SSF Guidelines promote in Paragraph 6.1, which says that: "All parties should consider integrated, ecosystem and holistic approaches to small-scale fisheries management and development that take the complexity of livelihoods into account. Due attention to social and economic development may be needed to ensure that small-scale fisheries communities are empowered and can enjoy their human rights."

Based on the situation of Hormozgan small-scale fisheries, which this chapter has described, we conclude that a holistic approach is essential for combatting IUU fishing. MCS may be necessary but insufficient for this endeavor, if what drives people to break rules is poverty. As a result, there is 'community failure', a situation where the social fabric of fishing communities is eroding so that communities cannot enforce their own rules and norms on their fishers (McCay and Jentoft 1998). The way to handle such irregularities in small-scale fisheries is to implement the SSF Guidelines to the full rather than just concentrating on those paragraphs that particularly mention the need for MCS. When the SSF Guidelines talk about the need to empower fishers and fishing communities, they also address indirectly the enabling conditions for addressing IUU fishing. When they talk about poverty and food security, they also address the reasons that people give for non-compliance vis-à-vis IUU fishing. When small-scale fishing communities thrive, their tenure rights are secured. When people can live a life of dignity, without poverty and marginalization, and when their active participation in management decision-making is called for, then their inclination to fish beyond the limits that are determined by management authorities will reduce. MCS should be there only as a last resort, and not as the only egg in the basket.

## References


Al-Abdulrazzak, D., & Pauly, D. (2013). Managing fisheries from space: Google Earth improves estimates of distant fish catches. *ICES Journal of Marine Science, 25*, 1–5.

Baird, R. J. (Ed.). (2006). *Aspects of illegal, unreported and unregulated fishing in the Southern Ocean*. Netherlands: Springer.

Cho, D. O. (2012). Eliminating illegal bottom trawl fishing in the coastal waters of Korea. *Marine Policy, 36*, 321–326.

Daliri, M. (2016). *Illegal, unreported and unregulated (IUU) small-scale fishing in the northern Persian Gulf (Hormozgan waters)* (Doctoral dissertation). Fisheries Department (Marine), Hormozgan University, Hormozgan, Iran.

Daliri, M., Kamrani, E., & Paighambari, S. Y. (2015). Illegal shrimp fishing in Hormozgan inshore waters of the Persian Gulf. *The Egyptian Journal of Aquatic Research, 41*, 345–352.

Daliri, M., Kamrani, E., Jentoft, S., & Paighambari, S. Y. (2016). Why is illegal fishing occurring in the Persian Gulf? A case study from the Hormozgan province of Iran. *Ocean & Coastal Management, 120*, 127–134.

Erceg, D. (2006). Deterring IUU fishing through state control over nationals. *Marine Policy, 30*, 173–179.

FAO. (2011). *Review of the state of world marine fishery resources*. Rome: Food and Agriculture Organization of the United Nations.

Hormozgan Governorship. (2013). *Statistical yearbook of Hormozgan province*. Ministry of Interior: Islamic Repulic of Iran.

Hosseini, S. A., Daliri, M., Raeisi, H., Paighambari, S. Y., & Kamrani, E. (2015). Destructive effects of small-scale shrimp trawl fisheries on bycatch fish assemblage in Hormozgan coastal waters. *Journal of Fisheries, 68*, 61–78.

IFO. (2014). *Fisheries yearbook*. Iran: Planning and Development Department.

IRNA (Islamic Republic News Agency). (2016). Fisheries and aquaculture: As axis of Hormozgan development. http://www8.irna.ir/fa/News/81774440/. Accessed 12 Nov 2016.

Jagers, S., Berlin, D., & Jentoft, S. (2012). Why comply? Attitudes towards harvesting regulations among Swedish fishers. *Marine Policy, 36*, 969–976.

Jensen, F., & Vestergaard, N. (2002). Moral hazard problems in fisheries regulation: The case of illegal landings and discard. *Resource and Energy Economics, 24*, 281–299.

Jentoft, S. (2014). Walking the talk: Implementing the international voluntary guidelines for securing sustainable small-scale fisheries. *Maritime Studies, 13*, 1–15.

Kannan, K. P. (1999). *Poverty alleviation as advancing basic human capabilities: Kerala's achievements compared*. Thiruvananthapuram: Centre for Development Studies.

Kurien, J. (2015). *Voluntary guidelines for securing sustainable small-scale fisheries in the context of food security and poverty eradication: Summary*. Chennai: International Collective in Support of Fishworkers.

Luomba, J., Chuenpagdee, R., & Song, A. M. (2016). A bottom-up understanding of illegal, unreported, and unregulated fishing in Lake Victoria. *Sustainability, 8*, 1062. doi:10.3390/8101062.

McCay, B., & Jentoft, S. (1998). Market or community failure? Critical perspectives on common property research. *Human Organization, 57*(1), 21–29.

Momeni, M., Safaie, M., Kamrani, E., Kamali, E., Karimi, H., & Iran, A. (2004). *Reproduction biology of silver pomfret (Pampus argenteus) in Hormozgan coastal waters*. Persian Gulf & Oman Sea Ecological Research Institute: Iran Fisheries Science Research Institute.

Pitcher, T., Kalikoski, D., & Pramod, G. (2012). *Evaluations of compliance with the FAO code of conduct for responsible fisheries*. Fisheries centre research reports. Vancouver: Fisheries Centre, University of British Columbia.

Pramod, G., & Pitcher, T. J. (2006). An estimation of compliance of the fisheries of Ireland with Article 7 (Fisheries Management) of the UN code of conduct for responsible fishing. In M. Safaie (Ed.), *Evaluations of compliance with the UN code of conduct for responsible fisheries* (pp. 1–23). Fisheries centre research reports 14(2). Vancouver: Fisheries Centre, University of British Columbia

Salari Shahr Babaki, M. M. (1999). *Law of protection and axploitation of the fisheries resources of Iran (Islamic Republic of Iran)*. Tehran: Noorbakhsh Press.

Sodik, D. M. (2007). *Combating illegal, unreported and unregulated fishing in Indonesian waters: The need for fisheries legislative reform* (Doctoral dissertation). University of Wollongong, Wollongong.

Stewart, J., & Walshe, K. (2008). Compliance costs and the small fisher: A study of exiters from the New Zealand fishery. *Marine Policy, 32*, 120–131.

Tavakol, M. (1991). *Socio-economical structure of small-scale fishing communities in Hormozgan province*. Iran Fisheries Organization, Planning and Studies Department.

Tghavi, S. A. A. (1999). *Fisheries monitoring, control and surveillance in the Islamic Republic of Iran. Muscat, Sultanate of Oman*. Proceeding report of a regional workshop on Fisheries Monitoring, Control and Surveillance.

# Chapter 27
# The Role of the Small-Scale Fisheries Guidelines in Reclaiming Human Rights for Small-Scale Fishing People in Colombia

**Lina María Saavedra-Díaz and Svein Jentoft**

**Abstract** For more than five decades, small-scale fisheries in Colombia have felt the devastating consequences of armed conflict and human rights violations. There is now a hope that the peace process will give the country a new start, and help to improve the well-being of small-scale fisheries communities and the sustainability of their natural resources. After the civil war and the drug violence, the government now has the opportunity to focus more on people's welfare and livelihood needs. With the Voluntary Guidelines for Securing Sustainable Small-Scale Fisheries (SSF Guidelines) endorsed by FAO member states in 2014, a new direction is outlined. Fisheries in Colombia suffer from the lack of a firm institutional foundation and a dysfunctional governance system, which has resulted in poor coordination of policies and actions targeting small-scale fisheries. This may also be a problem for the implementation of the broad agenda of the SSF Guidelines, which must engage governmental, non-governmental, private, and public institutions at the national, regional, and local levels alike. This chapter argues that there is a need for governance reform to facilitate the incorporation and implementation of international agreements such as the SSF Guidelines and related instruments. Its mandate should be to convert these commitments into national policies, management strategies, and regulations in accordance with the human rights and good governance principles and ambitions of the SSF Guidelines. This would also be an opportunity to involve all stakeholders and bring them under the same umbrella.

**Keywords** Colombia • Small-Scale Fisheries Guidelines • Armed conflict • Institutional fragmentation

L.M. Saavedra-Díaz (✉)
Programas de Biología e Ingeniería Pesquera, Universidad del Magdalena, Santa Marta, Magdalena, Colombia
e-mail: lsaavedra@unimagdalena.edu.co

S. Jentoft
Norwegian College of Fishery Science, UiT – The Arctic University of Norway, Tromsø, Norway
e-mail: svein.jentoft@uit.no



S. Jentoft et al. (eds.), *The Small-Scale Fisheries Guidelines*, MARE Publication Series 14, DOI 10.1007/978-3-319-55074-9_27

## Introduction

Colombia's 50 years of armed conflict has heavily affected small-scale fisheries communities. Fishers and their families have been victims of human rights violations, including violence and forced displacement (Rincón et al. 2013). Repeated peace negotiations between the government and armed groups have largely been unsuccessful, until recently when a peace agreement was signed between the goverment and the FARC (Revolutionary Armed Forces of Colombia), which was first rejected by a referendum and afterwards renegotiated and signed. The negotiation agreement included a new institutional structure for protecting the environment, participatory processes at the local level, and ways to promote a better quality of life, food security, information and knowledge sharing. This agenda resonates well with the main goals proposed by the Voluntary Guidelines for Securing Sustainable Small-Scale Fisheries (SSF Guidelines) endorsed by FAO member states in 2014.

The current peace agreement provides an opportunity for the implementation of the SSF Guidelines. The Guidelines prioritize the realization of human rights and the security and livelihood needs of vulnerable and marginalized groups such as Colombian small-scale coastal fishing communities, which are poor and, to a large extent, exist without social security, education, or medical services. In general, neither the national government nor the Colombian people recognize the important role of small-scale fisheries for providing food security, employment, and economic and social development on the coast. Being a fisher is not seen as a formal occupation in Colombia, which makes small-scale fisheries precarious and puts small-scale fisheries communities, families, and individuals at risk.

With the peace process and SSF Guidelines, this may now change, as they provide both an opportunity and an incentive to make small-scale fisheries figure more prominently in Colombian development policy, particularly for rural areas. However, we argue that for such a new policy to be successful, governance reform would be needed in the new context of peace. Until now, the institutional structure has been fragmented and dispersed, with overlapping mandates and poor coordination, which would hamper the implementation of the SSF Guidelines. The governing system must allow for the involvement of small-scale fisheries stakeholders to become more effective participants and partners in the governance process, as the SSF Guidelines point out.

The chapter provides an overview of the Colombian small-scale fisheries governance system and an invitation for Colombian fisheries stakeholders to reflect on the opportunities that the peace give for implementing the human rights-based approach that the SSF Guidelines promote. The next section explores the new context created by the peace, followed by an overview of small-scale fisheries in Colombia, illustrated by the situation in nine selected local communities in the Caribbean and Pacific coasts. After this, the institutional structure of the Colombian governance system is outlined, with particular emphasis on small-scale fisheries. The subsequent section presents how the Colombian small-scale fisheries governance system is structured and the characteristics of the regulatory framework. Finally, we reflect on how receptive this system is to the SSF Guidelines and the kinds of governance principles that they advocate in relation to Colombian fisheries governance.

## The SSF Guidelines and the Peace Process

Due to the situation described above, the small-scale fisheries sector in Colombia urgently need a governance policy such as the one advanced by the SSF Guidelines. This necessity is accentuated in developing countries such as Colombia. For the particular case of this country, in paragraphs 6.9 and 6.18, the SSF Guidelines recognize the effects of armed conflict on small-scale fishing communities:

Paragraph 6.18: "All parties should protect the human rights and dignity of small-scale fisheries stakeholders in situations of armed conflict in accordance with international humanitarian law to allow them to pursue their traditional livelihoods, to have access to customary fishing grounds and to preserve their culture and way of life..."

Paragraph 6.9: "All parties should create conditions for men and women of small-scale fishing communities to fish and to carry out fisheries-related activities in an environment free from crime, violence, organized crime activities, piracy, theft, sexual abuse, corruption and abuse of authority. All parties should take steps to institute measures that aim to eliminate violence and to protect women exposed to such violence in small-scale fishing communities. States should ensure access to justice for victims of inter alia violence and abuse, including within the household or community."

These two paragraphs speak directly to the situation in Colombia. Even though the confrontations have somewhat decreased in the last years due to peace talks, small-scale fishing communities have been heavily affected by armed confrontations, particularly those on the Pacific coast due to the presence of different criminal bands (Urabeños, Rastrojos, and Erpac) and guerrillas groups (FARC and ELN). While biologically diverse and rich, the Pacific coast is Colombia's poorest region. Due to the armed conflict, the region has seen massacres, kidnappings and massive displacements (Escobar 2003; Al 2008). On the Caribbean coast, there are also records of fishers being massacred due to the presence of different armed groups (i.e. Ciénaga Grande de Santa Marta), and this armed conflict has affected socio-ecological dynamics where violence and terror have undermined the ability of communities to enforce their own resource management rules (Vilardy and Renán-Rodríguez 2011). As armed conflict has subsided, the illegal crops have increased affecting the culture of the fishing communities (Rincón 2014). In specific, in Afro-fishing communities on the Pacific coast such crops have increased by more than 50% from 2014 to 2015 (UNODC 2016).

Parallel with the peace process, the Colombian government has designed a policy called 'Colombian Post-conflict Strategy' that recognizes how the armed conflict has created severe inequities between the urban and rural areas (CONPES 2016). The rural municipalities and regions, which host the small-scale fishing communities, are generally poor and show profound social, economic, and environmental decay after years of armed conflict and being exposed to confrontations with drug traffickers which are disruptive to the social cohesion of their communities. They are also characterized by insecure tenure rights, low self-governing and institutional capacity, and little state presence. These are problems all featured in the SSF Guidelines. Although, it is clear that small-scale fishers are located in zones where the armed conflict has taken place, the final peace agreement document with the FARC does not identify fishers as one of the social groups being conflict victims.

The SSF Guidelines recommend that "All parties should recognize that responsible governance of tenure of land, fisheries and forests applicable in small-scale fisheries is central for the realization of human rights, food security, poverty eradication, sustainable livelihoods, social stability, housing security, economic growth and rural and social development" (paragraph 5.2). For this to become a reality in Colombia, there is still a considerable way to go. However, the peace process now provides a window of opportunity for all relevant parties, not the least of which being the Colombian government, to prove their willingness to act upon this commitment. It would also have to be the first step in building the governance capacity at community level that is now missing after decades of armed conflict. The SSF Guidelines offer concrete ideas of how this could take place: "Government authorities and agencies at all levels should work to develop knowledge and skills to support sustainable small-scale fisheries development and successful co-management arrangements, as appropriate. Particular attention should be given to decentralized and local government structures directly involved in governance and development processes together with small scale fishing communities, including the area of research" (paragraph 12.4).

However, should such governance reform be implemented, it would represent a break with the established order of governance in Colombia, where small-scale fisheries policies and administration traditionally followed a top-down approach – as in many other places in Latin American countries (Salas et al. 2007; Sánchez and Moreno 2009; Saavedra et al. 2015). Although the first decade of the twenty-first century saw isolated efforts to decentralize the fisheries administration by involving fisheries communities in decision-making, fisheries management is still overwhelmingly centralized (Candelo et al. 2002; CORPOURABA 2005; Squalus 2008; Delgado et al. 2010; Ramírez-Luna 2013).

## Small-Scale Fisheries Case Studies

The study comprised 184 interviews with fishers (see detailed description of qualitative methods in Saavedra-Díaz et al. 2015). Questions targeted the problems that fishers experience in their community and sector and which they have to cope with on a regular basis. Data presented here were collected from August 2008 to June 2009 in nine communities selected, one for each Coastal and Marine Ecoregion (CME), of which there are five on the Caribbean and four on the Pacific coast. The Caribbean fishing communities are Ahuyama in the Guajira CME, Taganga in the Sierra Nevada de Santa Marta CME, Las Flores in the Magdalena CME, San Antero in the Morrosquillo and Sinú CME, and El Roto in the Darién CME. The Pacific fishing communities are Bahía Solano in the Alto Chocó CME, Pizarro in the Baudó CME, Juanchaco in the Málaga-Buenaventura CME, Tumaco in the Llanura Aluvial del Sur CME (Fig. 27.1).

The chosen communities exhibit heterogeneity. The number of fishers per community varied from 80 (in a village with 400 inhabitants) to 6000 (municipality with 160,034) (Saavedra-Díaz 2012). The number of fishers reflects the variety of economic alternatives offered in or around the community. Economic alternatives also influence the number of part time fishers who combine fishing with other sources of income. Consequently, a full range of economic strategies may be found among the fisheries population within and between communities. Nevertheless, the studied communities can be organized broadly in three categories based on the most common economic strategy: (a) Communities in which almost all active workers rely only on fishing (i.e. El Roto and Ahuyama on the Caribbean), (b) Communities in which fishing is a major economic activity among a number of alternatives (i.e., Pizarro, Juanchaco and Tumaco on the Pacific, and Las Flores on the Caribbean), and (c). Communities in which economic options have a greater number of part-time fishers. In the San Antero community, fishers cut mangrove during certain months and fish the rest of the year; others alternate fishing with agriculture. In Taganga and Bahía Solano, fishing is alternated with tourism activities (Saavedra-Díaz 2012).

Small-scale fishing activity occurs in all possible coastal environments contiguous to the community, including the mouths of rivers, estuaries, mangroves, sea grass beds, coral-reefs, swamps, littorals, and the open sea (see Fig. 27.2). This heterogeneity exacerbates the difficulties of efficiently monitoring and controlling fishing activities. Fishers who travel seasonally outside their local region to harvest lucrative fishing grounds aggravate the situation. Yet fishing is of fundamental importance to all these communities, providing protein-rich food as well as work to the inhabitants. The fishing communities participating in the present study are made up of a variety of racial and ethnic groups, including indigenous and people of Spanish and African descent, and people of mixed heritage. Some fishing communities are totally or partially indigenous, while some consist almost entirely of Afro-Colombians. The latter communities are particularly common on the Pacific coast. On the Caribbean side, most communities are of mixed ethnicity, with a lesser

**Fig. 27.1** Selected fishing communities (*white dots*) in Coastal and Marine Ecoregions (CMEs) on the Colombian Caribbean coast and Pacific coast. The spaced lines show the limits between each CME (Taken from Saavedra-Díaz et al. 2016)

number of Afro-Colombian Fishers. The ethnic diversity of all small-scale fishing communities is reflected in the fishing cultures found in this sector. Such differences need to be taken into account in when implementing the SSF Guidelines. An approach that might appeal to an indigenous fisher might fail to win over a fisher of African descent.

A. B.

C. D.

**Fig. 27.2** (**a**) *Atarraya* used by two fishers in *San Antero* in mangroves ecosystems. (**b**) *Chinchorro* used on *Tumaco* community from the beach. (**c**) A fisher checks out his *trasmallo* at the mouth of *Atrato river*. (**d**) Fisher from *Las Flores* using the *cometa* method on the Western cutwater at *Bocas de Ceniza*. (Photos taken by Lina María Saavedra-Díaz)

Fishing grounds are common property, as custom does not allow exclusive fishing regions to be established by community or restrictions imposed on the type of gear employed. Consequently, the same fishing territory is often shared by many small-scale fishing communities as well as migrant fishers. In all studied communities on the Caribbean and Pacific coasts, fishers complained about foreign fishers moving freely into their areas.

Caribbean fishers interviewed in this study employ 26 different small-scale fishing methods to catch marine species. Fifteen methods are used on the Pacific coast. While some are present on both coasts, others are unique to one region. In total, participants identified 30 different fishing methods (see Figure 3.5 and Table 3.2 in Saavedra-Díaz 2012). The Caribbean coast hosts a greater variety of fishing methods than the Pacific, due to its greater number of fishers and wide variety of marine ecosystems and fishing resources. In other words, intensive fishing competition within a region of highly diverse marine ecosystems and natural resources generally seems to promote a high diversity of fishing methods. Most fishing methods are used on both coasts. This great assortment of fishing methods is typical of countries located in the tropics, due to ecosystem variety and high fish diversity (Raakjaer et al. 2007).

Fishers in the nine communities described that they are victims of corruption by politicians and local administration officials, and that relief support directed towards their communities are lost before they reach their communities. Fishers also talked about how they have been displaced from their fishing areas by drug traffic activities. Some fisheries communities are located in remote areas used by paramilitary, guerrillas, and other illegal armed groups. These groups have established internal fishing rules that are violently enforced in these communities. Killings of fishers have also occurred because they were fishing in areas where and when armed confrontations have taken place. When violent events occur, fishers are forced to leave and move to urban areas where they feel better protected by the military.

Some fishers have themselves been involved in transporting drugs or cultivating illegal crops, and have incurred jail-time as a result. Fishers also say that illegal groups own fishing equipment (boats or fishing methods) and use it for the drug trade, thus tarnishing the image of small-scale fishers. Local fisheries culture is then affected by drug trafficking activities. Family and personal values are affected in communities where drugs are sold. Some fishers cultivate coca plants. Also, fishers sometimes find a 'paca' containing cocaine with a value of some USD $7000 each, which drug traffickers release wrapped in packs when they are being chased by the police. A 'paca' can bring a fisher out of poverty and into sudden affluence with the ability to purchase a new house, consumer goods, and luxury items. If not caught, fishers can earn more money this way than they do from fishing. Money earned can also be reinvested into more fishing equipment or encourage them to quit fishing altogether. When fishers leave their home community and move, pressure on fishing resources near urban areas increases. In some communities that once depended on tourism, violence has caused tourists to stay away and displaced employees are now involved in the fishery sector. Attacks on oil pipelines has caused spills that polluted rivers, mangroves and estuarine zones where fishers used to fish. Women are mostly affected since they used to collect 'piangua' from the mangrove roots. It is affecting not only their economic activity, but also the ecosystem, the fishery resource, and consumer health.

Fishing communities are also recipients of massive migrations due to violence. As a consequence, the number of fishers has increased dramatically. Displaced people who, without any previous experience, take up fishing often use destructive gear and other fishing methods that are increasing fishing pressure, which also affects the resources. The violent environment has changed fishers' mindsets, and resolving conflicts violently has become more common. Communities that have tried to implement internal fisheries management rules have been exposed to violence, especially in those communities that have many migrants.

Although the studied communities differ in particular aspects, for instance with regard to infrastructure services and living conditions, in all communities, family economy and food security are based on small-scale fisheries activities. Consequently, since these fishing communities are located where illegal and armed conflicts have taken place, the peace process brings a hope to reclaim a small-scale fisheries policy that deal with their social and environmental challenges under the Code of Conduct for Responsible Fisheries (FAO 1995) and the SSF Guidelines (FAO 2015). Currently, the small-scale fisheries sector is poorly organized and governed. Thus, there is a risk of overexploitation of fisheries resources (Díaz et al. 2011; Puentes et al. 2014), which can bring new conflicts in coastal communities. To improve the quality of life in small-scale fisheries communities, access to basic services (housing, water, electricity, education, health, and social rights) and decent jobs would be important.

## The Small-Scale Fisheries Governance System

For almost half a century, the fisheries administration has been led sequentially by six institutions: Renewable Natural Resources and Environmental National Institute (INDERENA), National Institute of Fisheries and Aquaculture (INPA), Colombian Rural Development Institute (INCODER), Colombian Agricultural Institute (ICA), INCODER again, and at present the National Authority of Aquaculture and Fisheries (AUNAP). In the last decade, these six institutions were reduced to four. Many other institutions are also involved in fisheries governance at local, regional, and national levels (see Fig. 27.3). In addition to the Ministry of Rural Development and Agriculture Ministry (MADR), five additional ministries are directly and indirectly responsible, with different agencies participating in overall management. These are the Ministry of Environment, Housing, and Development (MADVT), the Ministry of Education (ME), the Ministry of Social Protection (MSP), the Ministry of Defense (MD), and the Ministry of Interior and Justice (MIJ). Similar to the ministries, there are national state control agencies that oversee the state apparatus, among them the Office of the Comptroller, which supervises the Agriculture sector and, by extension, the fishery sector as well. In addition to these governmental institutions, private institutions also work directly with the fishery sector.

Figure 27.3 illustrates Colombia's fragmented small-scale fisheries governing system, which consists of multiple ministries and sub-units, including non-governmental organizations and industry actors. This diagram leaves the question open as to where in this governing system the implementation of the SSF Guidelines should be located. Given their broad agenda, and the institutional complexity described here, it is hard to imagine that it would be the responsibility of one single institution. Rather, the implementation of the SSF Guidelines would require a coordinating unit with a general overview.

Currently, MADR, through AUNAP, is formally in charge of the marine small-scale fisheries sector at the national level. Through this Ministry, INDERENA was active from 1968 to 1993 (Fig. 27.4). However, in 1990 the same Ministry created INPA, which became responsible for fishery resources until it was shut down by an administrative reform in 2003 (Decree 1293 May 21, 2003), although other hydro-biological resources continued to be the responsibility of INDERENA until 1993.

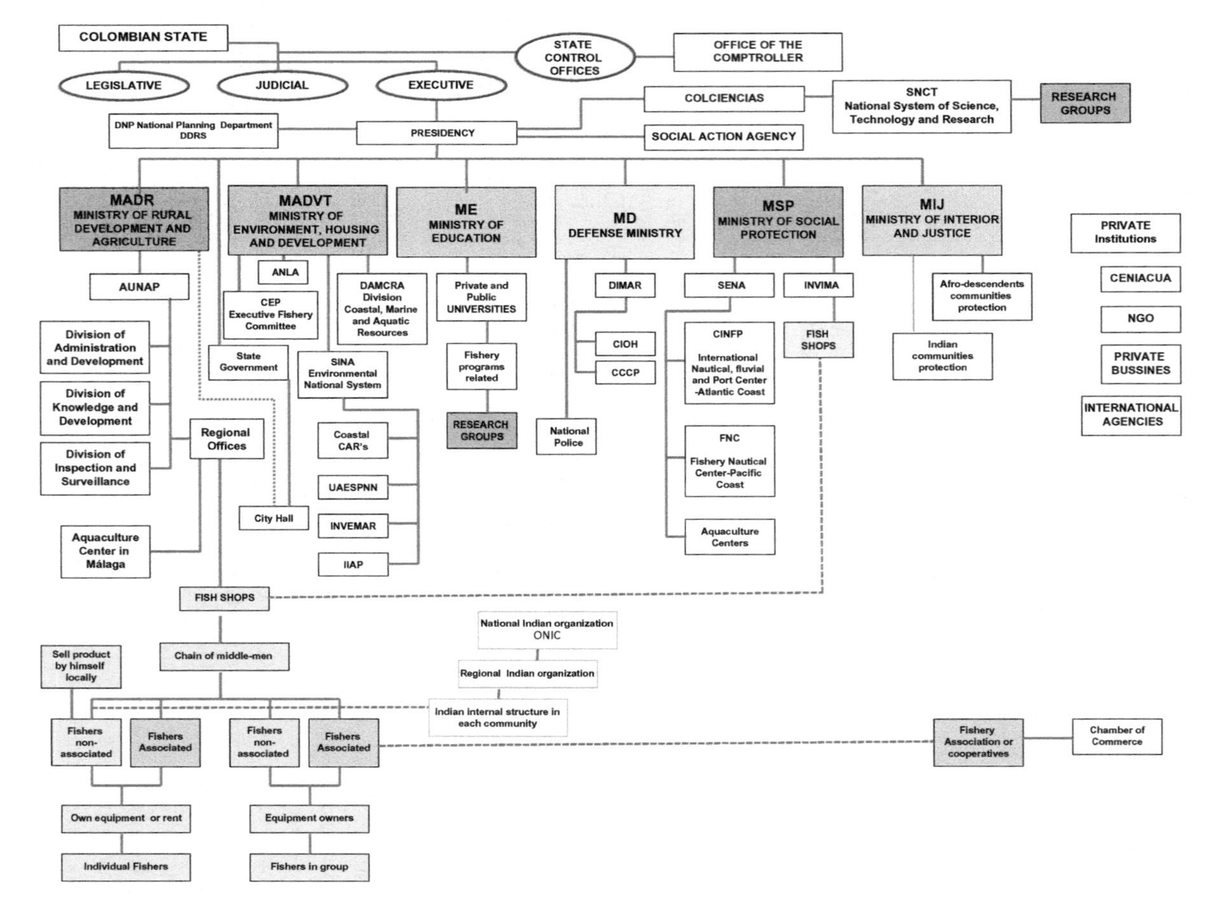


**Fig. 27.3** Present stakeholders that intervene in the small-scale fishery sector in Colombia

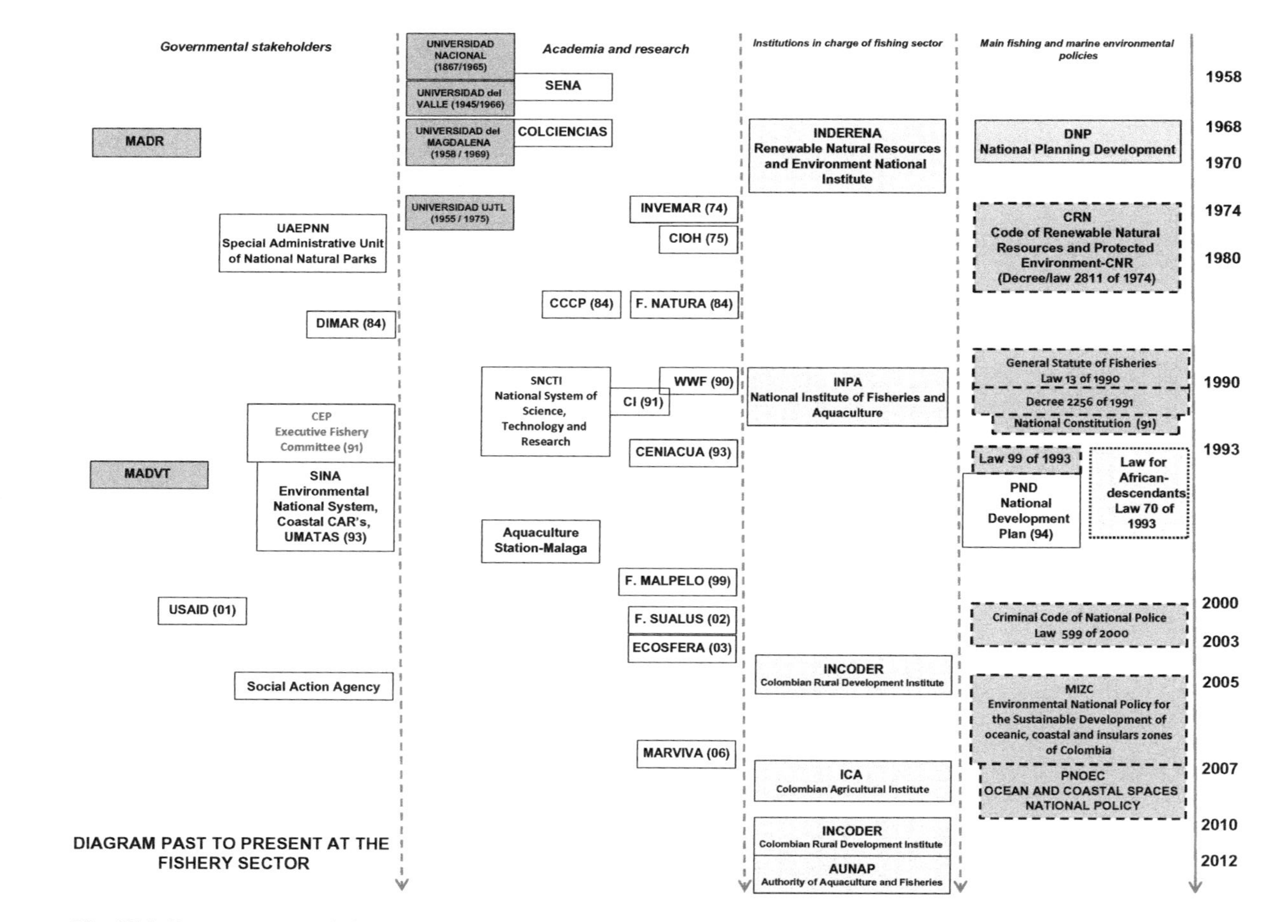


**Fig. 27.4** Past to present stakeholders at the small-scale fishery sector in Colombia

Due to this division of responsibilities, the fisheries sector faces uncertainty over which institution is in charge of it. Consequently, the environmental authority and fishery authority have always been at odds. After INDERENA was closed, all its responsibilities were transferred to the Environmental Ministry, today MADVT, which was created in 1993. Since then, fisheries have been administered primarily through both Ministries, with MADR as the fishery authority and MAVDT as the environmental authority. Since both shared oversight over the fishery sector, and created the Executive Fishery Committee (CEP) under Decree 2256 of 1991. CEP, MARD, and MAVDT jointly decide which species can be caught, set minimum fish size and catch quotas, among other issues related to fishery resources.

In 2003, the government enacted the Decree/law 1300 and consolidated INPA and other national institutes into INCODER. As a result, INPA's responsibilities were transferred to INCODER. However, the fisheries sector did not have a clear institutional representation from 2003 to 2004. Consequently, INCODER was only able to assume its administrative role in this sector from 2005 to 2007, when Law 1152 of 2007 shifted fisheries responsibilities mainly to ICA even though INCODER continued to be partially in charge. After Law 1152 of 2007, INCODER once again took charge of the fisheries sector and has remained so from 2010 to 2012. Finally, in 2012 AUNAP was created and is currently still in charge of the sector. AUNAP works through three technical offices centralized in the capital: Division of Administration and Development (DTAF), Division of Inspection and Surveillance (DTIV) and Division of Knowledge and Information (OGCI), and seven regional offices distributed along the national territory. Two of them are directly in charge of the coastal and marine small-scale fisheries sector: the Barranquilla office, with a responsibility for the Caribbean coast, and the Cali office for the Pacific coast.

At the time INDERENA closed in 1993, its environmental responsibilities were not only transferred to MAVDT but also to the Regional Autonomous Corporations (CAR), formed through the creation of the Environmental National System (SINA) by Law 99 of 1993. As part of SINA, there are 33 CARs at the national level whereas 12 of them are located in coastal states and act at the state level. Even though each corporation is the main environmental authority in each state, only since 2014 (Law 1092) the 11 coastal CAR's were recognized as coastal and marine environmental authorities. The only CAR before with legal authority was CORALINA that was able to make decisions and take action over the marine territory of San Andrés, Providencia, and Santa Catalina. As part of the same Ministry, in 2012 the Division of Coastal Marine Affairs and Aquatic Resources (DAMCRA) was created with the aim to preserve and sustainably use marine diversity and maintain ecosystem services.

A fundamental part of SINA is an important system of Protected Areas administrated by the Special Administrative Unit of National Natural Parks (UAESPNN) under MAVDT (Law 2811 of 1974). UAESPNN is organized in six Territorial Divisions with 59 parks at the national level. Ten coastal parks (1.379.751 hectares, an area which is equivalent to 1.4% of protected area in the country) are in constant interaction with small-scale fishing communities living in or fishing around park areas. Fishing activity long predated the establishment in 1974 of these protected areas, and conflicts between fishers and park administrators arise because of fishing activity in and around these areas. The Decree 622 of 1977 (Article 30/numeral 10, Chapter IX) prohibits: "...conduct which could have as a consequence the disturbance of the natural environment in areas of the Natural National System of Parks... Any act of fishing is prohibited, and only fishing activity for scientific purposes..., sport fishing, and subsistence fishing... are allowed". Subsistence is not well defined in the law, and ambiguities cause conflict. Fishers can only fish for subsistence purposes, but if they want to survive, they must sell some of their catch in order to pay for other needs. However, selling even a few fish makes fishing commercial. Consequently, most fishing that occurs in parks is illegal.

Decree 1753 of 1994 established the requirement for Environmental Impact Assessments. Private businesses must hire consulting groups in order to acquire permits from MADVT before they can alter any ecosystem, and in particular since 2011 through the creation of the National Authority of Environmental Licensees (ANLA). Impact assessments must be done before or after businesses interfere with the marine environment. Businesses manage their own studies for their own uses, but restrictions to this information limit outside access. Due to the nationwide epidemic of violence, poverty, drug-trafficking, and forced displacement, among other issues, the Presidency established the Agency for Social Action and International Cooperation-Acción Social (created by Decree 2467 of 2005), which provides many social programs at the national level. This agency and others have interacted with vulnerable communities, including coastal fishing communities. Although the majority of these programs work with small-scale fishers, some social programs were enhanced by interviewed fishers, such as the Families in Action and Forest Ranger Families programs. The United Nations Office to counteract Drugs and Crime (UNODC) and the United States Agency for International Development (USAID) have been working since 2001 in Colombia. Most of these programs are focused on Colombians in 1st, 2nd, and 3rd strata, in which fishers are situated. Through the Sisbén system, the National Planning Department (DNP) establishes which Colombian citizens can have access to economic support based on census information collected by the National Department of Statistics (DANE). Sisbén was created in order to invest governmental funds into poor and vulnerable populations via different social programs. A program of the Colombian Institute of Family Welfare (ICBF) provides children from 6 months to 5 years old with breakfast. Senior citizens are covered by social protection programs and food supplements are available for elders. Even though these social programs are not yet implemented in all municipalities, many coastal communities are covered by them.

Within this administrative structure, state governments and municipalities are very important due to the interactions of these offices with the marine fisheries sector. These interactions depend on the willingness of governors or mayors to support local fisheries; as a result, the person in office largely determines whether or not fishing communities receive support. Fishers fit into the governing system as members of fishers' associations or through their relationship with the chain of middlemen or fish shops directly. On one hand, once these associations are formed, they have to be registered with the local Chamber of Commerce, a private organization that does not belong to any ministry. On the other hand, non-associated and associated fishers engage with chains of middlemen that vary in number until the product gets to fish shops in large towns. These fish shops are registered in the new Rural Development Agency which is part of MADR, but are also overseen by the National Institute of Vigilance of Medicaments and Food (INVIMA) in MSP.

A host of governmental, non-governmental, private, and public institutions are directly involved with the marine fisheries sector at national, regional, and local levels (see diagram in Fig. 27.3). Each performs its own mission independently and simultaneously without any general framework to prevent overlapping jurisdictions. This lack of coordinating infrastructure creates a dysfunctional and unproductive system. Some small-scale fishing communities are constantly influenced by many agencies while others receive no support or do not interact with them. Furthermore, only fishers who fit the profile of an organization's study or agenda receive such temporary support. Interaction between fishing communities and governmental and non-governmental agencies is inconsistent, intermittent, and random, which confuse and sometimes overwhelm the fishers. As a result, many programs fail to yield results, and support given to communities is highly uncertain. Such disorganized, contradictory situations negatively affect any attempt at fisheries management. Many agencies interact with fishing communities on a host of fisheries issues, but since none of them is directly in charge of the fisheries sector, fishers often do not recognize any outside authority. Again, this might pose problems for the implementation of the SSF Guidelines.

## Small-Scale Fisheries Data and Knowledge

Even though government institutions in charge of the fisheries sector at the national level have generated important fisheries data through different research projects or studies, private and public institutions have also contributed substantially to fisheries research in Colombia. The Colombian Institute for the Development of Science and Technology 'Francisco José de Caldas' (COLCIENCIAS) was created in 1968 and has been in charge of research policies that promote knowledge, encourage development and improve the well-being of Colombian citizens. Fisheries research is addressed through the National Master Plan of Science and Sea Technology. There are currently around 123 groups carrying out research on fisheries, showing the diversity of groups that are generating knowledge under their own interest, and some could respond to the fishery sector priorities as other does not respond to it. Other master plans in SNCTI related to social and economic concerns also pertain to the fisheries sector. Groups from other disciplines (anthropology, education, history, economy, and law, among others) generate fisheries knowledge from a community perspective, even though their research is not directly connected with fisheries. Besides these groups, research potential exists at the national level to investigate a wide variety of knowledge areas pertaining to the coastal oceans.

Consequently, other ministries besides Agriculture and Environment support the fishery sector, including the Ministry of Education. Seventeen private and public universities on the nine coastal states offer academic undergraduate and graduate programs that are related to the marine fishery sector. However, the National University, Magdalena University, Jorge Tadeo Lozano University and Valle University have been foremost in marine fishery research. Besides the main programs in basic science observed in the Universities listed before, most large universities support research institutes focused on environmental science.

MSP within the Health Ministry includes a very important player in fishery training known as the National Service of Learning (SENA), created under the Decree/Law 118 of 1957. SENA has two centers in the fisheries sector, the International Nautical, Fluvial and Port Center (CINFP) in Cartagena on the Caribbean coast, and the Fishery Nautical Center (FNC) in Buenaventura on the Pacific coast. Besides these two centers, three Aquaculture centers provide education at the technical degree level. INVIMA also belongs to the MSP and controls the quality of fish sold in fish shops.

Among marine research institutes at the National level, the Marine and Coastal Research 'José Benito Vives De Andreis' (INVEMAR), created in 1974, belongs to MADR. The Oceanographic and Hydrographic Caribbean Research Center (CIOH), created in 1975, and the Center of Pacific Contamination Control (CCCP), created in 1984 belong to the MD. The Decree 2324 of 1984 placed them in the General Maritime Direction (DIMAR), the National Maritime Authority that regulates maritime security and provided mobilization permits to fishing boats through 13 Coast Guard Ports, nine on the Caribbean side and four on the Pacific side. The MD involves also the National Police, is part of the same Defense Ministry, and Criminal Code establishes penalties for crimes against natural resources and the environment.

Another institute in MADR, the Pacific Environmental Research Institute 'Jhon Von Neumann' (IIAP) is not focused on the marine environment, but works with coastal communities. When a researcher or research group studies African descendants or indigenous communities, they must get permission from the Ministry of Interior and Justice, following consultation processes since these minorities groups should be involve in any activity that affect their territory. Consequently, this Ministry is another important stakeholder affecting fishing communities since most small-scale fishers are of African or indigenous descendant.

Marine aquaculture is supported mainly by the Colombian Aquaculture Research Center (CENIACUA) (created in 1993), a private institution located on the Caribbean coast, and the ICA Aquaculture Station (2002) in Málaga Bay located on the Pacific. Other aquaculture stations belong to SENA and the Los Llanos University. Although they mainly work on freshwater fish species, SENA stations on the coast also work with marine fish species.

National and international NGOs (Non-Governmental Organizations), private foundations and corporations are very important actors that interact powerfully with small-scale fishery communities. Most of them base community projects and actions on community participation such as WWF, CI, Tropenbos, and Marviva, among others. At the same time, centers of research, such as the Colombian Caribbean Observatory and the Observatory of Racial Discrimination work directly with fishery communities of African descent. Recently, the AUNAP, in alliance with the University of Magdalena, has created the Colombian Fisheries Statistical System (SEPEC), which despite some limitations is now the most trusted source of fishing data. Consequently, this section shows the potential of different institutions that could help to provide a more elaborate fisheries knowledge-base. It is important to combine research from different disciplines (social, economic and environmental) and the knowledge of stakeholders in order to respond to the fishery crisis. Then, it would be for the government to apply this knowledge in management planning and decision-making.

## Towards a National Fisheries Policy

The legal framework that regulates fisheries resources at the national level is based on the Code of Renewable Natural Resources and Protected Environment (CNR) (created by the Decree/law 2811 of 1974) which delegated regulatory responsibility to the INDERENA. In 1990, Law 13 created the General Statute of Fisheries, and so far, this is the only regulation for the sector and requires urgently to be updated. The regulatory framework needs to be updated, based on solid science, policy planning, and management practices that recognize the need for conservation to achieve sustainability (ECOVERSA 2007).

The following concepts pertaining to marine fisheries resources are legally defined under Article 249 of Decree/law 2811 of 1974, Decree/law 1681 of 1978, Decree/ Law 13 of 1990, and Decree/ Law 2256 of 1991 (ECOVERSA 2007) are as follows: Fisheries resources are 'a part of hydro-biological resources that are able to be extracted or partially extracted without affecting their renewal capacity for purposes of consumption, processing, study, or for the purpose of other benefits'. Hydro-biological resources are 'all organisms that belong to the animal and plant kingdoms that have their life cycle totally in the aquatic environment'. Fishing is defined as "to make the most of any hydro-biological resource, through their caught products, extraction or collection". Even though the hydro-biological resources are wild fauna, Article 249 excludes 'fish and all species that have life cycle in water' from being so defined. Consequently, this overlapping among the concepts creates confusion of which institution should manage what resources and this misunderstanding needs to be resolved since the MAVDT is directly in charge of the hydro biological resources and the MADR is directly in charge of fisheries resources.

At the same time, the framework of fisheries law has been influenced by the Law of Reorganization of the Maritime and Ports Authority under Decree Law 2324 of 1984, the General Law of Agricultural and Livestock Development under Law 101 of 1993, the Food sanitary requirements under Decree 3075 of 1997, the System and

Implementation of Risks and Critical control points over fishing and aquaculture products (HACCP) through Resolution 730 of 1998 and Decree 60 of 2002, and the Law of Fishery and Aquiculture organizations chains under Law 811 of 2003.

Important resolutions and agreements that promote control and organization of small-scale fishing activity have included establishing closed seasons (shrimp), regulating fishing methods (marlin, sailfish, and swordfish), setting catch quota (queen conch, lobster, and Colombia's small-scale fishing catch quota), establishing minimum catch sizes (jaiba, lobster, and piangua), prohibitions against trade (shark fin traffic), regulating the features of fishing boats, and establishing exclusive areas for small-scale fishing. In relation to the designation of small-scale fishing catch quotas, it is not clear who has the right to use these quotas, how they are overseen or controlled, and who actually uses them (INCODER 2006). However, many of these regulations have been imposed from above without sounding out the opinion of fishers. Consequently, fishers reject their implementation. A reaction as the one from a fisher from Tumaco is common:

> If you tell me that I do not have to go to a certain place, I will be there right away. It is the same when it is the close season for shrimp. We were told that there are two months of closed season, we could not fish but at night, you will see 10, 20, 30 gillnets fishing. There is no order here in Tumaco and Borbur.

The General Statute of Fisheries (Law 13 of 1990) is the only Fishery Policy that has been regulated as Law and it urgently needs updating. Institutions have different policy documents related with fisheries and coastal marine ecosystems, but none of them have been converted into Law. In 2000, the MADVT and INVEMAR took over Leadership of the Environmental National Policy for the Sustainable Development of Oceanic, Coastal, and Insulars zones of Colombia (PNAOCI). The PNAOICI aims for sustainable use of oceanic spaces and coastal zones that, through their integrated management, improve the quality of life of the Colombian population, promote harmonious development among productive activities, and conserve marine resources and ecosystems (MMA 2000). In addition to PNAOCI, the Ocean Colombian Commission (CCO), in association with many institutions at the national level, created the Ocean and Coastal Spaces National Policy (PNOEC) in 2007, and recently in 2015 published an updated version. The National Presidency, through the National Planning Department-DNP, creates a National Development Plan every 4 years after each Presidential election. The directive informs a program of Agriculture, Livestock, Forestry, Fishing and Hunting through the Sustainable Rural Development Department-DDRS, and through the National Council of Economic and Social Policy-CONPES (established by Law 19 of 1958) advises the government every year in relation to policies that aid economic and social development. The social and economic portfolios within CONPES are the jurisdictions of different ministries. Some CONPES documents are concerned with fishery resources or fishing communities. For instance, CONPES 3164 (2002), the National Environmental Policy of Sustainable Development over marine, insular and coastal zones recognizes the importance of generating information about fisheries from the perspectives of zoning, planning, management, and sustainable use. Other documents focus on areas with marine fishing communities. Recently (2015), the new Fishery Authority (AUNAP) with the Ministry of Rural Development and Agriculture (MADR) and FAO created the Integral Fishery Sustainable Development National Policy (MADR 2015). All these Policies have defined and prioritized programs that respond to the small-scale fisheries sector but are not implemented under a legal framework. At the same time, there are governmental programs that donate boats and gears to fishing communities but they do not necessarily respond to the fishing community's needs. As stated by a fisher from *San Antero*:

> However, it is contradictory because the environmental institution -X- the first thing they do is to find money to buy gillnets to fish at the swamps that are breading places. I think that there are more gillnets than fishers. The swamp has many gillnets with small mesh size, and there are people who have even 1000 meters of them. This is too bad because they are closing the mouths of rivers and they do not let fish to get in. It is a big problem.

So far, there is evidence of efforts (technical documents such as Puentes Granada 2014) from different sectors, from Ministries to Fisheries Associations, governmental and non-governmental institutions that have tried to promote reforms of the Law 13, but so far with little success. There are only isolated reforms, such as on illegal fishing (Law 117 of 2015) that have been introduced. Even lately, the fisheries sector has received attention from different political parties. Some, parties, like MIRA and Partido Centro Democrático to promote legal change. It is interesting to observe that some of these Law projects are consistent with the SSF Guidelines principles. At the present time, it is known that the AUNAP is in the process to revive the Law project that will reform the Law 13 and it emphasizes the aquaculture sector as an important part of this new law.

Unfortunately, national planning documents also exist that are in conflict with rational and sustainable use of marine fisheries resources. One example is the 'Colombian Vision 2019' of DNP (2007), which expects an increase of as much as 30% in fisheries captures for 2019. Some fisheries experts consider these goals unrealistic, expecting such an increase to place critical pressure on fisheries resources that are already overfished (ECOVERSA 2007). On the other hand, interviewed fishers witness an incremental decrease in fishing effort. Consequently, they recommended the urgent need of implementing regulations and rule of law, since they believe it is not enough fish anymore and not enough for that many fishers (Saavedra-Diaz 2012). MADR submitted a recent law project (No. 117 of 2015 House and No. 162 of 2016 Senate) to combat illegal fishing activities. This law project is a top-down intervention that has not been subject to any consultation with the fishing population. Colombia is now in the process of joining OECD (Organization for Economic Co-operation and Development). This organization has released a document about the fisheries and aquaculture sector where they call for management plans with long-term objectives, built on the consensus with fishers (OECD 2016).

In this spirit, Law 70 of 1993 recognized the rights of African descendants to their territorial homes as collective property, regions where they have engaged in traditional activities such as agriculture, mining, and fishing, among others. It also aims to protect their culture identity by defining black communities as ethnic groups, and promotes socio-economic development in order to guarantee equal opportunities in Colombian society. This would also be a move that would find support in the SSF Guidelines, including in paragraphs that discuss tenure rights, but also ethnic minorities and customary practices.

## Conclusion

The small-scale fisheries governing system, as described above, is scattered among several agencies. Overall the system reveals complexity, fragmentation, and instability. Agencies have overlapping mandates and jurisdictions, and the coordination between them is poor. No permanent agency has the responsibility of incorporating international fisheries treaties that reflect the latest policy positions on international fisheries issues into national regulations, policies or management strategies, such as with the SSF Guidelines. Different government institutions and agencies have attempted to organize the fisheries sector and articulate policy through the publication of different sets of national documents. None of these policies reflect small-scale fishers' and fish workers' points of view. As a result, Colombia does not have a national fisheries policy that responds to the real problems of small-scale fisheries communities and fishing resources threats to them. The SSF Guidelines are therefore a timely reminder that the government and other parties need to get their act together, given that this institutional instability has been, and still is a major hazard which comes in addition to the problems described in the previous sections. Thus, the situation is characterized by a 'system-to-be governed' in disarray, a governing system that is very complex, and governing interactions which leaves small-scale fisheries communities at the receiving end of policies that have proven ineffective in addressing the poverty and marginalization problems that small-scale fishing people are experiencing in a context. With the peace arrival, there is time to do something with the obvious lack of a common platform: to create an institutional framework at national, regional, and local levels that integrates fisheries stakeholders in policy and management development.

In the Colombian context, the SSF Guidelines touch down in a country where small-scale fisheries communities have for a long time been victims of armed conflict and drug related violence. Small-scale fisheries have also been negatively affected by a largely dysfunctional and unproductive governing system, which has led to overfishing and environmental degradation, as well as a low quality of life due to poverty and lack of education and health services. Thus, the problems that the SSF Guidelines are aiming to address by invoking government, civil society and other stakeholders are certainly present in Colombian small-scale fisheries, but the governance systems and mechanisms that are needed to implement the initiatives that the SSF Guidelines are advocating, are underperforming at the outset. In this chapter, we argue, therefore, that in order for the SSF Guidelines to reach those people and communities they are meant to serve, not only is peace a necessary condition, it is now within reach. Also important is the need for institutional reform that can work for the benefit of small-scale fisheries in securing the human rights of small-scale fishers and fish workers. This process involves securing the food supply and tenure rights, but also at a very basic level the right to live in security, shielded from violence and other forms of abuse such as corruption. Obviously, the implementation of the SSF Guidelines in Colombia need less institutional fragmentation and more cooperation and coordination within various levels of government and with local communities and other stakeholders. For the implementation of the SSF Guidelines, an institutional overseer would be in order. The SSF Guidelines address other institutions that just those that specialize of fisheries, but they need an institutional 'home', one which takes responsibility for their implementation by working with other government agencies who have responsibilities for other functions that affect the well-being of small-scale fisheries communities, like public infrastructure, health, and education.

We argue that there is a need to incorporate international agreements that reflect the latest policy positions on international fisheries issues such as the SSF Guidelines and related instruments. This could, for instance, be a responsibility of the AUNAP. The aim should be to convert these commitments into national policies, management strategies, and regulations in accordance with the human rights and good governance principles and ambitions of the SSF Guidelines. This would also be an opportunity to involve all stakeholders and bring them under the same umbrella.

The implementation of the SSF Guidelines also would hinge on governance capacities and capabilities at the community level that can make local communities more self-reliant and resilient. That would imply a break with the traditional top-down governance structure that has dominated Colombia and its fisheries governance mechanism in the past. How to facilitate a governing system that works for small-scale fisheries communities is a question that relates directly to the prospects that the SSF Guidelines have to take effect. The good news is that the guiding governance principles as stated in the SSF Guidelines are highly consistent with those promoted in the Colombian peace process. Combining those two instruments should, therefore, bode well for the future of small-scale fishing communities in this country. Implementing the SSF Guidelines would hence be a useful tool for the Colombian government to reclaim and repair the fishing communities who have long been victims of the armed conflict but who now have a chance to experience peace.

## References


AI (Amnesty International). (2008). *'Leave us in Peace!' Targeting civilians in Colombia's internal armed confilct*. Amnesty International Publications.

Candelo, C., Cárdenas, J. C., Escobar, M., López, M. C., Maya, D. L., & Roldan, A. M. (2002). *Manglares comunidad y cooperación: Memorias de unos talleres y juegos en la Costa del Pacifico Nariñense (Cartilla)*. WWF Colombia.

CCO (Comisión Colombiana del Océano). (2015). *Política nacional del océano y de los espacios costeros-PNOEC*. Gobierno de Colombia.

CONPES (Consejo Nacional de Política Económica y Social). (2002). *Política nacional ambiental para el desarrollo sostenible de los espacios oceánicos y las zonas costeras e insulares de Colombia, Plan de Acción 2002–2004*. CONPES 3164. Departamento Nacional de Planeación-DNP y Ministerio del Medio Ambiente.

CONPES (Consejo Nacional de Política Económica y Social). (2016). *Estrategia de Posconflicto en Colombia*. Departamento Nacional de Planeación.

CORPOURABA. (2005). *Plan de manejo integral de los manglares del Golfo de Urabá y mar Caribe Antioqueño*. Municipio de Turbo: Consejo comunitario de Bocas del Atrato y Leoncito.

Delgado, M. F., Gualteros, W., Espinosa, S., Lucero, C., Roldan, A. M., Zapata, L. A., Cantera, J. R., Candelo, C., Palacio, C., Muñoz, O., Mayor, G., & Gil-Agudelo, D. L. (2010). *Pianguando – Estrategias para el manejo de la piangua (Cartilla). Serie de publicaciones generales INVEMAR no. 45*. Cali: Instituto de Investigaciones Marinas y Costeras.

Díaz, J. M., Vieira, C. A., & Melo, G. J. (Eds.). (2011). *Diagnóstico de las principales pesquerías del Pacífico Colombiano*. Bogotá: Fundación Marviva –Colombia.

DNP (Departamento Nacional de Planeación). (2007). *Documento Visión Colombia Segundo Centenario: 2019. Cuarta Edición*. http://www.dnp.gov.co/paginas_detalle.aspx?idp=366. Accessed 11 Dec 2016.

ECOVERSA. (2007). *Justificación sobre la necesidad de una nueva estructura para el manejo y ordenación de la pesca y acuicultura a nivel nacional (Draft version)*. Corporación ECOVERSA.

Escobar, A. (2003). Displacement, development, and modernity in the Colombian Pacific. *International Social Science Journal, 55*, 157–167.

INCODER (Instituto Colombiano de Desarrollo Rural). (2006). Producción histórica nacional de los recursos pesqueros en Colombia en el período 1975–2005. In *ECOVERSA, Justificación sobre la necesidad de una nueva estructura para el manejo y ordenación de la pesca y acuicultura a nivel nacional (Versión borrador)*. Corporación ECOVERSA.

FAO. (1995). *Code of conduct for responsible fisheries*. Rome: Food and Agriculture Organization of the United Nations.

FAO. (2015). *Voluntary guidelines for securing sustainable small-scale fisheries in the context of food security and poverty eradication*. Rome: Food and Agriculture Organization of the United Nations.

MADR (Ministerio de Agricultura y Desarrollo Rural). (2015). *Política integral para el desarrollo de la pesca sostenible en Colombia*. Food and Agriculture Organization of the United Nations y Ministerio de Agricultura y Desarrollo Rural.

MMA (Ministerio del Medio Ambiente). (2000). *Política nacional ambiental para el desarrollo sostenible de los espacios oceánicos y las zonas costeras e insulares de Colombia*. Ministerio del Medio Ambiente.

OECD (Organization for Economic Co-operation and Development). (2016). *Fisheries and aquaculture in Colombia*. https://www.oecd.org/tad/fisheries/Fisheries_Colombia_2016.pdf. Accessed 18 Mar 2017.

Puentes Granada, V. (2014). *Elementos para la construccion de una nueva ley de acuicultura y pesca en Colombia*. http://en.calameo.com/books/003960412ecd7b97c07e1. Accessed 10 Dec 2016.

Puentes, V., Escobar, F. D., Polo, C. J., & Alonso, J. C. (Eds.). (2014). Estado de los Principales Recursos Pesqueros de Colombia – 2014. Serie Recursos Pesqueros de Colombia – AUNAP. Oficina de Generación del Conocimiento y la Información, Autoridad Nacional de Acuicultura y Pesca – AUNAP.

Raakjaer, J., Son, D. M., Staehr, K.-J., Hovgard, H., Dieu Thuy, N. T., Ellegaard, K., Riget, F., Thi, D. V., & Hai, P. G. (2007). Adaptive fisheries management in Vietnam, the use of indicators and the introduction of a multi-disciplinary, marine fisheries specialist team to support implementation. *Marine Policy, 31*, 143–152.

Ramírez-Luna, V. (2013). *The exclusive fishing zone for the artisanal fishery in Chocó Colombia: Origins, development, and consequences for artisanal fisheries and food security* (Master thesis). Memorial University of Newfoundland, St. John's, NL.

Rincón Ruiz, A. (2014). Can common property regimes in Colombia curb the expansion of coca crops and the deforestation? In D. D. McBain (Ed.), *Power, justice and citizenship: The relationships of power* (pp. 47–62). Oxford: Interdisciplinary Press.

Rincón Ruiz, A., Pascual, U., & Romero, M. (2013). An exploratory spatial analysis of illegal coca cultivation in Colombia using local indicators of spatial association and socioecological variables. *Ecological Indicators, 34*, 103–112.

Saavedra-Díaz, L. M. (2012). *Towards Colombian small-scale marine fisheries management: Hacia un manejo de la pesca marina artesanal en Colombia*. (Doctoral dissertation). University of New Hampshire: Natural Resources and Earth System Science.

Saavedra-Díaz, L. M., Rosenberg, A. A., & Martín-López, B. (2015). Social perceptions of Colombian small marine fisheries conflicts: Insights for management. *Marine Policy, 56*, 61–70. doi:10.1016/j.marpol.2014.11.026.

Saavedra-Díaz, L. M., Pomeroy, R., & Rosenberg, A. (2016). Managing small-scale fisheries in Colombia. *Maritime Studies, 15*(6), 1–22. doi:10.1186/s40152-016-0047-z.

Salas, S., Chuenpagdee, R., Seijo, J. C., & Charles, A. (2007). Challenges in the assessment and management of small-scale fisheries in Latin America and the Caribbean. *Fisheries Research, 87*, 5–16.

Sánchez, O., & Moreno, C. (2009). *La política del recurso pesquero y acuícola en Colombia*. Contraloría General de la Republica, sector Agropecuario.

Squalus. (2008). *Pesquería artesanal de la zona norte del Pacífico Colombiano: Aportando herramientas para su co-manejo. Fase I. Caracterización espacial de las pesquerías artesanales*. Informe Técnico Final. Fundación Squalus e ICA.

UNODC (Oficina de las Naciones Unidas Contra la Droga y el Delito). (2016). *Colombia – Monitoreo de territorios afectados por cultivos ilícitos 2015*. UNODC y Gobierno de Colombia.

Vilardy, S., & Renán-Rodríguez, W. (2011). La influencia del conflicto armado en las dinámicas socioecológicas de la ecorregión Ciénaga Grande de Santa Marta. In S. Vilardy & J. A. González (Eds.), *Repensando la Ciénaga: Nuevas miradas y estrategias para la sostenibilidad en la Ciénaga Grande de Santa Marta* (pp. 152–171). Universidad del Magdalena y Universidad Autónoma de Madrid.

# Part VIII
# Building Capacity

Empowerment of small-scale fisheries and their people takes place along many dimensions of governance: legally, organizationally, and politically. Small-scale fishing people need to have security and control of their own destiny. But they also need knowledge and skills, and access to communication. This need, which is also stressed in the SSF Guidelines, is explored in the chapters of this part of the book. In Chap. 28, Kate Kincaid focuses on the potential and challenges of the implementation of the SSF Guidelines in the Bahamas, relating particularly to an integration of fishers' knowledge in sustainable resource management. The government is currently creating a network of marine protected areas (MPAs) in order to provide food security and sustainable livelihoods while conserving marine ecosystems. She argues for the need to collect fisheries and livelihood data within rural communities and to develop a national database on the extent, scale, and dependence of small-scale fishing. This problem is also the issue in the case study from Senegal reported in Chap. 29, where Aliou Sall and Cornelia Nauen critically examine conventional top-down research about small-scale fisheries and their communities. As an alternative, they aim to inform public policy making and resource management through understanding local knowledge and communicating from the ground up. Chapter 30 takes us further down along the coast of West Africa to Nigeria. Shehu Latunji Akintola, Kafayat Adetoun Fakoya, and Olufemi Olabode Joseph present a traditional small-scale fishery that, although successfully managed, faces issues such as policy inconsistency, inadequate communication, and limited participation of stakeholders. They argue that information, research, and communication should attempt to integrate the knowledge of fishers in order to improve planning processes. In Chap. 31, Zahidah Afrin Nisa takes us back to the Caribbean and Pacific regions. She stresses the need for cooperation and network-building between different sectors and disciplines, including the importance of dialogue between multiple layers of governance and the building of public–private partnerships at the local and national level.

# Chapter 28
# Challenges and Opportunities in Implementing the Small-Scale Fisheries Guidelines in the Family Islands, Bahamas

**Kate Kincaid**

**Abstract** This chapter discusses the challenges and opportunities for the Bahamas to implement the Voluntary Guidelines for Securing Sustainable Small-Scale Fisheries in the context of Food Security and Poverty Eradication (SSF Guidelines), developed through the Food and Agriculture Organization of the United Nations (FAO). Specifically, this chapter focuses on the implementation potential and challenges of the SSF Guidelines that relate to sustainable resource management and the need and opportunity to integrate fishers' knowledge. The Bahamas has a long fishing history based in small-scale fisheries. The Family Islands of the Bahamas are sparsely populated, largely undeveloped and fairly remote. Here, local populations rely on fishing as a primary source of food and income. The challenges for implementing the SSF Guidelines are particularly difficult in these islands. In addition, the Bahamian Government is implementing a wide network of marine protected areas (MPAs) to replenish fisheries and provide food security and sustainable livelihoods. Recognizing the importance of small-scale fishers and their knowledge contribution through the implementation of the SSF Guidelines in the Family Islands of the Bahamas is particularly timely as the Bahamas expands their MPA network. To assist with implementation in the Bahamas, there is a need to focus on the needs of small-scale fishers, to collect fisheries and livelihood data within rural communities and to develop a national database on the extent, scale, and dependence of small-scale fishing in the Family Islands of the Bahamas.

**Keywords** Artisanal • Bahamas • Resource management • Small-scale fisheries • SSF Guidelines

K. Kincaid (✉)
Department of Biology, Memorial University of Newfoundland, St. John's, NL, Canada

Cape Eleuthera Institute, Eleuthera, Bahamas
e-mail: kgb242@mun.ca



S. Jentoft et al. (eds.), *The Small-Scale Fisheries Guidelines*, MARE Publication Series 14, DOI 10.1007/978-3-319-55074-9_28

## Introduction

The small-scale fisheries sector and its communities include millions of rural people that are dependant on fisheries for their livelihoods worldwide (Jentoft 2014). However, there is often a lack of data, knowledge, and decision-making in small-scale fisheries compared to the large-scale fisheries sector. The Voluntary Guidelines for Securing Sustainable Small-Scale Fisheries in the context of Food Security and Poverty Eradication (SSF Guidelines) were adopted in 2014 by the United Nations' (UN) Food and Agriculture Organization (FAO) member states to address this problem (FAO 2015). These are the first internationally agreed guidelines designed specifically for the small-scale fisheries sector and are the result of extensive planning, consultations, and negotiations (Jentoft 2014). Implementing these Guidelines will present unique challenges for different countries. The SSF Guidelines includes three parts: an introduction that includes a set of guiding principles based on human rights, responsible fisheries and sustainable development. This is followed by descriptions of the SSF Guidelines for responsible fisheries and sustainable development and supporting implementation (FAO 2015). The SSF Guidelines include many aspects important to small-scale fisheries including governance, tenure rights, gender equality, disaster risks, capacity development, and monitoring.

In data-poor regions, fishers' knowledge is often the only source of information pertaining to an area. Local knowledge and natural history may be used in lieu of empirical data (Aswani and Hamilton 2004; Ban et al. 2009). In addition, the failure to understand fishers' needs will not benefit fisheries or conservation (Grafton et al. 2009). As primary stakeholders, there is a need to incorporate fishers' knowledge and perceptions into fisheries governance (Heck et al. 2011). There has been notable success in many parts of the world with community-based management (Johannes 1978, 2002; Mills et al. 2011). Fishers are able to provide bio-ecological data including habitat and site information alongside social economic data (Kincaid and Rose 2014). However, it is important to note that fishers cannot be included as a single group, since fishers' perceptions differ between different communities. This can be partially attributed to their individual fishing histories and backgrounds (Kincaid et al. 2014). Implementing the SSF Guidelines successfully will require the involvement and integration of local knowledge within rural fishing communities.

The Bahamas is an archipelago covering 259,000 km$^2$, and a Small Island Developing State (Smith and Zeller 2016). Land area covers just 13,935 km$^2$ that includes 700 islands, 29 of which are inhabited, with 90% of the population living on three main islands (Moultrie et al. 2016; Cox et al. 2005). These three islands are the economic centres for tourism, banking, and development industries. In contrast, the remaining islands are known as the Family Islands (or Out Islands) of the Bahamas. These are sparsely populated, largely undeveloped and fairly remote with small settlements (communities) located throughout. With limited employment opportunities, many local populations rely on subsistence fishing as a primary source of food. In many settlements on the Family Islands, fishing is a family tradition and is an important part of the community. Broad and Sanchirico (2008) interviewed local residents in the Family Islands on their support for a marine reserve in their local area. They found that respondents were less likely to support marine reserves if they were reliant on the area for fishing as their livelihood (Broad and Sanchirico 2008). In general, there is a lack of information and data on the fishing communities and the ecosystem including information on their needs and records on the status of subsistence fishing in the Bahamas (Moultrie et al. 2016; Smith and Zeller 2013).

The Bahamas have not yet implemented the SSF Guidelines however, in 2014 the Bahamas adopted a resolution to promote their implementation (Moultrie et al. 2016).This chapter aims to discuss the need and potential for their implementation, along with the associated challenges in doing so. Specifically, this question is approached through a focus on articles in Sections 5b (sustainable resource management) and 11 (information, research, and communication) of the SSF Guidelines. This chapter provides background to small-scale fisheries within the Bahamas and focuses on the rural communities within the more remote Family Islands and discusses how the use of fishers' knowledge and fisheries information could lead to successful implementation.

## Fisheries in the Bahamas

Historically in the Bahamas, fishing vessels were locally built sailboats (Bahamian sloop) and fishing gears were fish traps, hand lines, and nets (Adderley 1883). The largest fishing industry was the sponge trade, employing one third of the Bahamian workforce until 1940, when a fungal disease killed 90% of harvestable sponge within 2 years (Campbell 1978; Buchan 2000). In recent years, the sponge (*Porifera* spp.) fishery has returned to a few places. Sea turtles, particularly the green turtle (*Chelonia mydas*) were a commercial fishery until the 1990s and an important local food source until 2009, when the government issued a turtle fishing ban for conservation reasons. In the Bahamas today, there are around 4000 commercial fishing vessels with around 12,600 fishers, although the last fisheries census took place in 1998. The marine environment is productive, with a mixed fishery for shellfish and finfish (Fig. 28.1). Many fishing gears used are similar to those reported in 1883 (handlines, traps, and nets) along with dive compressor (with a license for conch and lobster), free diving, spears, and lobster hooks with condominiums (metal sheets on concrete blocks to aggregate lobsters – see Fig. 28.1). The government lists three categories of fishing: commercial, recreational (valuable for the tourism industry), and subsistence. Commercial fishing requires a license and is defined as fishing from any vessel over 20 ft. However, the majority of commercial vessels in use are often less than 20 ft and work from a large mothership. Only Bahamian citizens can own and operate commercials fishing vessels (Personal communication, E. Deleveaux, February 10th 2014).

**Fig. 28.1** Photographs illustrating fisheries in the Family Islands of the Bahamas. (**a**) Grouper, parrotfish and snapper mixed catch using spear and traps from one fisher's catch. (**b**) Local fisher at a typical landing site. (**c**) Fish trap in shallow patch reef habitat location. (**d**) Local fisher breath-holding and using a pull-spear to target snapper and spiny lobster in shallow patch reef habitat. (**e**) Mixed fish species (caught using nets) at the landing site ready to go direct to restaurants. Photo credits: (**a**, **b**, **c**, **e**), Kate Kincaid and (**d**), The Cape Eleuthera Island School

## Implementation Potential and Challenges

The diverse nature of the fisheries, habitats, variety of users, and diverging values and interests all add to difficulties in sustainable fisheries management for the Bahamas and to the future implementation of the SSF Guidelines (Holdschlag and Ratter 2013). Fisheries regulations and closed seasons are difficult to enforce and control (Holdschlag and Ratter 2013). Landings data reported to the FAO are from the main commercial landing sites and processing plants in New Providence, Grand Bahama, Abaco, Eleuthera (Spanish Wells), and Long Island (FAO 2009). It is not mandatory that fishers report landings data. A recent study by Smith and Zeller (2016) found reconstructed landings were 2.6 times greater than FAO figures that are based only on commercial landings supplied by the Department of Marine Resources. In 2014, the highest value ($USD) fishery species were spiny lobster (*Panulirus argus*), followed by conch (*Strombus gigas*), snapper (*Lutjanus* spp.), stone crab (*Menippe mercenaria*), and Nassau grouper (*Epinephelus striatus*) (Deleveaux 2016). Stocks of lobster, conch, snapper and grouper are thought to be fully, or overexploited (Moultrie et al. 2016, Deleveaux 2016; Hayes et al. 2015; Gascoigne 2002). In addition to these issues, the number of subsistence fishers is unknown, as well as the locations of landing sites, number of fishing vessels, and fishers. The following sections cover examples of how specific parts of Sections 5b and 11 of the SSF Guidelines relate to the Bahamas and the Family Islands, illustrate how they could be implemented, and discuss associated challenges in the rural communities of the Bahamas.

## SSF Guidelines Section 5b: Sustainable Resource Management

*5.13* States and all those engaged in fisheries management should adopt measures for the long-term conservation and sustainable use of fisheries resources and to secure the ecological foundation for food production. They should promote and implement appropriate management systems, consistent with their existing obligations under national and international law and voluntary commitments, including the Code, that give due recognition to the requirements and opportunities of small-scale fisheries.

There are a variety of fisheries regulations in place to sustainably manage marine resources in the Bahamas. The Department of Marine Resources (DMR) is responsible for fisheries management including enforcement and seafood processing regulations. Specifically, their objectives include sustainable fisheries for the benefit of the Bahamian people, local fishery development, and the improvement of the well-being of local fishermen (Government of the Bahamas 2011). They work closely with influential non-governmental organizations (NGOs), particularly the Bahamas National Trust (BNT). The aim of the BNT is the conservation and preservation of the natural environment and is the major organization that influences environmental policy in the Bahamas (Buchan 2000). A major focus of the BNT is the creation and implementation of a Protected Area network. In addition, the BNT works on environmental policy with the Bahamas Environment Science and Technology (BEST) commission. One other influential NGO is the Bahamas Reef Environment Education Foundation (BREEF), which focuses on educational training and programs.

The DMR manages 4 marine reserves, recognising traditional fishing rights in the South Berry Islands, Jewfish Cay, No Name Cay and Crab Cay. In addition, the DMR and The Nature Conservancy manage the South Berry Islands Marine Reserve (Moultrie et al. 2016). The government is implementing and expanding a wide network of marine protected areas (MPAs) managed by the BNT and in partnership with The Nature Conservancy. Through MPAs, the government aims to replenish fisheries and provide food security and sustainable livelihoods. In 2015, the Minister of the Environment and Housing announced the creation of 24 MPAs and the expansion of three existing MPAs adding three million hectares to the MPA network (UNEP 2015). Many of these areas are in the Family Islands and, while providing longer term biological benefits, would impact small-scale fishers through their displacement from fishing grounds if the MPA included closed areas.

*5.14* All parties should recognize that rights and responsibilities come together; tenure rights are balanced by duties, and support the long-term conservation and sustainable use of resources and the maintenance of the ecological foundation for food production. Small-scale fisheries should utilize fishing practices that minimize harm to the aquatic environment and associated species and support the sustainability of the resource.

Fishing regulations include gear restrictions (e.g. minimum mesh size for nets, hookah diving only in lobster closed season with a license, no spearguns), size regulations (specifically for target species: Nassau grouper (*Epinephelus striatus*), spiny lobster (*Panulirus argus*), conch (*Stombus gigas*), and stone crab (*Menippe mercenaria*)) and closed seasons (Nassau grouper, lobster, and stone crab). The Nassau grouper is both an endangered species according to the Convention on International Trade in Endangered Species of Wild Fauna and Flora (CITES) and is an important fishery. However, it is also a fully to over-exploited fishery and is protected through a three-month closed season (December – February) with a minimum landing weight of 3 lbs (1.36 kg) for al grouper species (BREEF 2013; FAO 2016).

*5.15* States should facilitate, train and support small-scale fishing communities to participate in and take responsibility for, taking into consideration their legitimate tenure rights and systems, the management of the resources on which they depend for their well-being and that are traditionally used for their livelihoods. Accordingly, States should involve small-scale fishing communities – with special attention to equitable participation of women, vulnerable and marginalized groups – in the design, planning and, as appropriate, implementation of management measures, including protected areas, affecting their livelihood options. Participatory management systems, such as co-management, should be promoted in accordance with national law.

Recognizing the importance of small-scale fishers and their contribution through the implementation of the SSF Guidelines and in the Family Islands of the Bahamas is particularly timely as the Bahamas expands their MPA network. Thus the fisheries and biodiversity conservation sectors will overlap and many of the designated and proposed MPAs are on the Family Islands (Bahamas Protected Areas Fund 2013). This would provide a platform for fishers' needs and rights to be considered and addressed. Local fishers have expressed their concerns about their displacement from their fishing grounds and reduced livelihoods due to MPAs (Wise 2014).

When looking at implementing Section 5 of the SSF Guidelines, it would be advisable to consider if local communities could be given responsibility to manage sustainable fishing practices and MPAs in a bottom-up approach. Such approaches have been successful in many parts of the world including the Pacific, as Locally Managed Marine Areas, and in Madagascar (Jupiter and Egli 2010; Harris 2007). It would, however, be prudent in some areas to implement new measures incrementally. For instance, in Madagascar, small closed areas resulted in an increased octopus fishery; local communities saw a benefit and supported further closed areas resulting in a successful large MPA network combining fisheries sustainability and marine conservation (Harris 2007). It is possible that the implementation of the SSF Guidelines could be a tool and an opportunity to build trust between the government and local fishers, utilize their local knowledge, and work together towards a common goal.

*5.17* States should ensure that the roles and responsibilities within the context of co-management arrangements of concerned parties and stakeholders are clarified and agreed through a participatory and legally supported process. All parties are responsible for assuming the management roles agreed to. All endeavours should be made so that small-scale fisheries are represented in relevant local and national professional associations and fisheries bodies and actively take part in relevant decision-making and fisheries policymaking processes.

A further challenge for the Bahamas is in documenting and integrating information nationally to support local food security, sustainable livelihoods, and conservation. The DMR has hosted meetings in local settlements but have found that it is hard to get participation if there is no perceived problem and the majority of fishers will only attend if there is a crisis at hand (E. Deleveaux, February 10th 2014, personal communication). This indicates a need to define the problems and develop an agenda based around needs that best support the local populations. There is also a need for a collective fishers' organisation to represent small-scale – subsistence fishers in the Family Islands (the already existing Bahamas Commercial Fishers Alliance represents the commercial sector).

## SSF Guidelines Section 11: Information, Research, and Communication

*11.1* States should establish systems of collecting fisheries data, including bioecological, social, cultural and economic data relevant for decision-making on sustainable management of small-scale fisheries with a view to ensuring sustainability of ecosystems, including fish stocks, in a transparent manner. Efforts should be made to also produce gender-disaggregated data in official statistics, as well as data allowing for an improved understanding and visibility of the importance of small-scale fisheries and its different components, including socioeconomic aspects.

There is a need to recognize and document levels of responsible and sustainable use in these areas, either through social surveys, in person interviews, or community meetings. This is fundamental to recognize human rights aspects and understand the importance of fisheries within these rural communities. This includes respecting local fishing practices and existing forms of organization within the communities. The DMR describes fishing communities in the Family Islands as crisis orientated conservative fishers that see no urgent reason to change (Personal communication E. Deleveaux, February 10th 2014). In addition, there are conflicts between fisher groups and gear users and many local communities are unhappy with the low level of consultation and outreach currently available (Waugh et al. 2010). This suggests that they may attend meetings, yet the low attendance recorded at such meetings is a cause for concern (Personal communication E. Deleveaux, February 10th 2014). This inconsistency is a problem that needs to be addressed before successful implementation of the SSF Guidelines. Fisheries data are difficult to assess because fish are largely sold directly to businesses leaving a lack of capacity to collect landing site data (Buchan 2000).

*11.2* All stakeholders and small-scale fisheries communities should recognize the importance of communication and information, which are necessary for effective decision-making. The government of the Bahamas does not know what the extent of the small-scale fishing is in the Family Islands and the DMR does not currently have the capacity to collect this information (Personal communication E. Deleveaux, February 10th 2014). This area should become a priority in the implementation of the SSF Guidelines, otherwise other issues and concerns of the SSF Guidelines cannot be properly addressed.

*11.5* States should ensure that the information necessary for responsible small-scale fisheries and sustainable development is available, including on illegal, unreported and unregulated (IUU) fishing. It should relate to, inter alia, disaster risks, climate change, livelihoods and food security with particular attention to the situation of vulnerable and marginalized groups. Information systems with low data requirements should be developed for data-poor situations.

There is a lack of biological, economic, and social data on fisheries in the Bahamas and the DMR lacks the resources to conduct such data collection and analysis (Waugh et al. 2010). Fishers' participation and inclusion is a necessary precursor to the success and support of such areas, otherwise they may not be willing to assist and provide data necessary to successful implementation. In data-poor areas, local fishers and small-scale fishing communities can provide extensive ecological and social knowledge (if they feel secure that this information will not be used against them). To fully implement the SSF Guidelines in the Bahamas, a focus on the Family Island fishing communities is needed including the rationalities that drive the fishers (understanding what they would gain and lose, particularly in the context of the expanding MPA network) and the data services they could provide.

*11.6* All parties should ensure that the knowledge, culture, traditions and practices of small-scale fishing communities, including indigenous peoples, are recognized and, as appropriate, supported, and that they inform responsible local governance and sustainable development processes. The specific knowledge of women fishers and fish workers must be recognized and supported. States should investigate and document traditional fisheries knowledge and technologies in order to assess their application to sustainable fisheries conservation, management and development.

Implementation will be a particular challenge within the Family Islands because of the lack of data available, as well as the low capacity to record data coupled with the remoteness of many of these island communities. It is unknown how many landing sites there are in each of the Family Islands. Hence, it is clear that there is a pressing need to establish a system to collect even basic fisheries data in the Family Islands, including information on the number of fishers, boats and fishing methods. A central database on fishers' knowledge data for the Family Islands would be a valuable asset and would recognize and include small-scale subsistence fishers that have been largely left out of marine resource management in the Bahamas. The DMR has a system in place that they aim to expand to the family islands in the future, this would greatly increase the documenting of traditional fisheries knowledge in the family islands (Moultrie et al. 2016).

*11.9* States and other parties should, to the extent possible, ensure that funds are available for small-scale fisheries research, and collaborative and participatory data collection, analyses and research should be encouraged. States and other parties should endeavour to integrate this research knowledge into their decision-making processes. Research organizations and institutions should support capacity development to allow small-scale fishing communities to participate in research and in the utilization of research findings. Research priorities should be agreed upon through a consultative process focusing on the role of small-scale fisheries in sustainable resource utilization, food security and nutrition, poverty eradication, and equitable development, including also DRM and CCA considerations.

Lastly, access to funding is an issue. The Bahamas has a high per-capita gross domestic product (GDP), yet the distribution of wealth is uneven, from high wealth to poverty and subsistence levels, particularly in the Family Islands. Due to this, the Bahamas is often not eligible for development funding to aid rural communities, a cause for concern for many Bahamians (Cox et al. 2005).

## Conclusions: Moving Forward

This chapter focused on the potential and challenges for the Bahamas to implement the SSF Guidelines with an emphasis on the need to integrate fishers' knowledge and fisheries information in sustainable resource management. This refers specifically to Chapter 5b and 11 of the SSF Guidelines and, within this context, provides examples relevant to some of the specific articles within. The Bahamas already has systems in place that apply to parts of the implementation of the SSF Guidelines (e.g. fisheries management measures for long-term conservation and sustainable use). The specific challenges for the Bahamas are in the Family Islands. These islands are data-poor and many small-scale fishers and their communities rely on fishing for their income and food. The extent of this reliance is unknown and there are no current data available on demographics, fisheries landings or the status of fisheries in the Family Islands. In addition to this knowledge gap for the Family Islands, the general status of fish stocks in the Bahamas is unknown.

The following are recommendations towards the successful implementation of the SSF Guidelines sections within the contexts of sustainable resource management and data collection:

1. There is a need to establish a consultative two-way process on the role of small-scale fisheries in the Family Islands in broad aspects of sustainable resources, conservation, food security, and poverty eradiation
2. This should include the development of a national database to collect traditional fisheries knowledge and fisheries information to assist with sustainable resource management (this is particularly timely as the Bahamas expands its MPA network). Through this, the integration of biological, ecological, social, and economic small-scale fisheries and information on the extent, scale, and dependence of the small-scale fishing sector for the Family Islands is vital
3. Alongside this, food security and livelihood support need to be taken into consideration for the Family Islands including isolation, poverty issues, and limited access to resources that make it harder for such fishers to be heard

Finally, to successfully implement the SSF Guidelines, the Department of Marine Resources will need to work closely with local organizations, the BNT, and BREEF, particularly in the areas of community education and supporting community decision making for fisheries management. These organizations already work together in fisheries management and marine protection. One priority should be to improve the relationship between local fishers and government through the formation of a fisher's organization that fully represents local fishers within rural communities, and to provide a spokesperson to speak to the government on individuals behalf. However, the challenges in this may be in allocating the time, personnel, and finances dedicated to such endeavours. One final point, is the important and recent report by Moultrie et al. (2016) reviewing fisheries and aquaculture in the Bahamas. This discussed the upcoming draft Strategic Plan for Fisheries and Aquaculture Development and Management in The Bahamas 2017-2022. In this Strategic Plan, the need for tenure rights for fishers is discussed alongside proposals to include artisinal fishers in the decision-making process and calls for their inclusion within a Fisheries Advisory Council. These are positive steps towards the implemetation of the SSF Guidelines, including Family Island fishers, and achieving sustainable fishery goals for the Bahamas.

## References

Adderley, A. (1883). The fisheries of The Bahamas. *International fisheries exhibition*. London: William Clowes and Sons.

Aswani, S., & Hamilton, R. J. (2004). Integrating indigenous ecological knowledge and customary sea tenure with marine and social science for conservation of Bumphead parrotfish (*Bolbometopon muricatum*) in the Roviana Lagoon, Solomon Islands. *Environmental Conservation, 31*(1), 69–83.

Bahamas Protected Areas Fund. (2013). *Information on protected areas in the Bahamas.*www.bahamasprotected.com. Accessed 28 Oct 2016.

Ban, N. C., Hansen, G. J., Jones, M., & Vincent, A. C. (2009). Systematic marine conservation planning in data-poor regions: Socioeconomic data is essential. *Marine Policy, 33*(5), 794–800.

BREEF (Bahamas Reef Environmental Foundation). (2013). *A quick guide to the fisheries regulations of the Bahamas*. www.breef.org. Accessed 3 Nov 2016.

Broad, K., & Sanchirico, J. N. (2008). Local perspectives on marine reserve creation in the Bahamas. *Ocean and Coastal Management, 51*(11), 763–771.

Buchan, K. C. (2000). The Bahamas. *Marine Pollution Bulletin, 41*, 94–111.

Campbell, D. G. (1978). *The ephemeral Islands: A natural history of The Bahamas*. London/Basingstoke: MacMillan Education Ltd..

Cox, L., Hammerton, J. L., & Wilchcombe, N. (Eds.) (2005). *GEO Bahamas - The Bahamas state of the environment report*. Global Environment Outlook, UNEP, & Bahamas Environment, Science, and Technology Commission (BEST).

Deleveaux, E. (2016). *Bahamas: An overview on the fisheries and marine resources inventory*. WECAFC-FIRMS data workshop. Barbados, 19–21 January 2016.

FAO. (2009). *Fishery and aquaculture country profiles. The Commonwealth of the Bahamas. Fisheries and Aquaculture Department*. Rome: Food and Agriculture Organization of the United Nations.

FAO. (2015). *Voluntary guidelines for securing sustainable small-scale fisheries in the context of food security and poverty eradication*. Rome: Food and Agriculture Organization of the United Nations. ISBN 978–92–5-108704-6.

FAO. (2016). *Fishery resources report: Bahamas Nassau grouper fishery*. Rome: Food and Agriculture Organization of the United Nations.

Gascoigne, J. (2002). *Nassau grouper and queen conch in The Bahamans. Status and management options*. London: MacAlister Elliott and Partners Ltd..

Government of the Bahamas. (2011). *About the Department of Marine Resources, Ministry of Agriculture, Marine Resources and Local Government*. http://www.bahamas.gov.bs. Accessed 28 Oct 2016.

Grafton, R. Q., Hilborn, R., Squires, D., Tait, M., & Williams, M. (2009). *Handbook of marine fisheries conservation and management*. Oxford: Oxford University Press.

Harris, A. (2007). 'To live with the sea': Development of the Velondriake community-managed protected area network, Southwest Madagascar. *Madagascar Conservation and Development, 2*(1), 43–49.

Hayes, M. C., Peterson, M. N., Heinen-Kay, J. L., & Langerhans, R. B. (2015). Tourism-related drivers of support for protection of fisheries resources on Andros Island, The Bahamas. *Ocean and Coastal Management, 106*, 118–123.

Heck, N., Dearden, P., & McDonald, A. (2011). Stakeholders' expectations towards a proposed marine protected area: A multi-criteria analysis of MPA performance criteria. *Ocean and Coastal Management, 54*(9), 687–695.

Holdschlag, A., & Ratter, B. M. (2013). Multiscale system dynamics of humans and nature in The Bahamas: Perturbation, knowledge, panarchy and resilience. *Sustainability Science, 8*(3), 407–421.

Jentoft, S. (2014). Walking the talk: Implementing the international Voluntary Guidelines for securing sustainable small-scale fisheries. *Maritime Studies, 13*(1), 1–15.

Johannes, R. E. (1978). Traditional marine conservation methods in Oceania and their demise. *Annual Review of Ecology and Systematics, 9*(1), 349–364.

Johannes, R. E. (2002). The renaissance of community-based marine resource management in Oceania. *Annual Review of Ecology and Systematics, 33*(1), 317–340.

Jupiter, S. D., & Egli, D. P. (2010). Ecosystem-based management in Fiji: Successes and challenges after five years of implementation. *Journal of Marine Biology*. doi:10.1155/2011/940765.

Kincaid, K. B., & Rose, G. A. (2014). Why fishers want a closed area in their fishing grounds: Exploring perceptions and attitudes to sustainable fisheries and conservation 10 years post closure in Labrador, Canada. *Marine Policy, 46*, 84–90.

Kincaid, K. B., Rose, G., & Mahudi, H. (2014). Fishers' perception of a multiple-use marine protected area: Why communities and gear users differ at mafia island, Tanzania. *Marine Policy, 43*, 226–235.

Mills, M., Jupiter, S. D., Pressey, R. L., Ban, N. C., & Comley, J. (2011). Incorporating effectiveness of community-based management in a national marine gap analysis for Fiji. *Conservation Biology, 25*(6), 1155–1164.

Moultrie, S., Deleveaux, E., Bethel, G., Laurent, Y., Maycock, V., Moss-Hackett, S., van Anrooy, R. (2016). Fisheries and Aquaculture in The Bahamas: A Review. Food and Agriculture Organisation of the United Nations/Department of Marine Resources, Nassau, The Bahamas.

Smith, N. S., & Zeller, D. (2013). *Bahamas catch reconstruction: Fisheries trends in a tourism-driven economy (1950–2010)* (Working paper #2013–08). Vancouver: University of British Columbia/Fisheries Centre.

Smith, N. S., & Zeller, D. (2016). Unreported catch and tourist demand on local fisheries of small island states: the case of The Bahamas, 1950–2010. *Fishery Bulletin, 114*(1).

UNEP. (2015). *Bahamas expand marine protected areas by more than 3 million hectares*. UNEP News Centre. http://www.unep.org/NEWSCENTRE. Accessed 21 Oct 2016.

Waugh, G. T., Braynen, M. T., Bethel, W. G., & Gittens, L. (2010). *The ministry of agriculture and marine resources: Department of marine resources five year sector strategic plan 2010–2014*. Nassau: The Government of the Bahamas.

Wise, S. P. (2014). Learning through experience: Non-implementation and the challenges of protected area conservation in the Bahamas. *Marine Policy, 46*, 111–118.

# Chapter 29
# Supporting the Small-Scale Fisheries Guidelines Implementation in Senegal: Alternatives to Top-Down Research

**Aliou Sall and Cornelia E. Nauen**

**Abstract** The Voluntary Guidelines for Securing Sustainable Small-scale Fisheries in the Context of Food Security and Poverty Eradication (SSF Guidelines) provide a generally accepted framework that emerged from 3 years of intense bottom-up consultations and that were finally endorsed by the FAO's Fisheries Committee (COFI) in 2014. Chapter 11 emphasizes the importance of information, research, and communication that is accessible to and involves small-scale fishers and their communities. Research questions in conventional research tend to meet the needs of donors or partners while the needs of the researched are only purportedly addressed. In this chapter, we explore types of research that attempt to understand local knowledge produced and seek ways to facilitate communication and exchange between different modes of knowledge production that overcome the sense of exclusion currently experienced by small-scale operators. The ability to access other sources of knowledge is important to their coping strategies, especially under conditions of advanced globalization of markets for fisheries products that result in changing working and living conditions. Field research conducted with fishers and summarized here focus on fishers' knowledge and perceptions. Implementing the SSF Guidelines will require a sustained process of building trust, mutual understanding, and institutional arrangements that enable respectful multi-directional communication and exchange. Our research suggests that where such conditions are created and sustained, they form a robust base for participatory and legitimate management and thus retain many of the positive social and distributional dimensions that could make small-scale fisheries central to food security and sustainable livelihoods.

A. Sall (✉)
Mundus maris – Sciences and Arts for Sustainability, Dakar, Senegal
e-mail: aliou@mundusmaris.org

C.E. Nauen
Mundus maris – Sciences and Arts for Sustainability, Brussels, Belgium
e-mail: ce.nauen@mundusmaris.org



S. Jentoft et al. (eds.), *The Small-Scale Fisheries Guidelines*, MARE Publication Series 14, DOI 10.1007/978-3-319-55074-9_29

**Keywords** Small-Scale fisheries • Endogenous knowledge • Participatory research • Critically engaged science

## Introduction

The authors of this chapter have throughout their working lives in Africa and Europe, and in the context of different types of international cooperation, experienced situations where research questions were primarily driven by either generic research curiosity, government or donor needs, or some other motivation than the perceived needs and challenges faced by small-scale fishers. Recent catch reconstructions by country show that global catches of small-scale fisheries increased between 1950 and 2010 from 8 to 22 million tons per year whereas industrial catches, which receive the bulk of attention from governments, while bigger, have declined since the mid-1990s (Pauly and Zeller 2016).

Indeed, we argue that the lack of attention to small-scale fishers as a whole and to their specific research needs in particular, aggravates their political marginalization and prevents recognizing their true socio-economic and environmental importance. Their importance is in terms of food production and local food security, employment and positive distributional effects across societies, the wider economy, and *de facto* management of coastal and marine territories and their ecosystems. The general lack of attention paid to small-scale fishers is compounded by a deficit of interest in effective valuations of the empirical knowledge present with fishers while conducting conventional research. Relatively few studies have been conducted explicitly to compare and, if appropriate, integrate fishers' knowledge into the research process (Christensen and Thi 2004; Wilson et al. 2006). A critically engaged use of fishers local knowledge would reflect the fact that scientists treat it as a legitimate source of knowledge. Simultaneously, recognizing fishers' knowledge would reduce their open defiance to conventional management efforts.

The combination of these two factors (marginalization of fishers and a deficit in the valuation of their local knowledge) has gradually led to a gap in many places between researchers on the one hand and fishing communities on the other. This results in turn in a lack of interest and recognition by fishing communities and the major social groups within them in conventional research and its importance. Searching for paths and means to establish bridges between these two modes of making sense of fisheries has potential to enrich both in many cases and ensure the sustainability of small-scale fisheries in their biological, economic, social, and cultural dimensions. The dynamism and high adaptability of many small-scale fisheries in the face of globalized markets of fisheries products make the development of new forms of collaboration between researchers and actors in the small-scale sector even more important and urgent, particularly for those charged with management.

## The Changing Face of Small-Scale vs. Industrial Fisheries – Critically Engaged Research and Exchange Is a Must

Good management uses effective and participatory monitoring and critically engaged research to inform all stakeholders in decision-making processes. That also entails keeping track of management outcomes and learning from experience. It should have the ability to identify and distinguish between different groups, and provide gendered perspectives that will exist amongst stakeholders. The speed of change and new dynamics arising in small-scale fisheries in response to changes in markets and environmental conditions, in terms of the constraints in access to resources as a result of these changes, and as a result of social and economic requirements of fishers, are often much faster than captured by conventional research (Sall 2012a; Meltzoff 2013).

During a workshop coorganized in 2012 by the National Collective of Artisanal Fishers in Senegal (CNPS) and its support NGO, Centre de Recherche sur les Technologies Intermédiaires de Pêche (CREDETIP), in Mbour, several changes in the country's fisheries were discussed that would have required adaptation of research protocols for collecting official statistical information. By way of example, we refer here to the organizational changes in the fisheries for high value bottom species. Formerly, fishing units within this segment of the fishery included a combination of a boat, prevalent fishing technique, target species, and a fixed landing place ('port'). However, as international market penetration and number of landing places increased, the opportunistic behavior of crews also increased. Catch was landed in ports other than the home port and often was not properly accounted for. This was at first true in the case of line fisheries with ice boxes, but gradually also occurred in other types of fisheries.

Another significant change in the country's fisheries was the growing importance of variable technological choices in small-scale fisheries for demersal species as compared to earlier stable specializations. This was a response to lower densities of traditionally fished species and market demand elasticity accepting substitute species or simply expanding overall.

Despite these fast and far-reaching changes in the fisheries, the data collection protocols of research centers operating in the seven member countries of the Sub-Regional Fisheries Commission (SRFC) headquartered in Dakar, Senegal, have not been adapted to reflect these developments. The gap between the assumptions underlying scientific data collection and evolving ground realities, in fact, is widening and may well lead to mistaken interpretations or weak management advice – see examples below (Sall 2012a) (Fig. 29.1).

Moreover, despite more attention being given to the role of women in small-scale fisheries in the last two decades, namely with regard to production and service activities, overall studies on gendered differences within fisheries are under-resourced in most countries. Even in Asia, where the vast majority of small-scale fishers are located, and where women are traditionally at least as important as men in management and postharvest activities (Williams et al. 2005; Brugère 2014), researchers

**Fig. 29.1** Awa Seye (standing, in *blue*) is a leader of fisher women in Guet Ndar, St. Louis and Senegal as a whole, fighting for their rights to health, education and working spaces (Photo credit: Mor Talla Ndione)

and activists working on gender issues, though growing, are still comparatively small in number. This is evident, for example, in the huge biannual Asian Fisheries Forum where presentations on gender in fisheries are limited to a comparatively small symposium taking place alongside (Gender in Aquaculture and Fisheries n.d.)

In addition to reliable information, access to information is important too. As we are seeing in Senegal, small-scale fishers are constantly seeking strategies to cope with constraints and opportunities. These strategies include getting involved in new multi-functional fisheries, migration, and the adoption of new coping strategies by women fish processors and marketers in the face of increasing competition to access the raw material. To illustrate increasing diversification among e.g. fishers focused on small pelagics we observe two main types: (a) processing on board the now bigger-sized pirogues and (b) prolonging fishing trips up to 20 h to more distant grounds and using large quantities of ice to ensure product quality, a practice previously confined to line fisheries for high-value demersal species.

A recent development is the apparent decapitalization in the pelagic fishery. Fishers used to deploy encircling nets for small pelagics. Pelagic landings formed a major part of the total fish processed and marketed by women. When the local CFA Franc currency was tied to the Euro, this resulted in certain inputs being priced out of the range of fishers. Hence, access to resources became a big problem and boat owners reduced the size of the pirogues, the planked undecked fishing boats typical for Senegalese small-scale fisheries. New pirogue constructions may now measure 14 or 18 m rather than 20 m in length. Boat owners also started downsizing the number of pirogues they had, from two to one. This could create the impression of a reduction of fishing capacity and effort. But a closer look at the ground reality showed that the reduced number of boats fished more intensely, with more fishing trips and additional use of monofilament driftnets, thus primarily reducing fixed cost, but not necessarily effort.

Not taking into account the fishers' commercial adaptation strategies in economic, social, and biological analyses leads to misleading conclusions about the adaptability of small-scale fishers in their continual search for survival and new opportunities. This change in deployment of boats and gear might thus amount to an actual increase of fishing effort rather than a decrease. At the same time, the change in target species might have major repercussions on the local postharvest economy, such as on women in processing and marketing, especially if uncommon species are caught for foreign markets. However, if monitoring and research efforts are not tracking these changes in a timely manner, the statistical data and analyses will not reflect the evolving realities.

Overall, the growth and permutations of small-scale fisheries around the world (Pauly and Zeller 2016) are in stark contrast to the predominantly industrial model underlying many national sector policies. There seems to exist a widespread 'social Darwinian' assumption in development aid and policy reform efforts that artisanal fisheries under competition from industrial fleets will evolve into a form of modern industry by way of an intermediate semi-industrial state. This has been largely falsified by empirical evidence (Sall 2012a, b, 2013).

## Why Participatory and More Fine-Grained Information and Research Are Particularly Important for Policy

At the same time, policy objectives that previously focused mostly on production and balance-of-payments are being modulated by new criteria and macro-objectives, such as food security, sustainability, poverty reduction, and employment. It is therefore important not only for the fishers themselves, but also for those setting the rules and defining policies, to gain realistic quantitative and qualitative insights into the developments of all types of fishing and postharvest activities and the social groups associated with these activities.

Formats of monitoring and statistical data collection in Senegal seem not to have evolved enough to capture those multiple changes reliably. Thus, a cursory glance of official documents also cited by donors, the FAO, and NGOs, show the same figure of approximately 600,000 jobs (between 63,000 and 100,000 in primary activities, depending on the source) created by the fisheries, mentioned for the last 10 or more years (Deme and Kebe 2000; FAO 2008; World Bank 2009) even though the above-mentioned changes within fisheries have certainly affected the size and structure of employment.

Even the classical distinction between industrial and small-scale fisheries may not be adequate in all cases. This is particularly the case in the artisanal fishing villages of Hann, Yoff, Mbour, and Joal in Senegal, where major quantities of fish are transshipped at sea – an officially illegal operation – from industrial vessels to pirogues to be landed as 'artisanal' catches. Entire value chains from fishmongers to women fish processors and micro-vendors have specialized in this raw material and supplied markets in the interior of the country (Sall 2012a, b).

The value chains identified during a field enquiry in 2009 of the transshipment of industrial vessels to small-scale pirogues in Senegal revealed that species concerned were mostly greater forkhead (*Phycis blennoides*), smoothmouth sea catfish (*Carlarius heudelotii*), lesser African threadfin (*Galeoides decadactylus*), law croaker (*Pseudotolithus senegallus*), and Guinean grunt (*Parapristipoma humile*). All these species were not accounted for in official documents (Sall 2009).

Some small-scale fisheries elsewhere in the West African region highlight similar situations. In Guinea, the fish smoking women in Bonfi and Taminataye rely heavily on industrial catches of smoothmouth sea catfish and barracuda (*Sphyraena barracuda*) transshipped to small-scale boats and landed as small-scale catches. While experienced scientists at the Marine Research Center in Boussoura are aware of the practice taking place, these catches are not reflected in national statistics (Sall 2008).

Moreover, the women managing the local traditional smoking operations of small pelagics in some Gulf of Guinea countries will have difficulties keeping up this activity without the pelagic freezer-trawlers providing the raw material. In fact, in the smoking villages of Gbesse Chorkor, Elmina, and Winiba in Ghana and in the port town of Cotonou, among other sites, industrial fisheries deliver the bulk of the mackerel (*Scomber japonicus*), horse mackerels (*Trachurus trecae*, *Caranx* spp.), and sardinella (*Sardinella* spp.) these women smoke. Unless the recording systems are well designed and implemented these landings might be attributed wrongly to small-scale fishers.

The fluid distinction between small-scale and industrial fisheries along the value chain suggests the need to review the way research data and national statistics are collected and processed. The wider problem for dramatically under-resourced national accounting systems in Africa has been analyzed and remedial action proposed (Jerven 2013). For the fisheries sector in particular, one way to distinguish small-scale and industrial catches is to assess frozen landings of industrial origin, irrespective of whether further handling or processing is done through traditional channels or not. In other cases, small-scale catches are grossly under-estimated. This and other developments underscore the importance of involving fishers in the research process.

The recent reconstructions of catches in the exclusive economic zones (EEZ) of all countries and of tuna and other catches of large pelagics in the high seas between 1950 and 2010 illustrate the seriousness of misrepresentations caused by an exclusive or at least predominant emphasis placed by a majority of countries on industrial fisheries landings (Zeller et al. 2016). This leads to a massive underestimation of the contribution of small-scale and subsistence fisheries to global catches. It also leads to serious distortions of statistics and other sources of input into sector policies, investment decisions, and the way fisheries as a whole are perceived. Overall, global catches were 53% higher than official, FAO-compiled, national statistics have us believe. The decline after peak production in 1996 was steeper than reported by the FAO. The decline was primarily in industrial fisheries; small-scale fisheries are still expanding (Pauly and Zeller 2016).

## Increasing the Benefits of Research – A Socio-economic and Ethical Necessity

Chapter 11 of the SSF Guidelines pertinently emphasizes the importance of making information, research, and communication accessible to small-scale fishers and involving them and their communities more in decision-making processes.

Thus, while the different research questions and drivers mentioned above are perfectly legitimate, we argue that the considerable effort and the resources channeled into such research could yield much higher benefits, including for small-scale fishing communities, if they are more involved. Among the factors that can be considered key to generating such additional benefits, we emphasize the need for greater attention being paid to making the process and resulting scientifically validated knowledge freely available. This should be done in different ways for different types of audiences so as to enable much more widespread access to this knowledge. This may take the form of executive summaries for decision makers, context-adapted narratives for small-scale fishers, visual representation of results, and dialogue among stakeholders to facilitate appropriation (Nauen 2002, 2005).

Moreover, specific research efforts are warranted to address the need for information and understanding of small-scale fishers beyond their own observations and experiences. This is best developed and conducted with fishermen, fisherwomen, and traders themselves in order to make sure that the questions and findings are framed in an understandable way and take account of local community knowledge. Recognizing explicitly the legitimacy and even need for different perspectives and approaches to research is among the most important ways to address interconnected issues in the complex societal and environmental systems in which small-scale fisheries operate (Nauen 1999; Sall 2012a, 2013).

In order to illustrate the desirability of doing research in a more critically engaged manner, we offer several examples. During multiple field studies over the last few years, the first author, as a fisheries socio-anthropologist, sought to enable some form of 'communication' between different perspectives on the research subject. This way, fisher communities gained access to scientific knowledge validated and recognized by conventional research and were able to compare it with their own knowledge.

Estimating the biomass of a stock and the part that can be safely harvested is a challenge for researchers as recently developed methods for data-poor situations are not yet widely adopted (Froese et al. 2016). Fishermen have developed their own methods of estimating the size of a school of small pelagics before paying out the encircling net, as it constitutes a heavy workload. They derive an estimate from the way the fish move at the surface (belly pointing down or sideways) and thus decide on whether it is profitable to deploy the net (Sall 2008). As fuel costs account for between 78 and 83% for encircling seines and gillnets respectively, the two main fishing gears for pelagic catches (Dème 2012), this is a vital assessment.

The empirical knowledge of fishermen about the environmental conditions their target species need and what gear type is best to catch that species is documented in other fisheries as well, both marine (e.g. observations along the coast from Mauritania to Guinea Bissau) and inland waters (e.g. in France) (Barthélémy 2005).

During a study of the biology and ecology of small pelagics in 2008, the first author discovered specialized knowledge among two distinct categories of female postharvest actors: women descaling and cleaning the landed catches and others having open-air fast food stands at the landing places in Hann, Joal, Mbour, Kayar, Guet Ndar, and Gokhou Mbathi in Senegal. The same profile was found in Guinea, e.g. in Boulbinet, Taminataye, and Bonfi. As the women clean raw fish daily, they can distinguish as to whether migration patterns of the round sardinella (*Sardinella aurita*) are due to reproduction or feeding purposes. Their experience allows them to identify egg-bearing fish at a glance. If their knowledge were recognized, these women could be of support to biological research, even though they appear only marginally involved with the fishery itself (Sall 2008).

Moreover, the women in these socio-professional categories could contribute in other ways to research, notably through their remarkable knowledge about the relationship between environmental conditions and the quality of eggs. They are able to distinguish when eggs mature during evisceration whether they will mature into liquid, viscous, or firm form, something worth validating through rigorous comparative observation.

We note that the involvement of fishers can have other significant advantages for different reasons as well. The communities' full involvement gives them a sense of responsibility throughout the process, from the field work up to the sharing of results. There is also potential for a substantial reduction in the transaction costs of the research activities thanks to their willingness to shoulder certain costs.

The second author studied 500 years of self-organization history of small-scale fishers in a fjord on the Baltic coast of Germany and used biological data from several decades of research to construct a model of the ecosystem that had supported the fishery during recent times (Nauen 1984). Her reporting back to the heads of the fishermen's cooperative was welcomed as she was the only scientist to have done so. It was also fruitful to both sides. The fishers were extremely pleased with a fresh interpretation of their empirical observations over a period of 40 years. They were also able to add more nuance to the ecosystem model of the scientist based on their intimate empirical knowledge.

## Collecting Recent First Hand Evidence in Senegal About Complaints and Perceived Needs of Small-Scale Fishers

In recognizing the challenge of exploring alternatives or complements to 'top-down' ways of characterizing and trying to manage fisheries we carried out field research in Senegal in 2014–2015. The trigger for searching for alternatives were the serious difficulties faced in implementing fisheries policy reform in Senegal, attempted since the late 1990, despite substantial support by major donors such as the World Bank. A declaration at the end of an Inter-ministerial Council on Fisheries in 2013 stated: 'In relation to the artisanal fishery the weak regulation of access to the resource is linked to the following factors: (i) ineffective implementation of fishing licenses and their unsuitability as the principal mean to reduce artisanal fishing effort, (ii) insufficient monitoring of pirogue registration, and (iii) weak implementation of the Fisheries Code resulting from insufficient personnel, weak deterrent effect of sanctions foreseen in the Fisheries Code and insufficient awareness and extension of the legal provisions.' (free translation) (République du Sénégal 2013). Resistance by coastal communities to the implementation of control measures supported by the Regional Program for West Africa of the World Bank (PRAO) was effective (République du Sénégal 2015).

Field research consisted of semi-structured interviews. We administered 270 questionnaires in a period of 3 months, from November 2014 to January 2015. The focus of our questions was on the obstacles to fisheries policy reform in Senegal. 234 questionnaires were detailed enough for further analysis. They produced some interesting leads of direct relevance to this chapter. Given these interesting leads, the first author followed up with three sessions of focus group discussions in 2015 involving 625 boat owners/fishermen, 340 women fish processors and retail vendors, and 263 fish mongers in seven major landing sites of the small-scale fisheries. These were Guet Ndar in Saint Louis, Kayar, Yoff, Hann, Mbour, Joal, and Ngaparou. The sites were chosen because together they account for the majority of small-scale landings and the major types of small-scale fisheries ranging from mostly localized fishing to fishing that required wide-ranging migration (Thiao and Ngom-Sow 2014). The sites also had encountered different experiences with regard to conservation initiatives that provided us a diversity of perspectives on the on-going policy reform efforts (Fig. 29.2).

The principal types of fisheries practiced in each of the seven sites and other important characteristics are summarized in Table 29.1.

Conversations were conducted in Wolof, a mostly oral language. Care was taken to remain faithful to the spirit of the expressions used during the conversations when translating them into English. We highlight those comments that were shared by the largest numbers of people in the respective categories and that we also encountered in other circumstances as of broad relevance beyond the specific Senegalese context (Wiber et al. 2004) (Fig. 29.3, Table 29.2).

Women fish processors, who are less publicly articulate, shared many of the concerns expressed by the fishermen and boat owners. There were too few of them

**Fig. 29.2** Caption: Smaller boats are beached, but bigger pirogues are moored past the surf in Kayar (Photo credit: Paolo Bottoni)

with clearly diverging views in the context of this inquiry to tease out strong gender-specific concerns.

Beyond gathering complaints about the predominant state of play between researchers and small-scale fishers in Senegal right now, focus group conversations also generated insights about what the participants' aspirations and expectations are of the type of research that would be most useful for them and how they hope to access such research. The demand for more reliable statistical information in easily accessible formats was articulated by participants across all groups. More specifically, women fish processors expressed a strong need for price information from major interior and regional markets in coastal Senegal where buyers come to buy produce. 86 women (23 from Joal, 11 from Mbour, 20 from Kayar, and 32 from Saint Louis) expressed such a view (Fig. 29.4).

Another noteworthy demand in line with the general framework of the SSF Guidelines concerns the production and sharing of data and analyses about good local governance practices. Altogether 72 informers (55 boat owners/fishermen, 5 fish mongers, and 12 women fish processors and vendors) were keen to learn about experiences regarding regenerating certain ecological niches as a result of establishing particular protection zones or marine protected areas (MPAs) in Ngaparou and Ouakam (Table 29.3).

The suggestions for improvements of access to information collected in field research have two principal aims: (a) get help to access the resource; and (b) increase trust among operators and managers through greater transparency, particularly with

**Table 29.1** Main characteristics of the seven landing sites in Senegal where qualitative field research was conducted in 2014 and 2015

| Landing site | Principal fisheries | Other remarks |
|---|---|---|
| Guet Ndar (Saint Louis), as identified in scientific publications and even official documents, is the name used here as well. However, on the ground Guet Ndar is subdivided further (see middle column). | Fisheries differ by neighborhood within the fishing village:<br>(a) Dakka is the original nucleus of the traditional village and predominantly still practices purse seine fisheries for small pelagics derived from an earlier gill net fishery<br>(b) Guet Ndar with its subdivisions of Gokhou Mbathj and Pondo Kholé practices mostly line fisheries for fresh fish landings, either for one day or for longer trips with ice box pirogues. | Due to the combined pressure of space limitations on the Barbarie peninsula, demography, and limited local market, the fishers of Saint Louis have seasonally migrated north and southwards for long periods of time. Operations are not confined to Senegal, but extend into Mauritania and other neighboring countries with entire families moving seasonally. These fishers are often involved in conflicts over access to resources within Senegal and elsewhere, where they often operate without licence. |
| Kayar | The submerged delta in front of Kayar is an ideal space for line fisheries for valuable fresh fish. A few encircling nets for small pelagics are also being operated more recently.<br>For the past 3–4 years a growing number of line fishermen from Kayar are migrating to Mauritania for the seasonal fishery of fresh largehead hairtail which is in demand from the factories serving mostly Asian markets. | Kayar does not only support a local line fishery alternating with vegetable growing in the off-season, but also hosts migrant fishers from Saint Louis in the line fishing season. Recently a largehead hairtail (*Trichiurus lepturus*), a circumtropical and temperate species much appreciated in Asian cuisines, was developed. |
| Yoff | Predominantly a daily line fishery for fresh fish, with few pirogues using encircling nets. | Yoff line fishermen going mostly on day-trips have traditionally migrated seasonally to Gambia and Petite Côte (Mbour, Ngaparou, and Saly) in Senegal. The boats using encircling nets have tried to be accepted as 'locals' in Gambia, but face new difficulties because local authorities have tried to constrain them since 2015 to land catches there instead of in Senegal. |
| Hann | Though officially prohibited, the dominant gear types practiced in Hann are gillnets and driftnets. | Pirogues for line fishing with ice on board owned by people from Saint Louis have mostly relocated to Bissau and Guinea Conakry in the last 3 years, reducing local effort.<br>Weak *Sardinella* catches in the same period have led to relocation of most pirogues operating encircling nets to Kafountine (Casamance) and Tanji (Gambia). Some line fishers from Yoff and Soumbédioune (Dakar) going for red pandora (*Pagellus bellottii*) now operate out of Hann. |

(continued)

**Table 29.1** (continued)

| Landing site | Principal fisheries | Other remarks |
|---|---|---|
| Mbour | Line fishing during day trips is predominant in Mbour. Encircling nets are also frequently used according to season. | Conversely, line fishers from Mbour participate in the frenzy for largehead hairtail off Kayar, while others, line fishers and encircling nets, move south to Tanji (Gambia) and Kafountine (Casamance). |
| | Depending on the seasonality of species Mbour also hosts fishers from other communities specialized in gillnets, lines, or shell fish. | |
| Joal | Though prohibited, encircling gillnets and driftnets are popular for catching small pelagics. Large quantities of yêêt (Wolof, Neptun's volute – FAO) (*Cymbium* sp.) and Toufa (Wolof, Crowned rock shell – FAO) (*Stramonita rustica*) are (gill)netted seasonally. | Some of the fishing effort for *Ethmalosa* has been displaced to the Saloum Islands as a result of gear restrictions and access to processing areas for inland markets (Burkina Faso). |
| Ngaparou | Primarily line fishing. The combination of strong fisher participation, active traditional authorities regulating access and long-term support from NGOs for local MPAs ensured enforcement of rules that keep destructive gears out of area. Conflicts with driftnet fishers from Thiaroye and Rufisque flared up nevertheless in 2013 and 2015 and have started to taint the perceptions of the policy reform process. | The frenzy for largehead hairtail led to seasonal migration of local fishers to Kayar and Yoff in 2014 and 2015. |

regard to knowledge about real catches. At the same time, the suggestions place institutional demands, notably the need to recognize fishers as stakeholders and fishers' rights to participate in official processes. Moreover, willingness of the fishers to participate actively in the research, including to reduce public costs, have been voiced.

These points tally well with what has also been found elsewhere (Wiber et al. 2004) and concrete efforts underway to produce such information. We do still observe, however, a great deficit in accessibility of results because comparatively little effort has been made to 'translate' important research results into formats that can be readily understood and used by the fishers themselves, especially when they may be illiterate in the country's official language or not have access to electricity (and hence sources of information running on electricity) on a regular basis.

**Fig. 29.3** Boats coming to the landing area in Hann, Senegal (Photo credit: Aliou Sall)

## What Can We Learn About Information and Communication Challenges and Opportunities Available from the Research So Far?

Several of the generic provisions of Section 11 of the SSF Guidelines, particularly Paragraphs 11.1. on fisheries data acquisition, 11.3 on transparency, 11.7 on support to, 11.8 on availability, flow and exchange of information, and 11.10 on promotion of research into the conditions of work of small-scale fishers, including those of migrant fishers, fish workers, and processors, many of whom are women, capture well the concrete expectations of fishers in Senegal about what type of research information would be most relevant for them (FAO 2015). The last of these, namely research into the conditions of small-scale fishers, is much under-researched (Salas et al. 2011).

The expectation of fishers are also echoed in the voices that emerge from field research elsewhere. There is a recurrent emphasis on the importance of fishers' and fisher communities' concrete abilities to actively participate (Berkes et al. 2001; Salas et al. 2007; Franks and Small 2016; Jounot 2016). Three areas of priority action were elaborated upon during the Southeast Asia Regional Consultation Workshop, a region of the world where a large percentage of the global population of small-scale fishers are located, on the implementation of the SSF Guidelines (Franz et al. 2015). The three areas were grouped around (a) governance of tenure

**Table 29.2** The most broadly shared comments and perspectives of small-scale boat owners/fishermen, women fish processors, and vendors and fish mongers with regard to research and access to its results

| Type of comment/perspective | Participants expressing views by category | Repartition by place [total] | |
|---|---|---|---|
| The data researchers produce serve primarily to accuse us of wrong-doings: fishing effort and increase of the number of boats. These data serve to justify restrictions. | Boat owners/Fishermen: 124 | Saint Louis: 28 | [Total: 124] |
| | Fish mongers: 0 | Kayar: 13 | |
| | Women fish processors and vendors: 0 | Yoff: 18 | |
| | | Hann: 21 | |
| | | Mbour: 25 | |
| | | Joal: 16 | |
| | | Ngaparou: 3 | |
| The researchers are more interested in fish and the money made in the fishery, but not in our living conditions. Research on our livelihoods and customs is absent. | Boat owners/Fishermen: 160 | Saint Louis: 60 | [Total: 160] |
| | Fish mongers: 0 | Kayar: 18 | |
| | Women fish processors and vendors: 0 | Yoff: 11 | |
| | | Hann: 12 | |
| | | Mbour: 34 | |
| | | Joal: 25 | |
| | | Ngaparou: 0 | |
| The management of MPAs, with the restoration of certain ecological niches, are based on the judgement of the biologists of the Dept. of Marine Fisheries (DPM) and of CRODT[a], but also on the expertise of WAMER-WWF[b]. The fishermen highlighted the case of thiof (white grouper) and spiny lobster, which recovered in record time in MPAs like Ngaparou. But speaking about Ngaparou the fishermen attribute the results there to a specific context – fully participatory community-based processes – where the local knowledge of fishermen has also been considered. | Boat owners/Fishermen: 74 | Saint Louis: 12 | [Total: 105] |
| | Fish mongers: 26 | Kayar: 3 | |
| | Women fish processors and vendors: 5 | Yoff: 11 | |
| | | Hann: 11 | |
| | | Mbour: 13 | |
| | | Joal: 13 | |
| | | Ngaparou: 42 | |
| Research is removed from daily practice in the fishery. (The interviewees argue that the dynamic developments in the fisheries cast doubt even on the inquiry sheets of the CRODT and DPM researchers, which are ill-adapted to track these changes. In Yoff e.g. fishermen challenge the traditional distinction between pirogues and decked boats). | Boat owners/Fishermen: 76 | Saint Louis: 25 | [Total: 107] |
| | Fish mongers: 21 | Kayar: 16 | |
| | Women fish processors and vendors: 10 | Yoff: 11 | |
| | | Hann: 23 | |
| | | Mbour: 23 | |
| | | Joal: 12 | |
| | | Ngaparou: 9 | |

(continued)

**Table 29.2** (continued)

| Type of comment/perspective | Participants expressing views by category | Repartition by place [total] | |
|---|---|---|---|
| The intellectuals of CRODT (researchers) make hasty judgements about certain species. When the expected results are not borne out by field observations, this is a problem for us as it casts doubt on the usefulness of the research (most interviewees mentioned octopus as an example where researchers frequently misjudged). | Boat owners/Fishermen: 89 | Saint Louis: 46 | [Total: 145] |
| | Fish mongers: 43 | Kayar: 25 | |
| | Women fish processors and vendors: 13 | Yoff: 15 | |
| | | Hann: 11 | |
| | | Mbour: 21 | |
| | | Joal: 13 | |
| | | Ngaparou: 14 | |
| The research reports serve to justify project applications and funding requests to donors. | Boat owners/Fishermen: 80 | Saint Louis: 21 | [Total: 134] |
| | Fish mongers: 29 | Kayar: 36 | |
| | Women fish processors and vendors: 25 | Yoff: 0 | |
| | | Hann: 0 | |
| | | Mbour: 33 | |
| | | Joal: 32 | |
| | | Ngaparou: 12 | |
| Rarely do scientists return to us after collecting data to share results – the majority return to validate their data because they work for the donors of projects: French Cooperation, African Development Bank, World Bank etc. – while explanatory feedback is rare. | Boat owners/Fishermen: 110 | Saint Louis: 42 | [Total: 176] |
| | Fish mongers: 34 | Kayar: 21 | |
| | Women fish processors and vendors: 32 | Yoff: 16 | |
| | | Hann: 10 | |
| | | Mbour: 35 | |
| | | Joal: 33 | |
| | | Ngaparou: 19 | |
| The research effort should be more directed at the small pelagics which are essential stocks on which the fishers depend (this alludes to the exploitation and disappearance of bottom fish species). | Boat owners/Fishermen: 45 | Saint Louis: 20 | [Total: 106] |
| | Fish mongers: 30 | Kayar: 21 | |
| | Women fish processors and vendors: 31 | Yoff: 0 | |
| | | Hann: 21 | |
| | | Mbour: 24 | |
| | | Joal: 20 | |
| | | Ngaparou: 0 | |
| We hear scientists talk about studies on small pelagics by the FAO in collaboration with our authorities, but these seem to be secret. Or they are only available on the internet or are difficult to access. The use of electronic media is not as widespread as many think. | Boat owners/Fishermen: 95 | Saint Louis: 26 | [Total: 123] |
| | Fish mongers: 25 | Kayar: 14 | |
| | Women fish processors and vendors: 3 | Yoff: 18 | |
| | | Hann: 11 | |
| | | Mbour: 25 | |
| | | Joal: 17 | |
| | | Ngaparou: 12 | |

(continued)

**Table 29.2** (continued)

| Type of comment/perspective | Participants expressing views by category | Repartition by place [total] | |
|---|---|---|---|
| Scientific discourse without illustrations is unsuitable in convincing fishermen. There are many seminars and conferences on climate change and / or El Niño, but they are not matched by attempts to explain the content in ways that fishers can understand. | Boat owners/Fishermen: 130 | Saint Louis: 56 | [Total: 168] |
| | Fish mongers: 21 | Kayar: 43 | |
| | Women fish processors and vendors: 17 | Yoff: 21 | |
| | | Hann: 7 | |
| | | Mbour: 17 | |
| | | Joal: 11 | |
| | | Ngaparou: 13 | |
| In addition to producing excessively long reports, no effort is made to make results more readily understandable, e.g. with the help of facilitators (most suggested that NGOs and their own associations would have a role to play). | Boat owners/Fishermen: 110 | Saint Louis: 25 | [Total: 132] |
| | Fish mongers: 13 | Kayar: 17 | |
| | Women fish processors and vendors: 9 | Yoff: 20 | |
| | | Hann: 17 | |
| | | Mbour: 27 | |
| | | Joal: 18 | |
| | | Ngaparou: 8 | |
| Until the end of the 1980s, we had information on the number of pirogues, catches, and even prices of fish at first point of sale and further down the marketing chain. The market at Gueule tapée is an example of this and CRODT published a red-colored bulletin about it. Through the CRODT research data collectors on the beaches everybody (including fishers) had access to this information. This important service has been discontinued since. | Boat owners/Fishermen: 110 | Saint Louis: 41 | [Total: 110] |
| | Fish mongers: 0 | Kayar: 0 | |
| | Women fish processors and vendors: 0 | Yoff: 10 | |
| | | Hann: 33 | |
| | | Mbour: 17 | |
| | | Joal: 4 | |
| | | Ngaparou: 5 | |
| Paper publications with data are now replaced by electronic communication which we cannot access. Data are only useful if we have access. | Boat owners/Fishermen: 118 | Saint Louis: 38 | [Total: 138] |
| | Fish mongers: 20 | Kayar: 13 | |
| | Women fish processors and vendors: 0 | Yoff: 0 | |
| | | Hann: 31 | |
| | | Mbour: 29 | |
| | | Joal: 27 | |
| | | Ngaparou: 0 | |

(continued)

**Table 29.2** (continued)

| Type of comment/perspective | Participants expressing views by category | Repartition by place [total] | |
|---|---|---|---|
| Unlike in the case of farmers and cattle breeders, we cannot see and touch a concrete research achievement (examples of farmers doing so are that of groundnuts, maize, and legumes). | Boat owners/Fishermen: 125 | Saint Louis: 47 | [Total: 125] |
| | Fish mongers: 0 | Kayar: 33 | |
| | Women fish processors and vendors: 0 | Yoff: 11 | |
| | | Hann: 8 | |
| | | Mbour: 8 | |
| | | Joal: 10 | |
| | | Ngaparou: 8 | |
| The meteorological service is a demonstration of the immense utility that research can play. It has contributed a lot to the improvement of our working and living conditions (deaths at sea have been reduced significantly). | Boat owners/Fishermen: 134 | Saint Louis: 42 | [Total: 164] |
| | Fish mongers: 23 | Kayar: 12 | |
| | Women fish processors and vendors: 7 | Yoff: 17 | |
| | | Hann: 30 | |
| | | Mbour: 31 | |
| | | Joal: 23 | |
| | | Ngaparou: 9 | |

[a]*CRODT* Centre de recherches océanographiques de Dakar-Thiaroye (Oceanographic Research Centre Dakar-Thiaroye)
[b]*WAMER-WWF* West African Marine Ecoregion Project of WWF

**Fig. 29.4** Refrigerated trucks from Senegal and Burkina Faso await loading of freshly landed sardinella in Guet Ndar (Photo credit: Paolo Bottoni)

**Table 29.3** Most widely shared proposals for types of desired research and improved access to research results as expressed by small-scale fishermen, boat owners, and fish mongers and women fish processors in seven major small-scale landing sites in Senegal

| Types of expected research information | Number of persons by category and site | Alternative approaches proposed |
|---|---|---|
| International fish price information (for fresh fish) | **132 fish mongers**, of whom 26 are in Saint Louis, 19 in Kayar, 46 in Hann, 19 in Mbour, and 22 in Joal | We need information about business practices and prices in major export markets, such as Spain, France, Italy, and Greece. The central fish market in Dakar as well as those fish landing sites with internet access could be focal points to relay this information. |
| Information about the biology and ecology of the major commercial species, particularly the small pelagics | **173 fishermen** (active and non-active boat owners and captains), of whom 42 are in Saint Louis, 41 in Kayar, 15 in Yoff, 23 in Hann, 14 in Mbour, 21 in Joal, and 17 in Ngaporou | Preparation and access to visual supports about the biology and ecology of small pelagic species (similar to what has been done for yêêt (snails), particularly round and flat sardinella and *Ethmalosa* (of particular interest to Nyominka fishers in Joal). Information should be made available as close as possible to actual working spaces, not in meetings and seminars in faraway places. |
| Information on visual supports about the effects of climate change on marine ecosystems, in particular on the health status of valuable demersal species, but also small pelagics | **118 boat owners/fishermen**, of whom 22 are in Yoff, 11 in Hann, 32 in Mbour, 20 in Joal, and 33 in Ngaparou | Production of posters and other visual supports that can be made available locally (fish landing places). |
| Statistical information collected by CRODT (the marine research center) should be available unconditionally at the beginning of the fishing season. | **108 boat owners/fishermen**, of whom 21 are in Saint Louis, 16 in Yoff, 12 in Hann, 21 in Mbour, 13 in Joal, and 25 in Ngaparou,<br>**42 fish mongers**, of whom 3 are in Kayar, 6 in Yoff, 18 in Hann (but based at Dakar central fish market), 8 in Mbour, 5 in Joal, and 2 in Ngaparou<br>**13 women fish processors and vendors**, of whom 5 are in Hann, 3 in Mbour, and 5 in Joal | A return to previous information policy of CRODT recommended – consisting of printed stats booklets freely available in the style of the 1970s. Current statistical information is deemed too complicated. All insist that such information should not only comprise quantities caught and landed, but also price information, and information about marketing and social aspects of fishing. |

(continued)

**Table 29.3** (continued)

| Types of expected research information | Number of persons by category and site | Alternative approaches proposed |
|---|---|---|
| Seasonal data about small pelagics | **100 boat owners/fishermen** operating encircling nets for small pelagics, of whom 36 are in Saint Louis, 22 in Yoff, 22 in Mbour, and 20 in Joal | Adapt the research programs to enable documenting seasonal and short-term changes in biomass of small pelagics. |
| Regular information about development of the Programme National d'Immatriculation des Pirogues (PNI) and the introduction of permits for small-scale fisheries (the fishermen commenting on this deplore what they perceive as a lack of transparency in the management of these two pillars of the reform. Such information could help restore trust and reduce suspicions about misuse of funds by the people in charge of local small-scale fisheries committees (CLPA) colluding with government officials. | **142 boat owners/fishermen** of whom 48 are in Saint Louis, 20 in Kayar, 21 in Yoff, 15 in Hann, 27 in Joal, and 11 in Ngaparou | Connect and integrate the information about the Programme National d'Immatriculation des Pirogues and the introduction of fishing licenses and make data collection a task for CRODT (in combination with re-issuing their information bulletin referred to above). |

(Chapter 5 of the SSF Guidelines), (b) social development, employment and decent work, and gender (Chapter 6 and 8), and (c) value chains, post-harvest, and trade (Chapter 7), all of which have significant unmet information and communication needs.

This is particularly true for the establishment of marine protected areas (MPAs), which have been shown to have beneficial effects on resource restoration, but have often met with opposition by fishers because they were excluded from the initial process defining and setting up MPAs. The general trend of rating a perceived loss much higher than a potential gain (Kahneman 2012) comes into play here. Moreover, the establishment of an MPA is certainly more than a simple technical management measure and should rather be seen as a socio-political enterprise (Chuenpagdee et al. 2013). This was also underscored by earlier field research that suggested that mistrust between fishers and scientists, which emerged again in the focus group discussions in 2015, can lead to limited engagement or even outright rejection of MPAs by fishers who have different representations of the variability of the environment. Fishers also tend to have little appreciation of scientific uncertainty, and more specifically oppose measures, when they are barely or not at all involved in the establishment process (Sall 2007; Meltzoff 2013).

Aichi Target 11 adopted in 2010 by the Conference of Parties to the Convention of Biological Diversity demands that 10% of marine and coastal areas, "especially areas of particular importance for biodiversity and ecosystem services, are conserved through effectively and equitably managed, ecologically representative and well connected systems of protected areas" by 2020 (CBD 2010). The specification of equitable management in setting up such protected areas needs rethinking, participatory processes, and particularly careful consideration of communication and information needs to succeed.

Our field research shows that the participatory establishment and long-term support by both government and NGOs of the locally managed MPA in Ngaparou has led to a high degree of acceptance of the MPA by local fishers. Having said that, the high demand for fish emerging from the American and Asian markets amongst others, as reflected in increasing exports and high fishing capacity, both of which have expanded significantly over the last decade (ANSD 2015), has meant that fishers from other communities have increasingly challenged the restrictions of the MPA accepted by local fishers (see Table 29.1). While the arrival of migrant fishers introducing greater diversity often has positive effects, it can also lower social consensus, as is the case in Ngaparou. This illustrates the need for continued investment in wider communication and information exchange, in order to build and rebuild consensus around at least a few shared key objectives. This is an essential precondition for managing greater complexity successfully (Page 2007).

Several researchers have suggested that traditional knowledge of fishers should play a more prominent role in filling knowledge gaps for effective fisheries management, and that this can be most successfully achieved by establishing two-way exchanges (Ruddle 1997; Wilson 1999; Berkes et al. 2001; Johnson 2010; Rosa et al. 2014; Bevilacqua et al. 2016). The need for critical engagement in this context can be seen in the progressive degradation of traditional knowledge where colonialism, technological change, urbanization and other factors bring about a weakening of knowledge transfer in traditional societies (Ruddle 1993). This is compounded by the need for updating such traditional knowledge with scientific knowledge as also demanded by the fishers in our field research, especially when market demand for unfamiliar species or overfishing requires rapid adaptations of fishing, processing and marketing strategies.

Gender issues – the elucidation of the social roles of women and men – have traditionally not attracted the attention they require. Despite the role women play in small scale fisheries, their contribution to the family, community and even national economy is frequently passed over in silence. Yet our research revealed that women often are managers running family businesses and are typically in charge of pre- and post-harvest activities, such as net mending, fish processing, including curing, drying and smoking, as well as marketing. They also glean seafood in tidal pools, harvest oysters in mangroves and even go fishing themselves with boats if needed (Williams et al. 2005). Growing and harvesting marine algae is another typical activity, particularly in the Asia-Pacific region (Fig. 29.5).

The SSF Guidelines envisage systematic gendered inquiries and special efforts to enhance information and communication with and about women's social roles. This is a pre-requisite to engage them at all levels of planning and decision making. The beneficial effects of such gendered attention, in conjunction with targeted

**Fig. 29.5** Algae drying by women in the Pacific (Photo credit: Dr. Mechthild Kronen)

capacity building and adaptation of rules to allow women to take leadership roles, have been documented (e.g. by Meltzoff 1995) in relation to barefoot shellfish gleaners (*mariscadoras*) in Galicia, Spain. Analyzing how to correct gender-blind approaches to projects, Brugere (2014) suggests the application of change theory with particular attention to small-scale fisheries.

A very significant step forward was the launch of a framework for bottom-up collection of actual production data about small-scale fisheries (Chuenpagdee et al. 2006). The painstaking work on global catch reconstructions orchestrated country by country, island by island, by the Sea Around Us Project led by Daniel Pauly, is a case of improving the reliability of and access to information, including information specifically about small-scale fisheries. This recent global summary builds on hundreds of individual reconstructions and concludes that global catches are on average 53% higher than reported by the FAO on behalf of national governments. Global production is dominated by industrial catches though these have been declining since 1996, particularly in those regions where industrial fisheries were first established and which have suffered from overfishing for extended periods of time. As scarcity is felt, discards at sea seem to decrease as well, but illegal, and otherwise doubtful catches continue to form a significant part of overall extractions. Small-scale fisheries, on the other hand, in different parts of the globe have increased their catches and are still expanding (Pauly and Zeller 2016).

Subsistence fisheries are even more poorly covered in national statistical systems than commercial small-scale fisheries. Their worldwide contribution to production was estimated at 3.8 million tons per year between 2000 and 2010 (Pauly and Zeller 2016). They are particularly important for local food security. As subsistence fisheries are predominantly the purview of women and girls, they have particularly suffered from gender blindness within fisheries.

The full implications of these major corrections to official mis-reportings by governments to the FAO are still emerging, but the data warrant digestion and sharing with small-scale fishers and all fishery stakeholders to review engrained perceptions about different fisheries, their share in production, and their roles in food security. One message is already clear and fully in tune with the SSF Guidelines: catch statistics need to be much improved and made available freely.

One technical approach to improving small-scale catch recordings has been developed by Stephen Box and his team in collaboration with fishers and traders in coastal communities in Honduras, Central America. Fishers have been given an ID card which gave them some recognition, which they did not have before and which motivated them to collaborate. The ID card facilitates recording their catches through a combination of a cheap and robust vessel tracking device and an App running on smart phones that keeps transaction records and helps buyers and associations with a business management tool (Mundus maris 2015). The fact that several Asian countries have asked for this approach to be presented to them is an indication of its relatively simple and cheap mode of operation.

## Conclusions

What is noticeable from the different research results referenced in this chapter is that content and processes with regard to fisheries management are highly site- and fisheries-specific. That is a significant finding. It underlines that while the principles enshrined in the SSF Guidelines are widely accepted, their concrete articulation and communication should be adapted to context. This way, the Guidelines can fulfill the needs and expectations of small-scale fishers and their communities and thus contribute to enabling their active participation in the governance of their fisheries.

Nevertheless, some useful generalizations can be made with regard to managing fisheries. Perhaps foremost is the urgent need for stepping up efforts to collect data on small-scale and subsistence fisheries around the world and refine the methods already developed and applied to global catch reconstructions cited above. Involving fishers in such collection efforts has become a real possibility with cheap technical approaches becoming available for that purpose and within reach of local communities.

There are still relatively few structured ways in which fishers can affirm and share their empirical knowledge to influence research and particularly public policy, despite the efforts of small-scale fishers' organizations to be represented in different fora, such as the FAO Fisheries Committee, the Fisheries Transparency Initiative

(FITI), and others. All too often small-scale fishers' relationships with governments have been and often still are more conflictual than cooperative (Nayak 2002; Sall 2004) as recently illustrated by protests of small-scale fishers in Chile against sector legislation in favor of industrial fishing interests (Díaz Medina 2016).

The SSF Guidelines, put together after long and intense exchanges and communications between a wide range of interested parties, can be an important instrument to help change perceptions about small-scale fisheries so that they are seen increasingly as opportunities for improved food security and lifting people out of poverty.

Approaches such as using local names in different languages together with scientific names in biodiversity information systems can also build bridges between different epistemologies (Palomares and Pauly 1993). The recent efforts at catch reconstructions by types of fisheries are an important way to recognize the real role of small-scale fisheries. Comparatively low internet penetration in many fishing communities together with language barriers still reduce fishers' access to such resources, something that was confirmed by our fieldwork. Migration and rapid expansion of hand-held communication devices are improving access to a wider array of information sources, including research. The channels of communication have certainly diversified and fishers and traders make ample use of mobile phones and information flows in networks of migrants within national borders and well beyond. Implementing the SSF Guidelines will require a sustained process of building up trust, better mutual understanding, and institutional arrangements that enable respectful multi-way communication and exchange.

Our research suggests that where such conditions are created and sustained, they form a robust underpinning of participatory and legitimate management as observed, for example, in the local marine protected area of Ngaparou in Senegal. Such multi-actor institutional arrangements can be quite effective in avoiding the widespread trend of overcapitalization and resource overuse that eventually destroys many of the positive social and distributional dimensions that could make small-scale fisheries role models for food security and sustainable livelihoods.

As women play such a crucial role in pre- and postharvest activities and are shown over and over to invest resources into maintaining and uplifting their families (Williams et al. 2005), the Guidelines rightly recognize the importance of gendering research, information, and communication for maximum effectiveness.

The SSF Guidelines point to the significant and sustained investment required to realize this potential. That it can be done and needs to be done is, for example, borne out by the rise of women shellfish growers in Galicia and others briefly mentioned. The demand on the part of fishermen, fisherwomen, and traders is clearly articulated. Acting on the research, information, and communication advice in the Guidelines will bundle and reinforce the resources and experience in currently too often separated knowledge communities.

Meeting the increased demands of understanding the social, economic, and ecological contexts in which men, women, and children in small-scale fisheries operate has excellent potential to support more adequate policies. Their implementation will need greater efforts – including human, institutional, and financial resources – at different levels, from local to national. These efforts may appear cumbersome and not expedient, but should pay off through much improved collective outcomes, not only in relation to food security and poverty reduction, but also in terms of other sustained benefits at community and national levels.

There is ample scope to make more top-down approaches to research useful to fishers by specific efforts to summarize, digest, and present results in formats that are accessible to them. But the implementation of the SSF Guidelines will require much more actor-centered and critically engaged research that addresses the specific information needs of small-scale fishers themselves and recognizes them explicitly as experts and legitimate stakeholders in their own right.

## References

ANSD. (2015). *Situation économique et sociale du Sénégal en 2012*. Chapitre 12 (pp. 201–211). Pêche maritime. Dakar: Agence nationale de la statistique et de la démographie.

Barthélémy, C. (2005). Les savoirs locaus: entre connaissances et reconnaissance. *VertigO – La revue électronique en sciences de l'environnement, 6*(1). https://vertigo.revues.org/2997. Accessed 15 Sept 2016.

Berkes, F., Mahon, R., McConney, P., Pollnac, R., & Pomeroy, R. (2001). *Managing small-scale fisheries. Alternative directions and methods*. Ottawa: International Development Research Centre.

Bevilacqua, A. H. V., Carvalho, A. R., Angelini, R., & Christensen, V. (2016). More than anecdotes: Fishers ecological knowledge can fill gaps for ecosystem modeling. *PLoS ONE, 11*(5), e0155655. doi:10.1371/journal.pone.0155655.

Brugere, C. (2014). Mainstreaming gender in transboundary natural resources projects – The experience of the Bay of Bengal Large Marine Ecosystem (BOBLME) project. *Environmental Development, 11*, 84–97. doi:10.1016/j.envdev.2014.05.003.

CBD. (2010). *Aichi biodiversity targets. Convention on Biological Diversity*. https://www.cbd.int/sp/targets/. Accessed 2 May 2016.

Christensen, S. & Thi, D. V. (2004). *Biological evaluation of local ecological knowledge regarding ecosystem health and exploitation status in the shrimp fishery in South Vietnam*. In 7th Asian Fisheries Forum, Penang, Malaysia, November 30–December 4.

Chuenpagdee, R., Liguori, L., Palomares, M. L. D., & Pauly, D. (2006). Bottom-up, global estimates of small-scale fisheries catches. *Fisheries Centre Research Reports, 14*(8), 105.

Chuenpagdee, R., Pascual-Fernández, J. J., Szeliánszky, E., Alegret, J. L., Fraga, J., & Jentoft, S. (2013). Marine protected areas: Rethinking their inception. *Marine Policy, 39*, 234–240.

Deme, M. (2012). *Étude des connaissances socio-économiques des pêcheries de petits pélagiques au Sénégal*. Commission Sous Régionale des Pêches.

Deme, M., & Kebe, M. (2000). *Révue sectorielle de la pêche au Sénégal: Aspects socio-économiques*. Dakar: Centre de recherches océanographiques.

Díaz Medina, I. (2016). Denuncia director de Ecoceanos: Ley de Pesca Longueira es hija de la corrupción. *El Clarín*. http://www.elclarin.cl/web/noticias/medios-y-periodismo/22-medio-ambiente/15617-denuncia-director-de-ecoceanos-ley-de-pesca-longueira-es-hija-de-la-corrupcion.html. Accessed 11 Dec 2016.

FAO. (2008). Vue générale du secteur des pêches national – La République du Sénégal. In *Profil des pêches et de l'aquaculture par pays*. Rome: Food and Agriculture Organization of the United Nations. ftp://ftp.fao.org/Fi/DOCUMENT/fcp/fr/FI_CP_SN.pdf. Accessed 11 Dec 2016.

FAO. (2015). *Voluntary guidelines for securing sustainable small-scale fisheries in the context of food security and poverty eradication*. Rome: Food and Agriculture Organization of the United Nations.

Franks, P., & Small, R. (2016). *Understanding the social impact of protected areas: A community perspective*. London: IIED Research Report. ISBN 978-1-78431-310-4.
Franz, N., Funge-Smith, S., Siar, S., & Westlund, L. (2015). Towards the implementation of the SSF Guidelines in the Southeast Asia region. In *Proceedings of the Southeast Asia regional consultation workshop on the implementation of the voluntary guidelines for Securing Sustainable Small-Scale Fisheries in the Context of Food Security and poverty eradication*. Rome: Food and Agriculture Organization of the United Nations.
Froese, R., Demirel, N., Gianpaolo, C., Kleisner, K. M., & Winker, H. (2016). Estimating fisheries reference points from catch and resilience. *Fish and Fisheries*. doi:10.1111/faf.12190.
Gender in Aquaculture and Fisheries. (n.d.). Homepage. https://genderaquafish.org/. Accessed 13 Dec 2016.
Jerven, M. (2013). *Poor numbers. How we are misled by African development statistics and what to do about it*. Cornell University Press, 187 p.
Johnson, T. R. (2010). Cooperative research and knowledge flow in the marine commons: Lessons from the Northeast United States. *International Journal of the Commons, 4*(1), 251–272.
Jounot, M. (2016). *Les océans: La voix des invisibles*. Film documentaire. Portfolio Production, France/2016/52'. https://vimeo.com/ondemand/oceanslavoix. Accessed 10 Apr 2016.
Kahneman, D. (2012). *Thinking fast and slow*. London: Penguin Books.
Meltzoff, S. K. (1995). Marisquadoras of the shellfish revolution: The rise of women in co-management on Illa de Arousa, Galicia. *Journal of Political Economy, 2*, 20–38.
Meltzoff, S. K. (2013). *Listening to sea lions. Currents of change from Galapagos to Patagonia*. Altamira Press.
Mundus maris. (2015). *New technology for small scale fisheries data collection*. http://mundus-maris.org/index.php/en/review/other/1821-ss-fisheries-data. Accessed 30 Apr 2016.
Nauen, C. E. (1984). The artisanal fishery in Schlei Fjord, eastern Schleswig Holstein, Federal Republic of Germany. *Studies and Reviews of the General Fisheries Council for the Mediterranean, 1*(61), 403–427.
Nauen, C. E. (1999). New players make a mark in in ocean governance. *International Journal of Sustainable Development, 2*(3), 328–337.
Nauen, C. E. (2002). How can collaborative research be more useful to fisheries management in developing countries? In J. M. McGlade, P. Cury, K. A. Koranteng, & N. J. Hardman-Mountford (Eds.), *The Gulf of Guinea large marine ecosystem. Environmental forcing and sustainable development of marine resources* (pp. 357–364). Amsterdam: Elsevier.
Nauen, C. E. (Ed.). (2005). *Increasing impact of the EU's international S&T cooperation for the transition towards sustainable development*. Luxembourg: Office for Official Publications of the European Communities.
Nayak, N. (2002). An essay on the fishworkers' movement in India. In: A. Sall, A. Belliveau, & N. Nayak (Eds.), *Conversations. A trialogue on power, intervention and organization in fisheries* (pp. 289–341). Chennai: International Collective in Support of Fishworkers.
Page, S. E. (2007). Making the difference: Applying a logic of diversity. *Academy of Management Perspectives, 21*(4), 6–20. doi:10.5465/AMP.2007.27895335.
Palomares, M. L. D. & Pauly, D. (1993). FishBase as a repository of ethno-ichthyology or indigenous knowledge of fishes. In *International Symposium on Indigenous Knowledge (IK) and Sustainable Development*, Philippines.
Pauly, D., & Zeller, D. (2016). Catch reconstructions reveal that global marine fisheries catches are higher than reported and declining. *Nature Communications, 7*. DOI: 10.1038/ncomms10244.
République du Sénégal. (2013). *Document introductif du Conseil interministériel sur la pêche*. http://www.aprapam.org/wp-content/uploads/2013/07/Document-introductif-du-conseil-intrministériel-sur-la-pêche_version-final.pdf. Accessed 11 Dec 2016.
République du Sénégal. (2015). Direction des pêches maritimes: Projet Régional des pêches en Afrique de l'Ouest (PRAO). Bulletin d'information trimestriel no. 3.
Rosa, R., Rosa Carvalho, A., & Angelini, R. (2014). Integrating fishermen knowledge and scientific analysis to assess changes in fish diversity and food web structure. *Ocean and Coastal Management, 102*, 258–268.

Ruddle, K. (1993). External forces and change in traditional community-based fishery management systems in the Asia-Pacific region. *MAST, 6*(1/2), 1–37.
Ruddle, K. (1997). The role of local management and knowledge systems in ICAM in the Pacific region: A review (Annex 10). In *Anon. ACP-EU Fisheries Research Initiative*. Proceedings of the third dialogue meeting Caribbean, Pacific and the European Union (pp. 121–130).
Salas, S., Chuenpagdee, R., Seijo, J. C., & Charles, A. (2007). Challenges in the assessment and management of small-scale fisheries in Latin America and the Caribbean. *Fisheries Research, 87*(1), 5–16.
Salas, R. C., Charles, S., & Seijo, J. C. (Eds.). (2011). *Coastal fisheries of Latin America and the Caribbean* (FAO Fisheries and Aquaculture Technical Paper 544). Rome: Food and Agriculture Organization of the United Nations.
Sall, A. (2004). Les pirogues font de la résistance: Histoire du Collectif national des pêcheurs artisans du Sénégal (CNPS). In A. Sall & N. Nayak (Eds.), *Les pêcheurs artisans du Sud s'organisent: Inde et Sénégal*. (pp. 59–120). Paris: Réseau des centres de documentation pour le développement et la solidarité internationale (RITIMO).
Sall, A. (2007). Loss of biodiversity: Representation and valuation processes of fishing communities. *Social Science Information, 46*(1), 153–187.
Sall, A. (2008). *Etude sur le savoir écologique des pêcheurs artisans des petits pélagiques en Afrique du Nord-Ouest*. CSRP – Projet BBI 13286: Pêche durable des petits pélagiques en Afrique Nord Ouest.
Sall, A. (2009). Relations entre pêche industrielle et pêche artisanale: Des frontières très vagues. In *Capitalisation des expériences des pêcheurs et femmes de la pêche en matière de complémentarité entre secteurs industriel et artisanal dans quelques pêcheries ouest africaines*. Document interne de travail du CREDETIP.
Sall, A. (2012a). *Contraintes d'accès aux ressources et nouvelles stratégies et tactiques développées par les pêcheurs d'espèces destinées à l'exportation*. Présentation à l'Atelier co-organisé par le CNPS et le CREDETIP, 6 janvier 2012, Mbour.
Sall, A. (2012b). Job satisfaction in the coastal pelagic fisheries of Senegal. *Social Indicators Research, 109*(1), 25–38. doi:10.1007/s11205-012-0053-5.
Sall, A. (2013). La force et la persistance des répères identitaires. In A. Fontana, & A. Samba (Eds.), *Artisans de la mer: Une histoire de la pêche maritime sénégalaise* (pp. 50–63). Dakar: Autopublié, Imprimerie Rochette. ISBN 978-2-7466-5677-2.
Thiao, D., & Ngom-Sow, F. (2014). Statistiques de la pêche maritime sénégalaise en 2013: Pêche artisanale et Pêche thonière. *Archives Scientifiques du Centre de Recherches Océanographiques de Dakar-Thiaroye, 223*, 36 p.
Wiber, M., Berkes, F., Charles, A., & Kearney, J. (2004). Participatory research supporting community-based fishery management. *Marine Policy, 28*(6), 459–468.
Williams, S. B., Hochet-Kibongui, A. M., & Nauen, C. E. (Eds.) (2005). *Gender, fisheries and aquaculture: Social capital and knowledge for the transition towards sustainable use of aquatic ecosystems*. Brussels: *ACP-EU Fish Res Rep*, (16), 128 p. ISSN 1025-3971/EUR 20432.
Wilson, D. C. (1999). *Fisheries science collaborations: The critical role of the community*. Keynote presentation at the Conference on Holistic Management and the role of Fisheries and Mariculture in the Coastal Community, 11–12 November, Tjärnö Marine Biological Laboratory, Sweden. http://www.oceandocs.org/bitstream/handle/1834/620/Co-Man45.pdf?sequence=1. Accessed 11 Dec 2016.
Wilson, D. C., Raakjaer, J., & Degnbol, P. (2006). Local ecological knowledge and practical fisheries management in the tropics: A policy brief. *Marine Policy, 30*(6), 794–801.
World Bank. (2009). *Senegal*. http://www.worldfishing.net/news101/regional-focus/senegal. Accessed 11 Dec 2016.
Zeller, D., Palomares, M. L. D., Tavakolie, A., Ang, M., Belhabib, D., Cheung, W. W. L., Lam, V. W. Y., Sy, E., Tsui, G., Zylich, K., & Pauly, D. (2016). Still catching attention: Sea around us reconstructed global catch data, their spatial expression and public accessibility. *Marine Policy, 70*, 145–152.

# Chapter 30
# Applying the Small-Scale Fisheries Guidelines in Nigeria: Status and Strategies for Badagry Coastal and Creek Fisheries

**Shehu Latunji Akintola, Kafayat Adetoun Fakoya, and Olufemi Olabode Joseph**

**Abstract** The essence of the Voluntary Guidelines for Securing Sustainable Small-Scale Fisheries (SSF Guidelines) juxtaposed with the Guidelines' voluntary nature has resulted in mixed outcomes. On the one hand, the Guidelines galvanise a willing state into action but on the other they are like a rusty tool unable to adequately ensure the principles and objectives are followed. This chapter presents a situation report of a traditional fishery that has the recipe for a successful management of a common pool resource. It explores the issues that require attention within the fishery so that the provisions of the SSF Guidelines are complied with. It is worthwhile to know the trajectory in which small-scale fisheries in Nigeria have moved pre- and post - SSF Guidelines, especially with regard to tenure rights. The chapter also examines the social development goals of the state with regard to provisioning of services that reflect the needs of fisheries- dependent communities and its attempts to deepen economic and social developments. Policy inconsistency is a common challenge which slows down progress and ultimately impacts negatively on the development of management plans, governance and the socio-economic interests of small-scale fishers. Inadequate communication coupled with limited participation of important stakeholders, especially small-scale fishers, in formal governance exacerbates slow progress. Areas for future research and data generation for fisheries management are spelled out. Information, research, and communication should also unpackage the indigenous knowledge of fishers. Such knowledge will help planning.

**Keywords** Small-Scale Fisheries Guidelines • Badagry Lagos State • Governance • Social Development • Gender • Policy Coherence • Information

---

S.L. Akintola (✉) • K.A. Fakoya
Fisheries Department, Lagos State University, Ojo, Lagos State, Nigeria
e-mail: shehu.akintola@lasu.edu.ng; shehu.akintola2@gmail.com; kafayat.fakoya@lasu.edu.ng

O.O. Joseph
Agriculture Department, Adeniran Ogunsanya College of Education, Ijanikin, Lagos State, Nigeria
e-mail: femifish64@gmail.com



S. Jentoft et al. (eds.), *The Small-Scale Fisheries Guidelines*, MARE Publication Series 14, DOI 10.1007/978-3-319-55074-9_30

## Introduction

There is increasing use of guidelines, codes, and other tools to foster a common blue print for the management of common pool resources (CPRs) and create a common standard for their operation. Hence, the eventual adoption of the Voluntary Guidelines for Securing Sustainable Small-Scale Fisheries in the Context of Food Security and Poverty Eradication (SSF Guidelines) (FAO 2015a) by the Committee of Fisheries (COFI) of the FAO in 2014, two decades after the adoption of the Code of Conduct for Responsible Fisheries in 1995 (CCRF) (FAO 2011), was heart-warming.

The SSF Guidelines recognize the great diversity of small-scale fisheries and that there is no single agreed definition of small-scale fisheries. In the context of this chapter, small-scale fisheries are fisheries in which people earn their livelihood catching fish using traditional or improved medium-sized crafts and traditional or modern gears, and operating inland or within 5 (9.13 km) nautical miles from the coast. In Nigeria small-scale fisheries account for 80% of total fish production and employ about 6 million coastal and riverine artisanal fisher folks fishing 46,300 km$^2$ of maritime area and 125,470.82 km$^2$ of inland waters (Ikenweiwe et al. 2011). However, in spite of this immense contribution, small-scale fishers in Nigeria face certain constraints (Aderounmu 1986), similar to those encountered by fishers globally. The SSF Guidelines, if adequately implemented, will ameliorate many constraints that small-scale fishers face which keep them poor despite their major contribution to the economy.

Nevertheless, expectations that the SSF Guidelines will rewrite the fortunes of small-scale fishers are dampened by the fear of non-implementation. Although formulating a consistent strategy is a difficult task for any management team, making that strategy work – implementing it throughout an organization – is even more difficult (Hrebiniak 2006). Often, a country adopts an international accord without a clear plan for putting the commitments into practice (Raustiala and Victor 1998). The main challenge with regards to the SSF Guidelines is whether states will really 'walk the talk' (Jentoft 2014). Many factors influence the success of strategy implementation, ranging from the people who communicate or implement the strategy to the systems or mechanisms in place for co-ordination and control (Yang et al. 2008). Strategic implementation largely will depend on the quality of the actors involved in the implementation of the SSF Guidelines and the will to comply to the Guidelines. Compliance depends on all stakeholders mentioned in the document, but especially the state as it is the key driver of the implementation process.

Nigeria's experience with regard to compliance with the CCRF was poor (Pitcher et al. 2009). It returned an abysmal 18% code compliance score. The country's profile report indicated that small-scale fisheries are considered only to a very limited extent in management plans, and only a few initiatives exist for consultation with small-scale fishers (Article 7: Management Objective) (Pramod and Pitcher 2006). As Nigeria is a member of Committee of Fisheries (COFI) of FAO and a signatory to CCRF, the SSF Guidelines provide a strategic policy framework for Nigeria to use; however, the country will face challenges when it comes to implementation.

In this chapter, anticipated challenges of implementing the SSF Guidelines in Nigeria are examined. Descriptions of institutional, social, political, and economic conditions conducive to the implementation of the SSF Guidelines are discussed. The chapter also explores extant regulatory frameworks as they relate to fisheries policies and institutions supporting small-scale fisheries and how they might facilitate or complicate implementation of the Guidelines. Recommendations are made for strategic implementation of the provisions of SSF Guidelines 5, 6, 10 and 11 vis-a-vis governance of tenure in small-scale fisheries and resource management; social development, employment and decent work; policy coherence, institutional coordination and collaboration; and information, research and communication towards improving the conditions of the small-scale fisheries.

## Data and Methods

### *Study Site*

Badagry coastal and creek fisheries, where this study was conducted, lies within longitude 2°42'E and 3°42'E and latitude 6°22'N and 6°42'N, and borders the Republic of Benin (Fig. 30.1). It is endowed with a lagoon system, deltaic distributaries, floodplains, and mangrove swamps and is directly connected to Nigeria's 960 km of coastline bordering the Atlantic Ocean in the Gulf of Guinea. The total

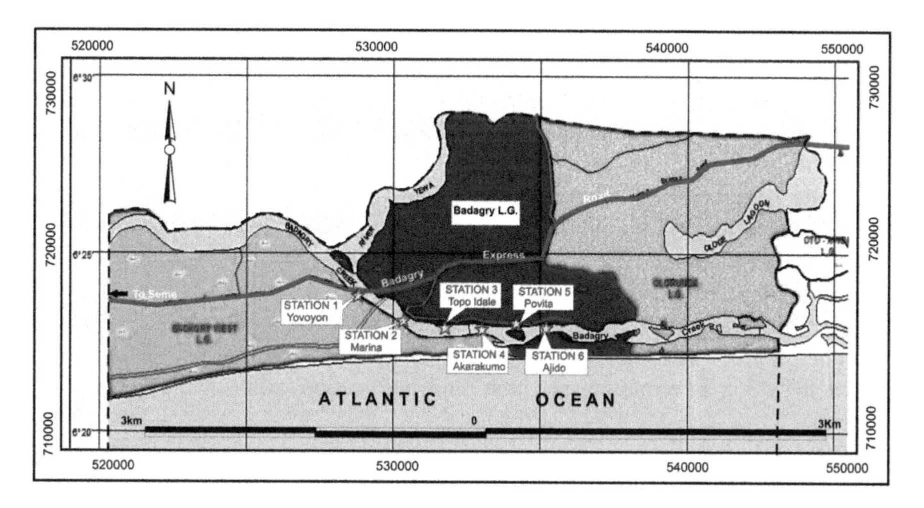


**Fig. 30.1** Some important fishing communities in badagry coastal and creek fisheries Lagos, Nigeria

area is 46,500 km$^2$ with a depth of up to 50 meters and an exclusive economic zone (EEZ) of 210,900 km$^2$.

The climate is dominated by a rainy season from April to October (Akintola et al. 2011), marked by heavier rainfall during the first period in May to July which results in serious flash floods that are aggravated by the poor surface drainage conditions of the coastal lowlands. September to October is another peak rainy period (Akintola et al. 2011). In terms of species diversity and distribution in the creek and coastal fisheries of Badagry there are 76 species across 47 families/orders of fin and shell fishes (Akintola and Fakoya 2017).

There are two main fishing regions within Badagry: the creek, which consists of fishers fishing in the main channel of the creek, and the coast which is made up of the marine fishermen of Yovoyan, Moba and Olomometa. There are a variety of fishing crafts including Ghanaian dugout canoes with planked free boards, smaller local dugout canoes, and local planked canoes. Outboard engines are of 5 to 55 HP. Gillnets are the most common gear and are surface or bottom, drifting and/or encircling, and trap.

## *Data Collection*

Regular visits were made to major fishing sites at Topo, Akarakumo, Marina, Yovoyon, and Moba and to other less important landing sites- Povita and Idale between July and December 2015. Data collection involved in-depth interviewing of identified key informants and follow up discussions. Focus group discussions were organized with tenure holders, experienced fisher folks, and community leaders to elicit responses on responsible governance of tenure and sustainable resource management. Information relating to management structures, plans, and arrangements aimed at conservation and sustainable use of fisheries resources of the creek were obtained from government documents.

Interactive sessions were also held with key informants (fisher elders and fisheries officers) relating to policy coherence, institutional coordination, and collaboration. Efforts were made to examine documentary evidence of violations of policies or failure in institutional coordination. Extensive literature reviews of important local and international publications on social and human resource development, fisheries policies, institutional coordination and collaboration, health, gender, and education were accessed from databases and internet websites using the Boolean search string Badagry, Lagos State, and Nigeria. Authors' knowledge and engagement in fisheries spanning over a decade were largely used to draw lessons about the extent to which information, research, and publication exist on the fisheries of Nigeria.

## SSF Guideline 5A. Responsible Governance of Tenure

### *Historical Analysis of the Aquatic Tenurial System in Badagry's Coastal and Creek Fisheries*

Evolution of small-scale fisheries property rights in Badagry's coastal and creek fisheries is synonymous with the historical development of Badagry. Furthermore, Badagry's position as an Atlantic and lagoon side port enabled the sub-region to emerge as an important commercial and political force in coastal affairs (Sorensen-Gilmour 1995). Three distinct periods (pre- colonial, colonial and post-colonial) are important with regards to management of the fisheries of Badagry (Akintola and Fakoya 2017). The pre-colonial period witnessed the emergence of a common property regime in which restrictions on fishing on days when deities had to be respected and the prevention of outsiders entering the fisheries system were the only barriers. In the colonial period, high chiefs, in a system of indirect rule, mandated craft registration in lieu of taxes. Open-access was re-introduced in the post-colonial period with the abolishment of taxes on fishers, a development ordered by the king, De WhenoAholu Menu-Toyi I.

#### Aquatic and Land Tenure

In the creek, fishing and fishing grounds are highly differentiated. There are differences of gear. Fishers use hooks, *acadjas,*[1] nets- gill and cast nets, and traps for shrimp operations. Access to specific fishing grounds is according to gear. Thus, fishing grounds are co-owned by those operating similar fish gear technologies and are fiercely protected from intruders. Land tenure is primarily customary and based on patrimonial inheritance of the male lineage, unlike in aquatic tenure where fishers hold common property rights. However, the 1978 Land Use Act enacted by the federal government delegated authority over land allocation to the states and local governments to ensure accessible and secured tenure to land by all.

Across fishing communities, the aforementioned land tenure systems prevail both among earlier settlers and recent immigrants. There is also gender neutrality. Perceived high costs and relatively long time lags to procure a Certificate of Occupancy (C of O), coupled with the burden that land holders face by being put on a compulsory tax role has meant that customary law governs land tenure among fishers more often than not. Conflicts are generally resolved through the traditional customary system though individuals are free to seek redress in court over land disputes.

[1]Acadja or brush park is a simple form of traditional fish enclosure culture or fish aggregating device used in West African lagoons.

## Types, Sources, and Governance of Fisheries

While there were two parallel governing systems, the traditional (customary) and the modern (state) systems, the former has remained the most popular, effective, and successful (Ruddle and Johannes 1985; Runge 1985; Ostrom 1986; Wade 1987; Olomola 1993; FAO 2008). A form of meta-governance of fisheries exists in Nigeria whereby according to Schedule II Part I, Item 29, of the 1999 Constitution (Nigeria's Constitution 1999), all issues relating to inland small-scale fisheries are within the purview of the state government to legislate on. The Nigerian Constitution permits legal pluralism but the governance of the small-scale fisheries is largely through traditional norms. The overall governance pattern can be described as one in which the state recognises other systems of governance and there is a measure of reciprocal adaptation, but yet little institutional or jurisdictional integration (Bavinck et al. 2013). State authorities perceive traditional systems as valid and useful and thus have accommodated them; however, the interconnections between customary and state law are weak. Conflicts generally are resolved by traditional institutions such as family, lineage, and chieftaincy.

Intra- fishing conflicts rarely take place since fishing crafts and fishing grounds are well differentiated to preclude any form of illegal incursion (Fig. 30.2). When they do occur, it is because of encroachments into fishing grounds by unentitled users, destruction of fishing gears by fishers and non-fishers, and marketing issues (over credit sales and remittances). These conflicts are resolved largely within traditional fishing resolution mechanisms in which case fish leaders intervened and find amicable solutions. Recourse to the traditional family and/or traditional kingship systems is an option rarely adopted. In the case of disputes with non-fishers, formal systems of State intervention are more commonly utilized. Though these types of disputes are rare, they are very much indicative of the changing scenario and influence of modernization. Conflicts between small-scale fishers and large-scale fishers are now more of an issue. Trawlers are known to make incursions into the 5 nautical miles of the continental shelf statutorily reserved for small-scale fishers, causing destruction to the latter's nets and higher fish mortalities.

## Immigrant Fishers

Immigrant fishers across the Gulf of Guinea from Benin, Ghana, Togo, and Cameroon abound in the coastal fishing villages of Badagry. A serene social order exists as fishing rights of individual fishers are respected regardless of whether they are immigrants or settlers. Both immigrants and settlers can switch from one line of fishing to another if they are properly introduced by a known and respected fisher to the other fishing type. Many years of intermarriage between immigrants and settlers have promoted harmonious relationships among fishers. Conflict resolution follows the pattern practiced by the indigenous fishers.

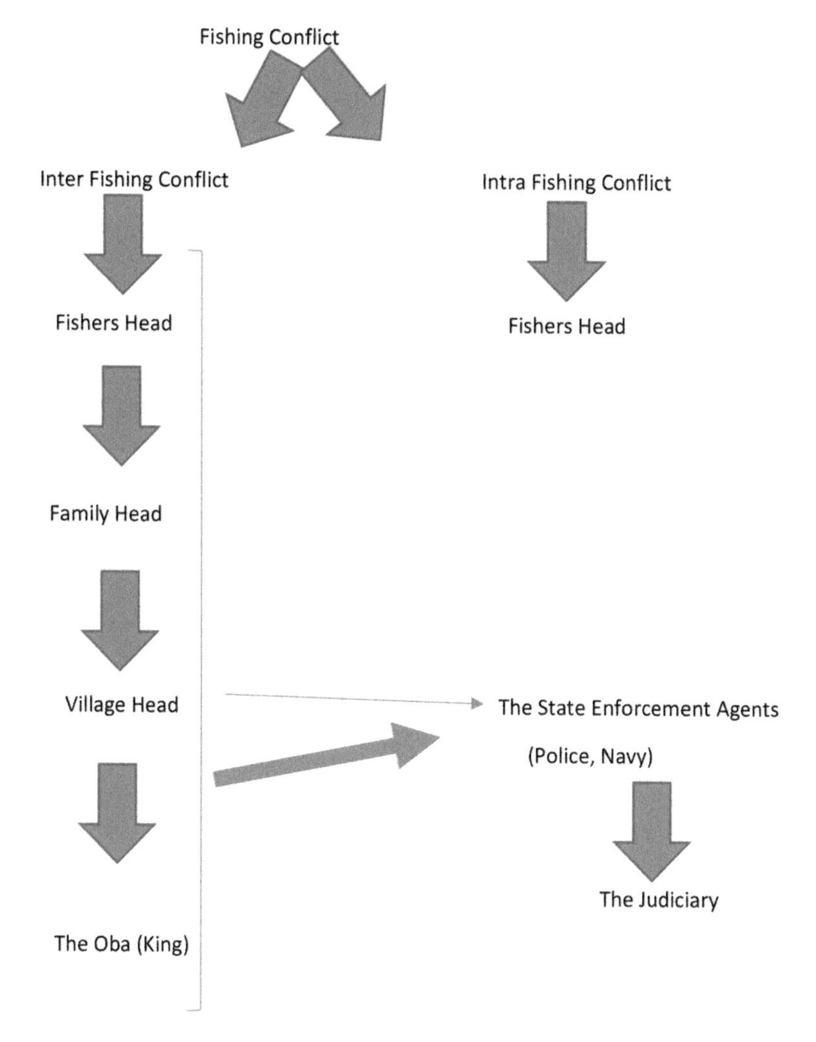


**Fig. 30.2** Model of fisheries governance and conflict resolution mechanisms in the badagry fishing communities

## Gender Equality

Women in the study area are primary, secondary, and tertiary users of the fisheries. They assume leadership roles among fishers using traps to harvest both fin fish and premium shrimps and prawns, but do not go for sea-fishing. Female fishers do not feel discriminated against and therefore have no cultural, economic, and social inhibition which may hinder their ability to contribute maximally to their livelihoods. They derive equal economic benefits to men from their engagement in fisheries. However, to a large extent, women do not have gender equity with male fishers in decision-making processes primarily due to their peripheral role in harvesting which limits their involvement in mainstream fisheries associations. The principle of gender equality is not yet enshrined in fisheries or agricultural policies in Nigeria. In fact, the country is still awaiting the birth of a national gender policy within the fisheries or agricultural sector to empower voiceless and underrepresented women in fisheries. Thus, women's inability to influence decisions aimed at increasing food security and food production at all levels still poses a serious challenge to the gender equity as addressed in the SSF Guidelines.

Women's roles in activities that may be considered downstream of the small-scale fisheries value chain, especially in fish processing, are well enshrined. They either sell fish to generate a supplementary income for the family, or consume it themselves (FAO 2015b). However, some fishers are not too happy with the role women play in credit and fish pricing. Market relationships, go sour when some women do not cooperate and fulfil payment agreement on credit lines. Non fulfilments of credit terms frustrate effective market credit systems in creek and coastal fisheries. To avoid losses, male fishers would rather sell to their wives or women mongers who are able to pay up. Wealthy women fishers who provide credit lines and finance less endowed male fishers are at times held in contempt because they act royal.

## SSF Guideline 5B: Sustainable Resource Management

Collective action by fishers in the fisheries of Badagry is a common way to promote conservation and guarantee sustainable livelihoods. Fishers have good knowledge of the fishing ground, fisheries ecology, tide movements, and other relevant technical and socio-economic knowledge. Their collective action, furthermore, has been guided by a number of management measures that are aimed at ensuring sustainable fishing: a closed season in September–October (reopened on 25th November) coinciding with the ebb tide when intrusion of seawater into the creek is common, sacrifices to appease the gods so that fish harvest is abundant, restrictions on nets with a mesh size of 0.5 inches, and a ban on harmful fishing are all geared toward resource management by prognosis.

### *State Management and Conservation Efforts*

A detailed examination of the full copy of the Lagos State Development Plan 2012–2025 (Lagos State Government 2013) and the Lagos State Ministry of Agriculture and Cooperatives, 2013–2015 Medium--Term Sector Strategy (MTSS) highlights the huge potential of fisheries, along with agriculture, to provide employment and food security in rural areas. The MTSS document, however, indicates the low priority given to small-scale coastal and lagoon fisheries despite the fact that water covers 22% of the total area of the State (Lagos state is often mentioned as the state of aquatic splendour) and is tagged as a fishing state by the federal government. In spite of the fact that fisheries at present contribute poorly to overall GDP (merely 0.36%), Lagos State has allocated a significant budget, though less than those to agriculture and aquaculture, mainly for a fisheries inputs subsidy scheme and plans to ensure conservation of endangered flora and fauna species as well as ecosystems in the state.

The hierarchical approach by both federal and state governments permits them to issue regulations in the form of decrees, laws, acts, and edicts (in the case of state governments). Federal laws/decrees which touch on activities of small-scale fisheries include the Inland Fisheries Decree, 1992 (No. 108 of 1992), the Coastal and Inland Shipping (Cabotage) Act (No. 5 of 2003), the Sea Fisheries (Fishing) Regulations, 1972 (L.N. No. 99 of 1971), the Sea Fisheries Decree, 1992 (No. 71 of 1992), and the National Inland Waterways Act, 1997 (No. 13 of 1997). Lagos State relies on regulations of the Inland Fisheries Decree, 1992 (No. 108 of 1992) to manage its inland fishery stocks. It is yet to conclude legislative processes and produce its own edict. By all indications, the extant decree is grossly violated and outdated with respect to present realities. It is mainly penal in nature and lacks provisions for stock management plans, data collection, environmental protection, and stakeholder participation.

## *Monitoring Control and Surveillance*

The state does not have an operational monitoring control and surveillance (MCS) unit. The need for MCS is often thought to be exclusive to industrial fisheries, ignoring the fact that small-scale fisheries are also carried out at sea. Illegal fishing by fishing trawlers is a major problem facing small-scale fishers. The regularity of trawler incursions into the non-trawling zone of 5 nautical miles has exacerbated the call for increased use of the MCS system in small-scale fishing zones. At present, fishers have to report intruders themselves and this has had no success.

## *Large-Scale Development Project*

The physical growth and urban development of Lagos is tied to its expanding economic and political roles and aided by rapid population growth. By 2025, Lagos will be ranked as the third largest city in the world after Tokyo and Mumbai. The state government, through the Lagos State Development Plan (LSDP) 2012–2025 document envisioned that by 2025 Lagos will be Africa's model megacity and global, economic, and financial Hub. Managing the growth and spread of Lagos Metropolitan Area is critical to the small-scale fisheries of Badagry creek and coast in terms of the future of sustainable livelihoods.

The Badagry sub-region extends over 80 km from the Republic of Benin border in the west to Lagos' port facilities at Apapa in the east. The Badagry fisheries is fronting onto the Atlantic Ocean, and a 'stretch of largely unspoilt and attractive sandy beaches, palm coves and a string of riverine islands inaccessible by land,' suggesting that it has to date not lost its pristine status. A major development that will directly impact this sub region of the state, especially the creek and coastal fisheries is the Badagry Mega Port and Free Trade Zone project (Badagry Port and Free Zone 2015). We have unfortunately not been able to lay our hands on social, economic, and environmental impact assessment studies that are said to have been conducted by the handlers of the project. However, fishers in the creek's channel are optimistic that once the project is brought to completion, it will open a passage for the inward movement of fishes from the sea that will lead once again to abundant fish availability for fishers. Studies of fish species in the Badagry fisheries highlight that many native species of the creek are diadromous, i.e. that they have to migrate between fresh and salt water as a necessary part of their life cycle (Solarin and Kusemiju 1991; Agboola et al. 2008; Akintola et al. 2009; Soyinka et al. 2010).

## SSF Guideline 6: Social Development, Employment and Decent Work

The state has made commendable strides with regard to providing an enabling environment for social development, employment, and decent work. There are a couple of public primary health centers, a general hospital which is a public secondary health care facility, and some private health care facilities. It is however important to note that inadequate funding of facilities and payment to health care providers continue to adversely affect all.

Presently, fishing communities have access to primary and secondary school education. Yet, fishers' children do not have the best opportunities that allow them to compete for placements in secondary school. Many fishers feel that access to quality education will help them find alternatives to the difficult profession of fishing. Moreover, young educated fishers will be better positioned to resolve issues within fisheries too. Fishing communities have access to vocational training centres for skill acquisition in many areas provided by the Lagos State Ministry of Women Affairs and Poverty Alleviation (WAPA). According to the 2013–2015 MTSS, the state has made concrete efforts to address gender dimensions within small-scale fisheries developmental programmes. Generally, WAPA is in favor of mainstreaming gender equality and youth empowerment. The center also runs computer literacy programmes to help bridge the gap in digital inclusion. Though not specific to small-scale fisheries, it provides real opportunities for livelihood diversification to cushion effects of declining income from fisheries. It is, however, regrettable to note that the Fishermen Vocational Training School in Yovoyan, Badagry, established by the state government for small-scale fishers is no longer functional.

At present, there are no viable schemes for fishers with regard to savings, credit, and insurance. Fisheries operations are largely financed through personal savings and credit from fisheries cooperative societies, which are limited. Therefore there is very little room for expansion. Currently, fishers are not particularly satisfied with the state's subsidy scheme and prefer to purchase from the open market where they can make informed choices with regard to inputs, based on their perceptions of the quality. Other pockets of finance exist in the shape of agricultural finance schemes offered by other financial institutions such as the National Agricultural Bank, previously known as the Nigerian Agricultural Cooperative and Rural Development Bank (NACRDB). However, fishers from the creek and coastal areas rarely utilise any of these schemes and are generally unhappy with the credit products offered.

WAPA may need to help make other government agencies provide infrastructure that will support women's roles in post-harvest operations. Women dominate the post-harvest part of the value chain. Section 7.2 of the SSF Guidelines emphasise the need to recognize women's roles in the post-harvest subsector and support improvements to facilitate women's participation. States should ensure that amenities and services appropriate for women are available as required to enable women retain and enhance their livelihoods in the post-harvest subsector.

WAPA is also the vehicle through which the state can address issues of importance to children. Child Rights Law and Domestic Violence Law address the issue of child abuse in whatever form it comes, but especially with regard to labor. Similarly, fishers are proud that they fare well with regard to the International Labor Organization's (ILO n.d.) conventions and regulations relevant to work and young persons. Incidence of school age children being out of school are not common place across the fishing community though fishers do take wards on fishing expeditions early in life as a form of training. They do not consider this child abuse. Many immigrant fishers who are used as labor enjoy favourable working conditions. A semblance of regional cooperation has been worked out where the embassy of the country to which immigrant fish workers belong facilitates registration and act as witness to agreed terms of operation and cooperation. The employer (fisher) only feeds the immigrant fish worker. Payment and settlement by the Nigerian employer is channelled through the embassy. This mechanism has thinned out cases of immigrant fish workers stealing fishing nets as witnessed in the past.

Fishing from time immemorial has been a risky business. In contrast to safety on industrial trawl fishing vessels and the recommendations in the SSF Guidelines, small-scale fishers lack well-developed safety programmes (Udolisa et al. 2013). In West Africa, the fatality rates of fishers in different countries of the region between 1991 and 1994 varied from 300 to 1,000 per 100,000 fishers (FAO 2000). The Lagos State Emergency Management Agency (LASEMA) is responsible for dealing with all hazards and risks. Presently, however, the activities of this agency remain largely inaccessible to fishers. Small-scale fishers face many occupational hazards. In the creek, they face hazards from waves, leaking boats and superior boats, all of which can lead to fatalities. Accidents in the coastal waters are often caused by trawl fishers. The stormy nature of the sea is sometimes overwhelming and leads to serious accidents. Fishers resort to unsafe and crude methods of saving their lives. To prevent loss of life, fishers often tie the legs of children to an anchor so that the waves will not sweep them away. The use of safety vests is rare and fishers complain of the poor quality when available.

## SSF Guideline 10: Policy Coherence, Institutional Coordination and Collaboration

Lack of policy coherence, institutional coordination, and collaboration results in power struggles between the Lagos State government and the federal government vis-a-vis governance and ownership rights over CPRs such as waterways and fisheries. Various ministries, agencies, and parastatals are embroiled in these struggles. The Lagos State Emergency Management Agency Law and National Emergency Management Agency Act have often clashed on matters pertaining to jurisdiction of land. Loss of jurisdiction or control to Federal Government on matters of regulation of natural resource is a usual issue with state governments.

As reflected in the LSDP document, fisheries have a huge potential. However, as fisheries are under the jurisdiction of the Fisheries Services Department in the Ministry of Agriculture and Cooperatives, it sits alongside other sectors including aquaculture. Aquaculture receives a greater share of programme contents, investments, funding, and support at the detriment of small-scale fisheries.

Another bottle neck with policy especially with reference to small-scale fisheries is mandate on conservation of fauna. Conservation activities including fisheries, is under the Ministry of Environment. However, there are overlaps in institutions because the Fisheries Department is also to address the conservation aspects of fisheries. Poor collaboration between agencies of the federal and state governments does not provide room for harmonious policies.

On the positive note, the state government is engaged with formal planning and uses GIS and remote sensing applications in guiding its plans. Spatial planning could be used to mark out fishing areas between fishers and non-fishers who compete for the same space within the same common pool resource. These technologies could be deplored to help monitor activities of small-scale sand mining, which also constitutes a threat to the survival of fishers' livelihoods.

## SSF Guideline 11: Information, Research, and Communication

Presently, the level of quantitative and qualitative information, research, and communication available on small-scale fisheries is inadequate to be useful in decision making. Many fisheries have no information about them at all while others require knowledge at different scale levels. The following challenges will have to be overcome in order to generate information with regard to governance, tenure, gender, livelihoods, climate change impacts, the economics of small-scale fish production, and spatial mapping of fishing grounds:

Limited funding of research activities in the country often affects the scale of research. Interested researchers are not able to access funding for regular seminars, conferences, workshops, and congresses in fisheries, especially small-scale fisheries. Only narrow network platforms exist for constructive engagement of stakeholders in small-scale fisheries to engage in discussions that will lead to better defining and understanding of the research required in small-scale fisheries. Moreover, subjects of the research (fishers, fish workers etc.) are excluded at the conceptualisation of research problems and planning, and execution of the research.

Little evidence of collaborative studies exists among researchers. The prevailing situation is, therefore, one of narrow scientific, social, and economic views being presented. There are no mechanisms in place to tap into indigenous knowledge of experienced and older fishers vis-a-vis fishing, fisheries, and governance. Younger generations lose local knowledge as they either opt for part-time fishing or even abandon fishing. The great and varied phroneses for which forebearers of the fishing community earned their reputation is increasingly being lost as more of the fishers drop out from fishing. The fishing world continues to lose part of its heritage. This situation is rather pathetic for a state and people whose traditional occupation is fishing.

Research into the trans boundary characteristics of the creek both within the country and regionally across the two coastal nations (i.e. Nigeria and Republic of Benin) of West Africa has not been conducted. Research at a broader scale, therefore, is required to understand the globalised nature of small-scale fisheries systems.

The state government's document Towards a Lagos State Climate Change Adaptation Strategy (LAS-CCAS) does not adequately address Disaster Risk Management (DRM) and Climate Change Adaption (CCA) relative to the needs of small-scale fishers. Climate change (CC) will negatively impact small-scale fisheries by placing additional costs on it and reducing profitability. The worst case scenario is that of fishing being completely abandoned (Mustapha 2013).

## Implementing the SSF Guidelines

Implementation is an iterative process in which ideas, expressed as policy, are transformed into behaviour, expressed as social action (Ottoson and Green 1987). Social action transformed from policy is typically aimed at social betterment and most frequently manifests itself as programs, procedures, regulations, or practices (DeGroff and Cargo 2009). The operationalization of the SSF Guidelines is based on a hybrid of meta-approaches consisting of rational-empirical, normative –reductive and coercive -power to effect planned change.

In the rational-empirical approach, the prerogative to disseminate and inform fishers about the SSF Guidelines lies with the state. Fishers, fish workers, and fishing communities are rational and will embrace the provisions of the SSF Guidelines once they are convinced their interests are addressed. The normative reductive approach leverages on the sharing of organizational power through the active involvement of significant stakeholders in problem definition and solution generation to motivate change. The fishers will be found to be willing partner and are waiting, this development is apparent from interactions with a broad group of fishers in the course of this study. They are actually desirous of the change and look forward to be partner in progress.

The coercive-power approach emphasizes the use of political and economic sanctions by the state as the principle way to bring about change. Policies, laws, and other legal agreements in support of the SSF Guidelines are examples of political measures that may be deplored towards ensuring compliance. Rewards (and incentives) or sanctions which focus on the provision (or withholding) of financial incentives so as to ensure compliance with the Guidelines are alternative economic strategies to effect change. These three meta-approaches are vital to the implementation of change aimed at addressing the objectives highlighted in Part 1, 1.1 (a–f) of the SSF Guidelines, namely to enhance the contribution of small-scale fisheries to global food security and nutrition and to support the progressive realization of the right to adequate food.

Nigeria often does not shy away at adopting international and regional instruments, and conventions but has problems with implementations. This might be partly due to lack of capacity to align pertinent provisions with national instruments or to the fact that these instruments and conventions are substantially non-reflective of the local or domestic reality. The voluntary nature of the Guidelines is to ensure that they are home grown and amendable to suit the reality of stakeholders' experiences and not cast in a rigid format nor imposed. In the quest towards implementation, legislative, bureaucratic, and consensus-building hurdles need to be scaled (Rein and Rabinovitz 1987; Jentoft 2014). All three hurdles are expected to thin out once stakeholders understand and agree that a shift in paradigm of fisheries management to accommodate small-scale fishers will ensure rapid reduction in the import bill for fish and guarantee food and nutritional security to Nigeria. Though implementation is discussed in the context of the case study, the SSF Guidelines will have to be addressed generally keeping in mind the perceived challenges in the context in which the Guideline is being implemented.

## Stakeholders and Actions Taken in Implementing the SSF Guidelines

To ensure a speedy attainment of the six objectives of the Guidelines, it is imperative that stakeholders own the document. The following groups, according to the Guidelines, are the stakeholders: governments, small-scale fishers, fish workers and their organizations (FWOs), civil society organizations (CSOs), researchers and academia, the private sector, and the donor community. All these groups define the context in which the fisheries are formed. The tendency is that each stakeholder pursues its own interests and is cocooned, thereby undermining the possibility to develop a strong, interactive win-win system of participation and decision- making. For stakeholders, the right to participate is a prerequisite for them accepting the SSF Guidelines.

There is a need for capacity building of FWOs so that they become credible, legitimate, and democratically accountable (ICSF 2014). ICSF noted that it was important to help FWOs gain recognition as representative bodies in the implementation process from their governments. FWOs are organized and coalesced as fisheries cooperative societies (FCSs). FCSs, however, lack the statutory status to partner state governments in fisheries governance and hence there is a long way to go before the relationship between CSOs and FWOs is that of a win-win situation. There is a need that CSOs partner with the government and local NGOs, as does the Fisheries Society of Nigeria (FISON), to develop small-scale fisheries in such a way that sustainable development is promoted. Likewise, FWOs should also partner with CSOs in articulating demands, particularly the demand that they are statutorily recognized as part of fisheries governance. Academia's relationship with fishers is also not necessarily based on mutual respect. Fishers often carry over their distrust of the state to academia. This is partly because researchers do not often come up with research problems that address the interests of fishers.

## *Consensus and Legislative Hurdles*

Consensus is the building block of implementation efforts. In a non-interactive governance system, it is heuristic to first seek consensus among stakeholders in order to instill the idea of interactive governance. The state aided by CSOs, the private sector, academia, and the donor community will need to take the initiative of bringing together all-policy makers and representatives of FWOs and conduct a dissemination workshop on the SSF Guidelines. The main objectives of the workshop should be to:

(i) Initiate and continue the participatory process of creating awareness about the SSF Guidelines and their applicability to the local context of Lagos State. It is important to note that both fisheries officers and fishers are not familiar with the SSF Guidelines;
(ii) Explore how extant legislation or pending legislation when passed into law reflects the goals of the SSF Guidelines and where there might be room for improvement;
(iii) Strengthen local fishers' (coastal and inland) awareness of the breadth of their rights both in the national and international context;
(iv) Identify roles and actions for stakeholders in implementing the SSF Guidelines on the ground;
(v) Shift the mode of governance to co-governance; and
(vi) Improve private sector participation.

Expected output from this workshop will be a brand new and localized document that will drive the sustainable development of small-scale fisheries in Lagos State with the aim of addressing food security and poverty eradication at the desirable scale. This high expectation is based on the first meta-theory, namely of an empirical-rational strategy that assumes people are rational. The implication is that once presented with information that demonstrates that a particular change is in their interest, small-scale fishers will support change as a means of achieving that interest (Miles et al. 2002), and be willing to make sacrifices such as sharing of power, information, and knowledge.

Under the present democratic disposition, support from the legislature is very much required and fundamental to effect the change being sought. Planned change must be premised on knowledge. The government (i.e. the executive and legislature), policy makers, the private sector, and CSOs will need appropriate data supporting the case for change. Basic data will need to be gathered and analyses undertaken on all facets of small-scale fisheries of the state to make wise judgements and decisions that will lead to the expected transformation. Efforts by researchers on their own will not suffice. There is need for a commissioned study on small-scale fisheries which will be adequate in terms of scale and quality. The government, policy makers, the private sector, and CSOs will have to fund the research. Research output should be processed to aid general understanding of the current fisheries context keeping in mind the objectives of the SSF Guidelines and should be made available to wider audiences through traditional and social media.

The model proposed in this chapter (Fig. 30.3) includes the third tier of governance: local governments. Involvement of local government in matters of inland fisheries will further deepen and make robust the principle of co-management. While constitutionally delegating matters related to inland fisheries to state governments, approval of the federal government will have to be sought. This will involve making a law or bye law that allows for an interim or ad-hoc arrangement that includes local government areas (LGAs) in the governance process, as many small-scale fisheries systems are within the geographical jurisdiction of LGAs. To make this more permanent, states may have to push for a review of the constitution so that matters relating to inland and coastal fisheries can be dealt with under the concurrent list.

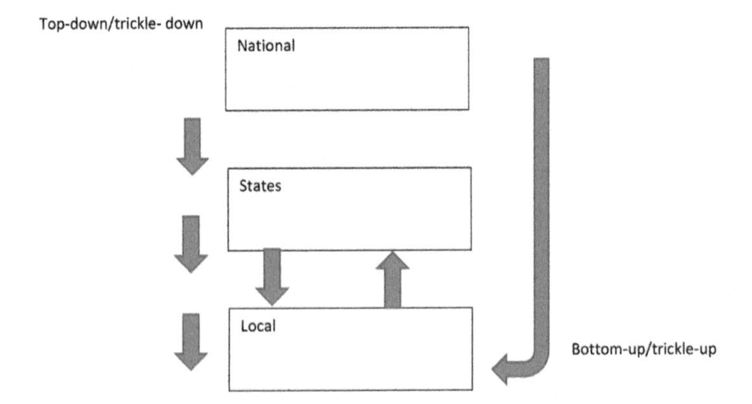


**Fig. 30.3** Model of implementation plan for Nigeria

## *Bureaucratic Hurdles*

Governance of small-scale fisheries is subject to both specific and general demands, which are not always easily harmonized, and which add to the governability challenge (Jentoft and Chuenpagdee 2015). Stakeholders will have to engage with the bureaucratic system. Already, FWOs have been interacting formally and informally with this system, namely with fisheries officers. Though fisheries officers are largely aware of the challenges facing small-scale fishers, they are handicapped by current policy and poor funding. Furthermore, the fisheries officers are constrained by the layers of reporting and communication necessary in fisheries governance. The hierarchical approach presently excludes consultation with and participation of non-state stakeholders in decision-making. Unfortunately, as mentioned in earlier sections of this chapter, the present bureaucratic emphasis is on aquaculture. Non-state stakeholders, specifically CSOs and local NGOs, will have to engage the bureaucracy in consultation given existing governability challenges.

The perspectives, vision, and interests of the bureaucracy may antagonise small-scale fishers. The bureaucracy may not have requisite fisheries background and understanding of the issues involved. It may, therefore, not appreciate the need to change existing structures and consequently slow down both consensus building and legislative efforts that are necessary for the anticipated change in policy direction towards co-governance. A meta-strategy that aims at normative re-education is most appropriate. This would involve harmonizing the values of the SSF Guidelines and those of the bureaucracy. This could lead to a shift in attitudes, values, norms, and relationships that help overcome distrust and build trust between FWOs and the bureaucracy. A sub-goal of the partnership between FWOs and CSOs is improvement in overall skills of the bureaucracy so as to direct its own change processes in the future. Lobbying may be an option of last resort and in this respect CSOs will have to take the lead. They will have to encourage FWOs, academics, and the private sector to lobby the executive and legislature.

Finally, there is the need to have a State Fisheries Commission to drive the process of integration of small-scale fisheries within wider socio-economic policies of the state. The commission will implement the localised SSF Guidelines. However, non-state stakeholders will have to be watchdogs. There is the possibility that the state's implementing agency may misfire if it does not have good information or get drawn into internal power struggles (Jentoft 2014).

# Discussion

There is little information on the socio-economic and governance dynamics of Badagry creek and coastal fisheries. The information required to move towards sustainable development is inadequate. Hence, this study is timely in providing this much needed information. As pointed out at the outset, the aquatic tenure system is a common property regime and differs significantly from the land tenure system which is governed by both customary law and the Land Use Act of 1978. Sections 5.3 and 5.4 of the Guidelines emphasize the need for the state to recognize customary or other forms of preferential access to fisheries resources in accordance with local norms and practices. This is all the more important because the state has not been able to address the socio-economic challenges faced by fishers.

Top-down hierarchical governance has proven to be inadequate and the traditional system too is beginning to show signs of weakness. Co-governance, as advocated in the SSF Guidelines, remains the only way out. In other words, the traditional ethos of fishers will have to be complemented at a larger scale. Therefore, the state will have to put in place mechanism by which the norms, beliefs and phroneses of fishers are adequately made provisions for and integrated into a formal co-governance regime.

The pillar of the SSF Guidelines is inclusive governance. This study shares the optimism of the SSF Guidelines and suggests that both non-state and state actors feel the Guidelines can lead to co-governance of a type that helps with economic rebuilding and growth of small-scale fisheries. Consequently food security and poverty eradication concerns can be addressed. Co-management will also allow for the institutionalization of modern fisheries management practices including that of data collection and analysis, MCS, and alongside strong local indigenous conservation practices that have contributed towards sustainable management in the past.

The state's effort at pursuing social development, as shown in the LSDP document, provides a veritable opportunity to push for the adoption and implementation of the SSF Guidelines. Fisheries policies developed in line with the SSF Guidelines also result in the private sector getting involved. There is need for the state government to encourage local private investors in boat construction, net production, and other inputs needs of small-scale fishers. The Lagos State Government is advised to pull funds presently invested in subsidy into a Capital Construction Fund (CCF) and Fisheries Loan Fund (FLF) to be accessed by private sector involved in production of fishing materials. Also state should divert present subsidy to fund fisheries infrastructure, collection of scientific data and strengthen capacity of the fisheries cooperative societies in terms of joint ownership of asset capital such as ice production and smoking kiln. Efforts of the state should be directed to capacity building towards fishers being able to manage their affairs.

However, to promote social goals and principles inherent in the Guidelines, precautionary measures to discourage price manipulations and dumping of substandard products which are vices that often accompany privatisation, have to be implemented. It will be important to implement a price control system approved in consultation with all stakeholders and also mandate regulatory bodies to ensure product quality compliance with international standards.

In many countries, explicit policy goals for fisheries development are still framed in terms of production targets, even if implicit goals – increasing government revenue, bolstering local or national food security, maintaining employment in a diverse coastal economy, or maintaining social stability - are more varied (Ratner and Allison 2012). Over time, and partly due to internal power struggles, goal displacement is to be expected causing disappointment among those who initially had high expectations of the Guidelines and for whom they were primarily intended (Jentoft 2014). Most of the fishers and fisheries officers interviewed agreed that implementing the Guidelines must bring mutual benefits to all and avoid certain stakeholders from unduly dominating.

If implementation is to be successful, non-state stakeholders must embolden the state to adopt a strong regulatory and enforcement power approach to manage the fisheries- a carrot and stick approach. The Kick Against Indiscipline (KAI) should be trained in fisheries regulation and enforcement and a specialized unit created to see to enforcement of fisheries regulations is envisaged. State governments could build on the existing monitoring control surveillance architecture to enforce compliance of policy streamlined by the SSF Guidelines and a plethora of fishing regulations and laws which explicitly prohibit fishing within 5 nautical miles by industrial fishing vessels. Violation of conservation rules and regulations must be viewed seriously and offenders brought to justice in accordance with the provisions of international and local legal instruments. It is also vital that through legitimate co-management, and a participatory process, all stakeholders agree on how to implement the Guidelines and develop a monitoring system which can give metrics and indicators to gauge both the success and shortcomings of the implementation process in relation to small-scale fisheries

## Conclusion

An in-depth assessment of the SSF Guidelines suggests its applicability to any form of fisheries governance. The strategy of the SSF Guidelines to integrate both phronesis and cultural values of small-scale fishers makes it adaptable to a multi-cultural nation such as Nigeria. The strength of the SSF Guidelines is vested in its motivational and voluntary nature to promote social well- being of fishers, fish workers and fishing communities and also to guarantee food and nutrition security. The Guidelines can accommodate the diverse, multitude of small-scale fisheries scattered across the landscape of Nigeria and the governance of these fisheries can align with the social, economic, and biological objectives of the SSF Guidelines.

This case study highlighted the status of Badagry creek and coastal fisheries clearly and how the SSF Guidelines could be brought to bear in this context. While the role of small-scale fisheries and fishers are well acknowledged in Badagry, both are marginalized in the hierarchical capture fisheries governance policy. Customary tenure based common property systems exist but are also weakening. Simultaneously, the state's existing regulatory framework, namely the Inland Fisheries Decree is certainly obsolete and MCS for small-scale fisheries absent. Co-governance can create a win-win for both fishers and the state. Policies and laws establishing the co-governance structure must allow for significant devolution of power in favor of small-scale fisheries so as to evolve community-based approaches and incentivize compliance to the guiding principles of the Guidelines.

Examination of the state's sustainable development program illustrated the encouraging efforts that have been made to put in place social amenities that will cater to the need of fishing communities. Formal schooling and vocational programs will further deepen the social development of small-scale fisheries and fishers. The SSF Guidelines can translate the state's intention to develop capture fisheries as a growth sector into reality. A fisheries commission can help implement the Guidelines once consensus has been reached and hurdles overcome. State and non-state actors will need to quickly work towards a consensus on many fronts, especially with regard to a new fisheries policy that is well engrained in the ethos and philosophy of the SSF Guidelines. Key to the operationalization of the SSF Guidelines will be the availability of basic fisheries data.

**Acknowledgments** Senior Evangelist Moses A.Y. Ashade provided many historical documents indicating that helped this study tremendously and the documentation of illegal trawling activities. Mr. Akpari provided us contact to many fishers. Both of them facilitated our gathering of data on Badagry coastal and creek fisheries. Efforts of the coordinators of the SSF Guideline research cluster of the Too Big To Ignore (TBTI) are deeply appreciated for driving the implementation of the SSF Guidelines. Finally, thanks are due to those who reviewed the chapter.

## References

Aderounmu, A. A. (1986). Small-scale fisheries development in Nigeria: Status, prospects, constraints/recommended solutions. *3rd Annual conference of the fisheries society of Nigeria* (FISON), 22–25 February, 1983 (pp. 30–39). Maiduguri, Nigeria.

Agboola, J. I., Anetekhai, M. A., & Denloye, A. A. A. (2008). Aspects of ecology and fishes of Badagry creek (Nigeria). *Journal of Fisheries and Aquatic Science, 3*(3), 184–194.

Akintola, S. L., & Fakoya, K. A. (2017). Inter-sectoral governance of inland fisheries: A case study of Badagry creek, Lagos State, Nigeria. In A. M. Song, S. D. Bower, P. Onyango, S. J. Cooke, & R. Chuenpagdee (Eds.), *Inter-sectoral governance of inland fisheries*. Too Big To Ignore e-book, available at http://toobigtoignore.net/research-highlights-1/e-book-inter-sectoral-governance-of-inland-fisheries/ accessed March 3, 2017

Akintola, S. L., Olusoji-Bello, O. A., Anetekhai, M. A., Clark, E. O., Ajibade, M. A., & Kumolu-Johnson, C. A. (2009). Catch composition and seasonal distribution of genera *Penaeus* and *Macrobrachium* in Badagry creek, Lagos Nigeria. *African Journal of Ecology, 48*, 828–830.

Akintola, S. L., Anetekhai, M. A., & Lawson, E. O. (2011). Some physicochemical characteristics of Badagry creek, Nigeria. *West African Journal of Applied Ecology, 18*, 96–107.

Badagry Port and Free Zone. (2015). *Badagry megaport and free zone: Bringing global growth to Nigerian waters*. http://www.badagry-port.com/downloads/Badagry240415.pdf. Accessed 7 Feb 2016.

Bavinck, M., Johnson, D., Amarasinghe, O., Rubinoff, J., Southwold, S., & Thomson, K. T. (2013). From indifference to mutual support – A comparative analysis of legal pluralism in the governing of South Asian fisheries. *European Journal of Development Research, 25*, 621–640.

DeGroff, A., & Cargo, M. (2009). Policy implementation: Implications for evaluation. *New Directions for Evaluation, 124*, 47–60.

FAO. (2000). Fishers' Safety. *The state of the world fisheries and aquaculture*. Rome: Food and Agriculture Organization of the United Nations.

FAO. (2008). *Customary water rights and contemporary legislation: Mapping out the interface* (FAO Legal Paper no. 76). Rome: Food and Agriculture Organization of the United Nations.

FAO. (2011). *Code of conduct for responsible fisheries, Special Edition 2011*. Rome: Food and Agriculture Organization of the United Nations. http://www.fao.org/docrep/013/i1900e/i1900e00.htm. Accessed 3 Feb 2016.

FAO. (2015a). Voluntary guidelines for securing sustainable small-scale fisheries in the context of food security and poverty eradication. Rome: Food and Agriculture Organization of the United Nations. http://www.fao.org/fishery/ssf/guidelines/en. Accessed 16 Jan 2016.

FAO. (2015b). A review of women's access to fish in small-scale fisheries. In A. Lentisco & U. Robert Lee (Eds.), Fisheries and aquaculture circular no. 1098. Rome: Food and Agriculture Organization of the United Nations.

Hrebiniak, L. G. (2006). Obstacles to effective strategy implementation. *Organizational Dynamics, 35*(1), 12–31.

ICSF. (2014). Reporting of group discussions in plenary. In K. G. Kumar (Ed.), *Towards socially just and sustainable fisheries: ICSF workshop on implementing the FAO voluntary guidelines for securing sustainable small-scale fisheries in the context of food security and poverty eradication (SSF Guidelines) report*. Puducherry, India.

Ikenweiwe, N. B., Idowu, A. A., Nathanael, A., Bamidele, O. S., & Fadipe, E. O. (2011). Effect of socio-economic factors on fish catch in Lower Ogun River, Isheri-olofin and Ihsasi, Ogun State, Nigeria. *International Journal of Agricultural Management and Development, 1*(4), 247–257.

ILO (International Labour Organization). (n.d.). *Conventions and recommendations relevant to work and young persons*.http://www.ilo.org/wcmsp5/groups/public/@dgreports/@dcomm/@inform/documents/genericdocument/wcms_345414.pdf. Accessed 3 Jan 2016.

Jentoft, S. (2014). Walking the talk: Implementing the international voluntary Guidelines for securing sustainable small-scale fisheries. *Maritime Studies, 13*(16), 1–15.

Jentoft, S., & Chuenpagdee, R. (2015). Assessing governability of small-scale fisheries. In S. Jentoft & R. Chuenpagdee (Eds.), *Interactive governance for small-scale fisheries*, MARE Publication Series, 13.

Lagos State Government. (2013). *Ministry of economic planning and budget Lagos State development plan 2012–2025*.http://documents.mx/documents/lagos-state-development-plan-2012-2025.html. Accessed 23 Mar 2016.

Miles, M., Thangaraj, A., Dawei, W., & Huiqin M. (2002). *Classic theories – Contemporary applications: A comparative study of the implementation of innovation in Canadian and Chinese public sector environments*.http://www.innovation.cc/scholarly-style/classic-theories.pdf. Accessed 4 Mar 2016.

Mustapha, M. K. (2013). Potential impacts of climate change on artisanal fisheries of Nigeria. *Journal of Earth Science & Climatic Change*. doi:10.4172/2157-7617.1000130.

Nigeria's Constitution. (1999). *Nigeria's Constitution of 1999*.https://www.constituteproject.org/constitution/Nigeria_1999.pdf?lang=en. Accessed 24 May 2016.

Olomola, A. S. (1993). The traditional approach towards sustainable management of common property fishery resources in Nigeria. *Maritime Anthropological Studies, 6*, 92–109.

Ostrom, E. (1986). Institutional arrangements for resolving the commons dilemma: Some contending approaches. In B. McCay & J. Acheson (Eds.), *The question of the commons* (pp. 250–265). Tuscon: University of Arizona Press.

Ottoson, J. M., & Green, L. W. (1987). Reconciling concept and context: A theory of implementation. *Advances in Health Education and Promotion, 2*, 353–382.

Pitcher, T., Kalikoski, D., Pramod, G., & Short, K. (2009). Not honouring the code. *Nature, 457*, 658–659.

Pramod, G., & Pitcher, T. J. (2006). An estimation of compliance of the fisheries of Nigeria with article 7 (Fisheries Management) of the FAO code of conduct for responsible fishing. In T. J. Pitcher, D. Kalikoski, & G. Pramod (Eds.), *Evaluations of compliance with the FAO code of conduct for responsible fisheries*. Food and Agriculture Organization of the United Nations: Rome.

Ratner, B. D., & Allison, E. H. (2012). Wealth, rights, and resilience: An agenda for governance reform in small-scale fisheries. *Development Policy Review, 30*(4), 371–398.

Raustiala, K., & Victor, D. G. (1998). Conclusions. In D. G. Victor, K. Raustiala, & B. Skolikoff (Eds.), *The implementation and effectiveness of international environmental commitments: Theory and practice*. Cambridge: The MIT Press.

Rein, M., & Rabinovitz, F. F. (1987). Implementation: A theoretical perspective. In W. D. Burnham & M. Wagner Weinberg (Eds.), *American politics and public policy* (pp. 659–708). Cambridge: The MIT Press.

Ruddle, K., & Johannes, R.E. (1985). The traditional knowledge and management of coastal systems in Asia and the Pacific. In *UNESCO regional office for science and technology for Southeast Asia*. Jakarta: UNESCO.

Runge, C. F. (1985). Common property and collective action in economic development. *World Development, 14*(5), 623–635.

Solarin, B. B., & Kusemiju, K. (1991). Day and night variations in beach seine catches in Badagry Creeks, Nigeria. *Journal of West African Fisheries, 5*, 241–248.

Sorensen-Gilmour, C. (1995). *Badagry 1784–1863: The political and lagoon-side community commercial history of a pre-colonial lagoon-side in South West Nigeria* (Doctoral dissertation). Department of History, University of Stirling, United Kingdom.

Soyinka, O. O., Kuton, M. P., & Ayo-Olalusi, C. I. (2010). Seasonal distribution and richness of fish species in the Badagry Lagoon, South-west Nigeria. *Estonian Journal of Ecology, 59*(2), 147–157.

Udolisa, R. E. K., Akinyemi, A. A., & Olaoye, O. J. (2013). Occupational and health hazards in Nigerian coastal artisanal fisheries. *Journal of Fisheries and Aquatic Science, 8*, 14–20.

Wade, R. (1987). The management of common property resources: Finding a cooperative solution. *The World Bank Research Observer, 2*(2), 219–234.

Yang, L., Sun, G., & Martin, J.E. (2008). *Making strategy work: A literature review on the factors influencing strategy implementation* (ICA Working Paper 2/2008).

# Chapter 31
# Building Capacity for Implementing the Small-Scale Fisheries Guidelines: Examples from the Pacific and the Caribbean Small Island Developing States

**Zahidah Afrin Nisa**

**Abstract** In June 2014, the United Nations (UN) Food and Agriculture Organization's (FAO) member states adopted the International Voluntary Guidelines for Securing Sustainable Small-Scale Fisheries in the context of Food Security and Poverty Eradication (SSF Guidelines). The current task is now for FAO member states and all partners to implement these SSF Guidelines. While many developing countries have in principle adopted the SSF Guidelines, their implementation may not unfold if the business of reforming ocean governance at the national and local levels does not occur. Governance reform outlined in the SSF Guidelines urgently calls for greater cooperation in developing countries between different sectors and disciplines. In particular, these SSF Guidelines require multiple layers of governance dialogue so that the articles of the Guidelines are communicated to those concerned and concrete solutions for small-scale fisheries' actors are negotiated. Articles from Part 3 of the SSF Guidelines give provisions for countries to establish local governance structures and promote networks necessary to achieve policy coherence and cross-sectoral collaboration, and implement holistic and inclusive ecosystem approaches in the fisheries sector. The policy provisions for establishing networks within the SSF Guidelines also provide provision to address the special concerns of women and vulnerable and marginalized people in small-scale fisheries governance reform at the national and local levels. This chapter gives insight into the negotiating and consensus seeking process of the SSF Guidelines and highlights existing communication gaps. Based on this analysis, the chapter suggests necessary future action in capacity building of communication and negotiation skills.

**Keywords** Networks • Negotiation • Reforms

Z.A. Nisa (✉)
Coastal Resource Center, University of Rhode Island, Kingston, RI, USA
e-mail: zaidy.oceans@gmail.com



S. Jentoft et al. (eds.), *The Small-Scale Fisheries Guidelines*, MARE Publication Series 14, DOI 10.1007/978-3-319-55074-9_31

## Introduction

Small-scale fisheries contribute to poverty alleviation and food security in rural coastal communities whereas large-scale industrial fisheries contribute mostly to national gross domestic product (GDP). More fish is harvested via large-scale fisheries in developing states; on the other hand, small-scale fisheries produce more fish for domestic human consumption and are a primary source of livelihood for rural populations in developing countries (FAO 2014). According to the 2014 Coasts at Risk report (UNU-EHS et al. 2014), 34% of average daily protein intake in the Pacific comes from fish whereas only 11.78% comes from fish in North America, Central America, and the Caribbean together. Intake for average daily protein from fish in South America is 11.02%, 9.92% in Europe, 16.91% in Asia and 20.96% in Africa.

Based on the nature of Small-Scale Fisheries (SSF) operations, fish caught and processed by small-scale fishers frequently goes unreported in official fisheries statistics. Lack of economic data implies that the GDP contribution of the SSF sector remains hidden from official view and government budgets (Garcia and FAO 2008; Kolding et al. 2014). As a result the role of SSF in poverty reduction, public health, and food security is less explored by fisheries administrations.

The Voluntary Guidelines for Securing Sustainable Small-Scale Fisheries in the Context of Food Security and Poverty Eradication (SSF Guidelines), adopted in 2014 by the United Nations (UN) Food and Agriculture Organization's (FAO) Member States, are the first-ever international instrument dedicated to promoting and defending small-scale fisheries. The Guidelines were formulated as a result of extensive consultation between FAO Member States, and through active participation by several civil society organizations (CSOs). Negotiations led to consensus on governance principles and guidelines addressing small-scale fisheries in the context of the world fisheries crisis.

The SSF Guidelines are intended to guide governance dialogue, policy processes, and actions at all levels, as well as help the fisheries sector realize its full contribution to food security and poverty eradication. The Guidelines follow international human rights standards, responsible fisheries governance, and also other commitments on poverty alleviation, food and nutritional security, and economic growth. The SSF Guidelines are a consequence of the reaffirmation of the UN Head of States Rio + 20 outcomes. These outcomes recognize that states are the frontline implementers of much needed small-scale fisheries governance reforms. States have recognized that national level policy and legal reforms gain legitimacy when they are linked and guided by global and regional agreements, conventions, and adopted articles (FAO 2008).

The challenge for small-scale fisheries stakeholders lies at the national level where windows of opportunity need to be created to influence national fisheries and rural development policy agendas that align with the principles of the SSF Guidelines. National policies, political will, dedicated funding, national capacity, and community empowerment issues provide further challenges to aligning national policy with the principles of the SSF Guidelines.

Today, the main driving forces underpinning FAO's efforts to reduce poverty and food security are the lack of institutional empowerment, networking, and linkages of small-scale producers that allow them to participate in the rural development agenda (Herbel et al. 2012). Over the last decades, small-scale producers have been continuously marginalized and isolated by growing large-scale industries, and are unable to seize economic opportunities in the marketplace or influence trade policies. In particular, asymmetric asset endowment and unbalanced power relations, as well as a lack of access to information on the rights of small-scale fisheries actors, are factors that exclude small-scale producers from markets and development opportunities. Experts have continuously highlighted that the primary reasons for the global fisheries crisis are inadequate legal and institutional frameworks, uncertain tenure and user rights, and a participation deficit of small-scale fisheries actors in ocean and fisheries reforms at the national and local levels (Burget 2013; FAO 2014; HLPE 2014).

The most promising solution for the above challenges is for small-scale rural producers to be more informed about their rights and latest governance reforms, and to be organized into producer cooperatives and networks. Small-scale fisheries actors can become active in shaping their path out of poverty through information-based tools such as the SSF Guidelines and the Voluntary Guidelines on the Responsible Governance of Tenure (Nisa 2014). An initial step towards implementation is to develop partnerships with well-functioning organizations, cooperatives, and networks in creating awareness and providing communication and development support. However, incentives are needed for these cooperatives and networks to make the necessary changes.

Harmonized funding efforts are needed for developing countries in financing governance reforms. The cost of reforming fisheries policy and legal frameworks to incorporate small-scale fisheries has to be well investigated and presented to policy makers in developing countries. Innovative financing mechanisms, enterprise support and market development to implement the SSF Guidelines, and consequently making the Guidelines a reality given the economic drivers of fisheries, have yet to be negotiated at all levels. These actions all require intensive communication and governance dialogue with the right actors and leaders to craft the way towards recognizing small-scale fisheries' contribution to food security at both the national and local levels in developing countries.

This chapter focuses on Article 12 of the SSF Guidelines, which is concerned with capacity development within small-scale fisher organizations and their networks. Part 3 of the SSF Guidelines is concerned with the enabling environment and implementation of the SSF Guidelines. Articles 10.5 and 10.6 specifically highlight the importance of countries establishing and promoting institutional linkages and networks necessary to achieve policy coherence and cross-sectoral collaboration in the fisheries sector. The chapter also discusses the 3-year period of negotiation that led to the SSF Guidelines and the knowledge gained from developing countries. Emphasis is placed on the need for future capacity development initiatives to focus on strengthening communication gaps and negotiation skills for small-scale fisheries development. These crucial skills are a prerequisite for initiating policy and legal reforms at the local and national levels.

## Results and Learning from the SSF Guidelines Consultative Process

The study titled '*Sharing the Responsibility of the Implementation of New International Voluntary Guidelines for Small-Scale Fisheries via Regional and National Networks*' conducted under the Nippon Foundation Fellowship program of the Division for Ocean Affairs and the Law of Sea, analysed the 3 year SSF Guidelines' consultation process of the FAO Small Island Developing States (SIDS) members. The study mapped the consultative process using social network analysis to understand the multi-layered small-scale fisheries information exchange and points of contact in government agencies working with small-scale fishing communities. The three consultation meetings (2010 in Rome, 2012 in the Pacific and Caribbean SIDS, and 2013 via electronic discussions) were used for the NetDraw software for social network visualization analysis shown in Fig. 31.1. The analysis reviewed the consultation meeting reports so as to assess representativeness and positions of the SIDS. The regions were grouped as Pacific and Caribbean SIDS. The analysis also mapped actors that were missing in the consultations. Figure 31.1 illustrates the visual NetDraw map of the SSF Guidelines consultative process for the SIDS (Nisa 2014).

Through the NetDraw analysis that visualized the 2010, 2012, and 2013 meetings (Fig. 31.1), points of contact in the government, are shown as red dots and clustered as Pacific SIDS in blue and the Caribbean SIDS in pink. Government officials who were present are labeled by their official surnames. Inter-government organizations' officials are labeled as IGO, private sector officials are labeled as P, Non-Government Organizations officials are labeled as NGO, academia institutions labeled as Acad, fisher organizations representative labeled as FO and their network, the Caribbean Network of Fisher Organizations, as CNFO.

The Pacific SIDS FAO consultations noted that the definition of small-scale fisheries in the Pacific SIDS context differs from other regions. Pacific SIDS indicated a preference for using the term community-based coastal and fisheries management approaches, which are already in, place, and highlighted that these should be the mechanism to build upon. Doing so would ensure that the SSF Guidelines avoid duplication, and that development efforts in the region are not fragmented (FAO 2012b).

The participants highlighted that details of the SSF Guidelines development process needed to be disseminated appropriately to governments and private sectors. Mention should also be made that community and private sectors stakeholders should have been allowed to be engaged in the FAO consultations but were not (FAO 2012b). Participants upon noticing the absence of CSOs also stressed the need to ensure that the voices of primary stakeholders of CSOs were included. They also urged that the FAO process be brought to the attention of those at the highest levels of national and regional authorities and not be confined to the corridors of inter-government organizations.

The sessions on policy coherence at the Pacific consultations highlighted some challenges at the local level with multi-layer governance systems. Implementation responsibility for fisheries and coastal management is split between national ministries,

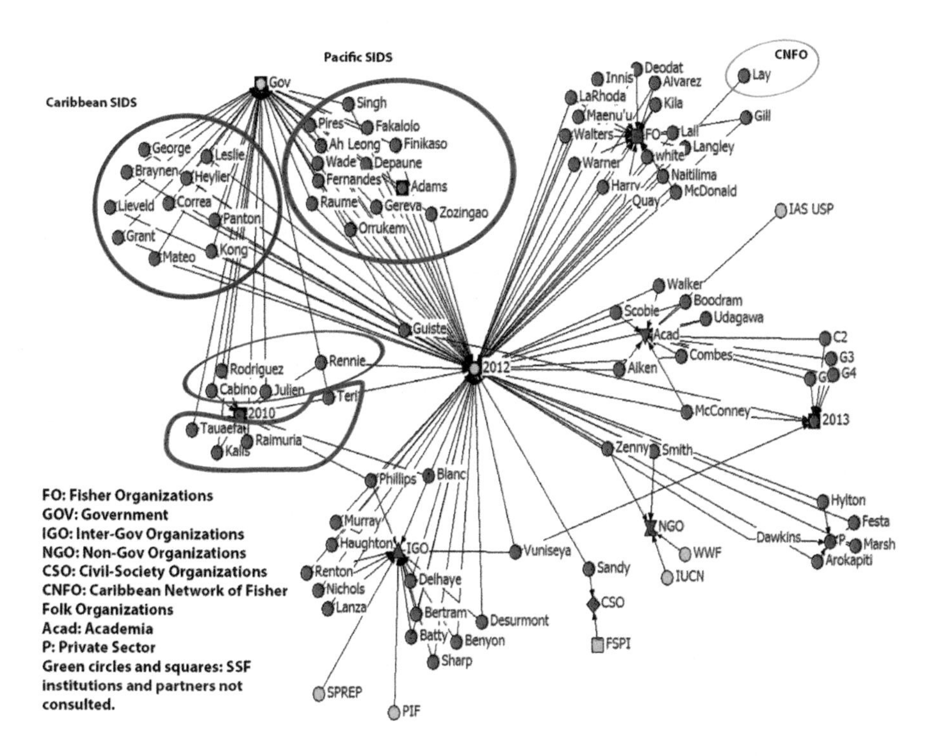


**Fig. 31.1** Visual NetDraw map of SSF Guidelines Consultations for SIDS Led by FAO in 2010, 2012, and 2013

provincial governments, and traditional resource owners. In some Pacific Island states it can be difficult to reconcile different layers of management decisions both horizontally (between coastal communities) and vertically (traditional leaders, provincial and national level authorities). Traditional institutions and structures are still paramount to the management of natural resources in Pacific Island communities. In Vanuatu devolution of power has already occurred at provincial, district, and community levels (Ibid). Memoranda of understanding between island councils and traditional owners to work together have been already been piloted in Pacific Island countries and the resultant governance arrangements have proven to be a success.

It was also noted at the Pacific consultations that the legislative process is often fragmented between different ministries (e.g. fisheries, environment, and agriculture). Fisheries and marine law enforcement is often constrained by the lack of resources and political will at the local level. Customary and culturally appropriate practices and tenure and customary rights of particular fisheries should be integrated within national and regional policies. Fisher cooperatives and associations have started to develop over the last few years but to date have often not been very effective (Ibid). All fisher associations present in the SSF Guidelines consultation process were industrial fisheries operators. Large-scale fish organizations were completely out of place given the agenda being discussed in 2014.

No regional civil society organization was part of the zero draft small-scale fisheries process and the 3-year consultation, according to reviews of the Pacific consultation, though CSOs were present in all three Caribbean consultations. Second, it is noted in the review of the Pacific consultation that the Department of the Pacific Center for Environment and Sustainable Development represented the University of the South Pacific (USP). This department is relatively new and undertakes teaching, training, and research for climate change and sustainable development in the Pacific Island region. It has not specialized in small-scale fisheries research, governance, and management. The Department of The Institute of Applied Sciences (IAS) of the USP is the most relevant institute for applied conservation science and small-scale fisheries resource management projects.

The consultation also did not involve the Locally Managed Marine Area Network (LMMA), a network that works with small-scale fisheries communities, though CNFO was present in all 3-year consultations. The LMMA Network is a network of marine management and conservation practitioners (community leaders/NGOs/government staff/researchers) working on LMMAs and small–scale fisheries for food security across Southeast Asia, Micronesia, Polynesia, and Melanesia. Practitioners from all levels collaborate for collective learning and understanding of best practices of local community-driven natural resource management (LMMA 2008). A LMMA by definition is an area of coastal and marine resources managed at a local level by coastal communities, using decentralized governance approaches (Ibid). The LMMA Network supports information sharing, learning, and development in strengthening community based adaptive management (Ibid). The IAS at the USP at this time had the most in-depth pilot case studies and expertise relating to small-scale fisheries, which it developed under the LMMA initiative over a decade in the Pacific region. IAS is also the Pacific focal point for the LMMA Network, and the learning portfolio of community-based marine conservation projects in Asia and the Pacific. Its vision is to support vibrant, resilient, and empowered communities who inherit and maintain healthy, well-managed, and sustainable marine resources and ecosystems.

Third, the Pacific SSF Guidelines consultations did not make reference to other relevant regional coastal fisheries policies or frameworks such as the framework for the Pacific Oceanscape. The recently adopted regional policy titled *Noumea Strategy 2015, a New Song for Coastal Fisheries and Community-based Ecosystem Approach to Fisheries Management Guidelines*, together with a vulnerability assessment on climate change and Pacific fisheries and aquaculture guidelines are critical linkages to the SSF Guidelines implementation process (Nisa 2014). Duplication of effort must be avoided. Meanwhile the consultation noted that two-way communication concerning policy development and implementation needs to be strengthened.

When FAO staffs were interviewed during the study regarding the lack of the right actors participating in the Pacific consultation, it was highlighted that inter-governmental organizations, in this case the South Pacific Commission, provided the list of names of stakeholders to the FAO headquarters. The Pacific consultations' study results indicate that due to FAO's weak regional relationships with small-scale fisheries actors, the consultative process did not include the 'right' actors from fisher networks and CSOs. The lack of right small-scale fisheries actors indicated that there was a major communication gap that could hinder the potential success of the SSF Guidelines in strengthening Pacific Islands coastal communities. To avoid such future problems, the FAO needs a better database of small-scale fisheries actors at local and national levels and cannot be dependent only on inter-government processes.

The FAO noted that CNFO was a key network present at the Caribbean Region SSF Guidelines' consultations. The CNFO's mission is to improve the quality of life for fisherfolk and develop a sustainable and profitable fishing industry through network representation and capacity building. The CNFO provided a CSO perspective on small-scale fisheries during the whole period of SSF Guidelines consultations for the Caribbean region (FAO 2012c).

It was noted during the session on policy coherence and institutional coordination that regional policies for small-scale fisheries need to be considered in the context of the existing Caribbean Community Common Fisheries Policy (CCCFP) (Ibid). The CCCFP of 2015 speaks of the SSF Guidelines. The consultation further noted that existing institutions and mechanisms should be used to promote policy coherence and collaboration to avoid institutional fatigue and financial burdens. In addition, the consultation pointed out that 'too much bureaucracy' hampered the implementation of regional policies.

What was also stressed at the Caribbean consultation was that fisheries management in the region had been largely top-down but that more recently there had been a shift to a bottom-up approach. For decentralized approaches to effectively work, appropriate governance mechanisms have to allow small-scale fisheries actors to be part of two-way communication systems that enable effective dialogue and policy processes. Many Caribbean Island States do not have functional and holistic fisheries management plans in place or are struggling with their implementation and enforcement. In cases where national management plans do exist, they are often not developed and implemented in a participatory manner. Responsibility for overall development often lies with different ministries. While there are ongoing discussions at various levels to explore opportunities for collaboration between fisheries and environmental institutions, the interest in working together is currently limited (Ibid). What can be learned from both the Pacific and Caribbean SSF Guidelines 3-year consultations is that implementers have to put enormous efforts in creating a more empowered communication environment at the country and local levels.

The study also reviewed the final negotiation process with regard to the SSF Guidelines in which a strong 37-member civil society delegation, with men and women from 18 countries participated. The CSO platform included members of the World Forum of Fish Harvesters and Fish Workers (WFF), the World Forum of Fisher Peoples (WFFP), and the International Collective in Support of Fish Workers (ICSF) (Sharma 2014). These three forums, between them, organized 20 national-level workshops spanning Asia, Africa, and Latin America, and two regional workshops in Africa to create awareness on the SSF Guidelines consultation process. There was no CSO delegation or expert from the Pacific, from the Caribbean region the CNFO coordinator was present at these meetings. Consultations with small scale fishers and fish workers were held in the EU and Canada, but none were held in the Pacific region in preparation for the final negotiation (Nisa 2014).

The March 2013 workshop on '*Strengthening Organizations and Collective Action in Fisheries: A Way forward in Implementing the International Guidelines for Securing Sustainable Small-scale Fisheries* 'explored the roles of different types of collective action and organizations in small-scale fisheries and proposed elements for a capacity development strategy to strengthen these. The types of collective action and organizational forms discussed included: customary community-based organizations, cooperatives and societies, and advocacy groups and networks.

The workshop recognized that bridging agencies such as the CSOs provide a platform through which small-scale fisheries stakeholders exercise their rights to organize and participate in development and decision-making processes. The workshop stressed the importance of supporting knowledge mobilization and transfer, leadership capabilities (of both men and women), research partnerships, use of effective communication tools (including new technologies and social media), and platforms and networks for experience sharing and collaboration. A Caribbean expert from academia was a resource person for this meeting and contributed the paper on*: 'Lessons Learned from Brazil and the Caribbean'* (FAO 2013).

No representation or expert was present from the Pacific SIDS region in this workshop. The most likely reason for this, based on information gathered from interviews with FAO staffs during the World Small-scale Fisheries Congress in September 2014 was that the Pacific experts were not invited since they had not been in the listing for the SSF Guidelines' consultation. By the time it was brought to the organizers' notice, it was too late both administratively and financially to facilitate new participation. It was also gathered from the Pacific SSF stakeholders that FAO had missed out on key small-scale fisheries leaders from the beginning of the zero draft. As a result, the SSF- Guidelines' negotiation process and the final workshops and SSF Guidelines document were unconnected to existing and ongoing coastal fisheries reform efforts in the Pacific region (Nisa 2014). Such oversight also meant that the SSF Guidelines were not forcefully recognized at regional and country levels in the Pacific. Major communication gaps continue to exist between international level actors and regional and country ones.

Key lessons can be learned from the analysis of the Pacific and the Caribbean SSF-Guidelines' consultations. The NetDraw visualization tool (Fig. 31.1) aided in evaluating the weak links in FAO's 3-year consultation and negotiation process with regard to the SSF Guidelines. The tool provided the basis to identify missing actors who in the future should be involved to champion the implementation of the SSF Guidelines implementation. The contacts can be used to follow-up with focal point actors who were nominated by various institutions and governments to represent their countries regionally. These points of contact or nodes are key focal points, especially in the fisheries sector through which to further the SSF Guidelines. It should also be noted that over time many contact people at the respective nodes had been replaced with different individuals. The task at the national level will, therefore, will be to identify new focal points.

Recognizing these gaps in the process can assist the FAO, small-scale fisheries knowledge brokers, and individual SIDS countries to position themselves better with regard to knowledge diffusion and development efforts pertaining to the SSF

Guidelines. Global assistance and UN funding programs can help support implementation. Involving actors not consulted also prevents duplication of on-going efforts.

## Recognition of the Communication and Negotiation Gaps at All Levels of Ocean Governance Reforms

While exploring the governance reforms called for at the international, regional and national levels, it is important to focus on three particular themes: human rights, sea and fisheries, and sustainable development. Each of these themes is addressed by multiple UN agreements, and a plethora of UN organizations are mandated to regulate sectoral activities. It is critical to note here that the recognition and urgent calls for integration of conventions and agreements, which developing states have signed, started decades ago and currently still faces bewildering proliferation and less viable national and local level transitioning pathways in its implementation (HLPE 2014). Therefore it is likely, that too much bureaucracy in relation to ocean governance and fisheries sector reforms will obstruct the SSF Guideline at all levels.

Serge Garcia, the former director of the FAO Fisheries Management Division, argues that cross-sectoral integration (or integrated approaches) is hindered by a number of factors which include: (i) old entrenched habits and cultural differences; (ii) unresolved differences in perceptions (world views); (iii) disagreement on a number of factors that guide action, including actual present and potential future levels of risks to ecological, economic, and social well-being; (iv) risk acceptance for failure to achieve environmental management objectives; and (v) risk tolerance for imposition of control measures that may constrain fishing more than the minimum necessary. Finally, there is also the question of what distribution of costs and benefits is deemed 'equitable' (Garcia et al. 2014). These challenges have been noted at the international level but need to be overcome at national and regional forums.

Another point worth noting is that while the SSF Guidelines call for multi-sectorial approaches in many of its clauses, the FAO SSF consultation and negotiation process, and meetings, where SIDS had been involved, did not fully display multi-sectoral participation or partnerships with other sectors of the governments. It is evident from the Pacific and Caribbean SSF Guidelines' consultations that only had representatives from fisheries divisions took part (Fig. 31.1) in the national consultation and negotiation processes (Nisa 2014). The FAO should have planned for multi-sectoral representation from fisheries, environment, and agriculture and rural development divisions. The current situation is one where the state fisheries divisions are handed to drive the process of small-scale fisheries governance reform and multi-sectoral approaches for the implementation of the Guidelines. Unless these shortcomings are overcome, it will be some time before the SSF Guidelines become a reality at the national level in this region.

## Hybrid and Networked Arrangements

FAO's March 2013 workshop on the *Strengthening Organizations and Collective Action in Fisheries: 'Way Forward in Implementing the International Guidelines'*, held in Rome, recognized that organizations and collective action in small-scale fisheries contribute to maximizing long–term community benefits and dealing with the threats to fisheries management. Organizations (customary, cooperatives and societies, associations and unions) provide a platform through which small-scale fisheries stakeholders exercise their rights to participate in development and decision-making processes and influence fisheries management outcomes. This workshop also discussed the role of what was described as 'hybrid and networked arrangements.'

FAO defines networks as a "system of the interlaced web of relationships in which control is loose, power is diffused, and centers of decision plural" (FAO 2012a). These hybrid arrangements are ones in which collective action is a mix of face to face and virtual organization aided by support groups and at times the state as well, both of which provide substantial information and communication technology for collective action and organizational management. It was concluded at the workshop that the future of small-scale fisheries governance lies in hybrid arrangements and organizations. Such hybrid arrangements and organizations provide a safe place for governance dialogue, debate, and communication platforms. Small-scale fisheries issues and management can be addressed appropriately and to the extent possible in a collective manner that results in a critical level of communication for advocacy and social change.

Based on the expert outcomes from the March 2013 workshop on the type of arrangements that would work best for collective action for SSF, the study reviewed the roles and functions of the CNFO and LMMA network. It was concluded that these networks in particular provide a well-established platform to exchange experiences and information among small-scale fisheries actors at the regional level and in some cases at the national level also. Also worth noting is that the CNFO and LMMA networks have a decade of work in process learning, legal and formal mechanisms to bolster policy and advocacy processes. Their regional and national efforts at doing so are a work in progress. What is clear, however, is that these mechanisms provide a strong basis for a partnership of networks within the FAO initiative to strength efforts to recognize small-scale fisheries at the national level. However, this will require FAO regional offices to recognize the work of existing regional and local networks and their bridging organizations and then build upon their work for knowledge transfer regarding the SSF Guidelines. It is also the responsibility of FAO, in its consultations, to find the right actors so that the process cannot be faulted in the name of misrepresentation (Nisa 2014).

Furthermore, networks have the potential to strengthen partnerships with like-minded networks at the international level, such as ICSF and WFFP with groups seeking recognition for indigenous peoples and women in fisheries. Small-scale fisheries representatives lack a platform within UN negotiations and larger consortiums that can help them gain momentum in fighting for a radical change to small-scale fisheries governance reforms at all levels. Donors supporting these networks can further strengthen the potential for international scale partnerships as they have the reach to promote information flow and community connections. Dialogues are necessary about the financial mechanisms required to start the necessary processes. However, one of limits to these networks is that they are dependent on philanthropic and international aid to scale up efforts. The FAO global assistance program for the implementation of the SSF Guidelines could be a source of support for building partnerships.

Existing networks (LMMA and CNFO) are not necessarily on the same wavelength as international bodies with regard to small-scale fisheries policy, advocacy and lobbying processes, and development demands. To address concerns of environmental justice, equitable fisheries management, tenure and indigenous rights, and human rights, an in-depth understanding of the policy and political process is necessary as these are contiguous and complicated issues. Networks need very substantial organizational capacity, communication and negotiation skills, appropriate legal basis, and funding to further their mandates, and in some cases, may need to change their strategies (Nisa 2014). The process of negotiating and finding a legal basis through which to address issues can be a lengthy process. The strength of current networks is that they can champion local and national issues, and provide more local network connections or nodes. Hence, the successful implementation of the SSF Guidelines greatly depends on recognizing the right actors, networks, and their bridging CSO partners with a wider range of communications skills. Integrated and adaptive approaches are critical to avoid duplication of effort and lack of harmonization.

## Way Forward for SSF Guidelines: Communicating Development for Collective Action

The SSF Guidelines, as highlighted in this chapter, can be meaningful to a wide range and variety of potential partnerships and networks such as the LMMA, CNFO, and WFFP. Small-scale fisheries governance reforms and multi-sectoral and multi-layer engagement can provide answers to multiple policy level and food security concerns. Concerns of importance include improving market access, social services and decent work environments of fishers. All of these require strong policy communication and advocacy building capacities so as to negotiate and drive change at local and national governance levels so as to meet the goals of the SSF Guidelines in developing countries.

Local and national level dialogues and negotiations on the implementation of the SSF Guidelines have to target a three-layered process: (i) multi–sectoral strategic level recognition of small-scale fisheries at local and national levels; (ii) revision and amendment of relevant legislation; and (iii) action and implementation that are supported by financial mechanisms and revenue streams that provide benefits to small-scale fisheries actors. All these changes will involve costly undertakings and have to be discussed and debated over time at the policy level if indeed the SSF Guidelines are to become a reality on the ground.

Funding opportunities to be explored to strengthen networks and small-scale fisheries organizations that will be critical to the implementation of the SSF Guidelines within countries will need to be sought from the FAO and other donors, have to be long-term, and aimed at addressing national and local sustainability concerns. Those who implement the SSF Guidelines will have to understand that it cannot be business as usual with regard to developing community-based or fisheries management plans. A shift away from a linear process that assumes that developing states will alter their practices in relation to small-scale fisheries on the basis of new information is required. Stakeholder dialogues and capacity building will have to be diversifies so as to be able to negotiate market-based solutions that bring economic opportunities and clear incentives for socio-economic change. Investments and funding is necessary to support capacity building of advocacy and negotiations skills. This will enable leaders, practitioners, and bridging organizations and networks to walk the talk in promoting small-scale fisheries in relation to national food security needs. Funding can also help small-scale fisheries bridging organizations, clusters, and networks develop their negotiating capacities.

Diverse funders and those who collaborate with the FAO at the international level should channel their support to capacity building of communication and negotiation skills as this will assist in developing new small-scale fisheries market opportunities, investments, and revenue schemes. Support for developing capacity in communication and negotiation skills will provide a means for seeking change in governance, power relations, social relations, attitudes, and even institutional functioning, all of which are necessary according to the SSF Guidelines. Furthermore, this will assist in the making of good policies and management decisions for small-scale fisheries that have been debated and negotiated and that bring good social returns. Such efforts and changes can ultimately reframe small-scale fisheries reform in ways that make them an investment opportunity generating greater economic opportunities for fishers, investors, and the wider community, as opposed to a costly affair not worth it.

**Acknowledgements** The research conducted for the present article was undertaken by the author while a Fellow of the United Nations - The Nippon Foundation of Japan Fellowship Programme. All views expressed in this article are those of the author only, and do not necessarily represent those of the United Nations nor The Nippon Foundation of Japan.

## References

Burget, N. V. D. (2013). *Legal aspects of sustainable development, The contribution of international fisheries law to human development: An analysis of multilateral and ACP-EU fisheries instruments*. Leiden: Martinus Nijhoff Publishers.

FAO. (2008). *Influencing policy processes: Lessons from experience. Policy assistance series 4*. Rome: Food and Agriculture Organization of the United Nations. http://www.fao.org/docs/up/easypol/756/influencing_policy_processes_202en.pdf. Accessed 10 June 2016.

FAO. (2012a). *Good practices in building innovation rural institutions to increase food security*. Rome: Food and Agriculture Organization of the United Nations. https://www.fao.org/docrep/015/i2258e/i2258e00.pdf. Accessed 9 July 2014.
FAO. (2012b). *Report of the FAO/SPC Pacific Regional Caribbean Regional Consultation on the development of international guidelines for securing sustainable small-scale fisheries, FAO fisheries and aquaculture report no. 1022*. Noumea: Food and Agriculture Organization of the United Nations.
FAO. (2012c). *Report of the FAO/CRFM/WECAFC Caribbean Regional Consultation on the development of international guidelines for securing sustainable small-scale fisheries, FAO fisheries and aquaculture report no. 1033*. Kingston: Food and Agriculture Organization of the United Nations.
FAO. (2013). *Strengthening organizations and collective action in fisheries – a way forward in implementing the international guideline for securing sustainable small-scale fisheries, FAO fisheries and aquaculture proceedings no. 32*. Rome: Food and Agriculture Organization of the United Nations.
FAO. (2014). *The state of world fisheries and aquaculture opportunities and challenges*. Rome: Food and Agriculture Organization of the United Nations. http://www.fao.org/3/a-i3720e.pdf. Accessed 17 April 2014.
Garcia, S.M., & FAO. (2008). *Towards integrated assessment and advice in small-scale fisheries. Principles and processes*. FAO Fisheries and Aquaculture Technical Paper no.515. Rome: Food and Agriculture Organization of the United Nations. http://www.fao.org/docrep/011/i0326e/i0326e00.htm. Accessed 20 September 2014.
Garcia, S. M., Rice, J., & Charles, A. (2014). *Governance for fisheries and marine conservation interaction and co-evolution synopsis*. New York: Wiley.
Herbel, D., Crowad, E., Haddad, O., & Lee, M. (2012). *Good practices in building innovative rural institutions to increase food security*. Rome: Food and Agriculture Organization of the United Nations. http://www.fao.org/docrep/015/i2258e/i2258e00.pdf. Accessed 20 April 2014.
HLPE. (2014). *Sustainable fisheries and aquaculture for food security and nutrition. A report by the High Level Panel of Experts on Food Security and Nutrition of the Committee on World Food Security*. Rome. http://www.fao.org/cfs/cfs-hlpe/en/. Accessed 10 May 2014.
Kolding, J., Béné, C., & Bavinck, M. (2014). Small-scale fisheries: Importance, vulnerability and deficient knowledge. In S. Garcia, J. Rice, & A. Charles (Eds.), *Governance of marine fisheries and biodiversity conservation* (pp. 317–331). London: Wiley.
LMMA. (2008). *LMMA Annual Report*. http://www.lmmanetwork.org/files/annual_reports/LMMA2008AR.pdf. Accessed 5 May 2014.
Nisa, A. Z. (2014). *Sharing the responsibility of the implementation of new International voluntary guidelines for small-scale fisheries via regional and national Networks*. The United Nations Nippon Foundation of Japan Fellowship Programme 2014 (Thesis). http://www.un.org/depts/los/nippon/unnff_programme_home/fellows_pages/grenada.html. Accessed 2 Feb 2015.
Sharma C. (2014). *Report of SSF – Guidelines sticky issues an update on the recent technical consultation on the international guidelines for securing sustainable small-scale fisheries*. International Collective in Support of Fishworkers (ICSF). http://www.icsf.net/en/page/588-About%20ICSF.html. Accessed 9 Sept 2014.
UNU-EHS (United Nations University-Institute for Environment and Human Security), TNC (The Nature Conservancy), & CRC (Coastal Resource Center). (2014). *Coasts at Risk: An assessment of coastal risks and the role of environment solutions*. http://www.crc.uri.edu/2014/07/crc-publishes-coasts-at-risk-report-launch-event-is-july-30/. Accessed 7 Sept 2014.

# Part IX
# Governing from Principles

Small-scale fishing people's poverty and marginalization spurred the process that resulted in the SSF Guidelines. The Guidelines are a strong moral statement about the need for policies and governance mechanisms that respect human rights, equity and justice. In Chap. 32, Ratana Chuenpagdee, Kungwan Juntarashote, Suvaluck Satumanatpan, Wichin Suebpala, Makamas Sutthacheep and Thamasak Yeemin discuss a new Royal Ordinance on Fisheries in Thailand, which, among its aims, seeks to protect and support small-scale and community-based fisheries in alignment with the SSF Guidelines. Proper operationalization could also mean fair and equitable benefits to small-scale fishing sectors. They present a case study from Trat province, which is one of the first provinces in the country to take steps towards implementing the new fisheries law, discussing how the process could benefit from the SSF Guidelines. Similarly, María José Barragán-Paladines talks about legal reform in Ecuador in Chap. 33, where a new constitution suggests an innovative development paradigm, based on the principle of *Buen Vivir* (good way of living). She finds that the governance of small-scale fisheries has been hampered by the rhetoric of the constitution and what is actually practiced by fisheries governing bodies. Chapter 34, by Milena Arias-Schreiber, Fillipa Säwe, Johan Hultman, and Sebastian Linke, addresses the situation of Swedish small-scale fisheries, which the authors find is unsustainable in biological, economic, and social terms. The authors explore the current process for stakeholder participation in the formulation of fishing policies and strategies in Sweden in the context of EU's Common Fisheries Policy. They believe that the SSF Guidelines may provide inspiration that could lead to positive change towards more sustainable coastal fisheries in their country. In Chap. 35, Danika Kleiber, Katia Frangoudes, Hunter T. Snyder, Afrina Choudhury, Steven M. Cole, Kumi Soejima, Cristina Pita, Anna Santos, Cynthia McDougall, Hajnalka Petrics, and Marilyn Porter examine the SSF Guidelines principle related to gender equity and equality, and the crucial role that women play throughout the value-chain in small-scale fisheries. They argue for the need to examine power relations as root causes of gender injustice and inequality, and stress the importance of capacity development for women and marginalized groups.

# Chapter 32
# Aligning with the Small-Scale Fisheries Guidelines: Policy Reform for Fisheries Sustainability in Thailand

**Ratana Chuenpagdee, Kungwan Juntarashote, Suvaluck Satumanatpan, Wichin Suebpala, Makamas Sutthacheep, and Thamasak Yeemin**

**Abstract** Global attention on issues affecting fisheries sustainability, particularly those related to illegal, unreported, and unregulated (IUU) fishing, has recently heightened. As one of the world's top seafood producers, Thailand is under immense pressure to illustrate commitment to address these issues. A new Royal Ordinance on Fisheries (2015) emerged as a result, replacing the Fisheries Act (2015). The Royal Ordinance includes several policies and regulations that aim to put a new order in the Thai fisheries, for instance, by addressing illegal fishing and promoting environmental protection and sustainable resource use. One of the main objectives of the new decree is to protect and assist or support small-scale and community-based fisheries, making it align well with the Voluntary Guidelines for Securing Sustainable Small-Scale Fisheries (SSF Guidelines). The Royal Ordinance could provide Thailand with a critical and timely opportunity to transform fisheries from an unsustainable and over-capacity situation to a well-balanced system. Proper operationalization of the new decree could also mean fair and equitable benefits to small-scale and large-scale fishing sectors, thus rectifying the existing economic and political imbalance. Similar to the prerequisites for the implementation of the SSF Guidelines, many conditions have to be met for this to happen, starting from having a common understanding about what the law says and what it implies in practice.

R. Chuenpagdee (✉)
Department of Geography, Memorial University of Newfoundland, St. John's, Newfoundland and Labrador, Canada
e-mail: ratanac@mun.ca

K. Juntarashote
Department of Fisheries Management, Kasetsart University, Bangkok, Thailand
e-mail: kungwan.j@ku.ac.th

S. Satumanatpan
Faculty of Environment and Resources Studies, Mahidol University, Bangkok, Thailand
e-mail: suvaluck.nat@mahidol.ac.th

W. Suebpala • M. Sutthacheep • T. Yeemin
Faculty of Sciences, Ramkhamhaeng University, Bangkok, Thailand
e-mail: wichin.s@gmail.com



S. Jentoft et al. (eds.), *The Small-Scale Fisheries Guidelines*, MARE Publication Series 14, DOI 10.1007/978-3-319-55074-9_32

This chapter presents the case of Trat province, which is one of the first provinces in the country to take the step towards operationalizing the new fisheries law.

**Keywords** Institutional analysis • Small-Scale Fisheries • Fisheries sustainability • SSF Guidelines • Fisheries royal ordinance • Thailand

## Introduction

Concerns about fisheries sustainability are global, as countries struggle with numerous management and governance challenges. The difficulty is enhanced when dealing with multi-species, multi-gear and multi-scale fisheries, as well as with multiple stakeholders with diverse interests. Global change, whether related to climate, market, or governance, adds complexity and uncertainty to the fisheries systems, constraining further management efforts. While some fisheries may be considered 'well-managed,' according to certain criteria (Hilborn 2007), the majority of the world's fisheries still face problems of overcapacity, resource competition, and use of destructive fishing practices, and illegal, unreported, and unregulated (IUU) fishing.

Ample evidence from research suggests that there is no 'one size fits all' solution to fisheries problems (Degnbol et al. 2005; Jentoft and Chuenpagdee 2009). Yet, some technical fixes, especially market-based approaches like individual transferable quotas and conservation-based approaches like marine protected areas, are still promoted, irrespective of the fisheries contexts that they are destined to deal with. Some of these are patchwork or Band-Aid solutions, which may do more harm than good since they often create confusion and uncertainty among stakeholders (Hilborn et al. 2004; Fulton et al. 2011). Moreover, these approaches can also result in making the problems more complex, for instance, by further marginalizing stakeholder groups that are already disadvantaged, leading consequently to social justice issues, among other things (Jentoft 2013). In such circumstances, major policy reform may be required to overhaul the entire fisheries system.

Reform may also be driven by external pressure. This was the case in Thailand, where pressure from the European Union (EU) and, to a lesser extent, the U.S., related to sustainable seafood trade and labor issues respectively. Thailand has been criticized for its weak law and poor enforcement, resulting in high levels of IUU fishing. This is despite the establishment of a Fisheries Patrolling Section within the Thai Department of Fisheries, equipped with patrol boats to perform monitoring, control, and surveillance activities and suppress illegal fishing within its exclusive economic zone (DoF 2015a). In April 2015, the EU issued Thailand with a so-called 'yellow card', barring it from exporting fisheries products to EU countries until certain conditions have been met. International human rights and labor organizations also take issue with the inadequate protection and callous treatment of foreign workers in the Thai fishing industries, and have launched media campaigns to advise customers against consumption of seafood products from Thailand. In the U.S. Trafficking in Person Report 2014, Thailand was lowered to 'Tier 3' for the lack of compliance with the minimum standards stipulated in Trafficking Victims Protection Act, and for not making significant effort to combat human trafficking.[1] It was noted that trafficking in the fishing industry played a key role in this downgrading. The Government of Thailand had no option but to take immediate measures to illustrate its commitment to addressing fisheries unsustainability and to improving working conditions, safety, and security for fishery workers.

One of the measures was the swift release of the Royal Ordinance on Fisheries (or Royal Ordinance in short) in 2015, superseding the Fisheries Act (2015), the making of which took a couple of years and involved consultation with many fisheries experts. The long-overdue Fisheries Act (2015) was a major amendment of the dated Fisheries Act (1947) but it was short-lived. The urgent need to comply with the EU demands resulted in the issuance of the Royal Ordinance, prepared mainly by a few legal experts.

The Royal Ordinance is a major institutional reform and can be operationalized in its own context. Yet, there is a clear advantage to align it with other international instruments, which is what was intended with respect to EU requirements. Given its relevance to small-scale fisheries, it is also imperative that the Royal Ordinance is seen in light of the Voluntary Guidelines for Securing Sustainable Small-Scale Fisheries (SSF Guidelines) (FAO 2015). It is argued here, following interactive governance theory, that the more coherent these two instruments are, the higher the chance that both could be effectively implemented (Kooiman et al. 2005). Further, any major reform is bound to create both intended and unintended consequences. Learning how such an institutional and legislative change affects small-scale fishing communities and what they do to cope with the situation is expected to provide valuable insights for future policy development.

This chapter presents a case study of small-scale fishing communities in Trat province on the east coast of Thailand, which is one of the first provinces in the country to attempt to operationalize the Royal Ordinance. In addition to documenting the process taken in the province and its outcomes, the chapter discusses how small-scale fishers in Trat may be able to draw on the SSF Guidelines when making the case to the government for special considerations. It also identifies gaps and missing elements in the current governance configuration that needs to be filled to achieve sustainable small-scale fisheries in Thailand. Information presented in the chapter came from literature review, participant observation, informal discussion with stakeholders, and previous studies.

[1] Thailand was moved to Tier 2 Watch List in 2016.

## Small-Scale Fisheries of Thailand: Status and Issues

Fishing is one of the most important economic activities in Thailand, with a long history of development. Major changes took place with the introduction of trawlers in the 1960s, which resulted in over-exploitation and the alteration of fisheries ecosystems (Pauly and Chuenpagdee 2003). During this period, fishing fleets were increasingly modernized, with the introduction of mechanized gears and growing sophistication in fishing technology (Butcher, 2004). Following expanded trawl fisheries was the development of purse seine fisheries in 1970s, targeting small pelagic species mainly in the Gulf of Thailand. Industrial fisheries started to dominate economically, with their high volume and value, particularly as export markets began to develop. It did not take long, however, before catches of both demersal and pelagic species started to plateau in the mid-1990s at around 2.8 million t, before descending to about 1.3 million t in 2014 (DoF 2015b). The early observation made by Panayotou and Jetanavanich (1987) about the mismatch between management capability and the fisheries exploitation capacity still holds today, posing challenges to the Department of Fisheries, the main government agency, under the Ministry of Agriculture and Cooperatives, responsible for sustainable fisheries management.

Many changes also occurred in small-scale fisheries. In one of the earlier studies on small-scale fisheries of Thailand, Juntarashote and Chuenpagdee (1987) showed that while the large-scale fishing sector had tripled in number from 1967 to 1985, small-scale fishery establishments increased by about 25%, from 38,321 to 47,938 during this period. According to the latest official census (DoF 2002), about 50,287 fishing households (87% of the total) were considered small-scale, the majority of which involved the use of outboard powered boat (about 82%). The other 12% fished without boats or with the use of non-powered boats, while the remaining 6% used inboard powered boats, with less than 5 gross registered tonnage (GRT) in size. In the most recent report by the Department of Fisheries, a total of 42,512 vessels were recorded, almost 70% of which were categorized as 'small artisanal' (i.e. less than 5 GRT, or mostly about 6 m in length) (DoF 2015a). Vessels between 5 and 10 GRT were considered 'large artisanal'(about 9%), which coincided with how some had considered them in the past (Pimoljinda 2002). The remaining of the fleet was commercial of various scales, from small, medium to large (Table 32.1). This categorization of fishing vessels reflects well what is observed on the ground. In the Royal Ordinance, no differentiation is made between small and large artisanal, as 10 GRT is used as an upper limit for small-scale fishing vessels. As will be later shown in the Trat case study, the new definition of small-scale fisheries is problematic since the majority of the small-scale fishing vessels are less than 5 GRT.

In the most recent Fishing Community Production Survey (2014), produced annually by the Department of Fisheries, catches from small-scale fisheries are associated with simple, unsophisticated gear, operated on a daily basis, mostly by household members (DoF, 2016). These include: (1) gill nets (for Indo-Pacific mackerel, crab, and shrimp); (2) mobile gears such as those using light luring net (e.g. squid, and anchovy falling nets), lift nets, push nets, and Acetes scoop net; (3)

**Table 32.1** Number of fishing gears by vessel size (2015)

| | Vessel tonnage | | | | | |
|---|---|---|---|---|---|---|
| Gear type | Small artisanal (<5 GRT) | Large artisanal (5–10 GRT) | Small commercial (10–20 GRT) | Medium commercial (20–60 GRT) | Large commercial (>60 GRT) | Total |
| Trawl | 225 | 304 | 517 | 1945 | 1096 | 4087 |
| Push net | 972 | 277 | 121 | 113 | 46 | 1529 |
| Gill net | 16,524 | 1282 | 356 | 229 | 24 | 18,415 |
| Trap | 3242 | 422 | 283 | 312 | 18 | 4277 |
| Hook/Lines | 2097 | 230 | 73 | 4 | 4 | 2447 |
| Falling net | 1452 | 673 | 672 | 638 | 35 | 3470 |
| Others | 2310 | 253 | 158 | 177 | 36 | 2934 |
| *Total demersal* | *26,822* | *3441* | *2180* | *3457* | *1259* | *37,159* |
| Anchovy purse seine | 13 | 49 | 58 | 129 | 190 | 439 |
| Anchovy falling net | 98 | 99 | 162 | 338 | 84 | 781 |
| Anchovy lift net | 2 | 0 | 4 | 7 | 0 | 13 |
| *Total anchovy* | *113* | *148* | *224* | *474* | *274* | *1233* |
| Surrounding net (incl. Purse seine) | 35 | 46 | 68 | 312 | 629 | 1090 |
| Gill net | 2185 | 329 | 184 | 189 | 42 | 2929 |
| Pound net | 68 | 16 | 8 | 9 | 0 | 101 |
| *Total other pelagic* | *2288* | *391* | *260* | *510* | *671* | *4120* |
| **GRAND TOTAL** | **29,223** | **3980** | **2664** | **4441** | **2204** | **42,512** |

Source: DoF (2015a)

stationary gears, particularly traps (for crab, squid, and fish), set bag net and bamboo stake trap; and (4) handline and hook and lines. Catches from gill nets constitute about 37% of the total production from small-scale fisheries (about 200,000 t), with another 17% coming from mobile gears. Table 32.1 shows, however, that a small portion of catches from small-scale fisheries comes from trawls, despite the fact that this gear is normally not considered part of small-scale fisheries.

Another feature of small-scale fisheries is the amount of hired labor. It is not often mentioned because of the lack of data, but it serves as an important indicator, suggesting the degree of involvement of family members in fisheries. The study by Juntarashote and Chuenpagdee (1987) showed, for instance, that among fishers using non-powered boats, only 7% hired labor. The percentage of hired labour increased with the type of boat, from 17% in outboard powered boats to 31% in small, inboard powered boats. Finally, distance from shore where fishing takes place is also used to delineate small-scale fisheries from large-scale sector.

Traditionally, small-scale fishing occurred in coastal areas within 3–5 km from shore. With the improvement of technology and boat engines, as well as the growing interest in small-scale coastal aquaculture, small-scale fishing takes place further from shore, beyond the traditional 3–5 km, creating problems with large-scale fisheries such as space competition and gear conflict. As will be seen in the Trat case study, spatial regulation and zoning is one of the most contentious issues in the new fisheries law.

Like in many of the world's fisheries context, despite being the majority, small-scale fisheries of Thailand are not properly recognized for the important roles they play to the local economy, how they generate goods and services, support livelihoods, and provide food security to the majority of the population. This could be because, when compared to large-scale fisheries, the production from small-scale fisheries constitutes less than 20% of the country total (Teh et al. 2015). Thus, many have argued for the inclusion of social, cultural, food provision, and other values associated with small-scale fisheries in policy and decision-making (Chuenpagdee 2011). It should be noted, however, that unlike many other countries, small-scale fishing communities in Thailand are well supported in terms of infrastructure and amenities, including roads, water, electricity, communication technology, schools and, healthcare. In other words, Thai small-scale fisheries are not remote or geographically detached from the attention of the state, market, and society, and thus are mostly neither poor nor marginalized, compared to, for example, Thai small-scale farmers (Chuenpagdee and Juntarashote 2011). Nevertheless, they are vulnerable to various types of change, such as those induced by the rising cost of fishing operations (Chuenpagdee and Juntarashote 2004), functional changes in the ecosystem (Pauly and Chuenpagdee 2003), and resource degradation brought about by other marine-based and coastal activities including destructive fishing, unsustainable aquaculture practices, rapid coastal development, market globalization, and global change (Satumanatpan 2011).

One of the issues affecting Thai small-scale fisheries is related to the EU's request to the Thai government to combat IUU fishing. Because measures to address IUU problems are designed to address large-scale, offshore fishing practices, their direct application to small-scale fisheries is highly problematic. For instance, the lack of data and information about small-scale fisheries, in terms of number of fishing people and boats, amount and values of the fisheries product, fishing locations, and landing sites, means that, by default, the majority of small-scale fisheries would be considered IUU. Such consideration goes against the very nature of small-scale fisheries, which normally involve both formal and informal activities. For instance, it is globally recognized that small-scale fisheries rely on family members, including women and children, to participate in various activities throughout the fish chain (FAO 2015). Fishers may carry more than one type of gear in their boats at one time, which means that they may target multiple species. Not all catches go through formal market channels, with some given to neighbors or friends, sometimes for free or in reciprocity, and some retained for household consumption. Finally, landing places are not always fixed, but may vary depending on weather or where the potential buyers are located, making it impossible to keep proper records. Not taking these formal and informal activities into account when discussing small-scale fisheries can result in inappropriate policies. Controlling and monitoring small-scale fisheries, for instance, through the issuing of fishing permits or licenses, as indicated in the Royal Ordinance, would likely meet with several challenges. Attempts to organize small-scale fisheries to resemble a formal economic sector would also go against what the SSF Guidelines suggest in Section 6.6, that formal and informal activities in small-scale fisheries should be promoted through decent work and social development and taken into account when discussing sustainability.

## Royal Ordinance on Fisheries and SSF Guidelines

The emphasis of the Royal Ordinance is on putting a new order in Thai fisheries, especially with respect to preventing overfishing and overcapacity, deterring IUU fishing, and improving welfare and working conditions of fishing labor, in order to achieve sustainability and to build confidence with Thailand's trade partners. The 12 objectives contained within the new law are comprehensive, covering similar elements specified in the FAO Code of Conduct for Responsible Fisheries and, to some extent, the SSF Guidelines. For instance, it follows good governance and precautionary principles and has a strong emphasis on information systems, ecosystem-based management, and the use of the best available science for sustainable development. All of these are the foundation of the SSF Guidelines.

While the majority of the new regulations are concerned with industrial fisheries, small-scale fisheries and local fishing communities are explicitly mentioned as one of the main objectives which avows to provide them with protection and support. Small-scale fisheries are defined as non-industrial fishing activities taking place in 'coastal seas', i.e. within 3 nautical miles from shoreline.[2] Given that industrialized fishing vessels are those greater than10 GRT, all fishing vessels less than 10 GRT are considered small-scale, by default. This has created several problems in practice, especially because some of the larger small-scale fishing vessels do operate beyond the 3 nautical mile limit. There is a provision in the law (Section 5), however, that allows an adjustment of 'coastal seas' boundary, to be between 1.5 and 12 miles, for resource management purpose. While small-scale fisheries are prohibited from fishing outside of the coastal seas (however defined), this particular clause, along with stipulations in other sections encouraging participation of small-scale fishers and community-based fisheries organizations in management (Section 25), opens up possibilities to discuss the appropriateness of some of the measures - and to propose alternatives. This particular provision aligns well with one of the guiding principles in the SSF Guidelines related to consultation and participation (Guiding Principle # 6). The Trat process illustrates the utility of this principle in formulating

[2] The terms used in the Royal Ordinance on Fisheries are artisanal and local fisheries. For consistency with the SSF Guidelines, the chapter uses the term small-scale fisheries to refer to this sector.

management measures that are context specific and deemed appropriate by local fishing communities, which is also promoted in the SSF Guidelines Section 5(b) related to sustainable resource management.

Although the importance of small-scale fisheries is well recognized in the Royal Ordinance and their participation in fisheries management is encouraged, several sections put restrictions on their operation. Under Section 32, small-scale fisheries using vessels must obtain a fishing license, which specifies type and number of gears used. This causes logistical problems and a financial burden to small-scale fishers who often use multiple gears on a year-round basis to target a wide range of species. A fishing license alone costs nearly USD 300 annually to acquire. They are also subjected to rules and regulations, under Section 33, that may or may not be suitable for their circumstance, like the use of logbooks to record catch amount by species and fishing areas. While information is essential for management and can support the implementation of the SSF Guidelines (as indicated in Section 11), some considerations are required, not only about the kind of data that are useful, but also about data collection methods suitable to the characteristics of small-scale fisheries. Otherwise, such efforts may lead to lack of compliance, in addition to the usual problem of data inefficiency and under-utilization.

One of the key elements of the Royal Ordinance is related to the restructuring of governing bodies and reforming fisheries institutions. At the national level, the 'National Fisheries Policy Committee' has been established, with broad membership from several government agencies, including Agriculture and Cooperatives, Foreign Affairs, Transportation, Natural Resources and Environment, Labour, Navy, Police, and Provincial Administration. Chaired by the Prime Minister, the committee also includes leaders of the National Farmers Federation, the Thai Chamber of Commerce, and the Federation of Thai Industries, as well as up to 10 experts and eminent persons, to be appointed by the Minister of Agriculture and Cooperatives for a two-year fixed term. Six of them are representatives from the following sectors: coastal fisheries, offshore fisheries, fisheries outside Thai waters, inland fisheries, aquaculture, and processing. Up to two other members are experts in natural resources and environment, and the remaining two members are fisheries academics. The mandate of this committee is to establish policies for all types of fisheries in order to achieve sustainability and a balance between fisheries resources and fishing capacity. This involves setting total allowable catches, determining fisheries development targets in accord with conservation and sustainability goals, promoting post-harvest industries, and preparing performance reports and dissemination to general public.

The Royal Ordinance further stipulates a set of management approaches that should be taken to achieve the objectives. Among them are safeguarding and protecting rights of Thai fishers, as well as developing and promoting their occupations and livelihoods. While these clauses are not specific to small-scale fisheries, they align well with the fundamental principles of the SSF Guidelines. For instance, the Royal Ordinance strongly promotes sustainable management through the use of the precautionary approach and best available science, with the aim to balance ecological, economic, and social considerations. This is the same goal promoted in Principle

10 (Economic, Social, and Environmental Sustainability) of the SSF Guidelines. Interestingly, a direct reference to small-scale fisheries in the management approaches is made only in the context of conflict resolution between small-scale and large-scale fisheries. This is one of the key considerations in the 'Fisheries Management Plan', to be prepared by the Department of Fisheries, which will be deliberated by the National Fisheries Policy Committee and recommended to the Council of Ministers for approval. This implies recognition of the vulnerable conditions of small-scale fisheries, which corresponds with the rationale for the SSF Guidelines. It is here in the fisheries management plan that other considerations relevant to small-scale fisheries, which are emphasized in the SSF Guidelines but absent in the law, can be added and elaborated. These include issues related to gender and gender equality, value chain, post-harvest and trade, and climate change and disaster risk.

The most important paragraphs in the Royal Ordinance on Fisheries for small-scale fisheries can be found in Section 25, about the participation of local fishing communities in resource management, and in Sections 26–28, related to the role of local fishing community organizations in governance, as members of the 'Provincial Fisheries Committee'. Section 25 aims to promote and support the participation of small-scale fishers by encouraging them to contribute to policy development. The Department of Fisheries is asked to facilitate their participation by providing knowledge and information about the management, maintenance, conservation, restoration, and utilization of aquatic resources, as well as by providing assistance and support to the implementation of communities' work, projects, or activities related to the above topics. Further, the new law encourages the establishment of associations and the registration of local fishing communities organizations. The registration needed to be completed, however, within 30 days from the date that the Royal Ordinance came into force (November 14, 2015) and the fact that a minimum of 30 members is required in order to form an organization would both serve as impediments to some fishing communities, and will have direct consequences for their formal participation in the decision-making process. As stipulated in the Royal Ordinance (Section 27), the 13 appointed experts of the 'Provincial Fisheries Committee' shall be representatives of the registered local fishing community organizations, covering a range of stakeholder groups from coastal and offshore fisheries, to inland fisheries, aquaculture, and processing. A member of the registered local fishing community organizations could also be appointed by the Director-General of the Department of Fisheries to assist fisheries officers in performing duties pursuant to the Royal Ordinance. The government, through its Public Relations Department, puts significant effort in communicating the information about this, clearly specifying the roles and opportunities that come with having such an organization. For instance, the announcement includes details about the rights of the local fishing community organizations to propose fisheries development plans or suggest pathways to address fisheries problems, including through legal measures such as area and seasonal closures and gear restrictions. As of September 2016, 662 local community fishing organizations have been registered, representing both inland and marine fisheries (Department of Fisheries, personal communication, October 2016).

Involving local people in decision-making and governance processes is not new in Thailand. The country started decentralizing governance in 1995 with the establishment of administration offices at the sub-district level and, two years later, at the provincial level. Members of the sub-district and the provincial administrative organizations are elected by local people, and serve in the position for a four-year fixed term. The sub-district administrative organization is responsible for providing basic infrastructure and amenities to people in the district. Its mandates also include promoting education, religion, and culture, supporting women, children, seniors, and people with disability, as well as protecting natural resources and environment (Chuenpagdee and Jentoft 2009). The provincial administrative organization has similar roles at the provincial level. Not only does this governance structure afford local people opportunities to influence decisions about natural resources through their elected members, it also allows problems and issues to be addressed in timely and efficient manners (Nasuchon and Charles 2009). While the sub-district administrative organization is not explicitly mentioned in the Royal Ordinance, the Chief Executive of the Provincial Administrative Organization, who serves as an *ex officio* member of the Provincial Fisheries Committee, can bring issues and concerns from the village and the sub-district to the decision-making table. The effectiveness of this decentralized system depends largely on two main factors, namely whether good governance principles such as transparency, inclusiveness, and accountability are followed, and whether there are proper channels and mechanisms for local people to participate in the governing process (Jones et al. 2015). The case study of the Trat province outlines some of these qualities and discusses what can be done to improve local participation and engagement.

## Trat Province Case Study

Trat is the eastern-most province of Thailand, about 300 km from Bangkok, with substantial coastline along the Gulf of Thailand bordering Cambodia (Fig. 32.1). It is a small province with about 225,000 people (2014 data), living on the mainland and the islands, particularly Koh Chang, which is the third largest island in the country. The province is organized into 7 districts, 38 sub-districts (20 of which are coastal), and 261 villages, and has one Provincial Administrative Organization and 29 Sub-District Administrative Organizations. Trat is famous for fruit orchards, gem mining, tourism, and fisheries. The province has 52 large and small islands, with Koh Chang as the most popular tourist destinations in the region among Thai and foreign tourists. Home to Mu Koh Chang National Park, established since 1982, the island went through major tourism development, with a steady increase in number of tourists (from about 250,000 in 2003 to more than one million in 2007) (Jentoft et al. 2011a). Along with this expansion, the number of island residents went up

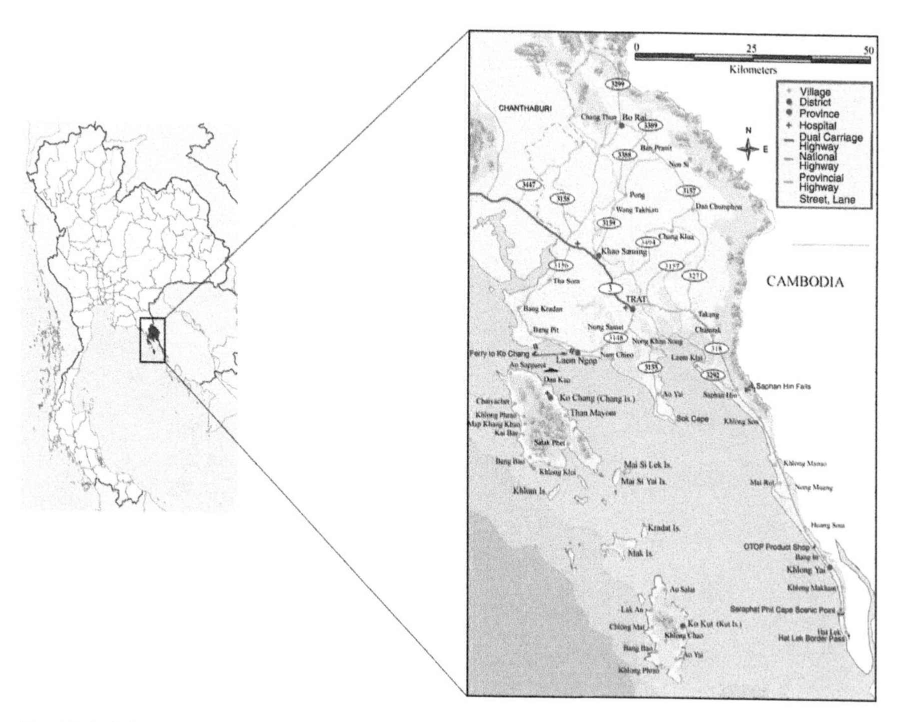


**Fig. 32.1** Map of Thailand, showing Trat Province and the Main Islands

significantly, through an influx of people from Bangkok and other areas looking for work in the tourism industry.

Fisheries are relative abundant in Trat province, compared to others in the region, mostly because of the healthy mangrove forests that line its coastlines. The coastal areas are also rich with coral reefs, serving as nursery and spawning grounds for marine life, and are high in biodiversity, including endangered species such as dolphin, dugong, and sea turtles. However, seagrass coverage has declined steadily and is no longer abundant in the area. A wide range of fish species, shellfish, and crustaceans are found in the areas, making them lucrative for both subsistence and commercial fishing. Similar to the national average, about 74% of the fishing vessels (or 2197 vessels) is small-scale, 1859 of which are less than 5 GRT, while the other 338 vessels are greater than 5 GRT but less than 10 GRT. The size distribution of the other 26% of the fleet (large-scale fishing vessels) is as follows: 26% between 10 and 15 GRT, 36% between 15 and 30 GRT, 25% between 30 and 60 GRT, and 13% greater than 60 GRT. About 90% of the large-scale fishing vessels have obtained fishing permits for commercial fisheries, and are dominated by anchovy falling nets (27%) and trawlers (22%). Small-scale fishing gears include shrimp trammel net, squid trap, crab trap, fish gill net, hook and lines, and reef fish trap (found mostly in Koh Chang), as well as "scoop net" (or "hand" push net, Fig. 32.2) used to catch Acetes shrimp (*Acetes spp.* and *Mesopodopsis spp.*), which is one of the "signature"

**Fig. 32.2** Scoop net used to catch Acetes shrimp in Trat province, Thailand (Photo credit: Wichin Suebpala)

species in the area (Lunn and Dearden 2006). Acetes shrimp are small crustaceans, with a thin soft shell, ranging in size from 6 to 32.9 mm in length. They are mostly found in shallow water up to about 50 m in depth. Acetes shrimp are the raw material for shrimp paste, which is an important ingredient in Thai cooking. Trat province is famous for producing good quality shrimp paste, and the Acetes scoop net fishery is considered a traditional fishing practice in the province. The differentiation between small-scale, family-based Acetes scoop net and the commercialized Acetes push net is necessary since all commercial push nets (whether for Acetes or other species) are prohibited from operating in coastal areas. The ban on push nets came about because the gear is non-selective, and its contact with the bottom can cause high levels of damage to the seafloor and coastal habitats. Push netters mostly argue that the fishing grounds they operate in are normally muddy and do not contain valuable habitats. In the case of Acetes push net, they also claim that their gear is highly selective, with little bycatch or discards.

The people of Trat province are known to have strong stewardship and conservation ethics. They have often been involved in collaborative efforts to address environmental concerns. For instance, coastal communities of Trat and nearby provinces were active participants in the discussion about ways to address coral bleaching problems, which occurred in the area since the late 1990s but more extensively in 2010 (Yeemin et al. 2012). They even agreed on allocating a provincial budget to initiate conservation projects to enhance resilience of coral reefs to climate change and other perturbation (Sutthacheep et al. 2012). The province also has a good track record of sustainable resource management through area protection and elimination of destructive fishing practices. For instance, zoning has long been used to successively demarcate areas protected from fishing activities. In addition to prohibiting the use of trawls and push nets within 3 km from shore (since 1972), use of light luring in purse seine fisheries is also extensively banned (in 1985). In 2000, trawls, push nets and dredges are either totally or seasonally restricted in areas between mainland and Koh Chang (Rogers 2015). The latest announcement was made in 2001 to prohibit lift net and other mobile gears using light luring, especially for anchovy, in certain areas, as well as to control mesh size. Restriction on anchovy fishing reflects the importance of the fisheries to the livelihoods of the people in the province. Along with shrimp paste from Acetes, the province is well known for high quality fish sauce made from anchovy caught in its waters. Therefore, it is not surprising that considerable efforts have been spent in management and research to achieve sustainable fisheries in the area.

The Eastern Marine Fisheries Research and Development Center is one of the key government divisions responsible for scientific research in the region. They conduct biological surveys and perform life history assessment of key marine species, including crab, Indian mackerel, and shrimp. According to their recent studies, catch per unit effort in large-scale fisheries remains low, most likely due to increasing fishing effort, suggesting an over-capacity situation. Things are looking better in small-scale fisheries where recovery and rebuilding efforts taking place along the coast seem to have resulted in sustainable yield, except in the case of Indian mackerel where number of juvenile fish is too low, which may suggest poor spawning conditions (Department of Fisheries, personal communication, September 2016). Positive signs can be found in small-scale crab fisheries since crabs caught by gill net and collapsible traps are larger than size at first maturity. Similar conditions are found in white shrimp. Data and information collected by this research center are well disseminated and used as a basis for discussion about fisheries management planning in the area, including at the main consultation forum about Trat fisheries, as detailed below.

## *The Trat Process*

Trat is one of the first provinces in the country to conduct a community-based consultative process to discuss the implication of the Royal Ordinance on small-scale fisheries. The Provincial Fisheries Committee, in collaboration with the Provincial Fisheries Office, initiated meetings first at the sub-district level, and later at the district level with all seven coastal districts in the province. At these meetings, fishers and community members were invited to discuss the appropriateness of fisheries policies affecting them, holding discussions on a gear by gear basis. The recommendations and propositions made during these meetings were compiled and presented at the provincial level meeting held on September 7, 2016. About 175 fishing people (men and women) and representatives of fishing groups and associations from around the province participated in this one-day meeting, co-hosted by the Department of Fisheries and the Department of Marine and Coastal Resources. The meeting began with presentations by the Fisheries Department about the status of the fisheries in Trat, followed by a summary from the district consultations. Participants were invited to comment on each recommendation and proposition before a conclusion was made about what to submit to the Provincial Fisheries Committee and later to the National Fisheries Policy Committee. The process was transparent and inclusive, with ample opportunities for fisher groups to provide additional information and/or explanation supporting their requests. The discussion included both viewpoints shared by the communities and those not reaching consensus at the district level meetings. During the provincial meeting, agreements were reached on almost all topics, and in the case of diverse opinions, notes were taken for further deliberation at future Provincial Fisheries Committee meetings.

While the process itself speaks volumes about the positive direction in which the province is going, which lends strong support to the implementation of the SSF Guidelines, three topics discussed in the provincial meeting are worth noting for their potential to enhance small-scale fisheries sustainability in Thailand.

First, much of the discussion evolved around issues related to conservation. Trat people expressed a lot of pride in their conservation efforts, and emphasized several measures that they would like to see strengthened. For instance, bestowed by the Royal Ordinance Section 56, the Provincial Fisheries Committee has already designated several areas in the province as sanctuaries for aquatic species. The communities generally agreed to this and, in some instances, demanded that more areas should be put under protection, such as mangrove areas and clam beds in certain districts. Communities elaborated on the protection of clam beds that it would have positive effects on tourism income. It was recognized, however, that the size and location of these protected areas should be determined at the district level so that they are suitable with the local ecological features. The emphasis on site-specific regulations acknowledges the local knowledge of fishers and reinforces their roles in fisheries governance, which follows the principle about 'respect of cultures' in the SSF Guidelines. Special considerations were given, for instance, about Koh Chang to enable small-scale fisheries to operate closer to shore due to its unique geography. The same request was not granted, however, to a nearby island of Koh Rang, due to a lack of evidence to suggest traditional use of the area.

The commitment of fishing communities to promoting conservation and protection of their fisheries resources was evident in their deliberation about impacts of having too many fishing gears in certain areas and in their desire to restrict people from other places from entering the area. One district, for instance, asked for the prohibition of fish gillnets and squid traps from operating within five nautical miles from shore. This request was not granted, however, on the basis that jurisdictional responsibility of the district extended only to three miles and not beyond. In another example, communities talked about how they communicated with each other and shared information about crab spawning period. They therefore argued for flexible seasonal closure in accord with the local knowledge.

Gear and seasonal restrictions were considered easier to implement if applied across the province. Prohibition of trawling, beach seining, and use of certain gears, such as gillnet, crab trap (smaller than 2.5 mesh size), and light luring for anchovy, within three nautical miles were generally agreed upon. An exemption was requested, however, for light luring for squid, practiced by small-scale fishing vessels, to be allowed within the three-mile limit, for safety reasons. This proposition was accepted, but only during the monsoon season (from October to March). Another request came from trawl fisheries for jellyfish in two districts. Arguing from the social wellbeing perspective (lack of alternative livelihoods), fishers asked that small trawlers (between 5 and 10 GRT) be allowed to fish in coastal areas (about 1.5 nautical miles from shore), from June to September. The request was granted, with the condition that the mesh size be greater than four inches. The compromises made at the meeting, especially when the requests were well justified, and the provision made to minimize fishing impacts on ecosystems, illustrate how responsible fisheries and sustainable development, the goals promoted in the SSF Guidelines, can potentially be achieved through a community-based, consultative process.

The discussion about self-imposed regulations was rather unusual, and went against the general belief about the need to deter fishers' behavior through heightened enforcement (Hatcher et al. 2000). Rather than being concerned about the restrictions, the fishing people of Trat, in some instances, asked for stricter regulations to be applied to their fishing areas. This corresponds with what Song and Chuenpagdee (2011) reveal in their study, that fishers and government officers may indeed agree about certain rules and regulations because they share fundamental conservation principles.

The second point that was exceptional about Trat is related to the strong stance that the fishing people took on the Acetes shrimp fishery. Lively discussion ensued at the meeting, with several fishers presenting their perspectives on the topic, arguing for the right to maintain this particular small-scale fishery as part of their traditional livelihoods. Clarification was provided by fishers about how the Acetes shrimp fishery operated, suggesting that it should not be included in 'push net' category, which is prohibited by law. Acetes shrimp fishery in Trat is done using a small scoop net that does not touch the seafloor when deployed. Fishers talked about the selectivity of their gear, that it has no bycatch, and that they knew when to scoop the shrimp (i.e. when the water is pink). Importantly, they argued that Acetes shrimp fishery is part of the traditional fishing practices, which must be preserved and passed on to younger generations. One fisher mentioned his involvement in the teaching of this particular fishery to schoolchildren as part of their curriculum. Fishers agreed to the current mesh size and seasonal regulations of the fishery, but did not consider area restriction to be reasonable. They requested that small-scale Acetes shrimp fishery (using vessels less than 10 GRT) be allowed to operate in shallow water, close to shore, where the shrimp can be found.

The management of Acetes shrimp fishery presents an interesting challenge to the Department of Fisheries, which is under pressure by the EU to eliminate push nets from operating in coastal waters. One push netter at the meeting spoke up, indicating how he was not able to benefit from the buyback program that the government established to help with the transition. It was later revealed that he was among the push netters who did not have a license to operate, and thus was not eligible to receive support. In the case of the small-scale Acetes shrimp push net fishery, fishers of Trat made an impressive case for maintaining their traditional fishing practice. Using local knowledge and scientific evidence to support their case, small-scale fishers argued that their fishing practice is non-destructive and thus should not be prohibited. The case emphasizes the importance of principles such as dexterity and subsidiarity in fisheries governance (Jentoft et al. 2011b). The former refers to management that is sensitive to details and recognizes differences from one location to another, while the latter posits that management should be vested at the lowest possible organization in order to be effective. Thus, the intervention by the EU in Thai fisheries may be necessary for reforming the large-scale, industrialized fisheries sector towards sustainability, but it is far from appropriate for small-scale fisheries.

The third and final point about the Trat process that invites further deliberation pertains to the presence and roles of various organizations in marine resource governance, which directly correspond with Part 3 of the SSF Guidelines about 'Ensuring an enabling environment and supporting implementation'. Trat province has attracted a host of research projects related mostly to biodiversity and conservation. For instance, Koh Chang was selected as a demonstration site for a coral restoration project in 2005 which was funded by the United Nations Environmental Program (UNEP) and the Global Environmental Fund (GEF) (UNEP 2007). In addition to enhancing scientific knowledge about marine ecosystems in the area, the project led to capacity development among local fishers and community members who participated in co-management efforts for restoring habitats and marine resources. Concerns still exist, however, about the health of coral reefs in the area due to both natural phenomena like coral bleaching and anthropocentric causes like rapid tourism development. Balancing between nature conservation and tourism development is the responsibility of DASTA (Designated Area for Sustainable Tourism Development Administration), established as a public organization in 2004 to encourage community participation in conservation projects while promoting tourism development (Rogers 2015). Recently, the Building Resilience to Climate Change Project, a four-year project launched in 2011, also selected Trat Province as one of its sites to study vulnerability and adaptability of communities to climate change. Through partnership between a non-profit organization, the Sustainable Development Foundation, and the Department of Marine and Coastal Resources, the project revealed a host of opportunities to engage local stakeholder groups in coastal planning, despite being identified as highly vulnerable to climate change (Morgan 2011). The Sustainable Development Foundation works closely with fishing communities in Trat Province, and has supported the establishment of local fishing community organizations in the area. Several fisher groups and other communities and non-government organizations, including the Sustainable Development

Foundation, were presented at the meeting, representing a wide range of coastal stakeholders interested in marine resource and ecosystem governance. Their active participation throughout the whole process bodes well about the potential roles they can play in the implementation of the SSF Guidelines.

The involvement of several government units is also important for resource governance in Trat province. The richness of coastal and marine ecosystems requires that all relevant units take part in collaborative planning for concerted actions. The fact that the meeting was hosted by the Department of Marine and Coastal Resources, held at the 'People's State' Center for Protection of the Sea,[3] sent a positive signal towards the development of cohesive policy and foster institutional coordination, which are imperative for the implementation of the SSF Guidelines (FAO 2015). Inputs from the representatives of the Department of Marine and Coastal Resources at the meeting, especially with respect to the conservation and protection of habitats and marine resources, were valuable in the discussion about zoning and spatial management. For instance, they were able to provide information about Koh Chang and Koh Rang, explaining the key ecosystem features in each system, thus justifying the different management rules.

Fisheries governance needs to consider a broad range of stakeholders and topics when dealing with transboundary issues. One of the key topics discussed at the meeting was illegal fishing and other fishing violations involving neighboring countries. Thai fishing vessels for anchovy were noted, for instance, to operate near the Cambodian border and may be landing their catches there, while other fishing violations occurred from fishing vessels of neighboring countries, such as Cambodia and Myanmar. The meeting discussed measures that should be implemented to prevent these from happening. Vessel monitoring systems installed in the larger vessels were considered useful in detecting violations. Some suggestions were made about improving signage to show the boundaries and increasing fines and penalties as deterrence. These jurisdictional issues are not easy to address and will require collaboration from other government agencies, including the Navy and the Coast Guard, as well as drawing from other international instruments. This is also an area where information availability and communication noted in the SSF Guidelines with respect to IUU fishing (paragraph 11.5) becomes highly relevant.

It should be noted that not all recommendations and suggestions discussed at the district level meetings met with approval. Likewise, not everyone was satisfied with the meeting outcomes. For instance, the request to allow collapsible crab traps of 1.5 inch mesh size to operate in one district was made, but not accepted. The regulation on the mesh size is that it must be no less than 2.5 inches. Further requests were made to the government to provide financial assistance to crab fishers in order to modify their gears in compliance with the rules. Another fisherwoman spoke about the unfair treatment that she and two other large-scale gillnetters received because they were prohibited from fishing in certain areas, but they were given no compensation. While nothing was done to rectify the situations at the meeting, the fact that

[3] The official name is 'Pracharath Center for Protection of the Sea'. Pracharath, or 'state of people,' is part of the government's new roadmap.

these issues were revisited and that the stakeholders were given opportunities to express their concerns marked a positive step towards relationship and trust building, which are imperative for co-governance (Jentoft 2000; Feeley et al. 2008).

## Thai Small-Scale Fisheries Moving Forward

The Royal Ordinance on Fisheries (2015) is a major reform that has generated a lot of discussion among fisheries stakeholders, civil society organizations, and researchers. The government sees the new legislation as the 'fundamental and comprehensive reform of legal framework governing Thai fisheries'.[4] However, its rapid emergence, and the motivation behind it, has drawn criticism from some stakeholders and raised concerns for others. There have also been many uncertainties about the implications of the new law, especially on small-scale fisheries. A close reading of the document allows for some optimism, but a lot depends on how the law will be interpreted and implemented and whether provisions can be made to bring clarification to certain sections as well as to accommodate stakeholders' concerns.

The most encouraging aspects of the new law are the establishment of the National Fisheries Policy Committee and the Provincial Fisheries Committee and their stipulated memberships, and the empowerment of the local governing bodies. This new governing structure offers ample opportunities for fishers and community members to participate in decision-making about fishing rules and regulations and in the planning of future fisheries development. Small-scale fishing people can take part in governance through local fishing community organizations. Even those without any representation can still provide inputs through other consultative processes, as seen in the Trat example. The capacity of local fishing people to participate in fisheries governance builds on a long tradition of community-based resource management promoted through many projects and initiatives (Juntarashote 1994; Ruangsivakul et al. 2002; Nasuchon and Charles 2009; Thiammueang et al. 2012).

The importance of Section 28, which bestows power and duties unto the Provincial Fisheries Committee, cannot be underestimated. The experience in Trat province elucidates clearly the need to consider local contexts in determining fisheries rules and regulations, which begins with the definition of small-scale fisheries. As argued by FAO and others (e.g. Johnson 2006), small-scale fisheries are not characterized only by capacity, vessel size, or engine power, but also by other features and values such as ownership and involvement of family members, utilization of catches, and the overall social and cultural importance of fisheries. Revision to the rules and regulations on small-scale fisheries proposed by Trat fishing communities reflect their knowledge about where and when to fish, what species, with what gear, as well as the impact of fishing on habitats and fisheries resources. Their desire

[4]According to the website of the Royal Thai Embassy to Belgium and Luxembourg/Mission of Thailand to the European Union, posted on 18 November 2015; http://www2.thaiembassy.be/highlights-of-thailands-new-fisheries-legislation/; accessed on September 25, 2016.

to preserve Acetes shrimp push net fishery illustrates traditional value that local communities place on their livelihoods and the importance of passing on local knowledge and tradition to future generations. While there were several requests from fishers with respect to area and gear restrictions, they were able to provide justification for modification and exemptions, supported in many instances by scientific research. It will be critical to see that the recommendations and proposals agreed upon in the meeting receive proper considerations at the provincial and national levels. Otherwise, the positive experience cultivated during the Trat process would be in vain and may initiate hostile relationship between fishers and government, especially with the uncertainty about what the new law may bring.

Moving Thai small-scale fisheries towards sustainability is not possible without addressing the many missing aspects specified in the SSF Guidelines. Primarily, gender and womens' issues do not receive any consideration in the Royal Ordinance. This is a major oversight given the prominent role that women play in Thai fisheries throughout the fish chain. More emphasis is also needed to improve post-harvest activities and the overall value chain, as suggested in the SSF Guidelines (Section 7). While occupational health and safety, hygiene, and wellbeing of fishers and fish workers are considered, the implications of climate change and disaster risk on small-scale fishing people and their communities are not. Finally and most importantly, the issue of rights is totally absent, which is highly problematic and could become a major impediment for the implementation of the SSF Guidelines, which are underlined by human rights-based approaches.

Further, in moving towards the implementation of the SSF Guidelines in Thailand, some emphasis should be placed on developing mechanisms that foster holistic and integrated approaches in fisheries governance, such as those that enhance coordination between different government agencies (Satumanatpan 2011). Such collaboration could also be extended to developing a common database, with an established baseline, and information systems that can be shared. This will then lead to concerted efforts and help enhance policy coherence. Opportunities to rectify these and fill other gaps identified in this study can be found with the process that the Southeast Asian Fisheries Development Center (SEAFDEC) is leading. SEAFDEC is an intergovernmental body, comprising 11 member countries, including Brunei Darussalam, Cambodia, Indonesia, Japan, Lao PDR, Malaysia, Myanmar, Philippines, Singapore, Thailand, and Vietnam. With its mandate to develop and manage fisheries in the region for food security and safety of the people, as well as for poverty alleviation, SEAFDEC has been working on the development of a regional plan for the implementation of the SSF Guidelines. Building from its experience and existing capacity, Thailand can play an active role in this process by, among other things, aligning its policy and regulations with the principles and directions posited in the SSF Guidelines. Doing so will help support small-scale fishing people in their negotiation with the government about the rules and legislation in the Royal Ordinance that prohibit their sustainability.

# References

Butcher, J. G. (2004). *The closing of the frontier – A history of the marine fisheries of Southeast Asia c. 1850–2000*. Leiden: KITLV Press.

Chuenpagdee, R. (2011). Thinking big on small-scale fisheries. In V. Christensen & J. Maclean (Eds.), *Ecosystem approaches to fisheries: A global perspective* (pp. 226–240). Cambridge, MA: Cambridge University Press.

Chuenpagdee, R., & Jentoft, S. (2009). Governability assessment for fisheries and coastal systems: A reality check. *Human Ecology, 37*, 109–120.

Chuenpagdee, R., & Juntarashote, K. (2004). *Regional overview of status and trend of "trash fish" from marine fisheries and their utilization, with special reference to aquaculture*. Report to FAO Regional Office for Asia and the Pacific. Bangkok.

Chuenpagdee, R., & Juntarashote, K. (2011). Learning from the experts: Attaining sufficiency in small-scale fishing communities in Thailand. In S. Jentoft & A. Eide (Eds.), *Poverty mosaics: Realities and prospects in small-scale fisheries* (pp. 309–331). Dordrecht: Springer.

Degnbol, P., Gislason, H., Hanna, S., Jentoft, S., Nielsen, J. R., Sverdrup-Jensen, S., & Wilson, D. C. (2005). Painting the floor with a hammer: Technical fixes in fisheries management. *Marine Policy, 30*, 534–543.

DoF. (2002). *Census of fishing household 2000*. Statistical unit, department of fisheries, ministry of agriculture and cooperatives. Bangkok.

DoF. (2015a). *Marine fisheries management plan of Thailand. A national policy for marine fisheries management, 2015–2019*. Department of fisheries, ministry of agriculture and cooperatives. Bangkok

DoF. (2015b). *Fisheries statistics of Thailand, 2013*. (No 7/2015). Fishery statistics analysis and research group, information and communication technology center, department of fisheries, ministry of agriculture and cooperatives. Bangkok.

DoF. (2016). *Fishing community production survey in 2014*. (No 4/2016). Fishery statistics analysis and research group, information and communication technology center, department of fisheries, ministry of agriculture and cooperatives. Bangkok.

FAO. (2015). *Voluntary guidelines for securing sustainable small-scale fisheries in the context of food security and poverty eradication*. Rome: United Nations food and agriculture organization, FAO.

Feeley, M. H., Pantoja, S. C., Agardy, T., Castilla, J. C., Farber, S. C., Hewawasam, I. V., Ibrahim, J., Lubchenco, J., McCay, B. J., Muthiga, N., Muthiga, S. B., Sathyendranath, S., Sissenwine, M. P., Suman, D. O., Tamayo, G., Roberts, S., Hall, F., & Bostrom, J. (2008). *Increasing capacity for stewardship of oceans and coasts: A priority for the 21st century*. Washington, DC: National Academies Press.

Fulton, E. A., Smith, A. D. M., Smith, D. C., & van Putten, I. E. (2011). Human behavior: The key source of uncertainty in fisheries management. *Fish and Fisheries, 12*, 2–17.

Hatcher, A., Jaffry, S., Thebaud, O., & Bennett, E. (2000). Normative and social influences affecting compliance with fishery regulations. *Land Economics, 76*, 448–461.

Hilborn, R. (2007). Moving to sustainability by learning from successful fisheries. *Ambio, 36*, 296–303.

Hilborn, R., Punt, A. E., & Orensanz, J. (2004). Beyond band-aids in fisheries management: Fixing world fisheries. *Bulletin of Marine Science, 74*, 493–507.

Jentoft, S. (2000). The community: A missing link of fisheries management. *Marine Policy, 24*(1), 53–59.

Jentoft, S. (2013). Social justice in the context of Fisheries – A governability challenge. In M. Bavinck, R. Chuenpagdee, S. Jentoft, & J. Kooiman (Eds.), *Governabiity of fisheries and aquculture: Theory and Applications* (pp. 45–65). Dordrecht: Springer.

Jentoft, S., & Chuenpagdee, R. (2009). Fisheries and coastal governance as a wicked problem. *Marine Policy, 33*, 553–560.

Jentoft, S., Chuenpagdee, R., & Pascual-Fernandez, J. (2011a). What are MPAs for? On goal formation and displacement. *Ocean and Coastal Management, 54*, 75–83.
Jentoft, S., Eide, A., Bavinck, M., Chuenpagdee, R., & Raakjær, J. (2011b). A better future: Prospects for small-scale fishing peoples. In S. Jentoft & A. Eide (Eds.), *Poverty mosaics: Realities and prospects in small-scale fisheries* (pp. 451–469). Dordrecht: Springer.
Johnson, D. S. (2006). Category, narrative, and value in the governance of small-scale fisheries. *Marine Policy, 30*, 747–756.
Jones, E., Schuttenberg, H., Gray, T., & Stead, S. (2015). The governability of mangrove ecosystems in Thailand: Comparative successes of different governance models. In S. Jentoft & R. Chuenpagdee (Eds.), *Interactive governance for small-scale fisheries: Global reflections* (pp. 413–432). Cham: Springer.
Juntarashote, K. (1994). Possible development of community-based management in Thailand with the advent of fishing right system. In *Proceedings of the 7th IIFET Conference, July 18–21 1994, Taipei, Taiwan*.
Juntarashote, K. & Chuenpagdee, R. (1987). Small-scale fisheries of Thailand. In Anon, *Research and small-scale fisheries* (pp. 555–573). IFREMER, France.
Kooiman, J., Bavinck, M., Jentoft, S., & Pullin, R. (Eds.). (2005). *Fish for life: Interactive governance for fisheries*. Amsterdam: Amsterdam University Press.
Lunn, K. E., & Dearden, P. (2006). Monitoring small-scale marine fisheries: An example from Thailand Koh Chang's archipelago. *Fisheries Research, 77*, 60–71.
Morgan, C. L. (2011). *Limits to adaptation: A review of limitation relevant to the project 'building resilience to climate change' – Coastal Southeast Asia*. Gland: IUCN.
Nasuchon, N., & Charles, A. (2009). Community involvement in fisheries management: Experiences in the Gulf of Thailand countries. *Marine Policy, 34*, 163–169.
Panayotou, T., & Jetanavanich, S. (1987). *The economics and management of Thai fisheries, ICLARM studies and reviews 14*. Manila/Arkansas: International center for aquatic living resources management, Winrock international institute for agricultural development.
Pauly, D., & Chuenpagdee, R. (2003). Development of fisheries in the Gulf of Thailand large marine ecosystem: Analysis of an unplanned experiment. In G. Hempel & K. Sherman (Eds.), *Large marine ecosystems of the world: Trends in exploitation, protection, and research* (pp. 337–354). Amsterdam: Elsevier Science.
Pimoljinda, J. (2002). Small-scale fisheries management in Thailand. In H. E. W. Seilert (Ed.), *Interactive mechanisms for small-scale fisheries management: Report of the regional consultation, RAP publication 2002/10* (pp. 80–91). Bangkok: FAO Regional Office for Asia and the Pacific.
Rogers, V.L. (2015). *Synergies for stewardship and governance of multiple-use coastal areas: A case study of Koh Chang, Thailand.* (Master's thesis). Department of Geography, Memorial University of Newfoundland, St. John's, NL, Canada.
Ruangsivakul, S., Yamao, M., & Kamhongsa, J. (2002). *People's groups and community-based arrangements in Tambol Pakklong, Pathew District, Chumporn Province, LBCRM-PD No. 4*. Bangkok: Collaborative project between Southeast Asian fisheries development center and department of fisheries.
Satumanatpan, S. (2011). *Coastal management: Integration to sustainability*. Nakorn Pathom: Mahidol University Press.
Song, A. M., & Chuenpagdee, R. (2011). Conservation principle: A normative imperative in addressing illegal fishing in Lake Malawi. *Maritime studies (MAST), 10*(1), 5–30.
Sutthacheep, M., Pengsakun, S., Chueliang, P., Suantha, P., Sakai, K., & Yeemin, T. (2012). Managing bleached coral reefs in the Gulf of Thailand. In Proceedings of the 12th International Coral Reef Symposium, Cairns, Australia, July 9–13 2012.
Teh, L. C. L., Zeller, D., & Pauly, D. (2015). *Preliminary reconstruction of Thailand's marine fisheries catches: 1950–2010, Fisheries centre working paper #2015–01*. Vancouver: University of British Columbia.

Thiammueang, D., Chuenpagdee, R., & Juntarashote, K. (2012). The "crab bank" project: Lessons from the voluntary fishery conservation initiative in Phetchaburi Province, Thailand. *Kasetsart Journal (Natural Science), 46*, 427–439.

UNEP. (2007). *National reports on coral reefs in the coastal waters of the South China Sea.* UNEP/GEF/SCS technical publication No. 11. UNEP.

Yeemin, T., Mantachitra, V., Plathong, S., Nuclear, P., Klinthong, W., & Sutthacheep, M. (2012). Impacts of coral bleaching, recovery and management in Thailand. In *Proceedings of the 12th International Coral Reef Symposium, Cairns, Australia*, July 9–13 2012.

# Chapter 33
# The *Buen Vivir* and the Small-Scale Fisheries Guidelines in Ecuador: A Comparison

**María José Barragán-Paladines**

**Abstract** The *Buen Vivir* paradigm, that inspired the Ecuadorian National Development Plan (PNBV) represents a shift in understanding development and articulates mother earth's and human's rights, both as subjects of legal protection. This chapter aims to disentangle contradictions between the PNBV and current practices at the small-scale fisheries sector. Theoretically grounded in the governability concept, we explore commonalities between the *Buen Vivir* objectives and the Voluntary Guidelines for Securing Sustainable Small-Scale Fisheries (SSF Guidelines) principles. The chapter (a) illustrates how are the PNBV objectives align to the guiding principles of the SSF Guidelines; (b) explores to what extent the actions taken by the state address issues desired to achieve small-scale fisheries sustainability; and (c) identifies what elements within the Ecuadorian fishing governance system are missing in order to enhance small-scale fisheries sustainability. The study involved a comprehensive literature review and empirical work using semi-structured interviews and participant observation. Commonalities on both instruments (e.g. diversity, sustainability, and human dignity) were found but also some unique elements such as in the PNBV (e.g. rights of nature and sovereignty) or in the SSF Guidelines (e.g. gender). Additionally, initiatives addressing some threats to the fisheries occur, but still are isolated practices with low connectivity to the entire fisheries systems. Finally, mismatches between the PNBV's rhetoric and the fisheries governance practices occur and have inordinately delayed the improvement of small-scale fisheries governability. Coherent theories and practices at the political and social realms, under innovative ontological and epistemological connotations of fisheries will help to achieve their sustainability under the *Buen Vivir* paradigm.

**Keywords** Small-scale fisheries • Ecuador • *Buen Vivir* • SSF Guidelines

---

M.J. Barragán-Paladines (✉)
Development and Knowledge Sociology Working Group, Leibniz Centre for Tropical Marine Research – ZMT, Bremen, Germany
e-mail: mariaj.barraganp@leibniz-zmt.de



S. Jentoft et al. (eds.), *The Small-Scale Fisheries Guidelines*, MARE Publication Series 14, DOI 10.1007/978-3-319-55074-9_33

## Introduction

Implementation is a process enabled by institutions – as collective or individual agents – that are operating at different dimensions and scales, in order to achieve particular aims. Implementation is certainly not a 'linear' set of steps to be followed. On the contrary, it can be described as a process deeply embedded within the complex structures that characterize most social and ecological systems, like the "systems-within-systems" format that is found in fisheries (Jentoft 2014).

In order to address the implementation of the Voluntary Guidelines for Securing Sustainable Small-Scale Fisheries in the Context of Food Security and Poverty Eradication (SSF Guidelines) in Ecuador, it is necessary to first understand how national development strategies have been designed during the last decade and what they contain. In 2008, for the first time ever, the Ecuadorian National Constitution (2008) granted inalienable rights to nature and recognized nature as a subject that enjoys juridical protection, at both Constitutional and Legislative levels (Berros 2015) under legally enforceable practices (Whittemore 2011). In the preamble of the Constitution, the Ecuadorian nation-state is defined to be one of "*constitutional rights and justice-based, social, democratic, sovereign, independent, unitary, intercultural, plurinational, and secular*." The Constitution was also meant to "[i]*ntegrate the diversity of peoples, cultures, notions* (i.e. Mother Earth or *Pachamama* and the *Sumak Kawsay*) *at all dimensions of National interest* (*e.g. economic, politic, financial, cultural, and environmental*)."

As an instrument of social bonding, the 2008 Constitution successfully recovered and integrated varied constituents of Ecuadorian society in such a way that enhanced Ecuadorian national pride, identity, and self-esteem. Since its approval, the Constitution has been symbolic of a return to the past in terms of recovering age old practices (e.g., traditional food and garments) that had been replaced by western-driven habits and practices over the last few centuries. The Constitution has also emphasized the need to not discriminate against traditionally marginalized groups. The making of the Constitution, therefore, can be seen as a successful example of a participatory process that "redeemed the past" of the Ecuadorian nation state (Acosta 2008; Acosta and Martínez 2009). Under the *Buen Vivir* approach of the Constitution, the existence of societal institutions – fostering reciprocity, cooperation, and solidarity – is envisioned as a key means to promote the good way of living that is embedded in this concept. Seen as an ancient ontological notion that has been recently recovered (Viveiro de Castro 2004; Haidar and Berros 2015), *Buen Vivir* constitutes an alternative approach to development, and as such it represents a potential response to the substantial critiques of post development (Acosta and Martínez 2009; Gudynas and Acosta 2011) that have populated the academic discourses in the last decades.

This chapter, therefore, undertakes a critical review of the implementation of the SSF Guidelines in the context of *Buen Vivir*. This un-orthodox perspective resembles to be an interesting standpoint that proposes a mindset shift in governing natural resources away from the dominance of western-science-based and technocratic-inspired approaches. In that light, it challenges the unique central position that political and managerial agenda has awarded to science while doing policymaking for resource governance (e.g., in the cases of oil palms, shrimps, and soy monocultures) (Escobar 2016).

Challenges in small-scale fisheries governance in Ecuador are varied (Barragán-Paladines 2015). They include legal and ethical concerns, including but not limited to, the accomplishment, promotion, and enforcement of both, rights of nature and fishers' rights.

Over the last few decades, some research has been conducted on environmental ethics (Johnson 2003; Saez-Marti and Weibull 2005), the jurisprudence of conservation (Stone 1972), new ways of development (Farah and Vasapollo 2011; Gudynas and Acosta 2011; Altmann 2013), ethics in fisheries (Lam and Pitcher 2012), and fisheries and human rights (Song 2015). This research bodies have collectively highlighted the need to integrate diverse epistemological notions and human expertise into natural resource governance.

In Ecuador, however, there has been limited research about environmental ethics in fisheries in general, and on the moral principles behind fisheries management in particular. Moreover, we find reasonable reasons to claim that little has been done in exploring and illustrating the foundational principles of Ecuador's National Development Plan (PNBV by its Spanish acronym), especially those central to marine resources governability, neither in Ecuador mainland nor in Galapagos Islands (Barragán-Paladines 2015). This scheme has influenced the inertia of governing bodies to tackle fisheries' problems appropriately and their consequent inability to govern this sector right from the time when fisheries became economically profitable, in the 1960s. In line with this idea, we call for the comprehensive integration of theories and practices that jointly address small-scale fisheries issues, beyond the purely managerial approach. This standpoint would help in avoiding the mistake of disdaining the institutional dimension of environmental change and the confusion between governance and government (Young et al. 2008).

This chapter searches for overlapping guiding principles in the concept of *Buen Vivir* and the SSF Guidelines as both could be conceived as normative instruments of critical relevance at governing fisheries resources. This study was an attempt to unravel commonalities between the two instruments mentioned above by paying special attention to the *rights of nature* and the *rights of fishers*. The methodology section details the type and origin of the data that has been used to inform this chapter, and the scales at which the implementation of the SSF Guidelines has taken place. In addition to that, varied dimensions of the implementation process under the current governing regime are described by analyzing, features that shaped the institutional performance and rules that regulate the small-scale fisheries sector in Ecuador. We suggest that that the apparent contradiction between the protection of the rights of nature and the defence of fishers' right, can be disentangled by looking at the former as the legal remedy to protect the rights of future generations of fishers. In line with that idea, the protection of nature – at the terrestrial and marine dimensions – and the rights of fishers to fish, is thus guaranteed.

We analyze the extent to which the *Buen Vivir* principles align with those from the Voluntary Guidelines for Securing Sustainable Small-scale Fisheries in the Context of Food Security and Poverty Eradication (SSF Guidelines). In doing so it, (a) illustrates at which level are the PNBV objectives aligned to the guiding principles of the SSF Guidelines; (b) explores the degree to which the actions taken by the state address critical issues identified by the SSF Guidelines as desirable to achieve small-scale fisheries sustainability; and (c) identifies what elements within the Ecuadorian fishing governance system are missing in order to enhance sustainability of the fisheries sector. Finally, reflections about how the *Buen Vivir* would promote small-scale fisheries sustainability – under current and future governing schemes – is discussed towards the end.

## Study Context – The *Buen Vivir* Principle

The *Sumak Kawsay* (in Quichua) or '*Buen Vivir*' (in Spanish) translated as 'good way of living' is not a new notion. This concept has been present in ancient Amerindian discourses and indigenous Andean cosmovisions (aka `non-dualist philosophies´) (Escobar 2016) that envisage a comprehensive way of understanding life. *Sumak Kawsay* retrieves and articulates varied ontologies and epistemologies about humans, animals, other beings, and the environment. These broader ontologies and epistemologies, as Berros (2015) has articulated, align nicely with newly-grounded ideas that currently belong to the fields of environmental – and animal – ethics that are also present in the juridical field (Haidar and Berros 2015). The *Sumak Kawsay* discourse circulates around the equilibrium and harmonic coexistence of beings, from both the social and natural realms, privileging the collective over the individual and solidarity over competition. *Buen Vivir* is a category in the Andean philosophy that has lost ground due to the dominance of religious beliefs, Western practices and rational theories (Viveiro de Castro 2004; Duarte and Belarde-Lewis 2015) that disregard and discredit these other ways of thinking (Haidar and Berros 2015).

Since 2008, when the new Constitution of Ecuador was approved, the *Buen Vivir* principle has become the center stage that contests the traditional notion of development focused on economic growth (Lind 2012). Contrary to that idea, the PNBV proposes the sustainable development not as the definitive desired outcome to pursue, but only as an interim goal in the way towards an alternative to develop, illustrated by the *Buen Vivir.* This proposal stresses the importance of a paradigmatic shift in the idea of development towards encompassing dimensions like happiness, freedom, and equal rights, in addition to sustainability (Gudynas 2011). In fact, several scholars (Escobar 1996; Acosta 2008; Acosta and Martínez 2009) propose *Buen Vivir* as the unconventional 'alternative for development' that contests the models for 'alternative development.' By prioritizing *Buen Vivir* over the last decade, the state has played a critical role in the drive towards achieving social wellbeing in Ecuador. Policies, programs, and practices within the political agenda have circulated around the *Buen Vivir*.

Despite the fact that the Ecuadorian Constitution describes 'nature' (i.e. Mother Earth or *Pachamama*) not just as an object of human ownership, but instead as an entity in-and-of itself, it still fails to acknowledge the implicit existence of the marine domain (i.e. Mother Sea) (Andreve-Díaz 2008) within the nature concept, traditionally linked to the terrestrial portion, only. By doing such, the role of the marine environment is underplayed. How *Buen Vivir* will be achieved in policies that address marine systems is unclear. The absence of a transparent vision for marine environments means that the small-scale fishing sector, fishing communities, and related marine systems are hardly integrated within the PNBV agenda. We thus argue that if the rights of nature, which are recognized by the Ecuadorian Constitution, are to be addressed it is necessary to explicitly acknowledge the marine dimension of nature and also the fisheries and fishing people that go along with it.

## The Obscurity of Small-Scale Fisheries Through the *Buen Vivir* Lens

In Ecuador, fishing has historically been an important cultural and social asset. Only lately (i.e. the 1960s) has fisheries, however, become an important economic feature. There is abundant evidence of fishing practices, consumption, and trading of fish produce, at a moderate scale, locally and regionally (Baumann 1978; Norton 1985; McEwan and Silva 1998) within pre-Hispanic Ecuadorian communities. There even are records of very complex fishing practices using special fishing strategies[1] and diverse gears (e.g. nets, lines, and hooks) (De Madariaga 1969). Until recently (i.e. the 1950s), small-scale fisheries in Ecuador were mainly a subsistence activity. In the 1960s, when its commercial potential was realized, it was developed with the support of international aid agencies (e.g. the FAO) (Allsopp 1985; Williams 1998). Since then, small-scale fisheries have been viewed as critical to economic growth, along with construction and tourism. Having said that, the importance of fish as a food source and its cultural and identity dimensions have not received adequate attention, though there has been reference to both dimensions in archaeological research (Norton 1985; McEwan and Silva 1998) that has described fishing in early Ecuadorian history (Baumann 1978; Rostworowski 2005).

Under the cultural construction of fisheries (Finley 2009) and as a consequence of the dominant and prevailing doctrine of free trade and markets that looks at fish only as a good to be traded, fish produced by small-scale fishers in Ecuador has

[1] Spanish conquerors recorded fishing practices of South American indigenous tribes who used a 'hunter fish' to catch bigger preys, even sharks. For a detailed description of these practices, see De Madariaga 1969, 116–130).

remained unnoticed, otherwise. This has meant also that historical, cultural, and even spiritual dimensions of fisheries within fishing communities have remained hidden in Ecuador under current governing practices that have been put in place in order to achieve *Buen Vivir*. Despite the appeal made by the PNBV to follow 'new ways to understand development', other notions have been disregarded.

We suggest that under the PNBV multiple dimensions of the fisheries sector have not been fully addressed by the principles enshrined in the new constitution and consequently have not received enough attention. Moreover, there is evidence of incongruent and dissonant approaches to governing fisheries and other marine resources that do not pay adequate attention to the ethical and/or moral dimensions of fisheries.

## Methodological Approach

This exploratory study starts with the premise that the quality of overall governance depends first and foremost on the inherent characteristics of the human and natural systems that are being governed and of the governing system (Chuenpagdee and Jentoft 2009, 2013). This chapter draws upon the interactive governance approach in the context of small-scale fisheries (Kooiman et al. 2005, 2008; Bavinck et al. 2013) and explores the extent to which the guiding principle (i.e. *Sumak Kawsay*) of the current governing system, represented by the Ecuadorian Constitution, enables the implementation of the SSF Guidelines. The variables used to perform the analysis were key words that illustrated critical human dimensions that were present in both the Constitution and the SSF Guidelines (e.g. rights, equity, dignity, etc.). We then elicited how the objectives proposed by the PNBV integrated (or failed to) elements that are present in the guiding principles of the SSF Guidelines.

Through an intensive literature review, which included perusing both published and grey literature concerning the new Constitution and documents related to it, non-structured in-depth interviews (Bernard 2011), e-mail-based communications with policy makers and practitioners, and participant observation (by the author in state and non-state policy meetings), empirical qualitative data was gathered to document the commonalities between the two instruments. Interviewees were first asked what their perception is concerning the articulation of the principles present in both instruments. Second, we 'tracked' evidence, through primary and secondary literature, of actions and practices that have enabled the implementation of any of the principles of the SSF Guidelines.

The key informants (n = 10) involved a range of actors from the fisheries sector, one of whom who actually took part in producing the new Constitution. We communicated with stakeholders throughout 2015 and counted fishers, (2) officials of the Undersecretary for Fisheries Resources (SRP), (3) politicians (1), scholars/researchers (2), and fisheries cooperative representatives (1). We also conducted one telephone conversation with a local fisheries manager and sent an email to a researcher who did not respond.

Finally, we undertook data analysis by following a categorization process which looked for themes patterns'. Our analysis was driven by a manual-based coding process that required separating qualitative data into theme-units', and setting theme categories which later contributed to the definition of discourse patterns (Anguera 1978). This coding process helped us in elucidating overlaps between the two instruments. However, we were unable to verify the data in the field, as suggested by (Aronson 1994) due to budget constraints.

## Main Findings

### *Contrasting the* Buen Vivir *Notion and the SSF Guidelines' Principles Using Interactive Governance*

The comparison of the guiding objectives of the PNBV and the guiding principles of the SSF Guidelines illustrates that some compatibilities and mismatches exist (e.g. diversity, sustainability, and human dignity), whereas others are at either the PNBV (e.g. rights of nature and sovereignty) or at the SSF Guidelines (e.g. gender issues). Coincident aspects found on both instruments are showed in Table 33.1, and can be read horizontally. These attributes fall under three thematic areas: social and environmental (global) sustainability; human rights, dignity, inclusion, people's participation, and peace; and diversity, multi-culturalism and multi-ethnicisms.

Some of the principles in the SSF Guidelines are incorporated in the twelve objectives of the PNBV. For example, Principle 11 of the SSF Guidelines, which calls for 'A comprehensive approach,' is clearly and explicitly covered by the *Buen Vivir* principle. The latter, as mentioned earlier, is a paradigmatic cosmovision that was recovered from traditional ancient practices and which can still quite often be encountered in present-day customary practices of indigenous communities in Ecuador.

Additionally, we looked at how the two instruments are articulated with the three dimensions of the interactive governance theory (Kooiman et al. 2005, 2008). First, we found that the twelve objectives of the PNBV (that were inspired by the *Buen Vivir* notion) address aspects falling under the three dimensions described by the interactive governance theory, namely: the governing system, the social- and natural-system to be governed, and the governing interactions.

Second, we also observed that the twelve objectives nicely articulate four aspects considered to be of relevance nationally, according to the initiatives promoted and taken by the state in the last decade: the productive matrix transformation through investment in technology and education; the strengthening of national identity through capacity building and the multi-cultural and pluri-national recognition; the social inclusion of traditionally marginalized groups; and the improvement of the quality of life of the population.

**Table 33.1** Integration of the 'Guiding Objectives of the PNBV' and the 'Guiding Principles of the SSF Guidelines' with the attributes of Interactive Governance Theory (IGT)

| | Ecuador's National *Buen Vivir* Plan (PNBV) | SSF Guidelines |
|---|---|---|
| | Guiding objectives | Guiding principles |
| Dimension (IGT) | | International human rights standards, responsible fisheries standards and practices, and sustainable development according to the United Nations Conference on Sustainable Development (Rio + 20) outcome document 'The Future We Want,' the Code and other relevant instruments, paying particular attention to vulnerable and marginalized groups and the need to support the progressive realization of the right to adequate food |
| Governing System | 1. The consolidation of the democratic state and the construction of people's power | 11. Holistic and integrated approaches |
| | 6. The consolidation of just transformation, strengthening integral security, and addressing human rights | |
| | 11. Sovereignty and efficiency of strategic sectors for industrial and technological transformation | |
| Social System-to-be-Governed | 3. Improvement of the population's quality of life | 1. Human rights and dignity |
| | 4. Strengthening of the citizen's capacities and potentialities | 2. Respect of cultures |
| | 5. Creation of common spaces to strength national identity, diverse identities, and pluri-national and inter-cultural society | 3. Non-discrimination |
| | 8. Sustainable consolidation of the social and solidarity economy | 4. Gender equality and equity |
| | 9. Warranty of of all jobs with dignity | 5. Equity and equality |
| | 10. Transformation of the productive matrix | 6. Consultation and participation |
| | 12. Warranty of sovereignty and peace, depth in strategic integration into the global system, and Latin American integration | 7. Rule of law |
| | | 8. Transparency |
| | | 9. Accountability |
| | | 12. Social responsibility |
| | | 13. Feasibility and social and economic viability |
| | | 10. Economic, social, and environmental sustainability |

(continued)

**Table 33.1** (continued)

| | Ecuador's National *Buen Vivir* Plan (PNBV) | SSF Guidelines |
|---|---|---|
| | Guiding objectives | Guiding principles |
| Natural System-to-be-Governed | 7. Warranty of the rights of nature and the promotion of territorial and global environmental sustainability | |
| Governing Interactions | 2. Equality, cohesion, inclusion, and social and territorial equity in a diverse environment | |

Source: SENPLADES (unpublished) and Guidelines for Small-scale Fisheries Sustainability (FAO 2015)

Third, the SSF Guidelines on the other hand only include two of the dimensions described by the interactive governance theory: the governing system and the social-system-to-be-governed. Governing interactions and the natural-system-to-be-governed attributes are not explicitly dealt with in the SSF Guidelines (Table 33.1). The objectives of PNBV and the guiding principles of the SSF Guidelines are linked to the extent that both instruments influence the governance of small-scale fisheries and thus the entire governability notion.

## Current Initiatives Addressing Small-Scale Fisheries Sustainability

After 2008 when the Ecuadorian Constitution was put into action, the Ecuadorian Fisheries Department initiated a series of measures in order to improve the status of the fisheries sector. In conjunction with our second research question aimed at highlighting interventions and actions, we suggest here that these measures were in four broad realms: social security for small-scale fishers, regularization of fishing through the clearing of fishing licences and permits, security at sea, and upgrading of fishing conditions by building or improving fishing harbor infrastructure, nationally. All these measures were ultimately aimed at promoting better labor conditions in the small-scale fisheries sector. Table 33.2 summarizes the measures taken since late 2009.

**Table 33.2** List of measures taken by the Ecuadorian government to improve the small-scale fisheries sector

| Government initiatives (according to the PNBV) (period 2006–2014)[a] | SSF guidelines -Guiding principles- | | | | | | | | | | | | |
|---|---|---|---|---|---|---|---|---|---|---|---|---|---|
| | 1 | 2 | 3 | 4 | 5 | 6 | 7 | 8 | 9 | 10 | 11 | 12 | 13 |
| *3.2.1. Improvement of fishers' life quality* | | | | | | | | | | | | | |
| (a) By creating a social security (i.e. health care and retirement pension) system for small-scale fishers | x | | x | x | x | | x | | x | x | x | x | x |
| (b) By increasing the access of small-scale fishers and fishing communities to state-based health and education services | x | | x | x | x | x | x | x | x | x | x | x | x |
| *3.2.2. Improvement of fishers' labour conditions* | | | | | | | | | | | | | |
| (a) By building/improving/modernizing the small-scale fishing harbor facilities | x | | x | x | x | | x | x | x | x | x | x | x |
| (b) By improving the security of fishers while at sea | x | | x | | x | x | x | x | x | x | x | x | x |
| *3.3.3. Improvement of environment quality* | | | | | | | | | | | | | |
| (a) By providing on-board equipment to reduce littering at sea | | | | | | | x | | x | x | x | x | x |
| (b) By establishing regulations against polluting the harbor waters | | | | | | | x | x | x | x | x | x | x |
| *3.3.4. Capacity and leadership building* | | | | | | | | | | | | | |
| (a) By including women in capacity building training initiatives | x | x | x | x | x | x | x | x | x | x | x | x | x |
| (b) By establishing formal training programs for fishers | x | x | x | x | x | x | x | x | x | x | x | x | x |

[a]The measures undertaken by the state have been communicated through different formal (reports) and informal (media) channels. People interviewed also mentioned these as current initiatives of the government that support the small-scale fisheries sector

Cells marked with "x" show the overlap of initiatives of the PNBV and the guiding principles of the SSF

## *What is Missing in the* Buen Vivir *Principle and Development Objectives to Enhance Small-Scale Fisheries' Sustainability?*

Concerning our third research question, namely analyzing what is necessary to achieve sustainability within small-scale fisheries, we find there are a number of issues that continue to threaten the viability of the sector. It is evident that despite existing 'shared common ground,' between the two instruments there are still incompatibilities between the rhetoric of *Buen Vivir* and the principles of the SSF

Guidelines aimed at sustainability and viability of small-scale fisheries. Table 33.3 illustrates the matches and mismatches between the two instruments and where it is desirable for further intersection between them. Interestingly, the 'wholeness' of the *Buen Vivir* principle, at least in theory, is entirely related to the 'Holistic and Integrated Approach' emphasized in Principle No.11 of the SSF Guidelines. This common ground can easily be used as a stepping stone to promote sustainability. However, as mentioned earlier, there are also incompatibilities between the two instruments. These mismatches, termed 'missing common grounds' here, become salient when one compares Objective 7 of the PNBV with Principle 1 of the SSF Guidelines. This gap represents a major absence within the common grounds and illustrates the aspect that is needed the most. If present, it would ideally integrate, the human's and the nature's rights, within one single holistic principle. Moreover, the presence of this element on both normative instruments would contribute to accomplish the superior aim of the *Buen Vivir* notion and would enhance the sustainability of the fisheries resources.

While the question of 'human rights' is present in both instruments, 'nature rights' is only explicitly presented in the PNBV. Having said that, there is scope to argue that the 'rights of nature' are indeed considered under the SSF Guidelines within the 'ecosystem approach' (i.e EAF) mentioned by Principle 11. The lack of explicit recognition and the apparent disconnection between human and natural systems, both considered elements of the system-to-be-governed according to interactive governance theory, underplays the necessity to address both systems under the same overarching governance perspective, put in place nationally.

## Discussion – Key Aspects to Address the Current Challenges

We have seen that some of the PNBV objectives overlap with some of the principles of the SSF Guidelines. But we also observed inconsistencies between the two instruments vis-à-vis the political, economic, and social realms. We, therefore, argue first of all that for the sustainability of the small-scale fisheries sector to become a reality, the implementation of the SSF Guidelines (or the willingness to do so) must be an explicit part of actions, activities, and programs developed by current and future fisheries-related departments and agencies[2] within Ecuador. As things stand, it seems that there is no real will neither motivation to follow the SSF Guidelines. Second, the implementation of the SSF Guidelines – or the intention to do so – would need to be aligned with the PNBV and thus the latter's normative rootings

[2] When this chapter is being revised (November 2016), the very first initiative from fisheries-related bodies (national and regional) based in Ecuador, have shown interest in the implementation of the SSF Guidelines. They have taken the initiative to ask for support in the implementation strategies, at national and regional scale and have developed the first workshop linked to the SSF Guidelines implementation (MAGAP et al. 2016).

**Table 33.3** Common elements identified between the principles of the SSF Guidelines and the objectives of the PNBV

| | PNBV guiding objectives[a, b] | | | | | | | | | | | |
|---|---|---|---|---|---|---|---|---|---|---|---|---|
| **SSF Guidelines Guiding Principles** | 1. Consolidation of Democratic State | 2. Social equality, cohesion, inclusion, social, and territorial equity | 3. Improvement of quality of life | 4. Strengthening citizen's capacities and potentialities | 5. Creation of common spaces, to strength the national identity, the diverse identities, the pluri-national and inter-cultural society | 6. Consolidation of a renewed justice and strengthening internal security by strictly addressing the human rights | 7. Warranty of the rights of nature and the promotion of the territorial and global environmental sustainability | 8. Sustainable consolidation of the social and solidarity-based economic system | 9. Warranty of the existence of all forms of jobs with dignity | 10. Transformation of the productive matrix | 11. Sovereignty and efficiency of the strategic sectors for industrial and technological transformation | 12. Warranty of the sovereignty and peace, the depth into the strategic integration in the global system and the Latin American integration |
| 1. Human rights and dignity | | O | O | O | O | O | * | | O | | O | O |
| 2. Respect of cultures | | O | | | O | O | O | | O | | | O |
| 3. Non-discrimination | O | O | O | O | O | O | O | O | O | | | O |
| 4. Gender equality and equity | O | O | O | O | O | O | | | O | | | O |
| 5. Equity and equality | O | O | O | O | O | | | | O | | | O |
| 6. Consultation and participation | O | | | O | | | | | | | | O |
| 7. Rule of law | O | O | O | | O | | O | O | O | O | O | O |
| 8. Transparency | O | O | | O | O | | O | | O | O | | O |
| 9. Accountability | O | O | | O | O | O | O | | | O | | O |

| | | | | | | | | | | | | |
|---|---|---|---|---|---|---|---|---|---|---|---|---|
| 10. Economic, social, and environmental sustainability | | | ○ | | | | ○ | ○ | ○ | ○ | ○ | ○ |
| 11. Holistic and integrated approaches | ○ | ○ | ○ | ○ | ○ | ○ | ○ | ○ | ○ | ○ | ○ | ○ |
| 12. Social responsibility | ○ | ○ | ○ | ○ | ○ | ○ | ○ | ○ | ○ | ○ | ○ | ○ |
| 13. Feasibility and social and economic viability | | | ○ | ○ | | | ○ | ○ | ○ | ○ | ○ | ○ |

○ Element common to both instruments * Elements desired to be present in the intersection of human- and nature-rights.

[a]The Ecuadorian Development objectives are based on the *Sumak Kawsay* principle which is a holistic perspective of development; therefore, all the twelve objectives implicitly address Principle No.11 of the SSF that calls for '*Holistic and Integrated Approaches*'

[b]It is assumed that all the twelve development objectives should tackle, either directly or indirectly, the small-scale fisheries sector. In this regard, we underlined the PNBV 'Guiding Objectives' (Objective 6 and 9) that explicitly mentions domains considered by the SSF Guidelines as critical to achieve the SSF sustainability

would have to be in accordance with existing institutional and legal frameworks around fisheries.

In order to assess the current status of the implementation of the SSF Guidelines in Ecuador, we looked at the three dimensions fostered by interactive governance theory. We argue that the implementation of the SSF Guidelines is in line with the needs of the governing system and the system-to-be-governed. In other words, achieving the sustainability of small-scale fisheries is less of a challenge than might have been expected, provided of course both instruments help strengthen the governing system and the social system-to be-governed. Moreover, it can be said that the common ground identified between both instruments already address aspects commonly recognized as priority areas (e.g. gender equity in fisheries, human dignity, and inclusiveness) that need of great attention from small-scale fisheries bodies (Barragán-Paladines 2015). However, there is limited intersection between the two instruments with regard to the natural system-to-be-governed and the governing interactions dimensions.

The incomplete integration of the principles of the two instruments requires critical examination. What seems necessary is to identify aspects of both instruments that need to be adapted to and synchronized with the national fisheries policy agenda. Only then will it be possible to improve the governability of small-scale fisheries with regard to the diversity, complexity, dynamics and scale of this complex system. We also argue that the SSF Guidelines are a good instrumental, analytical, and even methodological frame that, if implemented, would certainly support the sustainability and viability of the small-scale fisheries sector in Ecuador. For this to materialize, the new National Fisheries Policy needs to be updated, built upon, and sustained in the long-term. It would also require individuals, communities, organizations, authorities, practitioners, and institutions to feel the SSF Guidelines are in their self-interest, and that the benefits of the Guidelines outweigh the costs. Ideally the two instruments discussed in this chapter would be aligned and a high degree of coordination would exist between the executive agencies that operate under the current PNBV. Moreover, the national fisheries policy authority would have to take aboard the entire spectrum of practitioners and stakeholders involved in the small-scale fisheries sector.

Possible pitfalls that exist with regard to the sustainability of small-scale fisheries in the context of the PNBV lens include: the temporal scale of implementation, which remains unaddressed; the intergenerational notion in fisheries under the current fisheries policy, which is not evident; the still contested notions of sustainability' and 'development' in fisheries, driven by the tenets of the green (and blue) economy; the still needed (but absent) shift from a managerial approach to fisheries towards a more comprehensive fisheries governance approach; and finally, the gender issue, which is remarkably still left out of analyses of the small-scale fisheries sector.

## *Implementation as a Long-Lasting Process*

This chapter has highlighted thus far that while the two normative instruments have had different focuses, they also have shared key dimensions that have promoted small-scale fisheries sustainability and viability, at least in theory. As such, there is not 'one' obstacle that prevents the SSF Guidelines from being implemented. Constraints on higher governability of small-scale fisheries are varied and thus challenges to implement the Guidelines as well. In the course of this research, critical questions have arisen with regard '*to what extent*' and '*why, are (are not) institutional, social, political and economic conditions enabling the implementation of the SSF Guidelines in Ecuador*?'

The study revealed that current fisheries policy and institutions key to the small-scale fisheries sector in Ecuador have made important efforts to advance the sector's status and performance through the implementation of measures dedicated to improving fishers' working conditions (Table 33.2). These practices follow a state-based policy that calls for the general improvement of workers' labor conditions at the national scale, because this will lead to enhancing fishers' jobs, their quality of life and secure their dignity. This is in line with national government's objective of promoting *Buen Vivir* of all Ecuadorians and foreign residents. Yet fishers and fishing communities are still discontent due to the persistent unaddressed issues (e.g. poverty, marginalization, food insecurity, lack of food sovereignty, risks at sea, etc.) that represent a source of dissatisfaction and unwillingness to longer support the current ruling bodies policies and practices.

Poor implementation must be seen in the context of inappropriate actions and missed opportunities of former governing bodies. In fact, the political and economic priorities under neoliberal models of development ensured that critical issues threatening small-scale fisheries' sustainability (e.g., exclusion of fishers from fishing grounds through the concession of coastal zones for tourism industry benefit, lack of food sovereignty in fishing communities, etc.) were not addressed and solved. Lessons learnt from the past are that any action that can contribute to overall implementation of good policies are worthwhile. We argue that the awareness of the existence of the SSF Guidelines is a first step, which could enhance the knowledge and willingness of small-scale fishers, fishing communities, fisheries authorities, and practitioners, to commit to achieving increased governability, which would translate into *Buen Vivir*.[3]

[3] At the time of reviewing this chapter, the first workshop about the SSF Guidelines was developed in Ecuador, sponsored by the FAO Regional Office, the Agriculture, Cattle, Aquaculture and Fisheries Ministry, and the National Federation of Small-scale Fishers Cooperatives. The report that came out of this event has been produced only in Spanish, though (*Taller Nacional sobre la implementación de las Directrices Voluntarias para lograr la sostenibilidad de la pesca en pequeña escala, en el contexto de la seguridad alimentaria y la erradicación de la pobreza Guayaquil, Ecuador, 21–22 de Septiembre de 2016*).

Thus, based on the PNBV, the fisheries sector in Ecuador[4] (at least in the mainland) still follows a hierarchical-top-down governance model that best serves to adopt and impose urgent measures that may tackle symptoms but do not solve problems. The persistent and pressing topic of inclusiveness and legitimacy vis-à-vis representation of small-scale fishers is still a matter of disagreement. Whereas the rhetoric of the National Fisheries authority's discourse affirms their commitment to promoting participation of the small-scale fisheries sector, the claim of fisheries sector contradicts it. This highlights the existing gap between the 'theoretical' discourse of the PNBV, the actual policy and decision making, and the empirical evidence of current governing practices. Achieving small-scale fisheries sustainability in Ecuador ultimately depends on involving the main actors of the small-scale fisheries sector (i.e. fishers, fishing communities, authorities, and practitioners) in strategic lines of action.

## *The Sustainability of Small-Scale Fisheries: Different Views*

In order to situate the *Buen Vivir* paradigm in small-scale fisheries policy, a 'common-multiple practice driven by society at large, including among others, Civil Society Organizations (CSOs), fishers (men and women involved in fisheries), fisher's cooperatives and associations, environmental organizations, policy makers, the state, the market, etc. must be the focus. All these actors must be involved in the design and application of individual and collective practices. This will help move away from ecocentric and anthropocentric traditions, which only care about either nature or humans separately. As Haidar and Berros (2015) argue, there is also need for an equilibrated integration of 'other' (e.g. emotional, affective, or aesthetic) aspects that are equally relevant to the small-scale fisheries sector driven by *Sumak Kawsay* and *Pachamama*'s (i.e. Mother Earth) idea of holistic equilibrium. It is critical not only to find an equilibrium between the social and natural worlds, but also to take account of juridical and political dimensions, spiritual and affective attributes, and aesthetic components, in the National Fisheries Policy, as this would go some way towards implementing the SSF Guidelines.

To link *Sumak Kawsay* principle with the SSF Guidelines, effective mechanisms executed by individual countries (in the so-called undeveloped countries) at national level must be shared at global arenas within certain discussion fora and at conventions, declarations, manifests, and resolutions, internationally. Moreover, 'lessons learned' and 'steps-to-follow' from the 'good way of living' approach could serve as inspiration for future fisheries governance practices. Exporting notions, practices, and paradigms, from the 'Global South to the Global North' (De Souza-Santos 2012) would be the start of new era of knowledge circulation and mobilization, at global scale. It would also be the vindication of 'alternative' mindsets that have

[4] For a detailed description of existing governance modes of small-scale fisheries in Ecuador and particularly the Galapagos Islands, see Barragán-Paladines (2015).

been largely denied (or delayed) by dominant western practices aimed at ruling the natural and social world (Haidar and Berros 2015). The outcome would be governance of small-scale fisheries being improved as proposed by the SSF Guidelines.

Sustainable development, as a concept, is still understood as a 'friendly way of economic development' and is still a 'growth-oriented' notion inspired by a western-minded discourse (Forsythe 1997). If this is how sustainable development is imagined, then *Buen Vivir* must be seen a challenge to such 'civilizational models' of globalized development (Escobar 2016) proposal. Only alternatives to current notions of sustainable development can enable movement to good ways of living in fishing communities. The difficult is that *Sumak Kawsay* contests the idea of development as it is currently understood, and thus, could be seen as contradictory to the idea of sustainable fisheries proposed in the SSF Guidelines. However, if small-scale fisheries sustainability in Ecuador is viewed in terms of both, the rights of nature and the fishers´ human rights, it will strengthen and improve the mutual dependence of humans and nature.

Such an inclusive approach that integrates different ontologies and epistemologies within the 2008 Constitution (e.g. *Sumak Kawsay*) represents a great advance in the process of recognizing the diversity, complexity, dynamism, and scale of the natural and social systems involved in the small-scale fisheries. Furthermore, the recognition of "other" values that are central to the small-scale fisheries sector (e.g. the traditional knowledge and practices exercised by fishing communities, and the spiritual and aesthetic dimensions of fisheries), is exhorted, in line with Guiding Principle 11 (i.e. the call for a holistic and integrated approach" in dealing with small-scale fisheries)'of the SSF Guidelines. This is important to first, recognize fisheries not only as an important form of production (Galván-Tudela (2003) but also as a means, as Godelier (1989) argues, to recognize the overall importance of the entire fisheries system, including the fish, the fishers, their communities, and also the material instruments and technologies. Second, by recognizing fisheries as system that create, use, and control knowledge through certain territorial mechanisms that are utilized not only to access resources but also regulate them, will help put forth a more comprehensive assessment of these fisheries. Finally, doing so will also result in recognizing fisheries as important to the local identities of fishers and fishing communities (Galván-Tudela 2002).

In order to successfully achieve the *Buen Vivir* objectives and implement the SSF Guidelines´ guiding principles, it may require joint efforts and strategies at different scales. It is worth recalling that the dominant paradigm related to small-scale fisheries in Ecuador, in place since the 1950s (Barragán-Paladines 2015), has viewed the sector purely as an economic driving force for coastal communities as a whole. A shift of focus, should be aimed primarily at elevating the status of small-scale fisheries and thus, the SSF Guidelines can help improve the misery of small-scale fishers and their fishing communities, as well as overcome their marginalization, by enhancing their livelihoods and consequently their role as food providers.

## Conclusions

Achieving an improved governability of the small-scale fisheries sector in Ecuador has not happened thus far because of the narrow approach taken to the small-scale fisheries sector. It is argued in this chapter that a higher degree of coherence is needed in the fisheries governance agenda of Ecuador. Further, despite the relevance of the *Buen Vivir* principle in the National Development Plan, it needs to be embedded in the social, economic, and environmental dimensions of policy. Broader and more comprehensive approaches are needed that go beyond the purely managerial and economic focused model currently dominant in fisheries governance.

*Buen Vivir* has been termed as "the" paradigmatic and desired new approach to development and thus, to resource governance (e.g., fisheries). According to Farah and Vasapollo (2011), this notion resembles an ethic notion with more elevated objectives than the purely accumulation desire prayed by the developmentarist model. However, this paradigmatic shift needs a number of innovative strategies that enhance fisheries governance (i.e., creativity, imagination, innovation, and politic commitment) conducted by the governing systems. These ruling bodies must also not ignore the pluralist quality of the already existing structures that greatly enrich the spiritual, affective, and material dimensions of fisheries. The national fisheries policy must be reset in a way that translates rhetoric into action aimed at making the small-scale fisheries sector sustainable.

Finally, there are a number of other issues that need special attention. Gender issues have not been addressed though they align well with human rights discourse and equity and equality concerns. Despite women ostensibly being included in the vision of small-scale fisheries development, in practice they are not counted as active agents of change. An illustration of this is that women rarely become involved in strategies implemented for the fisheries sector (fishermen) which include skill upgradation related to business administration, leadership, accounting, techniques for better fish handling practices, etc. On the contrary, the capacity building program for women linked to small-scale fisheries (i.e. mainly fishers' wives) is limited to only two themes: bakery and jewelry. Unless women and men are treated equally within fisheries policy, meeting the goals of sustainability, achieving the principles of the SSF Guidelines, and the PNBV's objectives, will be largely delayed. Further research would certainly need to address issues on gender in fisheries, fish-as-food, and food sovereignty domains in small-scale fisheries, given the immense relevance that these topics have for the *Buen Vivir* principle in Ecuador.

**Acknowledgements** The author is grateful to the TBTI project and to the Social Sciences and Humanities Research Council (SSHR) of Canada for the funding provided to conduct this research. Thanks to Professor Svein Jentoft for his useful comments and feedback provided to early versions of this chapter. Thanks to the two anonymous reviewers whose comments greatly enhanced the quality of this publication. Thanks to the editorial team for helping me to improve the writing style of the chapter. Finally, my gratitude goes to the interviewees who agreed to participate in conversations and be part of interviews conducted for this study.

## References

Acosta, A. (2008). El Buen Vivir, una oportunidad por construir. *Ecuador Debate, 75*, 33–47.
Acosta, A., & Martínez, E. (2009). *El buen vivir. Una vía para el desarrollo*. Quito: AbyaYala.
Allsopp, W. H. L. (1985). *Fishery development experiences*. Farnham: Fishing New Books Ltd..
Altmann, P. (2013). Das Gute Leben als Alternative zum Wachstum?: Der Fall Ecuador. In *Sozialwissenschaften und Berufspraxis, 36*(1), 101–111. http://nbn-resolving.de/urn:nbn:de:0168-ssoar-406442.
Andreve-Díaz, J. L. (2008). Mother Earth, Mother Sea. *Samudra Report, 50*.
Anguera, M. T. (1978). *Metodología de la observación en las ciencias humanas*. Madrid: Ediciones Cátedra.
Aronson, J. (1994). A pragmatic view of the thematic analysis. *The Qualitative Report, 2*(1), Article 3.
Barragán-Paladines, M. J. (2015). Two rules for the same fish: Small-scale fisheries governance in mainland Ecuador and Galapagos Islands. In S. Jentoft & R. Chuenpagdee (Eds.), *Interactive governance for small-scale fisheries* (pp. 157–178). Cham: Springer.
Baumann, P. (1978). *Valdivia. El descubrimiento de la más antigua cultura de América*. Hamburg: Hoffmann und Campe Verlag.
Bavinck, M., Chuenpagdee, R., Jentoft, S., & Kooiman, J. (2013). *Governability of fisheries and Aquaculture. Theory and applications*. Dordrecht: Springer.
Bernard, H. R. (2011). *Anthropological research methods: Qualitative and quantitative approaches*. Lanham: Alta Mira Press.
Berros, V. (2015). Ética animal en diálogo con recientes reformas en la legislación de países latinoamericanos. *Revista de Bioética y Derecho, 33*, 82–93.
Chuenpagdee, R., & Jentoft, S. (2009). Governance assessment for fisheries and coastal systems: A reality check. *Human Ecology, 37*, 109–120.
Chuenpagdee, R., & Jentoft, S. (2013). Assessing governability: What's next. In M. Bavinck, R. Chuenpagdee, S. Jentoft, & J. Kooiman (Eds.), *Governability of fisheries and Aquaculture. Theory and applications* (pp. 335–349). Dordrecht: Springer.
De Madariaga, J. J. (1969). *La caza y la pesca al descubrirse América. Editorial*. Madrid: Prensa española.
De Souza-Santos, B. (2012). *De las dualidades a las ecologías*. La Paz: REMTE.
Duarte, M. E., & Belarde-Lewis, M. (2015). Imagining: Creating spaces for indigenous ontologies. *Cataloging & Classification Quarterly, 53*(5–6), 677–702.
Ecuadorian National Constitution. (2008). *Supra note 5, at Art.* 71, 72, 73, 74, 75.
Escobar, A. (1996). *La invención del Tercer Mundo. Construcción y deconstrucción del desarrollo*. Bogotá: Norma.
Escobar, A. (2016). Thinking-feeling with the Earth: Territorial struggles and the ontological dimension of the epistemologies of the South. *AIBR – Revista de Antropología Iberoamericana, 11*(1), 11–32.
FAO. (2015). *Voluntary guidelines for securing sustainable small-scale fisheries: In the context of food security and poverty eradication*. Rome: Food and Agriculture Organization of the United Nations.
Farah, I., & Vasapollo, L. (2011). *Vivir bien: ¿Paradigma no capitalista? Introduction*. La Paz/Bolivia: Plural Editores.
Finley, C. (2009). The social construction of fishing, 1949. *Ecology and Society, 14*(1), 6.
Forsythe, D. P. (1997). The United Nations, human rights, and development. *Human Rights Quarterly, 19*(2), 334–349.
Galván-Tudela, A. (2002). *Las Identidades pesqueras entre lo global y lo local. Reflexiones antropológicas*. Cádiz: Anuario Etnológico de Andalucía, 2000–2001.

Galván-Tudela, A. (2003). Sobre las culturas del mar. Prácticas y saberes de los pescadores de la Restinga. *El Pajar, Cuaderno de Etnografía Canaria*, 108–117.
Godelier, M. (1989). *Lo ideal y lo material. Pensamiento, economías, sociedades*. Madrid: Taurus.
Gudynas, E. (2011). Buen vivir: Germinando alternativas al desarrollo. *América Latina en Movimiento, 462*, 1–20.
Gudynas, E., & Acosta, A. (2011). La renovación de la crítica al desarrollo y el buen vivir como alternativa. Utopía y Praxis Latinoamericana. *Revista Internacional de Filosofía Iberoamericana y Teoría Social, 16*(53), 71–83.
Haidar, V., & Berros, V. (2015). Entre el Sumak Kawsay y la "vida en armonía con la naturaleza": Disputas en la circulación y traducción de perspectivas respecto de la regulación de la cuestión ecológica en el espacio global. *Revista Theomai Estudios Críticos sobre Sociedad y Desarrollo, 15*(32), 128.
Jentoft, S. (2014). Walking the talk: Implementing the international voluntary guidelines for securing sustainable small-scale fisheries. *Maritime Studies, 13*, 16.
Johnson, L. E. (2003). Future generations and contemporary ethics. *Environmental Values, 12*(4), 471–487.
Kooiman, J., Bavinck, M., Jentoft, S., & Pullin, R. (2005). *Fish for life: Interactive governance for fisheries*. Amsterdam: Amsterdam University Press.
Kooiman, J., Bavinck, M., Chuenpagdee, R., Mahon, R., & Pullin, R. (2008). Interactive governance and governability: An introduction. *The Journal of Transdisciplinary Environmental Studies, 7*(1), 171–190.
Lam, M. E., & Pitcher, T. J. (2012). The ethical dimensions of fisheries. *Current Opinion in Environmental Sustainability, 4*, 364–373.
Lind, A. (2012). Contradictions that endure: Family norms, social reproduction, and Rafael Correa's citizen revolution in Ecuador. *Politics and Gender, 8*(2), 254–260.
MAGAP (Ministerio Agricultura, Ganadería, Acuacultura y Pesca), FAO (Food and Agriculture Organization of the United Nations), & FENACOPEC (Federación Nacional de Cooperativas Pesqueras de Ecuador). (2016). *Informe Taller Nacional sobre la implementación de las Directrices Voluntarias para lograr la sostenibilidad de la pesca en pequeña escala, en el contexto de la seguridad alimentaria y la erradicación de la pobreza*. Workshop Proceedings, 21–22 September 2016, Guayaquil, Ecuador.
McEwan, C., & Silva, M. I. (1998). Arqueología y comunidad en el Parque Nacional Machalilla. In C. Josse & M. Iturralde (Eds.), *Compendio de Investigaciones en el Parque Nacional Machalilla*. Nuevo Arte: Quito.
Norton, P. (1985). *Boletín de los museos del Banco Central N° 6. Simposio 45 del Congreso Internacional de Americanistas*. Bogotá: Universidad de los Andes.
Rostworowski, M. (2005). *Recursos naturales renovables y pesca, siglos VXI-XVII: Curacas y sucesiones, costa norte*. Lima: Instituto de Estudios Peruanos.
Saez-Marti, M., & Weibull, J. W. (2005). Discounting and altruism to future decision-makers. *Journal of Economic Theory, 122*, 254–266.
Song, A. (2015). Human dignity: A fundamental guiding value for a human rights approach to fisheries? *Marine Policy, 61*, 164–170.
Stone, C. (1972). Should trees have standing? Toward legal rights for natural objects. *Southern California Law Review, 45*, 450.
Viveiro de Castro, E. (2004). The transformation of objects into subjects in Amerindian ontologies. Exchanging perspectives. Symposium: Talking Peace with Gods, Part1. *Common Knowledge, 10*(3), 463–484.
Whittemore, M. E. (2011). The problem of enforcing nature's rights under Ecuador's constitution: Why the 2008 environmental amendments have no bite. *Pacific Rim Law & Policy Journal*

*Association,* 659–691. https://digital.lib.washington.edu/dspace-law/bitstream/handle/1773.1/1032/20PacRimLPolyJ659.pdf?sequence=1. Accessed 6 Feb 2014.

Williams, M. (1998). Aquatic resources. Education for the development of world needs. In D. Symes (Ed.), *Fisheries dependent regions* (pp. 164–174). Oxford: Fishing New Books Blackwell Science.

Young, O. R., King, L. A., & Schroeder, H. (2008). *Institutional and environmental change. principal findings, applications, and research frontiers*. Cambridge: MIT Press.

# Chapter 34
# Addressing Social Sustainability for Small-Scale Fisheries in Sweden: Institutional Barriers for Implementing the Small-Scale Fisheries Guidelines

**Milena Arias-Schreiber, Filippa Säwe, Johan Hultman, and Sebastian Linke**

**Abstract** Swedish coastal fisheries are not sustainable in terms of the status of their main fish stocks, their economic profitability, and as source of regular employment. Social sustainability commitments in fisheries governance advocated by the Voluntary Guidelines for Securing Sustainable Small-Scale Fisheries (SSF Guidelines) have been so far mostly neglected. In this chapter, we bring attention to two institutional settings at different governance levels relevant for the implementation of the SSF Guidelines in the Swedish context. First, we look at the introduction of social goals under the perspective of the EU's *Common Fisheries Policy* (CFP). Second, we consider national tensions between forces advocating or opposing a further application of market-based economic instruments, often portrayed as an effective cure for all ills, in fisheries governance. Taking into account the logic on which the SSF Guidelines rest, we evaluate in both cases current processes for stakeholder participation in the formulation of fishing policies and strategies in Sweden. We conclude that the inclusion of a social dimension and stakeholder involvement at the EU level face procedural and institutional limitations that prevent the small-scale fisheries sector from exploiting opportunities for change. Further challenges to the implementation of the SSF Guidelines arise when central national authorities' interpretation of societal benefits opposes other interpretations, and consequently economic calculations take precedence over a participatory process-based, knowledge-accumulating approach to resource management. The SSF Guidelines, therefore, provide important material and intellectual resources to

---

M. Arias-Schreiber (✉)
Department of Philosophy, Linguistics and Theory of Science, Gothenburg University, Gothenburg, Sweden
e-mail: milena.schreiber@gu.se

F. Säwe • J. Hultman
Department of Service Management and Service Studies, Lund University, Lund, Sweden
e-mail: filippa.sawe@ism.lu.se; johan.hultman@ism.lu.se

S. Linke
School of Global Studies, Gothenburg University, Gothenburg, Sweden
e-mail: sebastian.linke@gu.se



S. Jentoft et al. (eds.), *The Small-Scale Fisheries Guidelines*, MARE Publication Series 14, DOI 10.1007/978-3-319-55074-9_34

make the most of new chances that can lead to an increased likelihood of change in the direction of sustainable coastal fisheries in Sweden.

**Keywords** Small-scale fisheries • Sweden • EU common fisheries policy • Market-based incentives • Social sustainability • Stakeholder participation

## Introduction

The FAO *Voluntary Guidelines for Securing Sustainable Small-Scale Fisheries* (hereafter SSF Guidelines) (FAO 2015) propose a suite of commitments that aim to improve the environmental, social, and economic conditions of small-scale fishers worldwide. Their implementation is still highly dependent on the context in which small-scale fishers operate. In the case of Sweden, coastal or inshore fishers (as they are more often referred to) are declining in number with very few prospects for them to continue working in these fisheries. Like in other small-scale fisheries in Europe, Swedish coastal fisheries are facing problems concerning the declining biomass of their main fish stocks, reduction in economic profitability and loss in their ability to provide regular employment (Macfadyen et al. 2011; STEFC 2014; Lade et al. 2015). Under a multilevel governance framework, the path to more sustainable Swedish coastal fisheries depends on regulations and the institutional functioning of the EU's *Common Fisheries Policy* (CFP), as well as on national fisheries policies of the Swedish government.

Ever since it was created in 1983, the CFP has failed in its most important task of reducing overfishing and halting the decline of key commercial fish stocks in EU coastal waters. The failure of the CFP has meant a continual revision by EU authorities of this particular policy approach and its implementation. It is the centralized and command-control decision-making system of the CFP and its reductionist technocratic approaches that rely only on expert knowledge which have been criticized eminently (Symes 2001; Daw and Gray 2005; Khalilian et al. 2010). During the last decade, therefore, different conceptualizations of what is meant by sustainable fisheries management and its main objectives have emerged, and new terms like 'governance', 'participatory decision making' and 'regionalization' have entered the language of the EU fisheries policy (Symes 2014). At the same time, the inclusion of social objectives in fisheries management worldwide is no longer limited to the establishment and enforcement of fishing rights and consultations with stakeholders in policy making (see Cochrane 2002). The new paradigm of fisheries management demands more substantial institutional reforms with the aim of taking into account not only the status of fishing stocks and the maximization of economic yields, but also how questions of employment, stakeholder knowledge and culture, social exclusion, legitimacy, and acceptability are addressed. When implementing the CFP, the inclusion of these, until now forgotten, social objectives is meant to balance environmental and economic objectives and improve sustainability by recognizing human-nature relations in the context of fisheries management (Berkes 2003; Lade et al. 2015;

Boonstra and Hentati-Sundberg 2016). In this chapter we ask: what are the main barriers in dealing with institutional reform demanded by the SSF Guidelines in the Swedish coastal fisheries context? Considering the logic on which the SSF Guidelines rest, we argue that an evaluation of the institutional structure and functioning of the CFP is necessary to foresee the challenges in implementing the Guidelines.

However, recent developments in Swedish fisheries policy suggest that the inclusion of social objectives in fisheries management is likely to face not only institutional barriers at the EU level but also opposition at the national level. This opposition arises from those authorities that prioritize only the economic objectives of fisheries (McGinley 1996; Sumaila 2010; Longo et al. 2015). Despite the United Nations World Commission on Environment and Development from 1987 setting forth ecologic, economic, and social dimensions of sustainability, sustainable development research and environmental policy including fisheries have so far mostly focused on bio-physical and economic aspects (Berkes 2003; Urquhart et al. 2011). Simultaneously, environmental policies have focused on economic modes of valuation to fix environmental problems (Alexander 2005; Fourcade 2011; Säwe and Hultman 2013). One consequence has been that sustainable development has become a subject for marketization (Redclift 2005).[1] Not surprisingly, therefore, the exposure of sustainable development to markets has been theorized as part of neoliberalization as evidenced by the enclosure of natural resources, and the privatization of ecosystem services (Banerjee 2003; Mansfield 2004; Heynen and Robbins 2005; Fairhead et al. 2012).

In this chapter we bring attention to two contextual settings relevant for the implementation of the SSF Guidelines under Swedish conditions. First, we look at the introduction of social/human goals and fisheries interactive governance under the perspective of the EU's CFP. Second, we call attention to the Swedish tensions at the national level between forces advocating or opposing a further application of market-based economic instruments to fisheries management. To justify our focus on these two settings, we start with a brief account of how the SSF Guidelines rest on a specific logic. Thereafter, our analysis of the preconditions necessary for the implementation of the SSF Guidelines in Sweden draws on two specific cases to illustrate the position of the Guidelines with regard to current social and economic objectives and consultative processes. In the first case, we evaluate the institutional structure and functioning of the CFP in terms of envisaging the integration of a human/social dimension in the EU's governance system. We examine two initiatives, as empirical objects for this analysis, which aim at the involvement and participation of stakeholders in the EU system. In the second case, we critically examine an on-going participatory process at the national level, where a new Swedish National Fishing Strategy – aimed at creating national sustainable fisheries through market-based resource management principles – is taking shape. Here we analyse how a stakeholders' perspective has different management logics. In this second case, our focus is on how relationships between different sustainability aspects are articulated and understood, and how this might complement or challenge the implementation of the SSF Guidelines.

[1] Marketization is used here as a synonym for commodification.

## The Logic of the SSF Guidelines in Context

The SSF Guidelines were built on a fully bottom-up participatory and consultative process involving 'more than 4000 representatives' and 'stakeholders from more than 120 countries' (FAO 2015, v). These processes of consultation were undertaken to assure credibility and approval. The SSF Guidelines have five clearly defined objectives:

- To safeguard global food security;
- to develop small-scale fishing communities and eradicate poverty;
- to conserve fisheries resources;
- to guide states and stakeholders to achieve sustainable fisheries; and
- to enhance public awareness of small-scale fisheries knowledge.

The nature of the SSF Guidelines is voluntary and global in scope. The fact that they are voluntary means that they have limited impact in deciding "how they should be applied in a national context" (FAO 2015, 2). Moreover, the SSF Guidelines recognize that it is difficult to come up with a standard definition of the small-scale sector. Instead, the SSF Guidelines advise each country to identify vulnerable and marginalized fishers through a multilevel, participatory, consultative, and objective-oriented manner. Similarly, they recommend that implementation occur in accordance to "national legal systems and their institutions" (FAO 2015, 2).

Thus, the SSF Guidelines recognize the wide-reaching existence of vulnerable people, who nevertheless deliver large quantities of a limited basic human need (food), and demand change and a transformation of their situation for the better. Similarly, they make a subtle but direct appeal to countries to foster this necessary transformation.[2] The main argument, therefore, can be pointed out already at this stage. This argument states that (a) relying on vulnerable fishing communities for access to safe and nutritious fish (food security) is too risky, (b) allowing that the fishers despite their suppliers' role live in poverty is distressing, and (c) ignoring the value and knowledge of the fishers is unjust.

The 13 general principles of the SSF Guidelines are 'based on international human rights standards' (FAO 2015, 2). They lay out the way forward in terms of changing behaviour towards small-scale fishers and their communities. The principles promote respect of cultures, non-discrimination, transparency, and accountability. Further, the principles seek to avoid overexploitation of fishery resources, not only because of the environmental impacts, but also because social and economic viability are necessary ingredients for secure and sustainable small-scale fisheries. The previous FAO Code of Conduct for Responsible Fisheries from which the SSF Guidelines were developed as a supplement (pages xi, 3) calls for a change of behaviour towards marine ecosystems. The FAO Code of Conduct appeals for a change on how marine ecosystems are managed with a specific focus on environmental justice.

[2] N.B.: At this point, the SSF Guidelines are voluntary and do not have to be implemented by states, the role of individual states (or countries) is therefore key.

The SSF Guidelines, on the other hand, emphasize a change directed at improving social justice "as an all-encompassing notion that affirms the value of life – all forms of life" (Cock 2011, 48).

In recent years, a human rights approach to fisheries has been gaining ground in fisheries governance discourse (Jentoft 2014; Ratner et al. 2014; Song 2015). This approach is distinct from, but acts as a counterpart to, the *rights-based management* approach, which advocates ecological stewardship through the allocation of property rights or fishing rights to individual fishers (Costello et al. 2008; Høst 2015). Rights-based management argues that perverse incentives to overexploit a common property resource like fish are overcome through privatizing access and conferring exclusive individual quotas or collective territorial rights to fishers. The 'human rights' approach, on the other hand, suggests that it is not enough to set technical rules to ensure responsible fishing and the protection of marine ecosystems. For the human rights approach, fishers' behaviour, values, and motivation to overexploit should be understood and generalizations or technical panaceas avoided (Boonstra and Hentati-Sundberg 2016). Hence, while the rights-based approach assumes that property right holders of fisheries resources will have an incentive to participate in the process of halting overexploitation and collapse of commercial fish stocks because their livelihoods depend on these stocks (Allison et al. 2012), the human rights approach presumes that small-scale fishers' motivations are not purely economic. Before agreeing to be part of conserving marine ecosystems, small-scale fishers very often demand recognition of their knowledge. They also demand recognition of the cultural and societal importance of fisheries to them, effective participation and non-discrimination in fisheries management and transparency and accountability in fisheries policy-making. All of this is aimed at making them feel less vulnerable.

Human rights in the case of fisheries include rights of fishing tenure. Part 2 of the SSF Guidelines refers to the necessity of tenure rights being addressed in the governance of small-scale fisheries and resource management (FAO 2015, 5). By highlighting the importance of tenure and effective ecological management through participation, the SSF Guidelines make a clear statement with regard to the balance of ecological, social, and economic expectations of responsible governance:

> *5.6 Where States own or control water (including fishery resources) and land resources, they should determine the use and tenure rights of these resources taking into consideration, inter alia, social, economic and environmental objectives* (FAO 2015, 5).

The social objectives that the SSF Guidelines aim at do not mean a denial or rejection of fishing property rights or an ecological approach to fisheries governance. The SSF Guidelines recognize the 'need to have secure tenure rights to the resources that form the basis for their social and cultural well-being, their livelihoods and sustainable development' (FAO 2015, 5). In this sense, a transformation that addresses tenure rights in small-scale fisheries should be respectful of and complemented by efforts to assure further human rights for fishing communities. To ensure that human rights are privileged the SSF Guidelines stress the need of states to "involve small-scale fishing communities... women, marginalized and vulnerable groups – in the design, planning and, as appropriate implementation of management measures" (FAO 2015, 7).

Part 3 of the SSF Guidelines (*Ensuring an enabling environment and supporting implementation*) presents a list of 'should do's' for smoothing over the transformation process. Out of the 29 paragraphs in Part 3 of the SSF Guidelines, 21 begin or contain the words 'states should.' While the SSF Guidelines as a whole demand participation in the governance of small-scale fisheries, Part 3 of the SSF Guidelines recognize the role of state governments in some particular subjects including, for example, policy coherence and harmonization, collection of fisheries data (including biological, social, cultural, and economic data), ensuring funds for small-scale fisheries research, and promotion of fish consumption.

In summary, the SSF Guidelines are relevant in the context of Swedish coastal fisheries because they introduce forgotten social principles and objectives in fisheries governance based on a human rights approach. Those social principles should lead to the democratization of the decision-making process in fisheries policy and the safeguard of tenure rights in accordance with human rights principles without overlooking environmental objectives and highlight the crucial role of the state in the inclusion of social/human dimensions in fisheries management. As we illustrate in the following sections, each of these tasks implies a need for changes in the way Swedish small-scale fisheries are currently managed.

## Swedish Coastal Fisheries as Part of the EU Fisheries Policy System

As part of the European Union (EU), the Swedish fisheries management system is governed via the CFP. This system was established and institutionalized through a path-dependent process, i.e. a co-developing institutional interdependence between science (giving advice) and management (preparing decisions) (Hegland and Raakjær 2008). This interface originated with some crucial discussions about finding appropriate management measures for international fisheries, carried out during the 1960s and 1970s in the two North Atlantic Fisheries Commissions.[3] As a consequence of these discussions fishing quotas emerged as the dominant measure for managing fisheries. The outcome of this was the TAC-based[4] fisheries management system, which needs to be seen as a consequence of the scientific and political circumstances of the time. These were those of an urgent need to connect fishing mortality with a feasible distribution of access rights among contracting members, i.e. coastal states (Gezelius 2008). A fish stock assessment tool called *Virtual Population Analysis* (VPA) allowed scientists to make catch forecasts and recommend suitable quotas, which in turn enabled politicians to make decisions based on TACs – in the form of annual fishing quotas for specific stocks. This tailored interplay between

[3] These are the *International Commission for the Northwest Atlantic Fisheries* (ICNAF) and the *North East Atlantic Fisheries Commission* (NEAFC). For further explications see Gezelius (2008).

[4] TAC: Total Allowable Catch is the total amount of fishing mortality to be imposed on a particular fish stock.

scientific assessments and TACs as political management tools fitted together like 'hand in glove' (Rozwadowski 2002). The arrangement served as the basic foundation for a standardized approach to international fisheries management, aptly described as the 'TAC Machine' by Holm and Nielsen (2004).

The 'TAC-machine' management approach has been applied in EU fisheries management under the CFP since its foundation. The system involves a linear division of labour: (fisheries) science measures the state of fish stocks, while decision-makers, for example EU authorities, negotiate and distribute quota shares in the form of TACs. While the TAC-machine still remains fundamental to Swedish fisheries science and management processes under today's CFP, it is not feasible to integrate broader management perspectives (such as those required for example by an Ecosystem Approach to Fisheries) into current ways of calculating and allocating TACs. This is the case especially for social and cultural issues as referred to in the SSF Guidelines, which have increasingly become stressed in in recent years (Urquhart et al. 2014). Because of the inextricable interplay between natural science on the one hand (in the form of biological advice on fish stocks and catch forecasts) and policy/politics on the other hand (in the form of TAC-based decisions and their distribution among member states), the EU fisheries policy system is caught in this 'institutional inertia' (Wilson 2009, 93). Strictly speaking the techno-scientific framing of the TAC-machine management system both requires taking notice primarily of biological information about fish, i.e. stock size and age distribution, and is also the result of it. Consequently, it ignores social, cultural, and other knowledge aspects of fisheries. In essence the CFP can be criticised for not managing fisheries, i.e. industries, people, and communities embedded in specific cultural contexts (St. Martin et al. 2007), but rather managing *fish* (in form of single fish stocks) (Wilson 2009). This way of institutionalizing fisheries has been interpreted as an attempt to keep politics out of EU fisheries management (Cardwell 2015). The resulting indifference in the CFP to the social conditions of fisheries (and to some extent also the economic conditions) adversely affects the small-scale fishing sector and local communities (Crilly and Esteban 2013) (Fig. 34.1).

## *ACs' and FLAGs' Potential for Addressing Social Issues in EU Fisheries*

### The Advisory Councils (ACs)

In response to increasing sustainability problems and accelerating legitimacy concerns with the science-based management approach of the TAC-machine, the CFP underwent a substantial policy reform in 2002 (EC 2002).[5] In 2002, a new rhetoric dominated the narratives and content of EU fisheries' policy documents. This rheto-

[5] The CFP is reformed every 10 years (Penas Lado 2016). Before the 2012/2013 CFP reform ACs were called *Regional Advisory Councils* (RACs).

**Fig. 34.1** A traditional coastal fishing Swedish boat as part of the EU system (Photo credit: Sebastian Linke)

ric, and the reform that went with it, recognized a need for '*more effective and participatory decision-making*' in order to cope with "*shortcomings and internal systemic weaknesses of the CFP*," identified as "*poor enforcement, lack of multi-annual management perspectives, fleet overcapacity and insufficient stakeholder involvement*" (COM 2002, 4). As a result of the reform attempts addressing legitimacy deficits and conflicts within the CFP, Advisory Councils (ACs) were established as new organizations to involve stakeholders in policy and management. As such the ACs could, at least in principle, also serve to raise various social and cultural issues for discussion and potentially include them more seriously in policy and management.

ACs consist of representatives from the fishing industry and so-called other interested groups (on a 60/40 allocation ratio)[6] such as environmental organizations, consumers, and recreational fishers (Long 2010; Linke et al. 2011; Hatchard and Gray 2014). During the last decade, ACs have progressed significantly and become an important new actor in the CFP (Ounanian and Hegland 2012). However, despite this new governance success, AC processes also expose new tensions and challenges when being confronted with the traditional science-policy interface of the institutionally inert TAC-machine (Griffin 2010, 2013; Linke and Jentoft 2013, 2014, 2016; Holm and Soma 2016). There is now a greater emphasis on including

[6] This has been changed under the last reform process in 2013, from a 2:1 representation ratio earlier.

social aspects in the management of the CPF, as shown in policy documents (EC 2013) and observed by the academic community (Urquhart et al. 2014). However, with respect to the issues mentioned above (social, cultural, knowledge, and value aspects) the success 'on the ground' has been rather limited. This is mainly due to the dominant linear approach of the TAC-machine's annual policy cycle of scientific and political procedures (i.e. scientific assessments & advice – policy-discussions – political decisions) (Linke et al. forthcoming). Moreover, despite the original idea that ACs should provide long-term consensual advice from all relevant stakeholders on regional fisheries management issues to the EU Commission, they also follow to a large extent the narrow, techno-scientific framing provided through the institutionalized annual TAC-machine cycle. This means they also discuss, interpret, and negotiate pre-formatted scientific advice and give their own recommendations in relation to it, mainly on annual TACs but also on other technical issues such as regulating fishing gear types.

In their capacity to integrate different stakeholders' perspectives in the EU fisheries governance system, ACs can be seen as a huge step towards more legitimate policy- and decision-making, and a substantial change away from the top-down, exclusively science-based management approach of the pre-2002 CFP (Ounanian and Hegland 2012; Griffin 2013; Linke and Jentoft 2016). Nevertheless, with regard to their principal ability to improve the democratic ideals they were created for, there remains substantial, but yet unrealized, potential for further developing their capabilities as new co-management actors (Linke and Bruckmeier 2015; Linke and Jentoft 2016). As shown in a study about the Baltic Sea (BS) AC's failing efforts to contribute effectively to a new salmon management plan for the Baltic Sea, the exclusion of social aspects – and of social science research, threatens the overall legitimacy of the CFP and leads to further conflicts, frustration, and fatigue among many fisheries stakeholders (Linke and Jentoft 2014, 2016). However, ACs should not be seen as final achievements on the road towards a new 'democracy heaven' in EU fisheries governance, but instead more realistically as work-in-progress. As expressed by Raakjær and Hegland (2012, 7), they can be seen as an "interim institutional stage…facilitating better information sharing and cultivating stakeholder relationships".

While high expectations rest on the ACs as the most important stakeholder organization to be integrated in EU fisheries (co-) management and policy-making at regional levels, analyses of their practical functioning reveal how they affect and institute new power relations both within and beyond the AC boundaries (Griffin 2013; Linke and Jentoft 2016). For example it has been observed that the Baltic AC (BSAC), in addition to providing short-term (annual) TAC recommendations, has often done so in a non-consensual manner that prioritized industry interests over those of other groups (Linke et al. 2011; Linke and Jentoft 2016). Griffin (2010, 2013) similarly concludes in the case of the North Sea AC by arguing that while it enables 'good governance' to some extent, it also exhibits a series of contradictory tendencies. ACs can, for example, hardly be seen as representative of "all the interests affected by the Common Fisheries Policy" as desired in the EU's foundational statements (e.g. EC 2004, 17), especially not of the small-scale fisheries sector of participating countries (Linke and Jentoft 2016). The issue of ACs' *representativeness* has been the key focus in a new CFP reform process (EC 2013), under which the small-scale sector has become more explicitly addressed, as reflected for example in a letter from the Fisheries Commissioner Maria Damanaki to the ACs which states:

> *In order to be able to carry out this role effectively ["proactive advisory role"], it is essential that the ACs have the necessary representativeness, i.e. that all legitimate stakeholders have a fair opportunity to participate and express their view. In this context, it is one of my priorities to ensure that small-scale fishermen, as well as all other legitimate stakeholders, have a real impact in the decision-making process through their effective participation in the Advisory Councils. It is particularly important for small-sale fishermen to organise themselves and become truly actively involved in the activities of the Advisory Councils … so that small-scale fishermen have a voice in Europe, and I am confident that progress can be made on this issue.*

The most recent CFP reform process, completed in 2013, took place in parallel to the development of the SSF Guidelines. The new basic regulation of the CFP states that "Member States should endeavour to give preferential access for small-scale, artisanal or coastal fishermen" (EC 2013, 24). In light of this new emphasis on the small-scale sector, apparent in both the EU's post-2013 CFP and the SSF Guidelines, statements such as those of Damanaki can be expected to have consequences for the EU's AC system and its representative function, if not immediately, at least in future reform processes.[7] How this in turn may change the capacity of the ACs to better address social and cultural issues, and take aboard additional knowledge and value perspectives from a broader variety of stakeholders, is an issue for further investigation.

## Fisheries Local Action Groups (FLAGs)

Another poorly represented stakeholder group in the ACs are local fishing communities who are affected by the policy- and decision-making procedures of the CFP. Yet, this stakeholder segment is by and large not considered in these organisations. Due to the institutional setup of ACs, it is mainly the large-scale national fishing organisations and NGOs acting as representatives in the ACs (Linke and Jentoft 2016). So-called Fisheries Local Action Groups (FLAGs) have more recently been established as community-level institutions throughout Europe. They aim to develop local fisheries areas primarily to the benefit of the small-scale and coastal fleet sectors (Linke and Bruckmeier 2015; Phillipson and Symes 2015). So far, however, there are no formal links or forms of cooperation between FLAGs and the

[7] Some ACs responded to the Damanaki letter. The North Sea AC has even surveyed its member organizations, stating that they all represent the small-scale sectors of their respective countries (NSAC 2013). However, we believe that a methodological problem exists in the asking of this question. By asking AC member organizations whether they ensure balanced representation, one does not speak to those who are not represented already in an AC.

regional ACs. Although FLAGs have been created primarily to support local development and as a means to apply for EU financing through the European Maritime and Fisheries Fund (EMFF), they too are an important interest group 'affected' by the CFP (EC 2002, 2004). Some FLAGs, for example the Bornholm FLAG in the BSAC, expressed their desire to be represented at higher policy and management levels. However, this has been a difficult issue to reach an agreement about both for the BSAC and the AC system in general. The BSAC administration contacted the EU Commission, which responded in a letter admitting that FLAGs may in fact be seen as a legitimate stakeholder though not generally allowed to participate in the AC. This was partly the case because of practical problems (there are for example ca. 150 FLAGs around the Baltic), and issues of parallel funding and overlapping functions within EU fisheries policy-making (i.e. national stakeholder representation) (BSAC 2012). In conclusion the EU Commission stated, somewhat ambiguously, "(R)AC membership does not seem a suitable nor workable solution for FLAGs, but participation in meetings as observers, on a case by case basis, and coordination and contact between FLAGs and (R)ACs who so wish, could be alternative solutions" (BSAC 2012).

Regardless of initiatives to involve local fishers and other small-scale fisheries actors in FLAGs and more generally in a regional-based management framework of the CFP, problems continue with regard to including social and cultural issues into the existing EU fisheries governance system. Despite these difficulties, FLAGs represent an innovative institutional arrangement at the community level that could, at least in principle, play an important role in the implementation of the SSF Guidelines if supported by national governments. For example, the support of individual EU governments to establish a new organization representing national FLAGs could be a possible solution to overcome the current exclusion of individual FLAGs in the regional ACs system. However, as we analyse in the following section, new barriers have arisen in terms of implementing the SSF Guidelines at the country (Sweden) level.

## Swedish Coastal Fisheries as Part of the National Fisheries Policy System

Garcia et al. (2008) argue that the absence of attention to small-scale fisheries at the national and supranational policy level in Europe might arise from a generalized lack of understanding and underestimation of the social and economic benefits from this subsector. Small-scale fisheries, both in developed and developing countries, have not merely been a source of employment and high quality fish for the market but also essential for the maintenance of traditional coastal communities for centuries (Symes et al. 2015). Besides conserving knowledge, values, and traditions that comprise the 'social glue' of coastal rural communities, small-scale fisheries enhance social capital that strengthens health delivery systems and the functioning of local governance within these communities (Grafton 2005; Béné et al. 2010; Symes 2014; Jentoft and Chuenpagdee 2015).

The tension illustrated within the supra-national TAC-machine of including social objectives in fisheries management – as emphasized in the SSF Guidelines – not only manifests itself in the AC/FLAG context. A similar struggle over the meaning of sustainability is also found in national legislations and policy strategies. In this section, we investigate one process in Swedish fisheries management as a way to highlight further tensions among interpretations of ecological, social, and economic sustainability. By focusing on a particular (but illustrative) moment within the process of articulating a new National Fishing Strategy (for 2015–2020) for Swedish professional fisheries, we illustrate how Sweden is currently moving towards a marketization of access rights to fish resources.

Swedish fisheries governance is under the jurisdiction of two central authorities, the Swedish Agency for Marine and Water Management (SwAM) and the Swedish Board of Agriculture (SBA). SwAM is responsible for fisheries control, environmental monitoring, and the allocation of fish resources. SBA is responsible for the promotion of coastal and small-scale fisheries, and coordination of fisheries management in relation to rural development. During 2015 and the beginning of 2016, the development of a National Fishing Strategy rested with SBA. SBA followed a bottom-up working mode and acted as the coordinating body for the process. Along with SBA, an advisory group was established. The advisory group consisted of a large number of fisheries stakeholders, regional authorities (county administrative boards), representatives of municipalities and researchers (mainly with bio-economic competence but also two of the co-authors Filippa Säwe and Johan Hultman who represented academia as action researchers). This advisory group regularly convened and discussed the strategy's text. Finally, there was also a small steering committee with members from SBA, SwAM and county administrative boards.

As work with the strategy's text progressed and a final version was being discussed in the advisory group, the consensus-driven process took a sudden top-down turn. SwAM unexpectedly stepped in and claimed shared responsibility for the strategy. This move resulted in a change of wording in the vision of the strategy and its resource allocation goal. It was a change that had neither been discussed in the large advisory group, nor presented in the steering committee. Nonetheless, the version of the National Fishing Strategy that was sent out for a final round of remittance opinions featured this vision and central goal that had been inserted top-down.

So what does the centrally decided upon vision look like, and how does the initial vision suggested by the county administrative boards, coastal municipalities, and social scientists compare to it? In the following section, we argue that the initial proposal integrated all three sustainability dimensions, (ecological, economic, and social) in line with the SSF Guidelines, whereas the top-down official proposal fails to do so (Fig. 34.2).

**Fig. 34.2** Different kinds of fishing boats in the harbor of Simrishamn in Sweden (Photo credit: Sebastian Linke)

## *Socio-Economic Profitability Assessments Versus Societal Benefits*

The initial suggestions for the text put forward by county administrative boards, coastal municipalities, and social scientists, had been discussed as mentioned earlier in the advisory group. As shown here, this first proposal focussed on the concept of societal benefits:

> *Swedish fish resources are distributed and utilized in an environmentally sustainable way that creates the greatest total societal benefits from social and economic perspectives.* (The Swedish National Strategy for Professional Fisheries, draft, February 2016, our translation).

The definition of societal benefits was as follows:

> *A well balanced fishing fleet with daily landings, which as far as possible strive to attain profitability from the quality of the catch rather than the prioritization of quantity, creates conditions for long-term regionalized management; a growing national service sector; rural development; consumer access to fresh fish; values associated with tourism and recreation; and a balance in the respective strengths offered by smaller and larger vessels. Such a fleet ensures a continuous environmental monitoring as well as national food security. Societal benefits can thus be understood as the way in which different activities in an industry become mutually reinforcing for the benefit of surrounding communities* (The Swedish National Strategy for Professional Fisheries, draft February 2016, our translation).

SwAM and SBA instead suggested the following wording and central resource allocation goal (SBA and SwAM 2016, our emphasis):

> *Swedish fish resources are managed, distributed and utilized in an environmentally sustainable way, and in a way that within these limits strive to attain the highest possible socio-economic gain/profit.*

Both proposals prioritize environmental sustainability, fully in accord with the reformed CFP. However, the SwAM/SBA suggestion embeds social and economic dimensions within the newly introduced concept of 'socio-economic profit'. This entails an interpretation of social and economic sustainability that is conditional upon computable parameters. "Socio-economic profit" was explained as follows:

> *A socio-economic profitability assessment means the attempt to measure welfare effects resulting from changes in the management of the fish resource, its distribution, and other measures. In other words, it is a socio-economic calculation that also includes assessments of relevant effects that has not been possible to quantify or value in monetary terms.*

The SFF Guidelines, as explained in this chapter, and the CFP are both explicit in their commitments and normative policies that fisheries management should implement and develop principles of inclusion, future orientation, and openness to a multitude of perspectives and knowledges. The post-2013 CFP states that:

> *When allocating the fishing opportunities available to them ... Member States shall use transparent and objective criteria including those of an environmental, social and economic nature* (EC 2013, article 17).

The final National Strategy version favours a maximized economic growth criterion. Social sustainability is implicitly expected to follow from this. A possible future scenario is that landings will become concentrated in a few large harbours and/or that the fish will be landed in foreign harbours associated with industrial processing plants and international markets. From this perspective, the Strategy's insistence (in sections other than the vision and central distribution goal) on the importance of fresh, high-quality fish and short supply-chains from sea to consumer seems problematic.

All these decisions, individually and collectively, have consequences for all sustainability dimensions. There is a bias towards economics in the kind of knowledge and logic the central authorities promote. Given this scenario, it is difficult to imagine the implementation of the SSF Guidelines in Swedish fisheries governance within the near future.

## Discussion and Conclusions

In the EU, small-scale fisheries account for about 9% of total volume and 30% of total value of marine capture, and employ ca. 100,000 fishers (Guyader et al. 2007). Possibly, due to their relatively limited productivity in terms of economic performance, small-scale fisheries in the EU have been largely ignored both within research and policy circles (Papaioannou et al. 2012; Guyader et al. 2013).

Consequently, small-scale fisheries are at present exposed to pressures from more powerful and economically relevant sectors. Small-scale fisheries are faced with pressures concerning large–scale and recreational fisheries, and have to compete with them for their share of the allowable catches and quotas, rights to fishing grounds, and market opportunities (Jacquet and Pauly 2008). These pressures seem to be minor in comparison with problems like poverty reduction, corruption of fisheries authorities, and lack of basic data for management that exist within small-scale fisheries in developing countries (Béné 2006; Purcell and Pomeroy 2015). However, as it has been illustrated in this chapter, Swedish small-scale fisheries also face analogous problems related to vulnerability due to the institutional functioning of the CFP, policy ignorance, limited sharing of decision-making and a failure to implement genuine participatory approaches.

Swedish fisheries under the CFP are inevitably impacted by 'institutional inertia' that is also familiar to other transnational bodies like the World Trade Organization (Wilkinson 2001). The TAC-machine still appears to be *the* dominant management perspective governing fisheries in the EU 'all the way down'. TACs, as fishing quotas, are distributed to individual fishing operators or vessels. The increasing calls for policy reforms towards new forms of stakeholder involvement and innovative change to address social and cultural issues on par with economic and ecological ones still face various kinds of institutional and procedural hindrances. Consequently possibilities to develop future co-management arrangements in EU fisheries have been slow to come (Linke and Bruckmeier 2015). Therefore the implementation of the SSF Guidelines, hanging on adaptive institutional structures and opportunities to include social and cultural concerns of local communities, faces serious challenges in the current framework of EU fisheries under the CFP. The techno-scientific logic of the TAC-machine instituted within the CFP substantially contradicts the social logic of the SSF Guidelines. A system that manages fisheries based solely on quotas to ensure the biological health of commercial marine resources and the allocation of these quotas without considering fishers' traditional tenure rights, culture, knowledge, and values is likely to fail in ensuring a biologically, socially, and economically sustainable fisheries. Thus, neglecting social objectives can be a critical mistake, especially given that fisheries management does not aim to manage resources but humans (Ommer and Paterson 2014). This mistake turns more visible when small-scale fisheries are the focus of attention. Despite the fact that small-scale fisheries are potentially more ecologically sustainable than large-scale fisheries (Jacquet and Pauly 2008), their cultural value and contribution to societal well-being has been largely ignored (Guyader et al. 2013). This has resulted in thousands of small-scale fishers living in poverty in developing countries (Jentoft 2014) and retreating from the occupation in the northern hemisphere (Ommer and Paterson 2014).

The impact of ignoring social sustainability – in terms of provision of basic needs and protection of ways of living – in the small-scale fisheries sector has become evident and hence too appeals for transformation as expressed in the SSF Guidelines. In the case of small-scale fisheries under the CFP, the techno-scientific and path-dependent institutional problems, in conjunction with lobbying attempts of groups (for example large-scale industrial fishers), serve to maintain the status quo and block transformation. Democratic innovations, like the implementation of participatory approaches in decision-making can be expected to be successful in delivering change (Newton 2012). However, as illustrated with the cases highlighted in this chapter, participation involves a process of adaptation and social learning that might be too slow in preventing the weakening and even disappearance of Swedish coastal fishing communities. A lack of mechanisms for systematic evaluation of the outputs of ACs and FLAGs complicates future effectiveness of these well-intentioned initiatives. Under these circumstances, EU policies predicating reforms in the governance of fisheries are not implemented and the sustainability focus falls short. The development of evaluation mechanisms and a reform of the quota system reserving seats for the coastal fisheries sector in any co-management fisheries initiative could improve the implementation of participatory principles as elucidated in the SSF Guidelines.

Regarding the development of a National Fishing Strategy in Sweden, different fisheries governance actors compete over two different versions and logics vis-a-vis sustainable social objectives. The adoption of a National Fishing Strategy in accordance with the social principles of the SSF Guidelines will depend on adaptation pressures and social learning processes successfully leading to the emergence of new norms and collective understandings. Social scientists and practitioners could play an important role as 'actors for change' in this process. In the Swedish context, the implementation of the SSF Guidelines will support the sustainability of small-scale fishers by assisting state decisions in fisheries governance. In that sense, while fisheries biology (fish stocks status) and economic outputs (productivity and efficiency) are relatively easily quantified, social benefits are mostly qualitative and should be monitored in accordance with defined 'human rights' standards.

We conclude that the challenge of including a social dimension in the governance of fisheries at the EU level confronts a path-dependent development that has a certain institutional inertia, which will seriously undermine the implementation of the SSF Guidelines. The societal value of Swedish coastal fisheries in this context could continue to be neglected. The current application of stakeholder involvement and participatory approaches at the EU level also faces procedural limitations in allowing the small-scale fisheries sector to exploit opportunities for sustainable change. Although participatory processes carried out at the national level can be translated into new fishing strategies and policies in accordance with the SSF Guidelines, problems exist with how central authorities interpret the meaning of societal benefits. Economic calculations and ad-hoc principles for resource allocation continue to take precedence which will act as hindrance to the implementation of the SSF Guidelines. Hence, for Swedish coastal fisheries, the SSF Guidelines have appeared at a crucial juncture. The SSF Guidelines can be used by the Swedish government and simultaneously hold the government accountable to the (voluntary) commitments enshrined in them. They provide important material and ideational resources that could and should be used to move in a new direction that increases the likelihood of change towards more sustainable coastal fisheries in Sweden. However, the continued emphasis on the performance of large-scale fisheries and market-based approaches for governing fisheries within the EU and especially northern European countries represents a severe challenge in terms of the immediate implementation of the SSF Guidelines. A better understanding of these negative impacts and the challenges they pose at national and EU levels, should help in refocusing priorities away from large-scale fisheries and institutionalizing a paradigm shift that focuses on the implementation of the SSF Guidelines' principles and goals.

**Acknowledgments** This research was partially founded by the main author's postdoctoral research grant from the Graduate School in Marine Environmental Research at the Centre for Sea and Society in Gothenburg University. FS and JH received funds from the Swedish Research Council FORMAS grant no. 250-2012-399 and SL from the Swedish Research Council FORMAS grant no. 211-2013-1282. Two peer reviewers are also thanked for valuable comments on an earlier version of the chapter.

## References

Alexander, C. (2005). Value: Economic valuations and environmental policy. In J. G. Carrier (Ed.), *A handbook of economic anthropology* (pp. 128–145). Cheltenham: Edward Elgar.

Allison, E. H., Ratner, B. D., Åsgård, B., Willmann, R., Pomeroy, R., & Kurien, J. (2012). Rights-based fisheries governance: From fishing rights to human rights. *Fish and Fisheries, 13*(1), 14–29.

Banerjee, S. B. (2003). Who sustains whose development? Sustainable development and the reinvention of nature. *Organization Studies, 24*(1), 143–180.

Béné, C. (2006). *Small-scale fisheries: Assessing their contribution to rural livelihoods in developing countries*. Rome: FAO Fisheries Circular. No. 108.

Béné, C., Hersoug, B., & Allison, E. H. (2010). Not by rent alone: Analyzing the pro-poor functions of small-scale fisheries in developing countries. *Development Policy Review, 28*, 325–358.

Berkes, F. (2003). Alternatives to conventional management: Lessons from small-scale fisheries. *Environments, 31*(1), 5–19.

Boonstra, W. J., & Hentati-Sundberg, J. (2016). Classifying fishers' behaviour. An invitation to fishing styles. *Fish and Fisheries, 17*(1), 78–100.

BSAC. (2012). *Letter from Ernesto Penas Lado*. European Commission, DG MARE to Baltic Sea AC on "Applications for membership by Bornholm FLAG and PAPF". Available on https://www.bsac.dk.

Cardwell, E. (2015). *Taming the ccean: Governmentality in the EU common fisheries policy* (Doctoral dissertation). Oxford University, United Kingdom.

Cochrane, K. L. (2002). *A fishery managers guidebook. Management measures and their application*. Rome: FAO Fisheries Technical Paper. No. 424.

Cock, J. (2011). Green capitalism' or environmental justice? A critique of the sustainability discourse. *Focus, 63*, 45–51.

COM. (2002). *Commission of the European Communities. Communication on the reform of the Common Fisheries Policy (roadmap)*. Brussels, COM, 181.

Costello, C., Gaines, S., & Lynham, J. (2008). Can catch shares prevent fisheries collapse? *Science, 32*, 1678–1681.

Crilly, R., & Esteban, A. (2013). Small versus large-scale, multi-fleet fisheries: The case for economic, social and environmental access criteria in European fisheries. *Marine Policy, 37*, 20–27.

Daw, T., & Gray, T. (2005). Fisheries science and sustainability in international policy: A study of failure in the European Union's Common Fisheries Policy. *Marine Policy, 29*(3), 189–197.

EC. (2002). Council regulation 2371/2002 on the conservation and sustainable exploitation of fisheries resources under the common fisheries policy. *Official Journal of the European Communities Legislation, 358*, 59–79.

EC. (2004). *Council decision of 19 July 2004 (585) establishing regional advisory councils under the common fisheries policy*. Brussels: European Commission.

EC. (2013). *Regulation (EU) No 1380/2013 of the European Parliament and of the council of 11 December 2013*. Strasbourg: European Parliament.

Fairhead, J., Leach, M., & Scoones, I. (2012). Green grabbing: A new appropriation of nature? *The Journal of Peasant Studies, 39*(2), 237–261.

FAO. (2015). *Voluntary guidelines for securing sustainable small-scale fisheries in the context of food security and poverty eradication*. Rome: FAO Food and Agriculture Organization of the United Nations.

Fourcade, M. (2011). Price and prejudice: On economics and the enchantement (and disenchantement) of nature. In J. Beckert & P. Aspers (Eds.), *The worth of goods: Valuation and pricing in the economy* (pp. 41–62). Oxford: Oxford University Press.

Garcia, S. M., Allison, E. H., Andrew, N., Béné, C., Bianchi, G., De Graaf, G., Kalikoski, D., Mahon, R., & Orensanz, L. (2008). *Towards integrated assessment and advice in small-scale fisheries: Principles and processes*. FAO Fisheries and Aquaculture Technical Paper No. 515.

Gezelius, S. (2008). The arrival of modern fisheries management in the North Atlantic: A historical overview. In S. S. Gezelius & J. Raakjær (Eds.), *Making fisheries management work* (pp. 27–40). Dordrecht: Springer.

Grafton, R. Q. (2005). Social capital and fisheries governance. *Ocean & Coastal Management, 48*, 753–766.

Griffin, L. (2010). The limits to good governance and the state of exception: A case study of North Sea fisheries. *Geoforum, 4*, 282–292.

Griffin, L. (2013). *Good governance, scale and power: A case study of North Sea fisheries*. New York: Routledge.

Guyader, O., Berthou, P., Koustikopoulos, C., Alban F., & Demaneche S. (2007). *Small-scale coastal fisheries in Europe*. Final report of the contract No FISH/2005/10. http://ec.europa.eu/fisheries/publications/studies_reports_en.htm.

Guyader, O., Berthou, P., Koutsikopoulos, C., Alban, F., Demanèche, S., Gaspar, M. B., Eschbaum, R., Fahy, E., Reynal, L., Curtil, O., Frangoudes, K., & Maynou, F. (2013). Small scale fisheries in Europe: A comparative analysis based on a selection of case studies. *Fisheries Research, 140*, 1–13.

Hatchard, J., & Gray, T. (2014). From RACs to advisory councils: Lessons from North Sea discourse for the 2014 reform of the European Common Fisheries Policy. *Marine Policy, 47*, 87–93.

Hegland, T. J., & Raakjær, J. (2008). Recovery plans and the balancing of fishing capacity and fishing possibilities: Path dependence in the common fisheries policy. In S. S. Gezelius & J. Raakjær (Eds.), *Making fisheries management work* (pp. 131–159). London: Springer.

Heynen, N., & Robbins, P. (2005). The neoliberalization of nature: Governance, privatization, enclosure and valuation. *Capitalism Nature Socialism, 16*(1), 5–8.

Holm, P., & Nielsen, K. N. (2004). *The TAC machine. Report of the working group on fishery systems, WGFS annual report*. Copenhagen: ICES.

Holm, P., & Soma, K. (2016). Fishers' information in governance: A matter of trust. *Current Opinion in Environmental Sustainability, 18*, 115–121.

Høst, J. (2015). *Market-based fisheries management – Private fish and captains of finance, Mare publication series 16*. Cham: Springer International Publishing.

Jacquet, J., & Pauly, D. (2008). Funding priorities: Big barriers to small-scale fisheries. *Conservation and Policy, 2*(4), 832–835.

Jentoft, S. (2014). Walking the talk: Implementing the international voluntary guidelines for securing sustainable small-scale fisheries. *Maritime Studies, 13*(1), 16.

Jentoft, R., & Chuenpagdee, R. (2015). *Interactive governance for small-scale fisheries: Global reflections, MARE publications series 13*. Cham: Springer.
Khalilian, S., Froese, R., Proelss, A., & Requate, T. (2010). Designed for failure: A critique of the common fisheries policy of the European Union. *Marine Policy, 34*(6), 1178–1182.
Lade, S. J., Niiranen, S., Hentati-Sundberg, J., Blenkner, T., Boonstra, W. J., Orach, K., Quaas, M. F., Österblom, H., & Schlütera, M. (2015). An empirical model of the Baltic Sea reveals the importance of social dynamics for ecological regime shifts. *Proceedings of the National Academy of Science, 112*(35), 11120–11125.
Linke, S., & Bruckmeier, K. (2015). Co-management in fisheries – Experiences and changing approaches in Europe. *Ocean & Coastal Management, 104*, 170–181.
Linke, S., & Jentoft, S. (2013). A communicative turnaround: Shifting the burden of proof in European fisheries governance. *Marine Policy, 38*, 337–345.
Linke, S., & Jentoft, S. (2014). Exploring the phronetic dimension of stakeholders' knowledge in EU fisheries governance. *Marine Policy, 47*, 153–161.
Linke, S., & Jentoft, S. (2016). Ideals, realities and paradoxes of stakeholder participation in EU fisheries governance. *Environmental Sociology, 2*(2), 144–154.
Linke, S., Dreyer, M., & Sellke, P. (2011). The Regional Advisory Councils: What is their potential to incorporate stakeholder knowledge into fisheries governance? *Ambio, 40*, 133–143.
Linke, S., Holm, P., & Hadjimichael, M. (forthcoming). Producing knowledge for fisheries governance: From linear to recursive management systems. In P. Holm, M. Hadjimichael, & S. Mackinson (Eds.), *Bridging the gap: Collaborative research practices in the fisheries*. Dordrecht: Springer.
Long, R. (2010). The role of Regional Advisory Councils in the European Common Fisheries Policy: Legal constraints and future options. *The International Journal of Marine Coastal Law, 25*, 289–346.
Longo, E., Clausen, R., & Clark, B. (2015). *The tragedy of the commodity*. New Brunswick: Rutgers University Press.
Macfadyen, G., Salz, P., & Cappell, R. (2011). *Characteristics of small-scale coastal fisheries in Europe*. Directorate General for Internal Policies. Policy Department B: Structural and Cohesion Policies. IP/B/PECH/IC/2010-158.
Mansfield, B. (2004). Neoliberalism in the oceans: "Rationalization," property rights, and the commons question. *Geoforum, 35*, 313–326.
McGinley, J. (1996). Are ITQs really a panacea? *Samudra, 18*, 13–16.
Newton, K. (2012). Curing the democratic malaise with democratic innovations. In B. Geissel & K. Newton (Eds.), *Evaluating democratic innovations – Curing the democratic malaise?* (pp. 3–20). London: Routledge.
NSAC. (2013). *Representation of small scale fishers*. Letter from NSRAC Secretariat to Lowri Evans, Director General DG Mare.
Ommer, R., & Paterson, B. (2014). Conclusions: Reframing the possibilities for natural and social science dialogue on the economic history of natural resources. *Ecology and Society, 19*(1), 17.
Ounanian, K., & Hegland, T. J. (2012). The regional advisory councils' current capacities and unforeseen benefits. *Maritime Studies, 11*(10), 1–20.
Papaioannou, E. A., Vafeidis, A. T., Quaas, M. F., & Schmidt, J. O. (2012). The development and use of a spatial database for the determination and characterization of the state of the German Baltic small-scale fishery sector. *ICES Journal of Marine Science, 69*, 1480–1490.
Penas Lado, E. (2016). *The common fisheries policy: The quest for sustainability*. Chichester: Wiley-Blackwell.
Phillipson, J., & Symes, D. (2015). Finding middle way to develop Europe's fisheries dependent areas: The role of fisheries local action groups. *Sociologia Ruralis, 55*(3), 344–359.
Purcell, S. W., & Pomeroy, R. S. (2015). Driving small-scale fisheries in developing countries. *Frontiers in Marine Science, 2*, 44.
Raakjær, J., & Hegland, T. J. (2012). Introduction: Regionalizing the common fisheries policy. *Maritime Studies, 11*(5), 1–7.

Ratner, B. D., Åsgård, B., & Allison, E. H. (2014). Fishing for justice: Human rights, development, and fisheries sector reform. *Global Environmental Change, 27*, 120–130.

Redclift, M. (2005). Sustainable development (1987–2005): An oxymoron comes of age. *Sustainable Development, 13*, 212–227.

Rozwadowski, H. M. (2002). *The sea knows no boundaries: A century of marine science under ICES*. Seattle/London: University of Washington Press.

Säwe, F., & Hultman, J. (2013). From moral to markets: The rhetoric of responsibility and resource management in EU fisheries policy. *Society & Natural Resources, 27*(5), 507–520.

SBA, & SwAM. (2016). *Swedish professional fisheries 2020 – Sustainable fishing and healthy food*. Available at https://www.jordbruksverket.se.

Song, A. M. (2015). Human dignity: A fundamental guiding value for a human rights approach to fisheries? *Marine Policy, 61*, 164–170.

St. Martin, K., McCay, B., Murry, G., Johnson, T., & Oles, B. (2007). Communities, knowledge, and fisheries of the future. *International Journal of Global Environmental Issues, 7*(2/3), 221–239.

STECF (Scientific, Technical and Economic Committee for Fisheries). (2014). *The 2014 Annual Economic Report on the EU Fishing Fleet (STECF-14-16). 2014*. Luxembourg: Publications Office of the European Union.

Sumaila, U. R. (2010). A cautionary note on individual transferable quotas. *Ecology and Society, 15*(3), 1–36.

Symes, D. (2001). The future of Europe's fisheries: Towards a 2020 vision. *Geography, 86*(4), 318–328.

Symes, D. (2014). Finding solutions: Resilience theory and Europe's small-scale fisheries. In J. Urquhart, T. Acott, D. Symes, & M. Zhao (Eds.), *Social sssues in sustainable fisheries management, MARE publication series 9* (pp. 23–38). Dordrecht: Springer.

Symes, D., Phillipson, J., & Salmi, P. (2015). Europe's coastal fisheries: Instability and the impacts of fisheries policy. *Sociologia Ruralis, 55*(3), 245–257.

Urquhart, J., Acott, T., Reed, M., & Courtney, P. (2011). Setting an agenda for social science research in fisheries policy in Northern Europe. *Fisheries Research, 108*, 240–247.

Urquhart, J., Acott, T., Symes, D., & Zhao, M. (2014). *Social sssues in sustainable fisheries management*. Dordrecht: Springer.

Wilkinson, R. (2001). The WTO in crisis: Exploring the dimensions of institutional inertia. *Journal of World Trade, 35*(3), 397–419.

Wilson, D. C. (2009). *The paradoxes of transparency – Science and the ecosystem approach to fisheries management in Europe*. Amsterdam: Amsterdam University Press.

# Chapter 35
# Promoting Gender Equity and Equality Through the Small-Scale Fisheries Guidelines: Experiences from Multiple Case Studies

**Danika Kleiber, Katia Frangoudes, Hunter T. Snyder, Afrina Choudhury, Steven M. Cole, Kumi Soejima, Cristina Pita, Anna Santos, Cynthia McDougall, Hajnalka Petrics, and Marilyn Porter**

**Abstract** Gender equity and equality is the fourth guiding principle of the Voluntary Guidelines for Securing Sustainable Small-Scale Fisheries (SSF Guidelines), and sits within its wider human rights framework. The SSF Guidelines contain acknowledgement of the roles of women in the small-scale fisheries value chain, the need for gender equity and equality in access to human well-being resources, and the need for equal gender participation in fisheries governance. While the inclusion of gender

D. Kleiber (✉)
Pacific Island Fisheries Science Centre, Joint Institute for Marine and Atmospheric Research, Honolulu, HI, USA

Sociology Department, Memorial University Newfoundland, St. John's, Canada
e-mail: danika.kleiber@noaa.gov

K. Frangoudes
Université de Bretagne Occidentale, Brest, France
e-mail: Katia.Frangoudes@univ-brest.fr

H.T. Snyder • H. Petrics
Dartmouth School of Graduate & Advanced Studies, Hanover, New Hampshire, USA
e-mail: huntertsnyder@gmail.com; Hajnalka.Petrics@fao.org

A. Choudhury
WorldFish, Dhaka, Bangladesh
e-mail: a.choudhury@cgiar.org

S.M. Cole
WorldFish, Lusaka, Zambia
e-mail: s.cole@cgiar.org

K. Soejima
National Fisheries University, Shimonoseki, Japan
e-mail: soejima@fish-u.ac.jp

C. Pita
University of Aveiro, Aveiro, Portugal
e-mail: c.pita@ua.pt



S. Jentoft et al. (eds.), *The Small-Scale Fisheries Guidelines*, MARE Publication Series 14, DOI 10.1007/978-3-319-55074-9_35

in the SSF Guidelines is unprecedented and encouraging, effective implementation is the critical next step. Part of the implementation process will include the creation of culturally and regionally-specific information that allows local agencies to recognize and prioritize gender needs. To provide an example of the diverse and interacting issues related to the implementation of the gender equity and equality principle, we use case studies and expertise from seven countries and regions. We examine the context-specific issues that should be considered in the implementation process and focus on the many barriers to gender equity and equality in small-scale fisheries. We conclude by outlining the many gender approaches that could be used to implement the SSF Guidelines, and suggest a gender transformative approach. Such an approach focuses on illuminating root causes of gender injustice and inequality, and requires on-going examination of power relationships as well as capacity development for women and marginalized groups.

**Keywords** Implementation • Gender equity • Equality • Small-scale fisheries • Value chain • Barriers • Opportunities

## Introduction

The presence of gender equity and equality in the Voluntary Guidelines for Securing Sustainable Small-Scale Fisheries (henceforth SSF Guidelines) is unprecedented in global fisheries policy. The SSF Guidelines include gender equity and equality as one of its 13 guiding principles and gender is also considered in the more detailed section on responsible fisheries and sustainable development (FAO 2015). As signatory countries begin to implement the SSF Guidelines, gender equity and equality discourse offers an important opportunity to introduce gender issues in small-scale fisheries contexts.

A. Santos
National Marine Fisheries Service, Alaska Fisheries Science Center,
Seattle, Washington, USA
e-mail: annasantos@email.tamu.edu

C. McDougall
WorldFish, Batu Maung, Malaysia
e-mail: C.McDougall@cgiar.org

M. Porter
Sociology Department, Memorial University Newfoundland, St. John's, Canada
e-mail: mporter2008@gmail.com

The inclusion of gender in the SSF Guidelines was not universally agreed upon. The SSF Guidelines were forged at Committee on Fisheries (COFI) meetings (2010–2013) which included representatives of each signatory country, and Civil Society Organizations (CSOs) who were present at COFI meetings as observers.[1] While country representatives were often reluctant to include a gender dimension, CSOs lobbied COFI decision makers between meetings to convince national delegates of the importance of women's contribution to small-scale fisheries and coastal communities, and the need to address gender equity and equality in these contexts. "CSOs outlined a strong gender agenda to ensure that the SSF Guidelines steer away from the mainstream approach of equating fisheries with fishing, with a focus on fishermen" (Sharma 2013, 9).

The current priority is to support the implementation of the gender equity and equality principle. The inclusion of gender in the SSF Guidelines is essential for three key reasons. First, it recognizes that women and men participate in all aspects of the small-scale fisheries value chain around the world, often in ecologically, economically, and culturally distinct ways (The WorldFish Center 2010; Kleiber et al. 2015). Women and men's fisheries labor are also often given different cultural and economic value, with women's work often going uncounted and not considered in fisheries governance, despite being vital to small-scale fisheries (Frangoudes 2013; Kleiber et al. 2014; Santos 2015). Second, it is essential to understanding the centrality of gender to other intersecting issues, particularly human rights and well-being, food security, and climate change (Badjeck et al. 2010). Small-scale fisheries sit within gendered social and cultural systems that perpetuate well-being disparities between men and women and introduce vulnerability within processes of ecological and social change (Gopal et al. 2015). Hence, gender is a key variable in understanding and enacting change to these systems. Lastly, it also highlights how gender differences in power and decision making exist in small-scale fisheries contexts and how those differences influence representative, fair, and sustainable small-scale fisheries governance (Ram-Bidesi 2015).

Our study examines the gender discourse in the SSF Guidelines to highlight what issues are being prioritized for implementation, and which areas may require more attention. We also explore issues specific to implementation including potential political, cultural, and institutional barriers that may overlook the gender concerns of the SSF Guidelines, and vice versa. Finally, we explore the different approaches that could be used to operationalize the gender discourse of the SSF Guidelines, and make some context-specific recommendations. To highlight the diversity of contexts that are considered when engaging with gender equity and equality in small-scale fisheries, we explore these concerns through examples that vary by discipline and location.

[1] These included the Collective in Support of Fishworkers (ICSF), the World Forum of Fish Harvesters and Fishworkers (WFF), the World Forum of Fisher Peoples (WFFP), and the International Planning Committee on Food Sovereignty (IPC), etc.

## Concepts, Analytical Framework and Methods

### *Concepts*

Gender is a social variable that permeates all aspects of human society and culture, and small-scale fisheries are no exception. Gender is fundamental to the organization of human institutions. Addressing issues of gender equity and equality is context dependent and requires working at multiple and intersecting scales and systems. Importantly, gender is internalized through constant social and cultural reinforcement, which can lead to false assumptions that gender roles are biologically-based and hence cannot be changed. Universalizing myths of women as saviours or victims deny their agency and belie the diversity and complexity of women's experiences, which may vary greatly within and between geographic contexts and by intersecting social categories of ethnicity, nationality, class, caste, age, education, among others (Cornwall et al. 2007).

Focusing only on policy changes or a limited list of inequality indicators is unlikely to create gender equity and equality (Cornwall et al. 2007). General support of women, without recognition of power difference among women, can lead to detrimental elite capture of development programs, without fundamentally changing gender relationships or addressing other systems of inequality such as poverty (Resurreccion 2008). Including gender in development policy can be a long-term and often challenging task, and one that is frequently made harder by the assumption that it is unnecessary, peripheral, or has 'been solved already' (Mukhopadhyay 2007). Meaningful change to gender equity and equality requires working with policy, society, and culture in context to engage with root causes of inequality.

### *Analytical Framework*

We use a feminist lens to explore the complexity of gender across national contexts and critically examine the discourse on gender equity in the SSF Guidelines. The theoretical grounding for work on gender and fisheries has followed a trajectory from women in development (WID) to gender and development (GAD) (Williams 2008), shifting from a focus on women-only projects and analyses to ones that examine and address larger issues of equity and equality and gendered power relationships at multiple interacting scales (Pearson and Jackson 1998). A feminist lens also supports more recent intersectional and gender transformative approaches in development work that acknowledge the diversity of experience among women and use collaborative research to openly catalyze pro-equity shifts in constraining gender and social norms (Cole et al. 2014a).

## *Methods*

In keeping with the use of case studies throughout this volume, we also illustrate many contexts in which the SSF Guidelines are being implemented. Examples are drawn from co-authors with expertise across seven countries or regions including Bangladesh,[2] Brazil,[3] the European Union,[4] Greenland,[5] Japan,[6] Portugal,[7] and Zambia.[8] Co-authors shared their expertise by answering a survey, which was developed to cover the topics related to gender found in the SSF Guidelines. Throughout the chapter the survey responses are used to illustrate the issues of gender and fisheries found in the SSF Guidelines. Secondary resources from the gender and fisheries literature are also used.

## The SSF Guidelines Gender Discourse

The discourse surrounding gender and women in the SSF Guidelines is found throughout the text, and often in tandem with 'equity', 'equality', and 'mainstreaming'. The meaning of these words in development contexts can have multiple and evolving interpretations (Reeves and Baden 2000). The flexibility of these definitions can allow for context-specific interpretation, but can also lead to loss of meaning and power to enact change (Cornwall and Rivas 2015). An important first step in an implementation process would be to agree on the interpretation of these words in small-scale fisheries contexts (See Table 35.1 for our definitions).

In the SSF Guidelines, gender equality is brought up in a variety of fisheries contexts and this was intentional (Sharma 2013). It begins very broadly in Part 1 (Introduction) that enumerates the 13 Guiding Principles of the SSF Guidelines. The 4th principle of the SSF Guidelines states: "Gender equality and equity is fundamental to any development. Recognizing the vital role of women in small-scale fisheries, equal rights and opportunities should be promoted." (FAO 2015, 2). Issues of gender are also brought up in the second principle "Respect of cultures". While this principle outlines a commitment to "respecting existing forms of organization, traditional and local knowledge and practices of small-scale fishing communities" it also encourages women's leadership and ends with the stipulation that respect of

[2] Choudhury 2016. Expert survey response (Bangladesh).

[3] Santos 2016. Expert survey response (Brazil).

[4] Frangoudes 2016. Expert survey response (European Union).

[5] Snyder 2016. Expert survey response (Greenland).

[6] Soejima 2016. Expert survey response (Japan).

[7] Pita 2016. Expert survey response (Portugal).

[8] Cole 2016. Expert survey response (Zambia).

**Table 35.1** Definitions

| |
|---|
| GENDER EQUALITY: Gender equality usually pertains to the creating of, or the outcome of equal opportunities for women and men by removing formal barriers (Reeves and Baden 2000). In a small-scale fisheries context this could mean changing policies that exclude (primarily) women from equal access to fisheries jobs, markets, or other resources. It can also be thought of as an outcome of efforts to create equal opportunities. |
| GENDER EQUITY: Gender equity is the process by which equality can be achieved. While equality and equity are often used synonymously, there are differences in emphasis, and hence operationalization. Equity works towards equality by acknowledging the different positions of women and men in society, and compensating for those differences (Reeves and Baden 2000). In small-scale fisheries context this could include capacity development aimed towards women, but also programs that incorporate elements of gender and power at several different levels. |
| GENDER MAINSTREAMING: The addition of gender considerations in policy-making, which necessitates addressing the implications of policy for women and men, and girls and boys. The aim is to ensure that gender is present in all aspects of a certain project or activity, with a larger goal of gender equality (UN Women 2016). It begins with an analysis of the context, capacities, attitudes, policies, and monitoring approaches, and when done properly is can be a very powerful tool to induce change. This is rarely done in small-scale fisheries contexts (but see Frangoudes 2015). Mainstreaming models that do not regard local context may fail to address the complexities of gender inequality (Subrahmanian 2007). |
| EMPOWERMENT: "The process by which those who have been denied the ability to make strategic life choices acquire such an ability" (Kabeer 2000, p. 435). According to Kabeer (2000) the focus should be on strategic life choices that can make a difference in a person's well-being, such as their choices related to livelihood, marriage, children, and living conditions. This includes having resources available that can be chosen between, but also the agency with which to make those choices. Empowerment can also be examined at broader scales to address power relationships between groups of people. The relational aspect of empowerment means that it is always shifting (Cornwall 2014). |

cultures must also consider the Convention on the Elimination of All Forms of Discrimination Against Women (CEDAW). The remaining principles do not mention women or gender, but the discourse in all other principles, such as the third principle of non-discrimination in policy and practice could easily be assumed to include dimensions of gender.

The discourse on gender in the principles is necessarily broad, but is more specific in the subsection "Gender equality" (8.1–8.4, Part 2). Part 2, section 8 details gender equality in terms of governance such as gender mainstreaming, and challenging gender discriminatory practices. There is also attention paid to gender in policy with calls for states to adapt legislation for gender equality, as well as compliance with CEDAW. This section also goes on to mention the representation of women in decision-making, leadership, and organizations, and in key personnel such as fisheries extension officers.

To help frame the different types of gender equality barriers and opportunities in small-scale fisheries we have grouped the discourse on gender into three broad interacting categories: (1) access to the fisheries value chain, (2) indicators of human well-being, and (3) governance (Table 35.2). Gender wording related to

**Table 35.2** Barriers to gender equity and equality in small-scale fisheries

| Barriers | SSF guidelines section | Major gender issues |
|---|---|---|
| Small-scale fisheries value chain | Tenure rights (5.3–5.4) | Fishing policy can deny women equal tenure rights. |
| | | Fishing policy can displace women fishers. |
| | | Women may be less likely to be granted lease or tenure over fishing resources. |
| | | Women may be denied membership to fisher groups that are given tenure rights. |
| | Access to fishing resources (6.4) | Women may not, or are less likely to own fishing gear. |
| | | Household owned fishing gear might not be available to women. |
| | Access to markets and marketing resources (7.6) | Fish markets may exclude or be dominated by women. |
| | | Women may have access to inferior product than men. |
| | | Women may have less access to credit or financial resources than men. |
| | | Women that can access credit may not have decision-making power over it. |
| | Recognition of and opportunities for fisheries labour (6.5) | "Gender neutral" policies that do not take unequal gender roles into account may give women fewer opportunities than men. |
| | Equal pay for fisheries labour (7.4) | Women's fisheries labour is often unpaid, or paid less. |
| Human well-being | Education (6.2) | Differences in access to education can impact women and men's fisheries labour. |
| | Food security (5.2 & 7.8) | Women's fishing often focused on small but reliable subsistence catch. |
| | | Women may have less access to food within households. |
| | Occupational health and safety (6.12) | Men and women are often exposed to different risks due to different roles in the fisheries value chain. |
| | Violence (6.9) | Shifting gender roles in fisheries related to changes in resource availability can also lead to increases in gender based domestic violence. |

(continued)

**Table 35.2** (continued)

| Barriers | SSF guidelines section | Major gender issues |
|---|---|---|
| Governance | Policy coherence (10.1) | Gender equity and equality cohere strongly with international CEDAW policy. |
| | | Major barriers may be in the will and capacity to implement existing policy. |
| | Capacity development (11.7, 12.1) | Lack of technical and formal fisheries training programs that are targeted to or include women. |
| | | Women are often not recognized as stakeholders and must contend with cultural barriers to their full participation in decision-making. |
| | | Capacity development should include increase training for gender work in fisheries institutions. |
| | Research and monitoring (11.1, 11.10, 13.3) | Lack of sex-disaggregated data collection. |
| | | Lack of prioritization, money, and training for gender research and gender researchers. |

access to the fisheries value chain, and indicators of human well-being are found throughout Part 2 (Responsible Fisheries and Sustainable Development). As stated above issues of governance are initially laid out in Part 2 (8.1–8.4), but are also specified in Part 3 (Ensuring and Enabling Environment and Supporting Implementation).

The inclusion of gender in these many and varied parts of the SSF Guidelines points to gender as a key cross-cutting issue. However, it is not universally present, and is missed in some other themes such as climate change (Djoudi and Brockhaus 2011). Other issues, such as reproductive health, which have been successfully integrated in small-scale fisheries and coastal management programs (Westerman et al. 2013), are not addressed in the SSF Guidelines, although they are often considered an important aspect of gender equity and equality (Singh et al. 2003).

## Barriers, Challenges, and Opportunities to Gender Equality in Small-Scale Fisheries Contexts

To understand why gender was included in many of the different themes of the SSF Guidelines, we will examine the barriers to gender equality in small-scale fisheries contexts. We will begin by discussing underlying cultural barriers to gender equity in access to the fisheries value chain, human well-being, and governance, and how those interact with policy (Table 35.2). We connect the main issues to specific sections in the SSF Guidelines in Table 35.2, and illustrate the issues with specific and diverse examples in the text.

## *Gender, Fisheries, and Culture*

Gender inequalities in all contexts are deeply embedded in cultural traditions and the organization of social institutions. The SSF Guidelines include principles of gender equity and equality and respect for cultures, while acknowledging that there is potential for tension between the two. The "Respect of Cultures" principle ends with deference to gender equality through the application of CEDAW. When discussing tenure rights, preference is given to traditional cultural practices with the stipulation that, "where constitutional or legal reforms strengthen the rights of women and place them in conflict with custom, all parties should cooperate to accommodate such changes in the customary tenure systems." In short, cultural practices should be respected, but when in conflict with gender equity, they should also be reconsidered. Successful efforts for change would not discount or disrespect cultural practices or impose others, but rather work from within existing systems. It is therefore important to directly examine cultural barriers that may prevent gender equity and equality in small-scale fisheries contexts (Onyango and Jentoft 2011).

Women and men often perform different roles in fisheries labour, and those roles are often given different cultural importance. Socially-proscribed gender roles can shape how, where, when, and what women and men fish, or what part of the value chain they predominately occupy. Women's participation in fisheries can often be limited by the domestic social obligations related to their gender roles. For example in Kiribati: "Gleaning shellfish is women's major fishing activity because it can be done close to home, takes relatively little time, requires no costly fishing equipment and may be done in the company of children" (Tekanene 2006). Many of the same social structures and gender roles that can limit women's equal participation in fisheries can also create gender difference in indicators of human-well-being, and access to full participation in governance. Differences in access to education, health care, and financial institutions can all be rooted in gender roles and interact with access to and roles in fisheries. For example in Greenland, current gender roles make women more likely to receive higher levels of education than men, and hence less likely to work directly in small-scale fisheries. The assumption that fisheries are the domain of men can limit women's full participation in fisheries governance, as can gender roles that restrain or stunt women's ability to participate in public spaces (Figs. 35.1 and 35.2)

**Fig. 35.1** Women shellfish gatherers of Cambados in Galicia, Spain

**Fig. 35.2** Fishers processing the morning catch of sea cucumber. Batasan Island, Bohol, The Philippines 2011 (Photo credit: Adam Cormier)

## *Barriers and Opportunities to Gender Equality in Fishing and the Value Chain*

### Tenure Rights (5.3–5.4)

One common form of fisheries policy is to determine who has access to fishing grounds, which falls within the domain of tenure rights. Discussions of tenure rights in fisheries relate to issues of access to fisheries resources, and rights-based approaches emphasize the need for inclusive regulations. A social justice framework goes further to include special attention to the needs of marginalized groups (Jentoft 2013). Women are often denied access to fishing either directly or indirectly through fisheries policy. In some cases, this is quite direct, such as Oman women, who are barred from obtaining fishing licenses (Anderson 2016, personal communication). In other cases, spatial management measures can have a disproportionate impact on women, such as the Bay of Bengal Large Marine Ecosystem in the Gulf of Mannar, India, where the creation of a Marine Protected Area (MPA) displaced women seaweed collectors (Rajagopalan 2007). In Bangladesh there is no formal barrier, but women's groups are simply unlikely to be given leases to fishing grounds while in Japan women are denied membership in the Japanese Fisheries Cooperative.

### Access to Fishing Resources (6.4), Markets, and Marketing Resources (7.6)

Women may also be limited in the types of fishing they can participate in because they do not have access to fishing gear. For example, in Zambia resource ownership within households is controlled by husbands, so women may not have unrestricted and regular access to gears needed to fish. In the Philippines, women fish in boats, but almost exclusively with their husbands or other male relatives, as women by themselves are unlikely to own boats (Kleiber et al. 2014). By contrast, in Brazil women can inherit fishing gear and often buy a boat specific to the needs of their mangrove fisheries. Similarly, in Ghana customary inheritance law favours women, which has led to some relatively affluent women being the sole owners of boats and fishing gear. However, the owners lease the gear exclusively to men, so this does not change women's direct participation in fishing (Walker 2001). In many European countries, wives of professional fishermen can inherit fishing boats, but may still be barred from fishing because they themselves do not have access to quotas that are only allotted to professional fishers. In some cases, this restriction has changed in countries such as Norway where, after many years of struggle, wives of fishers gained access to quotas along with inherited boats.

Participation in fishing markets is a key part of the small-scale fisheries value chain, and the gendered nature of this occupation can vary widely. In Bangladesh, markets are almost exclusively the domain of men, although poorer women are found marketing catch because poverty levels determine the strictness of gender norms. Poorer women face far more relaxed norms from necessity of survival. In

Kenya, both women and men participate in marketing fish, but men have access to the larger and more valuable catches (Matsue et al. 2014). In Ghana, elite women's ownership of boats and fishing gear allows them to dominate the marketing of the catch (Walker 2001), but this should not be assumed to be representative of all women's participation. In Southern Europe, including Portugal, selling fish at the market is often performed by women, especially in the marketing of small-scale fisheries catches (Frangoudes 2013).

Access to financial credit can also determine gender specific access to fishing gear, markets, and processing equipment. The ability of women and men to access credit is highly variable. In Japan, women are less likely than men to access credit for fishing gear, largely because they are not allowed to be members of the Fishing Cooperative Association. In Bangladesh, women often access microcredit loans precisely because they are usually denied formal loans through banks. However, while women bear the responsibility of repaying the loan, they often have little say in how the loan is used (Kabeer 1998). By contrast, in the Barotse fishery in western Zambia, while overall access to credit is low in rural areas, women participate more in village savings and lending groups, enabling them to gain access to small loans for buying and reselling fish, among other small-business ventures. In the European Union, women finance fish marketing through the European Fisheries structure fund. It is common that when men obtain bank loans for fishing gear and boats, their wives are always included as co-borrower.

## Recognition of and Opportunities for Fisheries Labor (6.5)

The SSF Guidelines state that "All [fisheries] activities should be considered: part-time, occasional and/or for subsistence" (FAO 2015, 22). This is particularly important for women since their labor in fisheries often falls into these three categories. In some cases, policies have been effectively changed to formally recognize women's fishing (Frangoudes et al. 2008), as well as women's fishing labor (Frangoudes and Keromnes 2008). In other cases, policies that are 'gender blind' impact women and men differently. These policies can also have broader gender-specific impacts on people in fishing communities and were implicated in changing gender roles in Norway, Iceland, and Newfoundland (Neis et al. 2013; Gerrard 2015). Furthermore, assumptions of equality can lead to lack of interest in gender-specific policies.

## Equal Pay for Fisheries Labor (7.4)

Women's labor in fisheries often goes unpaid because in many cases it is characterized as being part of women's household duties (Williams 2015). Labor can include pre-harvest activities such as gear manufacturing or maintenance, but also post-harvest activities such as marketing, processing, accounting, and cooking. In Portugal, women still carry out much of the unpaid fisheries work, including preparing and fixing gear, baiting fishing gear, as well as assisting their husbands and family members with other tasks. Since 1986, EU Members States have acknowledged women's labor contributions by giving them the legal status of 'assisting spouse', with corresponding social rights such as maternity leave and pensions (Frangoudes and Keromnes 2008; Frangoudes 2013). In other cases, women's labour is paid, but often less than that of men. In Brazil, women's wages earned as *marisquieras* or by participating in the shrimp fishery is reported as half that of men involved in the fishery, although some women shell-fishers may have income equal to fishermen. These gender differences are often overlooked because the income is seen as belonging to the household, not to the woman as an individual. However, viewing households as cooperative units overlooks within-household power dynamics that influence how resources are shared and distributed.

## *Barriers and Opportunities to Gender Equality in Human Well-Being*

### Education (6.2)

Gender differences in access to general education can impact the roles that women and men play in small-scale fisheries. In some cases, due to lack of education opportunities, women are more likely to perform lower-skill fisheries jobs. In other cases, such as Greenland, women attain higher levels of general education, while men are more likely to take vocational education training in fishing and hunting (women only comprise 5–7% of vocational students). The result is that men dominate the small-scale fisheries sector, while women have appreciably higher earning potential working as government employees.

### Food Security (5.2 & 7.8)

Small-scale fisheries are an important source of food security around the world and subsistence catch by women is often a key part of household food security strategies (Porter and Mbezi 2010; Béné et al. 2016). For example, in the Central Philippines men bring in the majority of the subsistence catch, but women are more likely to be solely subsistence fishers, and their catch can be relied on when other forms of fishing are not available (Kleiber 2014). In Eastern Brazil, women's catch concentrates on the daily consumption needs of their household, while men's catch (often larger but more variable) is distributed throughout the community and linked to systems of social capital (Santos 2015). Intra-household differences in how food gets distributed is noteworthy in Bangladesh, where there is a general belief that men work harder and hence should eat better and more. It is common for women to sacrifice some of their portions, which is especially harmful for pregnant and lactating women (D'Souza and Tandon 2015).

#### Occupational Health and Safety (6.12)

The gendered nature of occupational health and safety risks often relates to the division of labour in the fisheries value chain. The risks involved in fishing are high and can be fatal (Power 2008). For example, in 2014 in Japan 65 people were killed or went missing from fishing vessels accidents. Although the statistics are rarely disaggregated by sex, men are assumed to be at greater risk given their dominance as fishers. However there are also occupational health and safety risks in the processing and marketing of fisheries resources, which are often dominated by women. For example, in Indian shrimp processing plants, where all workers are women, occupational hazards put workers at greater risk of injury (Saha et al. 2006). In other cases such as Zambia, low access to financial resources has led to women fish processors and marketers exchanging sex for fish catch, putting them at higher risk of HIV infection (Béné and Merten 2008).

#### Violence (6.9)

Related to broader issues of health is a concern with gender-based violence in small-scale fishing communities. While in no way unique to small-scale fisheries contexts, violence against women was recognized by co-authors as a major issue in Bangladesh, Zambia, and Greenland. In Greenland, women who experience violence are often unable to move due to a housing shortage, which is a result of widespread employment-based housing where men who work in the fishing sector receive housing benefits as part of their remuneration. In other cases, such as the Philippines, increases in domestic violence have been linked to changing gender roles related to changes in availability of marine resources. Dwindling catch has led to shifting roles in fisheries where men are more likely to be at home, a sphere traditionally connected with women. Unhappiness with these shifting roles has led to an increase in domestic violence (Turgo 2014). It is also important to note that other forms of gendered violence, such a sexual harassment, may hinder women's participation in male-dominated parts of the fisheries value chain.

### *Barriers and Opportunities to Gender Equality in Small-Scale Fisheries Governance*

#### Policy Coherence (10.1)

The SSF Guidelines gender equity and equality principle may support or conflict with local, national, and international fisheries policy. The aims of gender equity and equality fit well with the international CEDAW policy, which has been a standard bearer of women's rights since the UN first adopted it in 1979. While the implementation from ratifying nations has been variable, its influence on national gender policy and women's rights is noticeable (Cho 2014), including recognition of the need for widespread cultural shifts to achieve social equality for women throughout the world. Both CEDAW and the SSF Guidelines agree on the need to include gender equity and equality in rights-based approaches. Recently the FAO recommended that the CEDAW Committee refer to the SSF Guidelines in the General Recommendations on the rights of rural women (CFS and FAO 2012; CEDAW 2016), in order for government to see how the two policy instruments could be mutually supportive. Hence having gender equity and equality discourse in the SSF Guidelines offers an opportunity to reinforce CEDAW policy in small-scale fisheries contexts.

While CEDAW and many other national gender equality policies are relevant to small-scale fishing communities, they often go unimplemented for a variety of reasons. For example, in Japan the national gender policy has not been broadly applied to the fisheries sector. It does mandate a quota of women represented on the boards of fishing collectives, but it does little to address broader gender labor disparities in the sector, and so cannot contribute to the resolution of gender equality issues. In other cases, broad gender equality policies that could include fishing communities were nonetheless beset by financial and capacity deficit barriers. In Zambia, the gender-specific policies have translated to women's greater access to and ownership of land, support for girls to return to school after pregnancy, and prevention of violence against women. However, few of these policies are adequately implemented or lead to widespread change at district and provincial levels – especially in fisheries contexts – due to lack of human and financial resources. By contrast, in the EU, gender policy has formalized women's previously unrecognized labor in the fisheries sector because all EU policies must include the principle of gender equality. In these cases, women are now eligible for social benefits that had previously been reserved for male fishers, although their role is still characterized as 'assisting spouse', rather than as fisheries participants in their own right.

## Capacity Development (11.7, 12.1)

Capacity development in the SSF Guidelines focuses on the inclusion of fishing communities in increasing their knowledge of fisheries through technical training and support (11.7), but also increasing their inclusion as stakeholders in the governance process and their ability to participate in decision making (12.1). In both cases the SSF Guidelines highlight the need to include women. To this definition of capacity development, we would also include the need for increasing capacity for gender work within fisheries institutions, and representation of women in more formal education programs necessary to become a fishery officer, researcher, or policy maker.

Technical training is not always available to women, and can reinforce deeply ingrained gender roles that exclude women from certain parts of the fisheries value chain, particularly fishing. For example, in Japan women's contributions to small-scale fishing activities have not been recognized and subsequently in the past women have not been included in training programs. More recently, women's groups have been targeted for fish processing and marketing training. In Bangladesh the recognition of gender differences in technical knowledge has led capacity development programs to target women. However, these programs have fallen short because they only address technical knowledge but do little to change the other social barriers that women face, such as the roles attributed to them by traditional or religious socio-cultural norms.

Fewer women are involved in small-scale fisheries decision-making institutions. Part of this disparity may be the perception that women are not equal actors, which can be tied to the devaluation of their contributions to fisheries value chains. In other cases, such as Bangladesh, women are perceived not to have the necessary knowledge and experience to participate in management decisions. In other cases, the labor associated with women's socially assigned roles can be a barrier to participation. For example, in Brazil women have multiple responsibilities as shellfish extractors, fishery processors, and domestic obligations, leaving little time to participate in fisheries council meetings.

The common absence of women from these fora has led to a great deal of emphasis on the representation of women in fisheries decision-making groups. While the presence of women has been found to influence decision making (Agarwal 2009), the presence of women does not assure that women's voices are being included. For example in Bangladesh, women in mixed-sex settings were less likely to speak (World Bank et al. 2009). In addition, the presence of women does not guarantee representation of the diversity of women's priorities. In cases where women are included specifically to fill a quota, women from elite groups (by wealth, education, caste) are often more likely to be chosen, and may not necessarily represent the needs of more vulnerable populations of women (Resurreccion 2008). The desire for women to have their needs prioritized may lead them to create women-only organizations where women's priorities could be addressed in a way that could not occur in either male-dominated, or mixed-sex settings (Agarwal 2001). For instance, in Japan women have been forming their own fisheries groups after they felt their needs and priorities were not being met by government-run groups. However, women-only groups do not necessarily guard against the marginalization of women's needs. In many cases women's groups are given responsibility over inferior resources, and receive less recognition and support than their male counterparts (Buchy and Rai 2008).

The capacity development of gender expertise or gender-responsive organizational culture within fisheries institutions also requires attention. In an effort to strengthen gender capacities, some institutions have a dedicated gender expert or gender team. While laudable, if the role of gender experts is not well understood or appreciated by all institutional divisions, gender issues may be easily compartmentalized, disregarded and under supported (Harrison 1997). These models may also mean that gender experts are over-burdened. For many co-authors who work as gender experts within their organizations, the existing gender expertise is thinly spread across a number of projects and too few people are responsible for gender inclusion. Gender training for fisheries officers in particular should be prioritized.

Fisheries officers are often the main way institutions connect with small-scale fishers and in fishing communities. They can be important for facilitating technical trainings, organizing programs, and ensuring stakeholder participation. Fisheries extension officers are also much more likely to be men, and may be less likely to include, consider, or address women's needs and priorities (Seniloli et al. 2002; Adeokun and Adereti 2003). In some cases women have stated that they would feel more comfortable talking to a female fisheries extension officer and this may be especially important in contexts where it is culturally inappropriate for unrelated men and women to speak to each other (Adeokun and Adereti 2003). Training and including women in fisheries management institutions is important, but it must be noted that simply adding women to the staff does not mean that programming will automatically be gender-sensitive. Nor should gender concerns be the sole burden of women. It is important to train both women and men in gender analysis and gender-sensitive design and delivery of the services of their institutions (Petrics et al. 2015).

## Research and Monitoring (11.1, 11.10, 13.3)

Research and monitoring is often used to inform policy decisions and is an important part of the governing process. However the lack of inclusion of gender in small-scale fisheries research limits and is limited by many of the issues already raised in this chapter. For example, lack of interest in or awareness of women's contribution to the fisheries value chain as well as limited gender expertise among fisheries researchers are some of the reasons gender and fisheries research (while growing) is still quite limited (Kleiber et al. 2015). Unfortunately, the lack of data only perpetuates the assumption that women's participation is either non-existent or unworthy of research notice.

The most basic form of data required for gender analysis is gender or sex-disaggregated data (Hill 2003). Sex-disaggregated data is rarely collected in natural resources research, and this is especially true of fisheries contexts. Many countries do not have regulations regarding the collection of gender-disaggregated data, but even in cases where it required, such as in Zambia, limitations of funding and training of fisheries extension officers results in sporadic collection. Women may also be left out because of narrow definitions of who counts as a fisher. In Japan, people who work on land are not considered to be working in fisheries, which leave out most of women's participation. As noted above, a deficit in training and time can also lead to gender data being sidelined. In Bangladesh, the collection of gender data required more time and training and was considered the job of the gender experts or social scientists. This was often done in separate studies or added in as an afterthought. In addition, data on women's involvement in fisheries value chains are challenging to collect because many of the tasks they carry out are either unpaid or less valued (at least by those conducting the research who inadvertently leave out questions that would capture the necessary data). Thus, greater attention by researchers working in small-scale fisheries contexts is needed to collect sex-disaggregated data to enhance our understanding of the complex settings where women and men, girls and boys reside and depend on for their livelihoods.

## Approaches to Implementing Gender Equality

In this section, we will highlight approaches that can be used to address the barriers and challenges to implementing gender equality in fisheries, while recognizing that there is no simple, short term, or "one size fits all" solution to achieving gender equality. A focus on development can lead to approaches that focus on measurable targets of human well-being and economic growth. However, the SSF Guidelines are very explicitly modeled on a human rights framework, which would also include approaches that recognize the aspiration for gender equity and equality, not only as a means to a material end, but simply because it is the right thing to do (Cole et al. 2015).

Many gender approaches begin with an understanding of historical and cultural contexts that highlights the gendered aspects of the passage of assets from one generation to the next, and what public and private spaces are available to women and men. This information can help to understand the underlying social structures that create the gendered distribution of resources (Zhang et al. 2008). Cultural understanding can also give greater depth to our understanding of the status of women and men in a society. For example, while women in Brazil are largely responsible for domestic tasks, certain women (Bahianas) have a reputable role in religious ceremonies that are highly regarded by communities. These types of data are typically found in ethnographic and anthropological research.

Other approaches, such as gender roles frameworks, examine the material realities of women and men by characterizing the gendered division of labor and access to resources (Razavi and Miller 1995). This information is often unavailable, so it is important to understand the local fisheries context by focusing on the roles of men and women inside the household, but also within the fisheries value chain. Livelihood approaches – which focus on material realities and adaptability – include gender as one social variable that can produce differences in access (Allison and Ellis 2001). Research on roles, labor, and material realities, would situate this focus in the realm of sociology and feminist economics.

Social relations frameworks and gender transformative approaches allow the inclusion of power relations (Cole et al. 2015). The use of these approaches allows the realization of a deeper analysis of the power differences that perpetuate differences in access in the first place (Razavi and Miller 1995). "Gender transformative research is informed by conceptual frameworks that recognize the influence social institutions have on creating and perpetuating gender inequalities" (Cole et al. 2015). These approaches help create greater gender equity and equality by addressing inequitable gender and social norms, differences in power, social expectations, and capacity to participate in civil society (Kantor et al. 2015). Engaging in a gender transformative approach means also acknowledging the diversity among women and among men and including intersecting power structures based on multiple social categories. This approach sits within the theoretical framework of intersectionality, and works to combat the homogenization of gendered experience (Walker and Robinson 2009).

The operationalization of gender transformative approaches can be quite similar to participatory action research in that it is a long-term process of collaboration between researchers and people in the communities they work in. WorldFish[9] has been using gender transformative approaches in their work in Zambia and Bangladesh (Cole et al. 2014b; Kantor et al. 2015). In western Zambia, projects designed to process high-quality fish with minimal loss also included a Gender Transformative Communication (GTC) tool of critical reflection sessions on the social limitations on women's participation in certain parts of the fisheries value chain. Through piloting these technical and social innovations together, the project aimed to reduce post-harvest losses and improve gender relations in the fishery value chain.

## Conclusion

The inclusion of gender equity and equality in the SSF Guidelines is the result of the hard work of many dedicated small-scale fisheries experts, practitioners, and CSOs. There has never been a global fisheries policy document that includes gender so broadly and thoroughly. Far from a call for small technical fixes, the SSF Guidelines has outlined the need for no less than gender equity and equality - a mighty aspiration for a policy document that is already ground-breaking in many other important ways. While the efforts to include gender equality and equity principles in the SSF Guidelines are commendable, the implementation phase that follows will demand even more resolve and attention.

Our regional review show that tenure rights, access to fishing resources and markets, recognition of and opportunities for fisheries labour, equal pay, education and food security, among several other themes emerge as gender concerns for many small-scale fisheries. Investigating small-scale fisheries through these themes provides direct evidence for sound policy design of the SSF Guidelines. It also illustrates the participation of women and men in all aspects of the small-scale fisheries value chain, that gender equality is inextricable linked to human rights and food security, and that women are often overlooked and undervalued by governing institutions leading to lack of fair gender representation in small-scale fisheries governance. Beyond presenting our findings in this chapter, our analyses also explain how and to what extent it may be possible for governments to address gender equity and equality concerns with the aid of the SSF Guidelines.

Unprecedented as it is to present gender within an instrument of this stature, our findings show potential for aligning global small-scale fisheries with effective and

[9] *WorldFish* is an international research organization of fisheries and aquaculture.

appropriate policy. Further research on how governments implement the SSF Guidelines will be needed, especially considering the ambitions of the SSF Guidelines and that lessons of implementation from one region may hold value for other countries with similar challenges.

## References

Adeokun, O. A., & Adereti, F. O. (2003). Agricultural extension and fisheries development: Training for women in fish industry in Lagos – State, Nigeria. *Journal of Agriculture and Social Research, 3*, 64–76.

Agarwal, B. (2001). Participatory exclusions, community forestry, and gender: An analysis for South Asia and a conceptual framework. *World Development, 29*, 1623–1648.

Agarwal, B. (2009). Rule making in community forestry institutions: The difference women make. *Ecological Economics, 68*, 2296–2308.

Allison, E. H., & Ellis, F. (2001). The livelihoods approach and management of small-scale fisheries. *Marine Policy, 25*, 377–388.

Badjeck, M. C., Allison, E. H., Halls, A. S., & Dulvy, N. K. (2010). Impacts of climate variability and change on fishery-based livelihoods. *Marine Policy, 34*, 375–383.

Béné, C., & Merten, S. (2008). Women and fish-for-sex: Transactional sex, HIV/AIDS and gender in African fisheries. *World Development, 36*, 875–899.

Béné, C., Arthur, R., Norbury, H., Allison, E. H., Beveridge, M., Bush, S., Campling, L., Leschen, W., Little, D., Squires, D., Thisted, S. H., Troell, M., & Williams, M. (2016). Contribution of fisheries and aquaculture to food security and poverty reduction: Assessing the current evidence. *World Development, 79*, 177–196.

Buchy, M., & Rai, B. (2008). Do women-only approaches to natural resource management help women? The case of community forestry in Nepal. In B. P. Resurreccion & R. Elmhirst (Eds.), *Gender and natural resource management* (pp. 127–150). London: Earthscan.

CEDAW. (2016). *General recommendation No. 34 on the rights of rural women*. New York: CEDAW (Committee on the Elimination of Descrimination Against Women), CEDAW/C/GC/34.

CFS, & FAO. (2012). *Voluntary guidelines on the responsible governance of tenure of land, fisheries and forests in the context of national food security*. Rome: FAO.

Cho, S. Y. (2014). International women's convention, democracy, and gender equality. *Social Science Quarterly, 95*, 719–739.

Cole, S. M., van Koppen, B., Puskur, R., Estrada, N., DeClerck, F., Baidu-Forson, J. J., Remans, R., Mapedza, E., Longley, C., Muyaule, C., & Zulu, F. (2014a). *Collaborative effort to operationalize the gender transformative approach in the Barotse Floodplain, Program brief: AAS-2014038*. Penang: CGIAR Research Program on Aquatic Agricultural Systems.

Cole, S. M., Kantor, P., Sarapura, S., & Rajaratnam, S. (2014b). *Gender-transformative approaches to address inequalities in food, nutrition and economic outcomes in aquatic agricultural systems, Working paper: AAS-2014-42*. Penang: CGIAR Research Program on Aquatic Agricultural Systems.

Cole, S. M., Puskur, R., Rajaratnam, S., & Zulu, F. (2015). Exploring the intricate relationship between povery, gender inequality and rural masculinity: A case study from an aquatic agricultural system in Zambia. *Culture, Society & Masculinities, 7*, 154–170.

Cornwall, A. (2014). *Women's empowerment: What works and why? WIDER working paper 2014/104*. Helsinki: UNU: WIDER.

Cornwall, A., & Rivas, A. M. (2015). From "gender equality and 'women's empowerment" to global justice: Reclaiming a transformative agenda for gender and development. *Third World Quarterly, 36*, 396–415.

Cornwall, A., Harrison, E., & Whitehead, A. (2007). Gender myths and feminist fables: The struggle for interpretive power in gender and development. *Development and Change, 38*, 1–173.
D'Souza, A., & Tandon, S. (2015). *Using household and intrahousehold data to assess food insecurity: Evidence from Bangladesh, Economic research report: 190*. Washington, DC: U.S. Department of Agriculture.
Djoudi, H., & Brockhaus, M. (2011). Is adaptation to climate change gender neutral? Lessons from communities dependent on livestock and forests in northern Mali. *International Forestry Review, 13*, 123–135.
FAO. (2015). *Voluntary guidelines for securing sustainable small-scale fisheries in the context of food security and poverty eradication*. Rome: Food and Agriculture Organization of the United Nations.
Frangoudes, K. (2013). *Women in fisheries: A European perspective*. European Union: Directorate-general for internal policies. Policy Department B: Structural and Cohesion Policies: BA-01-13-425-EN-C.
Frangoudes, K. (2015). *Fisheries policy, Policy-specific module: EIGE/2014/OPER/10*. Vilnius: European Institute for Gender Equality.
Frangoudes, K., & Keromnes, E. (2008). Women in artisanal fisheries in Brittany, France. *Development, 51*, 265–270.
Frangoudes, K., Marugán-Pintos, B., & Pascual-Fernández, J. J. (2008). From open access to co-governance and conservation: The case of women shellfish collectors in Galicia (Spain). *Marine Policy, 32*, 223–232.
Gerrard, S. (2015). Mobility practices and gender contracts in a fishery-related area. In S. M. Channa & M. Porter (Eds.), *Gender, livelihood and environment: How women manage resources* (pp. 113–140). Dehli: Orient Black Swan.
Gopal, N., Porter, M., Kusakabe, K., & Choo, P.-S. (2015). Guest editorial: Gender in aquaculture and fisheries – Navigating change. *Women in Fisheries Information Bulletin, 26*, 3–11.
Harrison, E. (1997). Fish, feminists and the FAO: Translating "gender" through differing institution in the development process. In A. M. Goetz (Ed.), *Getting institutions right for women in development* (pp. 61–76). London/New York: Zed Books.
Hill, C. L. M. (2003). *Gender-disaggregated data for agriculture and rural development: Guide for facilitators*. Rome: FAO Socio-economics and Gender Analysis (SEAGA) Programme.
Jentoft, S. (2013). Governing tenure in Norwegian and Sami small-scale fisheries: From common pool to common property? *Land Tenure Journal*, 91–115.
Kabeer, N. (1998). *"Can buy me love"? Re-evaluating the empowerment potential of loans to women in rural Bangladesh, IDS. Discussion paper: 363*. Sussex: IDS.
Kabeer, N. (2000). Resources, agency, achievements: Reflections on the measurement of women's empowerment. *Development and Change, 30*, 435–464.
Kantor, P., Morgan, M., & Choudhury, A. (2015). Amplifying outcomes by addressing inequality: The role of gender-transformative approaches in agricultural research for development. *Gender, Technology and Development, 19*, 292–319.
Kleiber, D. (2014). *Gender and small-scale fisheries in the Central Philippines*. Doctoral dissertation, The University of British Columbia, Canada.
Kleiber, D., Harris, L. M., & Vincent, A. C. J. (2014). Improving fisheries estimates by including women's catch in the Central Philippines. *Canadian Journal of Fisheries and Aquatic Sciences, 71*, 1–9.
Kleiber, D., Harris, L. M., & Vincent, A. C. J. (2015). Gender and small-scale fisheries: A case for counting women and beyond. *Fish and Fisheries, 16*, 547–562.
Matsue, N., Daw, T., & Garrett, L. (2014). Women fish traders on the Kenyan coast: Livelihoods, bargaining power, and participation in management. *Coastal Management, 42*, 531–554.
Mukhopadhyay, M. (2007). Mainstreaming gender or "streaming" gender away: Feminists marooned in the development business. In A. Cornwall, E. Harrison, & A. Whitehead (Eds.), *Feminisms in development: Contradictions, contestations and challenges* (pp. 135–149). London/New York: Zed Books.

Neis, B., Gerrard, S., & Power, N. G. (2013). Women and children first: The gendered and generational social-ecology of smaller-scale fisheries in Newfoundland and Labrador and Northern Norway. *Ecology and Society, 18*(4), 64.

Onyango, P. O., & Jentoft, S. (2011). Climbing the hill: Poverty alleviation, gender relationships, and women's social entrepreneurship in Lake Victoria, Tanzania. *Maritime Studies, 10*, 117–140.

Pearson, R., & Jackson, C. (1998). Introduction: Interrogating development: Feminism, gender and policy. In C. Jackson & R. Pearson (Eds.), *Feminist visions of development: Gender analysis and policy* (pp. 1–16). London/New York: Routledge.

Petrics, H., Blum, M., Kaaria, S., Tamma, P., & Barale, K. (2015). *Enhancing the potential of family farming for poverty reduction and food security: Through gender-sensistive rural advisory services, Occasional papers on innovation in family farming*. Rome: FAO.

Porter, M., & Mbezi, R. G. (2010). From hand to mouth: Fishery projects, women, men and household poverty. *Canadian Journal of Development Studies, 31*, 381–400.

Power, N. G. (2008). Occupational risks, safety and masculinity: Newfoundland fish harvesters' experiences and understandings of fishery risks. *Health Risk & Society, 10*, 565–583.

Rajagopalan, R. (2007). Uncertain future. *Yemaya, 24*, 3–5.

Ram-Bidesi, V. (2015). Recognizing the role of women in supporting marine stewardship in the Pacific Islands. *Marine Policy, 59*, 1–8.

Razavi, S., & Miller, C. (1995). *From WID to GAD: Conceptual shifts in the women and development discourse*. Geneva: United Nations Research Institute for Social Development. United Nations Development Programme.

Reeves, H., & Baden, S. (2000). *Gender and development: Concepts and definitions, Report: 55*. Brighton: University of Sussex.

Resurreccion, B. P. (2008). Gender, legitimacy and patronage-driven participation: Fisheries management in the Tonle Sap Great Lake, Cambodia. In B. P. Resurreccion & R. Elmhirst (Eds.), *Gender and natural resource management* (pp. 151–174). London: Earthscan.

Saha, A., Nag, A., & Nag, P. K. (2006). Occupational injury proneness in Indian women: A survey in fish processing industries. *Journal of Occupational Medicine and Toxicology, 1*, 23–27.

Santos, A. N. (2015). Fisheries as a way of life: Gendered livelihoods, identities and perspectives of artisanal fisheries in eastern Brazil. *Marine Policy, 62*, 279–288.

Seniloli, M., Taylor, L., & Fulivai, S. (2002). Gender issues in environmental sustainability and poverty reduction in the community: Social and community issues. *Development Bulletin, 58*, 96–98.

Sharma, C. (2013). Beyond lip service. *Yemaya, 42*, 9–11.

Singh, S., Darroch, J. E., Vlassoff, M., & Nadeau, J. (2003). *Adding it up: The benefits of investing in sexual and reproductive health care*. New York: UNFPA.

Subrahmanian, R. (2007). Making sense of gender in shifting institutional contexts: Some reflections on gender mainstreaming. In A. Cornwall, E. Harrison, & A. Whitehead (Eds.), *Feminisms in development: Contradictions, contestations and challenges* (pp. 112–121). London/New York: Zed Books.

Tekanene, M. (2006). The women fish traders of Tarawa, Kiribati. In P.-S. Choo, S. J. Hall, & M. J. Williams (Eds.), *Global symposium on gender and fisheries. Seventh Asian fisheries forum, 1–2 December 2004, Penang, Malaysia* (pp. 115–120). Penang: WorldFish Center.

The WorldFish Center. (2010). *Gender and fisheries: Do women support, complement, or subsidize small scale fishing activities, Issue brief: 2108*. Penang: The WorldFish Center.

Turgo, N. N. (2014). Redefining and experiencing masculinity in a Philippine fishing community. *Philippine Sociological Review, 62*, 7–38.

UN Women. (2016). *UN Women Training Centre glossary*. https://trainingcentre.unwomen.org/mod/glossary/view.php?id=36. Accessed 2 May 2016.

Walker, B. L. E. (2001). Sisterhood and seine-nets: Engendering development and conservation in Ghana's marine fishery. *The Professional Geographer, 53*, 160–177.

Walker, B. L. E., & Robinson, M. A. (2009). Economic development, marine protected areas and gendered access to fishing resources in a Polynesian lagoon. *Gender, Place & Culture, 16*, 467–484.
Westerman, K., Oleson, K. L. L., & Harris, A. R. (2013). Building socio-ecological resilience to climate change through community-based coastal conservation and development : Experiences in southern Madagascar. *Western Indian Ocean Journal of Marine Science, 11*(1), 87–97.
Williams, M. J. (2008). Why look at fisheries through a gender lens? *Development, 51*, 180–185.
Williams, M. J. (2015). Women in today's fisheries economy. *Yemaya, 50*, 2–4.
World Bank, FAO, International Fund for Agricultural Development. (2009). Gender in fisheries and aquaculture. In The World Bank, Food and Agriculture Orhganization, and International Fund for Agricultural Development (Ed.), *Gender in agriculture sourcebook* (pp. 561–600). Washington, DC: The World Bank.
Zhang, L., Liu, C., Liu, H., & Yu, L. (2008). Women's land rights in rural China: Current situation and likely trends. In B. P. Resurreccion & R. Elmhirst (Eds.), *Gender and natural resource management* (pp. 87–108). London: Earthscan.

# Part X
# Moving Forward

In the book's last section, we are returning to the Human Rights-Based Approach, which is the key guiding principle of the SSF Guidelines. In Chap. 36 Rolf Willmann, Nicole Franz, Carlos Fuentevilla, Thomas McInerney and Lena Westlund narrate the background and the legal and conceptual basis of this approach and discuss what legal and conceptual challenges are involved when trying to implement it in small-scale fisheries. In particular, they discuss how to facilitate the empowerment of small-scale fishing people. They argue that it is important to promote social development and decent work, including civil and political rights. Thereby, they illustrate the broad scope of the Human Rights-Based Approach. In Chap. 37, Svein Jentoft and Ratana Chuenpagdee bring it all together in a synthesis of findings and recommendations, drawing from the individual case studies that make up the major part of the book. They summarize what has happened so far with the SSF Guidelines and their implementation, and how they are received. It should not come as a surprise that the ground is unequally fertile for the SSF Guidelines. The chapter discusses which conditions are conducive to their successful implementation. One such condition, they hold, is funding for small-scale fisheries research.

# Chapter 36
# A Human Rights-Based Approach in Small-Scale Fisheries: Evolution and Challenges in Implementation

**Rolf Willmann, Nicole Franz, Carlos Fuentevilla, Thomas F. McInerney, and Lena Westlund**

*The ability of the poor and marginalized to break the veils of oppressive fear and injustice is the key to any process of socio–political and economic empowerment. This process of empowerment requires a rights–based perspective and the facilitating role of organizations and institutions to create an enabling environment for people to realise their own potential to change their lives. (ActionAid Bangladesh, Taking a Stand, Dhaka, November 2000).*

---

The text of this chapter reflects the views of the authors and should not be attributed in any form to their current or former employer.

R. Willmann (✉)
Independent Expert, Kressbronn, Germany
e-mail: rolf.willmann@gmail.com

N. Franz
Fisheries and Aquaculture Department, Food and Agriculture Organization of the United Nations (FAO), Rome, Italy
e-mail: Nicole.Franz@fao.org

C. Fuentevilla
Food and Agriculture Organization of the United Nations (FAO), Sub-Regional Office for the Caribbean, Bridgetown, Barbados
e-mail: carlos.fuentevilla@fao.org

T.F. McInerney
FAO Consultant, Rome, Italy
e-mail: tfmcinerney@me.com

L. Westlund
FAO Consultant, Rome, Italy

FAO Consultant, Stockholm, Sweden
e-mail: lena.m.westlund@telia.com



S. Jentoft et al. (eds.), *The Small-Scale Fisheries Guidelines*, MARE Publication Series 14, DOI 10.1007/978-3-319-55074-9_36

**Abstract** The Voluntary Guidelines for Securing Sustainable Small-Scale Fisheries in the Context of Food Security and Poverty Eradication (SSF Guidelines) call for realizing their six stated objectives through the promotion of a human rights-based approach (HRBA). This chapter will first present the foundations of such an approach and its specific guiding principles. It will then trace the evolution of the adoption of the HRBA in small-scale fisheries through the international community. The paper argues for the benefits of a HRBA in small-scale fisheries that is not confined to responsible fisheries management but also to furthering social development and decent work, gender equality and basic civil and political rights. The paper identifies some principal challenges in the implementation of a HRBA in small-scale fisheries and examines strategies and practical measures to overcome them.

**Keywords** Small-scale fisheries • Human rights-based approach • Human rights principles

## Introduction

The Voluntary Guidelines for Securing Sustainable Small-Scale Fisheries in the Context of Food Security and Poverty Eradication (SSF Guidelines) were endorsed by 147 FAO Members in the presence of a large number of civil society observers at the thirty-first session of the FAO Committee on Fisheries (COFI) in June 2014. They have been developed through a wide-ranging largely bottom-up consultation process and are the first international instrument dedicated entirely to the small-scale fisheries sector. The SSF Guidelines seek in the main to enhance the contribution of small-scale fisheries to food security and nutrition and support the realization of the right to adequate food; contribute to the equitable development of small-scale fishing communities and poverty eradication and improve the socio-economic situation of fishers and fishworkers; provide guidance for ecosystem-friendly and participatory policies, strategies and legal frameworks for responsible and sustainable small-scale fisheries; and enhance public awareness and promote the advancement of knowledge on the culture, role, and contribution of small-scale fisheries.

The guiding principles that underpin the SSF Guidelines are based on international human rights standards, the ecosystem approach to fisheries (EAF), responsible fisheries practices and standards, and sustainable development and other relevant instruments. These principles include: Human rights and dignity; respect of cultures; non-discrimination; gender equality and equity; equity and equality; consultation and participation; rule of law; transparency; accountability; economic, social and environmental sustainability; holistic and integrated approaches; social responsibility; feasibility; and social and economic viability (FAO 2015).

Vulnerability and marginalization of small-scale fishing communities are features that persist in spite of decade-old efforts to address them (Kurien 1995; FAO 2001; Béné 2003; Neiland and Béné 2004; Kurien and Willmann 2009; Chuenpagdee 2011; Jentoft and Eide 2011). Obstacles that have challenged these efforts include overexploitation of resources and threats to habitats and ecosystems, non-participatory and often centralized fisheries management regimes, unequal power relations and conflicts with large-scale fishing, and increasing interdependence or competition between small-scale fisheries and other sectors with stronger political or economic influence (FAO 2015).

This chapter seeks to provide guidance on the substantial new opportunities, as well as challenges that face the international community with the adoption of the SSF Guidelines, and the explicit recognition of human rights principles and the promotion of a human rights-based approach (HRBA) for small-scale fisheries. A better understanding of the concepts underlying the HRBA will help in the implementation of the principles and actions established by the SSF Guidelines.

## Legal and Conceptual Foundation of the Human Rights-Based Approach

A HRBA has been defined as a 'conceptual framework for the process of human development that is normatively based on international human rights standards and operationally directed to promoting and protecting human rights. It seeks to analyse inequalities which lie at the heart of development problems and redress discriminatory practices and unjust distributions of power that impede development progress.[1] The HRBA seeks to empower people to know and claim their rights and enhance the ability and accountability of duty-bearers of human rights. This means giving people greater opportunities to participate in shaping the decisions that impact on their lives and human rights. It also means increasing the ability of those with responsibility for fulfilling human rights to recognize and know how to respect those rights, and make sure they can be held to account (Table 36.1).

While human rights are held primarily by individuals, there are also collective human rights such as the right to self-determination and the rights of indigenous peoples. Article 1 of both the ICCPR and the ICESCR spell out the right to self-determination. Membership in a certain group such as ethnic and cultural minorities, such as indigenous peoples, for example, can give rise to collective rights for the protection of language, culture, and territory. This is reflected in the United Nations Declaration on the Rights of Indigenous Peoples (UNDRIP) (UN General Assembly 2007).

While international law and national legislation inspired by human rights are crucial pathways to promote and protect human rights, legislation is not a constitutive characteristic of human rights. Human rights should also be understood as ethical claims rather than just legal claims because their protection goes well beyond legislation and includes public recognition, advocacy, agitation and the monitoring of human rights violations. All of these can be done not just by state agents but by

[1] http://hrbaportal.org/ (accessed on 7 April 2016).

**Table 36.1** The core of international human rights law

| |
|---|
| The 1948 Universal Declaration of Human Rights (UDHR) laid the foundation for an international human rights system that now includes more than seventy adopted treaties and serves as a basis for many national constitutions and laws. The UDHR and its two implementing instruments, the International Covenant on Civil and Political Rights (ICCPR) and its two Optional Protocols and the International Covenant on Economic, Social and Cultural Rights (ICESCR) form together what is known as the International Bill of Rights which is commonly considered the main international reference point on human rights. |
| The civil and political rights of the ICCPR encompass rights to enjoy physical and spiritual freedom, fair treatment, and to participate meaningfully in the political process. They include the right to life and privacy, freedom from torture, slavery and arbitrary detention, the right to a fair trial, freedom of expression and religion, freedom of assembly, as well as the rights of minorities and freedom from discrimination. |
| The economic, social, and cultural rights of the ICESCR comprise rights to an adequate standard of living and health, rights to a fair wage and safe and healthy working conditions, the right to form and join trade unions, the right to education as well as the right to participate in cultural life), and freedom from discrimination in relation to the enjoyment of the Covenant's rights. |
| Other core UN human rights treaties address issues such as elimination of racial discrimination and discrimination against women, prohibitions against torture, and include conventions protecting the rights of children, migrants, and people with disabilities. |
| All human rights are considered to be universal, indivisible and interrelated (e.g. the Vienna Declaration and Programme of Action (UN General Assembly 1993, para. 5). As such, different types of human rights should not be seen to establish a ranking of the importance of one category of human right or one human right over another one. |

the public at large both nationally and internationally (Sen 2004). Human rights claims can be "addressed to all those who are in a position to help" (Sen 1999, 230). This should not, however, be seen to reduce the legal obligation on duty bearers, especially states. The duty of states entails a tripartite obligation, namely to respect, protect, and fulfil human rights. The obligation to respect requires states to refrain from interfering with the enjoyment of human rights. The obligation to protect requires states to prevent violations of such rights by third parties including business enterprises. The obligation to fulfil requires states to take appropriate legislative, administrative, budgetary, judicial, and other measures towards the full realization of such rights.

The emergence of the HRBA is the result of a convergence of the human rights and human development discourses both of which have been strongly influenced by economist, philosopher, and Nobel laureate Amartya Sen, especially his conceptual and philosophical underpinnings of the human development and related capability approach. The capability approach differs from other frameworks (for example the basic needs or neoliberal development approaches) by providing direct support for a broad characterization of fundamental freedoms and human rights that takes account of poverty, hunger and starvation as freedom-restricting conditions. Since wellbeing includes living with substantial freedoms, human development is integrally connected with enhancing certain functionings[2] and capabilities—the range of things a person can do and be in leading a life (Sen 1999; UNDP 2000; Fukuda-Parr 2003; Vizard 2005).

At the international level, the entry of development as a concept in human rights discourses occurred in the context of the movement seeking a new international economic order in the early 1970s. The call to add the right to development to the body of international human rights law came from developing countries in an attempt to underscore their demand for a better redistribution of resources and wealth between northern and southern countries (Uvin 2007).

In the 1990s, as increasing evidence showed the failure of structural adjustment policies to reduce and eventually eradicate poverty the international development community realized that alternative approaches were needed to more directly address poverty (World Bank 1990). New analyses and approaches were undertaken and developed during this period including UNDP's human development approach and its grounding in and complementarity with the human rights approach (Sen 1999; UNDP 2000).

A human rights perspective entered FAO's development discourse and practices through the process of developing and promoting the Voluntary Guidelines to Support the Progressive Realization of the Right to Adequate Food in the Context of National Food Security (Right to Food Guidelines) (FAO 2005a). Building on the human rights orientation of the Right to Food Guidelines and following a participative and wide-ranging consultation and negotiation approach, the 2012 Voluntary Guidelines on the Responsible Governance of Tenure of Land, Fisheries and Forests in the Context of National Food Security (Tenure Guidelines) and the 2014 SSF Guidelines broadened the human rights orientation in FAO's work. Both of these soft law instruments are based on human rights principles and explicitly (i.e. the SSF Guidelines) or implicitly (i.e. the Tenure Guidelines) promote a human rights-based approach for their implementation. Related work areas of FAO having human rights dimensions include nutrition, gender, livelihood and decent work, and indigenous peoples (Yeshanew 2014).

## Development of the SSF Guidelines as a Human Rights Oriented Instrument

During the first decade of the new millennium a human rights perspective emerged in respect to small-scale fisheries development. At the twenty-seventh session of COFI in March 2007, the FAO Secretariat tabled a paper on 'Social Issues in Small-Scale Fisheries' (FAO 2007). By that time, social aspects of small-scale fisheries including income and asset poverty, marginalization and vulnerability, exclusion

[2] In Sen's capability approach, functionings are the states and activities constitutive of a person's being, i.e. being healthy, working in a good job, having self-respect, and being happy. Capability entails the freedom to achieve valuable functionings (Sen 1999).

from decision-making, and others had become better known thanks to the publication of the FAO CCRF Technical Guidelines on Increasing the Contribution of Small-Scale Fisheries to Poverty Alleviation and Food Security (FAO 2005b) and a FAO technical paper on the same theme (Béné et al. 2007). The novelty of the 2007 COFI Secretariat paper was to discuss these issues in the context of human rights. Civil Society Organizations (CSOs) representing and supporting national fishworker organizations including the International Collective in Support of Fishworkers (ICSF), the World Forum of Fish Harvesters and Fish Workers (WFF), and the World Forum of Fisher Peoples (WFFP) supported the notion of a human rights-based approach in the sector, having themselves for many years used human rights language in their support of fishery sector workers (Allison et al. 2011). These three organizations had played and would continue to play a pivotal role in guiding the collaboration of civil society in the consultation and negotiation processes of the SSF Guidelines.

In 2008, FAO and the Royal Government of Thailand, in collaboration with the WorldFish Center and the Southeast Asian Fisheries Development Center (SEAFDEC), co-organized the first global multi-stakeholder conference focusing only on small-scale fisheries issues. Thematically and politically, the conference laid the ground work that would lead countries participating in COFI 2014 to embrace a human rights based approach in the SSF Guidelines. In her plenary presentation to the Conference, Chandrika Sharma, Executive Secretary of ICSF,[3] stated that "adopting a human rights approach for improving the life and livelihoods of fishing communities – and indeed all marginalized groups – was not really a matter of choice but an obligation (FAO 2009, 14; Sharma 2011)." Edward Allison, who helped draft a human rights perspective to small-scale fisheries in the FAO Secretariat document to COFI 2007, argued that the existing legal framework that supports the UDHR provides a potentially effective means of guiding investment and development action in securing sustainable small-scale fisheries. (FAO 2009, 15; Allison 2011; Allison et al. 2012).

A critical contribution to the global conference and the subsequent advocacy and lobbying by fishworkers' organizations around the world for the development of a human rights-based normative instrument on small-scale fisheries were a series of workshops organized by ICSF in Asia, Africa and Latin America on rights issues in small-scale fisheries. This also included the preparatory CSO workshop to the conference which had brought together more than hundred participants from 36 countries. In a statement to the global conference, participants of the preparatory workshop unanimously called upon FAO, Regional Fisheries Management Organizations, and national governments to secure access rights, post-harvest rights, and human rights of small-scale and indigenous fishing communities and include a specific chapter in the Code of Conduct for Responsible Fisheries (CCRF) on small-scale fisheries, recognizing the obligations of states towards them (ICSF 2007; FAO 2009; Sharma 2011).

[3] Chandrika Sharma tragically was aboard flight MH 370 that disappeared on 8 March 2014. By consensus, COFI Members dedicated the SSF Guidelines to her in respect of her tireless work for the betterment of the lives of fish workers all over the world, her deep care for people in general and for her invaluable contributions to the formulation of the SSF Guidelines.

The SSF Guidelines development process from 2010 onwards was based on wide-ranging consultations directly involving more than 4000 stakeholders from governments, CSOs, and the private sector in Africa, North and South America, Asia, the Near-East, the Caribbean, Europe and the South Pacific. In 2011–2012, the CSO consortium comprising ICSF, WFFP, WFF and the International Planning Committee (IPC) organized 20 national-level workshops spanning Asia, Africa, and Latin America, two regional workshops in Africa, as well as consultations among small-scale fishers and fishworkers in the European Union and Canada. More than 2300 people directly participated in these consultations, sharing their aspirations and making concrete proposals towards the content and principles of the SSF Guidelines (Sowman et al. 2012). Concurrently, and in partnership with ASEAN/SEAFDEC, BOBP-IGO, OSPESCA, NEPAD, Caribbean Regional Fisheries Mechanism (CRFM), and the Secretariat of the South Pacific Community (SPC), FAO convened six regional multi-stakeholder consultations and supported detailed national consultations in Cambodia and Malawi.

## Implementing a HRBA in Practice

A common understanding among UN agencies in 2003 on the human rights based approach to development cooperation resulted in a series of necessary and essential practices of a HRBA including that (i) people are recognized as key actors in their own development, rather than passive recipients of commodities and services; (ii) participation is both a means and a goal; (iii) development strategies are empowering, not disempowering; (iv) stakeholder analysis are inclusive; (v) development programs focus on marginalized, disadvantaged, and excluded groups, aim to reduce disparities and support accountability to all stakeholders; and (vi) the development process is locally owned and both top-down and bottom-up approaches are used in synergy (UN 2003).

Key features that differentiate conventional development programming and HRBA programming are summarized in Table 36.2.

A situation analysis based on HRBA must at a minimum address the following questions: (a) which rights of the individual or group are being violated and by whom?, (b) what are the immediate and underlying causes for rights violations and obstacles to right fulfilment?, (c) what are the views of the concerned people on rights and rights violations? are they aware of their rights and any violations and what are their priorities for action?, (d) who are the duty bearers responsible for upholding rights and preventing violations?, are they aware of their responsibilities and do they have the capacity to uphold them? (Harris-Curtis et al. 2005).

Especially in developing countries, resources constraints can limit the extent to which states are able to guarantee human rights such as the rights to housing, health, education and food. Resource constraints imply that the full realization of some human rights may have to occur over time in a progressive manner. This is recognized in Article 2 Paragraph 1 of the ICESCR which requires states to take steps, individually and through international assistance and co-operation, to the maximum

**Table 36.2** Conventional programming versus HRBA programming

| Conventional programming | HRBA programming |
|---|---|
| Successful development leads to respect for human rights | The realization of human rights is the central goal of development |
| Human rights activities are a distinct area of sectoral work | Human rights activities are an integral part of development |
| Respect for human rights is a useful tool to promote political stability and peaceful resolution of conflict | Development policies are guided by human rights both in terms of envisaged outcomes and the process of development |
| People cannot be developed. They must develop themselves | People have inherent rights |
| People, including the poorest, should be recognized as key actors in their own development rather than as passive beneficiaries | People, especially the poorest, should be empowered to recognize and claim their rights |
| Empowerment of stakeholders is important but not a strategy in itself: more a component of advocacy, capacity building, service delivery etc. | Empowerment of stakeholders is central. The role of outside agencies is to act as a support and catalyst for action as determined by the stakeholders |
| Role of stakeholder analysis is useful for social mobilization, program development and evaluation as it identifies accountability in the community and society | An analysis of the relationship between claim holders and duty bearers is essential for monitoring and accountability as well as to build capacity with the relevant (groups of) people |
| Programs should be developed on the basis of situation analysis that identifies problems and their immediate and underlying causes. These should be addressed either simultaneously or in sequence | Understanding causes at all levels: immediate, underlying and basic is essential. All causes must be addressed in respect to the indivisibility of human rights |

Source: Adapted from Jonsson (2003). Based on and adapted from Harris-Curtis et al. (2005, 22). The original by Jonsson can be found at: https://www.unicef.org/rightsresults/files/HRBDP_Urban_Jonsson_April_2003.pdf

of available resources. Strategies for the progressive realization of human rights should allow for the setting of priorities, time-bound targets, and benchmarks, and measures to address trade-offs. A HRBA imposes certain conditions on those features that the duty-bearers are required to respect. With regard to prioritization, a HRBA requires that this must involve effective participation of all stakeholders including the poor. With regard to trade-offs, a HRBA rules out any trade-offs which would result in or exacerbate unequal and discriminatory outcomes (OHCHR 2005).

Sunde and Sharma (2012) suggest that the unique location of small-scale fishing communities at the land-water interface and the 'common pool' nature of the resources they depend on results in a lower recognition of the spatial and resource components of their social identity. This in turn has caused the rights of fishers and their communities to their land and resource to go unrecognized. Thus they stress the requirement within the HRBA to increase awareness among fishers and fish-workers of the rights they are entitled to and the need to empower them to claim these rights through collective or other actions. Such mobilizations encourage fishers, fish workers and their communities to hold local and national governments accountable, realizing the full range of freedoms applicable to them as citizens and people (Sunde and Sharma 2012).

While the application of a HRBA might seem a daunting task, Allison et al. (2012), set out a practical strategy to integrate responsible fisheries, social development, and increased capacity for fishers to defend their fishing rights. Sharma (2011) presents policy and legislative measures on how to secure economic, social and cultural rights in fisheries, based on international fisheries instruments, including UNCLOS, UN Fish Stocks Agreement, and CCRF, as well as on national constitutions and laws. Ratner et al. (2014) prioritize, next to the application of a HRBA, the capacity to document, raise awareness of, and address specific human rights violations in fisheries, and human rights advocacy as a driver in fisheries sector reform.

## *Human Rights Principles in the SSF Guidelines*

The human rights-based focus of the SSF Guidelines is encapsulated in its guiding principles many of which are foundational for the responsible governance of fisheries, and other natural resources, as well as public affairs in general.

*Principle 1 Human Rights and Dignity* calls on all parties[4] to recognize, respect, promote, and protect human rights principles and their application to communities dependent on small-scale fisheries. It explicitly calls on states to respect and protect the rights of defenders of human rights in their work on small-scale fisheries. States are also asked to regulate the scope of activities in relation to small-scale fisheries of non-state actors to ensure their compliance with international human rights standards.

*Principle 2 Respect of Cultures* is of particular relevance to small-scale fishers and their communities, including indigenous peoples, and ethnic minorities which represent a vast assemblage of cultural traditions and practices that are fundamental for cultural identity and self-determination.

*Principle 3 Non-discrimination and Principle 5 Equity and Equality* are fundamental elements of international human rights law. The right to equality asserts that all persons are equal before the law. This requires that law is formulated in general terms applicable to every individual and that it be enforced in an equal manner. Further, all persons are entitled to equal protection under the law against arbitrary and discriminatory treatment. In order to guarantee non-discrimination and equality, it is necessary to look not only at the intention of legislation and policies but also at the effects they have in practice (Office of the UN High Commissioner on Human Rights (OHCHR 2005).

*Principle 4 Gender Equality and Equity* recognizes the vital role of women in small-scale fisheries. While the concept of gender, by definition, deals with both men and women, and boys and girls, and the socially, culturally and economically

[4] Parties involved/active/impacting small-scale fisheries.

established roles and relationships between them, women are often more disadvantaged than men. Gender equality efforts hence often mean supporting and empowering women whilst working with both men and women (Franz et al. 2015). When men and women enjoy equal rights and equitable benefits, poverty is reduced and development enhanced. In line with the statement by Landes (1999) that ‘the best clue to a nation's growth and development potential is the status and role of women’, it is imperative to ensure gender equality in order to secure sustainable small-scale fisheries. In agriculture, it is estimated that if women had the same access to productive resources as men, they could increase farm yields by 20–30%, which could reduce the number of hungry people in the world by 150 million people (FAO 2011). While specific data of this nature are not available in fisheries, it can be assumed that a similar gain in benefits could arise if there were greater resources available to women who work primarily in post-harvest fish processing, distribution and marketing.

The idea that women need to be afforded special attention and protection is found throughout the SSF Guidelines in respect to encouraging women leadership (Principle 2), eliminating discrimination against women in tenure rights (Articles 5.3 & 5.4), equitable participation of women in fisheries policy decision-making and management design, planning and implementation (Articles 5.15 & 8.2), preferential treatment of women in the provision of amenities and services such as savings, credit, and insurance schemes (Articles 6.2, 6.4 & 7.2), professional and organizational development including the ability of women to organize autonomously (Articles 6.5, 8.2 & 12.1), elimination of violence and protection of women exposed to violence (Article 6.9), comprehensive recognition of the role women play in the post-harvest subsector (Article 7.2), technology and capacity development (Articles 7.10 & 8.4) and recognition of knowledge held by women fishers and fishworkers (Article 11.6).

*Principle 6 Consultation and Participation* is not only consistent with but also demanded by a HRBA because the international human rights normative framework affirms the right to take part in the conduct of public affairs. It is not enough, for example, for small-scale fishers and fishworkers to participate in decision-making. They must be able to participate meaningfully and effectively. This may necessitate the promotion of a range of other human rights: they must be free to organize without restriction (right of association), to meet without impediment (right of assembly), to say what they want to without intimidation (freedom of expression), and to know the relevant facts (right to information). Furthermore, they must be allowed to receive support from sympathetic CSOs (including the media) that might be able to champion their cause. For this to be possible, states should create the necessary legal and institutional framework for an active independent civil society (OHCHR 2005, 14).

*Principle 7 Rule of Law, Principle 8 Transparency and Principle 9 Accountability* are the pillars of good governance at all levels and fundamental for a transparent rules-based approach to small-scale fisheries that makes duty-bearers of human rights – individuals, public agencies and non-state actors including business enterprises – answerable for their acts or omissions in relation to their duties. A rules-based approach requires that laws are applicable to all, equally enforced and independently adjudicated. Transparency requires that policies, laws, and procedures need to be clearly defined and widely publicized in applicable languages and accessible to all. An accountability procedure depends on monitoring but goes beyond it because it allows for explanations of conduct by duty-bearers and implies some form of remedy and reparation. The objective of monitoring is not just to allow a right-holder to hold a duty-bearer accountable but also to identify the areas on which duty-bearers may need to focus on in order to contribute to the realization of human rights in the most efficient and effective way (OHCHR 2005).

In order to ensure transparency and accountability, the SSF Guidelines note the importance of ascertaining which activities and operators are considered small-scale, and identifying vulnerable and marginalized groups needing greater attention. This should be undertaken at various levels (regional, sub-regional and national) and involve meaningful and substantive participation so that the voices of both men and women are heard (SSF Guidelines, Article 2.4). Further, the SSF Guidelines note the need for participatory monitoring in several instances (e.g. Articles 8.2, 13.4). Monitoring should use gender-sensitive approaches, indicators and data (Article 13.5).

Monitoring and accountability of the state can be strengthened by giving more power to parliaments, the decentralization and democratization of local-level governance, strengthening the legal framework to allow for independent monitoring through CSOs, better access to remedies and reparations for human rights violations and, where appropriate, punishment. States having ratified human rights treaties are also answerable to international treaty bodies and need to comply with established reporting, complaints, and inquiry procedures (OHCHR 2005).

Monitoring and accountability procedures should not be confined to states but also extend to global actor such as donors, intergovernmental organizations, international NGOs and transnational corporations (OHCHR 2005).

## *Indigenous Peoples*

Drawing on UNDRIP, the SSF Guidelines afford special recognition and attention to indigenous peoples not just in Principles 2 and 6 but also in a series of Articles (e.g. Articles 5.4; 5.5; 6.2; 9.2; 11.7). A higher standard of participation in decision-making is afforded to indigenous peoples through Article 19 of UNDRIP: "States shall consult and cooperate in good faith with the indigenous peoples concerned through their own representative institutions *in order to obtain their free, prior and informed consent* before adopting and implementing legislative or administrative measures that may affect them" (Italics by authors). In view of the importance of fisheries for many indigenous peoples, the SSF Guidelines reference UNDRIP in Principle 6, but the text falls short of applying the principle of free, prior and informed consent to all small-scale fishing communities. Instead, it requires the active, free, effective, meaningful, and informed participation of small-scale fishing communities, including indigenous peoples, taking into account UNDRIP in the decision-making process..., and taking power imbalances between different parties into consideration (FAO 2015). Power asymmetries are quite common where large-scale economic interests are concerned such as in mining, tourism and large-scale fisheries.

## Opportunities and Challenges in Adopting a HRBA in Small-Scale Fisheries

Above we have attempted to explain the specific meaning and features of implementing a HRBA in general and in small-scale fisheries specifically. In the following we examine some of the major opportunities and challenges to do so. As Jentoft (2014, 2) has argued '…the ultimate test [of the SSF Guidelines] is whether states will really 'walk the talk'.

As with other ideas for change, the foremost initial challenge is the necessary shift in mind sets, attitudes, and practices to transform common approaches applied today into a human rights-based approach. In the current climate, business as usual often entails a narrow focus on technological improvements in fishing, fish processing, and marketing to expand fish harvest and the valorization of catches in the value chain, alongside basic fisheries management and conservation that is often narrowly focused on a biological conception of sustainability.

This narrow approach, while also necessary, has not always resulted in a corresponding expansion of substantive human freedoms and the removal of deprivations in fishing communities. Development policies have rarely focused directly on the well-being and agency of fishing communities. Where fishers gained a political voice, it was most often as a result of them getting organized to battle injustice. The promotion of their civil and political rights and faculties to freely and effectively participate in decisions affecting their lives has rarely figured highly on government's development agendas.

Asserting a HRBA in small-scale fisheries requires a political process given that success will hinge on the extent of political power the fishing communities and their organizations are able to leverage. In democratic settings, political decision-makers are most responsive to demands from constituencies that can influence election outcomes.

While the SSF Guidelines have been negotiated and endorsed by a large number of countries, human rights concepts and language may be resisted because of the political or cultural traditions in a country or simply because powerful interests, for example at community level, might feel threatened by principles like gender equality or the preferential treatment of marginalized groups to achieve equitable outcomes.[5] In such contexts, a HRBA may need to be introduced in a gradual manner where initially only some of the principles could be applied and possibly only in a nuanced

[5] During the negotiations of the SSF Guidelines, a few delegations initially resisted the idea of giving preferential treatment to marginalized and vulnerable groups.

fashion. Some governments may also simply wish to continue their support to small-scale fisheries in the conventional way without reference to human rights.

Globally, the experience with a HRBA for development work is still limited even though the UN system and its specialized agencies have expressed a general commitment to it. The UN agency which has spear-headed the HRBA and can look back to an experience of over one decade is the United Nations Children's Fund (UNICEF). The findings of its first global evaluation are therefore of particular interest to our discussion in this chapter (UNICEF 2012). The objective of the evaluation was to examine whether there is adequate understanding of and commitment to HRBA throughout the organization and whether there is a proper enabling environment within the organization and within countries. Salient findings include that UNICEF staff's understanding of HRBA varies considerably and that more systematic training was needed. At the global level, UNICEF was able to integrate HRBA into the programming of various humanitarian and emergency frameworks. At the country level, there was considerable variation in the application of HRBA principles. For example, the principle of participation was applied in a mixed fashion because of the lack of a common understanding of this principle within the organization and due to external political and cultural constraint.[6] The application of non-discrimination was found to range from satisfactory to weak, with a lack of disaggregated data making it difficult to identify and thus target the most vulnerable and marginalized. The application of transparency was similarly between satisfactory and weak, reflecting positive efforts by UNICEF country offices in promoting the transparency of duty bearers but their lower level of success at ensuring transparency among rights holders. The application of the principle of accountability was constrained because of the lack of documentation on accountability mechanisms and of systems of complaint or redress within government or UNICEF programs.

## *Prioritization of Marginalized and Vulnerable Groups and Individuals*

The prioritization of support to meet the needs of those who are marginalized and vulnerable is a basic tenet of the HRBA. In poor countries, these may include most people living in small-scale fishing communities. The elders, women, children, disabled persons, orphans, and those who are most deprived of economic resources and social services are almost always among the marginalized and vulnerable. They often also include ethnic minorities, indigenous peoples, and migrant fishers. The challenge is to identify them, seek their views in the design and prioritization of development interventions, and provide them with access to legal recourse to address human rights violations. Remedying rights violations can occur through

[6] The principle of participation can mean different things to different people ranging from actively manipulating those whose participation is sought to participation as a human right in a process where those who participate can contribute to the decision-making process (Pretty 1994).

self-help means but having in place effective recourse/grievance mechanisms both judicial and non-judicial is an important factor in the efficacy of such efforts.

There are a wide range of techniques that have been applied in different contexts to help advance protection and provide opportunities to communities. For example, legal aid refers to a variety of approaches that provide legal assistance to poor and vulnerable groups. Assistance can be provided by the government, public interest law firms or NGOs, as pro bono services by private lawyers or by students, law clinics, or paralegals. A crucial component of legal assistance is to provide information on legal matters through channels and means that reach marginalized and vulnerable groups. Often, when confronted with issues concerning their livelihoods or rights, marginalized and vulnerable groups lack clarity on whether the issues are in fact legal and if so whether they can be resolved through legal means (McInerney 2013).

Vulnerable and marginalized groups must be included in monitoring and evaluation frameworks for development to ensure that their rights have indeed been strengthened and further marginalization has not occurred.

## *The Right to Food*

The first listed objective of the SSF Guidelines is to enhance the contribution of small-scale fisheries to global food security and nutrition and to support the progressive realization of the right to adequate food. Small-scale fisheries contribute about half of global fish catches and two-thirds of the fish destined for direct human consumption. Nearly the entire catch by small-scale inland fisheries is directed to human consumption (FAO 2015).

An added positive factor is the decentralized nature of the fish supplied through small-scale fisheries because of their geographically spread-out production structure. The distances between landing points and the points of final consumption are usually short, which lowers distribution costs. Fish produced by small-scale fisheries are, arguably more available and affordable to poorer consumers.

This availability is particularly important in countries where the staple crop is low in protein. This is the case of cassava and plantain. In such situations, as in many parts of Africa, a larger proportion of foods such as fish that are rich in proteins and fat may be essential for a healthy, robust population. This is especially true for the diets of young children, infants and pregnant women given the crucial role of fish in physical development (Kurien 2005).

A 2014 report on the contribution of fish to food security and nutrition from the High Level Panel of Experts on Food Security and Nutrition of the Committee of World Food Security noted that 'limited attention has been given so far to fish as a key element in food security and nutrition strategies at national level and in wider development discussions and interventions (HLPE 2014).' The SSF Guidelines are a formidable means to address this challenge of national and international neglect to recognize the critical role of fish and especially small-scale fisheries in food security and nutrition. They list a number of measures including secure tenure arrangements, sustainable resources management, good post-harvest practices, as well as avoiding the promotion of international fish trade and exports that would "… adversely affect the nutritional needs of people for whom fish is critical to a nutritious diet, their health and well-being, and for whom other comparable sources of food are not readily available or affordable" (FAO 2015, Art. 7.7).

In his October 2012 report to the General Assembly, the UN Special Rapporteur on the Right to Food, Olivier de Schutter (2008–2014) called upon countries and fisheries stakeholders to guide their actions on fisheries through a human rights prism, noting that a human rights approach to fisheries governance and policy is critical to achieve sustainable development in the sector, thus fulfilling its potential contribution to the realization of the right to food (De Schutter 2012).

In this context, the application of the SSF Guidelines through a HRBA to fisheries is a key step for the sector to fully contribute to the implementation of the Right to Food Guidelines which "aim to guarantee the availability of food in quantity and quality sufficient to satisfy the dietary needs of individuals; physical and economic accessibility for everyone, including vulnerable groups, to adequate food, free from unsafe substances and acceptable within a given culture; or the means of its procurement" (FAO 2005a).

The right to food is recognized in several international instruments.[7] Article 11 of ICESCR provides the most comprehensive formulation of the right to food as part of an adequate standard of living, and recognizes the fundamental right to be free from hunger (Skonhoft and Gobena 2009).

Facilitating the fulfilment of the right to food will require more far-reaching measures by states because vulnerable populations have to be actively identified and policies and programs have to be implemented to improve these people's access to food and their capacity to feed themselves. The obligation to fulfil also includes the obligation to ensure, as a minimum, that no one in a country suffers from hunger. The CESCR has considered that the obligation to fulfil also incorporates an obligation to promote human rights among its own agencies and private players (CESCR 2000, 2002; Skonhoft and Gobena 2009).

## *Natural Resources Access and Tenure*

One of the major infringements on the economic, social and cultural rights of small-scale fishers including their right to food is their gradual loss of access to traditional fishing grounds and fishery resources because of encroachments by large-scale

[7] Article 25 of the Universal Declaration of Human Rights protects the right to an adequate standard of living, including food. The Convention on the Rights of the Child (Art. 27(1)) and the Convention on the Elimination of Discrimination Against Women (Art. 12(2)) oblige states to combat child malnutrition and to ensure adequate nutrition for women during pregnancy and lactation, respectively. The International Convention on Civil and Political Rights (ICCPR) also outlaws deprivation of food and of means of subsistence in Article 1(2) (Skonhoft and Gobena 2009).

fishing fleets both domestic and foreign. Absurdly, many industrial fleets continue to receive annually billions of USD in subsidies (Schuhbauer and Sumaila 2016). Other reasons for loss of fishing opportunities include tourism development, coastal infrastructure, urban development, mining, energy generation, and others. The deterioration of the environmental quality of inland water bodies and oceans because of damming, pollution, acidification, damage to physical habitat, amongst other reasons has also infringed on the ability of fishers to make a living from fishing and exercise their cultural practices, a right of special importance to many indigenous peoples.

Infringements with the rights of fishing communities also happen because they commonly live on land that is customarily theirs but whose legal ownership is either with the state, or not well defined. They can, and often are, evicted at short notice and without adequate compensation. Around the world, small-scale fishing communities do not have secure tenure rights to either their fishing grounds, fishery resources or the lands on which they reside and privately use for processing or complementary activities such as livestock keeping, and cultivation and for their access to the sea or inland water bodies.

Common current practices ignore or outrightly deny traditional and customary tenure rights of fishing communities. When large-scale fishing vessels operate illegally in inshore waters, these are considered as infringements of, for example, zoning regulations where these exist, but not as violations of basic economic, social and cultural rights of fishing communities. Consequently, fines for such regulatory trespasses are minimal and rarely deter large-scale operators from continuing their illegal actions.

The SSF Guidelines call on all parties to recognize that responsible governance of tenure is "…central for the realization of human rights, food security, poverty eradication, sustainable livelihoods, social stability, housing security, economic growth and rural and social development (FAO 2015, paragraph 5.2)." States should ensure in accordance with their legislation that small-scale fishers, fish workers, and their communities have secure, equitable, and socially and culturally appropriate tenure rights not just to fishery resources in marine and inland waters but also to small-scale fishing grounds and adjacent land, with special attention paid to women with respect to tenure rights.

The SSF Guidelines list specific activities states and other parties should take to improve tenure arrangements for the benefit of small-scale fishers, fishworkers and their communities. These include the provision of legislation to secure customary tenure rights, granting SSF fishers preferential access to fishery resources, recognition of the role of SSF communities in the restoration, conservation and management of local aquatic and coastal ecosystem, redistributive reforms in favor of small-scale fishers in line with the Tenure Guidelines, avoidance of arbitrary evictions of small-scale fishing communities, and consultations with small-scale fishing communities alongside proper environmental impact assessments prior to large-scale developments that would affect them.

Moreover, the health of aquatic ecosystems and associated biodiversity are a fundamental basis for livelihoods and for the subsector's capacity to contribute to overall well-being (FAO 2015). As such, the granting of access rights within the context of human rights still requires proper management. As Charles (2011, 87) clearly explains, "while human rights are 'universal', a human rights perspective does not imply "universal" access to fisheries, and unlimited exploitation." Keeping fishery resources exploitation within sustainable limits is necessary for long-term food security and poverty alleviation and recognizes the rights of both present and future generations (Allison et al. 2012).

## *Decent Work*

People in poverty invariably lack adequate and secure livelihoods because of unemployment, underemployment, unreliable casual labor, poor wages, and unsafe working conditions. In rural areas (including small-scale fisheries) livelihoods are made precarious by multiple factors such as inadequate access to natural resources, deficient marketing and poor transportation facilities. Many people living in poverty are drawn into work that is dangerous or illegal such as bonded and forced labor and other slavery-like practices. On occasions they may become entrapped by human traffickers (OHCHR 2005). Children may work in small-scale fisheries in violation of international labor and human rights standards. Examples of such work can be found around the globe (FAO 2010; Mathew 2010). Occupational health and safety standards are also often poor within the small-scale sector.

Work as specified in international human rights law must be *decent work*, i.e. work in which human rights and the rights of workers, in terms of work safety and remuneration, are protected. The right to decent work is enshrined in a large number of international human rights and ILO labor conventions[8]. Article 6.6 of the SSF Guidelines calls on states to "promote decent work for all small-scale fisheries workers, including [those involved in] both the formal and informal sectors (FAO 2015, 15)." Access to decent work is instrumental in reducing poverty and in securing other rights such as the right to food, health and housing.

In the context of small-scale fisheries, it is important to note that the right to work encompasses self-employment, working at home, and other income-generating activities and thus is not confined to wage employment. Decent work is promoted through an enabling social, economic, and physical environment in which all people have fair and equal opportunities to make a living through their own endeavors and in a manner that is consistent with their dignity (OHCHR 2005, 24).

Small-scale fishing and other disadvantaged communities, mostly in least developed countries, increasingly serve as a reservoir of cheap labor supply for various economic sectors including the large-scale long distance fishing fleets and industrial fish processing of more prosperous countries. Workers from these communities are often recruited through specialized recruitment agencies which are frequently the

[8] Important ILO Conventions include the Declaration on Fundamental Principles and Rights at Work, and Conventions No. 138 on Minimum Age, No. 182 on Child Labour, No. 29 on Forced Labour, No. 105 on the Abolition of Forced Labour, No. 107 The Work in Fishing Convention, and No 155 Occupational Health and Safety.

first actors in a series of human rights violators all the way through to the ultimate employer. Forced labor, child labor, slave labor and human trafficking in fisheries have made headlines in recent years, primarily in relation to large-scale fishing and fish processing plants.[9] Investigative journalists rather than law enforcement agencies[10] have been responsible for bringing international attention to these violations.

Within the UN, the ILO is at the forefront of addressing forced labor and child labor. In fisheries specifically, ILO activities against forced labor include, inter alia, support for legal and policy reform, training of inspectors and promoting the ratification and implementation of the 2007 ILO Work in Fishing Convention. FAO and ILO are working to address child labor through various measures including the preparation and dissemination of guidance materials and assistance to countries in the elaboration and implementation of national programs of action against child labor, in particular the worst forms of child labor (FAO and ILO 2013; FAO 2010; Mathew 2010).

The SSF guidelines call on states to address occupational health and safety issues and unfair working conditions of all small-scale fishers and fish workers and eradicate forced labor, prevent debt-bondage of women, men and children, and adopt effective measures to protect fishers and fish workers, including migrants. They also highlight the importance of schooling and education to facilitate gainful and decent employment for youth (FAO 2015). Also important in the context of small-scale fisheries are social security and protection mechanisms in times when regular employment becomes unavailable because of economic, political crisis, and disasters. These are covered, for example, in article 6.3 of the SSF Guidelines. Social protection is widely recognized as an important means to reduce risks and adversaries among marginalized and vulnerable people including fishing communities and address poverty and food insecurity. Conditional cash transfers, for example, that require parents to send their children to school not only positively influence education levels but also contribute to reducing child labor. Social protection measures are also relevant to bridge times of reduced incomes caused by certain fisheries management measures such as seasonal and spatial closures and other restrictions on fishing.

## *Adequate Housing*

Poor housing conditions are constitutive of poverty and deprivation. Small-scale fishing communities are often characterized by precarious physical shelters, overcrowding, absence or inadequacies of infrastructure such as safe drinking water,

[9] Ratner et al. (2014) provide a review of case law and other documentation of human rights issues in fisheries including child labour, forced labour and unsafe working conditions.

[10] http://www.ilo.org/global/about-the-ilo/newsroom/features/WCMS_429031/lang--en/index.htm Accessed on 8 February 2016.

sanitation, power and decent access roads. Their habitats are also exposed to sea level rise and natural disasters resulting in insecurity of person and property. The right of adequate housing is critical for the enjoyment of other rights such as the right to health.

There are a number of actions that can be taken in support of the realization of the right to adequate housing including the (i) promotion of low-income and low cost housing programs including through formation of community-based and self-built housing schemes, subsidies, access to cheap credit, and others; (ii) strengthening and ensuring tenure security by prohibiting forced evictions, facilitating the conferring of tenure titles and others; and (iii) provision of infrastructure (e.g., roads, water and sanitation systems, drainage and lighting) for small-scale fishing communities (OHCHR n.d).

## *Disaster Risks and Climate Change*

Vulnerability of small-scale fishing communities is increasingly a function of susceptibility to disasters and climate change. Section 9 of the SSF Guidelines references the need for states to address disaster risks and climate change among small-scale fisheries, promote the livelihoods of small-scale fishing communities, and provide compensation to them when they are affected by disaster. Often individuals and communities that experience disasters have diminished rights and opportunities. It is thus important to both prevent harms associated with disaster, including those associated with climate change, and resolve the consequences of those disasters in ways that further their rights and increase their wellbeing.

## *Fishing in Foreign Waters*

During recent decades, small-scale fisheries have seen significant technological progress. This has greatly expanded their operational range and their ability to explore fishing opportunities in distant waters. Occasionally and at times inadvertently, this brings them into the waters of neighboring countries. As a consequence, there are a growing number of incidents where small-scale fishers are arrested and placed in prison, sometimes in contravention of international fisheries, maritime, and human rights law. A HRBA to this issue would require that foreign small-scale fishers are given due process and repatriated as expeditiously as possible. Neighboring coastal states should also be encouraged to conclude bilateral agreements and arrangements to address cross-border issues among their fishers in order to reduce hardship and improve joint fisheries management actions.

## *Organizational Development and Empowerment*

Fishers and their organizations must be at the forefront to mobilize and promote the respect, protection, and fulfilment of their rights as citizens and primary actors in the fisheries sector. Organizational and legal empowerment and capacity development are among the principal challenges that fishworker organizations and their supporters face in their struggle to realize human rights. Important aspects of organizational development include the promotion of strong and accountable leadership, transparent and comprehensive information flows between the leaders and their constituencies, as well as with relevant government agencies. Special attention is also required towards gender equality and equity in the functioning of the organizations to ensure that the voices of women are heard in decision-making. There are instances where distinct women's organizations may best ensure gender equality and equity.[11]

Institutional structures that allow small-scale fishing communities to take part in decision-making and policy processes, allowing them to fight their marginalization and influence their own development are an important ingredient towards the realization of human rights. A key strategy for implementing the SSF Guidelines and achieving successful outcomes in the context of human rights and equitable and sustainable development would hence be through a focus on empowerment through collective action.

Community-based organizational structures are an important building block for the effective co-management that is critical for securing sustainable development of small-scale fisheries. Fishers' organizations, both formal and informal, provide a platform through which small-scale fishers and fish workers exercise their right to organize, participate in development and decision-making processes and influence fisheries management outcomes (Jentoft 1989).

Fishworkers' organizations and support CSOs need to create alliances and solicit assistance from legal practitioners and human rights organizations to ensure that existing laws are used to empower communities to recognize and enjoy their rights. National human rights commissions exist in many countries but few have links with fishery sector actors. Specific human rights violations can be addressed through such commissions as well as by working with human rights NGOs, legal clinics, and pro bono lawyers to hold governments (and other actors) accountable and seek full compliance with obligations under national and international human rights law.

## *Policy Coherence*

The SSF Guidelines relate to a range of different development concerns ranging from environmental, labor, human rights, natural resource, and maritime affairs. In each of these fields, governments may undertake legislative or regulatory activities.

[11] On strategies to strengthen fishworkers' organizations and support CSOs, see ICSF (2014) and Kalikoski and Franz (2014).

Such efforts provide an opportunity to ensure that small-scale fisheries concerns are factored into any reform process. It is thus important to ensure that information is gathered horizontally on activities occurring outside of the small-scale fisheries context. Policy coherence and consistent and inclusive legal frameworks should be promoted taking the specific situation of small-scale fisheries into consideration.

The SSF Guidelines build on and hence overlap in some aspects with other existing international instruments and programs. This makes it desirable to build synergies between efforts to promote the SSF Guidelines and other initiatives. The Right to Food Guidelines and the Tenure Guidelines are obvious links (FAO 2005a, 2012), but many other complementary efforts exist, not least the SDGs (UN General Assembly 2015).

## Conclusions

The embracing of a HRBA in fisheries by FAO Members through their endorsement of the SSF Guidelines signals a fundamental re-orientation of efforts to promote sustainable small-scale fisheries development. The focus of this approach is on enhancing the capabilities, functionings, and agency of fishers, fishworkers, and their families and communities and empowering them to demand respect, protection and fulfilment of their human rights.

This shift in emphasis has been demanded by CSOs for some time and acknowledged by governments through the adoption of the SSF Guidelines. Now, when governments seek to introduce or make changes to a fishing rights regime they should examine the full array of human rights that might impact on the manner in which such a regime is being implemented: Are fishers and fishworkers being adequately consulted? Can they freely express their views without fear and shame? Are they able to assemble and organize to influence decisions that affect their lives and wellbeing? Have the most vulnerable fishers and communities, including minorities and indigenous peoples been identified and consulted? Is information being provided in forms and language understandable to them? Do fishers have the capacity and capability to deal with the administrative requirements to file their applications for fishing rights? Is there a mechanism in place to deal with grievances and are there options for recourse to appeals?

Apart from securing just and equitable tenure rights, there are several other areas of fundamental importance to the human development needs of small-scale fishing communities including health, education, housing, water, personal safety, and others. The conventional development approach talks about service delivery in catering to these basic needs. The HRBA considers these tasks as respect, protection and fulfilment of human rights, especially where vulnerable and marginalized people are concerned. It will ask us to consider the following: Who are the duty-bearers of these rights? How can their capacity and willingness to act be enhanced and accountability established? How can the protection and fulfilment of human rights be monitored, failings identified, and sanctions or remedies exercised?

A key concept that keeps reappearing when talking about promoting a HRBA in small-scale fisheries in order to achieve environmentally, economically, and socially sustainable development is empowerment and the inclusion of the marginalized and vulnerable – i.e. those that are in the most need of being empowered. Empowerment is a big word, encompassing a variety of circumstances and needs, which requires an enabling environment and favorable policy frameworks. Key aspects to support empowerment include organizational development – especially in the form of collective action – and access to legal systems.

The need for empowerment may not come as a surprise to the reader of this volume and she or he will be keenly aware of the obstacles that still exits. While anthropology, sociology, and other 'people' focused social sciences are increasingly recognized in fisheries policy making, more research is required on how to ensure the realization of human rights for small-scale fishing communities. Maybe more importantly, efforts to ensure that research results enter policy discourse should be strengthened.

There are important ongoing changes with regard to policy. These take place both at international and national scales. In the international arena the three international instruments – the Right to Food Guidelines, the Tenure Guidelines, and the SSF Guidelines – represent a shift in focus to an overall more holistic approach centered around human rights.

The authors strongly believe that advocacy, mobilization, and organizational development for the effective implementation of the SSF guidelines – grounded in the HRBA to development – is our best bet to ensure that small-scale fisheries are secured and sustained in the future. This entails greater economic, social, and cultural benefits to coastal and inland fishers and fishing communities at large but especially marginalized and vulnerable groups.

**Acknowledgments** The authors wish to thank Svein Jentoft, UiT – The Arctic University of Norway, Margret Vidar and Sisay Yeshanew, FAO Legal Office, and two anonymous referees for their valuable comments on an earlier draft of this chapter. Any remaining errors are the sole responsibility of the authors.

## References

Allison, E. H., Åsgård, B., & Willmann, R. (2011). Human rights approaches to governing fisheries. *Maritime Studies, 10*(2), 5–13.

Allison, E.H. (2011). Should states and international organizations adopt a human rights approach to fisheries policy? *Maritime Studies*, 10(2), 95-116.

Allison, E. H., Ratner, B. D., Åsgård, B., Willmann, R., Pomeroy, R., & Kurien, J. (2012). Rights-based fisheries governance: From fishing rights to human rights. *Fish and Fisheries, 13*, 14–29.

Béné, C. (2003). When fishery rhymes with poverty: A first step beyond the old paradigm on poverty in small-scale fisheries. *World Development, 31*(6), 949–975.

Béné, C., Macfadyen, G., & Allison, E. H. (2007). *Increasing the contribution of small-scale fisheries to poverty alleviation and food security* (Fisheries technical paper no. 481). Rome: Food and Agriculture Organization of the United Nations.

CESCR [UN Committee on Economic, Social and Cultural Rights]. (2000, August 11). *The right to the highest attainable standard of health (Art. 12)* (UN doc. E/C.12/2000/4). General Comment 14. Geneva, Switzerland.

CESCR. (2002). *The right to water (Arts.11 and 12)*. 20/01/2003 (UN doc. E/C.12/2002/11). General Comment 15. Geneva, Switzerland.

Charles, A. (2011). Small-scale fisheries: On rights, trade and subsidies. *Maritime Studies, 10*(2), 85–94.

Chuenpagdee, R. (Ed.). (2011). *World small-scale fisheries: Contemporary visions*. Delft: Eburon Academic Publishers.

De Schutter, O. (2012). *Fisheries and the right to food*. Report presented at the 67th Session of the United Nations General Assembly [A/67/268].

FAO. (2001). *Poverty in coastal fishing communities*. Report of the third session of the Advisory Committee on Fisheries Research (Fisheries report no. 639). Rome: Food and Agriculture Organization of the United Nations.

FAO. (2005a). *Voluntary Guidelines to support the progressive realization of the right to food in the context of national food security*. Rome: Food and Agriculture Organization of the United Nations.

FAO. (2005b). *Increasing the contribution of small-scale fisheries to poverty alleviation and food security, Technical guidelines for responsible fisheries no. 10*. Rome: Food and Agriculture Organization of the United Nations.

FAO. (2007). *Report of the twenty-seventh session of the Committee on Fisheries* (Fisheries and aquaculture report no. 830). Rome: Food and Agriculture Organization of the United Nations.

FAO. (2009). *Report of the Global conference on small-scale fisheries. Securing sustainable small-scale fisheries: Bringing together responsible fisheries and social development, Bangkok Thailand (2008)*. Rome: Food and Agriculture Organization of the United Nations.

FAO. (2010). *Report of the workshop on child labour in fisheries and aquaculture in cooperation with ILO* (Fisheries and aquaculture report no. 944). Rome: Food and Agriculture Organization of the United Nations.

FAO. (2011). *Women in agriculture. Closing the gender gap in development*. State of Food and Agriculture 2010–2011. Rome: Food and Agriculture Organization of the United Nations.

FAO. (2012). *Voluntary guidelines on the responsible governance of tenure of land, fisheries and forests in the context of national food security*. Rome: Food and Agriculture Organization of the United Nations.

FAO. (2015). *Voluntary guidelines for securing sustainable small-scale fisheries in the context of food security and poverty eradication*. Rome: Food and Agriculture Organization of the United Nations.

FAO, & ILO. (2013). *Guidance on addressing child labour in fisheries and aquaculture*. Rome/Geneva: Food and Agriculture Organization of the United Nations/International Labour Organization.

Franz, N., Fuentevilla, C., Westlund, L., & Willmann, R. (2015). A human rights-based approach to securing livelihoods depending on inland fisheries. In J. F. Craig (Ed.), *Freshwater fisheries ecology* (pp. 513–523). Chichester: Wiley & Sons.

Fukuda-Parr, S. (2003). The human development paradigm: Operationalizing Sen's ideas on capabilities. *Feminist Economics, 9*(2–3), 301–317.

Harris-Curtis, E., Marleyn, O., Bakewell, O. (2005). *The implications for northern NGOs of adopting rights-based approaches* (Occasional papers series, 41). International NGO Training Centre.

HLPE – High Level Panel of Experts on Food Security and Nutrition. (2014). *Sustainable fisheries and aquaculture for food security and nutrition*. Rome: Committee on World Food Security (CFS).

ICSF. (2007). *Asserting rights, defining responsibilities: Perspectives from small-scale fishing communities on coastal and fisheries management in Asia*. Reports the proceedings of the Asian workshop and symposium held in Siem Reap, Cambodia, Chennai.

ICSF. (2014, 21–24 July). *Report of the international workshop towards socially just and sustainable fisheries: Implementing the FAO Voluntary guidelines for securing sustainable small-scale fisheries in the context of food security and poverty eradication*, Puducherry, India.
Jentoft, S. (1989). Fisheries co-management – Delegating government responsibility to Fishermen's organisations. *Marine Policy, 13*(2), 137–154.
Jentoft, S. (2014). Walking the talk: Implementing the international voluntary guidelines for securing sustainable small-scale fisheries. *Maritime Studies, 13*(16), 1–15.
Jentoft, S., & Eide, A. (2011). *Poverty mosaics: realities and prospects in small-scale fisheries*. Dordrecht: Springer.
Jonsson, U. (2003). *Human rights approach to development programming*. UNICEF Eastern and Southern Africa Regional Office. https://www.unicef.org/rightsresults/files/HRBDP_Urban_Jonsson_April_2003.pdf. Accessed 20 Jan 2016.
Kalikoski, D., & Franz, N. (2014). *Strengthening organizations and collective action in fisheries – A way forward in implementing the international guidelines for securing sustainable small-scale fisheries* (Fisheries and aquaculture proceedings no. 32). Rome: Food and Agriculture Organization of the United Nations.
Kurien, J. (1995). The Kerala model: It's central tendency and the outlier. *Social Scientist, 23*, 1–3.
Kurien, J. (2005). *Responsible fish trade and food security* (FAO Fisheries technical paper no. 456). Rome: Food and Agriculture Organization of the United Nations.
Kurien, J., & Willmann, R. (2009). Special considerations for small-scale fisheries management in developing countries. In K. Cochrane & S. Garcia (Eds.), *A fishery manager's guidebook* (pp. 404–424). Oxford: Wiley-Blackwell.
Landes, D. S. (1999). *The wealth and poverty of nations: Why some are so rich and some so poor*. London: W.W. Norton.
Mathew, S. (2010). *Children's work and labour in fisheries: A note on principles and criteria for employing children and policies and action for progressively eliminating the worst forms of child labour in fisheries and aquaculture*. Background paper prepared for the Workshop on Child Labour in Fisheries and Aquaculture in cooperation with ILO, Rome.
McInerney, T. F. (2013). *Report on addressing human rights and legal empowerment in small scale fisheries*. Rome: Prepared for the UN Food and Agriculture Organization.
Neiland, A. E., & Bene, C. (2004). *Poverty and small-scale fisheries in West Africa*. Dordrecht: Springer.
OHCHR. (2005). *Frequently asked questions on a human rights-based approach to development cooperation*. Geneva: Office of the UN High Commissioner on Human Rights.
OHCHR. (n.d.). *Principles and guidelines for a human rights approach for poverty reduction strategies*. Geneva: Office of the UN High Commissioner on Human Rights.
Pretty, J. N. (1994). Alternative systems of inquiry for sustainable agriculture. *IDS Bulletin, 25*(2), 37–42.
Ratner, B. D., Asgard, B., & Allison, E. H. (2014). Fishing for justice: Human rights, development, and fisheries sector reform. *Global Environmental Change, 27*, 120–130.
Schuhbauer, A., & Sumaila, A. R. (2016). Economic viability and small-scale fisheries – A review. *Ecological Economics, 124*, 69–75.
Sen, A. K. (1999). *Development as Freedom*. Oxford: Oxford University Press.
Sen, A. K. (2004). Elements of a theory of human rights. *Philosophy and Public Affairs, 32*(4), 315–356.
Sharma, C. (2011). Securing economic, social and cultural rights of small-scale and artisanal fisher workers and fishing communities. *Maritime Studies, 10*(2), 41–62.
Skonhoft, A., & Gobena, A. (2009). *Fisheries and the right to food. Implementing the right to food in national fisheries legislation*. Rome: Food and Agriculture Organization of the United Nations.
Sowman, M., Sunde, J., Raemaekers, S., Mbatha, P., & Hara, C. (2012). *Towards international guidelines for sustainable small-scale fisheries*. Submission from Civil Society Organisations

to the FAO Consultative process on Small-scale Fisheries. Synthesis Document. Report developed for the CSO Co-ordinating Committee.
Sunde, J., & Sharma, C. (2012). Recognizing a rights-based approach to development in fisheries: Struggles of small-scale fishing communities to secure their human rights. In S. M. Suárez (Ed.), *The human rights framework in contemporary agrarian struggles* (pp. 239–290). *The Journal of Peasant Studies, 40*(1).
UN. (2003). *Statement on a common understanding of a human rights-based approach to development cooperation*. United Nations.
UN General Assembly. (1993, July 12). *Vienna declaration and programme of action* (A/CONF.157/23).
UN General Assembly. (2007). *United Nations declaration on the rights of indigenous peoples* (UNGA A/61/L.67 and Add.1).
UN General Assembly. (2015). *Transforming our world: The 2030 Agenda for Sustainable Development* (A/RES/70/1).
UNDP. (2000). *Human development report. Human rights and human development.* Oxford: Oxford University Press.
UNICEF. (2012). *Global evaluation of the application of the human rights-based approach to UNICEF programming, Final report, Volume I.* New York: United Nations Children's Fund.
Uvin, P. (2007). From the right to development to the rights-based approach: How 'human rights' entered development. *Development in Practice, 17*(4-5), 597–606.
Vizard, P. (2005). *The contributions of Professor Amartya Sen in the Field of Human Rights* (CASE Paper 91). Centre for Analysis of Social Exclusion. London School of Economics.
World Bank. (1990). *World development report 1990 – Poverty*. Oxford: Oxford University Press.
Yeshanew, S. A. (2014). Mainstreaming human rights in development programmes and projects: Experience from the work of a United Nations agency. *Nordic Journal of Human Rights, 32*(4), 372–386.

# Chapter 37
# From Rhetoric to Reality: Implementing the Voluntary Guidelines for Securing Sustainable Small-Scale Fisheries

**Svein Jentoft and Ratana Chuenpagdee**

**Abstract** In a historic decision in 2014, FAO member states endorsed the Voluntary Guidelines for Securing Sustainable Small-Scale Fisheries in the Context of Food Security and Poverty Eradication (SSF Guidelines). All around the world, initiatives to implement the SSF Guidelines are currently underway. In some instances, concrete measures have been taken. In other instances, the SSF Guidelines have yet to receive the attention. In some countries, the SSF Guidelines are brought in to support a policy process that is already on-going; in others they clearly contradict the current paradigms that dominate fisheries policies. In all instances described in this volume, interesting lessons can be drawn about the challenges and opportunities of implementation. This chapter summarizes what has happened so far with the SSF Guidelines and their implementation, according to the authors of this book. How are they received and perceived, and what challenges do they meet? Given the far-reaching change that the SSF Guidelines promote, this chapter discusses the conditions and opportunities for a successful implementation based on what the authors of the chapters of this volume have experienced and would predict.

**Keywords** FAO • Small-scale fisheries guidelines • Human rights approach • Implementation • Interactive governance

S. Jentoft (✉)
Norwegian College of Fishery Science, UiT – The Arctic University of Norway, Tromsø, Norway
e-mail: svein.jentoft@uit.no

R. Chuenpagdee
Department of Geography, Memorial University of Newfoundland, St. John's, Newfoundland and Labrador, Canada
e-mail: ratanac@mun.ca



S. Jentoft et al. (eds.), *The Small-Scale Fisheries Guidelines*, MARE Publication Series 14, DOI 10.1007/978-3-319-55074-9_37

## Introduction

After years of preparation, extensive consultation with civil society organizations (CSOs), and stakeholders ranging from the research community to governments, on June 9, 2014, the Committee of Fisheries (COFI) of FAO (Food and Agriculture Organization of the United Nations) adopted the Voluntary Guidelines for Securing Sustainable Small-Scale Fisheries in the Context of Food Security and Poverty Eradication (SSF Guidelines). This was indeed a remarkable achievement which has created hopes around the world that small-scale fisheries can advance from the marginalized and impoverished situation in which they often find themselves. Now it is time for governments, civil society, and the academic community to act on them. Many things need to take place, or be put in place, before the desired impacts of the SSF Guidelines can be fully realized. Even if FAO member states agreed to the SSF Guidelines, it remains to be seen how many of them will follow up, and to what extent, on them in practice. It will not be the first time that states do not act on what they have signed up for. As Raustiala and Victor (1989) conclude in a different context: 'Often, a country adopts an international accord without a clear plan for putting the commitments into practice' (660).

Many challenges now lie ahead to bring the SSF Guidelines home to those they are meant to serve: the millions of poor and marginalized fishers and fish workers around the world who expect that the SSF Guidelines will initiate the 'sea-change' that they need. Such change does not come easily, especially when other agendas compete for resources and attention from governments. Thus, as far as the SSF Guidelines are concerned, the ultimate test is whether states will really 'walk the talk' in a way that brings concrete change for people at the local community level, where the price of neglect is highest (Jentoft 2014).

The process of implementation has certainly started since the SSF Guidelines' adoption, with many initiatives and activities happening around the world to promote them and encourage their implementation. In Chap. 3, Franz and Barragán-Paladines provide an overview of what has transpired around the world as of November, 2016. They describe the implementation as a 'multi-directional, multi-scalar, and multi-temporal process', and argue that it is essential to make sure that implementation experiences are shared and can contribute to a collective learning process at all scales.

One of the most important features about the SSF Guidelines is their strong grounding in human rights standards and human dignity. 'Good governance' principles also thread through the entire document. These norms and principles are already globally accepted and legally codified. Still, they often meet resistance from many concerned parties, and even from the academic community, as Willmann et al. suggest in Chap. 2. While these norms may be new in the context of fisheries governance, they state the obvious: small-scale fishers and fish workers have had human rights, both as individuals and as collectives, long before the SSF Guidelines proclaimed it. These rights are about the fundamental freedom of fisheries communities and about securing sustainable small-scale fisheries by eliminating poverty, food insecurity, and political suppression. There is a long road from agreeing on general principles and objectives to acting upon them at regional, national, and local levels. State delegates were willing to compromise on principles and objectives during the negotiations. It remains to be seen, however, whether similar support exists at regional and local levels where the problems of small-scale fisheries are felt.

Seen in this light, and recognizing that the SSF Guidelines are here to stay, it is prudent to learn as much as possible about what challenges lie ahead and what can be done to address them, as governments at various levels and fishing people, groups, and communities begin their journey towards realizing the SSF Guidelines. The 32 case studies included in this book offer rich descriptions of small-scale fisheries contexts as the SSF Guidelines begin to unfold. In some instances, all the necessary conditions and factors are already in place to support implementation, and thus the SSF Guidelines can run unimpeded once they hit the ground. More commonly, however, a great deal of effort and reform is required before effective implementation can be expected. In some countries, concrete action is already taking place to implement the SSF Guidelines, initiated sometimes by government, sometimes by CSOs, and often together. In others, the SSF Guidelines have yet to receive notable recognition and stakeholders are still uninformed about them. In many instances, social innovation and governance transformation are likely imperative prior to implementation. In other cases, capacity needs to be developed, first and foremost, in order to facilitate the implementation.

On its own, each chapter offers valuable lessons about the SSF Guidelines that can be useful to others interested in implementing them in their own context, as presented below. Collectively, the chapters reflect the significance of this unique instrument in promoting sustainability, addressing poverty, and enhancing food security, and reaffirm that small-scale fisheries are indeed 'too big to ignore'.

## Consensus but Not Straightforward

Despite some gaps and imprecision in terminology and definition that chapter authors point out, the SSF Guidelines are the product of a hard-won consensus achieved among FAO member states. The language within the Guidelines is the result of intense negotiations over every single word in its hundred paragraphs. Indeed, what member states upheld was quite remarkable, and we should not think that the SSF Guidelines would land in a completely receptive environment when implemented. If implemented as intended, the SSF Guidelines will bring about social and ecological transformation of unprecedented proportions as far as this sector is concerned. Some chapters in this volume (for instance Chuenpagdee et al. Chap. 23 and McConney et al. Chap. 21) report on local enthusiasm in Newfoundland, Canada and in the Caribbean, respectively. However, putting the SSF Guidelines into practice will no doubt be challenging and local support alone will not be sufficient. One should not expect a linear, straightforward process. The SSF Guidelines do not enter a social, political, and cultural vacuum, but a complex policy landscape, as Coen et al. point out in Chap. 4 based on data from the Pacific Islands Countries and Territories. Rather, they will be integrated into functioning social and ecological systems that, despite commonalities, have unique histories and structural and cultural features, as suggested by Kleiber et al. in Chap. 35. These features, they argue, also regard gender roles and relations, issues which permeate throughout the SSF Guidelines. In some instances, like in Nigeria according to Akintola et al. (Chap. 30), women assume important leadership roles in fisheries and exert considerable bargaining power in the post-harvest value chain. Women, therefore, have an essential role to play in the implementation of the SSF Guidelines.

Given their visions and suggested institutional reforms, the SSF Guidelines are likely to meet resistance when they touch ground where they challenge established interests, ontologies, and paradigms (Jentoft and Søreng Chap. 13). This is especially the case where the current governing system is structured and operated from the top-down (Barragán-Paladines Chap. 33; Saavedra-Díaz and Jentoft Chap. 27). Arias Schreiber et al. (Chap. 34), whose case study involves reflection on the fate of the SSF Guidelines in Europe and their own country of Sweden, observe that neo-liberal approaches to fisheries governance, which favor large-scale fisheries, add severe challenges to the implementation of the SSF Guidelines. Sunde's observation from South Africa (Chap. 8) anticipates a similar paradigmatic gap, which she argues has become visible through the way in which the SSF Guidelines are being interpreted and implemented. In light of this interpretation, her chapter refers to customary tenure systems and governance as supported by the SSF Guidelines as being 'vulnerable to neo-liberal tendencies.'

Many stakeholders, including fisheries authorities, may not be all that enthusiastic when they learn about the SSF Guidelines, as they often contradict current ideologies and practices. Their chance of success is also restricted if they interfere with established power relations. This is an anticipation that can also be drawn from the SSF Guidelines preface (FAO 2015, X):

> Small-scale fishing communities also commonly suffer from unequal power relations. In many places, conflicts with large-scale fishing operations are an issue, and there is increasingly high interdependence or competition between small-scale fisheries and other sectors. These other sectors can often have stronger political or economic influence, and they include: tourism, aquaculture, agriculture, energy, mining, industry and infrastructure developments.

As other stakeholders, these sectors are not equally equipped and capable of securing their interests, and they do not always agree on issues being negotiated. Thus, it is difficult to imagine that they would easily yield to the concept of 'preferential treatment', mentioned in paragraph 5.4 (FAO 2015, 5):

> States should take appropriate measures to identify, record and respect legitimate tenure right holders and their rights. Local norms and practices, as well as customary or otherwise preferential access to fishery resources and land by small-scale fishing communities including indigenous peoples and ethnic minorities, should be recognized, respected and protected in ways that are consistent with international human rights law.

## Voluntary but Not Mechanically

The SSF Guidelines are, like other international 'soft law' instruments, voluntary, a term that appears in the document title and mentioned 18 times in the document. The Guidelines intend to "guide amendments and inspire new or supplementary legislative and regulatory provisions" (paragraph 4.2). Since the SSF Guidelines are not binding, states, CSOs, and other stakeholders can ignore them at will. Had the Guidelines been mandatory, and their legal status different, negotiations would most likely have been more cumbersome. It might, however, have led to a less complicated implementation process, as stakeholders would have been obliged to do things that they now can chose to disregard.

The voluntary nature of the SSF Guidelines makes the resulting implementation processes more interesting from an implementation research point of view - How will they be received? How is the 'as appropriate' clause being interpreted? Who will initiate and lead the process? Why is the implementation smooth in one country while bumpy in another? By explicitly linking the SSF Guidelines to 'hard' international law (such as the International Covenant on Civil and Political Rights, the International Covenant on Economic, Social and Cultural Rights, the Convention on the Elimination of All Forms of Discrimination against Women, the Convention on the Rights of the Child), the SSF Guidelines are less voluntary than they appear (see Willmann et al. Chap. 2). As Royo (2009) argues with regard to the implementation of the Declaration of the Rights of Indigenous Peoples, the distinction between soft and hard law does not need to be relevant in practice, and formally non-binding law has empirically proven to be even more effective in promoting respect for human rights than binding law. It is also an empirical question whether the SSF Guidelines will work in a similar way. Interestingly, and relevant to the SSF Guidelines, Royo (2009) refers to arguments that have been made that, given the comprehensive preparation and participation of indigenous peoples in the process, and the moral weight it therefore carries, although not legally binding, it is still binding 'as an instrument' (316).

Since the SSF Guidelines are voluntary and target issues that are politically contentious and institutionally demanding, they are likely to encounter resistance in many quarters when they meet reality on the ground, as observed by Jentoft and Søreng in the case of Norway and the indigenous Sami (Chap. 13). This is even the case when human rights principles are embedded in 'hard law', as with indigenous peoples, like in the Sami case. At the national and local level, the *status quo* tends to have strong defenders, while change often has only reluctant support. Interventions, like management regulations that somehow limit the action space of stakeholders, generally meet opposition.

However, states sometimes do not even implement their own laws. Nicaragua, which is the country in the chapter by González (Chap. 10), has progressive legislation to protect the tenure rights of indigenous peoples, but they do not always enforce them in practice. The big issue in that country now is the building of an inter-oceanic canal that will cut through titled indigenous land and fishing areas.

What the affected indigenous people believed were secure tenure rights do not seem to be so secure after all, especially when the government believes it is in the national interest to ignore them. The Nicaraguan case is a good illustration of how much weight the SSF Guidelines and the UN Declaration for the Rights of Indigenous peoples carry at the end of the day, and how strategic articulation with domestic and international legally binding norms and standards for the protection of the rights of indigenous peoples come into play. Both instruments constitute an important moral support for the causes of indigenous peoples, as well as non-indigenous communities, in securing sustainable fisheries, especially when their tenure rights are under siege.

The SSF Guidelines are a supplement to the Code of Conduct for Responsible Fisheries, which is also voluntary but still contains a set of norms that governments have accepted. The same holds true with other intergovernmental guidelines to which the SSF Guidelines frequently refer, including the Voluntary Guidelines to Support the Progressive Realization of the Right to Adequate Food in the Context of National Food Security and the Voluntary Guidelines on Responsible Governance of Tenure of Land, Fisheries, and Forests in the Context of National Food Security. All of these instruments are now well into their implementation and there is an overlap between them, which might create synergy. Text extracts from these other instruments found their way into the SSF Guidelines and helped secure their endorsement. Thus, the implementation of the SSF Guidelines is part of a more comprehensive effort, which should help facilitate their uptake. There are also lessons to learn from the implementation of each instrument which, although different in focus, draws on the same human rights and good governance principles, as previously mentioned. The fact that they largely talk to the same institutions and stakeholders should make them possible to implement.

## Flexibility Is Good but Also Problematic

The fact that the SSF Guidelines do not prescribe a standard definition of small-scale fisheries, or how it should be applied in a national context, creates flexibility as to their interpretation and, hence, their implementation. States may therefore decide for themselves for whom the SSF Guidelines are relevant – or, indeed, if they are relevant at all. It will thus be interesting to see how states choose to define small-scale fisheries in the context of the SSF Guidelines. In some instances, countries already have an official definition that is used for management and statistical purposes.

There will always be room for interpretation of key concepts of the SSF Guidelines. For instance, what does poverty mean, and are small-scale fishers poor? There must be a degree of consent on the term to know exactly what policies should be implemented and whom they should target. Clearly, the SSF Guidelines' conceptualization of poverty is broader than just about income, as it also includes education, health, and issues that are relevant for the viability and well-being of small-scale fisheries people, like tenure rights (Sunde Chap. 8; Nayak Chap. 9). These are also issues where human rights legislation apply (see Willman et al. Chap. 36). The SSF Guidelines focus on the causes of poverty, which involves the exposure of small-scale fishing communities to natural hazards and climate change – an issue that is the focus in the chapter by Islam and Jentoft (Chap. 24), which describes the situation of small-scale fishing people in Bangladesh. Poverty is also a likely outcome when small-scale fishers are displaced, such as when they lose access to fishing grounds or beaches, as illustrated by Soares (Chap. 7) in Jamaica and Saavedra-Díaz and Jentoft (Chap. 27) in Colombia. In the latter case, small-scale fishing communities have also been suffering from armed conflict, which the SSF Guidelines address (paragraphs 5.2, 6.18), and drug related violence.

Poverty is both absolute and relative. Small-scale fishers in industrialized countries may not be poor in an absolute sense, even if they are numerous, like in Japan where fisheries are overwhelmingly small-scale (Delany and Yagi Chap. 15). Neither are they as poor as small-scale fishers, fish workers, and their families in many countries in the Global South. But they are often poor and marginalized in comparison to people in other occupations in their own country (Jentoft and Eide 2011). The SSF Guidelines therefore also speak to small-scale fishers and fish workers in industrialized countries, as shown in Snyder et al. (Chap. 6) in Greenland, and Chuenpagdee et al. (Chap. 23) about Canada. Similarly, Arias-Schreiber et al. (Chap. 34) argue that small-scale fisheries in Sweden face many of the same problems as those in the Global South due to vulnerability and marginalization, which are rooted in ignorance about their situation and contribution to society, the limited sharing of decision-making, and a failure to implement genuine participatory approaches to governance.

Poverty is also relative from a gender perspective. The fourth guiding principle of the SSF Guidelines, which focuses on gender equality, states: "Gender equality and equity is fundamental to any development. Recognizing the vital role of women in small-scale fisheries, equal rights and opportunities should be promoted" (FAO 2015, 2). The implementation of this principle is the theme of Kleiber et al. in this volume (Chap. 35). Baragán-Paladines (Chap. 33) argues that, unless gender relations in small-scale fisheries are considered equally important as a driving force towards sustainability, the implementation of the SSF Guidelines will be largely delayed, and perhaps be implemented in a gender-biased manner.

The SSF Guidelines talk frequently about 'vulnerable' and 'marginalized' people. Again, what exactly do these terms refer to, and how are they linked? When are small-scale fisheries people sufficiently vulnerable and marginalized to deserve the 'special treatment' that the SSF Guidelines advocate in paragraph 3.5? This particular paragraph is about equity and equality, as well as the promotion of justice and fair treatment. These concepts relate to issues that are philosophically intricate, and therefore hard to define. Still, an agreement about what these terms mean would be needed for the SSF Guidelines' implementation.

One may think differently not only on what poverty, vulnerability, and marginalization are but also how they occur. Are small-scale fishers poor because they are marginalized, or is the opposite the case? Maybe they are poor and marginalized because they are vulnerable, or *vice versa*. Are people poor because they are marginalized materially (no secure access to resources), politically (no voice in the political process), or socially (outmaneuvered in the market place)? The combination of poverty, marginalization, and vulnerability of small-scale fisheries people is clearly demonstrated in the case of Bangladesh (see Islam and Jentoft Chap. 24). However, it is difficult to say what comes first of the three, and what causes of poverty leads to the others. In practice, they are compounding forces that interact to create poverty in different ways from case to case. What is clear though, is that together they lead to the destitution of small-scale fishing people, which the SSF Guidelines would need to address when implemented. Making people less vulnerable to climate change, natural hazards, and ecological disasters stemming from threats such as overfishing requires attention and political will to address their situation of poverty. People need to be secured in a social and economic sense, in addition to being safe in a physical sense. This requires concerted action on all paragraphs in the SSF Guidelines and not only those that deal with climate change and natural hazards. This is also an argument made by Luomba et al. (Chap. 25) and Daliri et al. (Chap. 26) in the context of IUU (Illegal, Unreported, and Unregulated) fishing. Therefore, Islam and Jentoft hold (in Chap. 24) argue that, as far as these two issues are concerned, and as long as they continue to have devastating effects on small-scale fisheries in Bangladesh, the SSF Guidelines are relevant in their totality.

## Similar Challenges but Different Contexts

Case studies like those presented in this volume are suitable for illustrating the diversity and complexity of the contexts in which the SSF Guidelines are introduced, and challenges that are associated with asymmetrical power relations within and outside the value chain (Gardner et al. Chap. 16). These particular contexts must be taken into account because they will contain both the obstacles and opportunities for the implementation of the SSF Guidelines (Soares Chap. 7). Therefore, Prescott and Steenbergen (Chap. 12) conclude that successful adoption of the SSF Guidelines across a wide spectrum of small-scale fisheries depends to a large extent on the compatibility with what is actually going on at the ground level. Prior to their implementation, small-scale fisheries systems and performances must be carefully analyzed through the 'holistic' approach that the SSF Guidelines advocate, which include both the natural and human dimensions of the social and ecological system-to-be-governed, as well as their governance arrangements (Kooiman et al. 2005; Jentoft and Chuenpagdee 2015).

Thus, the implementation approach must be attuned to the specific small-scale fisheries context in which it will ultimately be defined, or 'embedded,' as Gómez Mestres and Lloret phrase it in their case study of Cap de Creus, Spain (Chap. 19). Small-scale fisheries around the world contain too much variety in circumstances, problems, and needs for one formula to be effective. Rather, the approach must be adaptive and attentive to the situations on the ground as they exist where and when the SSF Guidelines enter. This requires a willingness to learn from the people that the Guidelines mean to benefit, which is why the document stresses the participatory approach. The Guidelines would need support from governments and CSOs, but imposing them from above may cause backlash.

As stressed throughout the document, the SSF Guidelines should be put into action in accordance with domestic law. This would have gone without saying if it had not been for the fact that the SSF Guidelines also underscore the need for legal reform in many instances, for instance with regard to tenure rights. For example, paragraph 5.4 states: "Where constitutional or legal reforms strengthen the rights of women and place them in conflict with custom, all parties should cooperate to accommodate such changes in the customary tenure systems." Thus, if a country insists that the SSF Guidelines should be implemented in conformity with existing domestic policies and legislation, they may end up confirming the *status quo*. This is not what the SSF Guidelines call for, at least not in those situations where important gaps exist between the current situation in small-scale fisheries and what the Guidelines promote. The SSF Guidelines call for substantive changes in the existing conditions that currently hold small-scale fishing people back. The firmness on the national legal lens as a given may suggest that member states are not eager to commit to meaningful reform. Perhaps they may even be unwilling to implement the SSF Guidelines if that means challenging the established order and the power relations that support it (Said Chap. 11). However, in some instances the state can be a supportive force of small-scale fisheries, like in Japan, as shown in Delany and Yagi (Chap. 15). The two authors hold that Japanese small-scale fisheries management has been overall a success in social, economic, and environmental terms due to the government's willingness to return to using traditional spatial and rights-based tenure systems, after flirting with Western style management approaches. Globally, Japan may be an exception to the general rule, but the case still demonstrates that apparent path dependencies are after all not irreversible. Legislation was established to empower small-scale fishers *vis-a-vis* powerful money-lenders and buyers.

Another example of how contexts matter is with respect to IUU, discussed in paragraph 11.5 of the SSF Guidelines, which is an issue in several chapters of this volume. In Chap. 26, Daliri et al. discuss many factors that cause IUU fishing in Hormozgan Province of Iran and the Persian Gulf, highlighting in particular the drivers related to community, culture, mismanagement, economic conditions, personal skills, and area features. They conclude that co-management can help address the problem of IUU fishing, if well implemented and when used in conjunction with other measures like monitoring, control, and surveillance. Co-management is, of course, the mode *a la mode* in Tanzania, but IUU fishing is persistent there, as shown by Luomba et al. (Chap. 25). Instead, these authors argue for a closer examination of why IUU fishing takes place and whether its persistence has anything to do with the characteristics of the ecological and the social systems of fisheries, and the capacity of the governing system. They also highlight that fishers may perceive and judge the severity of IUU fishing activities differently from government agencies, thus adding to the challenges of combatting IUU. In the case of Thailand, Chuenpagdee et al. (Chap. 32) suggest that turning IUU fishing towards small-scale fisheries may simply be a strategy employed by the government to meet other obligations (such as getting a 'pass' from the European Union for seafood trade). The consequence of such policies can be detrimental to the ability of small-scale fisheries to have gainful and viable fishing livelihoods.

## Implementation as an Ongoing Negotiation

As described in some chapters of this book, it happens that the SSF Guidelines are in accordance with policy reforms that are already underway in some countries, for instance, in Thailand as shown by Chuenpagdee et al. (Chap. 32). In all instances, the effect of the SSF Guidelines on the ground are still pending, given that a complex and ambitious undertaking like the implementation of the SSF Guidelines is bound to take years, or perhaps even decades. This is due to the fact that many of the problems and challenges that the SSF Guidelines focus on do not simply go away and cannot be solved once and for all. They are what Rittel and Webber (1973) called 'wicked problems.' For this reason, the SSF Guidelines' implementation is not a one-off action, but a process that is, and must be kept, alive. This is also because governance does not interfere in situations and systems that are static, but highly dynamic. The same point is made by Nayak (Chap. 9) about tenure systems, which he argues must therefore be understood as processes. This means that the implementation of the SSF Guidelines is a process about a process that must continue and must be iterative. Similarly, Gardner et al. argue in Chap. 16 that implementation will not achieve full impact without regularly revisiting and engaging the small-scale fisheries communities that it targets. The implementation of the SSF Guidelines is in itself a wicked problem, which requires political rather than technical solutions, also because the Guidelines are ethical, require sound judgement, involve human rights, and demand constant vigilance. These are issues that will stay with us throughout time. People may have different ideas what poverty and marginalization involve and what it would take to address them. Poverty, marginalization, and food security, issues that are key concerns of the SSF Guidelines, do not have a technical or scientific solution, but solutions that question our values and morality, as Garrett Hardin (1968) argued. This is also why it is relevant to talk about human rights and dignity, as the SSF Guidelines do (see Willmann et al. Chap. 36 and Song 2015), and why governance in fisheries should start at the meta level, as Kooiman (2003) points out. This is important because it is key to have a conscious perception of the basic norms and values that underpin the governance process.

The heated debate among state delegates and CSO representatives during the Technical Consultations in relation to the development of the SSF Guidelines held in 2013 and 2014 at FAO on concepts like gender, governance, co-management, redistribution, and small-scale fisheries as an informal sector may imply that states are hesitant to embrace the SSF Guidelines in full. Some state delegates also wanted more gender-neutral language, but lost the argument in the end. The gender perspective has a separate section in the SSF Guidelines (Section 8) and otherwise cuts through the entire document.

In Madagascar, as illustrated by Gardner et al. (Chap. 16), post-harvest actors, with government support, hold disproportionate negotiating power, with benefits from management initiatives accruing mainly to actors placed high in the value chain. Amarasinghe and Bavinck (Chap. 18) discuss the role of cooperatives in restructuring the way power works in the value chain so that small-scale fishers can gain more control of their working conditions, such as resource management, safety at sea, insurance problems, and marketing. They argue that more power would also be gained if individual cooperatives could form unions. Again, path dependencies can be changed.

Power relations in small-scale fisheries are also about gender, as Kleiber et al. (Chap. 35) point out. This and other chapters demonstrate why it is important to acknowledge the role and contributions of women within the family enterprise and the entire small-scale fisheries value chain, as Soares illustrates in the case of small-scale fisheries in Jamaica (Chap. 7). This is also important from a governance perspective, as women are often excluded from those arenas where collective action and decision-making take place (see Kleiber et al. Chap. 35 and Barragán-Paladines Chap. 33). When comparing the SSF Guidelines to existing policy documents for the small-scale fisheries in the Solomon Islands, Cohen et al. (Chap. 4) identify gender as a missing element that the SSF Guidelines may help repair.

State delegates' concerns about language in the SSF Guidelines suggest that they take them seriously. Words matter only if you care about what they refer to and attempt to convey. Nevertheless, it also indicates cautiousness or concern that the implementation process might become cumbersome, and that the negotiations over them will continue even after COFI's endorsement. The chapters of this volume confirm that this is a likely prospect, as the SSF Guidelines are to be implemented in concrete contexts where the same concept can take on different meanings. After all, implementation is the process by which "intent is translated into action"' (Rein and Rabinovitz 1987, 308). Now that FAO member states have endorsed the SSF Guidelines, the former is settled and the latter has just begun. The implementation of the SSF Guidelines will hardly be a straightforward transition, but a process where translation is a matter of negotiation. One should expect a cyclical, interactive, and iterative process, where objectives and concepts are subject to repeated questioning, debate, evaluation, and reformulation. Lessons learned in the course of implementation may even lead to reconsideration of the original intent and to subsequent reformulation of principles and goals and new interpretations of key concepts. From a research point of view, this would be something to look out for, also because stakeholders, even when they argue about principles, tend to do so from their particular interests and worldviews. This is the nature of politics, and also how the Technical Consultations on the SSF Guidelines worked.

Legislative reform and implementation are separate processes, but both are uphill battles. This is partly because the burden of proof for the need of reform rests with those who want change, not with those who defend the current order. The implementation of the SSF Guidelines is likely to experience the same, although the hindrance to reform is often structural and institutional. These barriers create a path dependency which is hard to turn around in a way that favors small-scale fisheries, as pointed out by Arias-Schreiber et al. in Chap. 34. Still, their endorsement may have shifted the burden of proof; it is now governments that must legitimize why they remain inactive, and defend why, after endorsing the SSF Guidelines, they are reluctant to implement them. Yet someone must raise the issue: someone must hold governments accountable and keep checking in on them. Here, CSOs have an important role to play.

## Dealing with Power

The SSF Guidelines promote norms and principles about issues that are social and ethical. They will intervene in situations where different interests are in conflict, even in armed conflict, as stressed in paragraphs 5.12 and 6.18, and where small-scale fisheries are the weaker party. Therefore, they will inevitably interfere with power. On this issue, the SSF Guidelines speak plainly. As stated in the Preface:

> Small-scale fishing communities also commonly suffer from unequal power relations. In many places, conflicts with large-scale fishing operations are an issue, and there is increasingly high interdependence or competition between small-scale fisheries and other sectors. These other sectors can often have stronger political or economic influence, and they include: tourism, aquaculture, agriculture, energy, mining, industry and infrastructure developments.

This is also an issue that the SSF Guidelines return to in paragraph 7.1:

> All parties should recognize the central role that the small-scale fisheries post-harvest subsector and its actors play in the value chain. All parties should ensure that postharvest actors are part of relevant decision-making processes, recognizing that there are sometimes unequal power relationships between value chain actors and that vulnerable and marginalized groups may require special support.

Chapter authors, like Gardner et al. (Chap. 16) in the case of Madagascar, argue that real change will require greater attention to the fundamental imbalances in bargaining power between actors. This bargaining power can be secured by facilitating fishers' representation at national fisheries policy decision-making arenas through the establishment and formalization of multi-stakeholder fisheries management platforms. Small-scale fishers are one of many stakeholders within the value chain and in their area. They have legitimate concerns and urgent needs, but do not always have the power they need to secure their interest relative to other stakeholders.

Small-scale fisheries are usually the weaker party, regardless of their legitimate concerns and urgent livelihood needs. They may well have conflicts with other stakeholders, like large-scale operators, as noted in the SSF Guidelines' preface and addressed in the chapter by Gunakar et al. in India (Chap. 14). However, they may also exist in a symbiotic relationship with other stakeholders, even large-scale fisheries (see Snyder et al. Chap. 6). Still, they are often in direct competition with more powerful stakeholders who have the government's ear more than small-scale fishers and fish workers. Whether the SSF Guidelines will change this, thus fulfilling one of their goals, remains to be seen.

In the case of Zanzibar, which Lindström and de la Torre-Castro investigate in Chap. 5, there is a clash caused by the lack of coherence between the recently established fisheries policies, which were funded by the World Bank, and the policies that follow from the strategies envisioned in the SSF Guidelines. Since the latter do not come with funding, the two authors anticipate that it will have to yield when there is conflict between the two. The authors point to a major conflict between the ideas of capitalization and commercialization, which are promoted heavily in the established policy framework, and the human rights approach that is fundamental to the SSF Guidelines. Zanzibar has a somewhat autonomous status within Tanzania, which extends to the governance of marine resources, but it is not a sovereign state and FAO member and thus was not invited to take part in the Technical Consultations.

There are usually political costs associated with intervening in established power relations, as they may easily prevent governments to act upon the SSF Guidelines, especially when their implementation challenges the established order. Still, as Sabau argues in her case study from Costa Rica (Chap. 17), while the situation is promising, actual change can only be brought about when real social inclusion takes place which goes beyond talk to the direct addressing of power relations and engagement in the transformation of social policies.

Governments may not have sufficient power to withstand powerful stakeholders who oppose the SSF Guidelines. During the Technical Consultations, some state delegates criticized the concept of 'redistributive reforms', but the term survived in paragraph 5.8. As illustrated by Gardner et al. in the case of Madagascar (Chap. 16), it is impossible to think of small-scale fisheries as a poverty alleviation strategy without addressing issues of distribution. Widespread poverty, food insecurity, low literacy, and other challenges, including disregard of small-scale fishing people by the authorities, makes small-scale fisheries more vulnerable to exploitation and the unfair distribution of benefits.

Efforts to level the playing field may therefore be a problem during the implementation process, as the effective implementation would require clout *vis-a-vis* powerful stakeholders that governments do not always have. Corruption is part of this problem, and may stand in the way of effective implementation of the SSF Guidelines (see Mattos et al. Chap. 22). Thus, paragraph 11.3 of the SSF Guidelines may therefore be hard to live up to in some instances: 'States should endeavor to prevent corruption, particularly through increasing transparency, holding decision-makers accountable, and ensuring that impartial decisions are delivered promptly and through appropriate participation and communication with small-scale fishing communities.'

For implementation to commence, someone must make the first move. Sitting on the fence may be a rational strategy, according to Mancur Olson (1971), who theorized about collective action, but if that is the position of all, the process becomes a quagmire before it really starts. The SSF Guidelines, however, talk primarily to the state, as governments have a major responsibility for their implementation but may also be the most effective initiator. Yet, in some instances, like in the Caribbean region (see Chap. 21 by McConney et al.), state institutions have been passive, whereas the CSOs have been the ones shouldering implementation by building networks. However, one should not necessarily expect a top-down implementation process initiated by state agencies if they are currently oblivious to the problems and potentials of small-scale fisheries, as Said argues is largely the case in Malta (Chap. 11). This is especially because of the current move 'from government to governance' in many countries (Bevir 2011), including fisheries (Kooiman et al. 2005; Bavinck et al. 2013). This transition transforms the role of the state from supreme governor to mediator, negotiator, and facilitator, which is a role more in line with the spirit of the SSF Guidelines. It results in a more open, inclusive and interactive form of governing, but may also lead to confusion about the division of labor.

Successful implementation depends as much on the messenger as the message. The SSF Guidelines may be met with skepticism just because the central government is the agent sponsoring them, especially if the state's track record in supporting small-scale fisheries is poor. This is an issue of trust. As Tsang et al. (2009) point out:

> Although government may be able to implement its own agenda without trust or exert absolute control over a population through the use of coercive resources, it is nowadays impossible to implement the programs of a modern state effectively without trust. A government has a great deal to gain by facilitating trust (101).

The implementation of the SSF-Guidelines will hinge on the trust that the state government enjoys among stakeholders. All too often, as Sall and Nauen point out in Chap. 29 based on their case study from Senegal, the relationship between government and SSF actors are more conflictual than cooperative. Building trustful relationships that can be sustained would therefore be essential for the SSF Guidelines to move from intent to action. This is particularly an issue when regulatory frameworks currently limit the freedom of fishers. Without trust, they will be inclined to break or ignore any rules that are imposed on them (Daliri et al. Chap. 26). For this reason as well, the implementation process must be interactive, taking the form of a functioning partnership between government, CSOs, and small-scale fisheries stakeholders. CSOs are often 'the first to spot deviations from the terms of consensus-based rights and rules' (Young 2006, 851), and may provide the arm that is needed to bring the SSF Guidelines into the policy area and keep them there (see McConney et al. Chap. 21). Therefore, the role of CSOs should not be restricted to filling the gaps left by states, as Espinosa-Romero et al. argue in Chap. 20. Their case study is situated in Mexico, where the federal government has legislated the right of CSOs to participate in forums and structures where planning, implementation, and follow-up occur. Their chapter also provides a vivid account of the wide-ranging role that CSO can play in the implementation process. Thus, the implementation process should be equally as interactive among all stakeholders and contributors to the implementation of the SSF Guidelines. The involvement of small-scale fisheries organizations and CSOs should therefore become part of the governance system, in the same way as they should become part of the solution to sustaining fisheries resources from an ecosystem perspective (see Prescott and Steenbergen Chap. 12). The implementation of the SSF Guidelines should not be an exception from the general rule.

## Reform and Transformation in Three Orders

Signing onto conventions, declarations, or in this case the SSF Guidelines, has symbolic value: it shows goodwill. However, states do not always implement what they have committed themselves to do, neither internationally nor domestically (FAO 2006). They may have had reservations to begin with, or they may meet resistance on the home front that make them think twice. Therefore, McConney et al. (Chap. 21) think that for their region (the Caribbean), it is unlikely that the SSF Guidelines will become a focus of policy due to the existing constraints.

According to Kooiman (2003), reform and transformation need to take place in all three orders of governance in order to become effective. The meta-order is where the basic values, principles, images – or problem definitions – are settled. In the SSF Guidelines, this happens in the Preface and Introduction sections, for instance where the human rights-based approach and the good governance ideals are declared. Second order governing is where values and principles are transformed into institutions and legal frameworks throughout the value chain, and which both restricts and enables the day-to-day governing that occurs at the first governing order. After opening with a section on the guiding principles, the SSF Guidelines in subsequent paragraphs provide concrete clues to how these principles would be operationalized at second and first governing order.

A critical analysis of the SSF Guidelines, which chapters of this book provide, would look for inconsistencies both between guiding principles and actions recommended, and between paragraphs as endorsed and as practiced. Are the recommendations at the second and first order sufficiently precise and robust as mechanisms for putting intent into action? How much room do they allow for manoeuvring? What happens if there are inconsistencies between governing orders when the SSF Guidelines hit the ground? One may anticipate two-way traffic, implying that institutions and action would be brought in line with the intent as established at the meta-level, but also that meta-order level principles, values, and images would be adapted to actual performance, thus perhaps lowering or raising the ambitions as expressed in the SSF Guidelines. Future evaluation and research will determine what the outcome will be. However, this mutual adjustment process may already have started.

Much in the same way as Kooiman talks about governing orders, Rein and Rabinovitz (1987) observe that implementation must surpass three 'imperatives': one legislative, one bureaucratic, and one related to consensus-building. These hurdles must be passed when the SSF Guidelines principles at the meta-order are converted into new institutions and legislation at the second order, and when stakeholders must be convinced about the actions taken by these institutions at the first order. Blockage may occur at each order. The mismatch between values and principles can easily lead to non-compliance, as demonstrated in the case of Lake Victoria by Luomba et al. (Chap. 25).

It should be noted that the orders do not necessarily correspond to different scales, such as global, national, and local. All three orders may be functioning at all scales, including at the local level where fisheries communities govern their own affairs. Implicitly or explicitly, local governing systems – such as customary tenure systems, which are frequently referred to in the SSF Guidelines and in chapters of this volume such as that of Akintola et al. (Nigeria), and Prescott and Steenbergen (Coral Triangle), and Sunde (South Africa) - have their meta-order values, images, and principles, as well as institutions that operate according to them and a governance practice that involves stakeholders. This would imply instances of 'legal pluralism', as discussed by Sunde in Chap. 8 and Jentoft and Søreng in Chap. 13. This is a situation where more than one normative system applies in the same situation, as when state law co-exists with customary law. There is, however, a risk that the former will be implemented in a way that overrides the latter. Therefore, paragraph 5.4 of the SSF Guidelines hold: 'States, in accordance with their legislation, and all other parties should recognize, respect and protect all forms of legitimate tenure rights, taking into account, where appropriate, customary rights to aquatic resources and land and small-scale fishing areas enjoyed by small-scale fishing communities.'

By supporting the concept of customary law, the SSF Guidelines' implementation should start by analyzing what formal or informal normative and legal systems exist, how they work, and whether they meet the general values and principles that underpin the SSF Guidelines, which in their own right represent a new global normative system. How do existing customary (and statutory) legal systems, such as in Nigeria (Akintola et al. Chap. 30), South Africa (Sunde Chap. 8), and Ecuador (Barragán-Paladines Chap. 33), hold up to human rights and gender equity as extolled in paragraph 5.4 of the Guidelines? Does the customary system fulfill human rights standards, and support them? If there is a gap between state law and customary law, a common ground should be sought. The opportunities that existing organizations such as the Spanish 'confraries' (Gomez Mestres and Lloret Chap. 19), or the community cooperatives in Sri Lanka (Amarasinghe and Bavinck Chap. 18), present for supporting the implementation of the SSF Guidelines must be recognized. The SSF Guidelines are there to help empower them.

What is legally required in order to bring about change for small-scale fisheries may vary from country to country. Therefore, it makes sense to add the phrase 'as appropriate', which happens frequently (19 times) in the SSF Guidelines text. The legal status of small-scale fisheries would need to be clarified in particular cases and, in some countries, as mentioned, new legislation may be required to achieve the aims of the SSF Guidelines. Legal processes and other governance reforms for the rights of small-scale fishing people are taking place in some countries, such as South Africa, as described by Sunde (Chap. 8), and in Thailand, as described by Chuenpagdee et al. (Chap. 32). In other countries, like Malta (Said Chap. 11), the SSF Guidelines can help trigger reforms which must take place in order to prevent small-scale fisheries from eradication but which so far do not seem to have inspired the government to act on them.

Even if the SSF Guidelines result in legislative reform at the second order, they might still not pass the bureaucratic hurdle. Bureaucratic practice is not always conducive to effective implementation at the first order. Bureaucrats are not supposed to foist their own meta-order values, images, and principles, as these are fundamentally political. However, they have knowledge about what is governable. In the case of small-scale fisheries, data are often scant (see Prescott and Steenbergen Chap. 12), which is a hindrance to the implementation of the SSF Guidelines. Therefore, the SSF Guidelines devote a whole chapter to the topic of information, research, and communication. Who are small-scale fishers, how many are there, where are they, and how do they operate under which conditions? Without such data, implementation would easily misfire. If statistically invisible, the poorest and most marginalized of small-scale fishers and fish workers may not be impacted by resulting policies. Therefore, as Kincaid argues in the case of the Bahamas (Chap. 28), there is a need to develop a small-scale fisheries database as part of the SSF Guidelines process. Other chapters make similar arguments, like by Akintola et al. in the case of Nigeria (Chap. 30). This is also in line with what the SSF Guidelines argue in paragraph 11.1:

> States should establish systems of collecting fisheries data, including bio- ecological, social, cultural and economic data relevant for decision-making on sustainable management of small-scale fisheries with a view to ensuring sustainability of ecosystems, including fish stocks, in a transparent manner. Efforts should be made to also produce gender-disaggregated data in official statistics, as well as data allowing for an improved understanding and visibility of the importance of small-scale fisheries and its different components, including socio-economic aspects.

Should the implementation process pass the second hurdle, it would still have to face the criticism of stakeholders. These actors may not agree with what the SSF Guidelines propose, even if they agree with the meta-order guiding principles and accept the legitimacy of second order institutions. The notion that small-scale fisheries deserve preferential treatment (paragraphs 3.5 and 6.2) may not sit well with other stakeholders, especially if they have problems distinguishing between equality and equity (as mentioned for example in paragraph 3.1.4 and 5). Thus, without consensus about practical measures, which would require a consultative process (Kincaid Chap. 28) at first governing order, implementation may come to a halt before it lands at the local level. There is also the risk of 'capture'; powerful stakeholders may bend the SSF Guidelines to their advantage or spin their arguments to fit the meta-level rhetoric. Therefore, the implementation must be sensitive to power differences at all governing orders, including those that exist between small-sale fisheries and other sectors, as well as those that exist within the small-scale fisheries sector between fishing leaders, boat owners, quota holders, and crew. Raustiala and Victor (1989), however, find that 'while regulatory capture is a risk, the capturing influence of target groups has been offset through informed participation by countervailing groups' (669). This calls for broad participation and transparent communication in the SSF Guidelines implementation process (e.g. paragraph 3.1.1, 3.1.6).

## Foster Collaboration and Build Partnerships

Despite this risk of capture, the broad scope of the SSF Guidelines requires an implementation process that fosters collaboration and builds partnerships, sometimes in the form of entirely new organizations, such as the Fjord Fishing Boards in Norway to enable participation of the Sami indigenous fishers in the management process (Jentoft and Søreng, Chap. 13). For the case of Bahamas, Kincaid argues in Chap. 28 that priority should be given to improve the relationship and interactions between local fishers, fish workers, and government through the formation of a fishers' organization that fully represents local communities. This is a recommendation that can easily be generalized for situations elsewhere where such organizations do not exist. Organizational development is about empowering fishers and fish workers, as Willmann et al. (Chap. 36) state. Such organizations must include other stakeholders, as they are both target groups as well as affected by implementation, and therefore have a right to have a voice.

Whether countries with customary governance institutions and a tradition of stakeholder participation at the grass roots level, such as Thailand (Chuenpagdee et al. Chap. 32), are particularly prone to the SSF Guidelines and to their implementation in an inclusive manner in accordance with the human rights-based approach, is a question worth pursuing. In democratic countries like Costa Rica, the implementation is also well underway, but democratic states may still be stuck in a centralized, bureaucratic approach, as Sabau points out (Chap. 17). A finding from studies of the implementation of environmental codes confirms that "participation during the negotiations of international commitments and the making of national implementing policy is high, but it has often proved difficult to expand participation at the implementation phase" (Victor et al. 1998, 23). Some countries may already have robust institutions in place at national and local levels (second order) that may help facilitate implementation at first order, while others may have to create them first. As the Guidelines state: "Accordingly, there should be support for the setting up and the development of cooperatives, professional organizations of the small-scale fisheries sector and other organizational structures, as well as marketing mechanisms, e.g. auctions, as appropriate" (paragraph 7.4). Institutional development is therefore an integral part of the implementation process in many situations. Many chapters of this volume illustrate what that process might imply and which organizational alternatives exist, such as Amarasinghe and Bavinck in the case of Sri Lanka (Chap. 18), and Sabau in Costa Rica (Chap. 17). Writing from Mexico, Espinosa-Romero et al. (Chap. 20) have a similar notion about the constructive role of cooperatives in the implementation of the SSF Guidelines.

It would perhaps help if there were a clear addressee for the SSF Guidelines at the national level that is more precise than just 'states'. In many instances, like in Jamaica (Soares Chap. 7), there is no single state institution with a fisheries mandate, like a dedicated Ministry. Soares therefore anticipates that Section 10 in the SSF Guidelines, which talks about policy coherence and institutional coordination, would be hard to implement, and that the creation of a more enabling institutional environment would be a necessary condition for the SSF Guidelines to take effect. In Jamaica, the Fisheries Act is still in progress after decades of development, and is currently in a stalemate. Whether the SSF Guidelines will help to lubricate the process would be interesting to follow.

The implementation process would need an overseer, like FAO. But there would also be a demand for a suitable monitoring and evaluation plan and a measurement instrument. The study conducted by Pitcher et al. (2009) is an example of what can be done. These authors compared country compliance with the Code of Conduct for Responsible Fisheries, and found substantial variation. However, this study measured conformity, not achievement. Countries may already have been living up to the principles of the Code at the time it was initiated, or their policies may have been initiated regardless thereof. This is also the case with the SSF Guidelines. However, there may still be policy incoherence between what the SSF Guidelines say and policies that currently dominate the policy context, like in Zanzibar, as pointed out by Lindström and de la Torre-Castro (Chap. 5), and in India by Gunakar et al. (Chap. 14). This is something a systematic comparison at country level with the SSF Guidelines may help to identify, as in Chap. 4 by Cohen et al. An example of this inquiry is found in McConney et al. (Chap. 21), who make a similar examination which found much overlap between existing policy documents and the SSF Guidelines, except for a focus on gender relations. A similar gap is identified both in the Brazilian case in Chap. 22 by Mattos et al. and in Chap. 33 by Barragán-Paladines whose case study is from Ecuador.

An overseer at national level would therefore be required, i.e. someone who has the full picture and takes responsibility for the implementation of the SSF Guidelines as a whole. Such an overseer is particularly needed in countries where the government structure is characterized by institutional fragmentation and instability, as described by Saavedra-Diaz and Jentoft for the case of Colombia (Chap. 27) and by Akintola et al. (Chap. 30) for Nigeria. A similar observation can be found in other chapters as well, like in McConney et al. (Chap. 21) and Nisa (Chap. 31), about the Caribbean region. They argue that both inter-agency and inter-sectoral interaction is "well below the ideal for an ecosystem approach to fisheries." The SSF Guidelines are broad and holistic, and consequently require engagement from a broad range of governmental and non-governmental organization. It is therefore a risk that the implementation of the SSF Guidelines may fall 'between the chairs', that no government institution sees the implementation of the SSF Guidelines as their responsibility. Colombia's complex and instable governance structure illustrates the problem well (Saavedra-Díaz and Jentoft Chap. 27). Policy coherence, coordination and cooperation would be essential, as pointed out in Part III of the SSF Guidelines, and illustrated in several other chapters of this book, like by Akintola et al. (Chap. 30) in the case of Nigeria and Chuenpagdee et al. (Chap. 32) on Thailand. For such reason, Amarasinghe and Bavinck talk about the need for a National Plan of Action (Chap. 18) for the implementation of the SSF Guidelines. Similarly, countries in

Southeast Asia have been discussing the Regional Plan of Action (cf. Chuenpagdee et al. Chap. 32). These plans would be needed in other countries and other regions as well, also because they would help place the SSF Guideline higher on the political agenda. Further, they will result in a more comprehensive approach than 'the purely normative and economic focus in fisheries governance' as Barragán-Paladines wants to see in the case of Ecuador (Chap. 33).

What the overseer and other sponsors of the SSF Guidelines would want to know is whether the SSF Guidelines are precipitating real policy change and whether they are making a positive difference to small-scale fisheries in their communities. Ideally, one would need to estimate how small-scale fisheries would have been without the SSF Guidelines, and be able to separate other factors that exert an influence. Small-scale fisheries are a dynamic sector and their sustainability is a moving target. Internal and external drivers are hard to control. The SSF Guidelines may therefore at best impact on how small-scale fisheries are developing, as they would never reach a steady state – as they are ridden with 'wicked' problems that do not go away.

## Develop Capacity and March on

As this volume demonstrates, small-scale fisheries are diverse and context specific. Despite some common features, like fishing close to shore, often being owner-operated, and using low-tech gear, they display features that are unique to a particular place and country. What is small in one country is big in another. A definition of small-scale fisheries that would apply everywhere is unlikely to work. Therefore, the SSF Guidelines do not contain such a definition. Rather, they start from the observation that fishers, fish workers, and fisheries communities are often poor and marginalized, with unsecure tenure rights, food insecurity, lack of basic education, health insurance, and social services. They are also usually excluded from political processes and are victims of unequal power relations, which in many instances amount to human rights violations. All these things hang together and keep small-scale fisheries trapped in a vicious circle of poverty, out of which it is difficult to break (Jentoft and Eide 2011). For these reasons, small-scale fisheries "rhyme with poverty" (Béné 2003). Rather than spending time on definitions, asking where they are and who the victims are would lead to fishers, fish workers, and fisheries communities that the SSF Guidelines mean to serve. The definition of small-scale fisheries will then emerge by itself.

Institutional contexts and capacities, policy agendas, and the preferences of key agents as they affect small-scale fisheries also differ in ways that will impact on the implementation of the SSF Guidelines. This is a point stressed by Prescott and Steenbergen in Chap. 12, where they compare the conditions between Australia and Indonesia as far as implementing the ecosystem approach to fisheries management, which is also an ambition of the SSF Guidelines, as enshrined in Guiding Principle

11. Similarly, Akintola et al. (Chap. 30) mention governing capacity problems or lack of awareness of domestic and local realities as some of the key impediments to the implementation of the SSF Guidelines in Nigeria.

Countries do not all have the same capacity to act. State governments may sometimes lack what Cappelli (2008) termed 'stateness', i.e. the resources, power, and control to implement public policies. Indeed, the SSF Guidelines target governments that in many instances would fit Myrdal's (1970) concept of 'soft states'.[1] Some would even qualify as 'failed states' (Thorpe et al. 2009). Nowhere are the SSF Guidelines as needed as in these places, but it is also here that the SSF Guidelines will fall on stony ground. CSOs may be called upon to fill the 'implementation gap', but they are usually short on economic resources (Hinds 2003; Béné et al. 2004). Therefore, as a necessary, enabling condition for implementation, capacities are needed at all levels, including at the local level in terms of community organizations, organizational networks, and education programs, as exemplified by Mattos et al. in the case of Brazil (Chap. 22). In some cases, it is easy to identify what capacities are needed and what enabling conditions are required. For instance, in the two case studies from India, by Nayak (Chap. 9) and Gunakar et al. (Chap. 14), the enabling conditions and capacity needs are clear. While in other cases, like in Jamaica (Soares Chap. 7), an analysis is required to detect the legal, institutional, knowledge, and policy gaps that need to be filled. Cohen et al. (Chap. 4) present a useful methodological approach for how such gaps can actually be analyzed.

A similar enabling environment should also be established for generating innovation in small-scale fisheries. One does not elevate the profile of small-scale fisheries, and/or generate the support they need, unless one can point to things that will not only address their problems but also achieve their opportunities. Implementing the SSF Guidelines certainly involves securing their rights, but it also should have a focus on strategies and means that can generate innovation. There are a myriad of things that can be done in order to remove bottlenecks that may help small-scale fisheries add value to their product and thus benefit their communities. Small-scale fisheries, including in industrialized countries like Japan, are based in local communities with strong ties, cultural links and solid fishing identity, as Delany and Yagi point out in Chap. 15. Chuenpagdee et al. (Chap. 23) illustrate similar traditional and cultural ties in their discussion about Newfoundland fisheries. The embedded nature of small-scale fisheries is a strength which makes them more robust and resilient than they would otherwise be. Small-scale fisheries need social capital, which is a resource that can tap also into for the implementation of the SSF Guidelines.

The chapter from Canada (Chuenpagdee et al. Chap. 23) also demonstrates what can be done in order to make small-scale fisheries 'rhyme' with modernity,

---

[1] To Myrdal, the concept refers to 'all the various types of social indiscipline which manifest themselves by deficiencies in legislation and, in particular, law observance and enforcement, a widespread disobedience by public officials and, often, their collusion with powerful persons and groups … whose conduct they should regulate. Within the concept of the soft states belongs also corruption' (Myrdal (1970, 208).

prosperity, and well-being. Social entrepreneurship and social innovation, especially in post-harvest and alternative livelihoods, are some of the examples. There are limits of relevance to what small-scale fisheries in the Global South can learn from their northern counterparts, and *vice versa*, or more generally from one context to another, but exchange of ideas and experiences that point to solutions through which stakeholders can associate positive values with small-scale fisheries, should always be welcome (see Chap. 2). Building regional networks and partnerships among countries that share similar challenges related to the implementation of the SSF Guidelines, as Nisa advocate in Chap. 31, is an idea that is relevant beyond the Caribbean and Pacific. This approach is valid given that the SSF Guidelines address principles and address challenges that are universal, and due to the fact that capacity building is something that must occur everywhere.

The news of the SSF Guidelines has yet to hit home in many parts of the world. People on the ground may still not have heard about them, as in Greenland, as described by Snyder et al. (Chap. 6), and Zanzibar described by Lindstõm and de la Torre-Castro (Chap. 5). Spreading the information to all stakeholders therefore must be a priority. However, the parties involved in implementation are in for a long haul. Consequently, evaluation and monitoring should be longitudinal, following the process as it unfolds at all levels of governance. People with experience in implementation research have relevant knowledge on how to set baselines and monitor progress. Their potential knowledge contribution to management and community development is particularly important in data poor regions, as Kincaid argues in Chap. 28. The relevance of local, traditional knowledge, stressed by Gunakar in Chap. 14, is also acknowledged in the SSF Guidelines, such as in paragraph 11.6, which partly reads: "All parties should ensure that the knowledge, culture, traditions and practices of small-scale fishing communities, including indigenous peoples, are recognized and, as appropriate, supported, and that they inform responsible local governance and sustainable development processes."

The literature points out that studies of implementation processes are in themselves costly and time-consuming (Goggin et al. 1990, 205). The member states that endorsed the SSF Guidelines have a responsibility to provide funding for their implementation, since the academic community also has an important role to play and a contribution to make. Paragraph 11.9 opens as follows: 'States and other parties should, to the extent possible, ensure that funds are available for small-scale fisheries research, and collaborative and participatory data collection, analyses and research should be encouraged. States and other parties should endeavor to integrate this research knowledge into their decision-making processes'.

This and other paragraphs in the SSF Guidelines that discuss research find support in Chap. 29 by Sall and Nauen. They stress the need for more data about small-scale fisheries and provide a long list of specific topics that interviewed stakeholders prioritized. Thus, they illustrate a research process that is bottom-up, like the development of the SSF Guidelines (see for example Chap. 31 by Nisa, where she describes the extensive consultations that occurred in the Caribbean and Pacific regions), by integrating the knowledge of small-scale fishers and fish workers and other relevant stakeholders. This is essential for them to become actively involved in the governance of their fisheries. Fisheries research should therefore be interactive and participatory, just as the governance process itself must be. The implementation of the SSF Guidelines should not be an exception from this rule, as was the case when they were developed through stakeholder consultations around the world (see Willmann et al. Chap. 36). These consultations, and the fact that 147 FAO member states reached a consensus on them, gave small-scale fisheries a legitimacy and momentum that make them more binding than they would otherwise be, and which may help their implementation. Future research would also need to be as holistic as the SSF Guidelines are: the focus must involve the social and ecological system in its entirety, and should include the drivers of change that come from outside the local community and nation in question.

Thus, research must be cross-scale and transdisciplinary, i.e. beyond science disciplines. It might integrate stakeholders in the research-making and its communication. Notably, research agendas should not be imposed from the top-down (Sall and Nauen Chap. 29), neither should research seek only to tap into the knowledge of local stakeholders in an instrumental way. Instead, knowledge should be co-produced by actively involving researchers and stakeholders in the research process. Such research should not narrowly focus on biological data and fishing effort, but also on social issues pertaining to livelihoods, communities, and governance systems. Willmann et al. (Chap. 36) call for more research on how to ensure the legalization of human rights for small-scale fishing communities. With the SSF Guidelines, the academic community has a challenge and also an opportunity, and indeed a responsibility, to become engaged in small-scale fisheries research plans and activities.

It is for this reason that the Too Big To Ignore global partnership and the cluster on the SSF Guidelines were formed. However, research needs the support of governments. The willingness of states to help fund research would be a sign of how serious and supportive they are. Together with FAO, the CSOs, and small-scale fisheries organizations, the academic community has a role to play in making sure that state governments live up to their commitments by endorsing the SSF Guidelines. By joining hands, they can make SSF Guidelines pass from rhetoric to creating new reality where they matter.

## References

Bavinck, M., Chuenpagdee, R., Jentoft, S., & Kooiman, J. (Eds.). (2013). *Governability of fisheries and aquaculture: Theory and applications*. Dordrecht: Springer.

Béné, C. (2003). When fishery rhymes with poverty: A first step beyond the old paradigm on poverty in small-scale fisheries. *World Development, 31*(6), 949–975.

Béné, C., Bennett, E., & Neiland, A. (2004). The challenge of managing small-scale fisheries with reference to poverty alleviation. In A. Neiland & C. Béné (Eds.), *Poverty and small-scale fisheries in West-Africa* (pp. 83–102). Dordrecht: Kluwer Academic Publisher.

Bevir, M. (2011). *The SAGE handbook of governance*. London: Sage Publications Inc..

Cappelli, O. (2008). Pre-modern state building in post-Soviet Russia. *Journal of Communist Studies and Transition Politics, 24*(4), 531–572.
FAO. (2006). *The right to food: Make it happen*. Rome: FAO.
FAO. (2015). *Voluntary guidelines for securing sustainable small-scale fisheries in the context of food security and poverty eradication*. Rome: FAO.
Goggin, M. L., Bowman, A. O.'. M., Lester, J. P., & O'Toole Jr., L. J. (1990). *Implementation theory and practice: Towards a third generation*. Glenview: Scott, Foreman/Little, Brown Higher Education.
Hardin, G. (1968). The tragedy of the commons. *Science, 162*, 1243–1248.
Hinds, L. (2003). Oceans governance and the implementation gap. *Marine Policy, 27*, 349–356.
Jentoft, S. (2014). Walking the talk: Implementing the international voluntary guidelines for securing sustainable small-scale fisheries. *Maritime Studies, 13*(16), 1–15.
Jentoft, S., & Chuenpagdee, R. (Eds.). (2015). *Interactive governance for small-scale fisheries: Global reflections*. Cham: Springer.
Jentoft, S., & Eide, A. (Eds.). (2011). *Poverty mosaics: Realities and prospects in small-scale fisheries*. Dordrecht: Springer Science.
Kooiman, J. (2003). *Governing as governance*. London: Sage Publications.
Kooiman, J., Bavinck, M., Jentoft, S., & Pullin, R. (Eds.). (2005). *Fish for life: Interactive governance for fisheries*. Amsterdam: Amsterdam University Press.
Myrdal, G. (1970). *The challenge of world poverty*. New York: Vintage Books.
Olson, M. (1971). *The logic of collective action. Public goods and the theory of groups*. Harvard: Harvard University Press.
Pitcher, T., Kalikoski, D., Pramod, G., & Short, K. (2009). Not honouring the code. *Nature, 457*, 658–659.
Raustiala, K., & Victor, D. G. (1989). Conclusions. In D. G. Victor, K. Raustiala, & B. Skolikoff (Eds.), *The implementation and effectiveness of international environmental commitments: Theory and practice*. Cambridge, MA: The MIT Press.
Rein, M., & Rabinovitz, F. F. (1987). Implementation: A theoretical perspective. In W. D. Burnham & M. Wagner Weinberg (Eds.), *American politics and public policy*. Cambridge, MA: The MIT Press.
Rittel, H. W. J., & Webber, M. M. (1973). Dilemmas in a general theory of planning. *Policy Sciences, 4*, 155–169.
Royo, L. R. -P. (2009). 'Where appropriate': Monitoring/implementing of indigenous peoples' rights under the Declaration. In C. Charters & R. Stavenhagen, (Eds.), *Making the declaration work: The United Nations declaration of the rights of indigenous peoples* (pp. 324–342). Copenhagen: Doc. 127, IWGIA.
Thorpe, A., Whitmarsh, D., Ndomahina, E., Baio, A., & Kemokai, M. (2009). Fisheries and failed states: The case of Sierra Leone. *Marine Policy, 33*, 393–400.
Tsang, S., Burnett, M., Hills, P., & Welford, R. (2009). Trust, public participation and environmental governance in Hong Kong. *Environmental Policy and Governance, 19*, 99–114.
Victor, D. G., Raustiala, K., & Skolikoff, B. (Eds.). (1998). *The implementation and effectiveness of international environmental commitments: Theory and practice*. Cambridge, MA: The MIT Press.
Young, O. (2006). Choosing governance systems. In M. Moran, M. Rein, & R. E. Goodin (Eds.), *The Oxford handbook of public policy*. Oxford: Oxford University Press.

# Appendix: List of Reviewers

Dedi S. Adhuri
Edward H. Allison
A. Minerva Arce Ibarra
María José Barragán-Paladines
Maarten Bavinck
Camilla Brattland
Mauricio Calderon
Leopoldo Cavaleri
Philippa J. Cohen
Moslem Daliri
Anthony Davis
Raquel de la Cruz Modino
Maricela de la Torre-Castro
Sérgio Macedo G. de Mattos
Cassandra de Young
Einar Eyþórsson
Julia Fraga
Katia Frangoudes
Carlos Fuentevilla
Jennifer Gee
Maria Hadjimichael
C. Emdad Haque
Mafaniso Hara
Bjørn Hersoug
Moenieba Isaacs
Mohammad Mahmudul Islam
Adam Jadhav
Jahn Petter Johnsen
Estellle Jones
Ronald W. Jones



S. Jentoft et al. (eds.), *The Small-Scale Fisheries Guidelines*, MARE Publication Series 14, DOI 10.1007/978-3-319-55074-9

Daniela Kalikoski
Kate Kincaid
Marloes Kraan
Lars Lindström
Sebastian Linke
Joseph Luomba
Patrick McConney
Sílvia Gómez Mestres
Knut H. Mikalsen
Iris Monnereau
Prateep Kumar Nayak
Friday Njaya
Paul Onyango
José J. Pascual-Fernández
Rodrigo Pereira Medeiros
Terrence Phillips
Evelyn Pinkerton
James Prescott
Carlos Pulgarin
Viviana Ramírez-Luna
Lina María Saavedra-Díaz
Gabriela Sabau
Alicia Said
Silvia Salas
Jyothis Sathyapalan
Suvaluck Satumanatpan
Joeri Scholtens
Milena Arias-Schreiber
Susana Siar
Andrew M. Song
Dirk J. Steenbergen
Johny Stephen
Jackie Sunde
Jan van Tatenhove
Rolf Willmann

# About the Authors

**Shehu Latunji Akintola** is presently an Associate Professor with the Department of Fisheries, Lagos State University (Nigeria). He holds a BA in Agricultural Technology (FUT, Akure), a M.Sc. (LASU), MBA (LAUTECH), and PhD (LASU). He has published widely in both local and international journals in areas of fisheries ecologies, fish as food, and fisheries governance. He has won many academic fellowships and grants and is a member of different academic societies including the Too Big To Ignore network. His research interests include coastal and inland fisheries governance and management.

**Oscar Amarasinghe** is a Professor, who was affiliated with the Department of Agricultural Economics of the University of Ruhuna, Sri Lanka. He has more than 30 years of experience in working in the field of small-scale fisheries in both teaching and research. He has been engaged in international projects focusing on a number of issues in small scale fisheries development, such as the Blue Revolution and its impacts, vulnerability and poverty, conflicts and their management, governance, resilience, and well-being. Dr. Amarasinghe has been working very closely with the government and was instrumental in drafting Sri Lanka's National Fisheries Policy in 2004.

**Milena Arias-Schreiber** is a Postdoctoral Researcher at the Graduate School in Marine Environmental Research at the Centre for Sea and Society at Gothenburg University. Her work focuses on the role of formal and informal institutions in sustainable fisheries governance. She has carried out research on the Peruvian pelagic fisheries and is currently studying the impacts of multilevel governance on small-scale fisheries in Sweden.

**Maarten Bavinck** is a Professor in the Department of Geography, Planning and International Development Studies (GPIO) of the University of Amsterdam and a member of the Governance and Inclusive Development programme group. He also holds a chair in coastal resource governance at the Norwegian Fisheries College of UiT – The Arctic University of Norway. Dr. Bavinck specializes in the governance of capture fisheries, particularly in the South, and is especially interested in the fate



S. Jentoft et al. (eds.), *The Small-Scale Fisheries Guidelines*, MARE Publication Series 14, DOI 10.1007/978-3-319-55074-9

of small-scale fisher peoples. His fieldwork is concentrated in South Asia (India and Sri Lanka). He is founder and co-director of the social-science Centre for Maritime Studies (MARE).

**Ramachandra Bhatta** is currently Emeritus Scientist (Economics) at the College of Fisheries, Mangalore, under the Indian Council of Agricultural Research. He previously served with the National Centre for Sustainable Coastal Management as Division Chair, Integrated Social Sciences and Economics, and as Professor of Fisheries Economics with the College of Fisheries, Mangalore. He was a Visiting Professor of ecological economics under the European Union Erasmus Mundus Program in Belgium in 2009, and also a SANDEE Fellow/Consultant in the Maldives in 2008 and 2009.

**David Bishop** is currently a Research Assistant with the Too Big To Ignore: Global Partnership for Small-Scale Fisheries Research project, working primarily with the Information System on Small-scale Fisheries. He obtained his Masters' degree in Marine Studies (Fisheries Resource Management) from the Fisheries and Marine Institute of Memorial University of Newfoundland in 2015. Prior to his Masters', he graduated with a B.Sc. majoring in Geography and minoring in Biology. During that time, his studies focused on topics such as marine biology and resource management.

**Meike Brauer** born in 1991 in Germany, is currently following the Nordic Master Aquatic Food Production – Quality and Safety at the Norwegian University of Life Science. In Summer 2016, she graduated from the B.Sc. Coastal and Marine Management program at the VHL University of Applied Sciences in the Netherlands. In early 2016, she conducted research for her Bachelors' thesis on the potential implementation of the SSF Guidelines in Newfoundland and Labrador, together with Sarah Pötter. Meike is highly interested in providing good quality seafood by sustainable fisheries and aquaculture.

**Joachim Carolsfeld** has a Masters' degree in Neurophysiology of invertebrates and a PhD in fish breeding from the University of Victoria, Canada. Since 1998, he has worked in international development, focusing on the interface between the natural and social sciences in order to find solutions to conserve aquatic resources and improve the lives of the most needy populations. He is currently the Executive Director of the Canadian charity organization World Fisheries Trust, teaches at Camosun College, and is Assistant Professor at the University of Victoria, both in Victoria, BC, Canada.

**Afrina Choudhury** works as Gender Specialist for WorldFish, Bangladesh, where she is responsible for the design and implementation of pro-poor gender sensitive strategies. Working in the field of aquatic-agriculture, her research has revolved around the integration of gender into technical interventions in ways that are sustainable and transformative. In particular, she works to understand how individuals negotiate and change the nuanced constraints brought about by social norms and relationships. She also co-chairs the National Gender Working Group, which brings together gender and equity work in Bangladesh. She holds a Masters' degree in Development Studies from BRAC University.

**Ratana Chuenpagdee** is Professor at Department of Geography, Memorial University of Newfoundland, St. John's, Canada. She held a position of Canada Research Chair in Natural Resource Sustainability and Community Development at Memorial University from 2006 to 2016. Her research emphasizes interdisciplinary approaches to coastal, fisheries, and ocean governance, focusing particularly on small-scale fisheries, marine protected areas, community-based management, and food security. She has worked in several countries including Cambodia, Malawi, Mexico, Spain, Thailand, and Canada. Dr. Chuenpagdee is a project director of the Too Big To Ignore (TBTI) global partnership, and has co-edited the book *Interactive governance for small-scale fisheries: Global reflections* (2015), with Professor Svein Jentoft for TBTI.

**Philippa J. Cohen** largely conducts research on understanding the roles of economic development and changing policy, culture, and demography on small-scale fisheries governance in the least developed countries of the Pacific. Her research employs an interdisciplinary approach to critically evaluate the contributions of community-based and cross-scale fisheries governance to ecological, food security, and livelihood outcomes. She received her PhD from the ARC Centre of Excellence for Coral Reef Studies at James Cook University in 2013, and since has continued there as an Adjunct Research Fellow. Dr. Cohen is a senior scientist in the small-scale fisheries research programme of WorldFish.

**Steven M. Cole** is a Gender Scientist for WorldFish. He obtained his PhD in Biological Anthropology and holds a M.Sc. degree in Agricultural Economics and a B.Sc. in Health and Nutrition. He leads the gender transformative research currently being carried out in small-scale fisheries contexts in Zambia (Barotse Floodplain and Lake Bangweulu wetlands) and in inland valley swamp areas in Sierra Leone. His research integrates and tests gender transformative approaches in development interventions that aim to achieve better, longer-lasting, and more gender-equitable development outcomes.

**Gustavo Henrique G. da Silva** graduated in Ecology with a Doctorate in Aquaculture in Continental Waters (São Paulo State University – UNESP, Brazil). He is currently Associate Professor at the Federal Rural University of the Semiarid (UFERSA-Brazil) and coordinator of Limnology Laboratory (LIMNOAQUA). Dr. Da Silva has experience in limnology, river and lake ecology, the ecology of aquatic organisms, and sustainable aquaculture. He was chief editor of the Bulletin of the Brazilian Association of Limnology, and is creator and coordinator of the graduate course in Ecology and postgraduate course in Ecology and Conservation (UFERSA).

**Moslem Daliri** is currently Assistant Professor of Fisheries at Hormozgan University of Iran. He earned his PhD in Fisheries Management from the Hormozgan University in 2016. For his PhD thesis, he worked on Illegal, Unreported, and Unregulated (IUU) small-scale fishing in the northern Persian Gulf (Hormozgan waters). Since

2010, Daliri has published 28 journal articles on fisheries management in the Persian Gulf region. He is a member of Iran's Elite Foundation, which he joined in 2015, and is also a member of the Too Big To Ignore small-scale fisheries research partnership. His recent research project is 'Destructive effects of small-scale shrimp trawl fisheries on by-catch fish assemblage in the northern Persian Gulf (Hormozgan waters)', which is funded by Gorgan University of Agricultural Sciences and Natural Resources.

**Maricela de la Torre-Castro** is Associate Professor at Stockholm University, from where she obtained her doctorate. She has a background in natural resource management and oceanology and a long history of cooperation with social scientists on interdisciplinary projects. Her research focuses on linked social-ecological systems in coastal communities. Her main interests are to understand human-nature interactions, sustainable development, and how to break poverty traps and increase human well-being. Her projects deal with ecology and the societal value of seagrasses, ecosystems goods and services, gender and management, small-scale fisheries, adaptation to climate change, livelihoods in the coastal zone and place-based development.

**Sérgio Macedo G. de Mattos** is a Public Manager at the Ministry of Planning, Brazil. He graduated in Fishing Engineering from the Federal Rural University of Pernambuco, Brazil, and holds a PhD in Marine Sciences from the Marine Sciences Institute of Barcelona. His work focuses on the management of small-scale fisheries and fisheries resources assessment, emphasizing fisheries bio-economics and issues related to public policy implementation. This approach highlights economic and biological sustainability and highlights participation of fishers and others stakeholders. He is also a coordinator of TBTI's 'Global Synthesis' cluster and a contributor to other TBTI clusters.

**Alyne Delaney** is an Associate Professor at Aalborg University's Innovative Fisheries Management program. Her research foci include resource rights, resilience, and social sustainability and she has worked on fisheries policy and coastal community–related projects around the world. She has served on the Scientific, Technical, and Economic Committee for Fisheries (EU advisory committee) and is an *ex officio* council member for the International Association for the Study of the Commons (IASC). Among her most recent publications includes 'The Neoliberal Reorganization of the Greenlandic Coastal Greenland Halibut Fishery in an Era of Climate and Governance Change' in Human Organization (2016).

**José Alberto Zepeda** works as a Consultant in fisheries management for NGOs, executive agencies and academic institutions. Since 2003, he has collaborated on more than 20 research and executive projects focused on marine and coastal resource management, mainly in Northwestern Mexico. He is a marine biologist with a B.Sc. in Sustainable Coastal Zone Management, M.Sc. in Marine Resources Management, and PhD in Marine Sciences. He is coauthor of nine scientific papers, has presented his work in more than 20 conferences and has received scholarships from academic and executive institutions. His main interest and expertise focus on fisheries management structures as socio-ecological systems.

**María José Espinosa-Romero** has a Master of Science in resource management and environmental studies from the University of British Columbia. Since 1999, she has collaborated with NGOs and governments in projects related to marine conservation and fisheries. Her research focuses on ecosystem-based management, fisheries management, and collective action for governing marine resources. Currently, she is the Director of Marine Conservation and Fisheries Programs at *Comunidad y Biodiversidad.*

**Kafayat Adetoun Fakoya** holds a B.Sc (LASU) M.Sc (Ibadan), and PhD (LASU), and is currently a senior lecturer at the Department of Fisheries, Lagos State University. Her professional interests include fisheries management and fisheries biology, where she is well published. A strong advocate of small-scale fisheries, gender, ecosystem-based management and traditional knowledge, her current activities include integrating traditional knowledge and scientific research in data-limited fisheries. Her professional affiliations include Too Big To Ignore, the International Institute of Fisheries Economics and Trade, and Ecosystem-Based Management Tools Network.

**Katia Frangoudes** is a Senior Researcher in UMR AMURE laboratory of the University of Brest, France. Her main research topics are women's roles in European fisheries, marine resource governance, and fisheries management. She is also the facilitator of AKTEA, a European Network of Women Organizations in Fisheries lobby for gender equality and women rights in the European Union.

**Nicole Franz** holds a Masters' degree in International Cooperation and Project Design from the University La Sapienza, Rome, as well as a Masters' in Economic and Cultural Cooperation and Human Rights in the Mediterranean Region. She started her career as a fisheries consultant with FAO, contributing to the publication 'The Sunken Billions'. Since 2011, Nicole has been a Fishery Planning Analyst at FAO. Since 2013, she has coordinated FAO's support in relation to the implementation of the SSF Guidelines. She has an interest in policy processes, stakeholder empowerment, and the promotion of the human rights based approach in fisheries.

**Carlos Fuentevilla** (MA University of Miami, USA) is a Fisheries Officer at FAO and is currently the manager of the project for Sustainable Management of Bycatch in Latin American and Caribbean trawl fisheries (REBYC-II LAC). He has diverse work experience, including as a research assistant at the University of Miami Experimental Fish Hatchery and in the Mexican government and NGOs. He helped facilitate the negotiation of the SSF Guidelines and continues to work on their implementation. He hopes to continue facilitating processes that build bridges between managers and fishers and increase attention to social and economic issues surrounding fisheries.

**Stuart Fulton** coordinates the national marine reserves program in Community and Biodiversity A.C. (COBI). Stuart has a Masters' degree in Oceanography from the

National Oceanography Centre at the University of Southampton in the UK. He is currently responsible for collaborative marine reserve projects in three priority marine ecosystems: the Mesoamerican Reef, Baja California's giant kelp forest, and the Gulf of California's rocky reef. Stuart works closely with coastal communities to establish, implement, and monitor the community marine reserves.

**Charlie J. Gardner** is an interdisciplinary conservation scientist and conservationist with a particular focus on Madagascar and the Western Indian Ocean. As a lecturer at the Durrell Institute for Conservation and Ecology (University of Kent, UK), his research interests center on the governance and management of protected areas, community-based natural resource management, small-scale fisheries, and the researcher-practitioner divide.

**Miguel González** (PhD, York University) is an Assistant Professor in the International Development Studies program at York University, Toronto, Canada. His research focuses on indigenous self-governance and territorial autonomous regimes in Latin America. On this question, he has published extensively (http://www.fygeditores.com/fgetnicidadynacion.htm) and co-edited a themed issue for a specialized academic journal in the field on indigenous studies (http://www.alternative.ac.nz). Miguel's second focus is the governance of small-scale fisheries in the global south, with a focus on the Nicaraguan Caribbean Coast. Miguel is a member of TBTI and the Centre for Research on Latin America and the Caribbean (CERLAC) at York University.

**Charlotte Gough** oversees the monitoring and evaluation of Blue Ventures' community conservation projects. A doctoral candidate at the University of Exeter, UK, Charlotte's research interests include the composition of small-scale fisheries catches in southwest Madagascar, as well as the social and economic contributions they make.

**Surathkal Gunakar** is Assistant Professor in the Department of Commerce at Pompei College, Aikala, Mangaluru in India. Dr. Gunakar has done extensive research on Karnataka fisheries socioeconomics. His 2013 doctoral dissertation – *Role of economic and social institutions in the fisheries post-harvest sector of coastal Karnataka* – focused on the quantification of social capital in the post-harvest fisheries sector. For many years, he has been actively participating in community-based fisheries management research, action, and outreach. He has published in peer-reviewed journals, participated in national and international conferences, and presented papers on coastal policies and small-scale fisheries.

**Alasdair Harris** is the founder and Executive Director of Blue Ventures and heads an interdisciplinary and international team of over 150 colleagues working worldwide. His work focuses on developing scalable solutions to marine environmental challenges, through approaches that make marine conservation make economic sense to coastal communities.

**Johan Hultman** is Professor in Human Geography at Lund University. His general research interest concerns different institutional understandings of sustainability.

Having previously published worked on waste narratives and value regimes, he now investigates natural resource management authorities in fisheries governance and their pragmatic implementation of sustainability policies.

**Mohammad Mahmudul Islam** is an Assistant Professor at Sylhet Agricultural University in Bangladesh. He received his PhD from the University of Bremen in Germany. His PhD research contextualized poverty and vulnerability in the livelihoods of coastal fishing communities in Bangladesh. His research interests include the migration trajectories of small-scale fishers, management, and conservation of hilsa fisheries, climate change impacts, livelihoods, and well-being analysis of small-scale fishers. He is currently involved in studies related to the establishment of co-management in hilsa sanctuaries of Bangladesh.

**Rikke Becker Jacobsen** is a Post-doctoral Fellow at Aalborg University's Innovative Fisheries Management and an associated researcher at Greenland's Climate Research Centre. Rikke has lived in Nuuk, Greenland from 2010 to 2015, where she has conducted ethnographic research on power and participation in Greenlandic fishery policy. Her focus is on problem-constructions, competing discourses, and the diverse cast of actors in Greenlandic fisheries governance. She has been engaged in climate change and adaptation projects under the Greenland Self Government and the Arctic Monitoring and Assessment Programme.

**Adam Jadhav** is a doctoral student in the Department of Geography at the University of California at Berkeley. His focuses include coastal and marine political ecology, political economy, and development, with a past emphasis on Karnataka fisheries. He has worked as researcher and consultant for multiple environment-related NGOs in India, including Dakshin Foundation, Panchabhuta Conservation Foundation, and the Centre for Environment Education. He also served as Fulbright-Nehru fellow working in the National Centre for Sustainable Coastal Management. He earned a master's in Global Environmental Policy from American University in Washington, DC in 2013.

**Svein Jentoft** is a sociologist and a Professor at the Norwegian College of Fishery Science at UiT – The Arctic University of Norway. Throughout his career, he has worked extensively on fisheries and coastal issues, including resource management, industrial organization and community development in his native Norway, as well as in many other countries. Jentoft has published more than 25 books and numerous journal articles on fisheries and coastal governance. He is a founding member and cluster leader within the Too Big To Ignore small-scale fisheries research partnership. His most recent book – *Interactive governance for small-scale fisheries: global reflections* – co-edited with Ratana Chuenpagdee, was published by Springer in 2015.

**Olufemi Olabode Joseph** holds OND, HND, and a B.Sc. He is currently an M.Sc. fisheries student at the Department of Fisheries, Lagos State University. His research interests include fisheries biology and small-scale fisheries, with an extensive interest and knowledge of fisheries of Lagos State and professional experience as a

Fisheries Officer. He is presently a Fisheries Technologist with the Department of Agriculture Education, Adeniran Ogunsanya College of Education Ijanikin, Lagos State Nigeria.

**Kungwan Juntarashote** is an Emeritus Professor at the Faculty of Fisheries at Kasetsart University, Bangkok, Thailand. He has 40 years of teaching and research experience in the field of fisheries management and small-scale fisheries development, both in Thailand and across Southeast Asia. He currently serves as an advisor to the Department of Fisheries, the Ministry of Agriculture and Cooperatives, and also to the Fisheries Association of Thailand. Dr. Juntarashote has recently been appointed as a Member of the National Fishery Policy Committee of Thailand.

**Ehsan Kamrani** is a Professor of Fisheries Ecology at Hormozgan University of Iran. He has a Master of Public Law and now he is a PhD student of International Public Law. Throughout his career, Kamrani has worked broadly on fisheries, marine biology, and environment, including fish stock assessment, fish population dynamics and management, and bio-socio-economy surveys in mangroves. He has published one book: *Biology and aquaculture of Sea cucumber,* in Persian, and 120 research articles in international and national journals. He has done a fellowship on Mangrove Ecology at ZMT in Bremen University, Germany in 2010. Kamrani has also supervised more than 60 M.Sc. and ten PhD students in fisheries and marine biology and also ten Masters' students in public law in Iranian universities.

**Vesna Kereži** is currently the Project Manager of the Too Big To Ignore: Global Partnership for Small-Scale Fisheries Research project. She holds a M.Sc. in the Human Dimensions of Wildlife Management from Memorial University of Newfoundland, Canada. In the past, she studied human-wildlife interactions with a focus on socio–cultural aspects of these interactions. Her interests lie in natural resource management, human dimensions of fisheries, and knowledge mobilization and transformation.

**Kate Kincaid** is a PhD Candidate at Memorial University of Newfoundland, Canada and a Research Department Head at the Cape Eleuthera Institute and Island School in Eleuthera, Bahamas.Kate has a B.Sc. in Zoology and Marine Zoology from Bangor University, UK and an M.Sc. in Marine Conservation from the University of Hull, UK. Her research focuses on the use of closures and marine protected areas within an Ecosystem-Based Management approach. Her research combines social and ecological methods and focuses on a global to local scale, addressing issues of maintaining both biodiversity conservation and sustainable livelihoods, particularly within island ecosystems.

**Danika Kleiber** will soon work for the Joint Institute for Marine and Atmospheric Research Socioeconomics Program, at the Pacific Islands Fisheries Science Center in Honolulu, Hawaii. Weaving together a background in biology and women's studies, her research focuses on the intersection of gender and small-scale fisheries. In particular, she has examined the gender dimensions of fishing practices and marine conservation. She holds a PhD in Resource Management and Environmental Studies from the University of British Columbia Institute for the Oceans and Fisheries.

**Mitchell Lay** is a professional small-scale fisherman of Antigua and Barbuda since 1988. His affiliations include being a member of the Antigua and Barbuda Fishers Alliance. Mitch is active both in leadership of fisherfolk organizations as well as in fishing operations. In 2007, he was elected Coordinator of the Caribbean Network of Network Fisherfolk Organisations (CNFO), Coordinating Committee. As Coordinator of the CNFO, he has done collaborative work in the development and implementation of national, regional, and international fisheries policy. In addition, he has done significant work in developing the capacity of the CNFO members.

**Adrian Levrel** oversees the development of Blue Ventures' fisheries work in Madagascar, working closely with the organization's field teams to ensure sound design for community-led fisheries monitoring and management approaches. A water engineer by training, Adrien holds an M.Sc. in biological oceanography.

**Lars Lindström** (PhD) is Associate Professor in Political Science at Stockholm University, Sweden. His research interests are in the areas of globalization, resistance, and democratization, and the governance and management of marine resources in poverty contexts. He is the co-founder of the Politics of Development Group at Stockholm University (PODSU). He has co-edited and co-authored several publications on the themes of globalization, governance, and resistance, as well as on small-scale fisheries.

**Sebastian Linke** is a Senior Lecturer in Human Ecology and Associate Professor in Environmental Social Science at the University of Gothenburg. He studies the interface between science and society for sustainable development with a focus on fisheries and marine management and the changing relations between scientists' and other stakeholders' knowledge within new forms of environmental governance.

**Josep Lloret** (PhD: Biology, 2000) is a marine and fisheries biologist who holds a Tenured Assistant Professor position at the University of Girona (UdG) in Catalonia, Spain. His research has been conducted at the UdG and the CSIC (Spain), the Thünen Institute (Germany) and the CNRS (France). He conducts research on the impact of fisheries and environmental factors on marine resources of the Mediterranean and the North Atlantic. His studies have focused particularly on artisanal and recreational fisheries in marine protected areas, with a view to safeguard vulnerable species and habitats and to guarantee the sustainability of coastal fisheries.

**Joseph Luomba** works with the Tanzania Fisheries Research Institute (TAFIRI). He has a MA in Geography from Memorial University of Newfoundland in Canada. His research interests are in fisheries governance, co-management, and fish marketing. Mr. Luomba has participated in several studies in Lake Victoria from 2007 to date. In addition, he has authored and co-authored a handful of publications in refereed journals. His currently working on fisheries tenure rights in Lake Victoria.

**Alison Elisabeth Macnaughton** has a BA in Geography (University of Victoria), an MA in Planning (University of British Columbia) and is currently a PhD student at the University of Victoria in the Department of Geography. Her research focuses on social assessment tools in small-scale fisheries, investigating a new fishery based on an introduced species (*Arapaima gigas*) in the Bolivian Amazon. Since 2004, she has worked with the World Fisheries Trust (www.worldfish.org) in Brazil and Bolivia with coastal and inland small-scale fishers on projects to improve livelihoods, empowerment, and conservation. She participates in Amazon Fish for Food (www.pecesvida.org) and the Community Conservation Research Network (www.communityconservation.net).

**Allyssandra Maria Lima R. Maia** graduated in Biological Sciences from the State University of Rio Grande do Norte (UERN). She is a specialist in Psychopedagogy (FVJ/CE), and has a Masters' degree and Doctorate in Animal Science by the Federal Rural University of the Semiarid (UFERSA-Brazil). She is a professor at the State University of Rio Grande do Norte/ UERN, at the Faculty of Health Sciences / Medicine, and is a member of the research Group of Studies in Collective Health (GESC). Her research focuses on the following topics: Collective Health, Epidemiology, Research Methodology, Population Dynamics, and Ecology.

**Patrick McConney** is a Senior Lecturer in the Centre for Resource Management and Environmental Studies (CERMES) at the University of the West Indies Cave Hill Campus in Barbados. His applied research includes resilience, socio-economics, and governance related to small-scale fisheries and marine protected areas in the Caribbean.

**Cynthia McDougall** is the Gender Leader for WorldFish and the CGIAR Research Program on Fish Agri-food Systems ('FISH'). She is an interdisciplinary social scientist with a background in systems thinking, community forestry, and food security. **She has a particular interest in transdisciplinary, mixed methods, and participatory action research approaches and how these can leverage constructive, locally-led, and scalable shifts towards empowerment, equality, poverty reduction, food and nutrition security, and sustainability. Cynthia** holds an MPhil (Geography) from Cambridge University in the UK and a PhD (Knowledge, Technology, and Innovation) from Wageningen University, The Netherlands.

**Thomas F. McInerney** is an international Corporate Lawyer and strategist with expertise in multilateral treaties and regulatory matters in the areas of agriculture, food, and natural resources. He advises numerous international organizations, including FAO, UNEP, Bioversity International, and WHO on innovative solutions to enhance the performance of global regulatory treaties. McInerney is the founder of the Treaty Effectiveness Initiative, which is dedicated to gathering, consolidating, and generating knowledge on practices related to improving treaty results. In addition, he is Distinguished Scholar in Residence at Loyola University Chicago School of Law, where he teaches international law, development finance, and institutional reform.

**Sílvia Gómez Mestres** (PhD: Social and Cultural Anthropology, 2003) is a researcher and Senior Lecturer at the Autonomous University of Barcelona. She has conducted research and lectures at the University of Barcelona, CNRS (France), EHSS (France), Université Paris 1 Phantéon-Sorbonne (France), and Oslo University (Norway). Her research focuses on the articulation between politico-legal provisions and social processes in different contexts, from an ethnographic and historical approach. Since 2004, her research has contributed to the understanding of artisanal fisheries in Marine Protected Areas of the Mediterranean, reinforcing and promote their value to cultural heritage and to ensure sustainable livelihoods.

**Tiffany H. Morrison** is a political geographer (PhD, University of Queensland) who draws on the disciplines of political science, public administration, geography, and sociology in order to understand and improve the design of complex and multi-scalar environmental governance regimes. Tiffany is currently an Associate Professor with the ARC Centre of Excellence for Coral Reef Studies at James Cook University. Prior to this, she was a tenured faculty member in the School of Geography, Planning and Environmental Management at the University of Queensland (2008–2015) and the School of Political and International Studies at Flinders University (2004–2008).

**Cornelia E. Nauen** holds a PhD in fisheries science/marine ecology from Kiel University, Germany. She worked in FAO's Fisheries Department from the late 1970s. Between 1985 and 2012, she served in the European Commission in development cooperation and in international science cooperation. Her major subject areas were aquatic resources management and restoration. Critically engaged science to support policy and action for social inclusion and living and being in sustainable ways were also a major focus of Cornelia's work. Since 2010, she has headed *Mundus maris* – Sciences and Arts for Sustainability, which seeks to combine scientific concepts with practice embedded in local and global cultural spaces.

**Prateep Kumar Nayak** is Assistant Professor in the School of Environment, Enterprise, and Development at the University of Waterloo. His academic training is in political science, environmental studies, and international development, and he holds a PhD in Natural Resources and Environmental Management from the University of Manitoba. Prateep's research focuses on the understanding of human-environment connections (or disconnections) with particular attention to social-ecological change in coastal-marine systems, its drivers, their influence and possible solutions. He is a past SSHRC Banting Fellow, Trudeau Scholar, a Harvard Giorgio Ruffolo Fellow in Sustainability Science and recipient of Canada's Governor General Academic Gold Medal.

**Nadine Nembhard** is the Administrative Assistant of the Caribbean Network of Fisherfolk Organisations (CNFO), and holds over ten years, which is almost her entire professional career, dedicated to looking after fisherfolk well-being. Her career working with fishers started in August 2005 and continued until July 2014 at the Belize Fishermen Cooperative Association (BFCA). During this time, she was elected to the position of secretary for the coordinating unit of the CNFO. In October

2014, the CNFO hired her as office assistant. In addition to this she is the Co-Chair of the World Forum of Fishers People (WFFP) 2014–2017.

**Zahidah Afrin Nisa** is a United Nations Nippon Fellow in the field of ocean affairs, the Law of the Sea, and related disciplines. She is a marine biologist from Fiji Islands with a specialization in environment management, development, and ocean governance reforms. Her work has been with both the private and public sector and civil society organizations in implementing locally managed marine area networks in the Pacific region. Her work in the Caribbean region has been on Marine Protected Area governance and the strengthening of fisher organizations and cooperatives. Currently, she is assisting the Grenada government in the development of its blue growth initiative.

**Kim Olson** has a passion for interdisciplinary approaches to fisheries governance and resource conservation. She is a member of the TBTI stewardship cluster, and has been working in areas related to small-scale fisheries, food security, governance, and fishery closures. She completed a Master's in Geography through the International Coastal Network at Memorial University, and was recognized as a fellow of the School of Graduate Studies for her research on voluntary fishery closures. Kim currently works with the Government of Newfoundland and Labrador, where she has worked to advance best practices for integrating stakeholder input into policy and decision-making processes.

**Paul Ochieng' Onyango** has a PhD in Fisheries Social Science from Tromso University in Norway. He teaches fisheries governance at the College of Natural and Aquatic Sciences (CONAS) of the University of Dar es Salaam. Prior to this, he worked with the Tanzania Fisheries Research Institute (TAFIRI) as a senior researcher. Dr. Paul has vast experience of more than 20 years in fisheries research in Tanzania. He has authored and co-authored various publications in refereed journals and books.

**María José Barragán-Paladines** obtained a PhD in Geography at Memorial University of Newfoundland, Canada. After her Bachelors' degree in Biological Sciences and a M.Sc. degree in Sustainable Resource Management (TUM – Germany), a shift in her research focus took her to Canada where she was immersed in interactive governance theory, which she applied to coastal and marine resources and protected area settings. She is currently a Post-doctoral Researcher at the Development and Knowledge Sociology Working Group in the Leibniz-Center for Tropical Marine Ecology-ZMT in Bremen, where she works primarily on fish-as-food in the context of food security in Latin America.

**Hajnalka Petrics** is a Gender and Development Officer with FAO. She supports mainstreaming gender into FAO operations, including institutional mechanisms, field programmes, and normative products. She leads the work on facilitating CEDAW implementation and on gender-sensitive rural advisory services. She also provides technical support to mainstreaming gender in small-scale fisheries and, in particular, to the formulation of the implementation guide of the gender equality principle of the Small-Scale Fisheries Guidelines. Hajnalka holds a PhD in International Cooperation and Sustainable Development Policies from the University of Bologna.

**Terrence Phillips** is a Senior Technical Officer at the Caribbean Natural Resources Institute, based in Trinidad and Tobago. He is responsible for planning and implementing the coastal and marine livelihoods and governance program, which is aimed at improving livelihoods and contributing to poverty reduction by promoting and facilitating sustainable use and governance of coastal and marine resources, building effective institutions and facilitating collaboration between key stakeholders (including fisherfolk and fisherfolk organisations), through participatory research, capacity building, and communication of lessons learned.

**Cristina Pita** holds a PhD in Social and Environmental Sustainability and is currently a Researcher at the Department of Environment and Planning & Centre for Environmental and Marine Studies (CESAM), at the University of Aveiro, Portugal. Her research focuses on coastal community development, interdisciplinary approaches to sustainable use of marine resources, and fisheries and coastal governance, especially in the context of small-scale fisheries.

**Joonas Plaan** is a PhD Candidate at Memorial University of Newfoundland. As an environmental anthropologist, he has been working with fisheries-based communities in Estonia and Newfoundland, Canada, studying various aspects of human-environment interactions. He is currently researching potential future effects of climate change to fishery dependent communities in Newfoundland. Joonas Plaan is a member of TBTI, an ISER Fellow and Kristjan Jaak Scholar.

**Marilyn Porter** is Professor Emerita at Memorial University in Newfoundland, Canada. Her work has focused on feminist methodology and women and development, especially women living and working in coastal contexts. As she ages, she has become increasingly interested in the challenges faced by older women and especially in their work as grandmothers.

**Sarah Pötter** born in 1992, is currently a Master's student of West Nordic Studies at the and the University of the Faroe Islands. She received her Bachelor's degree in Coastal and Marine Management from the VHL University of Applied Sciences in the Netherlands. In early 2016, she conducted research for her Bachelor's thesis about the potential implementation of the SSF Guidelines in Newfoundland and Labrador, together with Meike Brauer. Sarah is focused on international fisheries, both policy and management aspects, the Law of the Sea, as well as good governance.

**James Prescott** holds a Bachelor's degree in zoology from the University of Hawaii. His 40+ year career in fisheries has been divided between fisheries research and management in tropical developing country fisheries and temperate Australian fisheries. Jim has worked for the South Pacific Commission, the Papua New Guinea Marine Fisheries Service, the Forum Fisheries Agency, and the South Australian Research and Development Institute. He is currently employed at the Australian

Fisheries Management Authority. Jim's interest in governance is a consequence of his recent experiences while working with Indonesians who fish in Australian waters.

**Steve Rocliffe** is an interdisciplinary marine conservation scientist specializing in community-centered approaches to marine conservation and fisheries management. As Research and Learning Manager at Blue Ventures, Steve provides broad support to the organization's research program.

**Victoria Rogers** received a B.Sc. in Geography with a concentration in resource management from the University of Victoria in 2012, and an MA in Geography from Memorial University in 2015. Her research interests span a variety of resource management and governance topics, including parks and protected areas, marine conservation, sustainable tourism, and the role of small-scale fisheries in local food security. Following her studies, she completed a Global Affairs Canada internship with the WorldFish Centre in 2016. In this role, she joined a project that worked on assessing the welfare value of fish to agrarian households in Cambodia's Lower Mekong Basin.

**Lina María Saavedra-Díaz** is Associate Professor at the University of Magdalena (Biology Program and Fisheries Engineering) in Santa Marta, Colombia. Her research interests centre around understanding how Colombian small-scale fisheries could be managed on both coasts (Caribbean and Pacific) through a bottom-up approach, working with fishermen from different communities, and interacting at the same time with the National Fisheries Administration. Lina's research has facilitated a dialogue among government, fishermen, and academics partners through the co-production of different scenarios. She co-leads a research group in Social-Ecological Systems for Human Well-being in Colombia.

**Gabriela Sabau** holds a PhD in Economics from the Academy of Economic Studies in Romania. Since 2003, she has been Associate Professor of Economics and Environmental Studies at Grenfell Campus, Memorial University. Gabriela's research interests are sustainable development, fueled by scientific knowledge and value judgements; unjust un-economic growth; demand-side water management; and sustainable management of small-scale fisheries. Publications include: 'Whose fish is it anyway? Iceland's cod fishery rights', in *World small-scale fisheries. Contemporary visions* (Ed. Ratana Chuenpagdee), and 'From unjust uneconomic growth to sustainable fisheries in NL: The true cost of closing the inshore fishery of ground fish' (*Marine Policy*).

**Alicia Said** is a PhD Candidate in Biodiversity Management at the School of Anthropology and Conservation, University of Kent, UK. She holds a BA (Honours) and MA in Geography from the University of Malta. Her current research aims to understand the interacting processes of fisheries governance and management in Malta, with a specific focus on the fisheries and marine policy changes occurring since EU accession in 2004. She is particularly interested in investigating how

Maltese small-scale fishers are experiencing and responding to policy shifts and exploring how such changes affect the sustainability of the small-scale fishing fleet.

**Aliou Sall** is a Senegalese socio-anthropologist with university studies in Toulouse, France, and a PhD which he earned in Geneva, Switzerland. Co-founder of the NGO CREDETIP (Centre de recherche sur les technologies intermédiaires de pêche) in 1987, he has actively supported the social movements of small-scale fishers since then. He has led and carried out several studies and projects for organizations such as the Sub-regional Fisheries Commission (CSRP), the Program for the Protection of Marine and Coastal Resources (PRCM), the International Union for the Conservation of Nature (IUCN), the International Federation of Banc d'Arguin (FIBA), WWF, and the EU. He is co-founder and Vice President of *Mundus maris*.

**Anna Santos** holds a PhD in Geography and is currently working as a social science contractor for the Alaska Fisheries Science Center of NOAA fisheries. Her research has focused on governance and livelihoods of small-scale fisheries. She is currently working with a team developing social indicators of well-being of Alaska fishing communities.

**Suvaluck Satumanatpan** is a Lecturer in the faculty of Environment and Resource Studies, Mahidol University. She received the Fulbright U.S-ASEAN Visiting Scholars distinction in 2015. Her Thai textbook, *Coastal management: Integration to sustainability* received the Mahidol Excellent Textbook award in 2012, and has been used for teaching on coastal management subject in several Thai universities. She has contributed to several national policy papers for marine and coastal resources in Thailand. She is interested in coastal and fishery governance research (interactive governance), integrated coastal management, marine and environmental protected areas effectiveness, coastal and fishery resilience, and marine spatial planning.

**Filippa Säwe** is Senior Lecturer in sociology at Lund University. Her research covers a range of empirical fields to focus on issues of communication and communicative practices. Her current research seeks to develop the concept of transdisciplinarity within sustainability science, using qualitative methods to analyze the connection between the micro-level of social interaction and the macro-level of societal change.

**Rebecca L. Singleton** is responsible for monitoring and evaluating the social impacts of Blue Ventures' programs in Madagascar. She is a doctoral candidate at the University of British Columbia in Canada, where her interdisciplinary research focuses on the socioeconomic impacts of marine conservation programs in developing countries.

**Hunter T. Snyder** is a PhD student in the Program in Ecology, Evolution, Ecosystems and Society. He is committed to advancing the study of livelihoods that depend upon living marine resources. His ethnographic research has focused on fisheries, subsistence livelihoods, and social-ecological disruptions in the Arctic. He took his master's degree in anthropology at the University of Oxford and earned a graduate scholarship to read a second masters in fisheries resource management at the Marine Institute in St. John's, Newfoundland. Before beginning his PhD, he was a consultant at the Food and Agriculture Organization in the Fisheries and Aquaculture Department.

**Lisa K. Soares** completed her graduate studies (with distinction) in Global Affairs (focusing on International Business, Economics & Development) at New York University. She also holds an Investor Relations/Public Relations Certificate (CIR) from NYU. Lisa is currently a [3rd year] PhD Candidate in Politics and International Studies at the University of Warwick, UK. Current research interests include: Caribbean development and political economy and marine and fisheries politics and policy. Lisa is the 2015 recipient of the Government of Jamaica's Governor General's Achievement Award for Excellence (18–35 years) for the Jamaican Diaspora USA. Lisa is also an independent International Development consultant.

**Kumi Soejima** is a Lecturer at the National Fisheries University in Japan and holds a PhD in Fisheries Economics. Her work has focused on the role of women in the market and value chain, especially for small scale fisheries products. She and two other researchers have been trying to create networks of fisherwomen's groups in Japan since 2003.

**Francisco Javier Vergara Solana** provides links between academia and industry to facilitate the development of agro-industrial projects via business acceleration and the promotion of investment vehicles (enture.vc). Francisco is a marine biologist with an M.Sc. in *Marine Resources Management* and is currently a PhD student in Fisheries and Aquaculture Bioeconomics. His thesis evaluates the interdependency of the fishery and aquaculture of bluefin tuna in Mexico. Francisco has participated in research projects and authored articles relating to fisheries and aquaculture, bio-economics, ichtiology, and morphometrics, with complementary interests in R&D, econometrics and statistics, market analysis, and intelligence and technology-based business development.

**Andrew M. Song** is an early career social scientist interested in the geography and governance of natural resources with a particular focus on fisheries. He received his PhD from Memorial University of Newfoundland, Canada. He is currently a Research Fellow at the ARC Centre of Excellence for Coral Reef Studies at James Cook University, Australia, where his research focuses on the regional governance of small-scale fisheries in the Pacific Islands. Also affiliated with WorldFish, he holds familiarity with several marine and inland settings, including the Philippines, Malawi, South Korea, Atlantic Canada, and the Laurentian Great Lakes.

**Siri Ulfsdatter Søreng** is currently a Senior Advisor at the Norwegian Agriculture Agency. She has been working as a social scientist for several years, and has a PhD in fisheries management from the Norwegian College of Fishery Science at UiT – The Arctic University of Norway. Both as an academic and administrator, Søreng is concerned with interactive governance and legal pluralism in the management of natural resources, in particular with respect to challenges concerning the implementation of Sami/indigenous rights.

**Dirk J. Steenbergen** is a Postdoctoral Researcher at the Research Institute for Environment and Livelihoods at Charles Darwin University in Australia. Dirk's research focuses on co-management regimes around small scale fisheries and coastal natural resources, currently in eastern Indonesia and Timor-Leste. He is interested in interactions between technical agencies and resource user groups where continuous negotiations involving contesting interests, worldviews, and knowledge-bases occur, and which significantly affect local governance. His ongoing interest in local participation in common pool resource management builds on 10 years of working on research and applied projects with indigenous groups in Southeast Asia and Southern Africa.

**Wichin Suebpala** is a researcher at the Marine Biodiversity Research Group at Ramkhamhaeng University. He received B.Sc. and M.Sc. in Environmental Science from Ramkhamhaeng University and Kasetsart University, Thailand, respectively. He has been involved as a researcher in several projects regarding environmental science, integrated coastal management, and fisheries management. He is interested in applying integrated approaches to managing coastal fisheries and livelihoods to achieve sustainability. As a doctoral student of the Interdisciplinary Program of Environmental Science, Chulalongkorn University, he is currently conducting his PhD research project on assessing the ecological impacts of fisheries in Thailand.

**Jackie Sunde** is a Post-doctoral Researcher in the Department of Environmental and Geographical Science at the University of Cape Town, South Africa. Her PhD highlights issues of customary tenure and governance in small-scale fisheries in South Africa. She is a member of the International Collective in Support of Fishworkers (ICSF).

**Makamas Sutthacheep** is a Lecturer and Researcher at the Marine Biodiversity Research Group at Ramkhamhaeng University, Bangkok, Thailand. She accomplished her D.Sc. in Marine and Environmental Science from the University of the Ryukyus, Japan, and obtained her M.Sc. in Environmental Science from Chulalongkorn University. Her expertise includes marine ecology, management of marine and coastal resources, environmental impacts assessment, as well as applying remote sensing for coastal resource management. She is a project leader on the research project of assessing impacts of climate change on coral reef ecosystem in Thai waters.

**Jorge Torre** PhD is the General Director of *Comunidad y Biodiversidad, A.C.* His research has focused on the development of applied investigation in marine conservation and fisheries management, including both ecological and social aspects. He is particularly interested in promoting long-term evaluation and monitoring programs that incorporate traditional knowledge and community participation to measure the impact of conservation and human use on biodiversity. Jorge has collaborated in more than 40 publications. In 2012, he received the Conservation Award from the Desert Museum Arizona-Sonora.

**Xavier Vincke** formerly Blue Ventures' Fisheries Programme Manager, was responsible for supporting local fishers to manage southwest Madagascar's octopus fisheries. Xavier bade farewell to Madagascar in 2016 and moved to Belgium to take up a position with *Aviation Sans Frontières*.

**Lena Westlund** (M.Sc. University of Gothenburg, Sweden) has extensive experience in fisheries and development cooperation. She has lived and worked both long and short term in-country engaged in projects addressing poverty alleviation and food security in fishing communities. At the global level, she has been involved in policy development and has contributed to several FAO publications, e.g., on the ecosystem approach to fisheries (EAF) and marine protected areas (MPAs). She contributed to the SSF Guidelines development process and now supports their implementation. Ms. Westlund currently lives in Stockholm, Sweden, from where she continues to work as an FAO consultant.

**Rolf Willmann** (MA in Economics, University of Tübingen) is former Senior Fishery Planning Officer, Fisheries and Aquaculture Department, UN Food and Agriculture Organization. He has led FAO's normative work on small-scale fisheries having served as the secretary of the global conference on small-scale fisheries, Bangkok, Thailand, 13–17 October 2008, and coordinator of the team for the development of the international guidelines on securing sustainable small-scale fisheries.

**Matias John Wojciechowski** holds a MA in Local Economic Development from the University of Waterloo (2002) and an MA in Territorial Planning and Management from the Federal University of ABC (2014). He is currently working towards a PhD in Rural Development (thesis: *Territories of transition: Pathways from conventional agriculture to sustainable production systems*). Since 2006, he has worked with the World Fisheries Trust with inland and coastal fishing communities focusing on value-chain development, food safety, occupational health, economic development, and solidarity economy. From 2012 to 2014, he participated in the 'Peces Para la Vida' project in Bolivia, implementing participatory value-chain assessments and economic feasibility studies.

**Nobuyuki Yagi** is an Associate Professor at the University of Tokyo. His research focuses on the economics of ocean conservation. He received an MBA from the Wharton School of University of Pennsylvania, USA, and a PhD from the University of Tokyo, Japan. He has been a member of the Executive Committee of IIFET (International Institute of Fisheries Economics & Trade) and an Intergovernmental Science-Policy Platform on Biodiversity and Ecosystem Services expert panel since 2014. Publications include 'Applicability of ITQs in Japanese fisheries: a comparison of Rights Based Fisheries Management in Iceland, Japan, and United States' in *Marine Policy* (2012).

**Thamasak Yeemin** has worked at the Marine Biodiversity Research Group at Ramkhamhaeng University in Bangkok since 1992, after completing his D.Sc. in Biology from Kyushu University. He has experience in all aspects of coral reef and related ecosystem management, conservation, research, and administration based on over 20 years of field works in Thailand and other parts of East Asia. He has conducted research on ecology, biology, environmental science, and socio-economics of coastal ecosystem management and marine protected areas, and provided expertise to several management agencies to design and implement management plans for coastal resources and environment. Currently, he is the president of the Marine Science Association of Thailand.

# Index





S. Jentoft et al. (eds.), *The Small-Scale Fisheries Guidelines*, MARE Publication Series 14, DOI 10.1007/978-3-319-55074-9

## D

**E**

**G**

## H

## I

**M**

**N**

## O



## P

## Q

## R